How to access the supplemental web study guide

We are pleased to provide access to a web study guide that supplements *Physiology of Sport and Exercise, Fifth Edition*. This study guide offers an interactive learning experience that will allow you to practice, review, and develop knowledge and skills about the physiology of sport and exercise. You'll find interactive learning activities, practice quizzes, web links, career information, and more. We are certain you will enjoy this unique online learning experience.

Accessing the web study guide is easy!
Follow these steps if you purchased a new book:

1. Visit **www.HumanKinetics.com/PhysiologyOfSportAndExercise**.

2. Click the <u>fifth edition</u> link next to the corresponding fifth edition book cover.

3. Click the Sign In link on the left or top of the page. If you do not have an account with Human Kinetics, you will be prompted to create one.

4. If the online product you purchased does not appear in the Ancillary Items box on the left of the page, click the Enter Key Code option in that box. Enter the key code that is printed at the right, including all hyphens. Click the Submit button to unlock your online product.

5. After you have entered your key code the first time, you will never have to enter it again to access this product. Once unlocked, a link to your product will permanently appear in the menu on the left. For future visits, all you need to do is sign in to the textbook's website and follow the link that appears in the left menu!

→ Click the Need Help? button on the textbook's website if you need assistance along the way.

How to access the web study guide if you purchased a used book:

You may purchase access to the web study guide by visiting the text's website, **www.HumanKinetics.com/PhysiologyOfSportAndExercise**, or by calling the following:

800-747-4457 . . . . .U.S. customers
800-465-7301 . . . . .Canadian customers
+44 (0) 113 255 5665 . . . . European customers
08 8372 0999 . . . . Australian customers
0800 222 062 . . . . .New Zealand customers
217-351-5076 . . . . .International customers

For technical support, send an e-mail to:
support@hkusa.com . . . . U.S. and international customers
info@hkcanada.com . . . . Canadian customers
academic@hkeurope.com . . . . European customers
keycodesupport@hkaustralia.com . . . . . . Australian and New Zealand customers

HUMAN KINETICS
The Information Leader in Physical Activity & Health

04-2012

Product: Physiology of Sport and Exercise, Fifth Edition, web study guide

Key code: KENNEY-W7WMC7-OSG

This unique code allows you access to the web study guide.

Access is provided if you have purchased a new book. Once submitted, the code may not be entered for any other user.

Physiology of Sport and Exercise

FIFTH EDITION

W. Larry Kenney, PhD
Pennsylvania State University, University Park

Jack H. Wilmore, PhD
University of Texas, Austin

David L. Costill, PhD
Ball State University, Muncie, Indiana

Human
Kinetics

Library of Congress Cataloging-in-Publication Data

Kenney, W. Larry.
 Physiology of sport and exercise / W. Larry Kenney, Jack H. Wilmore, David L. Costill. -- 5th ed.
   p. ; cm.
 Rev. ed. of: Physiology of sport and exercise / Jack H. Wilmore, David L. Costill, W. Larry Kenney.
 Includes bibliographical references and index.
 ISBN-13: 978-0-7360-9409-2 (print)
 ISBN-10: 0-7360-9409-1 (print)
 I. Wilmore, Jack H., 1938- II. Costill, David L. III. Wilmore, Jack H., 1938- Physiology of sport and exercise. IV. Title.
 [DNLM: 1. Exercise--physiology. 2. Sports--physiology. 3. Physical Endurance. 4. Physical Fitness. QT 260]

 612'.044--dc23

                    2011035158

 ISBN-10: 0-7360-9409-1 (print)
 ISBN-13: 978-0-7360-9409-2 (print)

Permission notices and credits for photos in this book can be found on page xvii.

The web addresses cited in this text were current as of May 2011, unless otherwise noted.

Acquisitions Editor: Amy N. Tocco; **Developmental Editor:** Katherine Maurer; **Assistant Editor:** Steven Calderwood; **Copyeditor:** Joyce Sexton; **Indexer:** Susan Danzi Hernandez; **Permissions Manager:** Dalene Reeder; **Graphic Designer:** Nancy Rasmus; **Graphic Artists:** Nancy Rasmus and Denise Lowry; **Cover Designer:** Keith Blomberg; **Photographer (cover and title page):** PennTrackXC.com, a MileSplit U.S. Affiliate; **Cover Illustrator:** Jennifer Gibas; **Photo Asset Manager:** Laura Fitch; **Visual Production Assistant:** Joyce Brumfield; **Photo Production Manager:** Jason Allen; **Art Manager:** Kelly Hendren; **Associate Art Manager:** Alan L. Wilborn; **Art Style Development:** Joanne Brummett; **Illustrations:** © Human Kinetics; **Printer:** Courier Companies, Inc.

Printed in the United States of America   10 9 8 7 6 5 4 3

The paper in this book was manufactured using responsible forestry methods.

Human Kinetics
Website: www.HumanKinetics.com

United States: Human Kinetics, P.O. Box 5076, Champaign, IL 61825-5076
800-747-4457
e-mail: humank@hkusa.com

Canada: Human Kinetics, 475 Devonshire Road Unit 100, Windsor, ON N8Y 2L5
800-465-7301 (in Canada only)
e-mail: info@hkcanada.com

Europe: Human Kinetics, 107 Bradford Road, Stanningley, Leeds LS28 6AT, United Kingdom
+44 (0) 113 255 5665
e-mail: hk@hkeurope.com

Australia: Human Kinetics, 57A Price Avenue, Lower Mitcham, South Australia 5062
08 8372 0999
e-mail: info@hkaustralia.com

New Zealand: Human Kinetics, P.O. Box 80, Torrens Park, South Australia 5062
0800 222 062
e-mail: info@hknewzealand.com

E5149

First and foremost, to my wife, Patti, who takes care of all of the important things in our lives so that I can pursue academic endeavors such as research, teaching, and writing textbooks. To my grown-up children, Matthew, Alex, and Lauren, the three most important things in my life. It has been a joy watching you grow and achieve your own success in school, athletics, and life. Keep striving for new goals, continue to be good people who care about others, and stay happy along life's journey. To my parents, who inspired and supported me and continue to serve as role models. And to all my graduate students, past and present, who challenge me and teach me new things every day.

W. Larry Kenney

I dedicate this book to those who have had the greatest impact on my life. To my loving wife, Dottie, and our three wonderful daughters, Wendy, Kristi, and Melissa, for patience, understanding, and love. To our sons-in-law, Craig, Brian, and Randall, for being good husbands, fathers, and friends. To our grandchildren, who are a constant source of joy and amazement. To Mom and Dad for their love, sacrifice, direction, and encouragement. To my former students, who have been my friends and inspiration. And to my Lord, Jesus Christ, who provides for every one of my needs.

Jack H. Wilmore

To my grandchildren, Renee and David, who have added a new dimension to my life. To my wife, Judy, who gave me two loving daughters, Jill and Holly. To my college swimming coach, Bob Bartels, who "rescued my soul" on more than one occasion, and showed me the joys of research and teaching. To my former students, who taught me more than I taught them—their subsequent successes have been the highlight of my career.

David L. Costill

The body is an amazingly complex machine. At any given time there is intricate communication among various cells, tissues, organs, and systems that serve to coordinate its physiological functions. When you think of the numerous processes occurring within the body at any given time, it is truly remarkable that these physiological functions work so well together. Even at rest, the body is physiologically quite active. Imagine, then, how much more active all of these body systems become when you engage in exercise. During exercise, nerves excite muscles to contract. Exercising muscles are metabolically active and require more nutrients, more oxygen, and efficient clearance of waste products. How does the whole body respond to the increased physiological demands of exercise?

That is the key question when you study the physiology of sport and exercise. *Physiology of Sport and Exercise, Fifth Edition*, introduces you to the fields of sport and exercise physiology. Our goal is to build on the knowledge that you developed during basic course work in human anatomy and physiology and to apply those principles in studying how the body performs and responds to the added demands of physical activity.

What's New in the Fifth Edition

The fifth edition of *Physiology of Sport and Exercise* has been fully updated in both content and design. The illustrations, photos, and medical artwork are completely revamped for added detail, clarity, and realism, allowing both a greater insight into the body's response to activity and a better understanding of the underlying research. The layout is redesigned to focus on the new artwork and better present the content and numerous learning aids.

In addition to these visual changes, we have reorganized the chapters on metabolism and hormonal control during exercise—splitting the fourth edition's chapter 2 into two chapters, now chapters 2 and 4. Students often find information on exercise metabolism and bioenergetics to be challenging. To help students better understand these systems, chapter 2 has been completely updated and revised to offer a novel, big-picture perspective on how energy is derived from the foods we eat and used for muscle contraction and other physiological processes. All new figures help to make this information readily understandable. Likewise,

chapter 4 on hormonal control of exercise has been substantially revised and expanded in scope. We have also updated the text to include the latest research on important topics in the field, including the following:

- Updated principles of strength training based on the 2009 ACSM position stand and new sections on core strength, stability training, and high-intensity interval training (HIT)

- New content on lactic acid as a fuel source, muscle cramps, childhood obesity, substrate utilization and endocrine response to exercise, and vascular aging

- Updated coverage of central and peripheral cardiac functions, the female athlete triad, and the menstrual cycle

- New research on effects of physical activity on health, including the addition of international data on the incidence of cardiovascular disease and obesity

All of these changes are made with an emphasis on ease of reading and ease of understanding that have made this book the leading text for introducing students to this exciting field. Retained from the fourth edition are the reduced thickness and weight of the book compared to earlier editions and the overall structure and progression of the text. As with the fourth edition, our first focus is on muscle and how its needs are altered as an individual goes from a resting to an active state and how these needs are supported by—and interact with—other body systems. In later chapters we address principles of exercise training; considerations of environmental factors of heat, cold, and altitude; sport performance; and exercise for disease prevention.

Organization of the Fifth Edition

We begin in the introduction with a historical overview of sport and exercise physiology as they have emerged from the parent disciplines of anatomy and physiology, and we explain basic concepts that are used throughout the text. In parts I and II, we review the major physiological systems, focusing on their responses to acute bouts of exercise. In part I, we focus on how the muscular, metabolic, nervous, and endocrine systems interact to produce body movement. In part II, we look at how the cardiovascular and respiratory systems continue to deliver nutrients and oxygen to the active muscles and transport waste products away during physical activity. In part III, we consider how these systems adapt to chronic exposure to exercise (i.e., training).

We change perspective in part IV to examine the impact of the external environment on physical performance. We consider the body's response to heat and cold, and then we examine the impact of low atmospheric pressure experienced at altitude. In part V, we shift attention to how athletes can optimize physical performance. We evaluate the effects of different types and volumes of training. We consider the importance of appropriate body composition for optimal performance and examine athletes' special dietary needs and how nutrition can be used to enhance performance. Finally, we explore the use of ergogenic aids: substances purported to improve athletic ability.

In part VI, we examine unique considerations for specific populations. We look first at the processes of growth and development and how they affect the performance capabilities of young athletes. We evaluate changes that occur in physical performance as people age and explore the ways in which physical activity can help maintain health and independence. Finally, we examine issues and special physiological concerns of female athletes.

In the final part of the book, part VII, we turn our attention to the application of sport and exercise physiology to prevent and treat various diseases and the use of exercise for rehabilitation. We look at prescribing exercise for maintaining health and fitness, and then we close the book with a discussion of cardiovascular disease, obesity, and diabetes.

Special Features in the Fifth Edition

This fifth edition of *Physiology of Sport and Exercise* is designed with the goal of making learning easy and enjoyable. This text is comprehensive, but the many

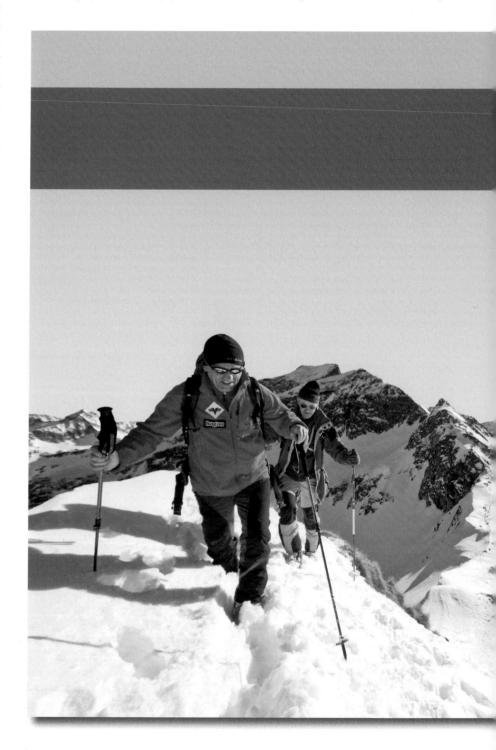

special features included will help you progress through the book without being overwhelmed by its scope. In addition to these features, the fully updated web study guide that accompanies this text provides opportunities for interactive learning and review.

Each chapter in the book begins with a chapter outline with page numbers to help you locate material.

Also noted in the chapter outline are the web study guide activities relating to each section of the chapter. Each chapter begins with a brief story that explores a real-world application of the concepts presented.

As you read through a chapter, you will find key points, labeled In Focus, that highlight important concepts and facts. At several points throughout each

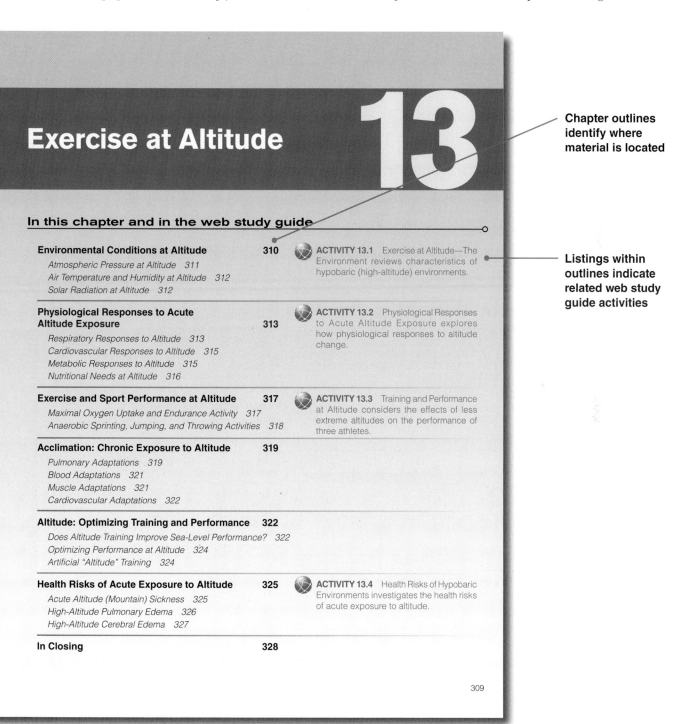

Exercise at Altitude

13

Chapter outlines identify where material is located

In this chapter and in the web study guide

ACTIVITY 13.1 Exercise at Altitude—The Environment reviews characteristics of hypobaric (high-altitude) environments.

ACTIVITY 13.2 Physiological Responses to Acute Altitude Exposure explores how physiological responses to altitude change.

ACTIVITY 13.3 Training and Performance at Altitude considers the effects of less extreme altitudes on the performance of three athletes.

ACTIVITY 13.4 Health Risks of Hypobaric Environments investigates the health risks of acute exposure to altitude.

Listings within outlines indicate related web study guide activities

309

chapter, longer In Review boxes take a moment to summarize the major points presented in the previous sections. And at the end of the chapter, the In Closing element wraps things up and notes how what you have learned sets the stage for the topics to come.

Key terms are highlighted in the text in red, are listed at the end of each chapter, and are defined in the glossary at the end of the book. At each chapter's end,

you will also find study questions to test your knowledge of the chapter's contents and a reminder of the study guide activities that are available, along with the web address of the online study guide.

At the end of the book is a comprehensive glossary that includes definitions of all key terms, a listing of numbered references for the sources cited in each chapter, and a thorough index. Finally, printed on the

Cardiovascular Disease and Physical Activity

In Review elements summarize major concepts

In review

➤ Risk factors for CHD that we cannot control are heredity (and family history), male sex, and advanced age. Those that we can control are abnormal blood lipids and lipoproteins, hypertension, smoking, physical inactivity, obesity, diabetes, and insulin resistance.

➤ Low-density lipoprotein cholesterol is thought to be responsible for depositing cholesterol in the arterial walls. Very low density lipoprotein cholesterol is also implicated in the development of atherosclerosis. However, HDL-C acts as a scavenger, removing cholesterol from the vessel walls. Thus, high HDL-C levels provide some degree of protection from CHD.

➤ The ratio of Total-C to HDL-C might be the best indicator of personal risk for CHD. Values below 3.0 reflect a low risk; values above 5.0 reflect a high risk.

➤ Risk factors for hypertension that can't be controlled include heredity, advanced age, and race. Those we can control are insulin resistance, obesity, diet (sodium and alcohol), use of tobacco products and oral contraceptives, stress, and physical inactivity.

In Focus elements present key points

In focus

Although the pathways are complex, it is becoming increasingly clear that hypertension, CHD, abnormal blood lipids, obesity, and diabetes might be linked through the common pathway of insulin resistance. It is also possible that obesity is the trigger that starts a cascade of events leading to the metabolic syndrome.

Reducing Risk Through Physical Activity

The role that physical activity might play in preventing or delaying the onset of CHD and hypertension has been of major interest to the medical community for many years. In the following sections, we try to unravel this mystery by examining the following areas:

- Epidemiological evidence
- Physiological adaptations with training that might reduce risk
- Risk factor reduction with exercise training

Reducing the Risk of Coronary Heart Disease

Physical activity has been proven effective in reducing the risk of CHD. In the following sections, we discover what is known about this topic and what physiological mechanisms are involved.

Epidemiological Evidence

Hundreds of research papers have dealt with the epidemiological relationship between physical inactivity and CHD. Generally, studies have shown the risk of heart attack in sedentary male populations to be about two to three times that of men who are physically active in either their jobs or their recreational pursuits. The early studies of Dr. J.N. Morris (see figure 21.7) and his colleagues in England in the 1950s were among the first to demonstrate this relationship.[37] In these studies, sedentary bus drivers were compared with active bus conductors who worked on double-decker buses, and sedentary postal workers were compared with active postal carriers who walked their routes. The death rate from CHD was about twice as high in the sedentary groups as in the active groups. Many studies published

Sidebars highlight current topics in exercise physiology

Metabolic Syndrome

Metabolic syndrome is a term that has been used to link CHD, hypertension, abnormal blood lipids, type 2 diabetes, and abdominal obesity to insulin resistance and hyperinsulinemia. This syndrome also has been referred to as syndrome X and the insulin resistance syndrome. It is not totally clear where the syndrome starts, but it has been observed that upper body obesity is associated with insulin resistance and that insulin resistance is highly correlated with increased risk for CHD, hypertension, and type 2 diabetes. It appears, however, that obesity or insulin resistance (or a combination of the two) is the trigger that initiates a cascade of events leading to the metabolic syndrome. Systemic inflammation has also been suggested as a causal factor. This became a major topic of research in the 1990s and continues to be today. The results of this research should help us better understand the pathophysiology of these diseases and their interrelationships.

Completely revamped photos, illustrations, and medical artwork offer a high level of detail and clarity

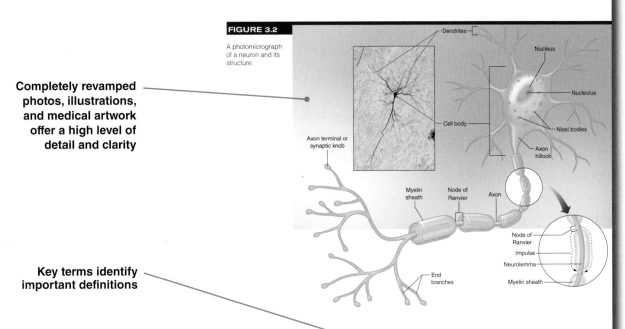

FIGURE 3.2

A photomicrograph of a neuron and its structure.

Dendrites

Nucleus

Nucleolus

Cell body

Nissl bodies

Axon terminal or synaptic knob

Axon hillock

Myelin sheath

Node of Ranvier

Axon

Node of Ranvier

Impulse

Neurolemma

Myelin sheath

End branches

Key terms identify important definitions

- The cell body, or soma
- The dendrites
- The axon

The cell body contains the nucleus. Radiating out from the cell body are the cell processes: the dendrites and the axon. On the side toward the axon, the cell body tapers into a cone-shaped region known as the **axon hillock**. The axon hillock has an important role in impulse conduction, as discussed later.

Most neurons contain only one axon but many dendrites. Dendrites are the neuron's receivers. Most impulses, or action potentials, that enter the neuron from sensory stimuli or from adjacent neurons typically enter the neuron via the dendrites. These processes then carry the impulses toward the cell body.

The axon is the neuron's transmitter and conducts impulses away from the cell body. Near its end, an axon splits into numerous **end branches**. The tips of these

branches are dilated into tiny bulbs known as **axon terminals** or synaptic knobs. These terminals or knobs house numerous vesicles (sacs) filled with chemicals known as **neurotransmitters** that are used for communication between a neuron and another cell. (This is discussed later in this chapter in more detail.) The structure of the neuron allows nerve impulses to enter the neuron through the dendrites, and to a lesser extent through the cell body, and to travel through the cell body and axon hillock, down the axon, and out through the end branches to the axon terminals. We next explain in more detail how this happens, including how these impulses travel from one neuron to another and from a somatic motor neuron to muscle fibers.

Nerve Impulse

Neurons are referred to as *excitable tissue* because they can respond to various types of stimuli and convert

71

(In)**closing**

In chapter 6, we discussed the role of the c: we looked at the role played by the respirat cardiovascular and respiratory systems resp

Key Terms

alveoli
arterial–mixed venous oxygen difference, or (
arterial–venous oxygen difference, or (a-v)O₂
Boyle's gas law
Dalton's law
expiration
external respiration
Fick's law
Henry's law
inspiration
internal respiration
myoglobin

spi........y
tidal volume
total lung capacity (TLC)
vital capacity (VC)

List of key terms provides check of vocabulary comprehension

Study Questions

1. Describe and differentiate between external and internal respiration.
2. Describe the mechanisms involved in inspiration and expiration.
3. What is a spirometer? Describe and define the lung volumes measured using spirometry.
4. Explain the concept of partial pressures of respiratory gases—oxygen, carbon dioxide, and nitrogen. What is the role of gas partial pressures in pulmonary diffusion?
5. Where in the lung does the exchange of gases occur with the blood? Describe the role of the respiratory membrane.
6. How are oxygen and carbon dioxide transported in the blood?
7. How is oxygen unloaded from the arterial blood to the muscle and carbon dioxide removed from the muscle into the venous blood?
8. What is meant by the arterial–mixed venous oxygen difference? How and why does this change from resting to exercise conditions?
9. Describe how pulmonary ventilation is regulated. What are the chemical stimuli that control the depth and rate of breathing? How do they control respiration during exercise?

Study questions test chapter knowledge

Study Guide Activities

In addition to the activities listed in the chapter opening outline on page 163, two other activities are available in the web study guide, located at

www.HumanKinetics.com/PhysiologyOfSportAndExercise

Additional reminder of web study guide activities promotes expanded learning opportunities

 The chapter **KEY TERMS** reviews important terms, and the end-of-chapter **QUIZ** tests your understanding of the material covered in the chapter.

179

inside front and back covers for easy reference are lists of common abbreviations and conversions.

Instructors using this text in their courses will find a wealth of updated ancillary materials available at www .HumanKinetics.com/PhysiologyOfSportAndExercise, including instructor guide, presentation package, and test package. New to this edition is an image bank, which contains art, content photos, and tables from the text as separate image files for reuse in custom presentations and class materials.

You might read this book only because it is a required text for a course. But we hope that the information will entice you to continue to study this relatively new and exciting area. We hope at the very least to further your interest in and understanding of your body's marvelous abilities to perform various types and intensities of exercise and sports, to adapt to stressful situations, and to improve its physiological capacities. This book is useful not only for anyone who pursues a career in exercise or sport science but also for anyone who wants to be active, healthy, and fit.

eBook
available at
your campus bookstore
or HumanKinetics.com

Student Resources

Students, visit the free **Web Study Guide** at www .HumanKinetics.com/PhysiologyOfSportAndExercise for dynamic and interactive learning activities, all of which can be conducted outside the lab or classroom. You'll be able to apply the key concepts learned by working with interactive versions of select artwork from the book, applying your knowledge to real-world scenarios, and more. The guide also includes key terms activities and quizzes to help test your knowledge of the material as you prepare for classroom quizzes or tests. You'll also have access to links to professional journals and information on organizations and careers in the field. The web study guide format conveniently allows you to practice, review, and develop knowledge and skills about the physiology of sport and exercise.

Instructor Resources

Instructor Guide

Specifically developed for instructors who have adopted *Physiology of Sport and Exercise, Fifth Edition,* the Instructor Guide includes sample lecture outlines, key points, and student assignments for every chapter in the text, along with sample laboratory exercises, and direct links to a wide range of detailed sources on the Internet.

Test Package

The Test Package includes a bank of over 1,600 questions created especially for *Physiology of Sport and Exercise,* *Fifth Edition.* Various question types are included, such as true-false, fill-in-the-blank, essay and short answer, and multiple choice. The Test Package is available for use through multiple formats, including a learning management system, Respondus, and rich text.

Presentation Package

The Presentation Package includes a comprehensive series of PowerPoint slides for each chapter. Learning objective slides present the major topics covered in each chapter; text slides list key points; and illustration and photo slides contain the outstanding graphics found in the text. The Presentation Package has more than 1,100 slides that can be used directly with PowerPoint and used to print transparencies or slides or make copies for distribution to students. Instructors can easily add to, modify, or rearrange the order of the slides, as well as search for slides based on keywords. You may access the Presentation Package by visiting www.HumanKinetics .com/PhysiologyOfSportAndExercise.

Image Bank

New for the fifth edition, the image bank includes all of the improved illustrations and artwork, as well as all of the tables from the text, sorted by chapter. These are provided as separate files for easy insertion into tests, quizzes, handouts, and other course materials and to provide instructors with greater flexibility when creating customized resources.

The instructor guide, test package, presentation package, and image bank are free to course adopters.

Acknowledgments

We would like to thank the staff at Human Kinetics for their continued support of the fifth edition of *Physiology of Sport and Exercise* and their dedication to publishing a high-quality product that meets the needs of instructors and students alike. Recognition goes to our capable developmental editors: Lori Garrett (first edition), Julie Rhoda (second and third editions), and Maggie Schwarzentraub (fourth edition). Amy Tocco (senior acquisitions editor) and Kate Maurer (developmental editor) took over the reins for the fifth edition and have worked tirelessly to bring this edition to completion, keeping all phases of the project on schedule with no concessions to quality. They have been a true pleasure to work with and their competence and skill are evident throughout the book. Special thanks as well go to Joanne Brummett for her artistic expertise and contributions to the enhanced art in the fifth edition.

For the fifth edition, special thanks go to a handful of Penn State colleagues who provided their valued expertise and time. In particular, direct feedback from Drs. Donna Korzick and Jim Pawelczyk at Penn State was invaluable in making substantive changes from an instructor's viewpoint. Dr. Korzick also graciously gave us access to several excellent cardiovascular figures. Special recognition goes to Dr. Lacy Alexander Holowatz at Penn State University for her hard work in helping us reorganize and revise selected chapters of the fifth edition. Her insights and skill in editing have made this a much better book. Dr. Mary Jane De Souza contributed her vast knowledge and expertise in the area of women in sport and exercise, and Dr. Pawelczyk did the same for the ergogenic aids chapter. Also, we would like to thank Dr. Caitlin Thompson-Torgerson and Dr. Bob Murray for their help with writing and updating the ancillaries for this edition.

Finally, we thank our families, who put up with our many long hours while we were writing, rewriting, editing, and proofing this book across all five editions. Their patience and support are sincerely appreciated.

W. Larry Kenney
Jack H. Wilmore
David L. Costill

Chapter and part opener photos

Introduction: © Icon Sports Media; **Part I:** © Human Kinetics; **Chapter 1:** © BSIP/Photoshot/Icon SMI; **Chapter 2:** © CNRI/Science Photo Library/Custom Medical Stock Photo; **Chapter 3:** Photo courtesy of Chuck Fong. studio2photo@yahoo.com.; **Chapter 4:** © Franck Faugere/DPPI/Icon SMI; **Chapter 5:** © Derick Hingle/Icon SMI; **Part II:** © Icon Sports Media; **Chapter 6:** © Collection CNRI/Phototake USA; **Chapter 7:** © BSIP/Age fotostock; **Chapter 8:** © Jonathan Larsen/Age fotostock; **Part III:** © Human Kinetics; **Chapter 9:** © Chai v.d. Laage/Imago/Icon SMI; **Chapter 10:** © Human Kinetics; **Chapter 11:** © Imago/Icon SMI; **Part IV:** © PA Photos; **Chapter 12:** © Gian Mattia D'Alberto/LaPresse/Icon SMI; **Chapter 13:** © Pritz/Age fotostock; **Part V:** © Michael Weber/Age fotostock; **Chapter 14:** © Human Kinetics; **Chapter 15:** © Chris Cheadle/All Canada Photos/Age fotostock; **Chapter 16:** © Ulrich Niehoff/Age Fotostock; **Part VI:** © George Shelley/Age fotostock; **Chapter 17:** © View Stock/Age fotostock; **Chapter 18:** © Rick Gomez/Age fotostock; **Chapter 19:** © Human Kinetics; **Part VII:** © Human Kinetics; **Chapter 20:** © Boccabella Debbie/Age fotostock; **Chapter 21:** © PA Photos; **Chapter 22:** © Frank Siteman/Age fotostock

Photos courtesy of the authors

Figures 0.1, 0.2, 0.3, 0.4*b-c*, 1.1*a-c*, 1.10, 1.11*a-c*, 5.9, 18.6, 20.1, 20.2, 21.7, and 22.8*a*; photos on p. 2, 12, 13, 21, 38, 41, 272, 491, 508, 620, and 621

Additional photos

Photo on p. x: © Pritz/Age fotostock; **photo on p. 2:** Photo courtesy of Dr. Larry Golding, University of Nevada, Las Vegas. Photographer Dr. Moh Youself; **photo on p. 5:** Photo courtesy of American College of Sports Medicine Archives. All rights reserved; **photo on p. 10:** Photo courtesy of American College of Sports Medicine Archives. All rights reserved; **photo on p. 10:** Photo courtesy of American College of Sports Medicine Archives. All rights reserved; **photo on p. 10:** Courtesy of Noll Laboratory, The Pennsylvania State University; **figure 0.4a:** Photo courtesy of American College of Sports Medicine Archives; **figure 0.5:** © Human Kinetics; **figure 0.6:** © Zuma Press/Icon SMI; **figure 1.2a:** © H.R. Bramaz/ISM/Phototake USA; **figure 1.4:** © Custom Medical Stock Photo; **photo in** figure 3.2: © Carolina Biological Supply Company/Phototake USA; **photo on p. 85:** © Human Kinetics; **figure 5.2a:** © Human Kinetics; **figure 5.2b:** © Panoramic/Imago/Icon SMI; **figure 6.7:** © Jochen Tack/age footstock; **photo on p. 157:** © MEIJER/DPPI-SIPA/ICON SMI; **photo in figure 6.16:** © B. Boissonnet/age footstock; **photo on p. 172:** © Phototake; **photo on p. 203:** © BSIP/Photoshot/Icon SMI; **figure 9.1:** © Human Kinetics; **photo on p. 217:** © Xinhua/IMAGO/Icon SMI; **figure 9.3:** © Human Kinetics; **figure 9.5:** © Human Kinetics; **figure 10.2:** Photos courtesy of Dr. Michael Deschene's laboratory; **photo on p. 236:** © Imago/Icon SMI; **figure 10.8:** Reprinted from *Physician and Sportsmedicine*, Vol. 12, R.C. Hagerman et al., "Muscle damage in marathon runners," pgs. 39-48, Copyright 1984, with permission from JTE Multimedia; **figure 10.9:** Reprinted from *Physician and Sportsmedicine*, Vol. 12, R.C. Hagerman et al., "Muscle damage in marathon runners," pgs. 39-48, Copyright 1984, with permission from JTE Multimedia; **photos on p. 251:** © Human Kinetics; **figure 12.2:** © Carolina Biological Supply Company/Phototake USA; **figure 12.3:** From Department of Health and Human Performance, Auburn University, Alabama. Courtesy of John Eric Smith, Joe Molloy, and David D. Pascoe. By permission of David Pascoe; **photo on p. 319:** © Norbert Eisele-Hein/Age fotostock; **photo on p. 348:** © Human Kinetics; **figure 15.2:** © Tom Pantages; **figure 15.3:** Photos courtesy of Hologic, Inc.; **figure 15.4:** © Zuma Press/Icon SMI; **figure 15.5:** © Human Kinetics; **figure 15.6:** © Human Kinetics; **photo on p. 363:** © Human Kinetics; **photo on p. 396:** © Franck Faugere/DPPI/Icon SMI; **photo on p. 402:** © Custom Medical Stock Photo; **photo on p. 436:** © Human Kinetics; **photo on p. 462:** © Human Kinetics; **figure 19.11a:** © H.R. Bramaz/ISM/Phototake USA; **figure 19.11b:** © ISM/Phototake USA; **figure 20.3:** © Human Kinetics; **photo on p. 516:** © Human Kinetics; **figure 21.3:** © 3D4Medical/Phototake; **photo on p. 540:** Photo courtesy of University of Arizona Sarver Heart Center; **photo on p. 546:** © AP Photo/NFL Photos; **figure 22.8b:** From J.C. Seidell et al., 1987, "Obesity and fat distribution in relation to health - Current insights and recommendations," *World Review of Nutrition and Dietetics* 50: 57-91; **figure 22.8c:** From J.C. Seidell et al., 1987, "Obesity and fat distribution in relation to health - Current insights and recommendations," *World Review of Nutrition and Dietetics* 50: 57-91; **photo on p. 560:** © Human Kinetics

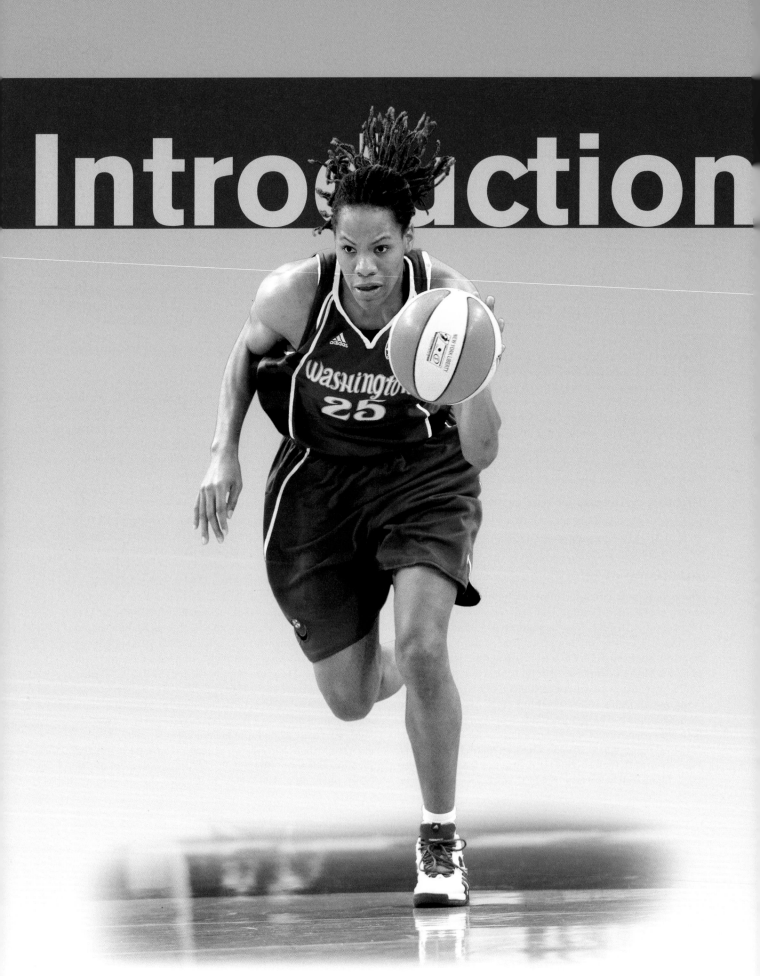

An Introduction to Exercise and Sport Physiology

In this chapter and in the web study guide

ACTIVITY 0.1 Timeline presents a historical perspective of the history of exercise physiology.

ACTIVITY 0.2 Interpreting Figures and Tables explains the components of charts, figures, and tables and how to interpret their data.

Much of the history of exercise physiology in the United States can be traced to the effort of a Kansas farm boy, David Bruce (D.B.) Dill, whose interest in physiology first led him to study the composition of crocodile blood. Fortunately for what would eventually grow into the discipline of exercise physiology, this young scientist redirected his research to humans when he became the first research director of the Harvard Fatigue Laboratory in 1927. Throughout his life he was intrigued by the physiology and adaptability of many animals that survive extreme exercise and environmental conditions; but he is best remembered for his research on *human* responses to exercise, heat, high altitude, and other environmental factors. Dr. Dill always served as one of the human "guinea pigs" in his own studies. During the Harvard Fatigue Laboratory's 20-year existence, he and his coworkers produced approximately 350 scientific papers along with a classic book titled *Life, Heat, and Altitude*.[8]

Dr. David Bruce (D.B.) Dill (*a*) at the beginning of his career; (*b*) as director of the Harvard Fatigue Laboratory at age 42; and (*c*) at age 92 just prior to his fourth retirement.

After the Harvard Fatigue Laboratory closed its doors in 1947, Dr. Dill began a second career as deputy director of medical research for the Army Chemical Corps, a position he held until his retirement from that post in 1961. Dr. Dill was then 70 years old—an age he considered too young for retirement—so he moved his research to Indiana University, where he served as a senior physiologist until 1966. In 1967 he obtained funding to establish the Desert Research Laboratory at the University of Nevada at Las Vegas. Dr. Dill used this laboratory as a base for his studies on human tolerance to exercise in the desert and at high altitude. He continued his research and writing until his final retirement at age 93, the same year he produced his last publication, a book titled *The Hot Life of Man and Beast*.[10]

The human body is an amazing machine. As you sit reading this introduction, countless perfectly coordinated and integrated events are occurring simultaneously in your body. These events allow complex functions, such as hearing, seeing, breathing, and information processing, to continue without any conscious effort. If you stand up, walk out the door, and jog around the block, almost all your body's systems will be called into action, enabling you to successfully shift from rest to exercise. If you continue this routine regularly for weeks or months and gradually increase the duration and intensity of your jogging, your body will adapt so that you can perform better. Therein lie the two basic components of the study of exercise physiology: the acute responses of the body to exercise in all its forms and the adaptation of those systems to repeated or chronic exercise, often called exercise training.

For example, as the point guard directs her team down the basketball court on a fast break, her body makes many adjustments that require a series of complex interactions involving many body systems. Adjustments occur even at the cellular and molecular levels. To enable the coordinated leg muscle actions as she moves rapidly down court, nerve cells from the brain, referred to as motor neurons, conduct electrical impulses down the spinal cord to the legs. On reaching the muscles, these neurons release chemical messengers that cross the gap between the nerve and muscle, each neuron exciting a number of individual muscle cells or fibers. Once the nerve impulses cross this gap, they spread along the length of each muscle fiber and attach to specialized receptors. Binding of the messenger to its receptor sets into motion a series of steps that activate the muscle fiber's contraction processes, which involve specific protein molecules—actin and myosin—and an elaborate energy system to provide the fuel necessary to sustain a single contraction and subsequent contractions. It is at this level that other molecules, such as adenosine triphosphate (ATP) and phosphocreatine (PCr), become critical for providing the energy necessary to fuel contraction.

In support of this sustained and rhythmic muscular contraction and relaxation, multiple additional systems are called into action, for example:

- The skeletal system provides the basic framework around which muscles act.
- The cardiovascular system delivers fuel to working muscle and to all of the cells of the body and removes waste products.

- The cardiovascular and respiratory systems work together to provide oxygen to the cells and remove carbon dioxide.
- The integumentary system (skin) helps maintain body temperature by allowing the exchange of heat between the body and its surroundings.
- The nervous and endocrine systems coordinate this activity, while helping to maintain fluid and electrolyte balance and assisting in the regulation of blood pressure.

For centuries, scientists have studied how the human body functions at rest during health and disease. During the past 100 years or so, a specialized group of physiologists have focused their studies on how the body functions during physical activity and sport. This introduction presents a historical overview of exercise and sport physiology and then explains some basic concepts that form the foundation for the chapters that follow.

Focus of Exercise and Sport Physiology

Exercise and sport physiology have evolved from the fundamental disciplines of anatomy and physiology. Anatomy is the study of an organism's structure, or morphology. While anatomy focuses on the basic structure of various body parts and their interrelationships, **physiology** is the study of body *function*. Physiologists study how the body's organ systems, tissues, cells, and the molecules within cells work and how their functions are integrated to regulate the body's internal environment, a process called **homeostasis**. Because physiology focuses on the functions of body structures, understanding anatomy is essential to learning physiology. Furthermore, both anatomy and physiology rely on a working knowledge of biology, chemistry, physics, and other basic sciences.

Exercise physiology is the study of how the body's functions are altered when we are exposed to exercise, a challenge to homeostasis. Because the environment in which one performs exercise has a large impact, **environmental physiology** has emerged as a subdiscipline of exercise physiology. **Sport physiology** further applies the concepts of exercise physiology to enhancing sport performance and optimally training the athlete. Thus, sport physiology derives its principles from exercise physiology. Because exercise physiology and sport physiology are so closely related and integrated, it is often hard to clearly distinguish between them. Because the same underlying scientific principles apply, exercise and sport physiology are often considered together, as they are in this text.

Acute and Chronic Responses to Exercise

The study of exercise and sport physiology involves learning the concepts associated with two distinct exercise patterns. First, exercise physiologists are often concerned with how the body responds to an individual bout of exercise, such as running on a treadmill for an hour or lifting weights. An individual bout of exercise is called **acute exercise**, and the responses to that exercise bout are referred to as acute responses. When examining the acute response to exercise, we are concerned with the body's immediate response to, and sometimes its recovery from, a single exercise bout.

The other major area of interest in exercise and sport physiology is how the body responds over time to the stress of repeated bouts of exercise, sometimes referred to as **chronic adaptation** or **training effects**. When one performs regular exercise over a period of days and weeks, the body adapts. The physiological adaptations that occur with chronic exposure to exercise or training improve both exercise capacity and efficiency. With resistance training, the muscles become stronger. With aerobic training, the heart and lungs become more efficient, and endurance capacity of the muscles increases. As discussed later in this introductory chapter, these adaptations are highly specific to the type of training the person does.

In focus

Exercise physiology evolved from its parent discipline, physiology. The two cornerstones of exercise physiology are (1) how the body responds to the acute stress of exercise, or physical activity, and (2) how it adapts to the chronic stress of repeated bouts of exercise, that is, exercise training. Some exercise physiologists use exercise or environmental conditions (heat, cold, altitude, etc.) to stress the body in ways that uncover basic physiological mechanisms. Others examine exercise training effects on health, disease, and well-being. Sport physiologists apply these concepts to athletes and sport performance.

The Evolution of Exercise Physiology

To students, contemporary exercise physiology may seem like a vast collection of new ideas never before studied with rigorous scientific scrutiny. On the contrary, however, the information in this book represents the lifelong efforts of hundreds of outstanding scientists who have collectively helped piece together what we currently know about the science of human movement. The theories and hypotheses of modern physiologists have been shaped by the efforts of scientists who may be long forgotten. What we consider original or new is most often an assimilation of previous findings or the application of basic science to problems in exercise physiology. As with every discipline, there are, of course, a number of key scientists and many pivotal scientific contributions that brought about significant advances in our knowledge of the physiological responses to exercise. The following section reflects on the history and on just a few of the people who shaped the field of exercise physiology. It is impossible in this short section to do justice to the hundreds of pioneering scientists who paved the way and laid the foundation for modern exercise physiology.

Beginnings of Anatomy and Physiology

One of the earliest descriptions of human anatomy and physiology was Claudius Galen's Greek text *De fascius*, published in the first century. As a physician to the gladiators, Galen had ample opportunity to study and experiment on human anatomy. His theories of anatomy and physiology were so widely accepted that they remained unchallenged for nearly 1,400 years. Not until the 1500s were any truly significant contributions made to the understanding of both the structure and function of the human body. A landmark text by Andreas Vesalius, titled *Fabrica Humani Corporis [Structure of the Human Body]*, presented his findings on human anatomy in 1543. Although Vesalius' book focused primarily on anatomical descriptions of various organs, he occasionally attempted to explain their functions as well. British historian Sir Michael Foster said, "This book is the beginning, not only of modern anatomy, but of modern physiology. It ended, for all time, the long reign of fourteen centuries of precedent and began in a true sense the renaissance of medicine" (p. 354).[13]

Most early attempts at explaining physiology were either incorrect or so vague that they could be considered no more than speculation. Attempts to explain how a muscle generates force, for example, were usually limited to a description of its change in size and shape during action because observations were limited to what could be seen with the naked eye. Yet from such observations, Hieronymus Fabricius (ca. 1574) suggested that a muscle's contractile power resided in its fibrous tendons, not in its "flesh." Anatomists did not discover the existence of individual muscle fibers until Dutch scientist Anton van Leeuwenhoek introduced the microscope (ca. 1660). How these fibers shortened and created force would remain a mystery until the middle of the 20th century, when the intricate workings of muscle proteins could be studied by electron microscopy.

Historical Aspects of Exercise Physiology

Exercise physiology is a relative newcomer to the world of science, although as early as 1793, a celebrated paper by Séguin and Lavoisier described the oxygen consumption of a young man as measured in the resting state and while he lifted a 7.3 kg (16 lb) weight numerous times for 15 min.[18] At rest the man used 24 L of oxygen per hour (L/h), which increased to 63 L/h during exercise. Lavoisier believed that the site of oxygen utilization and carbon dioxide production was in the lungs. Even though this concept was doubted by other physiologists of the time, it remained accepted doctrine until the middle of the 1800s, when several German physiologists demonstrated that combustion of oxygen occurred in tissues throughout the entire body.

Although many advances in the understanding of circulation and respiration occurred during the 1800s, few efforts were made to focus on the physiology of physical activity. However, in 1888, an apparatus was described that enabled scientists to study subjects during mountain climbing, even though the subjects had to carry a 7 kg (15.4 lb) "gasometer" on their backs.[21]

Arguably the first published textbook on exercise physiology, *Physiology of Bodily Exercise*, was written in French by Fernand LaGrange in 1889.[15] Considering the small amount of research on exercise that had been conducted up to that time, it is intriguing to read the author's accounts of such topics as "Muscular Work," "Fatigue," "Habituation to Work," and "The Office of the Brain in Exercise." This early attempt to explain the response to exercise was, in many ways, limited to speculation and theory. Although some basic concepts of exercise biochemistry were emerging at that time, LaGrange was quick to admit that many details were still in the formative stages. For example, he stated that "vital combustion [energy metabolism] has become very complicated of late; we may say that it is somewhat perplexed, and that it is difficult to give in a few words a clear and concise summary of it. It is a chapter of physiology which is being rewritten, and we cannot at this moment formulate our conclusions" (p. 395).[15]

Because the early text by LaGrange offered only limited physiological insights regarding bodily functions during physical activity, it might be argued that

A.V. Hill

October 16, 1923, was a significant milestone in the history of exercise physiology. A.V. Hill was inaugurated that day as Joddrell Professor of Physiology at University College, London. In his inaugural address he stated the principles that subsequently shaped the field of exercise physiology:

1921 Nobel Prize winner Archibald Hill (1927).

"It is strange how often a physiological truth discovered on an animal may be developed and amplified, and its bearings more truly found, by attempting to work it out on man. Man has proved, for example, far the best subject for experiments on respiration and on the carriage of gases by the blood, and an excellent subject for the study of kidney, muscular, cardiac and metabolic function. . . . Experiment on man is a special craft requiring a special understanding and skill, and 'human physiology,' as it may be called, deserves an equal place in the list of those main roads which are leading to the physiology of the future. The methods, of course, are those of biochemistry, of biophysics, of experimental physiology; but there is a special kind of art and knowledge required of those who wish to make experiments on themselves and their friends, the kind of skill that the athlete and the mountaineer must possess in realizing the limits to which it is wise and expedient to go.

Quite apart from direct physiological research on man, the study of instruments and methods applicable to man, their standardization, their description, their reduction to routine, together with the setting up of standards of normality in man are bound to prove of great advantage to medicine; and not only to medicine but to all those activities and arts where normal man is the object of study. Athletics, physical training, flying, working, submarines, or coalmines, all require a knowledge of the physiology of man, as does also the study of conditions in factories. The observation of sick men in hospitals is not the best training for the study of normal man at work. It is necessary to build up a sound body of trained scientific opinion versed in the study of normal man, for such trained opinion is likely to prove of the greatest service, not merely to medicine, but in our ordinary social and industrial life. Haldane's unsurpassed knowledge of the human physiology of respiration has often rendered immeasurable service to the nation in such activities as coal mining or diving; and what is true of the human physiology of respiration is likely also to be true of many other normal human functions."

During the late 1800s, many theories were proposed to explain the source of energy for muscle contraction. Muscles were known to generate much heat during exercise, so some theories suggested that this heat was used directly or indirectly to cause muscle fibers to shorten. After the turn of the century, Walter Fletcher and Sir Frederick Gowland Hopkins observed a close relation between muscle action and lactate formation.[11] This observation led to the realization that energy for muscle action is derived from the breakdown of muscle glycogen to lactic acid (see chapter 2), although the details of this reaction remained obscure.

Because of the high energy demands of exercising muscle, this tissue served as an ideal model to help unravel the mysteries of cellular metabolism. In 1921, Archibald V. (A.V.) Hill was awarded the Nobel Prize for his findings on energy metabolism. At that time, biochemistry was in its infancy, although it was rapidly gaining recognition through the research efforts of such other Nobel laureates as Albert Szent Gorgyi, Otto Meyerhof, August Krogh, and Hans Krebs, all of whom were actively studying how living cells generate and utilize energy.

Although much of Hill's research was conducted with isolated frog muscle, he also conducted some of the first physiological studies of runners. Such studies were made possible by the technical contributions of John S. Haldane, who developed the methods and equipment needed to measure oxygen use during exercise. These and other investigators provided the basic framework for our understanding of whole-body energy production, which became the focus of considerable research during the middle of the 20th century and is incorporated into the manual and computer-based systems that are used to measure oxygen uptake in exercise physiology laboratories throughout the world today.

the third edition of a text by F.A. Bainbridge titled *The Physiology of Muscular Exercise* should be considered the earliest scientific text on this subject.[3] Interestingly, that third edition was written by A.V. Bock and D.B. Dill, at the request of A.V. Hill, three key pioneers of exercise physiology discussed in this introductory chapter.

Era of Scientific Exchange and Interaction

From the early 1900s through the 1930s, the medical and scientific environment in the United States was changing. This was an era of revolution in the education of medical students, led by changes at Johns Hopkins. More medical and graduate programs based their educational endeavors on the European model of experimentation and development of scientific insights. There were important advances in physiology in areas such as bioenergetics, gas exchange, and blood chemistry that served as the basis for advances in the physiology of exercise. Building on collaborations forged in the late 1800s, interactions among laboratories and scientists were promoted, and international meetings of organizations such as the International Union of Physiological Sciences created an atmosphere for free scientific exchange, discussion, and debate.

Harvard Fatigue Laboratory

No laboratory has had more impact on the field of exercise physiology than the Harvard Fatigue Laboratory (HFL). A visit by A.V. Hill to Harvard University in 1926 had a significant impact on the founding and early activities of the HFL, which was established a year later in 1927. Interestingly, the early home of the HFL was the basement of Harvard's Business School, and its stated early mission was to conduct research on "fatigue" and other hazards in industry. Creation of this laboratory was due to the insightful planning of world-famous biochemist, Lawrence J. (L.J.) Henderson. A young biochemist from Stanford University, David Bruce D.B. Dill, was appointed as the first director of research, a title Dill held until the HFL closed in 1947.

As noted earlier, Dill had aided Arlen "Arlie" Bock in writing the third edition of Bainbridge's text on exercise physiology. Later in his career Dill credited the writing of that textbook with "shaping the program of the Fatigue Laboratory." Although he had little experience in applied human physiology, Dill's creative thinking and ability to surround himself with young, talented scientists created an environment that would lay the foundation for modern exercise and environmental physiology. For example, HFL personnel examined the physiology of endurance exercise and described the physical requirements for success in events such as distance running. Some of the most outstanding HFL investigations were conducted not in the laboratory but in the Nevada desert, on the Mississippi delta, and on White Mountain in California (with an altitude of 3,962 m, or 13,000 ft). These and other studies provided the foundation for future investigations on the effects of the environment on physical performance and in exercise and sport physiology.

In its early years, the HFL focused primarily on general problems of exercise, nutrition, and health. For example, the first studies on exercise and aging were conducted in 1939 by Sid Robinson (see figure 0.1), a student at the HFL. On the basis of his studies of subjects ranging in age from 6 to 91 years, Robinson described the effect of aging on maximal heart rate and oxygen uptake.[17] But with the onset of World War II, Henderson and Dill realized the HFL's potential contribution to the war effort, and research at the HFL took a different direction. Harvard Fatigue Lab scientists and support personnel were instrumental in forming new laboratories for the Army, Navy, and Army Air Corps (now the Air Force). They also published the methodologies necessary for relevant military research, methods that are still in use throughout the world.

Today's exercise physiology students would be amazed at the methods and devices used in the early days of the HFL and at the time and energy committed to conducting research projects in those days. What is now accomplished in milliseconds with the aid of computers and automatic analyzers literally demanded days of effort by HFL scientists. Measurements of oxygen uptake during exercise, for example, required collecting expired air in Douglas bags and analyzing it for oxygen and carbon dioxide by using a manually operated chemical analyzer, without the help of a computer, of course (see figure 0.2). The analysis of a single 1 min sample of expired air required 20 to 30 min of effort by one or more laboratory workers. Today, scientists make such measurements almost instantaneously and with little physical effort. One must marvel at the dedication, diligence, and hard work of the HFL's exercise physiology pioneers. Using the equipment and methods available at the time, HFL scientists published approximately 350 research papers over a 20-year period.

The HFL was an intellectual environment that attracted young physiologists and physiology doctoral students from all over the globe. Scholars from 15 countries worked in the HFL between 1927 and its closure in 1947. Most went on to develop their own laboratories and become noteworthy figures in

FIGURE 0.1   (*a*) Sid Robinson being tested by R.E. Johnson on the treadmill in the Harvard Fatigue Laboratory and (*b*) as a Harvard student and athlete in 1938.

FIGURE 0.2   (*a*) Early measurements of metabolic responses to exercise required the collection of expired air in a sealed bag known as a Douglas bag. (*b*) A sample of that gas then was measured for oxygen and carbon dioxide using a chemical gas analyzer, as illustrated by this photo of Nobel laureate August Krogh.

exercise physiology in the United States, including Sid Robinson, Henry Longstreet Taylor, Lawrence Morehouse, Robert E. Johnson, Ancel Keys, Steven Horvath, C. Frank Consolazio, and William H. Forbes. Notable international scientists who spent time at the HFL included August Krogh, Lucien Brouha, Edward Adolph, Walter B. Cannon, Peter Scholander, and Rudolfo Margaria, along with several other notable Scandinavian scientists discussed later. Thus, the HFL planted seeds of intellect at home and around the world that resulted in an explosion of knowledge and interest in this new field.

In focus

Founded by biochemist L.J. Henderson in 1927 and directed by D.B. Dill until its closure in 1947, the Harvard Fatigue Lab trained most of the scientists who became world leaders in exercise physiology during the 1950s and 1960s. Most contemporary exercise physiologists can trace their roots back to the HFL.

Scandinavian Influence

In 1909, Johannes Lindberg established a laboratory that became a fertile breeding ground for scientific contributions at the University of Copenhagen in Denmark. Lindberg and 1920 Nobel Prize winner August Krogh teamed up to conduct many classic experiments and published seminal papers on topics ranging from the metabolic fuels for muscle to gas exchange in the lungs. This work was continued from the 1930s into the 1970s by three young Danes, Erik Hohwü-Christensen, Erling Asmussen, and Marius Nielsen.

As a result of contacts between D.B. Dill and August Krogh, these three Danish physiologists came to the HFL in the 1930s, where they studied exercise in hot environments and at high altitude. After returning to Europe, each man established a separate line of research. Asmussen and Nielsen became professors at the University of Copenhagen, where Asmussen studied the mechanical properties of muscle and Nielsen conducted studies on control of body temperature. Both remained active at the University of Copenhagen's August Krogh Institute until their retirements.

In 1941, Hohwü-Christensen (see figure 0.3a) moved to Stockholm to become the first physiology professor at the College of Physical Education at Gymnastik-och Idrottshögskolan (GIH). In the late 1930s, he teamed with Ole Hansen to conduct and publish a series of five studies of carbohydrate and fat metabolism during exercise. These studies are still cited frequently and are considered to be among the first and most important sport nutrition studies. Hohwü-Christensen introduced Per-Olof Åstrand to the field of exercise physiology. Åstrand, who conducted numerous studies related to physical fitness and endurance capacity during the 1950s and 1960s, became the director of GIH after Hohwü-Christensen retired in 1960. During his tenure at GIH, Hohwü-Christensen mentored a number of outstanding scientists, including Bengt Saltin, who was the 2002 Olympic Prize winner for his many contributions to the field of exercise and clinical physiology (see figure 0.3b).

In addition to their work at GIH, both Hohwü-Christensen and Åstrand interacted with physiologists at the Karolinska Institute in Stockholm, Sweden, who studied clinical applications of exercise. It is hard to single out the most exceptional contributions from this institute, but Jonas Bergstrom's (figure 0.3c) reintroduction of the biopsy needle (ca. 1966) to sample muscle tissue was a pivotal point in the study of human muscle biochemistry and muscle nutrition. This technique, which involves withdrawing a tiny sample of muscle tissue with a needle inserted into the muscle through a small incision, was originally introduced in the early 1900s to study muscular dystrophy. The needle biopsy enabled physiologists to conduct histological and biochemical studies of human muscle before, during, and after exercise.

Other invasive studies of blood circulation were subsequently conducted by physiologists at GIH and at the Karolinska Institute. Just as the HFL had been the mecca of exercise physiology research between 1927 and 1947, the Scandinavian laboratories were equally noteworthy beginning in the late 1940s. Many leading investigations over the past 35 years were collaborations between American and Scandinavian exercise physiologists. Norwegian Per Scholander introduced a gas analyzer in 1947. Finn Martii Karvonen published a formula for calculating exercise heart rate that is still widely used today. (For a more detailed listing of the Scandinavian contributions to exercise physiology, consult Åstrand's review.[1])

Development of Contemporary Approaches

Much advancement in exercise physiology must be credited to improvements in technology. In the late 1950s, Henry L. Taylor and Elsworth R. Buskirk published two seminal papers[6, 19] describing the criteria for measuring maximal oxygen uptake and establishing that measure as the "gold standard" for cardiorespiratory fitness. In the 1960s, development of electronic analyzers to measure respiratory gases made studying energy metabolism much easier and more productive than before. This technology and radiotelemetry (which uses radio-transmitted signals), used to monitor heart rate and body temperature during exercise, were developed as a result of the U.S. space program. Although such instruments took much of the labor out of research, they did not alter the direction of scientific inquiry. Until the late 1960s, most exercise physiology studies focused on the whole body's response to exercise. The majority of investigations involved measurements of such variables as

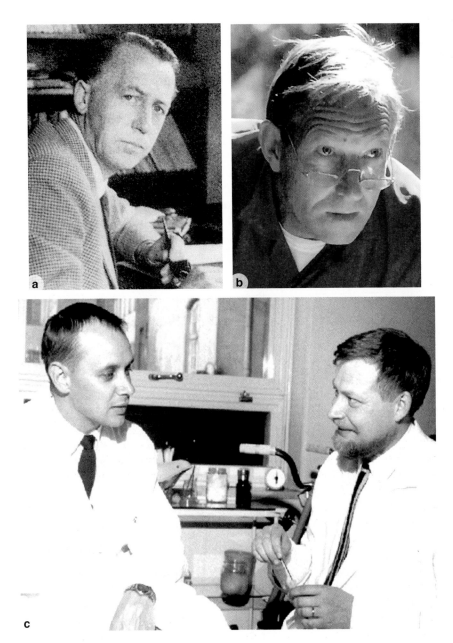

FIGURE 0.3 (a) Erik Hohwü-Christensen was the first physiology professor at the College of Physical Education at Gymnastik-och Idrottshögskolan in Stockholm, Sweden. (b) Bengt Saltin, winner of the 2002 Olympic Prize. (c) Jonas Bergstrom (left) and Eric Hultman (right) were the first to use muscle biopsy to study muscle glycogen use and restoration before, during, and after exercise.

oxygen uptake, heart rate, body temperature, and sweat rate. Cellular responses to exercise received little attention.

In the mid-1960s, three biochemists emerged who were to have a major impact on the field of exercise physiology. John Holloszy (figure 0.4a) at Washington University (St. Louis), Charles "Tip" Tipton (figure 0.4b) at the University of Iowa, and Phil Gollnick (figure 0.4c) at Washington State University first used rats and mice to study muscle metabolism and to examine factors related to fatigue. Their publications and their training of graduate and postdoctoral students have resulted in a more biochemical approach to exercise physiology research. Holloszy was ultimately awarded the 2000 Olympic Prize for his contributions to exercise physiology and health.

Exercise Physiology Milestones

Physiology has always been the basis for clinical medicine. In the same manner, exercise physiology has provided essential knowledge for many other areas, such as physical education, physical fitness, physical therapy, and health promotion. In the late 1800s and early 1900s, physicians such as Amherst College's Edward Hitchcock Jr. and Harvard's Dudley Sargent studied body proportions (anthropometry) and the effects of physical training on strength and endurance. Although a number of physical educators introduced science to the undergraduate physical education curriculum, Peter Karpovich, a Russian immigrant who had been briefly associated with the HFL, played a major role in introducing physiology to physical education. Karpovich established his own research facility and taught physiology at Springfield College (Massachusetts) from 1927 until his death in 1968.

Although he made numerous contributions to physical education and exercise physiology research, he is best remembered for the outstanding students he advised, including Charles Tipton and Loring Rowell, both recipients of the American College of Sports Medicine Honor and Citation Awards.

Another Springfield faculty member, swim coach T.K. Cureton, created an exercise physiology laboratory at the University of Illinois in 1941. He continued his research and taught many of today's leaders in physical fitness and exercise physiology until his retirement in 1971. Physical fitness programs developed by Cureton and his students, as well as Kenneth Cooper's 1968 book, *Aerobics*, established a physiological rationale for using exercise to promote a healthy lifestyle.[7]

(*a*) **Peter Karpovich introduced the field of exercise physiology during his tenure at Springfield College. (*b*) Thomas K. Cureton directed the exercise physiology laboratory at the University of Illinois at Urbana-Champaign from 1941 to 1971. (*c*) At Penn State, Elsworth Buskirk founded an intercollege graduate program focusing on applied physiology (1966) and constructed The Laboratory for Human Performance Research (1974).**

Another contributor to the establishment of exercise physiology as an academic endeavor was Elsworth R. "Buz" Buskirk. After holding positions as chief of the Environmental Physiology Section at the Quartermaster Research and Development Center in Natick, Massachusetts (1954-1957), and research physiologist at the National Institutes of Health (1957-1963), Buskirk moved to Pennsylvania State University, where he stayed for the remainder of his career. At Penn State, Buz founded the Intercollege Graduate Program in Physiology (1966) and constructed The Laboratory for Human Performance Research (1974), the nation's first freestanding research institute devoted to the study of human adaptation to exercise and environmental stress. He remained an active scholar until his death in April of 2010.

Although there was some awareness as early as the mid-1800s of a need for regular physical activity to maintain optimal health, this idea did not gain popular acceptance until the late 1960s. Subsequent research has continued to support the importance of exercise in slowing the physical decline associated with aging, preventing or mitigating the problems associated with chronic diseases, and rehabilitating injuries.

FIGURE 0.4 (*a*) John Holloszy was the winner of the 2000 Olympic Prize for scientific contributions in the field of exercise science. (*b*) Charles Tipton was a professor at the University of Iowa and the University of Arizona, and a mentor to many students who have become the leaders in molecular biology and genomics. (*c*) Phil Gollnick conducted muscle and biochemical research at Washington State University.

At about the time that Bergstrom reintroduced the needle biopsy procedure, exercise physiologists who were well trained as biochemists emerged. In Stockholm, Bengt Saltin realized the value of this procedure for studying human muscle structure and biochemistry. He first collaborated with Bergstrom in the late 1960s to study the effects of diet on muscle endurance and muscle nutrition. About the same time, Reggie Edgerton (University of California at Los Angeles) and Phil Gollnick were using rats to study the characteristics of individual muscle fibers and their responses to training. Saltin subsequently combined his knowledge of the biopsy procedure with Gollnick's biochemical talents. These researchers were responsible for many early studies on human muscle fiber characteristics and use during exercise. Although many biochemists have used exercise to study metabolism, few have had more impact on the current direction of human exercise physiology than Bergstrom, Saltin, Tipton, Holloszy, and Gollnick.

For more than 100 years, athletes have served as subjects for study of the upper limits of human endurance. Perhaps the first physiological studies on athletes occurred in 1871. Austin Flint studied one of the most celebrated athletes of that era, Edward Payson Weston, an endurance runner/walker. Flint's investigation involved measuring Weston's energy balance (i.e., food intake vs. energy expenditure) during Weston's attempt to walk 400 mi (644 km) in five days. Although the study resolved few questions about muscle metabolism during exercise, it did demonstrate that some body protein is lost during prolonged heavy exercise.[12]

Throughout the 20th century, athletes were used repeatedly to assess the physiological capabilities of human strength and endurance and to ascertain characteristics needed for record-setting performances. Some attempts have been made to use the technology and knowledge derived from exercise physiology to predict performance, to prescribe training, or to identify athletes with exceptional potential. In most cases, however, these applications of physiological testing are of little more than academic interest because few laboratory or field tests can accurately assess all the qualities required to become a champion.

The intent of this section has been to provide readers with an overview of the personalities and technologies that have helped to shape the field of exercise physiology. Naturally, a comprehensive review of all the scientists and research associated with this field is not possible in a text intended as an introduction to exercise physiology; but for those students who wish to take an in-depth look at the historical background in exercise physiology, there are several good sources.

Evolution of Exercise Physiology Tools and Techniques

The history of exercise physiology has, in some ways, been driven by advancements in technologies adapted from basic sciences. The early studies of energy metabolism during exercise were made possible by the invention of gas-collecting equipment and chemical analysis of oxygen and carbon dioxide. Chemical determination of blood lactic acid seemed to provide some insights regarding the aerobic and anaerobic aspects of muscular activity, but these data told us little regarding the production and removal of this by-product of exercise. Likewise, blood glucose measurements taken before, during, and after exhaustive exercise proved to be interesting data but were of limited value for understanding the energy exchange at the cellular level.

(a) Frank Booth and (b) Ken Baldwin.

Prior to the 1960s, there were few biochemical studies on the adaptations of muscle to training. Although the field of biochemistry can be traced to the early part of the 20th century, this special area of chemistry was not applied to human muscle until Bergstrom and Hultman reintroduced and popularized the needle biopsy procedure in 1966. Initially, this procedure was used to examine glycogen depletion during exhaustive exercise and its resynthesis during recovery. In the early 1970s, as noted earlier, a number of exercise physiologists used the muscle biopsy method, histological staining, and the light microscope to determine human muscle fiber types.

Over the last 30 years, muscle physiologists have used various chemical procedures to understand how muscles generate energy and adapt to training. Test tube experiments (in vitro) with muscle biopsy samples have been used to measure muscle proteins (enzymes) and to determine the muscle fiber's capacity to use oxygen. Although these studies provided a snapshot of the fiber's potential to generate energy, they often left more questions than answers. It was natural, therefore, for the sciences of cell biology to move to an even deeper level. It was apparent that the answers to those questions must lie within the fiber's molecular makeup.

Although not a new science, molecular biology has become a useful tool for exercise physiologists who wish to delve deeper into the cellular regulation of metabolism and adaptations to the stress of exercise. Physiologists like Frank Booth and Ken Baldwin have dedicated their careers to understanding the molecular regulation of muscle fiber characteristics and function and have laid the groundwork for our current understanding of the genetic controls of muscle growth and atrophy. The use of molecular biological techniques to study the contractile characteristics of single muscle fibers is discussed in chapter 1.

Well before James Watson and Francis Crick unraveled the structure of DNA (1953), scientists appreciated the importance of genetics in predetermining the structure and function of all living organisms. The newest frontier in exercise physiology combines the study of molecular biology and genetics. Since the early 1990s, scientists have attempted to explain how exercise causes signals that affect the expression of genes within skeletal muscle.

In retrospect, it is apparent that since the beginning of the 20th century, the field of exercise physiology has evolved from measuring whole-body function (i.e., oxygen consumption, respiration, and heart rate) to molecular studies of muscle fiber genetic expression. There is little doubt that exercise physiologists of the future will need to be well grounded in biochemistry, molecular biology, and genetics.

In 1968, D.B. Dill wrote a chapter, "History of the Physiology of Exercise," that detailed many of the events and scientists who contributed to this field before the founding of the HFL.[9] In that same year, Roscoe Brown Jr., the first African American exercise physiologist, coauthored *Classical Studies on Physical Activity*.[4] Although the authors subjectively selected those scientific studies they considered worthy of publi-

Women in Exercise Physiology

As in many areas of science, the contributions of female exercise physiologists have been slow to gain recognition. In 1954, Irma Rhyming collaborated with her future husband, P.-O. Åstrand, to publish a classic study that provided a means to predict aerobic capacity from submaximal heart rate.[2] Although this indirect method of assessing physical fitness has been challenged over the years, its basic concept is still in use today.

In the 1970s, two Swedish women, Birgitta Essen and Karen Piehl, gained international attention for their research on human muscle fiber composition and function. Essen, who collaborated with Bengt Saltin, was instrumental in adapting microbiochemical methods to study the small amounts of tissue obtained with the needle biopsy procedure. Her efforts enabled others to conduct studies on the muscle's use of carbohydrates and fats and to identify different muscle fiber types. Piehl published a number of studies that illustrated which muscle fiber types were activated during both aerobic and anaerobic exercise.

In the 1970s and 1980s, a third Scandinavian female physiologist, Bodil Nielsen, daughter of Marius Nielsen, actively conducted studies on human responses to environmental heat stress and dehydration. Her studies even encompassed measurements of body temperature during immersion in water. Interestingly, at about the same time an

(*a*) Birgitta Essen collaborated with Bengt Saltin and Phil Gollnick in publishing the earliest studies on muscle fiber types in human muscle. (*b*) Karen Piehl was among the first physiologists to demonstrate that the nervous system selectively recruits type I (slow-twitch) and type II (fast-twitch) fibers during exercise of differing intensities. (c) Barbara Drinkwater was among the first to conduct studies on female athletes and to address issues specifically related to the female athlete.

American exercise physiologist, Barbara Drinkwater, was doing similar work at the University of California at Santa Barbara. Her studies were often conducted in collaboration with Steven Horvath, D.B. Dill's son-in-law and director of UCSB's Environmental Physiology Laboratory. Drinkwater's contributions to environmental physiology and the physiological problems confronting the female athlete gained international recognition. In addition to their scientific contributions, the legacy of these and other women in physiology includes the credibility they earned and the roles they played in attracting other young women to the field of exercise physiology and medicine.

cation, the edited book provides an excellent sampling of important exercise physiological research from the early 1900s.

In the early 1970s, Dr. Steven and Betty Horvath (D.B. Dill's son-in-law and daughter) published a detailed history of the HFL, including the laboratory and field studies conducted by the key scientists of that era.[14] Although others have written different versions of exercise physiology history,[5, 20] most tend to provide the authors' views of important scientists and events, perhaps as we have done here. Finally, McArdle, Katch, and Katch[16] published one of the most comprehensive reviews of the evolution of exercise physiology. Their description of the early anatomists, physiologists, and exercise physiologists clearly illustrates the complexity and diversity of this field of science.

Now that we understand the historical basis for the discipline of exercise physiology, from which sport physiology emerged, we can explore some basic principles of, and tools utilized in, exercise and sport physiology.

Research: The Foundation for Understanding

Exercise and sport scientists actively engage in research to better understand the mechanisms that regulate the body's physiological responses to acute bouts of exercise as well as its adaptations to training and detraining. Most of this research is conducted at major research universities, medical centers, and specialized institutes using standardized research approaches and select tools of the exercise physiologist.

Research Settings

Research can be conducted either in the laboratory or in the field. Laboratory tests are usually more accurate because more specialized and sophisticated equipment can be used and conditions can be carefully controlled. As an example, the direct laboratory measurement of maximal oxygen uptake ($\dot{V}O_{2max}$) is considered the most accurate estimate of cardiorespiratory endurance capacity. However, some field tests, such as the 1.5 mi (2.4 km) run, are also used to estimate $\dot{V}O_{2max}$. These field tests, which measure the time it takes to run a set distance or the distance that can be covered in a fixed time, are not totally accurate; but they provide a reasonable estimate of $\dot{V}O_{2max}$, are inexpensive to conduct, and allow many people to be tested in a short time. Field tests can be conducted in the workplace, on a running track or in a swimming pool, or during athletic competitions. To measure $\dot{V}O_{2max}$ directly and accurately, one would need to go to a university or clinical laboratory.

Research Tools: Ergometers

When physiological responses to exercise are assessed in a laboratory setting, the participant's physical effort must be controlled to provide a measurable exercise intensity. This is generally accomplished through use of ergometers. An **ergometer** (*ergo* = work; *meter* = measure) is an exercise device that allows the intensity of exercise to be controlled (standardized) and measured.

Treadmills

Treadmills are the ergometers of choice for most researchers and clinicians, particularly in the United States. With these devices, a motor drives a large belt on which a subject can either walk or run; thus these ergometers are often called motor-driven treadmills (see figure 0.5). Belt length and width must accom-

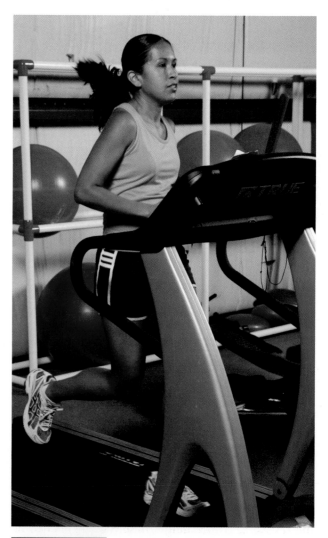

FIGURE 0.5 A motor-driven treadmill.

In focus

Treadmills generally produce higher peak values than other ergometers for almost all assessed physiological variables, such as heart rate, ventilation, and oxygen uptake.

modate the individual's body size and stride length. For example, it is nearly impossible to test elite athletes on treadmills that are too short, or obese subjects on treadmills that are too narrow or not sturdy enough.

Treadmills offer a number of advantages. Walking is a natural activity for almost everyone, so individuals normally adjust to the skill required for walking on a treadmill within a few minutes. Also, most people can achieve their peak physiological values on the treadmill, although some athletes (e.g., competitive cyclists) achieve higher values on ergometers that more closely match their mode of training or competition.

Treadmills do have some disadvantages. They are generally more expensive than simpler ergometers, like the cycle ergometers discussed next. They are also bulky, require electrical power, and are not very portable. Accurate measurement of blood pressure during treadmill exercise can be difficult because both the noise associated with normal treadmill operation and subject movement can make hearing through a stethoscope difficult.

Cycle Ergometers

For many years, the **cycle ergometer** was the primary testing device in use, and it is still used extensively in both research and clinical settings. Cycle ergometers can be designed to allow subjects to pedal either in the normal upright position (see figure 0.6) or in reclining or semireclining positions.

Cycle ergometers in a research setting generally use either mechanical friction or electrical resistance. With mechanical friction devices, a belt encompassing a flywheel is tightened or loosened to adjust the resistance against which the cyclist pedals. The power output depends on the combination of the resistance and the pedaling rate—the faster one pedals, the greater the power output. To maintain the same power output throughout the test, one must maintain the same pedaling rate, so pedaling rate must be constantly monitored.

With electrically braked cycle ergometers, the resistance to pedaling is provided by an electrical conductor that moves through a magnetic or electromagnetic field. The strength of the magnetic field determines the resistance to pedaling. These ergometers can be controlled so that the resistance increases automatically

as pedal rate decreases, and decreases as pedal rate increases, to provide a constant power output.

Similar to treadmills, cycle ergometers offer some advantages and disadvantages compared to other ergometers. Exercise intensity on a cycle ergometer does not depend on the subject's body weight. This is important when one is investigating physiological responses to a standard rate of work (power output). As an example, if someone lost 5 kg (11 lb), data derived from treadmill testing could not be compared with data obtained before the weight loss because physiological responses to a set speed and grade on the treadmill

FIGURE 0.6 A cycle ergometer.

vary with body weight. After the weight loss, the rate of work at the same speed and grade would be less than before. With the cycle ergometer, weight loss does not have as great an effect on physiological response to a standardized power output. Thus, walking or running is often referred to as weight-dependent exercise, while cycling is weight independent.

(In)focus

Cycle ergometers are the most appropriate devices for evaluating changes in submaximal physiological function before and after training in people whose weights have changed. Unlike the situation with treadmill exercise, cycle ergometer intensity is largely independent of body weight.

Cycle ergometers also have disadvantages. If the subject does not regularly engage in that form of exercise, the leg muscles will likely fatigue early in the exercise bout. This may prevent a subject from attaining a true maximal intensity. When exercise is limited in this way, responses are often referred to as peak exercise intensity rather than maximal exercise intensity. This limitation may be attributable to local leg fatigue, blood pooling in the legs (less blood returns to the heart), or the use of a smaller muscle mass during cycling than during treadmill exercise. Trained cyclists, however, tend to achieve their highest peak values on the cycle ergometer.

Other Ergometers

Other ergometers allow athletes who compete in specific sports or events to be tested in a manner that more closely approximates their training and competition. For example, an arm ergometer may be used to test athletes or nonathletes who use primarily their arms and shoulders in physical activity. Arm ergometry has also been used extensively to test and train athletes paralyzed below arm level. The rowing ergometer was devised to test competitive rowers.

Valuable research data have been obtained by instrumenting swimmers and monitoring them during swimming in a pool. However, the problems associated with turns and constant movement led to the use of two devices—tethered swimming and swimming flumes. In tethered swimming, the swimmer is attached to a harness connected to a rope, a series of pulleys, and counterbalancing weights and must swim against the pull of the apparatus to maintain a constant position in the pool. A swimming flume allows swimmers to more closely simulate their natural swimming strokes. The swimming flume operates by pumps that circulate

water past the swimmer, who attempts to maintain body position in the flume. The pump circulation can be increased or decreased to vary the speed at which the swimmer must swim. The swimming flume, which unfortunately is very expensive, has at least partially resolved the problems with tethered swimming and has created new opportunities to investigate the sport of swimming.

When one is choosing an ergometer, the concept of specificity is particularly important with highly trained athletes. The more specific the ergometer is to the actual pattern of movement used by the athlete in his or her sport, the more meaningful will be the test results.

Research Designs

In exercise physiology research, there are two basic types of research design: cross-sectional and longitudinal. With a **cross-sectional research design**, a cross section of the population of interest (that is, a representative sample) is tested at one specific time, and the differences between subgroups from that sample are compared. With a **longitudinal research design**, the same research subjects are retested periodically after initial testing to measure changes over time in variables of interest.

The differences between these two approaches are best understood through an example. The objective of a research study is to determine whether a regular program of distance running increases the concentration of cardioprotective high-density lipoprotein cholesterol (HDL-C) in the blood. High-density lipoprotein cholesterol is the desirable form of cholesterol; increased concentrations are associated with reduced risk for heart disease. Using the cross-sectional approach, one could, for example, test a large number of people who fall into the following categories:

- A group of subjects who do no training (the control group)
- A group of subjects who run 24 km (15 mi) per week
- A group of subjects who run 48 km (30 mi) per week
- A group of subjects who run 72 km (45 mi) per week
- A group of subjects who run 96 km (60 mi) per week

One would then compare the results from all the groups, basing one's conclusions on how much running was done. Using this approach, exercise scientists found that weekly running results in elevated HDL-C levels, suggesting a positive health benefit related to

FIGURE 0.7 The relation between distance run per week and average high-density lipoprotein cholesterol (HDL-C) concentrations across five groups: nontraining control (0 km/week), 24 km/week, 48 km/week, 72 km/week, and 96 km/week. This illustrates a cross-sectional study design.

FIGURE 0.8 The relation between months of distance-running training and average high-density lipoprotein cholesterol (HDL-C) concentrations in an experimental group (20 subjects, distance training) and a sedentary (20 subjects) control group. This illustrates a longitudinal study design.

running distance. Furthermore, as illustrated in figure 0.7, there was a **dose–response relation** between these variables—the higher the "dose" of exercise training, the higher the resulting concentration of HDL-C. It is important to remember, however, that with a cross-sectional design, these are different groups of runners, not the same runners at different training volumes.

Using the longitudinal approach to test the same question, one could design a study in which untrained people would be recruited to participate in a 12-month distance-running program. One could, for example, recruit 40 people willing to begin running and then randomly assign 20 to a training group and the remaining 20 to a control group. Both groups would be followed for 12 months. Blood samples would be tested at the beginning of the study and then at three-month intervals, concluding at 12 months when the program ended. With this design, both the running group and the control group would be followed over the entire period of the study, and changes in their HDL-C levels could be determined across each period. Actual studies have been conducted using this longitudinal design to examine changes in HDL-C with training, but their results have not been as clear as the results of the cross-sectional studies. See figure 0.8 as an example. Note that in this figure, in contrast to figure 0.7, there is only a small increase in HDL-C in the subjects who are training. The control group stays relatively stable, with only minor fluctuations in their HDL-C from one three-month period to the next.

In focus

Longitudinal research studies are the most accurate for studying changes in physiological variables over time. However, it is not always feasible to use a longitudinal design, and valuable information can be derived from cross-sectional studies.

A longitudinal research design is usually best suited to studying changes in variables over time. Too many factors that may taint results can influence cross-sectional designs. For example, genetic factors might interact so that those who run long distances are also those who have high HDL-C levels. Also, different populations might follow different diets; but in a longitudinal study, diet and other variables can be more easily controlled. However, longitudinal studies are time-consuming, are expensive to conduct, and are not always possible; and cross-sectional studies provide some insight into these questions.

Research Controls

When we conduct research, it is important to be as careful as possible in designing the study and collecting the data. We saw from figure 0.8 that changes in a variable over time resulting from an intervention such as exercise can be very small. Yet, even small changes in a variable such as HDL-C can mean a substantial reduction in risk for heart disease. Recognizing this, scientists design studies aimed at providing results that are both accurate and reproducible. This requires that studies be carefully controlled.

Research controls are applied at various levels. Starting with the design of the research project, the scientist must determine how to control for variation in the subjects used in the study. The scientist must determine if it is important to control for the subjects' sex, age, or body size. To use age as an example, for certain variables, the response to an exercise training program might be different for a child or an aged person compared with a young or middle-aged adult. Is it important to control for the subject's smoking or dietary status? Considerable thought and discussion are needed to make sure that the subjects used in a study are appropriate for the specific research question being asked.

For almost all studies, it is critical to have a control group. In the longitudinal research design for the cholesterol study described earlier, the **control group** acts as a comparison group to make certain that any changes observed in the running group are attributable solely to the training program and not to any other factors, such as the time of the year or aging of the subjects during the course of the study. Experimental designs often employ a **placebo group**. Thus, in a study in which a subject might expect to have a benefit from the proposed intervention, such as the use of a specific food or drug, a scientist might decide to use three groups of subjects: an intervention group that receives the actual food or drug, a placebo group that receives an inert substance that looks exactly like the actual food or drug, and a control group that receives nothing. (The last group often serves as a "time control," accounting for nonexperimentally induced changes that may occur over the course of the study period.) If the intervention and placebo groups improve their performance to the same level and the control group does not improve performance, then the improvement is likely the result of the "placebo effect," or the expectation that the substance will improve performance. If the intervention group improves performance and the placebo and control groups do not, then we can conclude that the intervention does improve performance.

One other way of controlling for the placebo effect is to conduct a study that uses a **crossover design**. In this case, each group undergoes both treatment and control trials at different times. For example, one group is administered the intervention for the first half of the study (e.g., 6 months of a 12-month study) and serves as a control during the last half of the study. The second group serves as a control during the first half of the study and receives the intervention during the second half. In some cases, a placebo can be used in the control phase of the study. Chapter 16, "Ergogenic Aids and Sport," provides further discussion of placebo groups.

It is equally important to control data collection. The equipment must be calibrated so the researcher knows that the values generated by a given piece of equipment are accurate, and the procedures used in collecting data must be standardized. For example, when using a scale to measure the weight of subjects, researchers need to calibrate that scale by using a set of calibrated weights (e.g., 10 kg, 20 kg, 30 kg, and 40 kg) that have been measured on a precision scale. These weights are placed on the weighing scale to be used in the study, individually and in combination, at least once a week to provide certainty that the scale is measuring the weights accurately. As another example, electronic analyzers used to measure respiratory gases need to be calibrated frequently with gases of known concentration to ensure the accuracy of these analyses.

Finally, it is important to know that all test results are reproducible. In the example illustrated in figure 0.8, the HDL-C of an individual is measured every three months. If that person is tested five days in a row before he or she starts the training program, one would expect the HDL-C results to be similar across all five days, providing diet, exercise, sleep, and time of day for testing remained the same. In figure 0.8, the values for the control group across 12 months varied from about 44 to 45 mg/dL, whereas the exercise group increased from 45 to 47 mg/dL. Over five consecutive days, the measurements should not vary by more than 1 mg/dL for any one person if the researcher is going to pick up this small change over time. To control for reproducibility of results, scientists generally take several measurements, sometimes on different days, and then average the results, before, during, and at the end of an intervention.

Confounding Factors in Exercise Research

Many factors can alter the body's acute response to a bout of exercise. For example, environmental conditions such as the temperature and humidity of the laboratory and the amount of light and noise in the test area can markedly affect physiological responses, both at rest and during exercise. Even the timing, volume, and content of the last meal and the quantity and quality of sleep the night before must be carefully controlled in research studies.

To illustrate this, table 0.1 shows how varying environmental and behavioral factors can alter heart rate at rest and during running on a treadmill at 14 km/h (9 mph). The subject's heart rate response during exercise differed by 25 beats/min when the air temperature was increased from 21 °C (70 °F) to 35 °C (95 °F). Most physiological variables that are normally measured during exercise are similarly influenced by

TABLE 0.1 **Heart Rate Responses to Running Differ With Variations in Environmental and Behavioral Conditions**

	HEART RATE (BEATS/MIN)	
Environmental and behavioral factors	Rest	Exercise
TEMPERATURE (50% HUMIDITY)		
21 °C (70 °F)	60	165
35 °C (95 °F)	70	190
HUMIDITY (21 °C)		
50%	60	165
90%	65	175
NOISE LEVEL (21 °C, 50% HUMIDITY)		
Low	60	165
High	70	165
FOOD INTAKE (21 °C, 50% HUMIDITY)		
Small meal 3 h before exercising	60	165
Large meal 30 min before exercising	70	175
SLEEP (21 °C, 50% HUMIDITY)		
8 h or more	60	165
6 h or less	65	175

environmental fluctuations. Whether one is comparing a person's exercise results from one day to another or comparing the responses of two different subjects, all of these factors must be controlled as carefully as possible.

Physiological responses, both at rest and during exercise, also vary throughout the day. The term **diurnal** **variation** refers to fluctuations that occur during a 24 h day. Table 0.2 illustrates the diurnal variation in heart rate at rest, during various levels of exercise, and during recovery. Body temperature shows similar fluctuations throughout the day. As seen in table 0.2, testing the same person in the morning on one day and in the

TABLE 0.2 **An Example of Diurnal Variations in Heart Rate at Rest and During Exercise**

	TIME OF DAY					
Condition	2 a.m.	6 a.m.	10 a.m.	2 p.m.	6 p.m.	10 p.m.
	HEART RATE (BEATS/MIN)					
Resting	65	69	73	74	72	69
Light exercise	100	103	109	109	105	104
Moderate exercise	130	131	138	139	135	135
Maximal exercise	179	179	183	184	181	181
Recovery, 3 min	118	122	129	128	128	125

Data from T. Reilly and G.A. Brooks (1990), "Selective persistence of circadian rhythms in physiological responses to exercise," *Chronobiology International*, 7: 59-67.

afternoon on the next can and will produce different results. Test times must be standardized to control for this diurnal effect.

At least one other physiological cycle must also be considered. The normal 28-day menstrual cycle often involves considerable variations in

- body weight,
- total body water and blood volume,
- body temperature,
- metabolic rate, and
- heart rate and stroke volume (the amount of blood leaving the heart with each contraction).

Exercise scientists must control for menstrual cycle phase or the use of oral contraceptives (which similarly alter hormonal status), or both, when testing women. When older women are being tested, testing strategies must take into account menopause and hormone replacement therapies.

In focus

Conditions under which research participants are monitored, at rest and during exercise, must be carefully controlled. Environmental factors, such as temperature, humidity, altitude, and noise, can affect the magnitude of response of all basic physiological systems, as can behavioral factors such as eating patterns and sleep. Likewise, physiological measurements must be well controlled for diurnal and menstrual cycle variations.

Units and Scientific Notation

A set of international standards for units and abbreviations (SI, Le Système International d'Unités) serves as the preferred units of measurement in exercise and sport physiology. In this text, alternate units in common use (such as weight in pounds) are often provided as well. Many of these units are provided on the inside cover of this text, and conversions between and among the SI units and other units in common use are found on the inside back cover.

In common writing and even in mathematics, the ratio between two numbers is typically written using a "slash" (/). For example, in dry air at 20 °C, the speed of sound is 343 m/s. That notation works well for simple fractions or ratios, and we have maintained it in this text. However, that notation gets confusing for relations among several, that is, more than two,

variables. Take, for instance, one of the cornerstone measurements in exercise physiology, an individual's maximal oxygen uptake or maximal aerobic capacity, abbreviated $\dot{V}O_{2max}$. This important physiological measurement is the maximal volume of oxygen that an individual can utilize during exhaustive aerobic exercise, and can be measured in liters per minute or L/min. However, because a large person can utilize more oxygen yet not be more aerobically fit, we often standardize this value to body weight in kilograms, that is, milliliters per kilogram per minute. Now the notation becomes a bit more complex and potentially more confusing. We could write the units as ml/kg/min, but what is being divided by what in this notation? Recall that L/min can also be written as $L \cdot min^{-1}$, just as the fraction $1/4 = 1 \cdot 4^{-1}$. To avoid errors and ambiguity, in exercise physiology we use the exponent notation any time more than two variables are involved. Therefore, milliliters per kilogram per minute is written as $ml \cdot kg^{-1} \cdot min^{-1}$ rather than ml/kg/min.

Reading and Interpreting Tables and Graphs

This book contains references to specific research studies that have had a major impact on our understanding of exercise and sport physiology. Once scientists complete a research project, they submit the results of their research to one of the many research journals in sport and exercise physiology. Some of the more widely used research journals appear on the list of selected readings and references at the back of this book, as well as in the study guide on the website www.HumanKinetics.com /PhysiologyOfSportAndExercise.

As in other areas of science, most of the quantitative research is presented in the form of tables and graphs. Tables and graphs provide an efficient way for researchers to communicate the results of their studies to other scientists. For the student in exercise and sport physiology, a working knowledge of how to read and interpret tables and graphs is critical.

Tables are usually used to convey a large number of data points or complex data that are affected by several factors. Take table 0.1 as an example. It is important to first look at the title of the table, which identifies what information is being presented. In this case, the table is designed to illustrate how various conditions affect heart rate, at rest and during exercise. The left-hand column, along with the horizontal subheadings (like "Humidity (21 °C)," specify the conditions under which the heart rate was measured. Columns 2 and 3 provide the mean heart rate values that correspond to each condition, with the middle column giving the resting value and the right-most column the exercise value. In every

Exercise Physiology Beyond Earth's Boundaries

An important segment of exercise physiology concerns the response and adaptation of people to extremes of heat, cold, depth, and altitude. Understanding and controlling the physiological stresses and adaptations that occur at these environmental limits have contributed directly to notable societal achievements such as construction of the Brooklyn Bridge, the Hoover Dam, pressurized aircraft, and underwater habitats for the commercial diving industry.

The next generation of environmental challenges will also require such physiological expertise. In January 2004, President George Bush announced the Vision for Space Exploration, a strategy to first return humans to the moon, then send explorers to the planet Mars, over the next 30 years. This ambitious plan to construct permanent human outposts on the moon beginning in 2017, followed by 2.5-year missions to the planet Mars, will require effective countermeasures to minimize the physiological changes that put space explorers at risk.

The continuous pull of gravity contributes to the growth and adaptation of postural skeletal muscles; loads bones, which increases their

Dr. James A. Pawelczyk.

size and density; and requires the cardiovascular system to maintain blood pressure and brain blood flow. In a microgravity environment (free fall around the earth or constant-velocity conditions in deep space), the reduction in loading leads to dramatic losses in muscle mass and strength, osteoporosis, and exercise intolerance at rates that mimic those seen in spinal cord–injured patients.

A series of dedicated space shuttle flights have studied these problems in detail. In 1983, the National Aeronautics and Space Administration (NASA) began flying the European Space Agency–developed Spacelab module, ushering a new era of internationally sponsored scientific research into low-earth orbit. The Spacelab Life Sciences (SLS-1, SLS-2) missions (STS-40 and STS-58) emphasized the study of cardiorespiratory, vestibular, and musculoskeletal adaptations to microgravity.

Subsequently the Federal German Aerospace Research Establishment (DLR) sponsored two missions (STS-61A and STS-68), perfecting a model of multidisciplinary, international investigation that was emulated by the Life and Microgravity Sciences Spacelab mission (STS-78), which concentrated on neuromuscular adaptation. The 1998 Neurolab Spacelab mission (STS-90), with an exclusive neuroscience theme, concluded flights of the Spacelab module. Dr. James A. Pawelczyk, a Penn State exercise physiologist and mission specialist on that flight, cotaught the first exercise physiology class from space! Even now, as of this writing, 250 mi (402 km) overhead, an active biomedical research program continues on the International Space Station.

For the exercise physiologist, the question is what combination of resistance and "aerobic" exercise training can prevent or diminish the changes that occur in space. At this time, the answer remains unclear. Furthermore, if physical conditioning is required before and during space exploration and as part of postflight rehabilitation, how should exercise prescriptions be individualized, evaluated, and updated? Without doubt, further research in exercise and environmental physiology will be essential to complete what is destined to be the largest exploration feat of the 21st century.

good table and graph, the units for each variable are clearly presented; in this table, heart rate is expressed in "beats/min," or beats per minute. Pay careful attention to the units of measure used when interpreting a table or graph. From this table—a relatively simple one by scientific standards—we see that both resting and exercise heart rate are increased by increased ambient temperature and humidity, while noise level only affected resting heart rate. Similarly, consuming a large meal or getting less than 6 h of sleep also raises heart rate. These data could not easily have been shown in graphical form.

Graphs can provide a better view of trends in data, response patterns, and comparisons of data collected from two or more groups of subjects. For some students, graphs can be more difficult to read and interpret; but graphs are, and will remain, a critical tool in the understanding of exercise physiology. First, every graph has a horizontal or *x*-axis for the **independent variable** and one (or sometimes two) vertical or *y*-axis for the **dependent variable** or variables. Independent variables are those factors that are manipulated or controlled by the researcher, while dependent variables are those that change with—that is, depend on—the independent variables.

In figure 0.9, time of day is the independent variable and is therefore placed along the *x*-axis of the graph, while heart rate is the dependent variable (since heart rate *depends on* the time of day) and is therefore plotted

on the *y*-axis. The units of measure for each variable are clearly displayed on the graph. Figure 0.9 is in the form of a line graph. Line graphs are useful in illustrating patterns or trends in data but should be used only to compare two variables that change in a continuous manner (for example, across time) and only if both the dependent and independent variables are numbers.

In a line graph, if the dependent variable goes up or down at a constant rate with the independent variable, the result will be a straight line. However, in physiology the response pattern between variables is often not a straight line but a curve of one shape or another. In such cases, pay close attention to the slope of various parts of the curve as it changes across the graph. For instance, figure 0.10 shows the concentration of lactate in the blood as subjects walk/run on a treadmill at various increasing speeds. At low treadmill speeds of 4 to 8 km/h, lactate increases very little. However, at about 8.5 km/h, a threshold is reached beyond which lactate increases more dramatically. In many physiological responses, both the threshold (onset of response) and the slope of the response beyond that threshold are important.

Data can also be plotted in the format of a bar graph. Bar graphs are commonly used when only the dependent variable is a number and the independent variable is a category. Bar graphs often show treatment effects, as in figure 0.7, which was previously discussed. Figure 0.7 shows the effect of distance run per week (a category) on HDL-C (a numerical response) in the bar graph format.

FIGURE 0.9 Understanding how to read and interpret a graph. This line graph depicts the relation between the time of day (on the *x*-axis, independent variable) and heart rate during low-intensity exercise (on the *y*-axis, dependent variable) that was measured at that time of day with no change in the exercise intensity.

FIGURE 0.10 A line graph showing the nonlinear nature of many physiological responses. This graph shows that, above a threshold (onset of response) of about 8.5 km/h, the slope of the blood lactate response increases sharply.

In closing

In this introduction we highlighted the historical roots and scientific underpinnings of exercise and sport physiology. We learned that the current state of knowledge in these fields builds on the past and is merely a bridge to the future—many questions remain unanswered. We briefly defined the acute responses to exercise bouts and chronic adaptations to long-term training. We concluded with an overview of the principles used in sport and exercise physiology research.

In part I, we begin examining physical activity the way exercise physiologists do as we explore the essentials of movement. In the next chapter, we examine the structure and function of skeletal muscle, how it produces movement, and how it responds during exercise.

Key Terms

acute exercise

chronic adaptation

control group

crossover design

cross-sectional research design

cycle ergometer

dependent variable

diurnal variation

dose–response relation

environmental physiology

ergometer

exercise physiology

homeostasis

independent variable

longitudinal research design

physiology

placebo group

sport physiology

training effect

treadmill

Study Questions

1. What is exercise physiology? How does sport physiology differ?
2. Provide an example of what is meant by studying acute responses to a single bout of exercise.
3. Describe what is meant by studying chronic adaptations to exercise training.
4. Describe the evolution of exercise physiology from the early studies of anatomy. Who were some of the key figures in the development of this field?
5. Describe the founding and the key areas of research emphasized by the Harvard Fatigue Laboratory. Who was the first research director of this laboratory?
6. Name the three Scandinavian physiologists who conducted research in the Harvard Fatigue Laboratory.
7. What is an ergometer? Name the two most commonly used ergometers and explain their advantages and disadvantages.
8. What factors must researchers consider when designing a research study to ensure that they get accurate and reproducible results?
9. List several environmental conditions that could affect one's response to an acute bout of exercise.
10. What are the advantages and disadvantages of a cross-sectional versus a longitudinal study design?
11. When should data be depicted as a bar graph as opposed to a line graph? What purpose does a line graph serve?

Study Guide Activities

In addition to the activities listed in the chapter opening outline on page 1, two other activities are available in the web study guide, located at

www.HumanKinetics.com/PhysiologyOfSportAndExercise

 The **KEY TERMS** activity reviews important terms, and the end-of-chapter **QUIZ** tests your understanding of the material covered in the chapter.

PART I

Exercising Muscle

In the introduction, we explored the foundations of exercise and sport physiology. We defined these fields of study, gained a historical perspective of their development, and established basic concepts that underlie the remainder of this book. We also examined the tools and research methods used by exercise physiologists. With this foundation, we can begin our mission to understand how the human body performs, and adapts to, exercise and physical activity. Because muscle is the true foundation of movement, we start with chapter 1, "Structure and Function of Exercising Muscle," where we focus on skeletal muscle, examining the structure and function of skeletal muscles and muscle fibers and how they produce body movement. We will learn how muscle fiber types differ and why these differences are important to specific types of activity. In chapter 2, "Fuel for Exercise," we study principles of metabolism, focusing on the primary source of energy, adenosine triphosphate (ATP), and how it is provided from the foods that we eat through three energy systems. In chapter 3, "Neural Control of Exercising Muscle," we discuss how the nervous system initiates and controls muscle action. Chapter 4, "Hormonal Control During Exercise," presents an overview of the endocrine system, then focuses on hormonal control of energy metabolism and of body fluid and electrolyte balance during exercise. Finally, chapter 5, "Energy Expenditure and Fatigue," discusses the measurement of energy expenditure, how energy expenditure changes from a state of rest to varying intensities of exercise, and the causes of fatigue that limits exercise performance.

Structure and Function of Exercising Muscle

In this chapter and in the web study guide

 ACTIVITY 1.1   Muscle Structure reviews the basic structures of muscle.

 ACTIVITY 1.2   Structure of a Skeletal Muscle Cell reviews the basic structures in a single muscle fiber.

 ACTIVITY 1.3   Structure of the Sarcomere reviews the basic structures in a sarcomere.

 ACTIVITY 1.4   Sliding Filament Theory describes this theory of muscle contraction and explores what happens at the cellular and gross motor movement levels.

 ACTIVITY 1.5   Fiber Types differentiates between type I and type II skeletal muscle fibers.

 ACTIVITY 1.6   Fiber Recruitment tests your understanding of the types of muscle fibers recruited and the order of recruitment based on the level of force demanded of the muscle.

 ACTIVITY 1.7   Generation of Force reviews the factors that influence the development of muscle force.

Liam Hoekstra possesses a physique and physical attributes like many professional athletes: rippled abdominal muscles, enough strength to perform feats like an iron cross and inverted sit-ups, and amazing speed and agility. However, Liam is 19 months old and weighs 22 lb! Liam has a rare genetic condition called myostatin-related muscle hypertrophy, a condition that was first described in an abnormally muscular breed of beef cattle in the late 1990s. Myostatin is a protein that inhibits the growth of skeletal muscles; myostatin-related muscle hypertrophy is a genetic mutation that blocks production of this inhibitory growth factor and thus promotes the rapid growth and development of skeletal muscles.

Liam's condition is extremely rare in humans, with fewer than 100 cases documented worldwide. However, studying this genetic phenomenon could help scientists unlock secrets of how skeletal muscles grow and deteriorate. Research on Liam's condition could lead to new treatments for debilitating muscular conditions such as muscular dystrophy. On the darker side, it could open up a whole new realm of abuse by athletes who are looking for shortcuts to develop muscle size and strength, not unlike the illicit and dangerous use of anabolic steroids.

When the heart beats, when partially digested food moves through the intestines, and when the body moves in any way, muscle is involved. These many and varied functions of the muscular system are performed by three distinct types of muscle (see figure 1.1): smooth muscle, cardiac muscle, and skeletal muscle.

Smooth muscle is sometimes called involuntary muscle because it is not under direct conscious control. It is found in the walls of most blood vessels, allowing them to constrict or dilate to regulate blood flow. It is also found in the walls of most internal organs, allowing them to contract and relax, for example to move food through the digestive tract, to expel urine, or to give birth.

Cardiac muscle is found only in the heart, composing the vast majority of the heart's structure. While it shares some characteristics with skeletal muscle, like smooth muscle it is not under conscious control. Cardiac muscle in essence controls itself, with some fine-tuning by the nervous and endocrine systems. Cardiac muscle will be discussed more fully in chapter 6.

Skeletal muscles are under conscious control and are so named because most attach to and move the

FIGURE 1.1 Microscopic photographs of the three types of muscle.

ⓐ Skeletal muscle

ⓒ Smooth muscle

ⓑ Cardiac muscle

skeleton. Together with the bones of the skeleton, they make up the **musculoskeletal system**. The names of many of these muscles have found their way into our everyday vocabulary—such as deltoids, pectorals (or "pecs"), and biceps—but the human body contains more than 600 skeletal muscles. The thumb alone is controlled by nine separate muscles!

Exercise requires movement of the body, which is accomplished through the action of skeletal muscles. Because exercise and sport physiology depends on human movement, the primary focus of this chapter is on the structure and function of skeletal muscle. Although the anatomical structures of smooth, cardiac, and skeletal muscle differ in some respects, their control mechanisms and principles of action are similar.

Functional Anatomy of Skeletal Muscle

When we think of muscles, we visualize each muscle as a whole, that is, as a single unit. This is natural because a skeletal muscle seems to act as a single entity. But skeletal muscles are far more complex than that.

If a person were to dissect a muscle, he or she would first cut through an outer connective tissue covering known as the **epimysium** (see figure 1.2). It surrounds the entire muscle and functions to hold it together. Once through the epimysium, one would see small bundles of fibers wrapped in a connective tissue sheath. These bundles are called fasciculi, and the connective tissue sheath surrounding each **fasciculus** (also called a fascicle) is the **perimysium**.

Finally, by cutting through the perimysium and using a microscope, one would see the individual **muscle fibers**, each of which is a muscle cell. Unlike most cells in the body, which have a single nucleus, muscle cells are multinucleated. A sheath of connective tissue, called the **endomysium**, also covers each muscle fiber. It is generally thought that muscle fibers extend from one end of the muscle to the other; but under the microscope, muscle bellies (the thick middle parts of muscles) often divide into compartments or more transverse fibrous bands (inscriptions).

Because of this compartmentalization, the longest human muscle fibers are about 12 cm (4.7 in.), which corresponds to about 500,000 sarcomeres, the basic functional unit of the myofibril. The number of fibers in

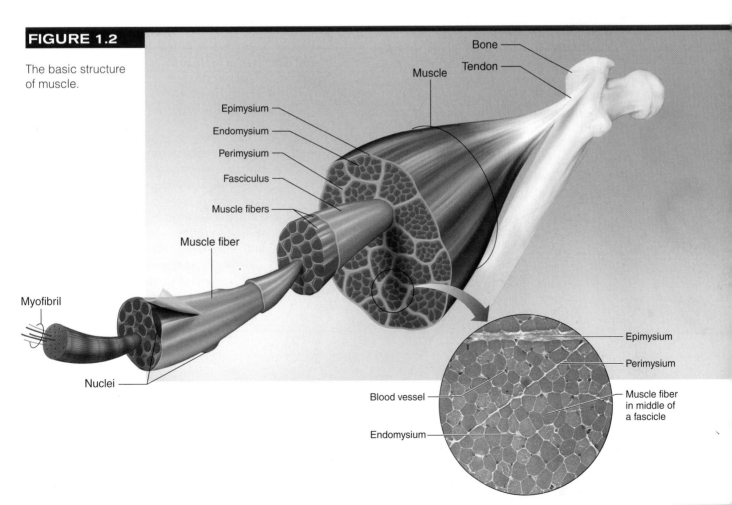

FIGURE 1.2

The basic structure of muscle.

different muscles ranges from several hundred (e.g., in the tensor pani, attached to the eardrum) to more than a million (e.g., in the medial gastrocnemius muscle).[6]

In focus

A single muscle cell is known as a muscle fiber. It has a cell membrane and the same organelles—mitochondria, lysosomes, and so on—as other cell types but is uniquely multinucleated.

Muscle Fibers

Muscle fibers range in diameter from 10 to 120 µm, so they are nearly invisible to the naked eye. The following sections describe the structure of the individual muscle fiber.

Plasmalemma

If one looked closely at an individual muscle fiber, one would see that it is surrounded by a plasma membrane, called the **plasmalemma** (figure 1.3). The plasmalemma is part of a larger unit referred to as the **sarcolemma**. The sarcolemma is composed of the plasmalemma and the basement membrane. (Some textbooks use the term sarcolemma to describe just the plasmalemma.[6]) At the end of each muscle fiber, its plasmalemma fuses with the tendon, which inserts into the bone. Tendons are made of fibrous cords of connective tissue that transmit the force generated by muscle fibers to the bones, thereby creating motion. So typically, individual muscle fibers are ultimately attached to bone via the tendon.

The plasmalemma has several unique features that are critical to muscle fiber function. It appears as a series of shallow folds along the surface of the fiber when the fiber is contracted or in a resting state, but these folds disappear when the fiber is stretched. This folding allows stretching of the muscle fiber without disrupting the plasmalemma. The plasmalemma also has junctional folds in the innervation zone at the motor end plate, which assists in the transmission of the action potential from the motor neuron to the muscle fiber as discussed later in this chapter. Finally, the plasmalemma helps to maintain acid–base balance and transport of metabolites from the capillary blood into the muscle fiber.[6]

Satellite cells are located between the plasmalemma and the basement membrane. These cells are involved in the growth and development of skeletal muscle and in muscle's adaptation to injury, immobilization, and training. This will be discussed in greater detail in subsequent chapters.

Sarcoplasm

Inside the plasmalemma, a muscle fiber contains successively smaller subunits, as shown in figure 1.3. The largest of these are myofibrils, the contractile element of the muscle, which are described later. A gelatin-like substance fills the spaces within and between the myofibrils. This is the **sarcoplasm**. It is the fluid part of the muscle fiber—its cytoplasm. The sarcoplasm contains mainly dissolved proteins, minerals, glycogen, fats, and necessary organelles. It differs from the cytoplasm of most cells because it contains a large quantity of stored glycogen as well as the oxygen-binding compound myoglobin, which is similar in structure and function to the hemoglobin found in red blood cells.

Transverse Tubules The sarcoplasm also houses an extensive network of **transverse tubules (T-tubules)**, which are extensions of the plasmalemma that pass laterally through the muscle fiber. These tubules are interconnected as they pass among the myofibrils, allowing nerve impulses received by the plasmalemma to be transmitted rapidly to individual myofibrils. The tubules also provide pathways from outside the fiber to its interior, enabling substances to enter the cell and waste products to leave the fibers.

Sarcoplasmic Reticulum A longitudinal network of tubules, known as the **sarcoplasmic reticulum (SR)**, is also found within the muscle fiber. These membranous channels parallel the myofibrils and loop around them. The SR serves as a storage site for calcium, which is essential for muscle contraction. Figure 1.3 depicts the T-tubules and the SR. Their functions are discussed in more detail later in this chapter when we describe the process of muscle contraction.

FIGURE 1.3 The structure of a single muscle fiber.

Transverse tubules — Sarcoplasmic reticulum — Opening into T-tubule — Nucleus — Mitochondria — Myofibril — Sarcoplasm — Plasmalemma — Terminal cisternae

Myofibrils

Each muscle fiber contains several hundred to several thousand **myofibrils**. These small fibers are made up of the basic contractile elements of skeletal muscle—the sarcomeres. Under the electron microscope, myofibrils appear as long strands of sarcomeres.

Sarcomeres

Under a light microscope, skeletal muscle fibers have a distinctive striped appearance. Because of these markings, or striations, skeletal muscle is also called striated muscle. This striation also is seen in cardiac muscle, so it too can be considered striated muscle.

Refer to figure 1.4, showing myofibrils within a single muscle fiber, and note the striations. Note that dark regions, known as A-bands, alternate with light regions, known as I-bands. Each dark A-band has a lighter region in its center, the H-zone, which is visible only when the myofibril is relaxed. There is a dark line in the middle of the H-zone called the M-line. The light I-bands are interrupted by a dark stripe referred to as the Z-disk, also known as the Z-line.

A **sarcomere** is the basic functional unit of a myofibril and the basic contractile unit of muscle. Each myofibril is composed of numerous sarcomeres joined end to end at the Z-disks. Each sarcomere includes what is found between each pair of Z-disks, in this sequence:

- An I-band (light zone)
- An A-band (dark zone)
- An H-zone (in the middle of the A-band)
- An M-line in the middle of the H-zone
- The rest of the A-band
- A second I-band

Myofibrils

FIGURE 1.4 An electron micrograph of myofibrils. Note the striping or striations. The blue regions are the A-bands, and the pink regions are the I-bands.

Looking at individual myofibrils through an electron microscope, one can differentiate two types of small protein filaments that are responsible for muscle contraction. The thinner filaments are composed primarily of **actin**, and the thicker filaments are primarily **myosin**. The striations seen in muscle fibers result from the alignment of these filaments, as illustrated in figure 1.4. The light I-band indicates the region of the sarcomere where there are only thin filaments. The dark A-band represents the regions that contain both thick and thin filaments. The H-zone is the central portion of the A-band and contains only thick filaments. The absence of thin filaments causes the H-zone to appear lighter than the adjacent A-band. In the center of the H-zone is the M-line, which is composed of proteins that serve as the attachment site for the thick filaments and assist in stabilizing the structure of the sarcomere. Z-disks, composed of proteins, appear at each end of the sarcomere. Along with two additional proteins, titin and nebulin, they provide points of attachment and stability for the thin filaments.

Thick Filaments About two-thirds of all skeletal muscle protein is myosin, the principal protein of the thick filament. Each myosin filament typically is formed by about 200 myosin molecules.

Each myosin molecule is composed of two protein strands twisted together (see figure 1.5). One end of each strand is folded into a globular head, called the myosin head. Each thick filament contains many such heads, which protrude from the thick filament to form cross-bridges that interact during muscle contraction with specialized active sites on the thin filaments. There is an array of fine filaments, composed of **titin**, that stabilizes the myosin filaments along their longitudinal axis (see figure 1.5). Titin filaments extend from the Z-disk to the M-line.

Myosin

Actin

Sarcomere

H-zone
Z-disk
I-band
M-line
A-band

Nebulin

Actin
Tropomyosin
Troponin

Thin filament:
Actin, troponin, tropomyosin

Thick filament:
Myosin

Z-disk

M-line

Myosin

Titin

Actin

Myosin
head

FIGURE 1.5 The sarcomere contains a specialized arrangement of actin (thin) and myosin (thick) filaments. The role of titin is to position the myosin filament to maintain equal spacing between the actin filaments. Nebulin is often referred to as an "anchoring protein" because it provides a framework that helps stabilize the position of actin.

Thin Filaments   Each thin filament, although often referred to simply as an actin filament, is actually composed of three different protein molecules—actin, **tropomyosin**, and **troponin**. Each thin filament has one end inserted into a Z-disk, with the opposite end extending toward the center of the sarcomere, lying in the space between the thick filaments. **Nebulin**, an anchoring protein for actin, coextends with actin and appears to play a regulatory role in mediating actin and myosin interactions (figure 1.5). Each thin filament contains active sites to which myosin heads can bind.

Actin forms the backbone of the filament. Individual actin molecules are globular proteins (G-actin) and join together to form strands of actin molecules. Two strands then twist into a helical pattern, much like two strands of pearls twisted together.

Tropomyosin is a tube-shaped protein that twists around the actin strands. Troponin is a more complex protein that is attached at regular intervals to both the actin strands and the tropomyosin. This arrangement is depicted in figure 1.5. Tropomyosin and troponin work together in an intricate manner along with calcium ions to maintain relaxation or initiate contraction of the myofibril, which we discuss later in this chapter.

In review

➤ Myofibrils are composed of sarcomeres, the smallest functional units of a muscle.

➤ A sarcomere is composed of two different-sized filaments, thick and thin filaments, which are responsible for muscle contraction.

➤ Myosin, the primary protein of the thick filament, is composed of two protein strands, each folded into a globular head at one end.

➤ The thin filament is composed of actin, tropomyosin, and troponin. One end of each thin filament is attached to a Z-disk.

Muscle Fiber Contraction

An **α-motor neuron** is a nerve cell that connects with and innervates many muscle fibers. A single α-motor neuron and all the muscle fibers it directly signals are collectively termed a **motor unit** (see figure 1.6). The synapse or gap between the α-motor neuron and a muscle fiber is referred to as a neuromuscular junction. This is where communication between the nervous and muscular systems occurs.

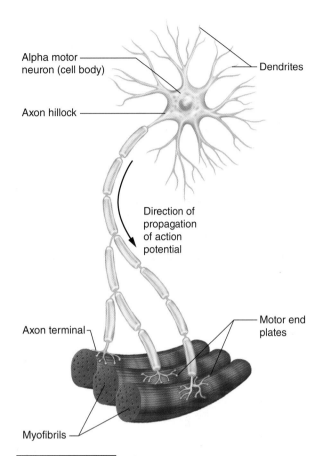

FIGURE 1.6   A motor unit includes one α-motor neuron and all of the muscle fibers it innervates.

In focus

When an α-motor neuron is activated, all of the muscle fibers in its motor unit are stimulated to contract.

Excitation–Contraction Coupling

The complex sequence of events that triggers a muscle fiber to contract is termed **excitation–contraction coupling** because it begins with the excitation of a motor nerve and results in contraction of the muscle fibers. The process, depicted in figure 1.7, is initiated by a nerve impulse, or **action potential**, from the brain or spinal cord to an α-motor neuron. The action potential arrives at the α-motor neuron's dendrites, specialized receptors on the neuron's cell body. From here, the action potential travels down the axon to the axon terminals, which are located very close to the plasmalemma.

When the action potential arrives at the axon terminals, these nerve endings release a signaling molecule or neurotransmitter called acetylcholine (ACh), which crosses the synaptic cleft and binds to receptors on the plasmalemma (see figure 1.7a). If enough ACh binds to the receptors, the action potential will be transmitted the full length of the muscle fiber as ion gates open in the muscle cell membrane and allow sodium to enter. This process is referred to as depolarization. An action potential must be generated in the muscle cell before the muscle cell can act. These neural events are discussed more fully in chapter 3.

Role of Calcium in the Muscle Fiber

In addition to depolarizing the fiber membrane, the action potential travels over the fiber's network of tubules (T-tubules) to the interior of the cell. The arrival of an electrical charge causes the adjacent SR to release large quantities of stored calcium ions (Ca^{2+}) into the sarcoplasm (see figure 1.7b).

In the resting state, tropomyosin molecules cover the myosin-binding sites on the actin molecules, preventing the binding of the myosin heads. Once calcium ions are released from the SR, they bind to the troponin on the actin molecules. Troponin, with its strong affinity for calcium ions, is believed to then initiate the contraction process by moving the tropomyosin molecules off the myosin-binding sites on the actin molecules. This is shown in figure 1.7c. Because tropomyosin normally covers the myosin-binding sites, it blocks the attraction between the **myosin cross-bridges** and actin molecules. However, once the tropomyosin has been lifted off the binding sites by troponin and calcium, the myosin heads can attach to the binding sites on the actin molecules.

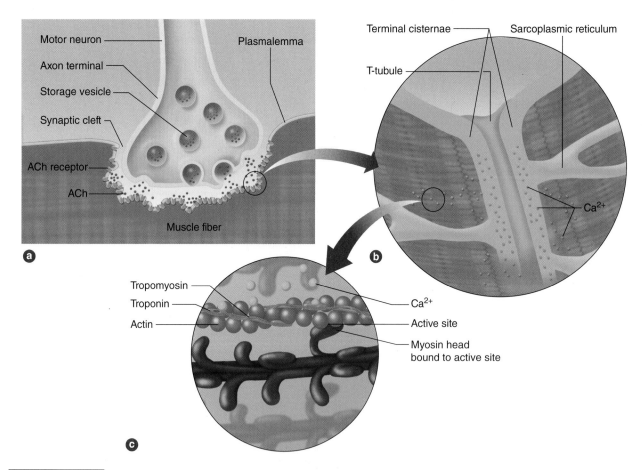

FIGURE 1.7 The sequence of events leading to muscle action, known as excitation–contraction coupling. (a) In response to an action potential, a motor neuron releases acetylcholine (ACh), which crosses the synaptic cleft and binds to receptors on the plasmalemma. If enough ACh binds, an action potential is generated in the muscle fiber. (b) The action potential triggers the release of calcium ions (Ca^{2+}) from the terminal cisternae of the sarcoplasmic reticulum into the sarcoplasm. (c) The Ca^{2+} binds to troponin on the actin filament, and the troponin pulls tropomyosin off the active sites, allowing myosin heads to attach to the actin filament.

The Sliding Filament Theory: How Muscles Create Movement

When muscle contracts, muscle fibers shorten. How do they shorten? The explanation for this phenomenon is termed the **sliding filament theory**. When the myosin cross-bridges are activated, they bind with actin, resulting in a conformational change in the cross-bridge, which causes the myosin head to tilt and to drag the thin filament toward the center of the sarcomere (see figures 1.8 and 1.9). This tilting of the head is referred to as the **power stroke**. The pulling of the thin filament past the thick filament shortens the sarcomere and generates force. When the fibers are not contracting, the myosin head remains in contact with the actin molecule, but the molecular bonding at the site is weakened or blocked by tropomyosin.

Immediately after the myosin head tilts, it breaks away from the active site, rotates back to its original position, and attaches to a new active site farther along the actin filament. Repeated attachments and power strokes cause the filaments to slide past one another, giving rise to the term *sliding filament theory*. This process continues until the ends of the myosin filaments reach the Z-disks, or until the Ca^{2+} is pumped back into the sarcoplasmic reticulum. During this sliding (contraction), the thin filaments move toward the center of the sarcomere and protrude into the H-zone, ultimately overlapping. When this occurs, the H-zone is no longer visible.

Recall that the sarcomeres are joined end to end within a myofibril. Because of this anatomical arrangement, as sarcomeres shorten, the myofibril shortens, and muscle fibers within a fascicle shorten.

The end result of many such fibers shortening is an organized muscle contraction.

Energy for Muscle Contraction

Muscle contraction is an active process, meaning that it requires energy. In addition to the binding site for actin, a myosin head contains a binding site for the molecule **adenosine triphosphate (ATP)**. The myosin molecule must bind with ATP for muscle contraction to occur because ATP supplies the needed energy.

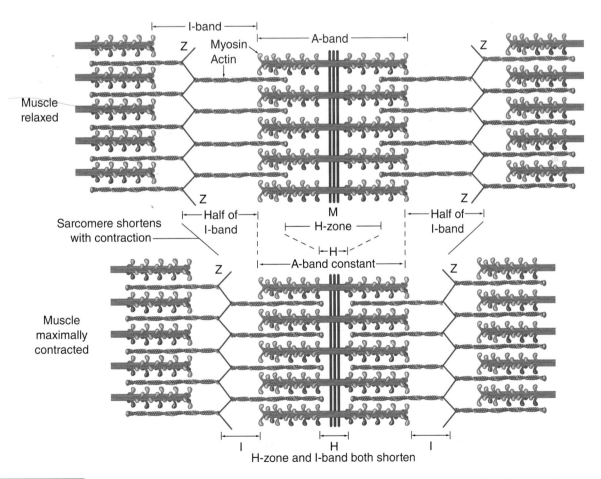

FIGURE 1.8 A sarcomere in its relaxed (top) and contracted (bottom) state, illustrating the sliding of the actin and myosin filaments with contraction.

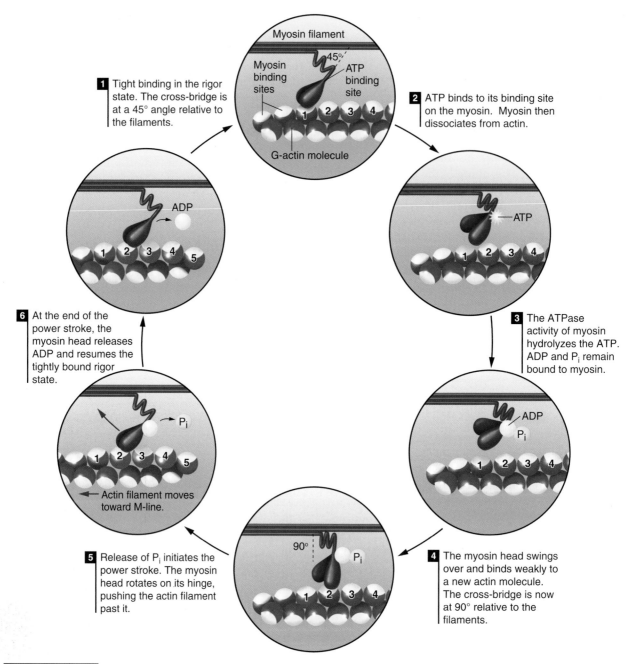

FIGURE 1.9 The molecular events of a contractile cycle illustrating the changes in the myosin head during various phases of the power stroke.

Fig. 12.9, p. 405 from HUMAN PHYSIOLOGY, 4th ed. By Dee Unglaub Silverthorn. Copyright © 2007 by Pearson Education, Inc. Adapted by permission.

The enzyme **adenosine triphosphatase (ATPase)**, which is located on the myosin head, splits the ATP to yield adenosine diphosphate (ADP), inorganic phosphate (P$_i$), and energy. The energy released from this breakdown of ATP is used to power the tilting of the myosin head. Thus, ATP is the chemical source of energy for muscle contraction. This process is discussed in much more detail in chapter 2.

Muscle Relaxation

Muscle contraction continues as long as calcium is available in the sarcoplasm. At the end of a muscle contraction, calcium is pumped back into the SR, where it is stored until a new action potential arrives at the muscle fiber membrane. Calcium is returned to the SR by an active calcium-pumping system. This is another energy-demanding process that also relies on ATP. Thus, energy is required for both the contraction and relaxation phases.

When the calcium is pumped back into the SR, troponin and tropomyosin return to the resting conformation. This blocks the linking of the myosin crossbridges and actin molecules and stops the use of ATP. As a result, the thick and thin filaments return to their original relaxed state.

In review

- The sequence of events that starts with a motor nerve impulse and results in muscle contraction is called excitation–contraction coupling.

- Muscle contraction is initiated by an α-motor neuron impulse or action potential. The motor neuron releases ACh, which opens up ion gates in the muscle cell membrane, allowing sodium to enter the muscle cell (depolarization). If the cell is sufficiently depolarized, an action potential is generated and muscle contraction occurs.

- The action potential travels along the plasmalemma, then moves through the T-tubule system, causing stored calcium ions to be released from the SR.

- Calcium ions bind with troponin. Then troponin moves the tropomyosin molecules off of the myosin-binding sites on the actin molecules, opening these sites to allow the myosin heads to bind to them.

- Once a strong binding state is established with actin, the myosin head tilts, pulling the thin filament past the thick filament. The tilting of the myosin head is the power stroke.

- Energy is required for muscle contraction to occur. The myosin head binds to ATP, and ATPase on the head splits ATP into ADP and P_i, releasing energy to fuel the contraction.

- The end of muscle contraction is signaled when neural activity ceases at the neuromuscular junction. Calcium is actively pumped out of the sarcoplasm and back into the SR for storage. Tropomyosin moves to cover active sites on actin molecules, leading to relaxation between the myosin heads and the binding sites.

- Like muscle contraction, muscle relaxation requires energy supplied by ATP.

Skeletal Muscle and Exercise

Having reviewed the overall structure of muscle and the process by which it develops force, we now look more specifically at how muscle functions during exercise. Strength, endurance, and speed depend largely on the muscle's ability to produce energy and force. This section examines how muscle accomplishes this task.

Muscle Fiber Types

Not all muscle fibers are alike. A single skeletal muscle contains fibers having different speeds of shortening and ability to generate maximal force: type I (also called slow or slow-twitch) fibers and type II (also called fast or fast-twitch) fibers. **Type I fibers** take approximately 110 ms to reach peak tension when stimulated. **Type II fibers**, on the other hand, can reach peak tension in about 50 ms. While the terms "slow twitch" and "fast twitch" continue to be used, scientists now prefer to use the terminology type I and type II, as is the case in this textbook.

Although only one form of type I fiber has been identified, type II fibers can be further classified. In humans, the two major forms of type II fibers are fast-twitch type a (type IIa) and fast-twitch type x (type IIx). Type IIx fibers in humans are approximately the equivalent of type IIb fibers in animals. Figure 1.10 is a micrograph of human muscle in which thinly sliced (10 µm) cross sections of a muscle sample have been chemically stained to differentiate the fiber types. The type I fibers are

FIGURE 1.10 A photomicrograph showing type I (black), type IIa (white), and type IIx (gray) muscle fibers.

The Muscle Biopsy Needle

It was once difficult to examine human muscle tissue from a living person. Most early (pre-1900) muscle research used muscle from laboratory animals or muscle from humans obtained by open incision surgery. In the early 1900s, a needle biopsy procedure was developed to study muscular dystrophy. In the 1960s, this technique was adapted to sample muscle for studies in exercise physiology.

Samples are removed by muscle biopsy, which involves removing a very small piece of muscle from the muscle belly for analysis. The area from which the sample is taken is first numbed with a local anesthetic, and then a small incision (approximately 1 cm, or 0.4 in.) is made with a scalpel through the skin, subcutaneous tissue, and connective tissue. A hollow needle is then inserted to the appropriate depth into the belly of the muscle. A small plunger is pushed through the center of the needle to snip off a very small sample of muscle. The biopsy needle is withdrawn, and the sample, weighing 10 to 100 mg, is removed, cleaned of blood, mounted, and quickly frozen. It is then thinly sliced, stained, and examined under a microscope.

This method allows us to study muscle fibers and gauge the effects of acute exercise and chronic training on fiber composition. Microscopic and biochemical analyses of the samples aid our understanding of the muscles' machinery for energy production.

(a) The use of a biopsy needle to obtain a sample from the leg muscle of an elite female runner. (b) A close-up view of a muscle biopsy needle and a small piece of muscle tissue.

---

stained black; type IIa fibers are unstained and appear white; and type IIx fibers appear gray. Although not apparent in this figure, a third subtype of fast-twitch fibers has also been identified: type IIc.

The differences among the type IIa, type IIx, and type IIc fibers are not fully understood, but type IIa fibers are believed to be the most frequently recruited. Only type I fibers are recruited more frequently than type IIa fibers. Type IIc fibers are the least often used. On average, most muscles are composed of roughly 50% type I fibers and 25% type IIa fibers. The remaining 25% are mostly type IIx, with type IIc fibers making up only 1% to 3% of the muscle. Because knowledge about type IIc fibers is limited, we will not discuss them further. The exact percentage of each of these fiber types varies greatly in various muscles and among individuals, so the numbers listed here are only averages. This extreme variation is most evident in athletes, as we will see later in this chapter when we compare fiber types in athletes across sports and events within sports.

Characteristics of Type I and Type II Fibers

Different muscle fiber types play different roles in physical activity. This is largely due to differences in their characteristics.

ATPase The type I and type II fibers differ in their speed of contraction. This difference results primarily from different forms of myosin ATPase. Recall that myosin ATPase is the enzyme that splits ATP to release energy to drive contraction. Type I fibers have a slow form of myosin ATPase, whereas type II fibers have a fast form. In response to neural stimulation, ATP is split more rapidly in type II fibers than in type I fibers. As a result, cross-bridges cycle more rapidly in type II fibers.

One of the methods used to classify muscle fibers is a chemical staining procedure applied to a thin slice of tissue. This staining technique measures the ATPase activity in the fibers. Thus, the type I, type IIa, and type

IIx fibers stain differently, as depicted in figure 1.10. This technique makes it appear that each muscle fiber has only one type of ATPase, but fibers can have a mixture of ATPase types. Some have a predominance of type I-ATPase, but others have mostly type II-ATPase. Their appearance in a stained slide preparation should be viewed as a continuum rather than as absolutely distinct types.

A newer method for identifying fiber types is to chemically separate the different types of myosin molecules (isoforms) by using a process called gel electrophoresis.

As shown in figure 1.11, the isoforms are separated by weight in an electric field to show the bands of protein (i.e., myosin) that characterize type I, type IIa, and type IIx fibers. Although our discussion here categorizes fiber types simply as slow-twitch (type I) and fast-twitch (type IIa and type IIx), scientists have further subdivided these fiber types. The use of electrophoresis has led to the detection of myosin hybrids or fibers that possess two or more forms of myosin. With this method of analysis, the fibers are classified as I; Ic (I/IIa); IIc (IIa/I); IIa; IIax; IIxa; and IIx.[6] In this book, we will use the histochemical method of identifying fibers by their primary isoforms, types I, IIa, and IIx.

Table 1.1 summarizes the characteristics of the different muscle fiber types. The table also includes alternative names that are used in other classification systems to refer to the various muscle fiber types.

Sarcoplasmic Reticulum Type II fibers have a more highly developed SR than do type I fibers. Thus, type II fibers are more adept at delivering calcium into the muscle cell when stimulated. This ability is thought to contribute to the faster speed of contraction (V_o) of type II fibers. On average, human type II fibers have a V_o that is five to six times faster than that of type I fibers. Although the amount of force (P_o) generated by type II and type I fibers having the same diameter is about the same, the calculated power ($\mu N \cdot fiber\ length^{-1} \cdot s^{-1}$) of a type II fiber is three to five times greater than that of a type I fiber because of a faster shortening velocity. This may explain in part why individuals who have a predominance of type II fibers in their leg muscles tend to be better sprinters than individuals who have a high percentage of type I fibers, all other things being equal.

Motor Units Recall that a motor unit is composed of a single α-motor neuron and the muscle fibers it innervates. The α-motor neuron appears to determine

FIGURE 1.11 Electrophoretic separation of myosin isoforms to identify type I, type IIa, and type IIx fibers. (*a*) Single fibers are isolated under a dissecting microscope. (*b*) The myosin isoforms are separated for each fiber using electrophoretic techniques. (*c*) The isoforms are then stained to show the myosin that indicates the fiber type.

TABLE 1.1 **Classification of Muscle Fiber Types**

	FIBER CLASSIFICATION		
System 1 (preferred)	Type I	Type IIa	Type IIx
System 2	Slow-twitch (ST)	Fast-twitch a (FTa)	Fast-twitch x (FTx)
System 3	Slow oxidative (SO)	Fast oxidative/glycolytic (FOG)	Fast glycolytic (FG)
	CHARACTERISTICS OF FIBER TYPES		
Oxidative capacity	High	Moderately high	Low
Glycolytic capacity	Low	High	Highest
Contractile speed	Slow	Fast	Fast
Fatigue resistance	High	Moderate	Low
Motor unit strength	Low	High	High

whether the fibers are type I or type II. The α-motor neuron in a type I motor unit has a smaller cell body and typically innervates a cluster of ≤300 muscle fibers. In contrast, the α-motor neuron in a type II motor unit has a larger cell body and innervates ≥300 muscle fibers. This difference in the size of motor units means that when a single type I α-motor neuron stimulates its fibers, far fewer muscle fibers contract than when a single type II α-motor neuron stimulates its fibers. Consequently, type II muscle fibers reach peak tension faster and collectively generate more force than type I fibers.[2]

In focus

The difference in maximal isometric force development between type II and type I motor units is attributable to two characteristics: the number of muscle fibers per individual motor unit and the difference in size of type II and type I fibers. Type I and type II fibers of the same diameter generate about the same force. On average, however, type II fibers tend to be larger than type I fibers and type II motor units tend to have more muscle fibers than do the type I motor units.

Distribution of Fiber Types

As mentioned earlier, the percentages of type I and type II fibers are not the same in all the muscles of the body. Generally, arm and leg muscles have similar fiber compositions within an individual. An endurance athlete with a predominance of type I fibers in his leg muscles will likely have a high percentage of type I fibers in his arm muscles as well. A similar relationship exists for

type II fibers. There are some exceptions, however. The soleus muscle (beneath the gastrocnemius in the calf), for example, is composed of a very high percentage of type I fibers in everyone.

Fiber Type and Exercise

Because of these differences in type I and type II fibers, one might expect that these fiber types would also have different functions when people are physically active. Indeed, this is the case.

Type I Fibers In general, type I muscle fibers have a high level of aerobic endurance. Aerobic means "in the presence of oxygen," so oxidation is an aerobic process. Type I fibers are very efficient at producing ATP from the oxidation of carbohydrate and fat, which will be discussed in chapter 2.

Recall that ATP is required to provide the energy needed for muscle fiber contraction and relaxation. As long as oxidation occurs, type I fibers continue producing ATP, allowing the fibers to remain active. The ability to maintain muscular activity for a prolonged period is known as muscular endurance, so type I fibers have high aerobic endurance. Because of this, they are recruited most often during low-intensity endurance events (e.g., marathon running) and during most daily activities for which the muscle force requirements are low (e.g., walking).

Type II Fibers Type II muscle fibers, on the other hand, have relatively poor aerobic endurance when compared to type I fibers. They are better suited to perform anaerobically (without oxygen). This means that in the absence of adequate oxygen, ATP is formed through anaerobic pathways, not oxidative pathways. (We discuss these pathways in detail in chapter 2.)

Single Muscle Fiber Physiology

One of the most advanced methods for the study of human muscle fibers is to dissect fibers out of a muscle biopsy sample, suspend a single fiber between force transducers, and measure its strength and **single-fiber contractile velocity (V_o)**.

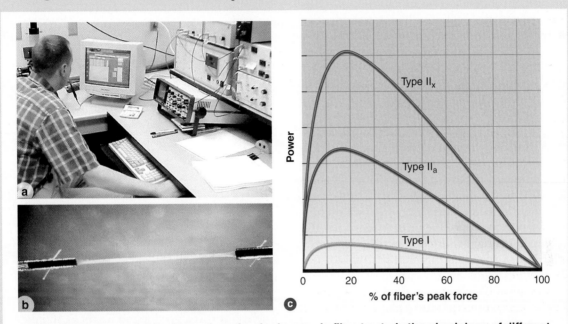

(*a*) The dissection and (*b*) suspension of a single muscle fiber to study the physiology of different fiber types. (*c*) Differences in peak power generated by each fiber type at various percentages of maximal force. Note that all of the fibers tend to reach their peak power when the fibers are generating only about 20% of their peak force. It is quite clear that the peak power of the type II fibers is considerably higher than that of the type I fibers.

Type IIa motor units generate considerably more force than do type I motor units, but type IIa motor units also fatigue more easily because of their limited endurance. Thus, type IIa fibers appear to be the primary fiber type used during shorter, higher-intensity endurance events, such as the mile run or the 400 m swim.

Although the significance of the type IIx fibers is not fully understood, they apparently are not easily activated by the nervous system. Thus they are used rather infrequently in normal, low-intensity activity but are predominantly used in highly explosive events such as the 100 m dash and the 50 m sprint swim. Characteristics of the various fiber types are summarized in table 1.2.

TABLE 1.2   Structural and Functional Characteristics of Muscle Fiber Types

Characteristic	FIBER TYPE		
	Type I	Type IIa	Type IIx
Fibers per motor neuron	≤300	≥300	≥300
Motor neuron size	Smaller	Larger	Larger
Motor neuron conduction velocity	Slower	Faster	Faster
Contraction speed (ms)	110	50	50
Type of myosin ATPase	Slow	Fast	Fast
Sarcoplasmic reticulum development	Low	High	High

Adapted from Close, 1967.

Determination of Fiber Type

The characteristics of muscle fibers appear to be determined early in life, perhaps within the first few years. Studies with identical twins have shown that muscle fiber type, for the most part, is genetically determined, changing little from childhood to middle age. These studies reveal that identical twins have nearly identical fiber types, whereas fraternal twins differ in their fiber type profiles. The genes we inherit from our parents likely determine which α-motor neurons innervate our individual muscle fibers. After innervation is established, muscle fibers differentiate (become specialized) according to the type of α-motor neuron that stimulates them. Some recent evidence, however, suggests that endurance training, strength training, and muscular inactivity may cause a shift in the myosin isoforms. Consequently, training may induce a small change, perhaps less than 10%, in the percentage of type I and type II fibers. Further, both endurance and resistance training have been shown to reduce the percentage of type IIx fibers while increasing the fraction of type IIa fibers.

Studies of older men and women have shown that aging may alter the distribution of type I and type II fibers. As we grow older, muscles tend to lose type II motor units, which increases the percentage of type I fibers.

In review

- Most skeletal muscles contain both type I and type II fibers.
- Different fiber types have different myosin ATPase activities. The ATPase in the type II fibers acts faster than the ATPase in type I fibers.
- Type II fibers have a more highly developed SR, enhancing the delivery of calcium needed for muscle contraction.
- α-Motor neurons innervating type II motor units are larger and innervate more fibers than do α-motor neurons for type I motor units. Thus, type II motor units have more (and larger) fibers to contract and can produce more force than type I motor units.
- The proportions of type I and type II fibers in a person's arm and leg muscles are usually similar.
- Type I fibers have higher aerobic endurance and are well suited to low-intensity endurance activities.
- Type II fibers are better suited for anaerobic activity. Type IIa fibers play a major role in high-intensity exercise. Type IIx fibers are activated when the force demanded of the muscle is high.

Muscle Fiber Recruitment

When an α-motor neuron carries an action potential to the muscle fibers in the motor unit, all fibers in the unit develop force. Activating more motor units is the way muscles produce more force. When little force is needed, only a few motor units are recruited. Recall from our earlier discussion that type IIa and type IIx motor units contain more muscle fibers than type I motor units do. Skeletal muscle contraction involves a progressive recruitment of type I and then type II motor units, depending on the requirements of the activity being performed. As the intensity of the activity increases, the number of fibers recruited increases in the following order, in an additive manner: type I → type IIa → type IIx.

Motor units are generally activated on the basis of a fixed order of fiber recruitment. This is known as the **principle of orderly recruitment**, in which the motor units within a given muscle appear to be ranked. Let's use the biceps brachii as an example: Assume a total of 200 motor units, which are ranked on a scale from 1 to 200. For an extremely fine muscle contraction requiring very little force production, the motor unit ranked number 1 would be recruited. As the requirements for force production increase, numbers 2, 3, 4, and so on would be recruited, up to a maximal muscle contraction that would activate most, if not all, of the motor units. For the production of a given force, the same motor units are usually recruited each time and in the same order.

A mechanism that may partially explain the principle of orderly recruitment is the **size principle**, which states that the order of recruitment of motor units is directly related to the size of their motor neuron. Motor units with smaller motor neurons will be recruited first. Because the type I motor units have smaller motor neurons, they are the first units recruited in graded movement (going from very low to very high rates of force production). The type II motor units then are recruited as the force needed to perform the movement increases. It is unclear at this time how the size principle relates to complex athletic movements.

During events that last several hours, exercise is performed at a submaximal pace, and the tension in the muscles is relatively low. As a result, the nervous system tends to recruit those muscle fibers best adapted to endurance activity: the type I and some type IIa fibers. As the exercise continues, these fibers become depleted of their primary fuel supply (glycogen), and the nervous system must recruit more type IIa fibers to maintain muscle tension. Finally, when the type I and type IIa fibers become exhausted, the type IIx fibers may be recruited to continue exercising.

This may explain why fatigue seems to come in stages during events such as the marathon, a 42 km (26.1 mi) run. It also may explain why it takes great conscious effort to maintain a given pace near the finish of the event. This conscious effort results in the activation of muscle fibers that are not easily recruited. Such information is of practical importance to our understanding of the specific requirements of training and performance.

In review

➤ Motor units give all-or-none responses. Activating more motor units produces more force.

➤ In low-intensity activity, most muscle force is generated by type I fibers. As the intensity increases, type IIa fibers are recruited, and at even higher intensities, the type IIx fibers are activated. The same pattern of recruitment is followed during events of long duration.

Fiber Type and Athletic Success

From what we have just discussed, it appears that athletes who have a high percentage of type I fibers might have an advantage in prolonged endurance events, whereas those with a predominance of type II fibers could be better suited for high-intensity, short-term, and explosive activities. But does the relative proportion of an athlete's muscle fiber types determine athletic success?

The muscle fiber makeup of successful athletes from a variety of athletic events and of nonathletes is shown in table 1.3. As anticipated, the leg muscles of distance runners, who rely on endurance, have a predominance of type I fibers.[3] Studies of elite male and female distance runners revealed that many of these athletes' gastrocnemius (calf) muscles may contain more than 90% type I fibers. Also, although muscle fiber cross-sectional area varies markedly among elite distance runners, type I fibers in their leg muscles average about 22% more cross-sectional area than type II fibers.

TABLE 1.3 Percentages and Cross-Sectional Areas of Type I and Type II Fibers in Selected Muscles of Male and Female Athletes

Athlete	Sex	Muscle	% type I	% type II	Cross-sectional area (µm²) Type I	Type II
Sprint runners	M	Gastrocnemius	24	76	5,878	6,034
	F	Gastrocnemius	27	73	3,752	3,930
Distance runners	M	Gastrocnemius	79	21	8,342	6,485
	F	Gastrocnemius	69	31	4,441	4,128
Cyclists	M	Vastus lateralis	57	43	6,333	6,116
	F	Vastus lateralis	51	49	5,487	5,216
Swimmers	M	Posterior deltoid	67	33	–	–
Weightlifters	M	Gastrocnemius	44	56	5,060	8,910
	M	Deltoid	53	47	5,010	8,450
Triathletes	M	Posterior deltoid	60	40	–	–
	M	Vastus lateralis	63	37	–	–
	M	Gastrocnemius	59	41	–	–
Canoeists	M	Posterior deltoid	71	29	4,920	7,040
Shot-putters	M	Gastrocnemius	38	62	6,367	6,441
Nonathletes	M	Vastus lateralis	47	53	4,722	4,709
	F	Gastrocnemius	52	48	3,501	3,141

In contrast, the gastrocnemius muscles are composed principally of type II fibers in sprint runners, who rely on speed and strength. Although swimmers tend to have higher percentages of type I fibers (60-65%) in their arm muscles than untrained subjects (45-55%), fiber type differences between good and elite swimmers are not apparent.[4, 5]

The fiber composition of muscles in distance runners and sprinters is markedly different. However, it may be a bit risky to think we can select champion distance runners and sprinters solely on the basis of predominant muscle fiber type. Other factors, such as cardiovascular function, motivation, training, and muscle size, also contribute to success in such events of endurance, speed, and strength. Thus, fiber composition alone is not a reliable predictor of athletic success.

Use of Muscles

We have examined the different muscle fiber types. We understand that all fibers in a motor unit, when stimulated, act at the same time and that different fiber types are recruited in stages, depending on the force required to perform an activity. Now we can move back to the whole-muscle level, turning our attention to how whole muscles work to produce movement.

Types of Muscle Contraction

Muscle movement generally can be categorized into three types of contractions—concentric, static, and eccentric. In many activities, such as running and jumping, all three types of contraction may occur in the execution of a smooth, coordinated movement. For the sake of clarity, though, we will examine each type separately.

A muscle's principal action, shortening, is referred to as a **concentric contraction**, the most familiar type of contraction. To understand muscle shortening, recall our earlier discussion of how the thin and thick filaments slide across each other. In a concentric contraction, the thin filaments are pulled toward the center of the sarcomere. Because joint movement is produced, concentric contractions are considered **dynamic contractions**.

Muscles can also act without moving. When this happens, the muscle generates force, but its length remains static (unchanged). This is called a **static**, or **isometric**, **muscle contraction**, because the joint angle does not change. A static contraction occurs, for example, when one tries to lift an object that is heavier than the force generated by the muscle, or when one supports the weight of an object by holding it steady with the elbow flexed. In both cases, the person feels the muscles tense, but there is no joint movement. In a static contraction, the myosin cross-bridges form and are recycled, producing force, but the external force is too great for the thin filaments to be moved. They remain in their normal position, so shortening can't occur. If enough motor units can be recruited to produce sufficient force to overcome the resistance, a static contraction can become a dynamic one.

Muscles can exert force even while lengthening. This movement is an **eccentric contraction**. Because joint movement occurs, this is also a dynamic contraction. An example of an eccentric contraction is the action of the biceps brachii when one extends the elbow to lower a heavy weight. In this case, the thin filaments are pulled farther away from the center of the sarcomere, essentially stretching it.

Generation of Force

Whenever muscles contract, whether the contraction is concentric, static, or eccentric, the force developed must be graded to meet the needs of the task or activity. Using golf as an example, the force needed to tap in a 1 m (~39 in.) putt is far less than that needed to drive the ball 250 m (273 yd) from the tee to the middle of the fairway. The amount of muscle force developed is dependent on the number and type of motor units activated, the frequency of stimulation of each motor unit, the size of the muscle, the muscle fiber and sarcomere length, and the muscle's speed of contraction.

Motor Units and Muscle Size More force can be generated when more motor units are activated. Type II motor units generate more force than type I motor units because a type II motor unit contains more muscle fibers than a type I motor unit. In a similar manner, larger muscles, having more muscle fibers, can produce more force than smaller muscles.

Frequency of Stimulation of the Motor Units: Rate Coding A single motor unit can exert varying levels of force dependent on the frequency at which it is stimulated. This is illustrated in figure 1.12.[1] The smallest contractile response of a muscle fiber or a motor unit to a single electrical stimulus is termed a twitch. A series of three stimuli in rapid sequence, prior to complete relaxation from the first stimulus, can elicit an even greater increase in force or tension. This is termed **summation**. Continued stimulation at higher frequencies can lead to the state of **tetanus**, resulting in the peak force or tension of the muscle fiber or motor unit. **Rate coding** is the term used to describe the process by which the tension of a given motor unit can vary from that of a twitch to that of tetanus by increasing the frequency of stimulation of that motor unit.

Muscle Fiber and Sarcomere Length

There is an optimal length of each muscle fiber relative to its ability to generate force. Recall that a given muscle fiber is composed of sarcomeres connected end to end and that these sarcomeres are composed of both thick and thin filaments. The optimal sarcomere length is defined as that length where there is optimal overlap of the thick and thin filaments, thus maximizing cross-bridge interaction. This is illustrated in figure 1.13.[6] When a sarcomere is fully stretched (1) or shortened (5), little or no force can be developed since there is little cross-bridge interaction.

Speed of Contraction

The ability to develop force also depends on the speed of muscle contraction. During concentric (shortening) contractions, maximal force development decreases progressively at higher speeds. When people try to lift a very heavy object, they tend to do it slowly, maximizing the force they can apply to it. If they grab it and quickly try to lift it, they will likely fail, if not injure themselves. However, with eccentric (lengthening) contractions, the opposite is true. Fast eccentric contractions allow maximal application

FIGURE 1.12 Variation in force or tension produced based on electrical stimulation frequency, illustrating the concepts of a twitch, summation, and tetanus.

Adapted, by permission, from G.A. Brooks, et al., 2005, *Exercise physiology: Human bioenergetics and its applications*, 4th ed. (New York: McGraw-Hill), 388. With permission of the McGraw-Hill Companies.

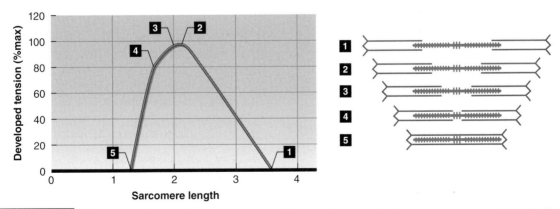

FIGURE 1.13 Variation in force or tension produced (% of maximum) with changes in sarcomere length, illustrating the concept of optimal length for force production.

Adapted, by permission, from B.R. MacIntosh, P.F. Gardiner, and A.J. McComas, 2006, *Skeletal muscle: Form and function*, 2nd ed. (Champaign, IL: Human Kinetics), 156.

of force. These relationships are depicted in figure 1.14. Eccentric contractions are shown on the left and concentric on the right.

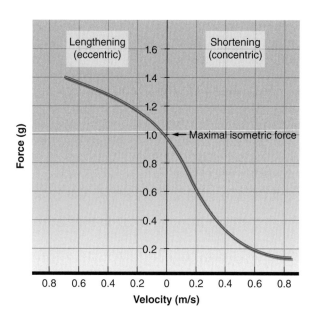

FIGURE 1.14 The relation between muscle lengthening and shortening velocity and force production. Note that the capacity for the muscle to generate force is greater during eccentric (lengthening) actions than during concentric (shortening) actions.

In review

➤ Among elite athletes, muscle fiber type composition differs by sport and event, with speed and strength events characterized by higher percentages of type II fibers and endurance events by higher percentages of type I fibers.

➤ The three main types of muscle contraction are concentric, in which the muscle shortens; static or isometric, in which the muscle acts but the joint angle is unchanged; and eccentric, in which the muscle lengthens.

➤ Force production can be increased both through the recruitment of more motor units and through an increase in the frequency of stimulation (rate coding) of the motor units.

➤ Force production is maximized at the muscle's optimal length. At this length, the amount of energy stored and the number of linked actin-myosin cross-bridges are optimal.

➤ Speed of contraction also affects the amount of force produced. For concentric contraction, maximal force is achieved with slower contractions. The closer to zero velocity (static), the more force can be generated. With eccentric contractions, however, faster movement allows more force production.

In closing

In this chapter, we reviewed the components of skeletal muscle. We considered the differences in fiber types and their impact on physical performance. We learned how muscles generate force and produce movement. Now that we understand how movement is produced, we turn our attention to how movement is fueled. In the next chapter, we focus on metabolism and energy production.

Key Terms

actin
action potential
adenosine triphosphatase (ATPase)
adenosine triphosphate (ATP)
α-motor neuron
concentric contraction
dynamic contraction
eccentric contraction
endomysium

epimysium
excitation–contraction coupling
fasciculus (fascicle)
motor unit
muscle fiber
musculoskeletal system
myofibril
myosin
myosin cross-bridge

nebulin

perimysium

plasmalemma

power stroke

principle of orderly recruitment

rate coding

sarcolemma

sarcomere

sarcoplasm

sarcoplasmic reticulum (SR)

satellite cells

single-fiber contractile velocity (V_o)

size principle

sliding filament theory

static (isometric) contraction

summation

tetanus

titin

transverse tubules (T-tubules)

tropomyosin

troponin

twitch

type I (slow-twitch) fiber

type II (fast-twitch) fiber

Study Questions

1. List and define the anatomical components that make up a muscle fiber.
2. List the components of a motor unit.
3. What are the steps in excitation–contraction coupling?
4. What is the role of calcium in muscle contraction?
5. Describe the sliding filament theory. How do muscle fibers shorten?
6. What are the basic characteristics that differ between type I and type II muscle fibers?
7. What is the role of genetics in determining the proportions of muscle fiber types and the potential for success in selected activities?
8. Describe the relationship between muscle force development and the recruitment of type I and type II motor units.
9. Differentiate among, and give examples of, concentric, static, and eccentric contractions.
10. What two mechanisms are used by the body to increase force production in a single muscle?
11. What is the optimal length of a muscle for maximal force development?
12. What is the relation between maximal force development and the speed of shortening (concentric) and lengthening (eccentric) contractions?

Study Guide Activities

In addition to the activities listed in the chapter opening outline on page 27, two other activities are available in the web study guide, located at

www.HumanKinetics.com/PhysiologyOfSportAndExercise

 The **KEY TERMS** activity reviews important terms, and the end-of-chapter **QUIZ** tests your understanding of the material covered in the chapter.

Fuel for Exercise: Bioenergetics and Muscle Metabolism

2

In this chapter and in the web study guide

ACTIVITY 2.1 ATP-PCr System reviews the stages in the ATP-PCr system.

ACTIVITY 2.2 Glycolytic System considers the main steps in the glycolytic system.

ACTIVITY 2.3 Glucose Oxidation describes how the complete oxidation of glucose produces ATP, heat, water, and carbon dioxide.

ACTIVITY 2.4 ATP Production explores three methods of ATP production, depending on the type, length, and intensity of an activity and the availability of oxygen.

"Hitting the wall" is a common expression heard among marathon runners, and more than half of all nonelite marathon runners report having "hit the wall" during a marathon regardless of how hard they trained. This phenomenon usually happens around mile 20 to 22. The runner's pace slows considerably and the legs feel like lead. Tingling and numbness are often felt in the legs and arms, and thinking often becomes fuzzy and confused. "Hitting the wall" is basically running out of available energy.

The runner's primary fuel sources during prolonged exercise are carbohydrates and fats. Fats might seem to be the logical first choice of fuel for endurance events—they are ideally designed to be energy dense, and stores are virtually unlimited. Unfortunately, fat metabolism requires a constant supply of oxygen, and delivery of energy is slower than that provided by carbohydrate metabolism.

Most runners are able to store 2,000 to 2,200 calories of glycogen in their liver and muscles, which is enough to provide energy for about 20 mi of moderate-pace running. Since the body is much less efficient at converting fat to energy, running pace slows and the runner suffers from fatigue. Furthermore, carbohydrates are the sole fuel source for brain function. Physiology, not coincidence, dictates why so many marathon runners hit the wall at around the 20 mi mark.

All energy originates from the sun as light energy. Chemical reactions in plants (photosynthesis) convert light into stored chemical energy. In turn, humans obtain energy either by eating plants or by eating animals that feed on plants. Nutrients from ingested foods are provided in the form of carbohydrates, fats, and proteins. These three basic fuels, or energy **substrates**, can ultimately be broken down to release the stored energy. Each cell contains chemical pathways that convert these substrates to energy that can then be used by that cell and other cells of the body, a process called **bioenergetics**. All of the chemical reactions in the body are collectively called **metabolism**.

Because all energy eventually degrades to heat, the amount of energy released in a biological reaction can be calculated from the amount of heat produced. Energy in biological systems is measured in calories. By definition, 1 calorie (cal) equals the amount of heat energy needed to raise 1 g of water 1 °C, from 14.5 °C to 15.5 °C. In humans, energy is expressed in **kilocalories (kcal)**, where 1 kcal is the equivalent of 1,000 cal. Sometimes the term *Calorie* (with a capital C) is used synonymously with kilocalorie, but *kilocalorie* is more technically and scientifically correct. Thus, when one reads that someone eats or expends 3,000 Cal per day, it really means the person is ingesting or expending 3,000 *kcal* per day.

Some free energy in the cells is used for growth and repair throughout the body. Such processes build muscle mass during training and repair muscle damage after exercise or injury. Energy also is needed for active transport of many substances, such as sodium, potassium, and calcium ions, across cell membranes. Active transport is critical to the survival of cells and the maintenance of homeostasis. Myofibrils also use some of the energy released in our bodies to cause sliding of the actin and myosin filaments, resulting in muscle action and force generation, as we saw in chapter 1.

Energy Substrates

Energy is released when chemical bonds—the bonds that hold elements together to form molecules—are broken. Substrates are composed primarily of carbon, hydrogen, oxygen, and (in the case of protein) nitrogen. The molecular bonds that hold these elements together are relatively weak and therefore provide little energy when broken. Consequently, food is not used directly for cellular operations. Rather, the energy in food molecular bonds is chemically released within our cells and then stored in the form of the high-energy compound introduced in chapter 1, adenosine triphosphate (ATP), which is discussed in detail later in this chapter.

At rest, the energy that the body needs is derived almost equally from the breakdown of carbohydrates and fats. Proteins serve important functions as enzymes that aid chemical reactions and as structural building blocks but usually provide little energy for metabolism. During intense, short-duration muscular effort, more carbohydrate is used, with less reliance on fat to generate ATP. Longer, less intense exercise utilizes carbohydrate and fat for sustained energy production.

Carbohydrate

The amount of **carbohydrate** utilized during exercise is related to both the carbohydrate availability and the muscles' well-developed system for carbohydrate metabolism. All carbohydrates are ultimately converted to the simple six-carbon sugar, **glucose** (figure 2.1), a monosaccharide (one-unit sugar) that is transported through the blood to all body tissues. Under resting conditions, ingested carbohydrate is stored in muscles and liver in the form of a more complex polysaccharide (multiple linked sugar molecules), **glycogen**. Glycogen is stored in the cytoplasm of muscle cells until those

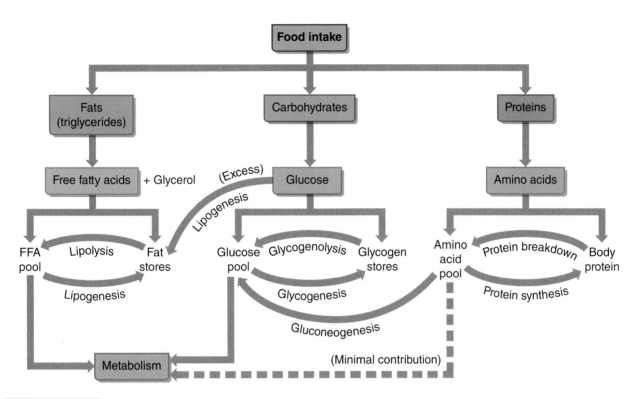

FIGURE 2.1 Cellular metabolism results from the breakdown of three fuel substrates provided by the diet. Once each is converted to its usable form, it either circulates in the blood as an available "pool" to be used for metabolism or is stored in the body.

cells use it to form ATP. The glycogen stored in the liver is converted back to glucose as needed and then transported by the blood to active tissues, where it is metabolized.

Liver and muscle glycogen stores are limited and can be depleted during prolonged, intense exercise, especially if the diet contains an insufficient amount of carbohydrate. Thus, we rely heavily on dietary sources of starches and sugars to continually replenish our carbohydrate reserves. Without adequate carbohydrate intake, muscles can be deprived of their primary energy source. Furthermore, carbohydrates are the only energy source utilized by brain tissue; therefore severe carbohydrate depletion results in negative cognitive effects.

(In)focus

Carbohydrate stores in the liver and skeletal muscle are limited to about 2,500 to 2,600 kcal of energy, or the equivalent of the energy needed for about 40 km (25 mi) of running. Fat stores can provide more than 70,000 kcal of energy.

Fat

Fats provide a large portion of the energy utilized during prolonged, less intense exercise. Body stores of potential energy in the form of fat are substantially larger than the reserves of carbohydrate, in terms of both weight and potential energy. Table 2.1 provides an indication of the total body stores of these two energy sources in a lean person (12% body fat). For the average middle-aged adult with more body fat (adipose tissue), the fat stores would be approximately twice as large, whereas the carbohydrate stores would be about the same. But fat is less readily available for cellular metabolism because it must first be reduced from its complex form, **triglyceride**, to its basic components, glycerol and **free fatty acids (FFAs)**. Only FFAs are used to form ATP (figure 2.1).

Substantially more energy is derived from breaking down a gram of fat (9.4 kcal/g) than from the same amount of carbohydrate (4.1 kcal/g). Nonetheless, the rate of energy release from fat is too slow to meet all of the energy demands of intense muscular activity.

Other types of fats found in the body serve non-energy-producing functions. Phospholipids are a

TABLE 2.1 Body Stores of Fuels and Associated Energy Availability

Location	g	kcal
CARBOHYDRATES		
Liver glycogen	110	451
Muscle glycogen	500	2,050
Glucose in body fluids	15	62
FAT		
Subcutaneous and visceral	7,800	73,320
Intramuscular	161	1,513
Total	**7,961**	**74,833**

Note. These estimates are based on a body weight of 65 kg (143 lb) with 12% body fat.

key structural component of all cell membranes and form protective sheaths around some large nerves. Steroids are found in cell membranes and also function as hormones or as building blocks of hormones such as estrogen and testosterone.

Protein

Protein also can be used as a minor energy source under some circumstances, but it must first be converted into glucose (figure 2.1). In the case of severe energy depletion or starvation, protein may even be used to generate FFAs for cellular energy. The process by which protein

or fat is converted into glucose is called **gluconeogenesis**. The process of converting protein into fatty acids is termed **lipogenesis**. Protein can supply up to 5% or 10% of the energy needed to sustain prolonged exercise. Only the most basic units of protein—the amino acids—can be used for energy. A gram of protein yields about 4.1 kcal.

Controlling the Rate of Energy Production

To be useful, free energy must be released from chemical compounds at a controlled rate. This rate is primarily determined by two things, the availability of the primary substrate and enzyme activity. The availability of large amounts of a substrate increases the activity of that particular pathway. An abundance of one particular fuel (e.g., carbohydrate) can cause cells to rely more on that source than on alternatives. This influence of substrate availability on the rate of metabolism is termed the *mass action effect*.

Specific protein molecules called **enzymes** also control the rate of free-energy release. Many of these enzymes speed up the breakdown (**catabolism**) of chemical compounds. Chemical reactions occur only when the reacting molecules have sufficient initial energy to start the reaction or chain of reactions. Enzymes do not cause a chemical reaction to occur and do not determine the amount of usable energy that is produced by these reactions. Rather, they speed up reactions by lowering the **activation energy** that is required to begin the reaction (figure 2.2).

FIGURE 2.2 Enzymes control the rate of chemical reactions by lowering the activation energy required to initiate the reaction. In this example, the enzyme creatine kinase binds to its substrate phosphocreatine to increase the rate of production of creatine.
Adapted from original figure provided by Dr. Martin Gibala, McMaster University, Hamilton, Ontario, Canada.

Although the enzyme names are quite complex, most end with the suffix *-ase*. For example, an important enzyme that acts to break down ATP and release stored energy is adenosine triphosphatase (ATPase).

Biochemical pathways that result in the production of a product from a substrate typically involve multiple steps. Each individual step is typically catalyzed by a specific enzyme. Therefore, increasing the amount of enzyme present or the activity of that enzyme (for example by changing the temperature or pH) results in an increased rate of product formation through that metabolic pathway. Additionally, many enzymes require other molecules called "cofactors" to function, so cofactor availability may also affect enzyme function and therefore the rate of metabolic reactions.

As illustrated in figure 2.3, typical metabolic pathways have one enzyme that is of particular importance in controlling the rate. This enzyme, usually located in an early step in the pathway, is known as the **rate-limiting enzyme**. The activity of a rate-limiting enzyme is determined by the accumulation of substances farther down the pathway that decrease enzyme activity through **negative feedback**.

One example of a substance that may accumulate and feed back to decrease enzyme activity would be the end product of the pathway; another would be ATP and its breakdown products, ADP and inorganic phosphate. If the goals of a metabolic pathway are to form a chemical product and release free energy in the form of ATP, it makes sense that an abundance of either that end product or ATP would feed back to slow further production and release, respectively.

In review

➤ Energy for cell metabolism is derived from three substrates in foods: carbohydrate, fat, and protein. Proteins provide little of the energy used for metabolism under normal conditions.

➤ Within cells, the usable storage form of the energy we derive from food is the high-energy compound adenosine triphosphate or ATP.

➤ Carbohydrate and protein each provide about 4.1 kcal of energy per gram, compared with about 9.4 kcal/g for fat.

➤ Carbohydrate, stored as glycogen in muscle and the liver, is more quickly accessible as an energy source than either protein or fat. Glucose,

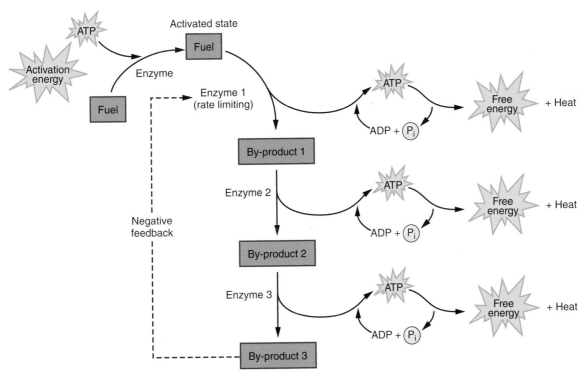

FIGURE 2.3 A typical metabolic pathway showing the important role of enzymes in controlling the rate of the reaction. An input of energy in the form of stored ATP is needed to begin the series of reactions (activation energy), but less initial energy is needed if one or more enzymes are involved in this activation step. As fuels are subsequently degraded into by-products along the metabolic pathway, ATP is formed. Utilization of the stored ATP results in the release of usable energy, heat, and the release of ADP and P_i.

- directly from food or broken down from stored glycogen, is the usable form of carbohydrate.

➤ Fat, stored as triglycerides in adipose tissue, is an ideal storage form of energy. Free fatty acids from the breakdown of triglycerides are converted to energy.

➤ Enzymes control the rate of metabolism and energy production. Enzymes can speed up the overall reaction by lowering the initial activation energy and by catalyzing various steps along the pathway.

➤ Enzymes can be inhibited through negative feedback of subsequent pathway by-products (or often ATP), slowing the overall rate of the reaction. This usually involves a particular enzyme located early in the pathway called the rate-limiting enzyme.

Storing Energy: High-Energy Phosphates

The immediately available source of energy for almost all metabolism including muscle contraction is adenosine triphosphate, or ATP. An ATP molecule (figure 2.4a) is composed of adenosine (a molecule of adenine joined to a molecule of ribose) combined with three inorganic phosphate (P_i) groups. Adenine is a nitrogen-containing base, and ribose is a five-carbon sugar. When an ATP molecule is combined with water (hydrolysis) and acted on by the enzyme ATPase, the last phosphate group splits away, rapidly releasing a large amount of free energy (approximately 7.3 kcal per mole of ATP under standard conditions, but possibly up to 10 kcal per mole of ATP or greater within the cell). This reduces the ATP to **adenosine diphosphate (ADP)** and P_i (figure 2.4b).

To generate ATP, a phosphate group is added to the relatively low-energy compound, ADP, in a process called **phosphorylation**. This process requires a considerable amount of energy. Some ATP is generated independent of oxygen availability, and such metabolism is called substrate-level phosphorylation. Other ATP-producing reactions (discussed later in the chapter) occur without oxygen, while still others occur with the aid of oxygen, a process called **oxidative phosphorylation**.

As shown in figure 2.3, ATP is formed from ADP and P_i via phosphorylation as fuels are broken down into fuel by-products at various steps along a metabolic pathway. The storage form of energy, ATP, can subsequently release free or usable energy when needed as it is once again broken down into ADP and P_i.

FIGURE 2.4 (a) The structure of an adenosine triphosphate (ATP) molecule, showing the high-energy phosphate bonds. (b) When the third phosphate on the ATP molecule is separated from adenosine by the action of adenosine triphosphatase (ATPase), energy is released.

The Basic Energy Systems

Cells can store only very limited amounts of ATP and must constantly generate new ATP to provide needed energy for all cellular metabolism including muscle contraction. Cells generate ATP through any one of (or a combination of) three metabolic pathways:

1. The ATP-PCr system
2. The glycolytic system (glycolysis)
3. The oxidative system (oxidative phosphorylation)

The first two systems can occur in the absence of oxygen and are jointly termed **anaerobic metabolism**. The third system requires oxygen and therefore comprises **aerobic metabolism**.

ATP-PCr System

The simplest of the energy systems is the **ATP-PCr system**, shown in figure 2.5. In addition to storing a very small amount of ATP directly, cells contain another high-energy phosphate molecule that stores energy called **phosphocreatine**, or PCr (sometimes called creatine phosphate). This simple pathway involves donation of a P_i from PCr to ADP to form ATP. Unlike the

limited freely available ATP in the cell, energy released by the breakdown of PCr is not directly used for cellular work. Instead, it regenerates ATP to maintain a relatively constant supply under resting conditions.

The release of energy from PCr is catalyzed by the enzyme **creatine kinase**, which acts on PCr to separate P_i from creatine. The energy released can then be used to add a P_i molecule to an ADP molecule, forming ATP. As energy is released from ATP by the splitting of a phosphate group, cells can prevent ATP depletion by breaking down PCr, providing energy and P_i to re-form ATP from ADP.

Following the principle of negative feedback and rate-limiting enzymes discussed earlier, creatine kinase activity is enhanced when concentrations of ADP or P_i increase, and is inhibited when ATP concentrations increase. When intense exercise is initiated, the small amount of available ATP in muscle cells is broken down for immediate energy, yielding ADP and P_i. The increased ADP concentration enhances creatine kinase activity, and PCr is catabolized to form additional ATP. As exercise progresses and additional ATP is generated by the other two energy systems—the glycolytic and oxidative systems—creatine kinase activity is inhibited.

This process of breaking down PCr to allow formation of ATP is rapid and can be accomplished without any special structures within the cell. The ATP-PCr system is classified as substrate-level metabolism. Although it can occur in the presence of oxygen, the process does not require oxygen.

During the first few seconds of intense muscular activity, such as sprinting, ATP is maintained at a relatively constant level, but PCr declines steadily as it is used to replenish the depleted ATP (see figure 2.6). At

FIGURE 2.6 Changes in type II (fast-twitch) skeletal muscle adenosine triphosphate (ATP) and phosphocreatine (PCr) during 14 s of maximal muscular effort (sprinting). Although ATP is being used at a very high rate, the energy from PCr is used to synthesize ATP, preventing the ATP level from decreasing. However, at exhaustion, both ATP and PCr levels are low.

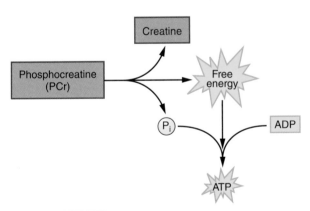

FIGURE 2.5 In the ATP-PCr system, adenosine triphosphate (ATP) can be re-created via the binding of an inorganic phosphate (P_i) to adenosine diphosphate (ADP) with the energy derived from the breakdown of phosphocreatine (PCr).

exhaustion, however, both ATP and PCr levels are low and are unable to provide energy for further muscle contraction and relaxation. Thus, the capacity to maintain ATP levels with the energy from PCr is limited. The combination of ATP and PCr stores can sustain the muscles' energy needs for only 3 to 15 s during an all-out sprint. Beyond that time, muscles must rely on other processes for ATP formation: glycolytic and oxidative pathways.

Glycolytic System

The ATP-PCr system has a limited capacity to generate ATP for energy, lasting only a few seconds. The second method of ATP production involves the liberation of energy through the breakdown ("lysis") of glucose. This system is called the glycolytic system because it entails **glycolysis**, the breakdown of glucose through a pathway that involves a sequence of glycolytic enzymes. Glycolysis is a more complex pathway than the ATP-PCr system, and the sequence of steps involved in this process is presented in figure 2.7.

Glucose accounts for about 99% of all sugars circulating in the blood. Blood glucose comes from the digestion of carbohydrate and the breakdown of liver glycogen. Glycogen is synthesized from glucose by a process called glycogenesis and is stored in the liver or in muscle until needed. At that time, the glycogen is broken down to glucose-1-phosphate, which enters the glycolysis pathway, a process termed **glycogenolysis**.

Before either glucose or glycogen can be used to generate energy, it must be converted to a compound called glucose-6-phosphate. Even though the goal of glycolysis is to release ATP, the conversion of a molecule of glucose to glucose-6-phosphate requires the expenditure or input of one ATP molecule. In the conversion of glycogen, glucose-6-phosphate is formed from glucose-1-phosphate without this energy expenditure. Glycolysis technically begins once the glucose-6-phosphate is formed.

Glycolysis requires 10 to 12 enzymatic reactions for the breakdown of glycogen to pyruvic acid, which is then converted to lactic acid. All steps in the pathway and all of the enzymes involved operate within the cell cytoplasm. The net gain from this process is 3 moles (mol) of ATP formed for each mole of glycogen broken down. If glucose is used instead of glycogen, the gain is only 2 mol of ATP because 1 mol was used for the conversion of glucose to glucose-6-phosphate.

This energy system obviously does not produce large amounts of ATP. Despite this limitation, the combined actions of the ATP-PCr and glycolytic systems allow the muscles to generate force even when the oxygen supply is limited. These two systems predominate during the early minutes of high-intensity exercise.

Another major limitation of anaerobic glycolysis is that it causes an accumulation of lactic acid in the muscles and body fluids. Glycolysis produces pyruvic acid. This process does not require oxygen, but the presence of oxygen determines the fate of the pyruvic acid. Without oxygen present, the pyruvic acid is converted directly to lactic acid, an acid with the chemical formula $C_3H_6O_3$. Anaerobic glycolysis produces lactic acid, but it quickly dissociates, and lactate is formed.

In focus

The terms "pyruvic acid" and "pyruvate," and "lactic acid" and "lactate," are often used interchangeably in exercise physiology. In each case, the acid form of the molecule is relatively unstable at normal body pH and rapidly loses a hydrogen ion. The remaining molecule is more correctly called pyruvate or lactate.

In all-out sprint events lasting 1 or 2 min, the demands on the glycolytic system are high, and muscle lactic acid concentrations can increase from a resting value of about 1 mmol/kg of muscle to more than 25 mmol/kg. This acidification of muscle fibers inhibits further glycogen breakdown because it impairs glycolytic enzyme function. In addition, the acid decreases the fibers' calcium-binding capacity and thus may impede muscle contraction.

The rate-limiting enzyme in the glycolytic pathway is **phosphofructokinase** or **PFK**. Like almost all rate-limiting enzymes, PFK catalyzes an early step in the pathway, the conversion of fructose-6-phosphate to fructose-1,6-biphosphate. Increasing ADP and P_i concentrations enhance PFK activity and therefore speed up glycolysis, while elevated ATP concentrations slow glycolysis by inhibiting PFK. Additionally, because the glycolytic pathway feeds into the Krebs cycle for additional energy production when oxygen is present (discussed later), products of the Krebs cycle, especially citrate and hydrogen ions, likewise feed back to inhibit PFK.

A muscle fiber's rate of energy use during exercise can be 200 times greater than at rest. The ATP-PCr and glycolytic systems alone cannot supply all the needed energy. Furthermore, these two systems are not capable of supplying all of the energy needs for all-out activity lasting more than 2 min or so. Prolonged exercise relies on the third energy system, the oxidative system.

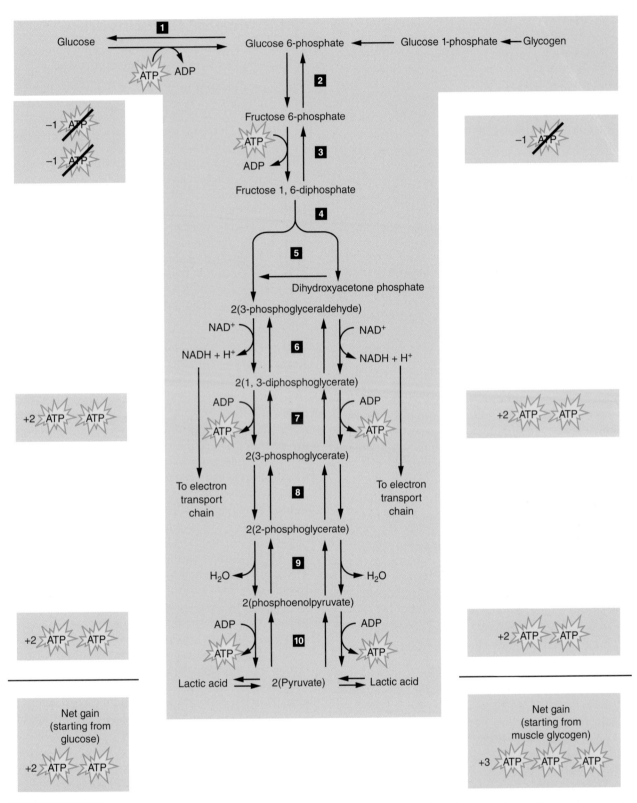

FIGURE 2.7 The derivation of energy (ATP) by glycolysis. Glycolysis involves the breakdown of one glucose (six-carbon) molecule to two three-carbon molecules of pyruvic acid. The process can begin with either glucose circulating in the blood or glycogen (a chain of glucose molecules, the storage form of glucose in muscle and liver). Note that there are roughly 10 separate steps in this anaerobic process, and the net result is the generation of either two or three ATP molecules depending on whether glucose or glycogen is the initial substrate.

In review

➤ Adenosine triphosphate is generated through three energy systems:
  1. The ATP-PCr system
  2. The glycolytic system
  3. The oxidative system

➤ In the ATP-PCr system, P_i is separated from PCr through the action of creatine kinase. The P_i can then combine with ADP to form ATP using the energy released from the breakdown of PCr. This system is anaerobic, and its main function is to maintain ATP levels early in exercise. The energy yield is 1 mol of ATP per 1 mol of PCr.

➤ The glycolytic system involves the process of glycolysis, through which glucose or glycogen is broken down to pyruvic acid. When glycolysis occurs without oxygen, the pyruvic acid is converted to lactic acid. One mol of glucose yields 2 mol of ATP, but 1 mol of glycogen yields 3 mol of ATP.

➤ The ATP-PCr and glycolytic systems are major contributors of energy during short-burst activities lasting up to 2 min and during the early minutes of longer high-intensity exercise.

Oxidative System

The final system of cellular energy production is the **oxidative system**. This is the most complex of the three energy systems, and only a brief overview is provided here. The process by which the body breaks down substrates with the aid of oxygen to generate energy is called cellular respiration. Because oxygen is required, this is an aerobic process. Unlike the anaerobic production of ATP that occurs in the cytoplasm of the cell, the oxidative production of ATP occurs within special cell organelles called **mitochondria**. In muscles, these are adjacent to the myofibrils and are also scattered throughout the sarcoplasm. (See figure 1.3, p. 30.)

Muscles need a steady supply of energy to continuously produce the force needed during long-term activity. Unlike anaerobic ATP production, the oxidative system is slow to turn on; but it has a much larger energy-producing capacity, so aerobic metabolism is the primary method of energy production during endurance activities. This places considerable demands on the cardiovascular and respiratory systems to deliver oxygen to the active muscles. Oxidative energy production can come from carbohydrates (starting with glycolysis) or fats.

Oxidation of Carbohydrate

As shown in figure 2.8, oxidative production of ATP from carbohydrates involves three processes:

• Glycolysis (figure 2.8a)
• The Krebs cycle (figure 2.8b)
• The electron transport chain (figure 2.8c)

Glycolysis In carbohydrate metabolism, glycolysis plays a role in *both* anaerobic and aerobic ATP production. The process of glycolysis is the same regardless of whether oxygen is present. The presence of oxygen determines only the fate of the end product, pyruvic acid. Recall that anaerobic glycolysis produces lactic acid and only three net mol of ATP per mol of glycogen, or two net mol of ATP per mol of glucose. In the presence of oxygen, however, the pyruvic acid is converted into a compound called **acetyl coenzyme A (acetyl CoA)**.

Krebs Cycle Once formed, acetyl CoA enters the **Krebs cycle** (also called the citric acid cycle or tricyclic acid cycle), a complex series of chemical reactions that permit the complete oxidation of acetyl CoA (shown in figure 2.9). Recall that, for every glucose molecule that enters the glycolytic pathway, two molecules of pyruvate are formed. Therefore, each glucose molecule that begins the energy-producing process in the presence of oxygen results in two complete Krebs cycles.

As depicted in 2.8b (and shown in more detail in figure 2.9), the conversion of succinyl CoA to succinate in the Krebs cycle results in the generation of guanosine triphosphate or GTP, a high-energy compound similar to ATP. GTP then transfers a P_i to ADP to form ATP. These two ATPs (per molecule of glucose) are formed by substrate-level phosphorylation. So at the end of the Krebs cycle, two additional mol of ATP have been formed directly, and the original carbohydrate has been broken down into carbon dioxide and hydrogen.

Like the other pathways involved in energy metabolism, Krebs cycle enzymes are regulated by negative feedback at several steps in the cycle. The rate-limiting enzyme in the Krebs cycle is isocitrate dehydrogenase, which, like PFK, is inhibited by ATP and activated by ADP and P_i, as is the electron transport chain. Because muscle contraction relies on the availability of calcium in the cell, excess calcium also stimulates the rate-limiting enzyme isocitrate dehydrogenase.

Electron Transport Chain During glycolysis, hydrogen ions are released when glucose is metabolized to pyruvic acid. Additional hydrogen ions are released in the conversion of pyruvate to acetyl CoA and at

FIGURE 2.8 In the presence of oxygen, after glucose (or glycogen) has been reduced to pyruvate, (*a*) the pyruvate is first catalyzed to acetyl coenzyme A (acetyl CoA), which can enter (*b*) the Krebs cycle, where oxidative phosphorylation occurs. Hydrogen ions released during the Krebs cycle then combine with coenzymes that carry the hydrogen ions to (*c*) the electron transport chain.

several steps in the Krebs cycle. If these hydrogen ions remained in the system, the inside of the cell would become too acidic. What happens to this hydrogen?

The Krebs cycle is coupled to a series of reactions known as the **electron transport chain** (figure 2.8*c*). The hydrogen ions released during glycolysis, during the conversion of pyruvic acid to acetyl CoA, and during the Krebs cycle combine with two coenzymes: nicotinamide adenine dinucleotide (NAD) and flavin adenine dinucleotide (FAD), converting each to its reduced form (NADH and $FADH_2$, respectively). During each Krebs cycle, three molecules of NADH and one molecule of $FADH_2$ are produced. These carry the hydrogen atoms (electrons) to the electron transport chain, a group of mitochondrial protein complexes located in the inner mitochondrial membrane. These protein complexes contain a series of enzymes and iron-containing proteins known as **cytochromes**. As high-energy electrons are passed from complex to complex along this chain,

some of the energy released by those reactions is used to pump H^+ from the mitochondrial matrix into the outer mitochondrial compartment. As these hydrogen ions move back across the membrane down their concentration gradient, energy is transferred to ADP, and ATP is formed. This final step requires an enzyme known as ATP synthase. At the end of the chain, the H^+ combines with oxygen to form water, thus preventing acidification of the cell. This is illustrated in figure 2.10. Because this overall process relies on oxygen as the final acceptor of electrons and H^+, it is referred to as **oxidative phosphorylation**.

For every pair of electrons transported to the electron transport chain by NADH, three molecules of ATP are formed, while the electrons passed through the electron transport chain by FADH yield only two molecules of ATP. However, because the NADH and FADH are outside the membrane of the mitochondria, the H^+ must be shuttled through the membrane, which

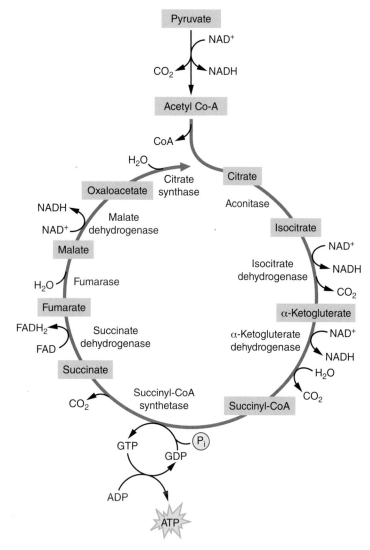

FIGURE 2.9 The series of reactions that take place during the Krebs cycle, showing the compounds formed and enzymes involved.

requires energy to be utilized. So in reality, the net yields are only 2.5 ATPs per NADH and 1.5 ATPs per FADH.

Energy Yield From Oxidation of Carbohydrate

The complete oxidation of glucose can generate 32 molecules of ATP, while 33 ATPs are produced from one molecule of muscle glycogen. The sites of ATP production are summarized in figure 2.11. The net production of ATP from substrate-level phosphorylation in the glycolytic pathway leading into the Krebs cycle results in a net gain of two ATPs (or three from glycogen). A total of 10 NADH molecules leading into the electron transport chain—two in glycolysis, two in the conversion of pyruvic acid to acetyl CoA, and six in the Krebs cycle—yields 25 net ATP molecules. Remember that while 30 ATPs are actually produced,

the energy cost of transporting ATP across membranes uses five of those ATPs. The two FAD molecules in the Krebs cycle that are involved in electron transport result in three additional net ATPs. And finally, substrate-level phosphorylation within the Krebs cycle involving the molecule GTP adds another two ATP molecules.

Accounting for the energy cost of shuttling electrons across the mitochondrial membrane is a relatively new concept in exercise physiology, and many textbooks still refer to net energy productions of 36-39 ATPs per molecule of glucose.

Oxidation of Fat

As noted earlier, fat also contributes importantly to muscles' energy needs. Muscle and liver glycogen stores can provide only ~2,500 kcal of energy, but the fat stored inside muscle fibers and in fat cells can supply at least 70,000 to 75,000 kcal, even in a lean adult. Although many chemical compounds (such as triglycerides, phospholipids, and cholesterol) are classified as fats, only triglycerides are major energy sources. Triglycerides are stored in fat cells and between and within skeletal muscle fibers. To be used for energy, a triglyceride must be broken down to its basic units: one molecule of glycerol and three FFA molecules. This process is called **lipolysis**, and it is carried out by enzymes known as lipases.

Free fatty acids are the primary energy source for fat metabolism. Once liberated from glycerol, FFAs can enter the blood and be transported throughout the body, entering muscle fibers by either simple diffusion or transporter-mediated (facilitated) diffusion. Their rate of entry into the muscle fibers depends on the concentration gradient. Increasing the concentration of FFAs in the blood increases the rate of their transport into muscle fibers.

β-Oxidation Recall that fats are stored in the body in two places, within muscle fibers and in adipose tissue cells called adipocytes. The storage form of fats is triglyceride, which is broken down into FFAs and

In focus

Although fat provides more kilocalories of energy per gram than carbohydrate, fat oxidation requires more oxygen than carbohydrate oxidation. The energy yield from fat is 5.6 ATP molecules per oxygen molecule used, compared with carbohydrate's yield of 6.3 ATP per oxygen molecule. Oxygen delivery is limited by the oxygen transport system, so carbohydrate is the preferred fuel during high-intensity exercise.

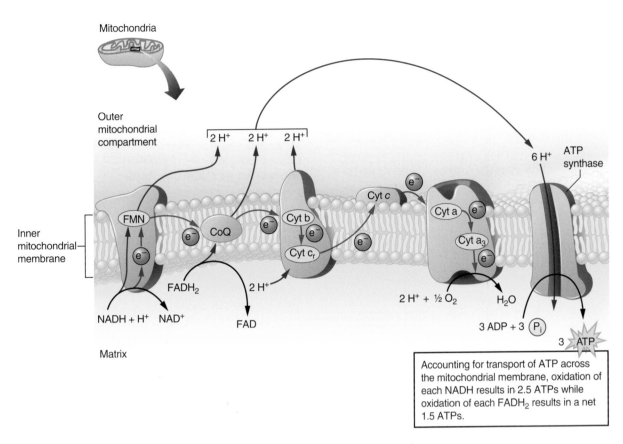

FIGURE 2.10 The final step in the aerobic production of ATP is the transfer of energy from the high-energy electrons of NADH and FADH$_2$ within the mitochondria, following a series of steps known as the electron transport chain.

FIGURE 2.11 The net energy production from the oxidation of one molecule of glucose is 32 molecules of ATP. Oxidation of glycogen as the original substrate would yield one additional ATP.

glycerol for energy metabolism. Before FFAs can be used for energy production, they must be converted to acetyl CoA in the mitochondria, a process called β-oxidation. Acetyl CoA is the common intermediate through which all substrates enter the Krebs cycle for oxidative metabolism.

β-Oxidation is a series of steps in which two-carbon acyl units are chopped off of the carbon chain of the FFA. The number of steps depends on the number of carbons in the FFA, usually between 14 and 24 carbons. For example, if an FFA originally has a 16-carbon chain, β-oxidation yields eight molecules of acetyl CoA. The acyl units become acetyl CoA, which then enters the Krebs cycle for the formation of ATP.

On entering the muscle fiber, FFAs must be enzymatically activated with energy from ATP, preparing them for catabolism (breakdown) within the mitochondria. Like glycolysis, β-oxidation requires an input energy of two ATPs for activation but, unlike glycolysis, produces no ATPs directly.

(In)focus

The maximum rate of ATP production from lipid oxidation is too low to match the rate of utilization of ATP during high-intensity exercise. This explains the reduction in an athlete's race pace when carbohydrate stores are depleted and fat, by default, becomes the predominant fuel source.

Krebs Cycle and the Electron Transport Chain After β-oxidation, fat metabolism follows the same path as oxidative carbohydrate metabolism. Acetyl CoA formed by β-oxidation enters the Krebs cycle. The Krebs cycle generates hydrogen, which is transported to the electron transport chain along with the hydrogen generated during β-oxidation to undergo oxidative phosphorylation. As in glucose metabolism, the byproducts of FFA oxidation are ATP, H_2O, and carbon dioxide (CO_2). However, the complete combustion of an FFA molecule requires more oxygen because an FFA molecule contains considerably more carbon molecules than a glucose molecule.

The advantage of having more carbon molecules in FFAs than in glucose is that more acetyl CoA is formed from the metabolism of a given amount of fat, so more acetyl CoA enters the Krebs cycle and more electrons are sent to the electron transport chain. This is why fat metabolism can generate much more energy than glucose metabolism. Unlike glucose or glycogen, fats are heterogeneous, and the amount of ATP produced depends on the specific fat oxidized.

Consider the example of palmitic acid, a rather abundant 16-carbon FFA. The combined reactions of oxidation, the Krebs cycle, and the electron transport chain produce 106 molecules of ATP from one molecule of palmitic acid (as shown in table 2.2), compared with only 32 molecules of ATP from glucose or 33 from glycogen.

TABLE 2.2 ATP Produced From One Molecule of Palmitic Acid

Stage of process	Direct (substrate level oxidation)	By oxidative phosphorylation
Fatty acid activation	0	–2
β-Oxidation (7 rounds)	0	28
Krebs cycle (8 rounds)	8	72
Subtotal	8	98
Total	**106**	

Oxidation of Protein

As noted earlier, carbohydrates and fatty acids are the preferred fuel substrates. But proteins, or rather the amino acids that compose proteins, are also used for energy under some circumstances. Some amino acids can be converted into glucose, a process called gluconeogenesis (see figure 2.1). Alternatively, some can be converted into various intermediates of oxidative metabolism (such as pyruvate or acetyl CoA) to enter the oxidative process.

Protein's energy yield is not as easily determined as that of carbohydrate or fat because protein also contains nitrogen. When amino acids are catabolized, some of the released nitrogen is used to form new amino acids, but the remaining nitrogen cannot be oxidized by the body. Instead it is converted into urea and then excreted, primarily in the urine. This conversion requires the use of ATP, so some energy is spent in this process.

When protein is broken down through combustion in the laboratory, the energy yield is 5.65 kcal/g. However, because of the energy expended in converting nitrogen to urea when protein is metabolized in the body, the energy yield is only about 4.1 kcal/g.

To accurately assess the rate of protein metabolism, the amount of nitrogen being eliminated from the body must be determined. This requires urine collection for 12 to 24 h periods, a time-consuming process. Because the healthy body uses little protein during rest and exercise (usually not more than 5-10% of total energy expended), estimates of total energy expenditure generally ignore protein metabolism.

Summary of Substrate Metabolism

As shown in figure 2.12, the ability to produce muscle contraction for exercise is a matter of energy supply and energy demand. Both the contraction of skeletal muscle fibers and their relaxation require energy. That energy comes from foodstuffs in the diet and stored energy in the body. The ATP-PCr system operates within the cytosol of the cell, as does glycolysis, and neither requires oxygen for ATP production. Oxidative phosphorylation takes place within the mitochondria. Note that under aerobic conditions, both major substrates—carbohydrates and fats—are reduced to the common intermediate acetyl CoA that enters the Krebs cycle.

FIGURE 2.12 The metabolism of carbohydrate, fat, and to a lesser extent protein share some common pathways within the muscle fiber. The ATPs generated by oxidative and nonoxidative metabolism are used by those steps in muscle contraction and relaxation that demand energy.

In review

➤ The oxidative system involves breakdown of substrates in the presence of oxygen. This system yields more energy than the ATP-PCr or the glycolytic system.

➤ Oxidation of carbohydrate involves glycolysis, the Krebs cycle, and the electron transport chain. The end result is H_2O, CO_2, and 32 or 33 ATP molecules per carbohydrate molecule.

➤ Fat oxidation begins with β-oxidation of FFAs and then follows the same path as carbohydrate oxidation: acetyl CoA moving into the Krebs cycle and the electron transport chain. The energy yield for fat oxidation is much higher than for carbohydrate oxidation, and it varies with the FFA being oxidized. However, the maximum rate of high-energy phosphate formation from lipid oxidation is too low to match the rate of utilization of high-energy phosphate during higher-intensity exercise, and the energy yield of fat per oxygen molecule used is much less than that for carbohydrate.

➤ Measurement of protein oxidation is more complex because amino acids contain nitrogen, which cannot be oxidized. Protein contributes relatively little to energy production, generally less than 5% to 10%, so its metabolism is often considered negligible.

Interaction Among the Energy Systems

The three energy systems do not work independently of one another, and no activity is 100% supported by a single energy system. When a person is exercising at the highest intensity possible, from the shortest sprints (less than 10 s) to endurance events (greater than 30 min), each of the energy systems is contributing to the total energy needs of the body. Generally one energy system dominates energy production, except when there is a transition from the predominance of one energy system to another. As an example, in a 10 s, 100 m sprint, the ATP-PCr system is the predominant energy system, but both the anaerobic glycolytic and the oxidative systems provide a small portion of the energy needed. At the other extreme, in a 30 min, 10,000 m (10,936 yd) run, the oxidative system is predominant, but both the ATP-PCr and anaerobic glycolytic systems contribute some energy as well.

Figure 2.13 shows the reciprocal relation among the energy systems with respect to power and capacity. The ATP-PCr energy system can provide energy at a fast rate but has a very low capacity for energy production.

(a) Maximal rate of ATP generation

(b) Maximal available energy

FIGURE 2.13 There is a reciprocal relationship among the various energy systems with respect to (a) the maximal rate at which energy can be produced and (b) the capacity to produce that energy.

Thus it supports exercise that is intense but of very short duration. By contrast, fat oxidation takes longer to gear up and produces energy at a slower rate; however, the amount of energy it can produce is unlimited.

The characteristics of the muscle fiber's energy systems are listed in table 2.3.

The Oxidative Capacity of Muscle

We have seen that the processes of oxidative metabolism have the highest energy yields. It would be ideal if these processes always functioned at peak capacity. But, as with all physiological systems, they operate within certain constraints. The oxidative capacity of muscle ($\dot{Q}O_2$) is a measure of its maximal capacity to use oxygen. This measurement is made in the laboratory, where a small amount of muscle tissue can be tested to determine its capacity to consume oxygen when chemically stimulated to generate ATP.

Enzyme Activity

The capacity of muscle fibers to oxidize carbohydrate and fat is difficult to determine. Numerous studies have shown a close relation between a muscle's ability to perform prolonged aerobic exercise and the activity of its oxidative enzymes. Because many different enzymes are required for oxidation, the enzyme activity of the muscle fibers provides a reasonable indication of their oxidative potential.

Measuring all the enzymes in muscles is impossible, so a few representative enzymes have been selected to reflect the aerobic capacity of the fibers. The enzymes most frequently measured are succinate dehydrogenase and citrate synthase, mitochondrial enzymes involved in the Krebs cycle (see figure 2.9). Figure 2.14 illustrates the close correlation between succinate dehydrogenase activity in the vastus lateralis muscle and that muscle's oxidative capacity. Endurance athletes' muscles have oxidative enzyme activities nearly two to four times greater than those of untrained men and women.

Fiber Type Composition and Endurance Training

A muscle's fiber type composition primarily determines its oxidative capacity. As noted in chapter 1, type I, or slow-twitch, fibers have a greater capacity for aerobic activity than type II, or fast-twitch, fibers because type I fibers have more mitochondria and higher concentrations of oxidative enzymes. Type II fibers are better suited for glycolytic energy production. Thus, in general, the more type I fibers in one's muscles, the greater

TABLE 2.3 **Characteristics of the Various Energy Supply Systems**

Energy system	Oxygen necessary?	Overall chemical reaction	Relative rate of ATP formed per second	ATP formed per molecule of substrate	Available capacity
ATP-PCr	No	PCr to Cr	10	1	<15 s
Glycolysis	No	Glucose or glycogen to lactate	5	2-3	~1 min
Oxidative (from carbohydrate)	Yes	Glucose or glycogen to CO_2 and H_2O	2.5	36-39*	~90 min
Oxidative (from fat)	Yes	FFA or triglycerides to CO_2 and H_2O	1.5	>100	days

*Production of 36-39 ATP per molecule of carbohydrate excludes energy cost of transport through membranes. The net production is slightly lower (see text).

Courtesy of Dr. Martin Gibala, McMaster University, Hamilton, Ontario, Canada.

FIGURE 2.14 The relation between muscle succinate dehydrogenase (SDH) activity and its oxidative capacity ($\dot{Q}O_2$), measured in a muscle biopsy sample taken from the vastus lateralis.

the oxidative capacity of those muscles. Elite distance runners, for example, possess more type I fibers, more mitochondria, and higher muscle oxidative enzyme activities than do untrained individuals.

Endurance training enhances the oxidative capacity of all fibers, especially type II fibers. Training that places demands on oxidative phosphorylation stimulates the muscle fibers to develop more mitochondria, larger mitochondria, and more oxidative enzymes per mitochondrion. By increasing the fibers' enzymes for β-oxidation, this training also enables the muscle to rely more on fat for aerobic ATP production. Thus, with endurance training, even people with large percentages of type II fibers can increase their muscles'

aerobic capacities. But it is generally agreed that an endurance-trained type II fiber will not develop the same high endurance capacity as a similarly trained type I fiber.

Oxygen Needs

Although the oxidative capacity of a muscle is determined by the number of mitochondria and the amount of oxidative enzymes present, oxidative metabolism ultimately depends on an adequate supply of oxygen. At rest, the need for ATP is relatively small, requiring minimal oxygen delivery. As exercise intensity increases, so do energy demands. To meet them, the rate of oxidative ATP production increases. In an effort to meet the muscles' need for oxygen, the rate and depth of respiration increase, improving gas exchange in the lungs, and the heart beats faster and more forcefully, pumping more oxygenated blood to the muscles. Arterioles dilate to facilitate delivery of arterial blood to muscle capillaries.

The human body stores little oxygen. Therefore, the amount of oxygen entering the blood as it passes through the lungs is directly proportional to the amount used by the tissues for oxidative metabolism. Consequently, a reasonably accurate estimate of aerobic energy production can be made by measuring the amount of oxygen consumed at the lungs (see chapter 5).

In focus

A muscle's oxidative capacity depends on its oxidative enzyme concentrations, fiber type composition, and oxygen availability.

In this chapter, we focused on energy metabolism and the synthesis of the storage form of energy in the body, ATP. We described in some detail the three basic energy systems used to generate ATP and their regulation and interaction. Finally, we highlighted the important role that oxygen plays in the sustained generation of ATP for continued muscle contraction and the three fiber types found in human skeletal muscle. We next look at the neural control of exercising muscle.

Key Terms

acetyl coenzyme A (acetyl CoA)

activation energy

adenosine diphosphate (ADP)

aerobic metabolism

anaerobic metabolism

ATP-PCr system

bioenergetics

β-oxidation

carbohydrate

catabolism

creatine kinase

cytochrome

electron transport chain

enzyme

free fatty acids (FFAs)

gluconeogenesis

glucose

glycogen

glycogenolysis

glycolysis

kilocalories

Krebs cycle

lipogenesis

lipolysis

metabolism

mitochondria

negative feedback

oxidative phosphorylation

oxidative system

phosphocreatine (PCr)

phosphofructokinase (PFK)

phosphorylation

rate-limiting enzyme

substrate

triglycerides

Study Questions

1. What is ATP and how is it of importance in metabolism?
2. What is the primary substrate used to provide energy at rest? During high-intensity exercise?
3. What is the role of PCr in energy production? Describe the relationship between muscle ATP and PCr during sprint exercise.
4. Describe the essential characteristics of the three energy systems.
5. Why are the ATP-PCr and glycolytic energy systems considered anaerobic?
6. What role does oxygen play in the process of aerobic metabolism?
7. Describe the by-products of energy production from ATP-PCr, glycolysis, and oxidation.
8. What is lactic acid and why is it important?
9. Discuss the interaction among the three energy systems with respect to the rate at which energy can be produced and the sustained capacity to produce that energy.
10. How do type I muscle fibers differ from type II fibers in their respective oxidative capacities? What accounts for those differences?

Study Guide Activities

In addition to the activities listed in the chapter opening outline on page 49, two other activities are available in the web study guide, located at

www.HumanKinetics.com/PhysiologyOfSportAndExercise

 The **KEY TERMS** activity reviews important terms, and the end-of-chapter **QUIZ** tests your understanding of the material covered in the chapter.

Neural Control of Exercising Muscle

3

In this chapter and in the web study guide

 ACTIVITY 3.1 The Neuron reviews the basic structure of a neuron.

 ACTIVITY 3.2 Action Potentials explores the sequence of events that occur during an action potential.

 ACTIVITY 3.3 Communication Among Components explores the way that neurons and muscle fibers communicate.

 ACTIVITY 3.4 Central Nervous System describes the components of the central nervous system.

 ACTIVITY 3.5 Higher Brain Center Function reviews the functions of the higher brain centers.

 ACTIVITY 3.6 Peripheral Nervous System identifies the functions and components of the peripheral nervous system.

In 1964, Jimmie Heuga and teammate Billy Kidd made history by winning the first U.S. Olympic medals in men's alpine skiing in Innsbruck, Austria. The two close friends and teammates made the cover of *Sports Illustrated* before the 1968 Olympics, but Heuga fell to a seventh-place finish in the slalom and to 10th place in the giant slalom at the 1968 Olympic Games in Grenoble, France. By that time, he was beginning to feel the early symptoms of what would later be diagnosed as multiple sclerosis (MS), a chronic neurological disorder that slows nerve signals to the muscles.

At that time, people with MS were told that physical activity would exacerbate their condition, so he was advised to rest and conserve his energy. Heuga followed that advice for a while but began feeling unhealthy, unmotivated, and less energetic. He began to deteriorate physically and mentally. Six years later, Heuga decided to defy medical convention. He developed a cardiovascular endurance exercise program and began stretching and strengthening exercises. He established realistic goals for his personal wellness program. Inspired by his own success, he created the Jimmie Heuga Center, a nonprofit organization (now known as Can Do Multiple Sclerosis) based in Colorado that served over 10,000 people in 2008 alone. The center's most important research contribution, published in the *Annals of Neurology* in 1996,[4] demonstrated that an exercise training program enhances physiological and psychological function and general quality of life in MS patients, countering the medical wisdom at that time, which dictated a life of restricted activity.

Heuga, Hall of Fame skier and physical activity activist, died on February 8, 2010, at the age of 66.

All functions within the human body are, or can be, influenced by the nervous system. Nerves provide the wiring through which electrical impulses are sent to and received from virtually all parts of the body. The brain acts as a central computer, integrating incoming information, selecting an appropriate response, and then signaling the involved organs and tissues to take appropriate action. Thus, the nervous system forms a vital network, allowing communication, coordination, and interaction among the various tissues and systems in the body and between the body and the external environment.

The nervous system is one of the body's most complex systems. Because this book is primarily concerned with neural control of muscle contraction and voluntary movement, we will limit our coverage of this complex system. We first review the structure and function of the nervous system and then focus on specific topics relevant to sport and exercise.

Before we examine the intricate details of the nervous system, it is important to step back and look at the big picture—how the nervous system is organized and how that organization functions to integrate and control movement. The nervous system as a whole is commonly divided into two parts: the **central nervous system (CNS)** and the **peripheral nervous system (PNS)**. The CNS is composed of the brain and spinal cord, while the PNS is further divided into **sensory** (or **afferent**) **nerves** and **motor** (or **efferent**) **nerves**. Sensory nerves are responsible for informing the CNS about what is going on within and outside the body. The motor nerves are responsible for sending information from the CNS to the various tissues, organs, and systems of the body in response to the signals coming in from the sensory division. The efferent nervous system is composed of two parts, the autonomic nervous system and the somatic nervous system. Figure 3.1 provides a schematic of these relationships. More detail concerning each of these individual units of the nervous system is presented later in this chapter.

FIGURE 3.1 Organization of the nervous system.

Structure and Function of the Nervous System

The **neuron** is the basic structural unit of the nervous system. We first review the anatomy of the neuron and then look at how it functions—allowing electrical impulses to be transmitted throughout the body.

Neuron

Individual nerve fibers (nerve cells), depicted in figure 3.2, are called neurons. A typical neuron is composed of three regions:

FIGURE 3.2

A photomicrograph of a neuron and its structure.

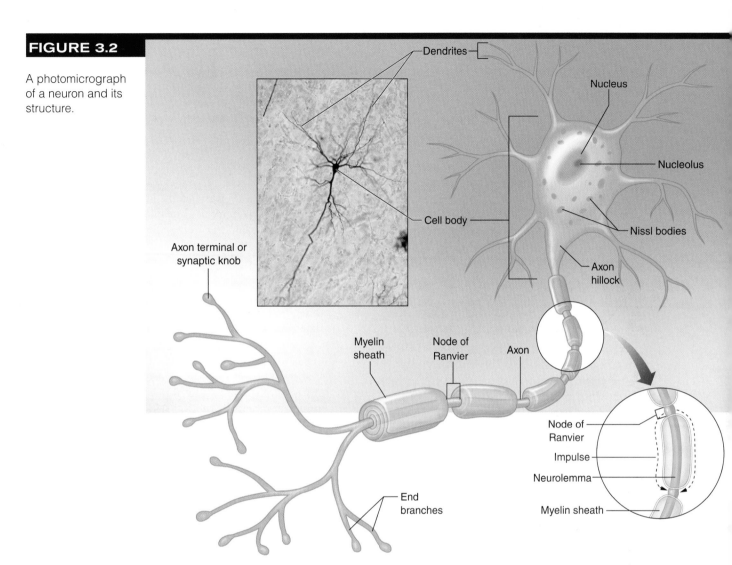

- The cell body, or soma
- The dendrites
- The axon

The cell body contains the nucleus. Radiating out from the cell body are the cell processes: the dendrites and the axon. On the side toward the axon, the cell body tapers into a cone-shaped region known as the **axon hillock**. The axon hillock has an important role in impulse conduction, as discussed later.

Most neurons contain only one axon but many dendrites. Dendrites are the neuron's receivers. Most impulses, or action potentials, that enter the neuron from sensory stimuli or from adjacent neurons typically enter the neuron via the dendrites. These processes then carry the impulses toward the cell body.

The axon is the neuron's transmitter and conducts impulses away from the cell body. Near its end, an axon splits into numerous **end branches**. The tips of these

branches are dilated into tiny bulbs known as **axon terminals** or synaptic knobs. These terminals or knobs house numerous vesicles (sacs) filled with chemicals known as **neurotransmitters** that are used for communication between a neuron and another cell. (This is discussed later in this chapter in more detail.) The structure of the neuron allows nerve impulses to enter the neuron through the dendrites, and to a lesser extent through the cell body, and to travel through the cell body and axon hillock, down the axon, and out through the end branches to the axon terminals. We next explain in more detail how this happens, including how these impulses travel from one neuron to another and from a somatic motor neuron to muscle fibers.

Nerve Impulse

Neurons are referred to as *excitable tissue* because they can respond to various types of stimuli and convert those messages to a nerve impulse. A **nerve impulse**—an electrical signal—arises when a stimulus is strong enough to substantially change the normal electrical charge of the neuron. That signal then moves along the neuron down the axon and toward an end organ, such as another neuron or a group of muscle fibers. For simplicity, think of the nerve impulse traveling through a neuron as electricity travels through the electrical wires in a home. This section describes how the electrical impulse is generated and how it travels through a neuron.

Resting Membrane Potential

The cell membrane of a typical neuron at rest has a negative electrical potential of about –70 mV. This means that if one were to insert a voltmeter probe inside the cell, the electrical charges found there and the charges found outside the cell would differ by 70 mV, and the inside would be negative relative to the outside. This electrical potential difference is known as the **resting membrane potential (RMP)**. It is caused by an uneven separation of charged ions across the membrane. When the charges across the membrane differ, the membrane is said to be polarized.

The neuron has a high concentration of potassium ions (K^+) on the inside of the membrane and a high concentration of sodium ions (Na^+) on the outside. The imbalance in the number of ions inside and outside the cell causes the RMP. This imbalance is maintained in two ways. First, the cell membrane is much more permeable to K^+ than to Na^+, so the K^+ can move more freely. Because ions tend to move to establish equilibrium, some of the K^+ will move to the area where they are less concentrated, outside the cell. The Na^+ cannot move to the inside as easily. Second, **sodium-potassium pumps** in the neuron membrane, which contain Na^+-K^+ adenosine triphosphatase (Na^+-K^+-ATPase), maintain the imbalance on each side of the membrane by actively transporting potassium ions in and sodium ions out. The sodium-potassium pump moves three Na^+ out of the cell for each two K^+ it brings in. The end result is that more positively charged ions are outside the cell than inside, creating the potential difference across the membrane. Maintenance of a constant RMP of about –70 mV is primarily a function of the sodium-potassium pump.

Depolarization and Hyperpolarization

If the inside of the cell becomes less negative relative to the outside, the potential difference across the membrane decreases. The membrane will be less polarized. When this happens, the membrane is said to be depolarized. Thus, **depolarization** occurs any time the charge difference becomes more positive than the RMP of –70 mV, moving closer to zero. This typically results from a change in the membrane's Na^+ permeability.

The opposite can also occur. If the charge difference across the membrane increases, moving from the RMP to an even more negative value, then the membrane becomes more polarized. This is known as **hyperpolarization**. Changes in the membrane potential control the signals used to receive, transmit, and integrate information within and between cells. These signals are of two types, graded potentials and action potentials. Both are electrical currents created by the movement of ions.

Graded Potentials

Graded potentials are localized changes in the membrane potential, either depolarization or hyperpolarization. The membrane contains ion channels with gates that act as doorways into and out of the neuron. These gates are usually closed, preventing a large number of ions from flowing into and out of the membrane, that is, above and beyond the constant movement of Na^+ and K^+ that maintain the RMP. However, with potent enough stimulation, the gates open, allowing more ions to move from the outside to the inside or vice versa. This ion flow alters the charge separation, changing the polarization of the membrane.

Graded potentials are triggered by a change in the neuron's local environment. Depending on the location and type of neuron involved, the ion gates may open in response to the transmission of an impulse from another neuron or in response to sensory stimuli such as changes in chemical concentrations, temperature, or pressure.

Recall that most neuron receptors are located on the dendrites (although some are on the cell body), yet the impulse is always transmitted from the axon terminals at the opposite end of the cell. For a neuron to transmit an impulse, the impulse must travel almost the entire length of the neuron. Although a graded potential may result in depolarization of the entire cell membrane, it is usually just a local event such that the depolarization does not spread very far along the neuron. To travel the full distance, an impulse must be sufficiently strong to generate an action potential.

In focus

Neurons are considered excitable tissues because they have the ability to respond to various types of stimuli and convert them to an electrical signal or nerve impulse.

Action Potentials

An action potential is a rapid and substantial depolarization of the neuron's membrane. It usually lasts only about 1 ms. Typically, the membrane potential changes from the RMP of about –70 mV to a value of about +30 mV and then rapidly returns to its resting value. This is illustrated in figure 3.3. How does this marked change in membrane potential occur?

All action potentials begin as graded potentials. When enough stimulation occurs to cause a depolarization of at least 15 to 20 mV, an action potential results. In other words, if the membrane depolarizes from the RMP of –70 mV to a value of –50 to –55 mV, an action potential will occur. The membrane voltage at which a graded potential becomes an action potential is called the depolarization **threshold**. Any depolarization that does not attain the threshold will not result in an action potential. For example, if the membrane potential changes from the RMP of –70 mV to –60 mV, the change is only 10 mV and does not reach the threshold; thus,

no action potential occurs. But any time depolarization reaches or exceeds the threshold, an action potential will result. This is commonly referred to as the *all-or-none principle*.

When a segment of an axon's sodium gates are open and it is in the process of generating an action potential, it is unable to respond to another stimulus. This is referred to as the *absolute refractory period*. When the sodium gates are closed, the potassium gates are open, and repolarization is occurring, that segment of the axon can potentially respond to a new stimulus, but the stimulus must be of substantially greater magnitude to evoke an action potential. This is referred to as the *relative refractory period*.

Propagation of the Action Potential

Now that we understand how a neural impulse, in the form of an action potential, is generated, we can look at how the impulse is propagated, that is, how it travels

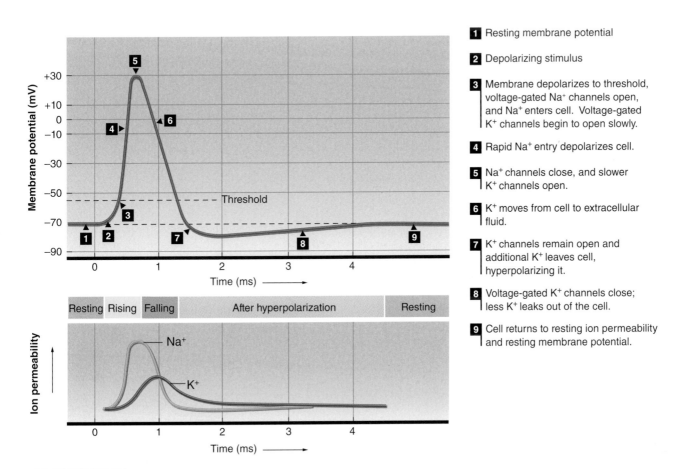

1 Resting membrane potential

2 Depolarizing stimulus

3 Membrane depolarizes to threshold, voltage-gated Na+ channels open, and Na+ enters cell. Voltage-gated K+ channels begin to open slowly.

4 Rapid Na+ entry depolarizes cell.

5 Na+ channels close, and slower K+ channels open.

6 K+ moves from cell to extracellular fluid.

7 K+ channels remain open and additional K+ leaves cell, hyperpolarizing it.

8 Voltage-gated K+ channels close; less K+ leaks out of the cell.

9 Cell returns to resting ion permeability and resting membrane potential.

FIGURE 3.3 Voltage and ion permeability changes during an action potential.

Fig. 8.9, p. 259 from HUMAN PHYSIOLOGY, 4th ed. By Dee Unglaub Silverthorn. Copyright © 2007 by Pearson Education, Inc. Reprinted by permission.

through the neuron. Two characteristics of the neuron determine how quickly an impulse can pass along the axon: myelination and diameter.

Myelination   The axons of many neurons, especially large neurons, are myelinated, meaning they are covered with a sheath formed by myelin, a fatty substance that insulates the cell membrane. This **myelin sheath** (see figure 3.2) is formed by specialized cells called Schwann cells.

The myelin sheath is not continuous. As it spans the length of the axon, the myelin sheath has gaps between adjacent Schwann cells, leaving the axon uninsulated at those points. These gaps are referred to as *nodes of Ranvier* (see figure 3.2). The action potential appears to jump from one node to the next as it traverses a myelinated fiber. This is referred to as **saltatory conduction**, a much faster type of conduction than occurs in unmyelinated fibers.

In focus

The velocity of nerve impulse transmission in large myelinated fibers can be as high as 100 m/s, or 5 to 50 times faster than that in unmyelinated fibers of the same size.

Myelination of peripheral motor neurons occurs over the first several years of life, partly explaining why children need time to develop coordinated movement. Individuals affected by certain neurological diseases, such as MS as discussed in our chapter opening, experience degeneration of the myelin sheath and a subsequent loss of coordination.

Diameter of the Neuron   The velocity of nerve impulse transmission is also determined by the neuron's size. Neurons of larger diameter conduct nerve impulses faster than neurons of smaller diameter because larger neurons present less resistance to local current flow.

Synapse

For a neuron to communicate with another neuron, first an action potential must occur and travel along the first neuron, ultimately reaching its axon terminals. How does the action potential then move from the neuron in which it starts to another neuron to continue transmitting the electrical signal?

Neurons communicate with each other at junctions called synapses. A **synapse** is the site of action potential transmission from the axon terminals of one neuron to the dendrites or soma of another. There are both chemical and mechanical synapses, but the most common type is the chemical synapse, which is our focus. It is important to note that the signal that is transmitted from one neuron to another changes from electrical to chemical, then back to electrical.

As seen in figure 3.4, a synapse between two neurons includes

- the axon terminals of the neuron sending the action potential,
- receptors on the neuron receiving the action potential, and
- the space between these structures.

The neuron sending the action potential across the synapse is called the presynaptic neuron, so axon ter-

In review

➤ A neuron's RMP of about –70 mV results from the uneven separation of charged sodium and potassium ions, with more potassium inside the membrane and more sodium on the outside.

➤ The RMP is maintained by actions of the sodium-potassium pump, coupled with low sodium permeability and high potassium permeability of the neuron membrane.

➤ Any change that makes the membrane potential less negative results in depolarization. Any change making this potential more negative is a hyperpolarization. These changes occur when ion gates in the membrane open, permitting more ions to move across the membrane.

➤ If the membrane is depolarized by 15 to 20 mV, the depolarization threshold is reached and an action potential results. Action potentials are not generated if the threshold is not met.

➤ In myelinated neurons, the impulse travels through the axon by jumping between nodes of Ranvier (gaps between the cells that form the myelin sheath). This process, saltatory conduction, results in nerve transmission rates 5 to 50 times faster than in unmyelinated fibers of the same size. Impulses also travel faster in neurons of larger diameter.

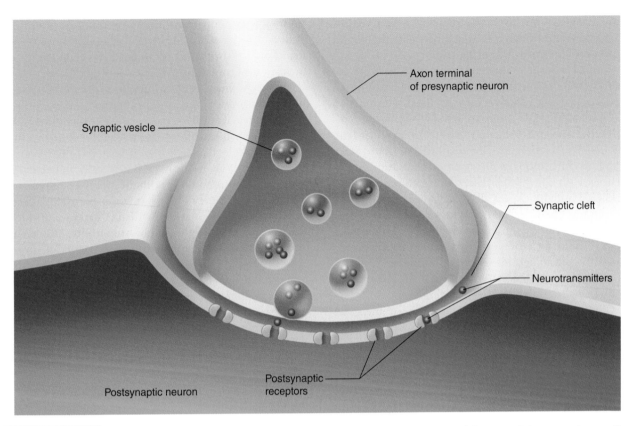

FIGURE 3.4 A chemical synapse between two neurons, showing the synaptic vesicles containing neurotransmitter molecules.

minals are presynaptic terminals. Similarly, the neuron receiving the action potential on the opposite side of the synapse is called the postsynaptic neuron, and it has postsynaptic receptors. The axon terminals and postsynaptic receptors are not physically in contact with each other. A narrow gap, the synaptic cleft, separates them.

The action potential can be transmitted across a synapse in only one direction: from the axon terminal of the presynaptic neuron to the postsynaptic receptors, about 80% to 95% of which are on the dendrites of the postsynaptic neuron. (The remaining 5% to 20% of the postsynaptic receptors are adjacent to the cell body instead of being located on the dendrites.[2]) Why can the action potential go in only one direction?

The presynaptic terminals of the axon contain a large number of saclike structures called synaptic (or storage) vesicles. These vesicles contain a variety of chemical compounds called neurotransmitters because they function to transmit the neural signal to the next neuron. When the impulse reaches the presynaptic axon terminals, the synaptic vesicles respond by releasing the neurotransmitters into the synaptic cleft. These neurotransmitters then diffuse across the synaptic cleft to the postsynaptic neuron's receptors. Each neurotransmitter then binds to its specialized postsynaptic receptors. When sufficient binding occurs, a series of

graded depolarizations occurs; and if the depolarization reaches the threshold, an action potential occurs and the impulse has been transmitted successfully to the next neuron. Depolarization of the second nerve depends on both the amount of neurotransmitter released and the number of available receptor binding sites on the postsynaptic neuron.

In focus

Nerve-to-nerve signal transmission occurs across synapses by the presynaptic release of neurotransmitters that diffuse across the synaptic cleft and bind with specific postsynaptic receptors.

Neuromuscular Junction

Recall from chapter 1 that a single α-motor neuron and all of the muscle fibers it innervates is called a motor unit. Whereas neurons communicate with other neurons at synapses, an α-motor neuron communicates with its muscle fibers at a site known as a **neuromuscular junction**, which functions in essentially the same manner as a synapse. In fact, the proximal part of the neuromuscular junction is the same: It starts with the axon terminals of the motor neuron, which release neurotransmitters into the space between the motor nerve and the muscle fiber in response to an action potential. However, in the neuromuscular junction, the axon terminals protrude into motor end plates, which are invaginated (folded to form cavities) segments on the plasmalemma of the muscle fiber (see figure 3.5).

Neurotransmitters—primarily acetylcholine (ACh)—released from the α-motor neuron axon terminals diffuse across the synaptic cleft and bind to receptors on the muscle fiber's plasmalemma. This binding typically causes depolarization by opening sodium ion channels, allowing more sodium to enter the muscle fiber. Again, if the depolarization reaches the threshold, an action potential is formed. It spreads across the plasmalemma into the T-tubules, initiating muscle fiber contraction. As in the neuron, the plasmalemma, once depolarized, must undergo repolarization. During the period of repolarization, the sodium gates are closed and the potassium gates are open; thus, like the neuron, the muscle fiber is unable to respond to any further stimulation during this refractory period. Once the resting membrane potential of the muscle fiber is restored, the fiber can respond to another stimulus. Thus, the refractory period limits the motor unit's firing frequency.

Now we know how the impulse is transmitted from nerve to nerve or nerve to muscle. But to understand what happens once the impulse is transmitted, we must first examine the chemical signals that accomplish this signal transmission.

Neurotransmitters

More than 50 neurotransmitters have been positively identified or are suspected as potential candidates. These can be categorized as either (a) small-molecule, rapid-acting neurotransmitters or (b) neuropeptide, slow-acting neurotransmitters. The small-molecule, rapid-acting transmitters, which are responsible for most neural transmissions, are our main focus.

Acetylcholine and norepinephrine are the two major neurotransmitters involved in regulating the multiple physiological responses to exercise. **Acetylcholine** is the primary neurotransmitter for the motor neurons that innervate skeletal muscle as well as for most parasympathetic autonomic neurons. It is generally an excitatory neurotransmitter in the somatic nervous system but can have inhibitory effects at some parasympathetic nerve endings, such as in the heart. **Norepinephrine** is the neurotransmitter for most sympathetic autonomic neurons, and it too can be either excitatory or inhibitory, depending on the receptors involved. Nerves that primarily release norepinephrine are called **adrenergic**, and those that have acetylcholine as their primary neurotransmitter are termed **cholinergic**. The sympathetic and parasympathetic branches of the autonomic nervous systems are discussed later in this chapter.

In focus

Receptors on the motor end plates of the neuromuscular junction are called cholinergic, meaning that they bind the primary neurotransmitter involved in excitation of muscle fibers, acetylcholine.

Once the neurotransmitter binds to the postsynaptic receptor, the nerve impulse has been successfully transmitted. The neurotransmitter is then either degraded by enzymes, actively transported back into the presynaptic terminals for reuse, or diffused away from the synapse.

Nerve fiber branches

Motor neuron fiber

Myofibril of muscle fiber

Synaptic vesicle

Synaptic cleft

Motor end plate

Neurotransmitters

FIGURE 3.5 The neuromuscular junction, illustrating the interaction between the α-motor neuron and the plasmalemma of a single muscle fiber.

Postsynaptic Response

Once the neurotransmitter binds to the receptors, the chemical signal that traversed the synaptic cleft once again becomes an electrical signal. The binding causes a graded potential in the postsynaptic membrane. An incoming impulse may be either excitatory or inhibitory. An excitatory impulse causes depolarization, known as an **excitatory postsynaptic potential (EPSP)**. An inhibitory impulse causes a hyperpolarization, known as an **inhibitory postsynaptic potential (IPSP)**.

The discharge of a single presynaptic terminal generally changes the postsynaptic potential less than 1 mV. Clearly this is not sufficient to generate an action potential, because reaching the threshold requires a change of at least 15 to 20 mV. But when a neuron transmits an impulse, several presynaptic terminals typically release their neurotransmitters so that they can diffuse to the postsynaptic receptors. Also, presynaptic terminals from numerous axons can converge on the dendrites and cell body of a single neuron. When multiple presynaptic terminals discharge at the same time, or when only a few fire in rapid succession, more neurotransmitter is released. With an excitatory neurotransmitter, the more that is bound, the greater the EPSP and the more likely an action potential will result.

Triggering an action potential at the postsynaptic neuron depends on the combined effects of all incoming impulses from these various presynaptic terminals. A number of impulses are needed to cause sufficient depolarization to generate an action potential. Specifically, the sum of all changes in the membrane potential must equal or exceed the threshold. This accumulation of the individual impulses' effects is called summation.

For summation, the postsynaptic neuron must keep a running total of the neuron's responses, both EPSPs and IPSPs, to all incoming impulses. This task is done at the axon hillock, which lies on the axon just past the cell body. Only when the sum of all individual graded potentials meets or exceeds the threshold can an action potential occur.

In focus

Summation refers to the cumulative effect of all individual graded potentials as processed by the axon hillock. Once the sum of all individual graded potentials meets or exceeds the depolarization threshold, an action potential occurs.

Individual neurons are grouped together into bundles. In the CNS (brain and spinal cord), these bundles are referred to as tracts, or pathways. Neuron bundles in the PNS are referred to simply as nerves.

In review

➤ Neurons communicate with each other across synapses composed of the axon terminals of the presynaptic neuron, the postsynaptic receptors on the dendrite or cell body of the postsynaptic neuron, and the synaptic cleft between the two neurons.

➤ A nerve impulse causes neurotransmitters to be released from the presynaptic axon terminal into the synaptic cleft.

➤ Neurotransmitters diffuse across the cleft and bind to the postsynaptic receptors.

➤ Once sufficient neurotransmitters are bound, the impulse is successfully transmitted and the neurotransmitter is then destroyed by enzymes, removed by reuptake into the presynaptic terminal for future use, or diffuses away from the synapse.

➤ Neurotransmitter binding at the postsynaptic receptors opens ion gates in that membrane and can cause depolarization (excitation) or hyperpolarization (inhibition), depending on the specific neurotransmitter and the receptors to which it binds.

➤ Neurons communicate with muscle fibers at neuromuscular junctions. A neuromuscular junction involves presynaptic axon terminals, the synaptic cleft, and motor end-plate receptors on the plasmalemma of the muscle fiber and functions much like a neural synapse.

➤ The neurotransmitters most important in regulating exercise responses are acetylcholine in the somatic nervous system and norepinephrine in the autonomic nervous system.

➤ Excitatory postsynaptic potentials are graded depolarizations of the postsynaptic membrane; inhibitory postsynaptic potentials are hyperpolarizations of that membrane.

➤ A single presynaptic terminal cannot generate enough of a depolarization to fire an action potential. Multiple signals are needed. These may come from numerous neurons or from a single neuron when numerous axon terminals release neurotransmitters repeatedly and rapidly.

➤ The axon hillock keeps a running total of all EPSPs and IPSPs. When their sum meets or exceeds the threshold for depolarization, an action potential occurs. This process of accumulating incoming signals is known as summation.

Central Nervous System

To comprehend how even the most basic stimulus can cause muscle activity, we next consider the complexity of the CNS. In this section, we present an overview of the components of the CNS and their functions.

In focus

The CNS contains more than 100 billion neurons.

Brain

The brain is a highly complex organ composed of numerous specialized areas. For our purposes, we subdivide it into the four major regions illustrated in figure 3.6: the cerebrum, diencephalon, cerebellum, and brain stem.

Cerebrum

The cerebrum is composed of the right and left cerebral hemispheres. These are connected to each other by fiber bundles (tracts) referred to as the *corpus callosum,* which allows the two hemispheres to communicate with each other. The cerebral cortex forms the outer portion of the cerebral hemispheres and has been referred to as the site of the mind and intellect. It is also called the gray matter, which simply reflects its distinctive color resulting from lack of myelin on the neurons located in

this area. The cerebral cortex is the conscious brain. It allows people to think, to be aware of sensory stimuli, and to voluntarily control their movements.

The cerebrum consists of five lobes—four outer lobes and the central insular lobe—having the following general functions (see figure 3.6):

- Frontal lobe: general intellect and motor control
- Temporal lobe: auditory input and interpretation
- Parietal lobe: general sensory input and interpretation
- Occipital lobe: visual input and interpretation
- Insular lobe: diverse functions usually linked to emotion and self-perception

The three areas in the cerebrum that are of primary concern to exercise physiology are the primary motor cortex, located in the frontal lobe; the basal ganglia, located in the white matter below the cerebral cortex; and the primary sensory cortex, located in the parietal lobe. In this section, the focus is on the primary motor cortex and basal ganglia, which work to control and coordinate movement.

Primary Motor Cortex The primary motor cortex is responsible for the control of fine and discrete muscle movements. It is located in the frontal lobe, specifically within the precentral gyrus. Neurons here, known as *pyramidal cells,* let us consciously control movement of

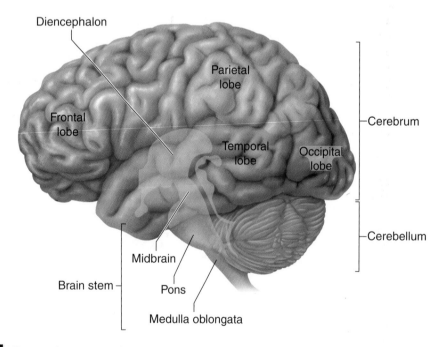

FIGURE 3.6 Four major regions of the brain and four outer lobes of the cerebrum (note that the insular lobe is not shown because it is folded deep within the cerebrum between the temporal lobe and the frontal lobe).

skeletal muscles. Think of the primary motor cortex as the part of the brain where decisions are made about what movement one wants to make. For example, in baseball, if a player is in the batter's box waiting for the next pitch, the decision to swing the bat is made in the primary motor cortex, where the entire body is carefully mapped out. The areas that require the finest motor control have a greater representation in the motor cortex; thus, more neural control is provided to them.

The cell bodies of the pyramidal cells are housed in the primary motor cortex, and their axons form the extrapyramidal tracts. These are also known as the corticospinal tracts because the nerve processes extend from the cerebral cortex down to the spinal cord. These tracts provide the major voluntary control of skeletal muscles.

In addition to the primary motor cortex, there is a premotor cortex just anterior to the precentral gyrus in the frontal lobe. Learned motor skills of a repetitious or patterned nature are stored here. This region can be thought of as the memory bank for skilled motor activities.[3]

Basal Ganglia The basal ganglia (nuclei) are not part of the cerebral cortex. Rather, they are in the cerebral white matter, deep in the cortex. These ganglia are clusters of nerve cell bodies. The complex functions of the basal ganglia are not well understood, but the ganglia are known to be important in initiating movements of a sustained and repetitive nature (such as arm swinging during walking), and thus they control complex movements such as walking and running. These cells also are involved in maintaining posture and muscle tone.

Diencephalon

The region of the brain known as the diencephalon (see figure 3.6) is composed mostly of the thalamus and the hypothalamus. The thalamus is an important sensory integration center. All sensory input (except smell) enters the thalamus and is relayed to the appropriate area of the cortex. The thalamus regulates what sensory input reaches the conscious brain and thus is very important for motor control.

The hypothalamus, directly below the thalamus, is responsible for maintaining homeostasis by regulating almost all processes that affect the body's internal environment. Neural centers here assist in the regulation of most physiological systems, including

- blood pressure, heart rate and contractility;
- respiration;
- digestion;
- body temperature;
- thirst and fluid balance;
- neuroendocrine control;
- appetite and food intake; and
- sleep–wake cycles.

Cerebellum

The cerebellum is located behind the brain stem. It is connected to numerous parts of the brain and has a crucial role in *coordinating* movement.

The cerebellum is crucial to the control of all rapid and complex muscular activities. It helps coordinate the timing of motor activities and the rapid progression from one movement to the next by monitoring and making corrective adjustments in the motor activities that are elicited by other parts of the brain. The cerebellum assists the functions of both the primary motor cortex and the basal ganglia. It facilitates movement patterns by smoothing out the movement, which would otherwise be jerky and uncontrolled.

The cerebellum acts as an integration system, comparing the programmed or intended activity with the actual changes occurring in the body and then initiating corrective adjustments through the motor system. It receives information from the cerebrum and other parts of the brain and also from sensory receptors (proprioceptors) in the muscles and joints that keep the cerebellum informed about the body's current position. The cerebellum also receives visual and equilibrium input. Thus, it notes all incoming information about the exact tension and position of all muscles, joints, and tendons and the body's current position relative to its surroundings; then it determines the best plan of action to produce the desired movement.

The primary motor cortex is the part of the brain that makes the decision to perform a movement. This decision is relayed to the cerebellum. The cerebellum notes the desired action and then compares the intended movement with the actual movement based on sensory feedback from the muscles and joints. If the action is different than planned, the cerebellum informs the higher centers of the discrepancy so corrective action can be initiated.

Brain Stem

The brain stem, composed of the midbrain, the pons, and the medulla oblongata (see figure 3.6), connects the brain and the spinal cord. Sensory and motor neurons pass through the brain stem as they relay information in both directions between the brain and the spinal cord. This is the site of origin for 10 of the 12 pairs of cranial nerves. The brain stem also contains the major autonomic centers that control the respiratory and cardiovascular systems.

A specialized collection of neurons in the brain stem, known as the *reticular formation*, is influenced by, and has an influence on, nearly all areas of the CNS. These neurons help

- coordinate skeletal muscle function,
- maintain muscle tone,
- control cardiovascular and respiratory functions, and
- determine state of consciousness (arousal and sleep).

The brain has a pain control system located in the reticular formation, a group of nerve fibers in the brain stem. Opiod substances such as enkephalins and β-endorphin act on the opiate receptors in this region to help modulate pain. Research has demonstrated that exercise of long duration increases the concentrations of these substances. While this has been interpreted as the mechanism causing the "endorphin calm" or "runner's high" experienced by some exercisers, the cause–effect association between these endogenous opioids and these sensations has not been substantiated.

Spinal Cord

The lowest part of the brain stem, the medulla oblongata, is continuous with the spinal cord below it. The spinal cord is composed of tracts of nerve fibers that allow two-way conduction of nerve impulses. The sensory (afferent) fibers carry neural signals from sensory receptors, such as those in the skin, muscles, and joints, to the upper levels of the CNS. Motor (efferent) fibers from the brain and upper spinal cord transmit action potentials to end organs (e.g., muscles, glands).

Peripheral Nervous System

The PNS contains 43 pairs of nerves: 12 pairs of cranial nerves that connect with the brain and 31 pairs of spinal nerves that connect with the spinal cord. Cranial and spinal nerves directly supply the skeletal muscles. Functionally, the PNS has two major divisions: the sensory division and the motor division.

Sensory Division

The sensory division of the PNS carries sensory information toward the CNS. Sensory (afferent) neurons originate in such areas as

- blood vessels,
- internal organs,
- special sense organs (taste, touch, smell, hearing, vision),
- the skin, and
- muscles and tendons.

Sensory neurons in the PNS end either in the spinal cord or in the brain, and they continuously convey information to the CNS concerning the body's constantly changing status, position, and internal and external environment. Sensory neurons within the CNS carry the sensory input to appropriate areas of the brain, where the information can be processed and integrated with other incoming information.

The sensory division receives information from five primary types of receptors:

In review

- ➤ The CNS includes the brain and the spinal cord.
- ➤ The four major divisions of the brain are the cerebrum, the diencephalon, the cerebellum, and the brain stem.
- ➤ The cerebral cortex is the conscious brain. The primary motor cortex, located in the frontal lobe, is the center of conscious motor control.
- ➤ The basal ganglia, in the cerebral white matter, help initiate some movements (sustained and repetitive ones) and help control posture and muscle tone.
- ➤ The diencephalon includes the thalamus, which receives all sensory input entering the brain, and the hypothalamus, which is a major control center for homeostasis.
- ➤ The cerebellum, which is connected to numerous parts of the brain, is critical for coordinating movement. It is an integration center that decides how to best execute the desired movement, given the body's current position and the muscles' current status.
- ➤ The brain stem is composed of the midbrain, the pons, and the medulla oblongata.
- ➤ The spinal cord contains both sensory and motor fibers that transmit action potentials between the brain and the periphery.

1. *Mechanoreceptors* that respond to mechanical forces such as pressure, touch, vibrations, or stretch
2. *Thermoreceptors* that respond to changes in temperature
3. *Nociceptors* that respond to painful stimuli
4. *Photoreceptors* that respond to electromagnetic radiation (light) to allow vision
5. *Chemoreceptors* that respond to chemical stimuli, such as from foods, odors, or changes in blood or tissue concentrations of substances such as oxygen, carbon dioxide, glucose, and electrolytes

Virtually all of these receptors are important in exercise and sport. Let's consider just a few. Free nerve endings detect crude touch, pressure, pain, heat, and cold. Thus, they function as mechanoreceptors, nociceptors, and thermoreceptors. These nerve endings are important for preventing injury during athletic performance. Special muscle and joint nerve endings are of many types and functions, and each type is sensitive to a specific stimulus. Here are some important examples:

- Joint kinesthetic receptors located in the joint capsules are sensitive to joint angles and rates of change in these angles. Thus, they sense the position and any movement of the joints.
- Muscle spindles sense muscle length and rate of change in length.
- Golgi tendon organs detect the tension applied by a muscle to its tendon, providing information about the strength of muscle contraction.

Muscle spindles and Golgi tendon organs are discussed later in this chapter.

Motor Division

The CNS transmits information to various parts of the body through the motor, or efferent, division of the PNS. Once the CNS has processed the information it receives from the sensory division, it determines how the body should respond to that input. From the brain and spinal cord, intricate networks of neurons go out to all parts of the body, providing detailed instructions to the target areas including—and central to exercise and sport physiology—muscles.

Autonomic Nervous System

The autonomic nervous system, often considered part of the motor division of the PNS, controls the body's involuntary internal functions. Some of these functions that are important to sport and activity are heart rate, blood pressure, blood distribution, and lung function.

The autonomic nervous system has two major divisions: the sympathetic nervous system and the parasympathetic nervous system. These originate from different sections of the spinal cord and from the base of the brain. The effects of the two systems are often antagonistic, but the systems always function together.

Sympathetic Nervous System

The sympathetic nervous system is sometimes called the fight-or-flight system: It prepares the body to face a crisis and sustains its function during the crisis. When fully engaged, the sympathetic nervous system can produce a massive discharge throughout the body, preparing it for action. A sudden loud noise, a life-threatening situation, or those last few seconds before the start of an athletic competition are examples of circumstances in which this massive sympathetic excitation may occur. The effects of sympathetic stimulation are important to the athlete:

- Heart rate and strength of cardiac contraction increase.
- Coronary vessels dilate, increasing the blood supply to the heart muscle to meet its increased demands.
- Peripheral vasodilation increases blood flow to active skeletal muscles.
- Vasoconstriction in most other tissues diverts blood away from them and to the active muscles.
- Blood pressure increases, allowing better perfusion of the muscles and improving the return of venous blood to the heart.
- Bronchodilation improves ventilation and effective gas exchange.
- Metabolic rate increases, reflecting the body's effort to meet the increased demands of physical activity.
- Mental activity increases, allowing better perception of sensory stimuli and more concentration on performance.
- Glucose is released from the liver into the blood as an energy source.
- Functions not directly needed at that time are slowed (e.g., renal function, digestion).

These basic alterations in bodily function facilitate motor responses, demonstrating the importance of the autonomic nervous system in preparing the body for and sustaining it during acute stress or physical activity.

Parasympathetic Nervous System

The parasympathetic nervous system is the body's housekeeping system. It has a major role in carrying out such processes as digestion, urination, glandular secretion, and conservation of energy. This system is more active when one is calm and at rest. Its effects tend to oppose those of the sympathetic system. The parasympathetic division causes decreased heart rate, constriction of coronary vessels, and bronchoconstriction.

The various effects of the sympathetic and parasympathetic divisions of the autonomic nervous system are summarized in table 3.1.

In review

➤ The PNS contains 43 pairs of nerves: 12 cranial and 31 spinal.

➤ The PNS can be subdivided into the sensory and motor divisions. The motor division also includes the autonomic nervous system.

➤ The sensory division carries information from sensory receptors to the CNS. The motor division carries motor impulses from the CNS to the muscles and other organs.

➤ The autonomic nervous system includes the sympathetic nervous system and the parasympathetic system. Although these systems often oppose each other, they always function together to create an appropriately balanced response.

TABLE 3.1 Effects of the Sympathetic and Parasympathetic Nervous Systems on Various Organs

Target organ or system	Sympathetic effects	Parasympathetic effects
Heart muscle	Increases rate and force of contraction	Decreases rate of contraction
Heart: coronary blood vessels	Cause vasodilation	Cause vasoconstriction
Lungs	Cause bronchodilation; mildly constrict blood vessels	Cause bronchoconstriction
Blood vessels	Increase blood pressure; cause vasoconstriction in abdominal viscera and skin to divert blood when necessary; cause vasodilation in the skeletal muscles and heart during exercise	Little or no effect
Liver	Stimulates glucose release	No effect
Cellular metabolism	Increases metabolic rate	No effect
Adipose tissue	Stimulates lipolysis[a]	No effect
Sweat glands	Increase sweating	No effect
Adrenal glands	Stimulate secretion of epinephrine and norepinephrine	No effect
Digestive system	Decreases activity of glands and muscles; constricts sphincters	Increases peristalsis and glandular secretion; relaxes sphincters
Kidney	Causes vasoconstriction; decreases urine formation	No effect

[a]Lipolysis is the process of breaking down triglyceride to its basic units to be used for energy.

Sensory-Motor Integration

Having discussed the components and divisions of the nervous system, we now discuss how a sensory stimulus gives rise to a motor response. How, for example, do the muscles in the hand know to pull one's finger away from a hot stove? When someone decides to run, how do the muscles in the legs coordinate while supporting weight and propelling the person forward? To accomplish these tasks, the sensory and motor systems must communicate with each other.

This process is called **sensory-motor integration**, and it is depicted in figure 3.7. For the body to respond to sensory stimuli, the sensory and motor divisions of the nervous system must function together in the following sequence of events:

1. A sensory stimulus is received by sensory receptors (e.g., pinprick).

2. The sensory action potential is transmitted along sensory neurons to the CNS.

3. The CNS interprets the incoming sensory information and determines which response is most appropriate, or reflexively initiates a motor response.

4. The action potentials for the response are transmitted from the CNS along α-motor neurons.

5. The motor action potential is transmitted to a muscle, and the response occurs.

Sensory Input

Recall that sensations and physiological status are detected by sensory receptors throughout the body. The action potentials resulting from sensory stimulation are transmitted via the sensory nerves to the spinal cord. When they reach the spinal cord, they can trigger a local reflex at that level, or they can travel to the upper regions of the spinal cord or to the brain. Sensory

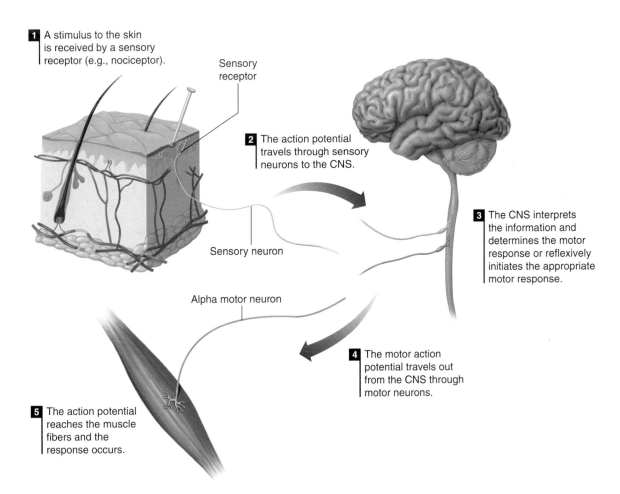

1 A stimulus to the skin is received by a sensory receptor (e.g., nociceptor).

Sensory receptor

2 The action potential travels through sensory neurons to the CNS.

3 The CNS interprets the information and determines the motor response or reflexively initiates the appropriate motor response.

Sensory neuron

Alpha motor neuron

4 The motor action potential travels out from the CNS through motor neurons.

5 The action potential reaches the muscle fibers and the response occurs.

FIGURE 3.7 The sequence of events in sensory-motor integration.

pathways to the brain can terminate in sensory areas of the brain stem, the cerebellum, the thalamus, or the cerebral cortex. An area in which the sensory impulses terminate is referred to as an integration center. This is where the sensory input is interpreted and linked to the motor system. Figure 3.8 illustrates various sensory receptors and their nerve pathways back to the spinal cord and up into various areas of the brain. The integration centers vary in function:

- Sensory impulses that terminate in the spinal cord are integrated there. The response is typically a simple motor reflex (discussed later), which is the simplest type of integration.

- Sensory signals that terminate in the lower brain stem result in subconscious motor reactions of a higher and more complex nature than simple spinal cord reflexes. Postural control during sitting, standing, or moving is an example of this level of sensory input.

- Sensory signals that terminate in the cerebellum also result in subconscious control of movement. The cerebellum appears to be the center of coordination, smoothing out movements by coordinating the actions of the various contracting muscle groups to perform the desired movement. Both fine and gross motor movements appear to be coordinated by the cerebellum in concert with the basal ganglia. Without the control exerted by the cerebellum, all movement would be uncontrolled and uncoordinated.

- Sensory signals that terminate at the thalamus begin to enter the level of consciousness, and the person begins to distinguish various sensations.

- Only when sensory signals enter the cerebral cortex can one discretely localize the signal. The primary sensory cortex, located in the postcentral gyrus (in the parietal lobe), receives general sensory input from receptors in the

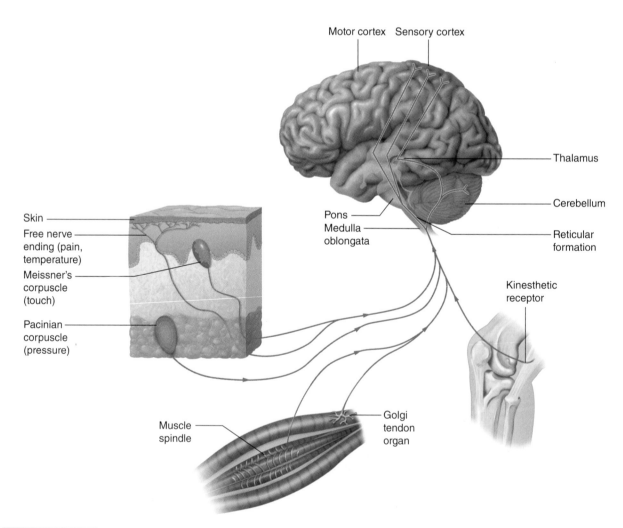

FIGURE 3.8 The sensory receptors and their pathways back to the spinal cord and brain.

skin and from proprioceptors in the muscles, tendons, and joints. This area has a "map" of the body. Stimulation in a specific area of the body is recognized, and its exact location is known instantly. Thus, this part of the conscious brain allows us to be constantly aware of our surroundings and our relationship to them.

Once a sensory impulse is received, it may evoke a motor response, regardless of the level at which the sensory impulse stops. This response can originate from any of three levels:

- The spinal cord
- The lower regions of the brain
- The motor area of the cerebral cortex

As the level of control moves from the spinal cord to the motor cortex, the degree of movement complexity increases from simple reflex control to complicated movements requiring basic thought processes. Motor responses for more complex movement patterns typically originate in the motor cortex of the brain, and the basal ganglia and cerebellum help to coordinate repetitive movements and to smooth out overall movement patterns. Sensory-motor integration is also assisted by reflex pathways for quick responses and specialized sensory organs within muscles.

Reflex Activity

What happens when one unknowingly puts one's hand on a hot stove? First, the stimuli of heat and pain are received by the thermoreceptors and nociceptors in the hand, and then sensory action potentials travel to the spinal cord, terminating at the level of entry. Once in the spinal cord, these action potentials are integrated instantly by interneurons that connect the sensory and motor neurons. The action potentials move to the motor neurons and travel to the effectors, the muscles controlling the withdrawal of the hand. The result is that the person reflexively withdraws the hand from the hot stove without giving the action any thought.

A **motor reflex** is a preprogrammed response; any time the sensory nerves transmit certain action potentials, the body responds instantly and identically. In examples like the one just used, whether one touches something that is too hot or too cold, thermoreceptors will elicit a reflex to withdraw the hand. Whether the pain arises from heat or from a sharp object, the nociceptors will also cause a withdrawal reflex. By the time one is consciously aware of the specific stimulus (after sensory action potentials also have been transmitted to the primary sensory cortex), the reflex activity is well under way, if not completed. All neural activity occurs extremely rapidly, but a reflex is the fastest mode of response because the impulse is not transmitted up the spinal cord to the brain before an action occurs. Only one response is possible; no options need to be considered.

The level of nervous system response to sensory input varies according to the complexity of movement necessary. Most simple reflexes are handled by the spinal cord, whereas complex reactions and movements require activation of higher centers in the brain.

Muscle Spindles

Now that we have covered the basics of reflex activity, we can look more closely at two specific reflexes that help control muscle function. The first involves a special structure: the muscle spindle (figure 3.9).

The **muscle spindle** is a group of specialized muscle fibers found between regular skeletal muscle fibers, referred to as *extrafusal* (outside the spindle) fibers. A muscle spindle consists of 4 to 20 small, specialized

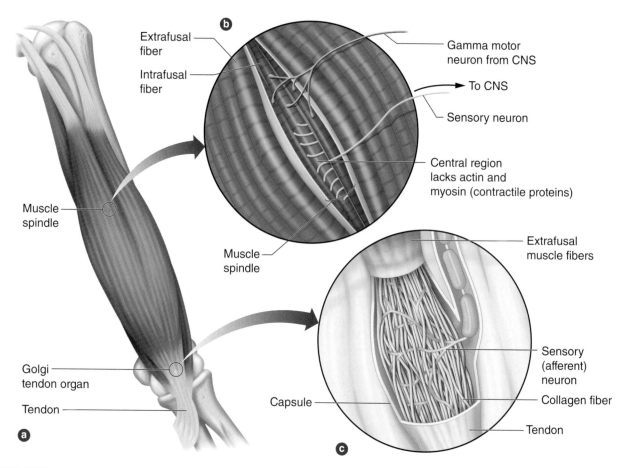

FIGURE 3.9 (*a*) A muscle belly showing (*b*) a muscle spindle and (*c*) a Golgi tendon organ.

intrafusal (inside the spindle) fibers and the nerve endings, sensory and motor, associated with these fibers. A connective tissue sheath surrounds the muscle spindle and attaches to the endomysium of the extrafusal fibers. The intrafusal fibers are controlled by specialized motor neurons, referred to as γ-*motor neurons* (or *gamma motor neurons*). In contrast, extrafusal fibers (the regular fibers) are controlled by α-*motor neurons.*

The central region of an intrafusal fiber cannot contract because it contains no or only a few actin and myosin filaments. This central region can only stretch. Because the muscle spindle is attached to the extrafusal fibers, any time those fibers are stretched, the central region of the muscle spindle is also stretched.

Sensory nerve endings wrapped around this central region of the muscle spindle transmit information to the spinal cord when this region is stretched, transmitting a signal to the CNS about the muscle's length. In the spinal cord, the sensory neuron synapses with an α-motor neuron, which triggers reflexive muscle contraction (in the extrafusal fibers) to resist further stretching.

Let's illustrate this action with an example. A person's arm is bent at the elbow, and the hand is extended, palm up. Suddenly someone places a heavy weight in the palm. The forearm starts to drop, which stretches the muscle fibers in the elbow flexors (e.g., biceps brachii), which in turn stretch the muscle spindles. In response to that stretch, the sensory neurons send action potentials to the spinal cord, which then activates the α-motor neurons of motor units in the same muscles. These cause the muscles to increase their force production, overcoming the stretch.

γ-Motor neurons excite the intrafusal fibers, prestretching them slightly. Although the midsection of the intrafusal fibers cannot contract, the ends can. The γ-motor neurons cause slight contraction of the ends of these fibers, which stretches the central region slightly. This prestretch makes the muscle spindle highly sensitive to even small degrees of stretch.

The muscle spindle also assists normal muscle action. It appears that when the α-motor neurons are stimulated to contract the extrafusal muscle fibers, the γ-motor neurons are also activated, contracting the ends of the intrafusal fibers. This stretches the central region of the muscle spindle, giving rise to sensory impulses that travel to the spinal cord and then to the α-motor neurons. In response, the muscle increases its force production. Thus, muscle force production is enhanced through this function of the muscle spindles.

Information brought into the spinal cord from the sensory neurons associated with muscle spindles does not merely end at that level. Impulses are also sent up to higher parts of the CNS, supplying the brain with continuous feedback on the exact length of the muscle and the rate at which that length is changing. This information is essential for maintaining muscle tone and posture and for executing movements. The muscle spindle functions as a servo-mechanism to continuously correct movements that do not proceed as planned. The brain is informed of errors in the intended movement at the same time that the error is being corrected at the spinal cord level.

Golgi Tendon Organs

Golgi tendon organs are encapsulated sensory receptors through which a small bundle of muscle tendon fibers pass. These organs are located just proximal to the tendon fibers' attachment to the muscle fibers, as shown in figure 3.9. Approximately 5 to 25 muscle fibers are usually connected with each Golgi tendon organ. Whereas muscle spindles monitor the length of a muscle, Golgi tendon organs are sensitive to tension in the muscle–tendon complex and operate like a strain gauge, a device that senses changes in tension. Their sensitivity is so great that they can respond to the contraction of a single muscle fiber. These sensory receptors are inhibitory in nature, performing a protective function by reducing the potential for injury. When stimulated, these receptors inhibit the contracting (agonist) muscles and excite the antagonist muscles.

Golgi tendon organs are important in resistance exercise. They function as safety devices that help prevent the muscle from developing excessive force during a contraction that may ultimately damage the muscle. Additionally, some researchers speculate that reducing the influence of Golgi tendon organs disinhibits the active muscles, allowing a more forceful muscle action. This mechanism may explain at least part of the gains in muscular strength that accompany strength training.

Motor Response

Now that we have discussed how sensory input is integrated to determine the appropriate motor response, the last step in the process is how muscles respond to motor action potentials once they reach the muscle fibers.

Once an action potential reaches an α-motor neuron, it travels the length of the neuron to the neuromuscular junction. From there, the action potential spreads to all muscle fibers innervated by that particular α-motor neuron. Recall that the α-motor neuron and all muscle fibers it innervates form a single motor unit. Each muscle fiber is innervated by only one α-motor neuron, but each α-motor neuron innervates up to several thousand muscle fibers, depending on the function of the muscle. Muscles controlling fine movements, such as those controlling the eyes, have only a small number of muscle fibers per α-motor neuron. Muscles with more general functions have many fibers per α-motor neuron.

The muscles that control eye movements (the extraocular muscles) have an innervation ratio of 1:15, meaning that one α-motor neuron controls only 15 muscle fibers. In contrast, the gastrocnemius and tibialis anterior muscles of the lower leg have innervation ratios of almost 1:2,000.

The muscle fibers in a specific motor unit are homogeneous with respect to fiber type. Thus, one will not find a motor unit that has both type II and type I fibers. In fact, as mentioned in chapter 1, it is generally believed that the characteristics of the α-motor neuron actually determine the fiber type in that motor unit.[1, 5]

In review

➤ Sensory-motor integration is the process by which the PNS relays sensory input to the CNS and the CNS interprets this information and then sends out the appropriate motor signal to elicit the desired motor response.

➤ Sensory input can terminate at various levels of the CNS. Not all of this information reaches the brain.

➤ Reflexes are the simplest form of motor control. These are not conscious responses. For a given sensory stimulus, the motor response is always identical and instantaneous.

➤ Muscle spindles trigger reflexive muscle action when stretched.

➤ Golgi tendon organs trigger a reflex that inhibits contraction if the tendon fibers are stretched from high muscle tension.

We have seen how muscles respond to neural stimulation, whether through reflexes or under complex control of the higher brain centers. We discussed how the individual motor units respond and how they are recruited in an orderly manner depending on the required force. Thus, we have learned how the body functions to allow people to move. In the next chapter, we examine the energy needs of the body at rest and during exercise.

Key Terms

acetylcholine

adrenergic

afferent nerves

axon hillock

axon terminal

central nervous system (CNS)

cholinergic

depolarization

efferent nerves

end branches

excitatory postsynaptic potential (EPSP)

Golgi tendon organ

graded potential

hyperpolarization

inhibitory postsynaptic potential (IPSP)

motor nerves

motor reflex

muscle spindle

myelin sheath

nerve impulse

neuromuscular junction

neuron

neurotransmitter

norepinephrine

peripheral nervous system (PNS)

resting membrane potential (RMP)

saltatory conduction

sensory nerves

sensory-motor integration

sodium-potassium pump

synapse

threshold

Study Questions

1. What are the major divisions of the nervous system? What are their major functions?
2. Name the different parts of a neuron.
3. Explain the resting membrane potential. What causes it? How is it maintained?
4. Describe an action potential. What is required before an action potential is activated?
5. Explain how an action potential is transmitted from a presynaptic neuron to a postsynaptic neuron. Describe a synapse and a neuromuscular junction.
6. What brain centers have major roles in controlling movement, and what are these roles?
7. How do the sympathetic and parasympathetic systems differ? What is their significance in performing physical activity?
8. Explain how reflex movement occurs in response to touching a hot object.
9. Describe the role of the muscle spindle in controlling muscle contraction.
10. Describe the role of the Golgi tendon organ in controlling muscle contraction.

Study Guide Activities

In addition to the activities listed in the chapter opening outline on page 69, two other activities are available in the web study guide, located at

www.HumanKinetics.com/PhysiologyOfSportAndExercise

 The **KEY TERMS** activity reviews important terms, and the end-of-chapter **QUIZ** tests your understanding of the material covered in the chapter.

Hormonal Control During Exercise

4

In this chapter and in the web study guide

ACTIVITY 4.1   Endocrine Glands reviews the body's major endocrine glands.

ACTIVITY 4.2   Hormones reviews hormones and their functions.

ACTIVITY 4.3   Hormones and Exercise considers hormones and their key roles in maintaining homeostasis during physical activity.

On May 22, 2010, a 13-year-old American boy became the youngest climber to reach the top of Mount Everest, a grueling trek to an altitude 29,035 ft above sea level. The climb was extremely controversial because of the boy's age. In fact, because the Nepalese government would not give the family permission to climb Everest from Nepal, the climbing team ascended from the more difficult Chinese side where there was no age restriction. To prepare for the climb, the boy and his father (and climbing partner) slept for months in a hypoxic tent to prepare their bodies for ascent to high altitude. One goal of high-altitude acclimation is to increase the concentration of oxygen-carrying red blood cells in the blood. Two important hormones facilitated this goal. An increase in the hormone erythropoietin signaled the bone marrow to produce more red blood cells, and a decrease in vasopressin (also called antidiuretic hormone) caused the kidneys to produce excess urine to better concentrate the red blood cells. Because of these adaptations, the climbers were able to summit Mount Everest with less time spent in the various base camps along the way.

During exercise and exposure to extreme environments, the body is faced with tremendous demands that require a multitude of physiological adjustments. Energy production must increase and metabolic by-products must be cleared. Cardiovascular and respiratory function must be constantly adjusted to match the demands placed upon these and other body systems, such as those regulating temperature. While the body's internal environment is in a constant state of flux even at rest, during exercise these well-orchestrated changes must occur rapidly and frequently.

The more rigorous the exercise, the more difficult it is to maintain homeostasis. Much of the regulation required during exercise is accomplished by the nervous system (chapter 3). But another physiological system affects virtually every cell, tissue, and organ in the body. It constantly monitors the body's internal environment, noting all changes that occur and responding quickly to ensure that homeostasis is not dramatically disrupted. It is the endocrine system that exerts this control through the hormones it releases. In this chapter, we focus on the importance of hormones in making adjustments and maintaining homeostasis amid all the internal processes that support physical activity. Because we cannot cover all aspects of endocrine control during exercise, the focus will be on hormonal control of metabolism and body fluids. Additional hormones, including those that regulate growth and development, muscle mass, and reproductive function, are covered in other chapters of this book.

The Endocrine System

As the body transitions from a resting to an active state, the rate of metabolism must increase to provide the necessary energy. This requires the coordinated integration of many physiological and biochemical systems. Such integration is possible only if all of the involved tissues, organs, and systems can efficiently communicate. Although the nervous system is responsible for much of this communication, fine-tuning the physi-

ological responses to any disturbance in homeostasis is primarily the responsibility of the endocrine system. The endocrine and nervous systems, often collectively called the neuroendocrine system, work in concert to initiate and control movement and all of the physiological processes that support exercise. The nervous system functions quickly, having short-lived, localized effects, whereas the endocrine system responds more slowly but has longer-lasting effects.

The endocrine system is defined as all tissues or glands that secrete **hormones**. The major endocrine glands are illustrated in figure 4.1. Endocrine glands typically secrete their hormones directly into the blood where they act as chemical signals throughout the body. When secreted by the specialized endocrine cells, hor-

- Hypothalamus
- Pituitary gland
- Thyroid gland
- Parathyroid glands
- Thymus gland
- Adrenal glands
- Pancreas
- Kidneys
- Testes (in male)
- Ovaries (in female)

FIGURE 4.1 Location of the major endocrine organs of the body.

mones are transported via the blood to specific **target cells**—cells that possess specific hormone receptors. On reaching their destinations, hormones can control the activity of the target tissue. Historically a hormone denoted a chemical made by a gland that traveled to a remote tissue in the body to exert its action. Now hormones are more broadly defined as any chemical that controls and regulates the activity of certain cells or organs. Some hormones affect many body tissues, whereas others target very specific cells of the body.

Hormones

Hormones are involved in most physiological processes, so their actions are relevant to many aspects of exercise and sport. Because hormones play key roles in almost every system of the body, total coverage of that topic is well beyond the scope of this book. In the following sections, the chemical nature of hormones and the general mechanisms through which they act are discussed. An overview of the major endocrine glands and their hormones is presented for completeness. With respect to exercise, the focus is on two major aspects of hormonal control, the control of exercise metabolism and the regulation of body fluids and electrolytes during exercise.

Chemical Classification of Hormones

Hormones can be categorized as two basic types: steroid hormones and nonsteroid hormones. **Steroid hormones** have a chemical structure similar to cholesterol, since most are derived from cholesterol. For this reason, they are soluble in lipids so they diffuse rather easily through cell membranes. This group includes the hormones secreted by

- the adrenal cortex (such as cortisol and aldosterone),
- the ovaries (estrogen and progesterone),
- the testes (testosterone), and
- the placenta (estrogen and progesterone).

Nonsteroid hormones are not lipid soluble, so they cannot easily cross cell membranes. The nonsteroid hormone group can be subdivided into two groups: protein or peptide hormones and amino acid–derived hormones. The two hormones produced by the thyroid gland (thyroxine and triiodothyronine) and the two from the adrenal medulla (epinephrine and norepinephrine) are amino acid hormones. All other nonsteroid hormones are protein or peptide hormones. The chemical structure of a hormone determines its mechanism of action on target cells and tissues.

Hormone Secretion and Plasma Concentration

Control of hormone secretion must be rapid in order to meet the demands of changing bodily functions. Hormones are not secreted constantly or uniformly, but often in a pulsitile manner, that is, in relatively brief bursts. Therefore, plasma concentrations of specific hormones fluctuate over short periods of an hour or less. But these concentrations also fluctuate over longer periods of time, showing daily or even monthly cycles (such as monthly menstrual cycles). How do endocrine glands know when to release their hormones and how much to release?

Most hormone secretion is regulated by a negative feedback system. Secretion of a hormone causes some change in the body, and this change in turn inhibits further hormone secretion. Consider how a home thermostat works. When the room temperature decreases below some preset level, the thermostat signals the furnace to produce heat. When the room temperature increases to the preset level, the thermostat's signal ends, and the furnace stops producing heat. When the temperature again falls below the preset level, the cycle begins anew. In the body, secretion of a specific hormone is similarly turned on or off (or up or down) by specific physiological changes.

Negative feedback is the primary mechanism through which the endocrine system maintains homeostasis. Using the example of plasma glucose concentrations and the hormone insulin, when the plasma glucose concentration is high, the pancreas releases insulin. Insulin increases cellular uptake of glucose, lowering plasma concentration of glucose. When plasma glucose concentration returns to normal, insulin release is inhibited until the plasma glucose level increases again.

The plasma concentration of a specific hormone is not always the best indicator of that hormone's activity because the number of receptors on target cells can be altered to increase or decrease that cell's sensitivity to the hormone. Most commonly, an increased volume of a specific hormone decreases the number of cell receptors available to it. When this happens, the cell becomes less sensitive to that hormone, because with fewer receptors, fewer hormone molecules can bind. This is referred to as **downregulation**, or desensitization. In some people with obesity, for example, the number of insulin receptors on their cells appears to be reduced. Their bodies respond by increasing insulin secretion from the pancreas, so their plasma insulin concentrations increase. To obtain the same degree of plasma glucose control as normal, healthy people, these individuals must release much more insulin.

In a few instances, a cell may respond to the prolonged presence of large amounts of a hormone by increasing its number of available receptors. When this happens, the cell becomes more sensitive to that hormone because more can be bound at one time. This is referred to as **upregulation**. In addition, one hormone occasionally can regulate the receptors for another hormone.

Hormone Actions

Because hormones travel in the blood, they contact virtually all body tissues. How, then, do they limit their effects to specific targets? This ability is attributable to the specific hormone receptors possessed by the target tissues that can bind only specific hormones. Each cell typically has from 2,000 to 10,000 receptors. The combination of a hormone bound to its receptor is referred to as a hormone–receptor complex.

Recall that steroid hormones are lipid soluble and can therefore pass through cell membranes whereas nonsteroid hormones cannot. Receptors for nonsteroid hormones are located on the cell membrane, while those for steroid hormones are found either in the cytoplasm or in the nucleus of the cell. Each hormone

is usually highly specific for a single type of receptor and binds only with its specific receptors, thus affecting only tissues that contain those specific receptors. Once hormones are bound to a receptor, numerous mechanisms allow them to control the actions of cells.

(In)focus

Hormones influence specific target tissues or cells through a unique interaction between the hormone and the specific receptors for that hormone on the cell membrane (nonsteroid hormones) or within the cytoplasm or nucleus of the cell (steroid hormones).

Steroid Hormones

The general mechanism of action of steroid hormones is illustrated in figure 4.2. Once inside the cell, a steroid hormone binds to its specific receptors. The hormone–receptor complex then enters the nucleus, binds to part of the cell's DNA, and activates certain genes. This process is referred to as **direct gene activation**. In response to this activation, mRNA is synthesized within

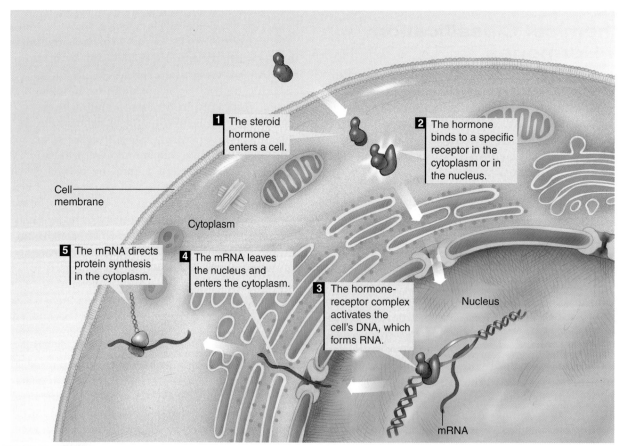

1 The steroid hormone enters a cell.

2 The hormone binds to a specific receptor in the cytoplasm or in the nucleus.

Cell membrane

Cytoplasm

5 The mRNA directs protein synthesis in the cytoplasm.

4 The mRNA leaves the nucleus and enters the cytoplasm.

3 The hormone-receptor complex activates the cell's DNA, which forms RNA.

Nucleus

mRNA

FIGURE 4.2 The general mechanism of action of a typical steroid hormone, leading to direct gene activation and protein synthesis.

response to this activation, mRNA is synthesized within the nucleus. The mRNA then enters the cytoplasm and promotes protein synthesis. These proteins may be

- enzymes that can have numerous effects on cellular processes,
- structural proteins for tissue growth and repair, or
- regulatory proteins that can alter enzyme function.

Nonsteroid Hormones

Because nonsteroid hormones cannot cross the cell membrane, they bind with specific receptors on the cell membrane. A nonsteroid hormone molecule binds to its receptor and triggers a series of reactions that lead to the formation of an intracellular **second messenger**. In addition to their job as a signal-relaying molecule, second messengers also help intensify the strength of the signal. One important second messenger that mediates a specific hormone–receptor response in many cells is **cyclic adenosine monophosphate** (cyclic AMP, or **cAMP**), and its mechanism of action is depicted in figure 4.3. In this case, attachment of the hormone

to the appropriate membrane receptor activates an enzyme, adenylate cyclase, situated within the cell membrane. This enzyme catalyzes the formation of cAMP from cellular adenosine triphosphate (ATP). Cyclic AMP then can produce specific physiological responses that can include

- activation of cellular enzymes,
- change in membrane permeability,
- promotion of protein synthesis,
- change in cellular metabolism, or
- stimulation of cellular secretions.

Some of the hormones that achieve their effects through cAMP as a second messenger are epinephrine, glucagon, and luteinizing hormone.

While nonsteroid hormones often activate the cAMP system of the cell, many other second messengers exist, including

- cyclic guanine monophosphate (cGMP),
- inositol trisphosphate (IP3) and diacylglycerol (DAG), and
- calcium ions (Ca^{2+}).

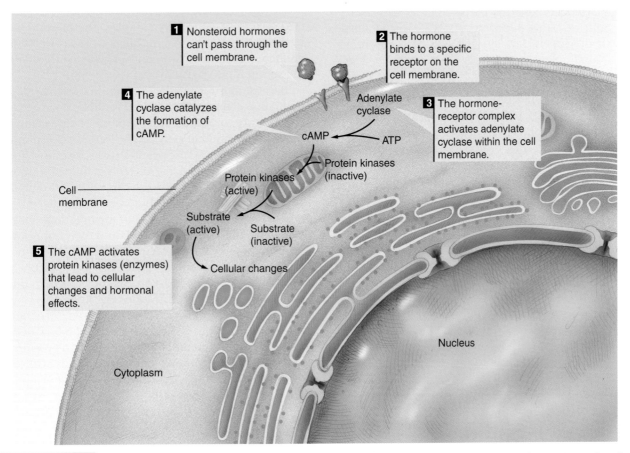

FIGURE 4.3 The mechanism of action of a nonsteroid hormone, in this case activating a second messenger (cyclic adenosine monophosphate, or cAMP) within the cell to activate cellular functions.

Prostaglandins

Prostaglandins, although by strict definition not hormones, are often considered to be a third class of hormones. These substances are derived from a fatty acid, arachidonic acid, and they are associated with the plasma membranes of almost all body cells. Prostaglandins typically act as *local* hormones, exerting their effects in the immediate area where they are produced. But some also survive long enough to circulate through the blood to affect distant tissues. Prostaglandin release can be triggered by many stimuli, such as other hormones or a local injury. Their functions are quite numerous because there are several different types of prostaglandins. They often mediate the effects of other hormones. They are also known to act directly on blood vessels, increasing vascular permeability (which promotes swelling) and vasodilation. In this capacity, they are important mediators of the inflammatory response. They also sensitize the nerve endings of pain fibers; thus, they promote both inflammation and pain.

In review

➤ Hormones are classified chemically as either steroid or nonsteroid. Steroid hormones are lipid soluble, and most are formed from cholesterol. Nonsteroid hormones are formed from proteins, peptides, or amino acids.

➤ Hormones generally are secreted nonuniformly into the blood and then circulate to target cells. They act by binding with specific receptors found only in target tissues.

➤ A negative feedback system regulates secretion of most hormones.

➤ The number of receptors for a specific hormone can be altered to meet the body's demands. Upregulation refers to an increase in available receptors, and downregulation refers to a decrease. These two processes change cell sensitivity to hormones.

➤ Steroid hormones pass through cell membranes and bind to receptors in the cytoplasm or nucleus of the cell. At the nucleus, they use a mechanism called direct gene activation to cause protein synthesis.

➤ Nonsteroid hormones cannot easily enter cells, so they bind to receptors on the cell membrane. This activates a second messenger within the cell, which in turn can trigger numerous cellular processes.

Endocrine Glands and Their Hormones: An Overview

The endocrine glands and their respective hormones are listed in table 4.1. This table also lists each hormone's primary target and actions. Because the endocrine system is extremely complex, the presentation here has been greatly simplified to focus on those endocrine glands and hormones of greatest importance to sport and physical activity.

Because hormones play such an important role in regulation of many physiological variables during exercise, it is not surprising that hormone release changes during acute bouts of activity. The hormonal responses to an acute bout of exercise and to exercise training are summarized in table 4.2. This table is limited to those hormones that play major roles in sport and physical activity. Further details of these exercise-induced hormonal responses are provided in the following discussion of specific endocrine glands and their hormones.

As mentioned earlier, a comprehensive description of neuroendocrine control is well beyond the scope of this textbook. Two important functions of the endocrine glands and their hormones are the regulation of metabolism during exercise and the regulation of body fluids and electrolytes. The sections that follow detail these two important functions. Each section provides a description of the primary endocrine glands involved, the hormones produced, and how those hormones serve that regulatory role.

TABLE 4.1 — The Endocrine Glands and Their Hormones, Target Organs, Controlling Factors, and Functions

Endocrine gland	Hormone	Target organ	Controlling factor	Major functions
Anterior pituitary	Growth hormone (GH)	All cells in the body	Hypothalamic GH-releasing hormone; GH-inhibiting hormone (somatostatin)	Promotes development and enlargement of all body tissues until maturation; increases rate of protein synthesis; increases mobilization of fats and use of fat as an energy source; decreases rate of carbohydrate use
	Thyrotropin (TSH)	Thyroid gland	Hypothalamic TSH-releasing hormone	Controls the amount of thyroxin and triiodothyronine produced and released by the thyroid gland
	Adrenocorticotropin (ACTH)	Adrenal cortex	Hypothalamic ACTH-releasing hormone	Controls the secretion of hormones from the adrenal cortex
	Prolactin	Breasts	Prolactin-releasing and -inhibiting hormones	Stimulates milk production by the breasts
	Follicle-stimulating hormone (FSH)	Ovaries, testes	Hypothalamic FSH-releasing hormone	Initiates growth of follicles in the ovaries and promotes secretion of estrogen from the ovaries; promotes development of the sperm in the testes
	Luteinizing hormone (LH)	Ovaries, testes	Hypothalamic FSH-releasing hormone	Promotes secretion of estrogen and progesterone and causes the follicle to rupture, releasing the ovum; causes testes to secrete testosterone
Posterior pituitary	Antidiuretic hormone (ADH or vasopressin)	Kidneys	Hypothalamic secretory neurons	Assists in controlling water excretion by the kidneys; elevates blood pressure by constricting blood vessels
	Oxytocin	Uterus, breasts	Hypothalamic secretory neurons	Controls contraction of uterus; milk secretion
Thyroid	Thyroxine (T_4) and triiodothyronine (T_3)	All cells in the body	TSH, T_3, and T_4 concentrations	Increase the rate of cellular metabolism; increase rate and contractility of the heart
	Calcitonin	Bones	Plasma calcium concentrations	Controls calcium ion concentration in the blood
Parathyroid	Parathyroid hormone (PTH or parathormone)	Bones, intestines, and kidneys	Plasma calcium concentrations	Controls calcium ion concentration in the extracellular fluid through its influence on bones, intestines, and kidneys

(continued)

TABLE 4.1 *(continued)*

Endocrine gland	Hormone	Target organ	Controlling factor	Major functions
Adrenal medulla	Epinephrine	Most cells in the body	Baroreceptors, glucose receptors, brain and spinal centers	Stimulates breakdown of glycogen in liver and muscle and lipolysis in adipose tissue and muscle; increases skeletal muscle blood flow; increases heart rate and contractility; increases oxygen consumption
	Norepinephrine	Most cells in the body	Baroreceptors, glucose receptors, brain and spinal centers	Stimulates lipolysis in adipose tissue and in muscle to a lesser extent; constricts arterioles and venules, thereby elevating blood pressure
Adrenal cortex	Mineralocorticoids (aldosterone)	Kidneys	Angiotensin and plasma potassium concentrations; renin	Increase sodium retention and potassium excretion through the kidneys
	Glucocorticoids (cortisol)	Most cells in the body	ACTH	Control metabolism of carbohydrates, fats, and proteins; exert an anti-inflammatory action
	Androgens and estrogens	Ovaries, breasts, and testes	ACTH	Assist in the development of female and male sex characteristics
Pancreas	Insulin	All cells in the body	Plasma glucose and amino acid concentrations	Controls blood glucose levels by lowering glucose levels; increases use of glucose and synthesis of fat
	Glucagon	All cells in the body	Plasma glucose and amino acid concentrations	Increases blood glucose; stimulates the breakdown of protein and fat
	Somatostatin	Islets of Langerhans and intestines	Plasma glucose, insulin, and glucagon concentrations	Depresses the secretion of both insulin and glucagon
Kidney	Renin	Adrenal cortex	Plasma sodium concentrations	Assists in blood pressure control
	Erythropoietin (EPO)	Bone marrow	Low tissue oxygen concentrations	Stimulates erythrocyte production
Testes	Testosterone	Sex organs, muscle	FSH and LH	Promotes development of male sex characteristics, including growth of testes, scrotum, and penis, facial hair, and change in voice; promotes muscle growth
Ovaries	Estrogens and progesterone	Sex organs and adipose tissue	FSH and LH	Promote development of female sex organs and characteristics; increase storage of fat; assist in regulating the menstrual cycle

TABLE 4.2   Hormone Response to Acute Exercise and Change in Response With Exercise Training

Endocrine gland	Hormone	Response to acute exercise (untrained)	Effect of exercise training
Anterior pituitary	Growth hormone (GH)	Increases with increasing rates of work	Attenuated response at same rate of work
	Thyrotropin (TSH)	Increases with increasing rates of work	No known effect
	Adrenocorticotropin (ACTH)	Increases with increasing rates of work and duration	Attenuated response at same rate of work
	Prolactin	Increases with exercise	No known effect
	Follicle-stimulating hormone (FSH)	Small or no change	No known effect
	Luteinizing hormone (LH)	Small or no change	No known effect
Posterior pituitary	Antidiuretic hormone (ADH or vasopressin)	Increases with increasing rates of work	Attenuated response at same rate of work
	Oxytocin	Unknown	Unknown
Thyroid	Thyroxine (T_4) and triiodothyronine (T_3)	Free T_3 and T_4 increase with increasing rates of work	Increased turnover of T_3 and T_4 at same rate of work
	Calcitonin	Unknown	Unknown
Parathyroid	Parathyroid hormone (PTH or parathormone)	Increases with prolonged exercise	Unknown
Adrenal medulla	Epinephrine	Increases with increasing rates of work, starting at about 75% of $\dot{V}O_2$max	Attenuated response at same rate of work
	Norepinephrine	Increases with increasing rates of work, starting at about 50% of $\dot{V}O_2$max	Attenuated response at same rate of work
Adrenal cortex	Aldosterone	Increases with increasing rates of work	Unchanged
	Cortisol	Increases only at high rates of work	Slightly higher values
Pancreas	Insulin	Decreases with increasing rates of work	Attenuated response at same rate of work
	Glucagon	Increases with increasing rates of work	Attenuated response at same rate of work
Kidney	Renin	Increases with increasing rates of work	Unchanged
	Erythropoietin (EPO)	Unknown	Unchanged
Testes	Testosterone	Small increases with exercise	Resting levels decreased in male runners
Ovaries	Estrogens and progesterone	Small increases with exercise	Resting levels might be decreased in highly trained women

Hormonal Regulation of Metabolism During Exercise

As noted in chapter 2, carbohydrate and fat metabolism are responsible for maintaining muscle ATP levels during prolonged exercise. Various hormones work to ensure glucose and free fatty acid (FFA) availability for muscle energy metabolism. In the next sections we examine (1) the major endocrine glands and hormones responsible for metabolic regulation, and (2) how the metabolism of glucose and fat is affected by these hormones during exercise.

Endocrine Glands Involved in Metabolic Regulation

While many complex systems interact to regulate metabolism at rest and during exercise, the major endocrine glands responsible are the anterior pituitary gland, the thyroid gland, the adrenal glands, and the pancreas.

Anterior Pituitary

The pituitary gland is a marble-sized gland attached to the hypothalamus at the base of the brain. It is composed of three lobes: anterior, intermediate, and posterior. The intermediate lobe is very small and is thought to play little or no role in humans, but both the anterior and posterior lobes serve major endocrine functions. The secretory action of the anterior pituitary is controlled by hormones secreted by the hypothalamus, while the posterior pituitary receives direct nerve signals from neural projections from the hypothalamus. Therefore, the pituitary gland can be thought of as the relay between central nervous system control centers and peripheral endocrine glands. The posterior pituitary is discussed later in the chapter.

The anterior pituitary, also called the adenohypophysis, secretes six hormones in response to **releasing factors** or **inhibiting factors** (which are also categorized as hormones) secreted by the hypothalamus. Communication between the hypothalamus and the anterior lobe of the pituitary occurs through a specialized circulatory system that transports the releasing and inhibiting factors from the hypothalamus to the anterior pituitary. The major functions of each of the anterior pituitary hormones, along with their releasing and inhibiting factors, are listed in table 4.1. Exercise appears to be a strong stimulus to the hypothalamus because exercise increases the release rate of all six anterior pituitary hormones (see table 4.2).

Of the six anterior pituitary hormones, four are tropic hormones, meaning they affect the functioning of other endocrine glands. The exceptions are growth hormone and prolactin. **Growth hormone (GH)** is a potent anabolic agent (a substance that builds up organs and tissues, producing growth and cell differentiation and an increase in size of tissues). It promotes muscle growth and hypertrophy by facilitating amino acid transport into the cells. In addition, GH directly stimulates fat metabolism (lipolysis) by increasing the synthesis of enzymes involved in that process. Growth hormone concentrations are elevated during aerobic exercise in proportion to the exercise intensity and typically remain elevated for some time after exercise.

Thyroid Gland

The thyroid gland is located along the midline of the neck, immediately below the larynx. It secretes two important nonsteroid hormones, **triiodothyronine (T_3)** and **thyroxine (T_4)**, which regulate metabolism in general, and an additional hormone, calcitonin, which assists in regulating calcium metabolism.

The two metabolic thyroid hormones share similar functions. Triiodothyronine and thyroxine increase the metabolic rate of almost all tissues and can increase the body's basal metabolic rate by as much as 60% to 100%. These hormones also

- increase protein synthesis (including enzymes);
- increase the size and number of mitochondria in most cells;
- promote rapid cellular uptake of glucose;
- enhance glycolysis and gluconeogenesis; and
- enhance lipid mobilization, increasing FFA availability for oxidation.

Release of **thyrotropin** (thyroid-stimulating hormone, or **TSH**) from the anterior pituitary increases during exercise. TSH controls the release of triiodothyronine and thyroxine, so the exercise-induced increase in TSH would be expected to stimulate the thyroid gland. Exercise increases plasma thyroxine concentrations, but a delay occurs between the increase in TSH concentrations during exercise and the increase in plasma thyroxine concentration. Furthermore, during prolonged submaximal exercise, thyroxine concentrations remain relatively constant after a sharp initial increase as exercise begins, and triiodothyronine concentrations tend to decrease.

Adrenal Glands

The adrenal glands are situated directly atop each kidney and are composed of the inner adrenal medulla and the outer adrenal cortex. The hormones secreted by these two parts are quite distinct. The adrenal medulla produces and releases two hormones, **epi-**

nephrine and norepinephrine, which are collectively referred to as **catecholamines**. Because of its origin in the adrenal gland, a synonym for epinephrine is **adrenaline**. When the adrenal medulla is stimulated by the sympathetic nervous system, approximately 80% of its secretion is epinephrine and 20% is norepinephrine, although these percentages vary with different physiological conditions. The catecholamines have powerful effects similar to those of the sympathetic nervous system. Recall that these same catecholamines function as neurotransmitters in the sympathetic nervous system; however, the hormones' effects last longer because these substances are removed from the blood relatively slowly compared to the quick reuptake and degradation of the neurotransmitters. These two hormones prepare a person for immediate action, often called the "fight-or-flight response."

Although some of the specific actions of these two hormones differ, the two work together. Their combined effects include

- increased heart rate and force of contraction,
- increased metabolic rate,
- increased glycogenolysis (breakdown of glycogen to glucose) in the liver and muscle,
- increased release of glucose and FFAs into the blood,
- redistribution of blood to the skeletal muscles,
- increased blood pressure, and
- increased respiration.

Release of epinephrine and norepinephrine is affected by a wide variety of factors, including changes in body position, psychological stress, and exercise. Plasma concentrations of these hormones increase as individuals gradually increase their exercise intensity. Plasma norepinephrine concentrations increase markedly at work rates above 50% of $\dot{V}O_2$max, but epinephrine concentrations do not increase significantly until the exercise intensity exceeds 60% to 70% of $\dot{V}O_2$max. During long-duration steady-state activity of moderate intensity, blood concentrations of both hormones increase. When the exercise bout ends, epinephrine returns to resting concentrations within only a few minutes of recovery, but norepinephrine can remain elevated for several hours.

The adrenal cortex secretes more than 30 different steroid hormones, referred to as corticosteroids. These generally are classified into three major types: mineralocorticoids (discussed later in the chapter), glucocorticoids, and gonadocorticoids (sex hormones).

The **glucocorticoids** are essential to the ability to adapt to exercise and other forms of stress. They also help maintain fairly consistent plasma glucose concentrations even during long periods without ingestion of food. **Cortisol**, also known as hydrocortisone, is the major corticosteroid. It is responsible for about 95% of all glucocorticoid activity in the body. Cortisol

- stimulates gluconeogenesis to ensure an adequate fuel supply;
- increases mobilization of FFAs, making them more available as an energy source;
- decreases glucose utilization, sparing it for the brain;
- stimulates protein catabolism to release amino acids for use in repair, enzyme synthesis, and energy production;
- acts as an anti-inflammatory agent;
- depresses immune reactions; and
- increases the vasoconstriction caused by epinephrine.

We discuss cortisol's important role in exercise later in this chapter when we consider the regulation of glucose and fat metabolism.

Pancreas

The pancreas is located behind and slightly below the stomach. Its two major hormones are insulin and glucagon. The balance of these two opposing hormones provides the major control of plasma glucose concentration. When plasma glucose is elevated (**hyperglycemia**), as occurs after a meal, the pancreas releases **insulin** into the blood. Among its actions, insulin

- facilitates glucose transport into the cells, especially muscle fibers;
- promotes glycogenesis; and
- inhibits gluconeogenesis.

Insulin's main function is to reduce the amount of glucose circulating in the blood. But it is also involved in protein and fat metabolism, promoting cellular uptake of amino acids and enhancing synthesis of protein and fat.

The pancreas secretes **glucagon** when the plasma glucose concentration falls below normal concentrations (**hypoglycemia**). Its effects generally oppose those of insulin. Glucagon promotes increased breakdown of liver glycogen to glucose (glycogenolysis) and increased gluconeogenesis, both of which increase plasma glucose levels.

During exercise lasting 30 min or longer, the body attempts to maintain plasma glucose concentrations; however, insulin concentrations tend to decline.

The ability of insulin to bind to its receptors on muscle cells increases during exercise, due in large part to increased blood flow to muscle. This increases the body's sensitivity to insulin and reduces the need to maintain high plasma insulin concentrations for transporting glucose into the muscle cells. Plasma glucagon, on the other hand, shows a gradual increase throughout exercise. Glucagon primarily maintains plasma glucose concentrations by stimulating liver glycogenolysis. This increases glucose availability to the cells, maintaining adequate plasma glucose concentrations to meet increased metabolic demands. The responses of these hormones are usually blunted in trained individuals, and those who are well trained are better able to maintain plasma glucose concentrations.

Regulation of Carbohydrate Metabolism During Exercise

As we learned in chapter 2, the heightened energy demands of exercise require that more glucose be made available to the muscles. Recall that glucose is stored in the body as glycogen, primarily in the muscles and the liver. Glucose must be freed from its storage form of glycogen, so glycogenolysis must increase. Glucose freed from the liver enters the blood to circulate throughout the body, allowing it access to active tissues. Plasma glucose concentration also can be increased through gluconeogenesis, the production of "new" glucose from noncarbohydrate sources like lactate, amino acids, and glycerol.

Regulation of Plasma Glucose Concentration

The plasma glucose concentration during exercise depends on a balance between glucose uptake by exercising muscles and its release by the liver. Four hormones work to increase the amount of circulating plasma glucose:

- Glucagon
- Epinephrine
- Norepinephrine
- Cortisol

At rest, glucose release from the liver is facilitated by glucagon, which promotes both liver glycogen breakdown and glucose formation from amino acids. During exercise, glucagon secretion increases. Muscular activity also increases the rate of catecholamine release from the adrenal medulla, and these hormones (epinephrine and norepinephrine) work in concert with glucagon

to further increase glycogenolysis. After a slight initial drop, cortisol concentration increases during the first 30 to 45 min of exercise. Cortisol increases protein catabolism, freeing amino acids to be used within the liver for gluconeogenesis. Thus, all four of these hormones can increase plasma glucose by enhancing the processes of glycogenolysis (breakdown of glycogen) and gluconeogenesis (making glucose from other substrates). In addition to the effects of the four major glucose-controlling hormones, growth hormone increases mobilization of FFAs and decreases cellular uptake of glucose, so less glucose is used by the cells and more remains in circulation. The thyroid hormones promote glucose catabolism and fat metabolism.

The amount of glucose released by the liver depends on both exercise intensity and duration. As intensity increases, so does the rate of catecholamine release. This can cause the liver to release more glucose than is being taken up by the active muscles. Consequently, during or shortly after an explosive, short-term sprint, blood glucose concentrations may be 40% to 50% above the resting value, as glucose is released by the liver at a greater rate than the rate of uptake by the muscles.

The greater the exercise intensity, the greater the catecholamine release, and thus the rate of glycogenolysis is significantly increased. This process occurs not only in the liver but also in the muscle. The glucose released from the liver enters the blood where it becomes available to the muscle fibers. But the muscle has a more readily available source of glucose: its own glycogen stores. The muscle uses its own glycogen stores before using the plasma glucose during explosive, short-term exercise. Glucose released from the liver is not used as readily, so it remains in the circulation, elevating the plasma glucose. Following exercise, plasma glucose concentration decreases as glucose enters the muscle to replenish glycogen stores that were depleted through glycogenolysis.

During exercise bouts that last for several hours, however, the rate of liver glucose release more closely matches the muscles' needs, keeping plasma glucose at or only slightly above the resting concentrations. As muscle uptake of glucose increases, the liver's rate of glucose release also increases. In most cases, plasma glucose does not begin to decline until late in the activity as liver glycogen stores become depleted, at which time the glucagon concentration increases significantly. Glucagon and cortisol together enhance gluconeogenesis, providing more fuel.

Figure 4.4 illustrates the changes in plasma concentrations of epinephrine, norepinephrine, glucagon, cortisol, and glucose during 3 h of cycling. Although the hormonal regulation of glucose remains intact throughout such long-term activities, the liver's glycogen supply may become limiting and the liver's rate of glucose release may be unable to keep pace with the muscles' rate of glucose uptake. Under this condition, plasma glucose may decline despite strong hormonal stimulation. Glucose ingestion during the activity can play a major role in maintaining plasma glucose concentrations.

In focus

Plasma glucose concentrations are increased by the actions of glucagon, epinephrine, norepinephrine, and cortisol. This is important during exercise, particularly long-duration or high-intensity exercise, during which blood glucose concentrations might otherwise decline. Glucose ingestion during exercise also helps maintain plasma glucose concentrations.

Glucose Uptake by Muscle

Merely releasing sufficient amounts of glucose into the blood does not ensure that the muscle cells will have enough glucose to meet their energy demands. Not only must the glucose be released and delivered to these cells; it also must be taken up by the cells. Transport of glucose through the cell membranes and into muscle cells is controlled by insulin. Once glucose is delivered to the muscle, insulin facilitates its transport into the fibers.

Surprisingly, as seen in figure 4.5, plasma insulin concentration tends to decrease during prolonged submaximal exercise, despite a slight increase in plasma glucose concentration and glucose uptake by muscle. This apparent contradiction between the plasma insulin concentrations and the muscles' need for glucose serves as a reminder that a hormone's activity is determined not only by its concentration in the blood but also by a cell's sensitivity to that hormone. In this case, the

FIGURE 4.4 Changes (as a percentage of preexercise values) in plasma concentrations of epinephrine, norepinephrine, glucagon, cortisol, and glucose during 3 h of cycling at 65% $\dot{V}O_{2max}$.

FIGURE 4.5 Changes in plasma concentrations of glucose and insulin during prolonged cycling at 65% to 70% of $\dot{V}O_{2max}$. Note the gradual decline in insulin throughout the exercise, suggesting an increased sensitivity to insulin during prolonged effort.

cell's sensitivity to insulin is at least as important as the concentration of circulating hormone. Exercise may enhance insulin's binding to receptors on the muscle fiber, thereby reducing the need for high concentrations of plasma insulin to transport glucose across the muscle cell membrane into the cell. This is important, because during exercise four hormones are working to release glucose from its storage sites and create new glucose. High insulin concentrations would oppose their action, preventing this needed increase in plasma glucose supply.

Regulation of Fat Metabolism During Exercise

Although fat generally contributes less than carbohydrate to muscles' energy needs during exercise, mobilization and oxidation of FFAs are critical to performance in endurance exercise. During such prolonged activity, carbohydrate reserves become depleted, and muscle must rely more heavily on the oxidation of fat for energy production. When carbohydrate reserves are low (low plasma glucose and low muscle glycogen), the endocrine system can accelerate the oxidation of fats (lipolysis), thus ensuring that muscles' energy needs can be met.

Free fatty acids are stored as triglycerides in adipose tissue and within muscle fibers. Adipose tissue triglycerides, however, must be broken down to release the FFAs, which are then transported to the muscle fibers. The rate of FFA uptake by active muscle correlates highly with the plasma FFA concentration. Increasing this concentration would increase cellular uptake of the FFA. Therefore, the rate of triglyceride breakdown may determine, in part, the rate at which muscles use fat as a fuel source during exercise.

The rate of lipolysis is controlled by at least five hormones:

- (Decreased) insulin
- Epinephrine
- Norepinephrine
- Cortisol
- Growth hormone

The major factor responsible for adipose tissue lipolysis during exercise is a fall in circulating insulin. Lipolysis is also enhanced through the elevation of epinephrine and norepinephrine. In addition to having a role in gluconeogenesis, cortisol accelerates the mobilization and use of FFAs for energy during exercise. Plasma cortisol concentration peaks after 30 to 45 min of exercise and then decreases to near-normal levels. But the plasma FFA concentration continues to increase throughout the activity, meaning that lipase continues to be activated by other hormones. The hormones that continue this process are the catecholamines and growth hormone. The thyroid hormones also contribute to the mobilization and metabolism of FFAs but to a much lesser degree.

In focus

Free fatty acids are a primary source of energy at rest and during prolonged endurance exercise. They are derived from triglycerides through the action of the enzyme lipase, which breaks down triglycerides into FFA and glycerol.

Thus, the endocrine system plays a critical role in regulating ATP production during exercise as well as controlling the balance between carbohydrate and fat metabolism.

In review

➤ Plasma glucose concentration is increased by the combined actions of glucagon, epinephrine, norepinephrine, and cortisol. These hormones promote glycogenolysis and gluconeogenesis, thus increasing the amount of glucose available for use as a fuel source.

➤ Insulin allows circulating glucose to enter the cells, where it can be used for energy production. But insulin concentrations decline during prolonged exercise, indicating that exercise increases cell sensitivity to insulin so that less of the hormone is required during exercise than at rest.

➤ When carbohydrate reserves are low, the body turns more to fat oxidation for energy, and lipolysis increases. This process is facilitated by a decreased insulin concentration and increased concentrations of epinephrine, norepinephrine, cortisol, and growth hormone.

Hormonal Regulation of Fluid and Electrolytes During Exercise

Fluid balance during exercise is critical for optimal metabolic, cardiovascular, and thermoregulatory function. At the onset of exercise, water is shifted from the plasma volume to the interstitial and intracellular spaces. This water shift is specific to the amount of muscle that is active and the intensity of effort. Metabolic by-products begin to accumulate in and around the muscle fibers, increasing the osmotic pressure there. Water is then drawn into these areas by diffusion. Also, increased muscular activity increases blood pressure, which in turn drives water out of the blood (hydrostatic forces). In addition, sweating increases during exercise. The combined effect of these actions is that the muscles and sweat glands gain water at the expense of plasma volume. For example, prolonged running at approximately 75% of $\dot{V}O_{2max}$ decreases plasma volume by 5% to 10%. Reduced plasma volume can decrease blood pressure and increase the strain on the heart to pump blood to the working muscles. Both of these effects can impede athletic performance.

Endocrine Glands Involved in Fluid and Electrolyte Homeostasis

The endocrine system plays a major role in monitoring fluid levels and correcting imbalances, along with regulating electrolyte balance, especially that of sodium. The two major endocrine glands involved in these processes are the posterior pituitary and the adrenal cortex. The kidneys are not only the primary target organ for hormones released by these glands but are also endocrine glands themselves.

Posterior Pituitary

The pituitary's posterior lobe is an outgrowth of neural tissue from the hypothalamus. For this reason, it is also referred to as the neurohypophysis. It secretes two hormones: **antidiuretic hormone** (**ADH**; also called vasopressin or arginine vasopressin) and oxytocin. Both of these hormones are actually produced in the hypothalamus. They travel through the neural tissue and are stored in vesicles within nerve endings in the posterior pituitary. These hormones are released into capillaries as needed in response to neural impulses from the hypothalamus.

Of the two posterior pituitary hormones, only ADH is known to play an important role during exercise. Antidiuretic hormone promotes water conservation by increasing water reabsorption by the kidneys. As a result, less water is excreted in the urine, creating an "antidiuresis."

Muscular activity and sweating cause electrolytes to become concentrated in the blood plasma as more fluid leaves the plasma compared to electrolytes. This is called **hemoconcentration**, and it increases the plasma **osmolality** (the ionic concentration of dissolved substances in the plasma). An increased plasma osmolality is the primary physiological stimulus for ADH release.

The increased osmolality is sensed by osmoreceptors in the hypothalamus. A second and related stimulus for ADH release is a low plasma volume sensed by baroreceptors in the cardiovascular system. In response to either stimulus, the hypothalamus sends neural impulses to the posterior pituitary, stimulating ADH release. The ADH enters the blood, travels to the kidneys, and promotes water retention in an effort to dilute the plasma electrolyte concentration back to normal levels. This hormone's role in conserving body water minimizes the extent of water loss and therefore the risk of severe dehydration during periods of heavy sweating and hard exercise. Figure 4.6 illustrates this process.

In focus

Loss of fluid (plasma) from the blood results in a concentration of the constituents of the blood, a phenomenon referred to as hemoconcentration. Conversely, a gain of fluid in the blood results in a dilution of the constituents of the blood, which is referred to as hemodilution.

Adrenal Cortex

A group of hormones called **mineralocorticoids**, secreted from the adrenal cortex, maintain electrolyte balance in the extracellular fluids, especially that of sodium (Na^+) and potassium (K^+). **Aldosterone** is the major mineralocorticoid, responsible for at least 95% of all mineralocorticoid activity. It works primarily by promoting renal reabsorption of sodium, thus causing the body to retain sodium. When sodium is retained, so is water; thus, aldosterone, like ADH, results in water retention. Sodium retention also enhances potassium excretion, so aldosterone plays a role in potassium balance as well. For these reasons, aldosterone secretion is stimulated by many factors, including decreased plasma sodium, decreased blood volume, decreased blood pressure, and increased plasma potassium concentration.

The Kidneys as Endocrine Organs

Although the kidneys are not typically considered major endocrine organs, they do release a hormone called erythropoietin. **Erythropoietin (EPO)** regulates red blood cell (erythrocyte) production by stimulating bone marrow cells. The red blood cells are essential for transporting oxygen to the tissues and removing carbon dioxide, so this hormone is extremely important in our adaptation to training and altitude.

The kidneys also play a role in determining the aldosterone concentration in the blood. While the

2 Sweating causes loss of blood plasma, resulting in hemo-concentration and increased blood osmolality.

1 Muscular activity promotes sweating.

3 Increased blood osmolality stimulates osmoreceptors in the hypothalamus.

Hypothalamus

4 The hypothalamus sends a neural signal to the posterior pituitary gland.

Pituitary gland

5 The posterior pituitary gland secretes ADH into the blood.

7 Decreased urine output aids in minimizing body fluid losses

6 ADH acts on the kidneys, increasing the water permeability of the renal tubules and collecting ducts, leading to increased resorption of water.

FIGURE 4.6 The mechanism by which antidiuretic hormone (ADH) helps to conserve body water.

primary regulator of aldosterone release is plasma potassium and sodium concentrations, a second set of hormones also determines aldosterone concentration and thus helps regulate body fluid balance. In response to a fall in blood pressure or plasma volume, blood flow to the kidneys decreases. Stimulated by activation of the sympathetic nervous system, the kidneys release **renin**. Renin is an enzyme that is released into the circulation, where it converts a molecule called angiotensinogen to angiotensin I. Angiotensin I is subsequently converted to its active form, angiotenin II, in the lungs with the aid of an enzyme, **angiotensin-converting enzyme**, or **ACE**. Angiotenin II stimulates aldosterone release from the adrenal cortex for sodium and water resorption at the kidneys. Figure 4.7 shows the mechanism involved in renal control of blood pressure, the **renin–angiotensin-aldosterone mechanism**.

1 Prolonged exercise without adequate fluid replacement can lead to dehydration.

5 Angiotensin-converting enzyme in the lungs converts angiotensin I to angiotensin II.

Angiotensin I

Liver

4 Renin converts the protein angiotensinogen, released by the liver, to angiotensin I.

Angiotensin II

Angiotensinogen

6 Angiotensin II stimulates the adrenal cortex to release aldosterone and vasoconstricts blood vessels, increasing blood pressure.

3 Renin is secreted by the kidneys.

Renin

⊖

Blood pressure

Adrenals

Aldosterone

2 Dehydration can, in turn, cause a decrease in blood pressure, which is sensed by the kidneys.

Kidney

7 Aldosterone acts on the kidneys to increase Na⁺ resorption.

8 Urine volume decreases.

FIGURE 4.7 Water loss from plasma during exercise leads to a sequence of events that promotes sodium (Na^+) and water reabsorption from the renal tubules, thereby reducing urine production. In the hours after exercise when fluids are consumed, the elevated aldosterone concentration causes an increase in the extracellular volume and an expansion of plasma volume.

In focus

In addition to stimulating aldosterone release from the adrenal cortex, angiotensin II causes blood vessels to constrict. Because ACE catalyzes the conversion of angiotensin I to angiotensin II, ACE inhibitors are sometimes prescribed for individuals with hypertension, since relaxation of the blood vessels lowers blood pressure.

Recall that aldosterone's primary action is to promote sodium reabsorption in the kidneys. Because water follows sodium, this renal conservation of sodium causes the kidneys to also retain water. The net effect is to conserve the body's fluid content, thereby minimizing the loss of plasma volume while keeping blood pressure near normal. Figure 4.8 illustrates the changes in plasma volume and aldosterone concentrations during 2 h of exercise.

Osmolality

Body fluids contain many dissolved molecules and minerals. The presence of these particles in various body fluid compartments (i.e., intracellular, plasma, and interstitial spaces) generates an osmotic pressure or attraction to retain water within that compartment. The amount of osmotic pressure exerted by a body fluid is proportional to the number of molecular particles (osmoles, or Osm) in solution. A solution that has 1 Osm of solute dissolved in each kilogram (the weight of a liter) of water is said to have an osmolality of 1 osmole per kilogram (1 Osm/kg), whereas a solution that has 0.001 Osm/kg has an osmolality of 1 milliosmole per kilogram (1 mOsm/kg). Normally, body fluids have an osmolality of 300 mOsm/kg. Increasing the osmolality of the solutions in one body compartment generally causes water to be drawn away from adjacent compartments that have a lower osmolality (i.e., more water).

FIGURE 4.8 Changes in plasma volume and aldosterone concentration during 2 h of cycling exercise. Note that plasma volume declines rapidly during the first few minutes of exercise and then shows a smaller rate of decline despite large sweat losses. Plasma aldosterone concentration, on the other hand, increases rather steadily throughout the exercise.

FIGURE 4.9 Changes in plasma volume during three days of repeated exercise and dehydration. Subjects exercised in heat on days 3 to 6. Note the sudden decline in plasma volume when the subjects stopped training (sixth day). The changes in hemoglobin and hematocrit reflect the expansion and contraction of plasma volume during and after the three-day training period.

The hormonal influences of ADH and aldosterone persist for up to 12 to 48 h after exercise, reducing urine production and protecting the body from further dehydration. In fact, aldosterone's prolonged enhancement of Na^+ reabsorption may cause the body's Na^+ concentration to increase above normal following an exercise bout. In an effort to compensate for this elevation in Na^+ concentrations, more of the water ingested shifts into the extracellular compartment.

As shown in figure 4.9, individuals who are subjected to three repeated days of exercise and dehydration show a significant increase in plasma volume that continues to increase throughout the period of activity. This increase in plasma volume appears to parallel the body's retention of dietary Na^+. When the daily bouts

of activity are terminated, the excess Na^+ and water are excreted in urine.

Most athletes involved in heavy training have an expanded plasma volume, which dilutes various blood constituents. The actual amount of proteins and electrolytes (solutes) within the blood remains unaltered, but the substances are dispersed throughout a greater volume of water (plasma), so they are diluted and their concentration decreases. This phenomenon is called **hemodilution**.

Hemoglobin is one of the substances diluted by plasma expansion. For this reason, some athletes who actually have normal hemoglobin concentrations may appear to be anemic as a consequence of Na$^+$-induced hemodilution. This condition, not to be confused with true anemia, can be remedied with a few days of rest, allowing time for aldosterone concentrations to return to normal and for the kidneys to unload the extra Na$^+$ and water.

In review

➤ The two primary hormones involved in the regulation of fluid balance are antidiuretic hormone (ADH) and aldosterone.

➤ ADH is released in response to increased plasma osmolality. When osmoreceptors in the hypothalamus sense this increase, the hypothalamus triggers ADH release from the posterior pituitary. Low blood volume is a secondary stimulus for ADH release.

➤ ADH acts on the kidneys, directly promoting water reabsorption and thus fluid conservation. As more fluid is resorbed, plasma volume increases and plasma osmolality decreases.

➤ When plasma volume or blood pressure decreases, the kidneys release an enzyme called renin that converts angiotensinogen into angiotensin I, which later becomes angiotensin II in the lung circulation. Angiotensin II is a powerful constrictor of blood vessels and increases peripheral resistance, increasing the blood pressure.

➤ Angiotensin II also triggers the release of aldosterone from the adrenal cortex. Aldosterone promotes sodium reabsorption in the kidneys, which in turn causes water retention, thus minimizing the loss of plasma volume.

In this chapter, we focused on the role of the endocrine system in regulating some of the many physiological processes that accompany exercise. We discussed the role of hormones in regulating the metabolism of glucose and fat for energy metabolism and the role of other hormones in maintaining fluid balance. We next look at the energy expenditure and fatigue during exercise.

Key Terms

adrenaline

aldosterone

angiotensin-converting enzyme (ACE)

antidiuretic hormone (ADH)

catecholamines

cortisol

cyclic adenosine monophosphate (cAMP)

direct gene activation

downregulation

epinephrine

erythropoietin (EPO)

glucagon

glucocorticoids

growth hormone

hemoconcentration

hemodilution

hormone

hyperglycemia

hypoglycemia

inhibiting factors

insulin

mineralocorticoids

nonsteroid hormones

osmolality

prostaglandins

releasing factors

renin

renin–angiotensin-aldosterone mechanism

second messenger

steroid hormones

target cells

thyrotropin (TSH)

thyroxine (T_4)

triiodothyronine (T_3)

upregulation

Study Questions

1. What is an endocrine gland, and what are the functions of hormones?
2. Explain the difference between steroid hormones and nonsteroid hormones in terms of their actions at target cells.
3. How can hormones have very specific functions when they reach nearly all parts of the body through the blood?
4. How are plasma concentrations of specific hormones controlled?
5. Define the terms upregulation and downregulation. How do target cells become more or less sensitive to hormones?
6. What are secondary messengers and what role do they play in hormonal control of cell function?
7. Briefly outline the major endocrine glands, their hormones, and the specific action of these hormones.
8. Which of the hormones outlined in question 7 are of major significance during exercise?
9. What hormones are involved in the regulation of metabolism during exercise? How do they influence the availability of carbohydrates and fats for energy during exercise lasting for several hours?
10. Describe the hormonal regulation of fluid balance during exercise.

Study Guide Activities

In addition to the activities listed in the chapter opening outline on page 91, two other activities are available in the web study guide, located at

www.HumanKinetics.com/PhysiologyOfSportAndExercise

 The **KEY TERMS** activity reviews important terms, and the end-of-chapter **QUIZ** tests your understanding of the material covered in the chapter.

Energy Expenditure and Fatigue

5

In this chapter and in the web study guide

 ACTIVITY 5.1 Evaluating Energy Use explores six methods of measuring energy use and the advantages and disadvantages of each.

 ACTIVITY 5.2 Energy Expenditure at Rest and During Exercise reviews the measurement of basal metabolic rate and the common terms used to refer to the best single measurement of cardiorespiratory endurance and aerobic fitness.

 ACTIVITY 5.3 Fatigue considers various sources of fatigue and how they affect athletic performance.

They call it the greatest football game ever played. On January 2, 1982, the Miami Dolphins and the San Diego Chargers battled in the hot, sticky night for over 4 hours. Players were carted off the field time and time again, only to return to the field each time. Hall of Fame tight end Kellen Winslow overcame severe fatigue and excruciating back spasms and became one of many heroes of this epic test of wills. As Rick Reilly's *Sports Illustrated* (October 25, 1999) story noted, "No player on either team would ever take himself that far or that high again." One player commented, "You hear coaches say 'leave everything on the field.' Well that actually happened that day. Both teams." Another player quipped, "Guys would refuse to come out of the game . . . just so they didn't have to run all the way to the sideline!" Perhaps no other account so vividly highlights the concepts of energy and fatigue, the topics discussed in this chapter.

One cannot understand exercise physiology without understanding some key concepts about energy expenditure at rest and during exercise. In chapter 2, we discussed the formation of adenosine triphosphate (ATP), the major form of chemical energy stored within cells. ATP is produced from substrates by a series of processes that are known collectively as metabolism. In the first half of this chapter we discuss various techniques for measuring the body's energy expenditure or metabolic rate; then we describe how energy expenditure varies from basal or resting conditions up to maximal exercise intensities. If exercise is sustained for a prolonged time, eventually muscular contraction cannot be sustained and performance will diminish. This inability to maintain muscle contractions is broadly called "fatigue." Fatigue is a complex, multidimensional phenomenon that may or may not result from an inability to maintain metabolism and expend energy. Because fatigue often has a metabolic component, it is discussed in the present chapter along with energy expenditure.

Measuring Energy Expenditure

The energy utilized by contracting muscle fibers during exercise cannot be directly measured. But numerous indirect laboratory methods can be used to calculate whole-body energy expenditure at rest and during exercise. Several of these methods have been in use since the early 1900s. Others are new and have only recently been used in exercise physiology.

Direct Calorimetry

Only about 40% of the energy liberated during the metabolism of glucose and fats is used to produce ATP. The remaining 60% is converted to heat, so one way to gauge the rate and quantity of energy production is to measure the body's heat production. This technique is called **direct calorimetry** ("measuring heat"), since the basic unit of heat is the **calorie (cal)**.

This approach was first described by Zuntz and Hagemann in the late 1800s.[10] They developed the

calorimeter illustrated in figure 5.1, which consists of an insulated, airtight chamber. The walls of the chamber contain copper tubing through which water is circulated. In the chamber, the heat produced by the body radiates to the walls and warms the water. The water temperature change is recorded, as are temperature changes of the air entering and leaving the chamber. These changes are caused by the heat the body generates. One's metabolism can be calculated from the resulting values.

Calorimeters are expensive to construct and to operate and are slow to generate results. Their only real advantage is that they measure heat directly, but they have several disadvantages for exercise physiology. Although a calorimeter can provide an accurate measure of total-body energy expenditure over time, it cannot follow rapid changes in energy expenditure. Therefore, while direct calorimetry is useful for measuring resting metabolism and energy expended during prolonged, steady-state aerobic exercise, energy metabolism during more typical exercise situations cannot be adequately studied with a direct calorimeter. Second, exercise equipment such as a motor-driven treadmill gives off its own heat that must be accounted for in the calculations. Third, not all heat is liberated from the body; some is stored and causes body temperature to rise. And finally, sweating affects the measurements and the constants used in the calculations of heat produced. Consequently, this method is seldom used today because it is easier and less expensive to quantify energy expenditure by measuring the exchange of oxygen and carbon dioxide that occurs during oxidative phosphorylation.

Indirect Calorimetry

As discussed in chapter 2, oxidative metabolism of glucose and fat—the main substrates for aerobic exercise—utilizes O_2 and produces CO_2 and water. The rate of O_2 and CO_2 exchanged in the lungs normally equals the rate of their usage and release by the body tissues. Based on this principle, energy expenditure can be determined by measuring respiratory gases. This method of estimating total-body energy expenditure is called **indirect calorimetry** because heat production

FIGURE 5.1 A direct calorimeter for the measurement of energy expenditure by an exercising human subject. The heat generated by the subject's body is transferred to the air and walls of the chamber (through conduction, convection, and evaporation). This heat produced by the subject—a measure of his or her metabolic rate—is measured by recording the temperature change in the air entering and leaving the calorimeter and in the water flowing through its walls.

is not measured directly. Rather, energy expenditure is calculated from the respiratory exchange of O_2 and CO_2.

In order for oxygen consumption to reflect energy metabolism accurately, energy production must be almost completely oxidative. If a large portion of energy is being produced anaerobically, respiratory gas measurements will not reflect all metabolic processes and will underestimate the total energy expenditure. Therefore, this technique is limited to steady-state aerobic activities lasting a minute or longer, which fortunately takes into account most daily activities including exercise.

Respiratory gas exchange is determined through measurement of the volume of O_2 and CO_2 that enters and leaves the lungs during a given period of time. Because O_2 is removed from the inspired air in the alveoli and CO_2 is added to the alveolar air, the expired O_2 concentration is less than the inspired, whereas the CO_2 concentration is higher in expired air than in inspired air. Consequently, the differences in the concentrations

of these gases between the inspired and the expired air tells us how much O_2 is being taken up and how much CO_2 is being produced by the body. Because the body has only limited O_2 storage, the amount taken up at the lungs accurately reflects the body's use of O_2. Although a number of sophisticated and expensive methods are available for measuring the respiratory exchange of O_2 and CO_2, the simplest and oldest methods (i.e., Douglas bag and chemical analysis of collected gas sample) are probably the most accurate, but they are relatively slow and permit only a few measurements during each session. Modern electronic computer systems for respiratory gas exchange measurements offer a large time savings and multiple measurements.

Notice in figure 5.2 that the gas expired by the subject passes through a hose into a mixing chamber. Note that the subject is wearing a nose clip so that all expired gas is collected from the mouth and none is lost to the air. From the mixing chamber, samples are pumped to electronic oxygen and carbon dioxide analyzers. In this setup, a computer uses the measurements of expired

FIGURE 5.2 *(a)* Typical equipment that is routinely used by exercise physiologists to measure O_2 consumption and CO_2 production. These values can be used to calculate $\dot{V}O_{2max}$ and the respiratory exchange ratio, and therefore energy expenditure. Although this equipment is cumbersome and limits movement, smaller versions have recently been adapted for use under a variety of conditions in the laboratory, on the playing field, in industry, and elsewhere. *(b)* Illustration of a portable metabolic gas analyzer being used to monitor the O_2 consumption of a subject performing aerobic exercise.

gas (air) volume and the fraction (percentage) of oxygen and carbon dioxide in a sample of that expired air to calculate O_2 uptake and CO_2 production. Sophisticated equipment can do these calculations breath by breath, but calculations are more typically done over discrete time periods lasting one to several minutes.

Calculating Oxygen Consumption and Carbon Dioxide Production

Using equipment like that illustrated in figure 5.2, exercise physiologists can measure the three variables needed to calculate the actual volume of oxygen consumed ($\dot{V}O_2$) and volume of CO_2 produced ($\dot{V}CO_2$). Generally, values are presented as oxygen consumed per minute ($\dot{V}O_2$) and CO_2 produced per minute ($\dot{V}CO_2$). The dot over the V ($\dot{V}$) indicates the rate of O_2 consumption or CO_2 production per minute.

In simplified form, $\dot{V}O_2$ is equal to the volume of O_2 inspired minus the volume of O_2 expired. To calculate the volume of O_2 inspired, we multiply the volume of air inspired by the fraction of that air that is composed of O_2; the volume of O_2 expired is equal to the volume of air expired multiplied by the fraction of the expired air that is composed of O_2. The same holds true for CO_2.

Thus, calculation of $\dot{V}O_2$ and $\dot{V}CO_2$ requires the following information:

- Volume of air inspired ($\dot{V}_I$)
- Volume of air expired ($\dot{V}_E$)
- Fraction of oxygen in the inspired air (F_IO_2)
- Fraction of CO_2 in the inspired air (F_ICO_2)
- Fraction of oxygen in the expired air (F_EO_2)
- Fraction of CO_2 in the expired air (F_ECO_2)

The oxygen consumption, in liters of oxygen consumed per minute, can then be calculated as follows:

$$\dot{V}O_2 = (\dot{V}_I \times F_IO_2) - (\dot{V}_E \times F_EO_2).$$

The CO_2 production is similarly calculated as

$$\dot{V}CO_2 = (\dot{V}_E \times F_ECO_2) - (\dot{V}_I \times F_ICO_2).$$

These equations provide reasonably good estimates of $\dot{V}O_2$ and $\dot{V}CO_2$. However, the equations are based on the fact that inspired air volume exactly equals expired air volume and there are no changes in gases stored within the body. Since there are differences in gas storage during exercise (discussed next), more accurate equations can be derived from the variables listed.

Haldane Transformation

Over the years, scientists have attempted to simplify the actual calculation of oxygen consumption and CO_2 production. Several of the measurements needed in the preceding equations are known and do not change. The gas concentrations of the three gases that make up inspired air are known: oxygen accounts for 20.93% (or 0.2093), CO_2 accounts for 0.03% (0.0003), and nitrogen accounts for 79.03% (0.7903) of the inspired air. What about the volume of inspired and expired air? Aren't they the same, such that we would need to measure only one of the two?

Inspired air volume equals expired air volume only when the volume of O_2 consumed equals the volume of CO_2 produced. When the volume of oxygen consumed is greater than the volume of CO_2 produced, $\dot{V}_I$ is greater than $\dot{V}_E$. Likewise, $\dot{V}_E$ is greater than $\dot{V}_I$ when the volume of CO_2 produced is greater than the volume of oxygen consumed. However, the one thing that is constant is that the volume of nitrogen inspired ($\dot{V}_I N_2$) is equal to the volume of nitrogen expired ($\dot{V}_E N_2$). Because $\dot{V}_I N_2 = \dot{V}_I \times F_I N_2$ and $\dot{V}_E N_2 = \dot{V}_E \times F_E N_2$, we can calculate $\dot{V}_I$ from $\dot{V}_E$ by using the following equation, which has been referred to as the **Haldane transformation**:

$$(1)\ \dot{V}_I \times F_I N_2 = \dot{V}_E \times F_E N_2,$$

which can be rewritten as

$$(2)\ \dot{V}_I = (\dot{V}_E \times F_E N_2)\ /\ F_I N_2.$$

Furthermore, because we are actually measuring the concentrations of O_2 and CO_2 in the expired gases, we can calculate $F_E N_2$ from the sum of $F_E O_2$ and $F_E CO_2$, or

$$(3)\ F_E N_2 = 1 - (F_E O_2 + F_E CO_2).$$

So, in pulling all of this information together, we can rewrite the equation for calculating $\dot{V}O_2$ as follows:

$$\dot{V}O_2 = (\dot{V}_I \times F_I O_2) - (\dot{V}_E \times F_E O_2).$$

By substituting equation 2, we get the following:

$$\dot{V}O_2 = [(\dot{V}_E \times F_E N_2)\ /\ (F_I N_2 \times F_I O_2)] - [(\dot{V}_E) \times (F_E O_2)].$$

By substituting known values for $F_I O_2$ of 0.2093 and for $F_I N_2$ of 0.7903, we get the following:

$$\dot{V}O_2 = [(\dot{V}_E \times F_E N_2)\ /\ 0.7903) \times 0.2093] - [(\dot{V}_E) \times (F_E O_2)].$$

By substituting equation 3, we get the following:

$$\dot{V}O_2 = [(\dot{V}_E) \times (1 - (F_E O_2 + F_E CO_2)) \times (0.2093\ /\ 0.7903)] - [(\dot{V}_E) \times (F_E O_2)]$$

or, simplified,

$$\dot{V}O_2 = (\dot{V}_E) \times [(1 - (F_E O_2 + F_E CO_2)) \times (0.265)] - [(\dot{V}_E) \times (F_E O_2)]$$

or, further simplified,

$$\dot{V}O_2 = (\dot{V}_E) \times \{[(1 - (F_E O_2 + F_E CO_2)) \times (0.265)] - (F_E O_2)\}.$$

This final equation is the one actually used in practice by exercise physiologists, although computers now do the calculating automatically in most laboratories.

One final correction is necessary. When air is expired, it is at body temperature (BT), is at the prevailing atmospheric or ambient pressure (P), and is saturated (S) with water vapor, or what are referred to as BTPS conditions. Each of these influences would not only add error to the measurement of $\dot{V}O_2$ and $\dot{V}CO_2$ but would also make it difficult to compare measurements made in laboratories at different altitudes, for example. For that reason, every gas volume is routinely converted to its standard temperature (ST: 0 °C or 273 °K) and pressure (P: 760 mmHg), dry equivalent (D), or STPD. This is accomplished by a series of correction equations.

Respiratory Exchange Ratio

To estimate the amount of energy used by the body, it is necessary to know the type of food substrate (combination of carbohydrate, fat, protein) being oxidized. The carbon and oxygen contents of glucose, free fatty acids (FFAs), and amino acids differ dramatically. As a result, the amount of oxygen used during metabolism depends on the type of fuel being oxidized. Indirect calorimetry measures the rate of CO_2 release ($\dot{V}CO_2$) and oxygen consumption ($\dot{V}O_2$). The ratio between these two values is termed the **respiratory exchange ratio (RER)**.

$$RER = \dot{V}CO_2\ /\ \dot{V}O_2.$$

In general, the amount of oxygen needed to completely oxidize a molecule of carbohydrate or fat is proportional to the amount of carbon in that fuel. For example, glucose ($C_6H_{12}O_6$) contains six carbon atoms. During glucose combustion, six molecules of oxygen are used to produce six CO_2 molecules, six H_2O molecules, and 32 ATP molecules:

$$6\ O_2 + C_6H_{12}O_6 \rightarrow 6\ CO_2 + 6\ H_2O + 32\ ATP.$$

By evaluating how much CO_2 is released compared with the amount of O_2 consumed, we find that the RER is 1.0:

$$RER = \dot{V}CO_2\ /\ \dot{V}O_2 = 6\ CO_2\ /\ 6\ O_2 = 1.0.$$

As shown in table 5.1, the RER value varies with the type of fuels being used for energy. Free fatty acids have considerably more carbon and hydrogen but less oxygen than glucose. Consider palmitic acid, $C_{16}H_{32}O_2$. To completely oxidize this molecule to CO_2 and H_2O requires 23 molecules of oxygen:

$$16\ C + 16\ O_2 \rightarrow 16\ CO_2$$
$$\underline{32\ H + 8\ O_2 \rightarrow 16\ H_2O}$$
Total = 24 O_2 needed
$$\underline{-1\ O_2 \text{ provided by the palmitic acid}}$$
23 O_2 must be added

Ultimately, this oxidation results in 16 molecules of CO_2, 16 molecules of H_2O, and 129 molecules of ATP:

$$C_{16}H_{32}O_2 + 23\ O_2 \rightarrow 16\ CO_2 + 16\ H_2O + 129\ ATP.$$

Combustion of this fat molecule requires significantly more oxygen than combustion of a carbohydrate molecule. During carbohydrate oxidation, approximately 6.3 molecules of ATP are produced for each molecule of O_2 used (32 ATP per 6 O_2), compared with 5.6 molecules of ATP per molecule of O_2 during palmitic acid metabolism (129 ATP per 23 O_2).

Although fat provides more energy than carbohydrate, more oxygen is needed to oxidize fat than carbohydrate. This means that the RER value for fat is substantially lower than for carbohydrate. For palmitic acid, the RER value is 0.70:

$$RER = \dot{V}CO_2 / \dot{V}O_2 = 16 / 23 = 0.70.$$

Once the RER value is determined from the calculated respiratory gas volumes, the value can be compared with a table (table 5.1) to determine the food

TABLE 5.1 Respiratory Exchange Ratio (RER) as a Function of Energy Derived from Various Fuel Mixtures

% KCAL FROM			
Carbohydrates	Fats	RER	Energy (kcal/L O_2)
0	100	0.71	4.69
16	84	0.75	4.74
33	67	0.80	4.80
51	49	0.85	4.86
68	32	0.90	4.92
84	16	0.95	4.99
100	0	1.00	5.05

mixture being oxidized. If, for example, the RER value is 1.0, the cells are using only glucose or glycogen, and each liter of oxygen consumed would generate 5.05 kcal. The oxidation of only fat would yield 4.69 kcal/L of O_2, and the oxidation of protein would yield 4.46 kcal/L of O_2 consumed. Thus, if the muscles were using only glucose and the body were consuming 2 L of O_2/min, then the rate of heat energy production would be 10.1 kcal/min (2 L/min · 5.05 kcal/L).

Limitations of Indirect Calorimetry

While indirect calorimetry is a common and extremely important tool of exercise physiologists, it has some limitations. Calculations of gas exchange assume that the body's O_2 content remains constant and that CO_2 exchange in the lung is proportional to its release from the cells. Arterial blood remains almost completely oxygen saturated (about 98%), even during intense effort. We can accurately assume that the oxygen being removed from the air we breathe is in proportion to its cellular uptake. Carbon dioxide exchange, however, is less constant. Body CO_2 pools are quite large and can be altered simply by deep breathing or by performance of highly intense exercise. Under these conditions, the amount of CO_2 released in the lung may not represent that being produced in the tissues, so calculations of carbohydrate and fat used based on gas measurements appear to be valid only at rest or during steady-state exercise.

Use of the RER can also lead to inaccuracies. Recall that protein is not completely oxidized in the body because nitrogen is not oxidizable. This makes it impossible to calculate the body's protein use from the RER. As a result, the RER is sometimes referred to as nonprotein RER because it simply ignores any protein oxidation.

Traditionally, protein was thought to contribute little to the energy used during exercise, so exercise physiologists felt justified in using the nonprotein RER when making calculations. But more recent evidence suggests that in exercise lasting for several hours, protein may contribute up to 5% of the total energy expended under certain circumstances.

The body normally uses a combination of fuels. Respiratory exchange ratio values vary depending on the specific mixture being oxidized. At rest, the RER value is typically in the range of 0.78 to 0.80. During exercise, though, muscles rely increasingly on carbohydrate for energy, resulting in a higher RER. As exercise intensity increases, the muscles' carbohydrate demand also increases. As more carbohydrate is used, the RER value approaches 1.0.

This increase in the RER value to 1.0 reflects the demands on blood glucose and muscle glycogen, but it also may indicate that more CO_2 is being unloaded from the blood than is being produced by the muscles. At or near exhaustion, lactate accumulates in the blood. The body tries to reverse this acidification by releasing more CO_2. Lactate accumulation increases CO_2 production because excess acid causes carbonic acid in the blood to be converted to CO_2. As a consequence, the excess CO_2 diffuses out of the blood and into the lungs for exhalation, increasing the amount of CO_2 released. For this reason, RER values approaching 1.0 may not accurately estimate the type of fuel being used by the muscles.

Another complication is that glucose production from the catabolism of amino acids and fats in the liver produces an RER below 0.70. Thus, calculations of carbohydrate oxidation from the RER value will be underestimated if energy is derived from this process.

Despite its shortcomings, indirect calorimetry still provides the best estimate of energy expenditure at rest and during aerobic exercise.

Isotopic Measurements of Energy Metabolism

In the past, determining an individual's total daily energy expenditure depended on recording food intake over several days and measuring body composition changes during that period. This method, although widely used, is limited by the individual's ability to keep accurate records and by the ability to match the individual's activities to accurate energy costs.

Fortunately, the use of isotopes has expanded our ability to investigate energy metabolism. Isotopes are elements with an atypical atomic weight. They can be either radioactive (radioisotopes) or nonradioactive (stable isotopes). As an example, carbon-12 (^{12}C) has a molecular weight of 12, is the most common natural form of carbon, and is nonradioactive. In contrast, carbon-14 (^{14}C) has two more neutrons than ^{12}C, giving it an atomic weight of 14. ^{14}C is created in the laboratory and is radioactive.

Carbon-13 (^{13}C) constitutes about 1% of the carbon in nature and is used frequently for studying energy metabolism. Because ^{13}C is nonradioactive, it is less easily traced within the body than ^{14}C. But although radioactive isotopes are easily detected in the body, they pose a hazard to body tissues and thus are used infrequently in human research.

^{13}C and other isotopes such as hydrogen-2 (deuterium, or ^2H) are used as tracers, meaning that they can be selectively followed in the body. Tracer techniques involve infusing isotopes into an individual and then following their distribution and movement.

Although the method was first described in the 1940s, studies that used doubly labeled water for monitoring energy expenditure during normal daily living

in humans were not conducted until the 1980s. The subject ingests a known amount of water labeled with two isotopes ($^2H_2\ ^{18}O$), hence the term doubly labeled water. The deuterium (^2H) diffuses throughout the body's water, and the oxygen-18 (^{18}O) diffuses throughout both the water and the bicarbonate stores (where much of the CO_2 derived from metabolism is stored). The rate at which the two isotopes leave the body can be determined by analysis of their presence in a series of urine, saliva, or blood samples. These turnover rates then can be used to calculate how much CO_2 is produced, and that value can be converted to energy expenditure through the use of calorimetric equations.

Because isotope turnover is relatively slow, energy metabolism must be measured for several weeks. Thus, this method is not well suited for measurements of acute exercise metabolism. However, its accuracy (more than 98%) and low risk make it well suited for determining day-to-day energy expenditure. Nutritionists have hailed the doubly labeled water method as the most significant technical advance of the past century in the field of energy metabolism.

(In) review

➤ Direct calorimetry involves using a large sophisticated chamber to directly measure heat produced by the body; while it can provide very accurate measures of resting metabolism, it is neither a common nor a useful tool for exercise physiologists.

➤ Indirect calorimetry involves measuring whole-body O_2 consumption and CO_2 production from expired gases. Since we know the fraction of O_2 and CO_2 in the inspired air, three additional measurements are needed: the volume of air inspired ($\dot{V}_I$) or expired ($\dot{V}_E$), the fraction of oxygen in the expired air (F_EO_2), and the fraction of CO_2 in the expired air (F_ECO_2).

➤ By calculating the RER value (the ratio of CO_2 production to O_2 consumption) and comparing the RER value with standard values to determine the metabolic substrates being oxidized, we can calculate the energy expended per liter of oxygen consumed in kilocalories.

➤ The RER value at rest is usually 0.78 to 0.80. The RER value for the oxidation of fat is 0.70 and is 1.00 for carbohydrates.

➤ Isotopes can be used to determine metabolic rate over long periods of time. They are injected or ingested into the body. The rates at which they are cleared can be used to calculate CO_2 production and then caloric expenditure.

Energy Expenditure at Rest and During Exercise

With the techniques described in the previous section, exercise physiologists can measure the amount of energy a person expends in a variety of conditions. This section deals with the body's rates of energy expenditure, or metabolic rates, under resting conditions, during submaximal and maximal exercise intensities, and during the period of recovery following an exercise bout.

Basal and Resting Metabolic Rates

The rate at which the body uses energy is termed the metabolic rate. Estimates of energy expenditure at rest and during exercise are often based on measurement of whole-body oxygen consumption ($\dot{V}O_2$) and its caloric equivalent. Under resting conditions, an average person consumes about 0.3 L of O_2/min. This equals 18 L of O_2/h or 432 L of O_2/day.

Knowing an individual's $\dot{V}O_2$ allows us to calculate that person's caloric expenditure. Recall that at rest, the body usually burns a mixture of carbohydrate and fat. An RER value of approximately 0.80 is fairly common for most resting individuals on a mixed diet. The caloric equivalent associated with an RER value of 0.80 is 4.80 kcal per liter of O_2 consumed (from table 5.1). Using these values, we can calculate this individual's caloric expenditure as follows:

$$\text{kcal/day} = \text{liters of } O_2 \text{ consumed per day}$$
$$\times \text{ kcal used per liter of } O_2$$
$$= 432 \text{ L } O_2/\text{day} \times 4.80 \text{ kcal/L } O_2$$
$$= 2,074 \text{ kcal/day}.$$

This value closely agrees with the average resting energy expenditure expected for a 70 kg (154 lb) man. Of course, it does not include the extra energy needed for normal daily activity or any excess energy used for exercise.

One standardized measure of energy expenditure at rest is the **basal metabolic rate (BMR)**. The BMR is the rate of energy expenditure for an individual at rest in a supine position, measured in a thermoneutral environment immediately after at least 8 h of sleep and at least 12 h of fasting. This value reflects the minimum amount of energy required to carry on essential physiological functions.

Because muscle has high metabolic activity, the BMR is directly related to an individual's fat-free mass and is generally reported in kilocalories per kilogram of fat-free mass per minute (kcal · kg FFM^{-1} · min^{-1}). The higher the fat-free mass, the more total calories expended in a day. Because women tend to have a lower fat-free mass and a greater fat mass than men, women tend to have a lower BMR than men of similar weight.

Body surface area also affects BMR. The higher the surface area, the more heat loss occurs from the skin, which raises the BMR because more energy is needed to maintain body temperature. For this reason, the BMR is sometimes reported in kilocalories per square meter of body surface area per hour (kcal · m^{-2} · h^{-1}). Because we are discussing daily energy expenditure, we've opted for a simpler unit: kcal/day.

Many other factors affect BMR, including these:

- Age: BMR gradually decreases with increasing age, generally because of a decrease in fat-free mass.

- Body temperature: BMR increases with increasing temperature.

- Psychological stress: Stress increases activity of the sympathetic nervous system, which increases the BMR.

- Hormones: For example, increased release of thyroxine from the thyroid gland or epinephrine from the adrenal medulla can both increase the BMR.

Instead of BMR, most researchers measure resting metabolic rate (RMR), which in practice is similar to BMR but does not require the stringent standardized conditions associated with a true BMR. Basal metabolic rate and RMR values are typically within 5% to 10% of each other, with BMR slightly lower, and range from 1,200 to 2,400 kcal/day. But the average total metabolic rate of an individual engaged in normal daily activity ranges from 1,800 to 3,000 kcal.

In focus

While the basal metabolic rate can be as low as 1,200 kcal/day, the energy expenditure for large athletes engaged in intense training, for example large football players in two-a-day practice sessions, can exceed 10,000 kcal/day!

Metabolic Rate During Submaximal Exercise

Exercise increases the energy requirement well in excess of RMR. Metabolism increases in direct proportion to the increase in exercise intensity, as shown in figure 5.3a. As this subject exercised on a cycle ergometer for 5 min at 50 watts (W), oxygen consumption ($\dot{V}O_2$) increased from its resting value to a steady-state value within 1 min or so. The same subject then cycled for 5 min at 100 W, and again a steady-state $\dot{V}O_2$ was reached in 1 to 2 min. In a similar manner, the subject cycled for 5 min at 150 W, 200 W, 250 W, and 300 W, respectively, and steady-state values were achieved at each power output. The steady-state $\dot{V}O_2$ value represents the energy cost for that specific power output.

The steady-state $\dot{V}O_2$ values were plotted against their respective power outputs (right portion of figure 5.3*a*), showing clearly that there is a linear increase in the $\dot{V}O_2$ with increases in power output.

From more recent studies, it is clear that the $\dot{V}O_2$ response at higher rates of work does not follow the steady-state response pattern shown in figure 5.3*a* but rather looks more like the graphs illustrated in figure 5.3*b*. At power outputs above the lactate threshold (the

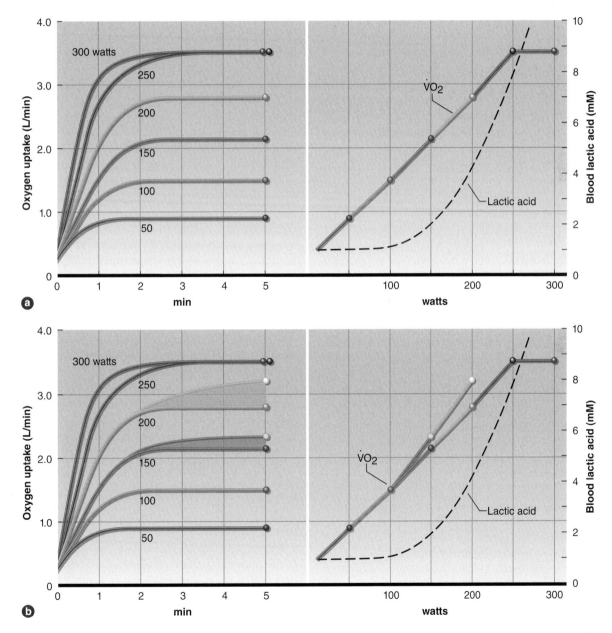

FIGURE 5.3 The increase in oxygen uptake with increasing power output *(a)* as originally proposed by P.-O. Åstrand and K. Rodahl (1986), *Textbook of work physiology: Physiological bases of exercise,* 3rd ed. (New York: McGraw-Hill), p. 300; and *(b)* as redrawn by Gaesser and Poole (1996, p. 36). See the text for a detailed explanation of this figure.

Reprinted, by permission, from G.A. Gaesser and D.C. Poole, 1996, "The slow component of oxygen uptake kinetics in humans," *Exercise and Sport Sciences Reviews* 24: 36.

lactate response is indicated by the dashed line in the right half of figure 5.3, *a* and *b*), the oxygen consumption continues to increase beyond the typical 1 to 2 min needed to reach a steady-state value. This increase has been called the slow component of oxygen uptake kinetics.[4] The most likely mechanism for this slow component is an alteration in muscle fiber recruitment patterns, with the recruitment of more type II muscle fibers, which are less efficient (i.e., they require a higher $\dot{V}O_2$ to achieve the same power output).[2, 4]

A similar, but unrelated, phenomenon is referred to as $\dot{V}O_2$ drift. $\dot{V}O_2$ **drift** is defined as a slow increase in $\dot{V}O_2$ during prolonged, submaximal, constant power output exercise. Unlike the slow component, $\dot{V}O_2$ drift is observed at power outputs well below lactate threshold, and the magnitude of the increase in $\dot{V}O_2$ drift is much less. Although not understood completely, $\dot{V}O_2$ drift is likely attributable to an increase in ventilation and effects of increased circulating catecholamines.

Maximal Capacity for Aerobic Exercise

In figure 5.3*a*, it is clear that when the subject cycled at 300 W, the $\dot{V}O_2$ response was not different from that achieved at 250 W. This indicates that the subject had reached the maximal limit of his ability to increase his $\dot{V}O_2$. This value is referred to as aerobic capacity, **maximal oxygen uptake**, or $\dot{V}O_{2max}$. $\dot{V}O_{2max}$ is widely regarded as the best single measurement of cardiorespiratory endurance or aerobic fitness. This concept is further illustrated in figure 5.4, which compares the $\dot{V}O_{2max}$ of a trained and an untrained man.

FIGURE 5.4 The relation between exercise intensity (running speed) and oxygen uptake, illustrating $\dot{V}O_{2max}$ in a trained and an untrained man.

In some exercise settings, as intensity increases, a subject reaches volitional fatigue before a plateau occurs in the $\dot{V}O_2$ response (the criterion for a true $\dot{V}O_{2max}$). In such cases, the highest oxygen uptake achieved is more correctly termed the **peak oxygen uptake** or $\dot{V}O_{2peak}$. For example, a highly trained marathon runner will almost always achieve a higher $\dot{V}O_2$ value ($\dot{V}O_{2max}$) on a treadmill than when he or she is tested to volitional fatigue on a cycle ergometer ($\dot{V}O_{2peak}$). In the latter case, fatigue of the quadriceps muscles is likely to occur before a true maximal oxygen uptake is achieved.

Although some exercise physiologists have suggested that $\dot{V}O_{2max}$ is a good predictor of success in endurance events, the winner of a marathon race cannot be predicted from the runner's laboratory-measured $\dot{V}O_{2max}$. Likewise, an endurance running performance test is only a modest predictor of one's $\dot{V}O_{2max}$. This suggests that while a relatively high $\dot{V}O_{2max}$ is a necessary attribute for elite endurance athletes, a stellar endurance performance requires more than a high $\dot{V}O_{2max}$, a concept discussed in chapters 11 and 14.

Also, research has documented that $\dot{V}O_{2max}$ increases with physical training for only 8 to 12 weeks and then plateaus despite continued higher-intensity training. Although $\dot{V}O_{2max}$ does not continue to increase, participants continue to improve their endurance performance. It appears that these individuals develop the ability to perform at a higher percentage of their $\dot{V}O_{2max}$. Most trained marathon runners, for example, can complete a 42 km (26.1 mi) marathon at an average pace that equals approximately 75% to 80% of their $\dot{V}O_{2max}$.

Consider the case of Alberto Salazar, arguably the premier marathon runner in the world in the 1980s. His measured $\dot{V}O_{2max}$ was 70 ml · kg^{-1} · min^{-1}. That is below the $\dot{V}O_{2max}$ one might expect based on his best marathon performance of 2 h 8 min. He was, however, able to run at a race pace in the marathon at 86% of his $\dot{V}O_{2max}$, a percentage considerably higher than that of other world-class runners. This may partly explain his world-class running ability.

Because individuals' energy requirements vary with body size, $\dot{V}O_{2max}$ generally is expressed relative to body weight, in milliliters of oxygen consumed per kilogram of body weight per minute (ml · kg^{-1} · min^{-1}). This allows a more accurate comparison of the cardiorespiratory endurance capacity of different-sized individuals who exercise in weight-bearing events, such as running. In non-weight-bearing activities, such as swimming and cycling, endurance is better reflected by $\dot{V}O_{2max}$ measured in liters per minute.

Normally active, but untrained, 18- to 22-year-old college students have average $\dot{V}O_{2max}$ values of about 38 to 42 ml · kg^{-1} · min^{-1} for women and 44 to 50 ml · kg^{-1} · min^{-1} for men. After the age of 25 to 30 years, the $\dot{V}O_{2max}$ of inactive individuals decreases at a rate of about 1% per year, attributable to the combination of biological aging and sedentary lifestyle. In addition, adult women generally have $\dot{V}O_{2max}$ values considerably below those of adult men. While some of this difference may be attributed to a more sedentary lifestyle across the population, some actual physiological differences may play a role as well (discussed further in chapter 19). Two reasons for the sex difference in equally trained men and women are body composition differences (women generally have less fat-free mass and more fat mass) and blood hemoglobin content (lower in women, thus they have a lower oxygen-carrying capacity).

(In)focus

Aerobic capacities of 80 to 84 ml · kg⁻¹ · min⁻¹ have been measured for elite male long-distance runners and cross-country skiers. The highest $\dot{V}O_{2max}$ value recorded for a man is that of a champion Norwegian cross-country skier who had a $\dot{V}O_{2max}$ of 94 ml · kg⁻¹ · min⁻¹. The highest value recorded for a woman is 77 ml · kg⁻¹ · min⁻¹ for a Russian cross-country skier. In contrast, poorly conditioned adults may have values below 20 ml · kg⁻¹ · min⁻¹.

Anaerobic Effort and Exercise Capacity

No exercise is 100% aerobic or 100% anaerobic. The methods we have discussed thus far ignore the anaerobic processes that accompany aerobic exercise. How can the interaction of the aerobic (oxidative) processes

and the anaerobic processes be evaluated? The most common methods for estimating anaerobic effort involve examination of either the excess postexercise oxygen consumption (EPOC) or the lactate threshold.

Postexercise Oxygen Consumption

The matching of energy requirements during exercise with oxygen delivery is not perfect. When aerobic exercise begins, the oxygen transport system (respiration and circulation) does not immediately supply the needed quantity of oxygen to the active muscles. Oxygen consumption requires several minutes to reach the required (steady state) level at which the aerobic processes are fully functional, even though the body's oxygen requirements increase the moment exercise begins.

Because oxygen needs and oxygen supply differ during the transition from rest to exercise, the body incurs an oxygen deficit, as shown in figure 5.5. This deficit accrues even at low exercise intensities. The oxygen deficit is calculated simply as the difference between the oxygen required for a given exercise intensity (steady state) and the actual oxygen consumption. Despite the insufficient oxygen delivery at the onset of exercise, the active muscles are able to generate the ATP needed through the anaerobic pathways described in chapter 2.

During the initial minutes of recovery, even though active muscle activity has stopped, oxygen consumption does not immediately decrease to a resting value. Rather, oxygen consumption remains temporarily elevated (figure 5.5). This excess oxygen consumption, which exceeds that required at rest, was traditionally

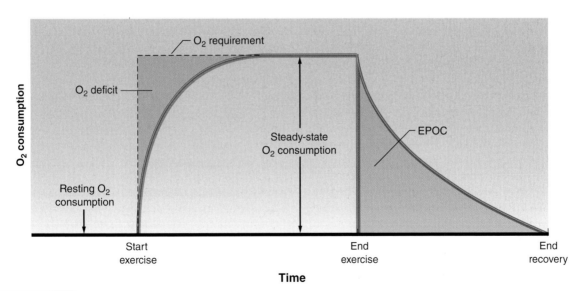

FIGURE 5.5 Oxygen requirement (dashed line) and oxygen consumption (red solid line) during exercise and recovery, illustrating the oxygen deficit and the concept of excess postexercise oxygen consumption (EPOC).

referred to as the "oxygen debt." The more common term today is **excess postexercise oxygen consumption (EPOC)**. The EPOC is the volume of oxygen consumed during the minutes immediately after exercise ceases that is above that normally consumed at rest. Everyone has experienced this phenomenon at the end of an intense exercise bout: A fast climb up several flights of stairs leaves one with a rapid pulse and breathing hard. These physiological adjustments are serving to support the EPOC. After several minutes of recovery, the pulse and breathing return to resting rates.

For many years, the EPOC curve was described as having two distinct components: an initial fast component and a secondary slow component. According to classical theory, the fast component of the curve represented the oxygen required to rebuild the ATP and phosphocreatine (PCr) used during the initial stages of exercise. Without sufficient oxygen available, the high-energy phosphate bonds in these compounds were broken to supply the required energy. During recovery, these bonds would need to be re-formed, via oxidative processes, to replenish the energy stores, or repay the debt. The slow component of the curve was thought to result from removal of accumulated lactate from the tissues, by either conversion to glycogen or oxidation to CO_2 and H_2O, thus providing the energy needed to restore glycogen stores.

According to this theory, both the fast and slow components of the curve reflected the anaerobic activity that had occurred during exercise. The belief was that by examining the postexercise oxygen consumption, one could estimate the amount of anaerobic activity that had occurred.

However, more recently researchers have concluded that the classical explanation of EPOC is too simplistic. For example, during the initial phase of exercise, some oxygen is borrowed from the oxygen stores (hemoglobin and myoglobin), and that oxygen must be replenished during early recovery as well. Also, respiration remains temporarily elevated following exercise partly in an effort to clear CO_2 that has accumulated in the tissues as a by-product of metabolism. Body temperature also is elevated, which keeps the metabolic and respiratory rates high, thus requiring more oxygen; and elevated concentrations of norepinephrine and epinephrine during exercise have similar effects.

Thus, the EPOC depends on many factors other than merely the rebuilding of ATP and PCr and the clearing of lactate produced by anaerobic metabolism.

Lactate Threshold

Many investigators consider the lactate threshold to be a good indicator of an athlete's potential for endurance exercise. The **lactate threshold** is defined as the point at which blood lactate begins to substantially accumulate above resting concentrations during exercise of increasing intensity. For example, a runner might be required to run on the treadmill at different speeds with a rest between each speed. After each run, a blood sample is taken from his or her fingertip, or from a catheter in one of the arm veins, from which blood lactate is measured. As illustrated in figure 5.6, the results of such testing can be used to plot the relationship between blood lactate and running velocity. At low running velocities, blood lactate concentrations remain at or near resting levels. But as running speed increases, the blood lactate concentration increases rapidly beyond some threshold velocity. The point at which blood lactate appears to increase disproportionately above resting values is termed the lactate threshold.

The lactate threshold has been thought to reflect the interaction of the aerobic and anaerobic energy systems. Some researchers have suggested that the lactate threshold represents a significant shift toward anaerobic glycolysis, which forms lactate from pyruvic acid. Consequently, the sudden increase in blood lactate with increasing effort has also been referred to as the "anaerobic threshold." However, blood lactate concentration is determined not only by the production of lactate in skeletal muscle or other tissues but also by the clearance of lactate from the blood by the liver, skeletal muscle, cardiac muscle, and other tissues in the body. Thus, lactate threshold is best defined as that point in time during exercise of increasing intensity when the rate of lactate production exceeds the rate of lactate clearance or removal.

FIGURE 5.6 The relationship between exercise intensity (running velocity) and blood lactate concentration. Blood samples were taken from a runner's arm vein and analyzed for lactate after the subject ran at each speed for 5 min. LT = lactate threshold.

In untrained people, the lactate threshold typically occurs at approximately 50% to 60% of their $\dot{V}O_{2max}$. Elite endurance athletes may not reach lactate threshold until closer to 70% or 80% of $\dot{V}O_{2max}$.

The lactate threshold is usually expressed as the percentage of maximal oxygen uptake ($\% \dot{V}O_{2max}$) at which it occurs. The ability to exercise at a high intensity without accumulating lactate is beneficial to the athlete because lactate accumulation contributes to fatigue. From the previous section, we learned that the major determinants of successful endurance performance are $\dot{V}O_{2max}$ and the percentage of $\dot{V}O_{2max}$ that an athlete can maintain for a prolonged period. The latter is probably related to the lactate threshold, because the lactate threshold is likely the major determinant of the pace that can be tolerated during a long-term endurance event. So the ability to perform at a higher percentage of $\dot{V}O_{2max}$ probably reflects a higher lactate threshold. Consequently, a lactate threshold at 80% $\dot{V}O_{2max}$ suggests a greater aerobic exercise tolerance than a threshold at 60% $\dot{V}O_{2max}$. Generally, in two individuals with the same maximal oxygen uptake, the person with the highest lactate threshold usually exhibits the best endurance performance, although other factors contribute as well.

Lactate threshold, when expressed as a percentage of $\dot{V}O_{2max}$, is one of the best determinants of an athlete's optimal pace in endurance events such as distance running and cycling.

Economy of Effort

As people become more skilled at performing an exercise, the energy demands during exercise at a given pace are reduced. In a sense, people become more economical. (Note that we avoid calling this "efficiency," which has a more stringent mechanical definition.) This is illustrated in figure 5.7 by the data from two distance runners. At all running speeds faster than 11.3 km/h (7.0 mph), runner B used significantly less oxygen than runner A. These men had similar $\dot{V}O_{2max}$ values (64-65 ml · kg^{-1} · min^{-1}), so runner B's lower submaximal energy use would be a decided advantage during competition.

FIGURE 5.7 The oxygen requirements for two distance runners running at various speeds. Although they had similar $\dot{V}O_{2max}$ values (64-65 ml · kg^{-1} · min^{-1}), runner B was more economical and therefore could run at a faster pace for a given oxygen cost.

Measuring Anaerobic Capacity

Unlike the situation with maximal aerobic capacity, which is measured as $\dot{V}O_{2max}$, there is no clear acceptable method for determining one's anaerobic capacity. Several methods have been described; but their validity has been challenged, and at best they offer only a crude estimate of anaerobic capacity. Early attempts to determine anaerobic capacity measured blood lactate after exhaustive exercise. Although it is generally agreed that lactate in the blood indicates increased anaerobic glycolysis, such measurements do not give a quantitative estimate of the anaerobic energy production. Maximal EPOC also has been proposed as an index of anaerobic capacity, but subsequent research has not supported this. In 1988, Medbø and coworkers[8] proposed the use of the maximal accumulated oxygen deficit as a measure of anaerobic capacity. Subsequently, a number of research studies confirmed that this test provides valid estimates of anaerobic capacity. Other tests that have shown great promise are the Wingate anaerobic test[1] and the critical power test.[7] Despite the limitations inherent in each of these methods, they remain our only indirect indicators of the metabolic potential of anaerobic capacity.

These two runners competed on numerous occasions. During marathon races, they ran at paces requiring them to use 85% of their $\dot{V}O_{2max}$. On the average, runner B beat runner A by 13 min in their competitions. Because their $\dot{V}O_{2max}$ values were so similar but their energy needs so different during these events, much of runner B's competitive advantage could be attributed to his greater running economy. Unfortunately, there is no single specific explanation for the underlying causes of these differences in economy, and they are likely due to several complex physiological and biomechanical factors.

Various studies with sprint, middle-distance, and distance runners have shown that marathon runners are generally the most economical. In general, these ultra-long-distance runners use 5% to 10% less energy than middle-distance runners and sprinters at a given pace. However, this economy of effort has been studied at only relatively slow speeds (paces of 10-19 km/h, or 6-12 mph). We can reasonably assume that distance runners are less economical at sprinting than runners who train specifically for short, faster races. It is probable that runners self-select their chosen events in part because they achieve early success, success achieved in part due to better running economy.

Variations in running form and the specificity of training for sprint and distance running may account for at least part of these differences in running economy. Film analyses reveal that middle-distance and sprint runners have significantly more vertical movement when running at 11 to 19 km/h (7-12 mph) than marathoners do. But such speeds are well below those required during middle-distance races and probably do

not accurately reflect the running economy of competitors in shorter events of 1,500 m (1 mi) or less.

Performance in other athletic events might be even more affected by economy of movement than is running. Part of the energy expended during swimming, for example, is used to support the body on the surface of the water and to generate enough force to overcome the water's resistance to motion. Although the energy needed for swimming depends on body size and buoyancy, the efficient application of force against the water is the major determinant of swimming economy.

Energy Cost of Various Activities

The amount of energy expended for different activities varies with the intensity and type of exercise involved. Despite subtle differences in economy, the *average* energy costs of many activities have been determined, usually through the measurement of oxygen consumption during the activity to determine an average oxygen uptake per unit of time. The amount of energy expended per minute (kcal/min) then can be calculated from this value.

These values typically ignore the anaerobic aspects of exercise and the EPOC. This omission is important because an activity that costs a total of 300 kcal during the actual exercise period may cost an additional 100 kcal during the recovery period. Thus, the total cost of that activity would be 400, not 300, kcal.

The body requires 0.16 to 0.35 L of oxygen per minute to satisfy its resting energy requirements. This would amount to 0.80 to 1.75 kcal/min, 48 to 105

Characteristics of Successful Athletes in Aerobic Endurance Events

From our discussion of the metabolic characteristics of aerobic endurance athletes in this chapter and of their muscle fiber type characteristics in chapter 1, it is clear that to be successful in aerobic endurance activities, one needs some combination of the following:

- High $\dot{V}O_{2max}$
- High lactate threshold when expressed as a percentage of $\dot{V}O_{2max}$
- High economy of effort, or a low $\dot{V}O_2$ for a given absolute exercise intensity
- High percentage of type I muscle fibers

From the limited data available, these four characteristics appear to be properly ranked in their order of importance. As an example, running velocity at lactate threshold and $\dot{V}O_{2max}$ are the best predictors of actual race pace among a group of elite distance runners. However, each of those runners already has a high $\dot{V}O_{2max}$. Although economy of effort is important, it does not vary much between elite runners. Finally, having a high percentage of type I muscle fibers is helpful but not essential. The bronze medal winner in one of the Olympic marathon races had only 50% type I muscle fibers in his gastrocnemius muscle, one of the primary muscles used in running.

kcal/h, or 1,152 to 2,520 kcal/day. Obviously, any activity above resting levels will add to the projected daily expenditure. The range for total daily caloric expenditure is highly variable. It depends on many factors, including

- activity level (by far the largest influence),
- age,
- sex,
- size,
- weight, and
- body composition.

The energy costs of sport activities also differ. Some, such as archery or bowling, require only slightly more energy than rest. Others, such as sprinting, require such a high rate of energy delivery that they can be maintained for only seconds. In addition to exercise intensity, the duration of the activity must be considered. For example, approximately 29 kcal/min are expended during running at 25 km/h (15.5 mph), but this pace can be endured for only brief periods. Jogging at 11 km/h (7 mph), on the other hand, expends only 14.5 kcal/min, half that of running at 25 km/h (15.5 mph). But jogging can be maintained for considerably longer, resulting in a greater total energy expenditure for an exercise session.

Table 5.2 provides an estimate of energy expenditure during various activities for average men and women. Remember that these values are merely averages. Most activities involve moving the body mass, so these figures may vary considerably with individual differences such as those previously listed and with individual skill (economy of movement).

TABLE 5.2 Average Values for Energy Expenditure During Various Physical Activities

Activity	Men (kcal/min)	Women (kcal/min)	Relative to body mass (kcal · kg^{-1} · min^{-1})
Basketball	8.6	6.8	0.123
Cycling 11.3 km/h (7.0 mph)	5.0	3.9	0.071
16.1 km/h (10.0 mph)	7.5	5.9	0.107
Handball	11.0	8.6	0.157
Running 12.1 km/h (7.5 mph)	14.0	11.0	0.200
16.1 km/h (10.0 mph)	18.2	14.3	0.260
Sitting	1.7	1.3	0.024
Sleeping	1.2	0.9	0.017
Standing	1.8	1.4	0.026
Swimming (crawl), 4.8 km/h (3.0 mph)	20.0	15.7	0.285
Tennis	7.1	5.5	0.101
Walking, 5.6 km/h (3.5 mph)	5.0	3.9	0.071
Weightlifting	8.2	6.4	0.117
Wrestling	13.1	10.3	0.187

Note. Values presented are for a 70 kg (154 lb) man and a 55 kg (121 lb) woman. These values will vary depending on individual differences.

In review

- ➤ The BMR is the minimum amount of energy required by the body to sustain basic cellular functions and is highly related to fat-free body mass and body surface area. It typically ranges from 1,100 to 2,500 kcal/day; but when daily activity is added, the typical daily caloric expenditure is 1,700 to 3,100 kcal/day.

- ➤ Metabolism increases with increased exercise intensity, but oxygen consumption is limited. Its maximal value is termed the $\dot{V}O_{2max}$. Successful aerobic performance is linked to a high $\dot{V}O_{2max}$, to the ability to perform for long periods at a high percentage of $\dot{V}O_{2max}$, and to the running velocity at lactate threshold.

- ➤ The EPOC is the elevated metabolic rate above resting levels that occurs during the recovery period immediately after exercise has ceased.

- ➤ Lactate threshold is that point at which blood lactate production begins to exceed the body's ability to clear or remove lactate, resulting in a rapid increase in blood lactate concentration during exercise of increasing intensity. Generally, individuals with higher lactate thresholds, expressed as a percentage of their $\dot{V}O_{2max}$, are capable of better endurance performances.

- ➤ A high endurance performance capacity is also associated with a high economy of effort, or a low $\dot{V}O_2$ for a given absolute exercise intensity.

Fatigue and Its Causes

What exactly is the meaning of the term **fatigue** during exercise? Sensations that individuals call fatigue are markedly different when a person is exercising to exhaustion in events lasting 45 to 60 s, such as the 400 m run, than during prolonged exhaustive muscular effort, such as marathon running. Therefore it is not surprising that the causes of fatigue are different in those two scenarios as well. We typically use the term *fatigue* to describe decrements in muscular performance with continued effort accompanied by general sensations of tiredness. An alternative definition is the inability to maintain the required power output to continue muscular work at a given intensity. To distinguish fatigue from muscle weakness or damage, one can think of fatigue as being reversible by rest.

Ask most exercisers what causes fatigue during exercise, and the most common two-word answer is "lactic acid." Not only is this common misconception an oversimplification, but there is mounting evidence that lactic acid may actually have beneficial effects on exercise performance!

Fatigue is an extremely complex phenomenon. Most efforts to describe underlying causes and sites of fatigue have focused on

- decreased rate of energy delivery (ATP-PCr, anaerobic glycolysis, and oxidative metabolism);
- accumulation of metabolic by-products, such as lactate and H⁺;

- failure of the muscle fiber's contractile mechanism; and
- alterations in neural control of muscle contraction.

The first three causes occur within the muscle itself and are often referred to as peripheral fatigue. In addition to alterations at the motor unit level, changes in the brain or central nervous system may also cause central fatigue. None of these alone can explain all aspects of fatigue, and several causes may act synergistically to bring about fatigue. Mechanisms of fatigue depend on the type and intensity of the exercise, the fiber type of the involved muscles, the subject's training status, and even his or her diet. Many questions about fatigue remain unanswered, especially about cellular sites of fatigue within the muscle fibers themselves. Keep in mind that fatigue is rarely caused by a single factor but typically by multiple factors acting at multiple sites. Potential sites of fatigue are discussed next.

Energy Systems and Fatigue

The energy systems are an obvious area to explore when considering possible causes of fatigue. When we feel fatigued, we often express this by saying "I have no energy." But this use of the term *energy* is far removed from its physiological meaning. What role does energy availability play in fatigue during exercise, in the true physiological sense of providing ATP from substrates?

Lactic Acid as a Source of Energy During Exercise

Lactic acid is in a state of constant turnover within cells, being produced by glycolysis and being removed from the cell, primarily through oxidation. Thus, despite its reputation as a cause of fatigue, lactic acid can be, and is, used as an actual fuel source during exercise. This occurs through several mechanisms.

First, we now know that lactate produced by glycolysis in the cytoplasm of a muscle fiber can be taken up by the mitochondria within that same fiber and directly oxidized. This occurs mostly in cells with a high density of mitochondria like type I (high oxidative) muscle fibers, cardiac muscle, and liver cells.

Second, lactate produced in a muscle fiber can be transported away from its site of production and used elsewhere, by a process called the lactate shuttle and first described by Dr. George Brooks. Lactate is produced primarily by type II muscle fibers but can be transported to adjacent type I fibers by diffusion or active transport. In that regard, most of the lactate produced in a muscle never leaves that muscle. It can also be transported through the circulation to sites where it can be directly oxidized. The lactate shuttle allows for glycolysis in one cell to supply fuel for use by another cell.

Finally, some of the lactic acid produced in the muscle is transported by the blood to the liver, where it is reconverted to pyruvic acid and back to glucose (gluconeogenesis) and transported back to the working muscle. This is called the Cori cycle. Without this recycling of lactate into glucose for use as an energy source, prolonged exercise would be severely limited.

PCr Depletion

Recall that PCr is used under anaerobic conditions, such as short-term high-intensity effort, to rebuild ATP as it is used and thus to maintain ATP stores within the muscle. Biopsy studies of human thigh muscles have shown that during repeated maximal contractions, fatigue coincides with PCr depletion. Although ATP is directly responsible for the energy used during such activities, it is depleted less rapidly than PCr during muscular effort because ATP is being produced by other systems (see figure 2.6, p. 55). But as PCr is depleted, the ability to quickly replace the spent ATP is hindered. Use of ATP continues, but the ATP-PCr system is less able to replace it. Thus, ATP concentration also decreases. At exhaustion, both ATP and PCr may be depleted. It now appears that P_i, which increases during intense short-term exercise because of the breakdown of PCr, is a potential cause of fatigue in this type of exercise.[9]

To delay fatigue, an athlete must control the rate of effort through proper pacing to ensure that PCr and ATP are not prematurely exhausted. This holds true even in endurance-type events. If the beginning pace is too rapid, available ATP and PCr concentrations will quickly decrease, leading to early fatigue and an inability to maintain the pace in the event's later stages. Training and experience allow the athlete to judge the optimal pace that permits the most efficient use of ATP and PCr for the entire event.

Glycogen Depletion

Muscle ATP concentrations are also maintained by the breakdown of muscle glycogen. In events lasting longer than a few seconds, muscle glycogen becomes the primary energy source for ATP synthesis. Unfortunately, glycogen reserves are limited and are depleted quickly. Since the muscle biopsy technique was first established, studies have shown a correlation between muscle glycogen depletion and fatigue during prolonged exercise.

As with PCr use, the rate of muscle glycogen depletion is controlled by the intensity of the activity. Increasing the intensity results in a disproportionate decrease in muscle glycogen. During sprint running, for example, muscle glycogen may be used 35 to 40 times faster than during walking. Muscle glycogen can be a limiting factor even during mild effort. The muscle depends on a constant supply of glycogen to meet the high energy demands of exercise.

Muscle glycogen is used more rapidly during the first few minutes of exercise than in the later stages, as seen in figure 5.8.[3] The illustration shows the change in muscle glycogen content in the subject's gastrocnemius (calf) muscle during the test. Although the subject ran the test at a steady pace, the rate of muscle glycogen

FIGURE 5.8 *(a)* The decline in gastrocnemius (calf) muscle glycogen during 3 h of treadmill running at 70% of $\dot{V}O_{2max}$, and *(b)* the subject's subjective rating of the effort. Note that the effort was rated as moderate for nearly 1.5 h of the run, although glycogen was decreasing steadily. Not until the muscle glycogen became quite low (less than 50 mmol/kg) did the rating of effort increase.

Adapted, by permission, from D.L. Costill, 1986, *Inside running: Basics of sports physiology* (Indianapolis: Benchmark Press). Copyright 1986 Cooper Publishing Group, Carmel, IN.

metabolized from the gastrocnemius was greatest during the first 75 min.

The subject also reported his perceived exertion (how difficult his effort seemed to be) at various times during the test. He felt only moderately stressed early in the run, when his glycogen stores were still high, even though he was using glycogen at a high rate. He did not perceive severe fatigue until his muscle glycogen levels were nearly depleted. Thus, the sensation of fatigue in long-term exercise coincides with a decreased concentration of muscle glycogen but not with its rate of depletion. Marathon runners commonly refer to the sudden onset of fatigue that they experience at 29 to 35 km (18-22 mi) as "hitting the wall." At least part of this sensation can be attributed to muscle glycogen depletion.

Glycogen Depletion in Different Fiber Types

Muscle fibers are recruited and deplete their energy reserves in selected patterns. The individual fibers most frequently recruited during exercise may become depleted of glycogen. This reduces the number of fibers capable of producing the muscular force needed for exercise.

This glycogen depletion is illustrated in figure 5.9, which shows a micrograph of muscle fibers taken from a runner after a 30 km (18.6 mi) run. Figure 5.9a has been stained to differentiate type I and type II fibers. One of the type II fibers is circled. Figure 5.9b shows a second sample from the same muscle, stained to show glycogen. The redder (darker) the stain, the more glycogen is present. Before the run, all fibers were full of glycogen and appeared red (not depicted). In figure 5.9b (after the run), the lighter type I fibers are almost completely depleted of glycogen. This suggests that type I fibers are used more heavily during endurance exercise that requires only moderate force development, such as the 30 km run.

The pattern of glycogen depletion from type I and type II fibers depends on the exercise intensity. Recall that type I fibers are the first fibers to be recruited during light exercise. As muscle tension requirements increase, type IIa fibers are added to the workforce. In exercise approaching maximal intensities, the type IIx fibers are added to the pool of recruited fibers.

Depletion in Different Muscle Groups

In addition to selectively depleting glycogen from type I or type II fibers, exercise may place unusually heavy demands on select muscle groups. In one study, subjects ran on a treadmill positioned for uphill, downhill, and level running for 2 h at 70% of $\dot{V}O_{2max}$. Figure 5.10 compares the resultant glycogen depletion in three muscles of the lower extremity: the vastus lateralis (knee extensor), the gastrocnemius (ankle extensor), and the soleus (another knee extensor).

The results show that whether one runs uphill, downhill, or on a level surface, the gastrocnemius uses more glycogen than does the vastus lateralis or the

FIGURE 5.9 (a) Histochemical staining for fiber type after a 30 km run; a type II (fast-twitch) fiber is circled. (b) Histochemical staining for muscle glycogen after the run. Note that a number of type II fibers still have glycogen, as noted by their darker stain, whereas most of the type I (slow-twitch) fibers are depleted of glycogen.

FIGURE 5.10 Muscle glycogen use in the vastus lateralis, gastrocnemius, and soleus muscles during 2 h of level, uphill, and downhill running on a treadmill at 70% of $\dot{V}O_{2max}$. Note that the greatest glycogen use is in the gastrocnemius during uphill and downhill running.

soleus. This suggests that the ankle extensor muscles are more likely to become depleted during distance running than are the thigh muscles, isolating the site of fatigue to the lower leg muscles.

Glycogen Depletion and Blood Glucose Muscle glycogen alone cannot provide enough carbohydrate for exercise lasting several hours. Glucose delivered by the blood to the muscles contributes a lot of energy during endurance exercise. The liver breaks down its stored glycogen to provide a constant supply of blood glucose. In the early stages of exercise, energy production requires relatively little blood glucose; but in the later stages of an endurance event, blood glucose may make a large contribution. To keep pace with the muscles' glucose uptake, the liver must break down increasingly more glycogen as exercise duration increases.

Liver glycogen stores are limited, and the liver cannot produce glucose rapidly from other substrates. Consequently, blood glucose concentration can decrease when muscle uptake exceeds the liver's glucose output. Unable to obtain sufficient glucose from the blood, the muscles must rely more heavily on their glycogen reserves, accelerating muscle glycogen depletion and leading to earlier exhaustion. On the other hand, most studies have shown no effect of carbohydrate ingestion on net muscle glycogen utilization during prolonged, strenuous exercise.

Not surprisingly, endurance performances improve when the muscle glycogen supply is elevated before the start of activity. The importance of muscle glycogen storage for endurance performance is discussed in chapter 15. For now, note that glycogen depletion and hypoglycemia (low blood sugar) limit performance in activities lasting longer than 60 to 90 min.[6]

Mechanisms of Fatigue With Glycogen Depletion It does not appear likely that glycogen depletion directly causes fatigue during endurance exercise performance. Rather, the depletion of muscle glycogen may be the first step in a series of events that leads to fatigue. A certain level of muscle glycogen metabolism is necessary to maintain oxidative metabolism of both carbohydrates and fats using the Krebs cycle. That is, we now know that a certain rate of glycogen breakdown is needed for the optimal production of reduced nicotinamide adenine dinucleotide (NADH) and to maintain the electron transport system.

Additionally, as glycogen is depleted, exercising muscle relies more heavily on the metabolism of FFAs. To accomplish this, more FFAs must be moved into the mitochondria, and the rate of transfer may limit FFA oxidation to the point where it can no longer keep up with the need for fat oxidation.

Metabolic By-Products and Fatigue

Various by-products of metabolism have been implicated as factors causing, or contributing to, fatigue. One example is P_i, which increases during intense short-term exercise as PCr and ATP are being broken down.[9] Additional metabolic by-products that have received the most attention in discussions of fatigue are heat, lactate, and hydrogen ions.

Heat, Muscle Temperature, and Fatigue

Recall that energy expenditure results in a relatively large heat production, some of which is retained in the body, causing core temperature to rise. Exercise in the heat can increase the rate of carbohydrate utilization and hasten glycogen depletion, effects that may be stimulated by the increased secretion of epinephrine. It is hypothesized that high muscle temperatures impair both skeletal muscle function and muscle metabolism.

The ability to continue moderate- to high-intensity cycle performance is affected by ambient temperature. Galloway and Maughan[5] studied performance time to exhaustion of male cyclists at four different air temperatures: 4 °C (38 °F), 11 °C (51 °F), 21 °C (70 °F), and 31 °C (87 °F). Results of that study are shown in figure 5.11. Time to exhaustion was longest when subjects

FIGURE 5.11 Time to exhaustion for a group of men performing cycle exercise at about 70% $\dot{V}O_{2max}$. The subjects were able to perform longer (delay fatigue longer) in a cool environment of 11 °C. Exercising in colder or warmer conditions hastened fatigue.

Adapted, by permission, from S.D.R. Galloway and R.J. Maughan, 1997, "Effects of ambient temperature on the capacity to perform prolonged cycle exercise in man," *Medicine and Science in Sports and Exercise*, 29: 1240-1249.

exercised at an air temperature of 11 °C and was shorter at colder and warmer temperatures. Fatigue set in earliest at 31 °C. Precooling of muscles similarly prolongs exercise, while preheating causes earlier fatigue. Heat acclimation, discussed in chapter 12, spares glycogen and reduces lactate accumulation.

Lactic Acid, Hydrogen Ions, and Fatigue

Recall that lactic acid is a by-product of anaerobic glycolysis. Although most people believe that lactic acid is responsible for fatigue in all types of exercise, lactic acid undergoes constant turnover and accumulates within the muscle fiber only during relatively brief, highly intense muscular effort. Marathon runners, for example, may have near-baseline lactate concentrations at the end of the race, despite their fatigue. As noted in the previous section, their fatigue is likely caused by inadequate energy supply, not excess lactic acid.

Short sprints in running, cycling, and swimming can all lead to large accumulations of lactic acid. But the presence of lactic acid should not be blamed for the feeling of fatigue in itself. When not cleared, the lactic acid dissociates, converting to lactate and causing an accumulation of hydrogen ions. This H^+ accumulation causes muscle acidification, resulting in a condition known as acidosis.

Activities of short duration and high intensity, such as sprint running and sprint swimming, depend heavily on anaerobic glycolysis and produce large amounts of lactate and H^+ within the muscles. Fortunately, the cells and body fluids possess buffers, such as bicarbonate (HCO_3), that minimize the disrupting influence of the H^+. Without these buffers, H^+ would lower the pH to about 1.5, killing the cells. Because of the body's buffering capacity, the H^+ concentration remains low even during the most severe exercise, allowing muscle pH to decrease from a resting value of 7.1 to no lower than 6.6 to 6.4 at exhaustion.

However, pH changes of this magnitude adversely affect energy production and muscle contraction. An intracellular pH below 6.9 inhibits the action of phosphofructokinase, an important glycolytic enzyme, slowing the rate of glycolysis and ATP production. At a pH of 6.4, the influence of H^+ stops any further glycogen breakdown, causing a rapid decrease in ATP and ultimately exhaustion. In addition, H^+ may displace calcium within the fiber, interfering with the coupling of the actin-myosin cross-bridges and decreasing the muscle's contractile force. Most researchers agree that low muscle pH is the major limiter of performance and the primary cause of fatigue during maximal, all-out exercise lasting more than 20 to 30 s.

As seen in figure 5.12, reestablishing the preexercise muscle pH after an exhaustive sprint bout requires

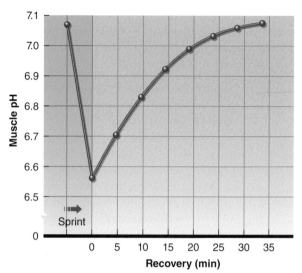

FIGURE 5.12 Changes in muscle pH during sprint exercise and recovery. Note the drastic decrease in muscle pH during the sprint and the gradual recovery to normal after the effort. Note that it took more than 30 min for pH to return to its preexercise level.

about 30 to 35 min of recovery. Even when normal pH is restored, blood and muscle lactate levels can remain quite elevated. However, experience has shown that an athlete can continue to exercise at relatively high intensities even with a muscle pH below 7.0 and a blood lactate level above 6 or 7 mmol/L, four to five times the resting value.

Neuromuscular Fatigue

Thus far we have considered only factors within the muscle that might be responsible for fatigue. Evidence also suggests that under some circumstances, fatigue may result from an inability to activate the muscle fibers, a function of the nervous system. As noted in chapter 3, the nerve impulse is transmitted across the neuromuscular junction to activate the fiber's membrane, and it causes the fiber's sarcoplasmic reticulum to release calcium. The calcium, in turn, binds with troponin to initiate muscle contraction, a process collectively called excitation–contraction coupling. Several possible neural mechanisms could disrupt this process and possibly contribute to fatigue, and two of those—one peripheral and one central—are discussed next.

Neural Transmission

Fatigue may occur at the neuromuscular junction, preventing nerve impulse transmission to the muscle fiber membrane. Studies in the early 1900s clearly established such a failure of nerve impulse transmission in fatigued muscle. This failure may involve one or more of the following processes:

- The release or synthesis of acetylcholine (ACh), the neurotransmitter that relays the nerve impulse from the motor nerve to the muscle membrane, might be reduced.

- Cholinesterase, the enzyme that breaks down ACh once it has relayed the impulse, might become hyperactive, preventing sufficient concentration of ACh to initiate an action potential.

- Cholinesterase activity might become hypoactive (inhibited), allowing ACh to accumulate excessively, inhibiting relaxation.

- The muscle fiber membrane might develop a higher threshold for stimulation by motor neurons.

- Some substance might compete with ACh for the receptors on the muscle membrane without activating the membrane.

- Potassium might leave the intracellular space of the contracting muscle, decreasing the membrane potential to half of its resting value.

Although most of these causes for a neuromuscular block have been associated with neuromuscular diseases (such as myasthenia gravis), they may also cause some forms of neuromuscular fatigue. Some evidence suggests that fatigue also may be attributable to calcium retention within the sarcoplasmic reticulum, which would decrease the calcium available for muscle contraction. In fact, depletion of PCr and lactate buildup might simply increase the rate of calcium accumulation within the sarcoplasmic reticulum. However, these theories of fatigue remain speculative.

Central Nervous System

The central nervous system (CNS) also might be a site of fatigue. Undoubtedly, there is some CNS involvement in most types of fatigue. When a subject's muscles appear to be nearly exhausted, verbal encouragement, shouting, playing of music, or even direct electrical stimulation of the muscle can increase the strength of muscle contraction. The precise mechanisms underlying the CNS role in causing, sensing, and even overriding fatigue are not fully understood.

The recruitment of muscle depends, in part, on conscious control. The stress of exhaustive exercise may lead to conscious or subconscious inhibition of the athlete's willingness to tolerate further pain. The CNS may slow the exercise pace to a tolerable level to protect the athlete. Indeed, researchers generally agree that the perceived discomfort of fatigue precedes the onset of a physiological limitation within the muscles. Unless they are highly motivated, most individuals terminate exercise before their muscles are physiologically exhausted. To achieve peak performance, athletes train to learn proper pacing and tolerance for fatigue.

In review

➤ Fatigue may result from depletion of PCr or glycogen; both situations impair ATP production.

➤ Lactic acid often has been blamed for fatigue in general, but it is generally not directly related to fatigue during prolonged endurance exercise.

➤ In short-duration exercise, like sprinting, it is actually the H^+ generated by dissociation of lactic acid that often contributes to fatigue. The accumulation of H^+ decreases muscle pH, which impairs the cellular processes that produce energy and muscle contraction.

➤ Failure of neural transmission may be a cause of some types of fatigue. Many mechanisms can lead to such failure, and further research is needed.

➤ The CNS plays a role in most types of fatigue, perhaps limiting exercise performance as a protective mechanism. Perceived fatigue usually precedes physiological fatigue, and athletes who feel exhausted can often be encouraged to continue by various cues that stimulate the CNS, such as listening to music.

In closing

In previous chapters, we discussed how muscles and the nervous system function together to produce movement. In this chapter we focused on energy expenditure during exercise and fatigue. We considered the energy needed for movement. We saw how energy is stored in the form of ATP and explored how energy production and availability can limit performance. We also learned that metabolic needs vary considerably. In the next chapter, we turn our attention to the cardiovascular system and its control.

Key Terms

basal metabolic rate (BMR)

calorie (cal)

calorimeter

direct calorimetry

excess postexercise oxygen consumption (EPOC)

fatigue

Haldane transformation

indirect calorimetry

lactate threshold

maximal oxygen uptake ($\dot{V}O_{2max}$)

peak oxygen uptake ($\dot{V}O_{2peak}$)

respiratory exchange ratio (RER)

resting metabolic rate (RMR)

$\dot{V}O_2$ drift

Study Questions

1. Define direct calorimetry and indirect calorimetry and describe how they are used to measure energy expenditure.
2. What is the respiratory exchange ratio (RER)? Explain how it is used to determine the oxidation of carbohydrate and fat.
3. What are basal metabolic rate and resting metabolic rate, and how do they differ?
4. What is maximal oxygen uptake? How is it measured? What is its relationship to sport performance?
5. Describe two possible markers of anaerobic capacity.
6. What is the lactate threshold? How is it measured? What is its relationship to sport performance?
7. What is economy of effort? How is it measured? What is its relationship to sport performance?
8. What is the relationship between oxygen consumption and energy production?
9. Why do athletes with high $\dot{V}O_{2max}$ values perform better in endurance events than those with lower values?
10. Why is oxygen consumption often expressed as milliliters of oxygen per kilogram of body weight per minute ($ml \cdot kg^{-1} \cdot min^{-1}$)?
11. Describe the possible causes of fatigue during exercise bouts lasting 15 to 30 s and 2 to 4 h.
12. Discuss three mechanisms through which lactate can be used as an energy source.

Study Guide Activities

In addition to the activities listed in the chapter opening outline on page 113, two other activities are available in the web study guide, located at

www.HumanKinetics.com/PhysiologyOfSportAndExercise

 The **KEY TERMS** activity reviews important terms, and the end-of-chapter **QUIZ** tests your understanding of the material covered in the chapter.

PART II

Cardiovascular and Respiratory Function

In part I of the book, we learned how skeletal muscle contracts in response to neural signaling and how the body produces energy through metabolism to fuel its movement. We also examined hormonal control of metabolism and of body fluid and electrolyte balance. Finally, we looked at how energy expenditure is measured and the causes of fatigue. Part II focuses on how the cardiovascular and respiratory systems provide oxygen and fuel to the active muscles, how they rid the body of carbon dioxide and metabolic wastes, and how these systems respond to exercise. In chapter 6, "The Cardiovascular System and Its Control," we look at the structure and function of the cardiovascular system: the heart, blood vessels, and blood. Our primary focus is on how this system provides all parts of the body with an adequate blood supply to meet their needs, especially during exercise. In chapter 7, "The Respiratory System and Its Regulation," we examine the mechanics and regulation of breathing, the process of gas exchange in the lungs and at the muscles, and the transportation of oxygen and carbon dioxide in the blood. We also see how this system regulates the body's pH within a very narrow range. In chapter 8, "Cardiorespiratory Responses to Acute Exercise," we concentrate on the cardiovascular and respiratory changes that occur in response to an acute bout of exercise.

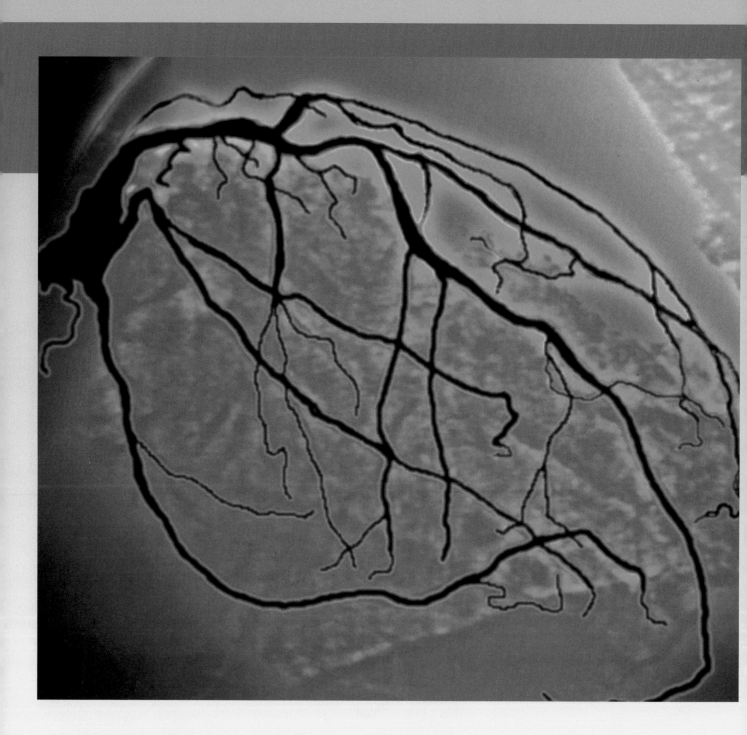

The Cardiovascular System and Its Control

6

In this chapter and in the web study guide

ACTIVITY 6.1 Anatomy of the Heart reviews the names and locations of the structures of the heart.

ACTIVITY 6.2 Functioning of the Heart describes blood flow through the heart and differentiates the heart's functions.

ACTIVITY 6.3 Cardiac Conduction explores the function of each of the components of the heart's conduction system.

ACTIVITY 6.4 Control of the Vascular System explains the role that parts of the vascular system play to guarantee an adequate blood supply where it is most needed.

On January 5, 1988, the sport world lost one of its greatest athletes. "Pistol Pete" Maravich, former National Basketball Association star, collapsed and died of cardiac arrest at 40 years of age during a pickup basketball game. His death came as a shock, and its cause surprised the medical experts. Maravich's heart was abnormally enlarged, primarily because he was born with only a single coronary artery on the right side of his heart—he was missing the two coronary arteries that supply the left side of the heart! The medical community was amazed that this single right coronary artery had taken over supplying the left side of Maravich's heart and that this adaptation had allowed him to compete for many years as one of the top players in the history of basketball. Although Maravich's death was a tragedy and shocked the sport world, he was able to perform at the highest level of competition for 10 years in one of the most physically demanding sports. More recently, a number of highly promising high school and college athletes have had their lives cut short by sudden cardiac death. Most of these deaths are attributable to hypertrophic cardiomyopathy, a disease characterized by an abnormally enlarged heart muscle mass, generally involving the left ventricle. In about half of the cases, the disease is inherited. Although this remains the major cause of sudden cardiac death in adolescent and young adult athletes (~36%), it is relatively rare, with an estimated occurrence of between 1 and 2 cases per 1 million athletes annually.

The cardiovascular system serves a number of important functions in the body and supports every other physiological system. The major cardiovascular functions can be grouped into six categories:

- Delivery of oxygen and other nutrients
- Removal of carbon dioxide and other metabolic waste products
- Transport of hormones and other molecules
- Support of thermoregulation and control of body fluid balance
- Maintenance of acid–base balance
- Regulation of immune function

The cardiovascular system delivers oxygen and nutrients to, and removes carbon dioxide and metabolic waste products from, every cell in the body. It transports hormones (chapter 4) from endocrine glands to their target receptors. The cardiovascular system supports body temperature regulation (chapter 12), and the blood's buffering capabilities help control the body's pH. The cardiovascular system maintains appropriate fluid balance across the fluid compartments of the body and helps prevent infection from invading organisms. Although this is just an abbreviated list of roles, the cardiovascular functions listed here are important for understanding the physiological basis of exercise and sport. Obviously these roles change and become even more critical with the challenges imposed by exercise.

All physiological functions and virtually every cell in the body depend in some way on the cardiovascular system. Any system of circulation requires three components:

- A pump (the heart)
- A system of channels or tubes (the blood vessels)
- A fluid medium (the blood)

In order to keep blood circulating, the heart must generate sufficient pressure to drive blood through the continuous network of blood vessels in a closed-loop system. Thus, the primary goal of the cardiovascular system is to ensure that there is adequate blood flow throughout the circulation to meet the metabolic demands of the tissues. We look first at the heart.

Heart

About the size of a fist and located in the center of the thoracic cavity, the heart is the primary pump that circulates blood through the entire cardiovascular system. As shown in figure 6.1, the heart has two atria that act as receiving chambers and two ventricles that serve as the pumping units. It is enclosed in a tough membranous sac called the **pericardium**. The thin cavity between the pericardium and the heart is filled with pericardial fluid, which reduces friction between the sac and the beating heart.

Blood Flow Through the Heart

The heart is sometimes considered to be two separate pumps, with the right side of the heart pumping deoxygenated blood to the lungs through the pulmonary circulation and the left side of the heart pumping oxygenated blood to all other tissues in the body through the systemic circulation. Blood that has circulated through the body, delivering oxygen and nutrients and picking up waste products, returns to the heart through the great veins—the superior vena cava and inferior vena cava—to the right atrium. This chamber receives all the deoxygenated blood from the systemic circulation.

From the right atrium, blood passes through the tricuspid valve into the right ventricle. This chamber pumps the blood through the pulmonary valve into

Superior vena cava

Right pulmonary veins

Pulmonary valve

Right atrium

Tricuspid valve

Chordae tendineae

Right ventricle

Interventricular septum

Inferior vena cava

Aorta

Right and left pulmonary arteries

Aortic valve

Left pulmonary veins

Left atrium

Mitral valve

Papillary muscles

Left ventricle

Descending aorta

FIGURE 6.1 An anterior (as if the person is facing you) cross-sectional view of the human heart.

the pulmonary artery, which carries the blood to the lungs. Thus, the right side of the heart is known as the pulmonary side, sending the blood that has circulated throughout the body into the lungs for reoxygenation.

After blood is oxygenated in the lungs, it is transported back to the heart through the pulmonary veins. All freshly oxygenated blood is received from the pulmonary veins by the left atrium. From the left atrium, the blood passes through the mitral valve into the left ventricle. Blood leaves the left ventricle by passing through the aortic valve into the aorta and is distributed to the systemic circulation. The left side of the heart is known as the systemic side. It receives the oxygenated blood from the lungs and then sends it out to supply all other body tissues.

Myocardium

Cardiac muscle is collectively called the **myocardium** or myocardial muscle. Myocardial thickness at various locations in the heart varies according to the amount of stress placed on it. The left ventricle is the most powerful pump of the four heart chambers because

it must generate sufficient pressure to pump blood through the entire body. When a person is sitting or standing, the left ventricle must contract with enough force to overcome the effect of gravity, which tends to pool blood in the lower extremities.

Because the left ventricle must generate considerable force to pump blood to the systemic circulation, it has the thickest muscular wall compared with the other heart chambers. This hypertrophy is the result of the pressure placed on the left ventricle at rest or under normal conditions of moderate activity. With more vigorous exercise—particularly intense aerobic activity, during which the working muscles' need for blood increases considerably—the demand on the left ventricle to deliver blood to the exercising muscles is higher. In response to both intense aerobic and resistance training, the left ventricle will hypertrophy. In contrast to the positive adaptations that occur as a result of exercise training, cardiac muscle also hypertrophies as a result of several diseases, such as high blood pressure or valvular heart disease. In response to either training or disease, over time the left ventricle adapts by increasing its size and pumping capacity, similar to the way skeletal muscle adapts to physical training. However, the mechanisms for adaptation and cardiac performance with disease are different from those observed with aerobic training.

Although striated in appearance, the myocardium differs from skeletal muscle in several important ways. First, because the myocardium has to contract as if it were a single unit, individual cardiac muscle fibers are anatomically interconnected end to end by dark-staining regions called **intercalated disks**. These disks have desmosomes, which are structures that anchor the individual cells together so that they do not pull apart during contraction, and gap junctions, which allow rapid transmission of the action potentials that signal the heart to contract as one unit. Secondly, the myocardial fibers are rather homogeneous in contrast to the mosaic of fiber types in skeletal muscle. The

myocardium contains only one fiber type, similar to type I fibers in skeletal muscle in that it is highly oxidative, has a high capillary density, and has a large number of mitochondria.

In addition to these differences, the mechanism of muscle contraction also differs between skeletal and cardiac muscle. Cardiac muscle contraction occurs by "calcium-induced calcium release" (figure 6.2). The action potential spreads rapidly along the myocardial sarcolemma from cell to cell via gap junctions and also to the inside of the cell through the T-tubules. Upon stimulation, calcium enters the cell by the dihydropyridine receptor in the T-tubules. Unlike what happens in skeletal muscle, the amount of calcium that enters the cell is not sufficient to directly cause the cardiac muscle to contract; but it serves as a trigger to another type of receptor, called the ryanodine receptor, to release calcium from the sarcoplasmic reticulum. Figure 6.3 summarizes some of the similarities and differences between cardiac and skeletal muscle.

Sarcolemma

Dihydropyridine receptors (DHPR): Ca^{2+} channels on the sarcolemma

T-tubule

Ryanodine receptors (RYR): Ca^{2+} release channels on the sarcoplasmic reticulum

ATPase pump

Cardiac myofibril

Ca^{2+}

Ca^{2+}

Ca^{2+}

FIGURE 6.2 The mechanism of contraction in a cardiac muscle fiber, termed calcium-induced calcium release.

Courtesy of Dr. Donna H. Korzick, Pennsylvania State University.

Muscle type	Location	Appearance	Type of activity	Stimulation
Skeletal ("striated" or "voluntary") muscle Striation Muscle fiber Nucleus	Named muscle (e.g., the biceps of the arm) attached to the skeleton and fascia of limbs, body wall, and head or neck	Large, long, unbranched, cylindrical fibers with transverse striations (stripes) arranged in parallel bundles; multiple, peripherally located nuclei	Strong, quick intermittent (phasic) contraction above a baseline tonus; acts primarily to produce movement or resist gravity	Voluntary (or reflexive) by the somatic nervous system
Cardiac muscle Nucleus Intercalated disk Striation Muscle fiber	Muscle of heart (myocardium) and adjacent portions of the great vessels (aorta, vena cava)	Branching and anastomosing shorter fibers with transverse striations (stripes) running parallel and connected end to end by complex junctions (intercalated disks); single, central nucleus	Strong, quick continuous rhythmic contraction; pumps blood from the heart	Involuntary; intrinsically (myogenically) stimulated and propagated; rate and strength of contraction modified by the autonomic nervous system

FIGURE 6.3 Functional and structural characteristics of skeletal and cardiac muscle.

Adapted, by permission, from K.L. Moore, and A.F. Dalley, 1999, *Clinically oriented anatomy,* 4th ed. (Baltimore, MD: Lippincott, Williams, and Wilkins), 27.

The myocardium, just like skeletal muscle, must have a blood supply to deliver oxygen and nutrients and remove waste products. Although blood courses through each chamber of the heart, little nourishment comes from the blood within the chambers. The primary blood supply to the heart is provided by the coronary arteries, which arise from the base of the aorta and encircle the outside of the myocardium (figure 6.4). The right coronary artery supplies the right side of the heart, dividing into two primary branches, the marginal artery and the posterior interventricular artery. The left coronary artery, also referred to as the left main coronary artery, also divides into two major branches, the circumflex artery and the anterior descending artery. The posterior interventricular artery and the anterior descending artery merge, or anastomose, in the lower posterior area of the heart, as does the circumflex. Blood flow increases through the coronary arteries when the heart is between contractions (during diastole).

The coronary arteries are very susceptible to atherosclerosis, or narrowing by the accumulation of plaque and inflammation, leading to coronary artery disease. This disease is discussed in greater detail in chapter 21. Anomalies—shortenings, blockages, or misdirections—sometimes occur in the coronary arteries, and

such congenital abnormalities are a common cause of sudden death in athletes.

In addition to its unique anatomical structure, the ability of the myocardium to contract as a single unit also depends on initiation and propagation of an electrical signal through the heart, the cardiac conduction system.

Cardiac Conduction System

Cardiac muscle has the unique ability to generate its own electrical signal, called spontaneous rhythmicity, which allows it to contract without any external stimulation. The contraction is rhythmical, in part because of the anatomical coupling of the myocardial cells through gap junctions. Without neural or hormonal stimulation, the intrinsic heart rate (HR) averages ~100 beats (contractions) per minute. This resting heart rate of about 100 beats/min can be observed in patients who have undergone cardiac transplant surgery, because their transplanted hearts lack neural innervation.

Even though all myocardial fibers have inherent rhythmicity, the heart has a series of specialized myocardial cells that function to coordinate the heart's excitation and contraction and maximize the efficient pumping of blood. Figure 6.5 illustrates the four main components of the cardiac conduction system:

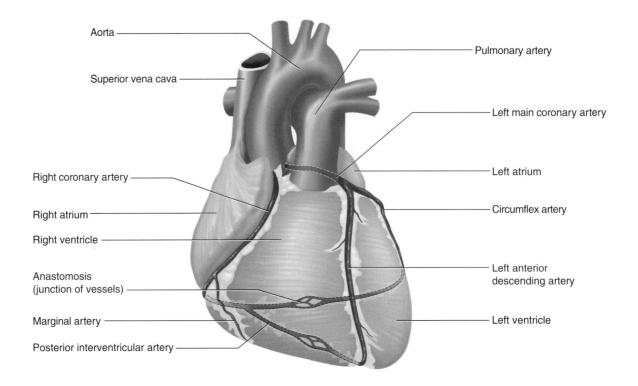

Aorta

Superior vena cava

Pulmonary artery

Left main coronary artery

Right coronary artery

Left atrium

Right atrium

Circumflex artery

Right ventricle

Anastomosis
(junction of vessels)

Left anterior
descending artery

Marginal artery

Left ventricle

Posterior interventricular artery

FIGURE 6.4   The coronary circulation, illustrating the right and left coronary arteries and their major branches.

- Sinoatrial (SA) node
- Atrioventricular (AV) node
- AV bundle (bundle of His)
- Purkinje fibers

The impulse for normal heart contractions is initiated in the **sinoatrial (SA) node**, a group of specialized cardiac muscle fibers located in the upper posterior wall of the right atrium. These specialized cells spontaneously depolarize at a faster rate than other myocardial muscle cells because they are especially leaky to sodium. Because this tissue has the fastest intrinsic firing rate, typically at a frequency of about 100 beats/min, the SA node is known as the heart's pacemaker, and the rhythm it establishes is called the sinus rhythm. The electrical impulse generated by the SA node spreads through both atria and reaches the **atrioventricular (AV) node**, located in the right atrial wall near the center of the heart. As the electrical impulse spreads through the atria, they are signaled to contract.

The AV node conducts the electrical impulse from the atria into the ventricles. The impulse is delayed by about 0.13 s as it passes through the AV node, and then it enters the AV bundle. This delay is important because it allows blood from the atria to completely empty into the ventricles to maximize ventricular filling before the ventricles contract. While most blood moves passively from the atria to the ventricles, active contraction of the atria (sometimes called the "atrial kick") completes the process. The AV bundle travels along the ventricular septum and then sends right and left bundle branches into both ventricles. These branches send the impulse toward the apex of the heart and then outward. Each bundle branch subdivides into many smaller ones that spread throughout the entire ventricular wall. These terminal branches of the AV bundle are the Purkinje fibers. They transmit the impulse through the ventricles approximately six times faster than through the rest of the cardiac conduction system. This rapid conduction allows all parts of the ventricle to contract at virtually the same time.

Extrinsic Control of Heart Activity

Although the heart initiates its own electrical impulses (intrinsic control), both the rate and force of contraction can be altered. Under normal conditions, this is accomplished primarily through three extrinsic systems:

SA node

Right atrium

AV node

Purkinje fibers

Left atrium

AV bundle

Bundle branches

Right bundle branch

FIGURE 6.5 The specialized conduction system of the heart.

- The parasympathetic nervous system
- The sympathetic nervous system
- The endocrine system (hormones)

Although an overview of these systems' effects is offered here, they are discussed in more detail in chapters 3 and 4.

The parasympathetic system, a branch of the autonomic nervous system, originates centrally in a region of the brain stem called the medulla oblongata and reaches the heart through the vagus nerve (cranial nerve X). The vagus nerve carries impulses to the SA and AV nodes, and when stimulated it releases acetylcholine, which causes hyperpolarization of the conduction cells. The result is a slower spontaneous depolarization and a decrease in heart rate. At rest, parasympathetic system activity predominates and the heart is said to have "vagal tone." Recall that, in the absence of vagal tone, intrinsic heart rate would be approximately 100 beats/min. The vagus nerve has a depressant effect on the heart: It slows impulse generation and conduction and thus decreases the heart rate. Maximal vagal stimulation can decrease the heart rate

to as low as 20 to 30 beats/min. The vagus nerve also decreases the force of cardiac muscle contraction.

The sympathetic nervous system, the other branch of the autonomic system, has opposite effects. Sympathetic stimulation increases the rate of depolarization and conduction speed, and thus heart rate. Maximal sympathetic stimulation can increase the heart rate to 250 beats/min. Sympathetic input also increases the force of contraction of the ventricles. Sympathetic control predominates during times of physical or emotional stress, when the heart rate is greater than 100 beats/min. The parasympathetic system dominates when heart rate is less than 100. Thus, when exercise begins, or if exercise is at a low intensity, heart rate first increases due to withdrawal of vagal tone, with further increases due to sympathetic activation, as shown in figure 6.6.

The third extrinsic influence, the endocrine system, exerts its effect through two hormones released by the adrenal medulla: norepinephrine and epinephrine (see chapter 4). These hormones are also known as catecholamines. Like the norepinephrine released as a neurotransmitter by the sympathetic nervous system,

FIGURE 6.6 Relative contribution of sympathetic and parasympathetic nervous systems to the rise in heart rate during exercise.

Adapted from L.B. Rowell, 1993, *Human cardiovascular control.* (Oxford, UK: Oxford University Press).

circulating norepinephrine and epinephrine stimulate the heart, increasing its rate and contractility. In fact, release of these hormones from the adrenal medulla is triggered by sympathetic stimulation during times of stress, and their actions prolong the sympathetic response.

Normal resting heart rate (RHR) typically varies between 60 and 100 beats/min. With extended periods of endurance training (months to years), the RHR can decrease to 35 beats/min or less. A RHR as low as 28 beats/min has been observed in a world-class, long-distance runner. These lower training-induced RHRs result from increased parasympathetic stimulation (vagal tone), with reduced sympathetic activity playing a lesser role.

In focus

Heart rate is established by the SA node, the heart's intrinsic pacemaker, but can be altered by the sympathetic and parasympathetic nervous systems as well as circulating catecholamines.

Electrocardiogram

The electrical activity of the heart can be recorded (figure 6.7) to monitor cardiac changes or diagnose potential cardiac problems. Because body fluids contain electrolytes, they are good electrical conductors. Electrical impulses generated in the heart are conducted through body fluids to the skin, where they can be amplified, detected, and printed out on an **electrocardiograph**. This printout is called an **electrocardiogram**, or **ECG**. A standard ECG is recorded from 10 electrodes placed in specific anatomical locations. These 10 electrodes correspond to 12 leads that represent different views of the heart. Three basic components of the ECG represent important aspects of cardiac function (figure 6.8):

- The P wave
- The QRS complex
- The T wave

The P wave represents atrial depolarization and occurs when the electrical impulse travels from the SA node through the atria to the AV node. The QRS complex represents ventricular depolarization and occurs as the impulse spreads from the AV bundle to the **Purkinje fibers** and through the ventricles. The T wave represents ventricular repolarization. Atrial repolarization cannot be seen, because it occurs during ventricular depolarization (QRS complex).

Artificial Pacemakers

Occasionally, chronic problems develop within the cardiac conduction system, hampering its ability to maintain appropriate sinus rhythm throughout the heart. In such cases, an artificial pacemaker can be surgically installed. This small, battery-operated electrical stimulator, usually implanted under the skin, has tiny electrodes attached to the right ventricle. An electrical stimulator is useful, for example, to treat a condition called AV block. With this disorder, the SA node creates an impulse, but the impulse is blocked at the AV node and cannot reach the ventricles, resulting in the heart rate being controlled by the intrinsic firing rate of the pacemaker cells in the ventricles (closer to 40 beats/min). The artificial pacemaker takes over the role of the disabled AV node, supplying the needed impulse and thus controlling ventricular contraction.

FIGURE 6.7 Recording an exercise electrocardiogram.

FIGURE 6.8 A graphic illustration of the various phases of the resting electrocardiogram.

Electrocardiograms are often obtained during exercise as clinical diagnostic tests of cardiac function. As exercise intensity increases, the heart must beat faster and work harder to deliver more blood to active muscles. Indications of coronary artery disease, not evident at rest, may show up on the ECG as the strain on the heart increases. Exercise ECGs are also invaluable tools for research in exercise physiology because they provide a convenient method for tracking heart rate and rhythm changes during acute exercise.

In focus

An ECG provides a graphic record of the electrical activity of the heart and can be used to aid clinical diagnoses, for example in someone who has had a myocardial infarction in the past or is at risk for one in the future. It is important to remember that the ECG provides no information about the pumping capacity of the heart, only its electrical activity.

> ➤ The atria serve primarily as filling chambers, receiving blood from the veins; the ventricles are the primary pumps that eject blood from the heart.

> ➤ Because the left ventricle must produce more force than other chambers to pump blood throughout the systemic circulation, its myocardial wall is thicker.

> ➤ Cardiac tissue is capable of spontaneous rhythmicity and has its own specialized conduction system made up of myocardial fibers that serve specialized functions.

> ➤ Because it has the fastest inherent rate of depolarization, the SA node is normally the heart's pacemaker.

> ➤ Heart rate and force of contraction can be altered by the autonomic nervous system (sympathetic and parasympathetic) and the endocrine system through circulating catecholamines (epinephrine and norepinephrine).

> ➤ The ECG is a surface recording of the heart's electrical activity. An exercise ECG can sometimes be used to detect underlying cardiac disorders.

Cardiac Arrhythmias

Occasionally, disturbances in the normal sequence of cardiac events can lead to an irregular heart rhythm, called an arrhythmia. These disturbances vary in degree of seriousness. Bradycardia and tachycardia are two types of arrhythmias. **Bradycardia** is defined as a RHR lower than 60 beats/min, whereas **tachycardia** is defined as a resting rate greater than 100 beats/min. With these arrhythmias, the sinus rhythm is normal, but the rate is altered. In extreme cases, bradycardia or tachycardia can affect maintenance of blood pressure. Symptoms of both arrhythmias include fatigue, dizziness, light-headedness, and fainting. Tachycardia can sometimes be sensed as palpitations or a "racing" pulse.

Other arrhythmias may also occur. For example, **premature ventricular contractions (PVCs)**, which result in the feeling of skipped or extra beats, are relatively common and result from impulses originating outside the SA node. Atrial flutter, in which the atria depolarize at rates of 200 to 400 beats/min, and atrial fibrillation, in which the atria depolarize in a rapid and uncoordinated manner, are more serious arrhythmias that may cause ventricular filling problems. **Ventricular tachycardia**, defined as three or more consecutive premature ventricular contractions, is a very serious arrhythmia that compromises the pumping capacity of the heart and can lead to **ventricular fibrillation**, in which depolarization of the ventricular tissue is random and uncoordinated. When this happens, the heart is extremely inefficient, and little or no blood is pumped out of the heart. Under such conditions, the use of a defibrillator to shock the heart back into a normal sinus rhythm must occur within minutes if the victim is to survive.

Interestingly, most highly trained endurance athletes develop resting bradycardias, an advantageous adaptation, as a result of training. Also, the heart rate naturally increases during physical activity to meet the increased demands of exercising muscle for blood flow. These adaptations should not be confused with pathological causes of bradycardia or tachycardia, which are abnormal alterations in the RHR that usually indicate underlying disease or dysfunction.

Terminology of Cardiac Function

The following terms are essential to understanding the work done by the heart and to our later discussions of the cardiac responses to exercise: cardiac cycle, stroke volume, ejection fraction, and cardiac output ($\dot{Q}$).

Cardiac Cycle

The **cardiac cycle** includes all the mechanical and electrical events that occur during one heartbeat. In mechanical terms, it consists of all heart chambers undergoing a relaxation phase (diastole) and a contraction phase (systole). During diastole, the chambers fill with blood. During systole, the ventricles contract and expel blood into the aorta and pulmonary arteries. The diastolic phase is approximately twice as long as the systolic phase. Consider an individual with a heart rate of 74 beats/min. At this heart rate, the entire cardiac cycle takes 0.81 s to complete (60 s divided by 74 beats). Of the total cardiac cycle at this rate, diastole accounts for 0.50 s, or 62% of the cycle, and systole accounts for 0.31 s, or 38%. As the heart rate increases, these time intervals shorten proportionately.

Refer to the normal ECG in figure 6.8. One cardiac cycle spans the time between one systole and the next. Ventricular contraction (systole) begins during the QRS complex and ends in the T wave. Ventricular relaxation (diastole) occurs during the T wave and continues until the next contraction. Although the heart is continually working, it spends slightly more time in the diastole (~2/3 of the cardiac cycle) than in systole (~1/3 of the cardiac cycle).

The pressure inside the heart chambers rises and falls during each cardiac cycle. When the atria are

Heart Murmur

The four heart valves prevent backflow of blood, ensuring one-way flow through the heart. These valves maximize the amount of blood pumped out of the heart during contraction. A heart murmur is a condition in which abnormal heart sounds are detected with the aid of a stethoscope. Normally, a heart valve makes a distinct clicking sound when it snaps shut. With a murmur, the click is replaced by a sound similar to blowing. This abnormal sound can indicate the turbulent flow of blood through a narrowed valve or backward (retrograde) flow back toward the atria through a leaky valve. It could also indicate errant blood flow through a hole in the wall (septum) separating the right and left sides of the heart (septal defect).

Mild heart murmurs are fairly common in growing children and adolescents. During growth periods, valve development does not always keep up with enlargement of the heart openings. Valves can leak as a result of disease, such as stenosis, in which the valve is narrowed and often thickened and rigid. This condition can require surgical replacement of the valve. With mitral valve prolapse, the mitral valve allows some blood to flow back into the left atrium during ventricular contraction. This disorder, relatively common in adults (6-17% of the population including athletes), usually has little clinical significance unless there is significant backflow.

Most murmurs in athletes are benign, affecting neither the heart's pumping nor the athlete's performance. Only when there is a functional consequence, such as light-headedness or dizziness, are murmurs a cause for immediate concern.

relaxed, blood from the venous circulation fills the atria. About 70% of the blood filling the atria during this time passively flows directly through the mitral and tricuspid valves into the ventricles. When the atria contract, the atria push the remaining 30% of their volume into the ventricles.

During ventricular diastole, the pressure inside the ventricles is low, allowing the ventricles to passively fill with blood. As atrial contraction provides the final filling volume of blood, the pressure inside the ventricles increases slightly. As the ventricles contract, pressure inside the ventricles rises sharply. This increase in ventricular pressure forces the atrioventricular valves (i.e., tricuspid and mitral valves) closed, preventing any backflow of blood from the ventricles to the atria. The closing of the atrioventricular valves results in the first heart sound. Furthermore, when ventricular pressure exceeds the pressure in the pulmonary artery and the aorta, the pulmonary and aortic valves open, allowing blood to flow into the pulmonary and systemic circulations, respectively. Following ventricular contraction, pressure inside the ventricles falls and the pulmonary and aortic valves close. The closing of these valves corresponds to the second heart sound. The two sounds together, the result of valves closing, results in the typical "lub, dub" heard through a stethoscope during each heartbeat.

The interactions of the various events that take place during one cardiac cycle are illustrated in figure 6.9,

called a Wiggers diagram after the physiologist who created it. The diagram integrates information from the electrical conduction signals (ECG), heart sounds from the heart valves, pressure changes within the heart chambers, and left ventricular volume.

Stroke Volume

During systole, most, but not all, of the blood in the ventricles is ejected. This volume of blood pumped during one beat (contraction) is the **stroke volume (SV)**. This is depicted in figure 6.10a. To understand stroke volume, consider the amount of blood in the ventricle before and after contraction. At the end of diastole, just before contraction, the ventricle has finished filling. The volume of blood it now contains is called the **end-diastolic volume (EDV)**. At rest in a normally healthy adult, this value is approximately 100 ml. At the end of systole, just after the contraction, the ventricle has completed its ejection phase, but not all the blood is pumped out of the heart. The volume of blood remaining in the ventricle is called the **end-systolic volume (ESV)** and is approximately 40 ml under resting conditions. Stroke volume is the volume of blood that was ejected and is merely the difference between the volume of the filled ventricle and the volume remaining in the ventricle after contraction. So stroke volume is simply the difference between EDV and ESV; that is, SV = EDV − ESV (example: SV = 100 ml − 40 ml = 60 ml).

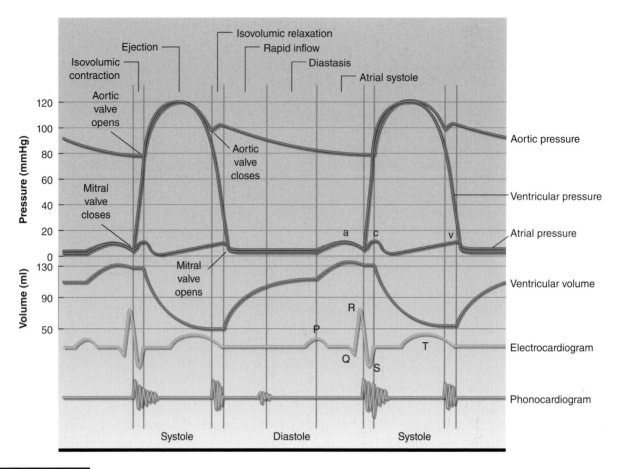

FIGURE 6.9 The Wiggers diagram, illustrating the events of the cardiac cycle for left ventricular function. Integrated into this diagram are the changes in left atrial and ventricular pressure, aortic pressure, ventricular volume, electrical activity (electrocardiogram), and heart sounds.

Ejection Fraction

The fraction of the blood pumped out of the left ventricle in relation to the amount of blood that was in the ventricle before contraction is called the **ejection fraction (EF)**. Ejection fraction is determined by dividing the stroke volume by EDV (60 ml / 100 ml = 60%), as shown in figure 6.10b. The EF, generally expressed as a percentage, averages about 60% at rest in healthy, active young adults. Thus, 60% of the blood in the ventricle at the end of diastole is ejected with the next contraction, and 40% remains in the ventricle. Ejection fraction is often used clinically as an index of the pumping ability of the heart.

Cardiac Output

Cardiac output ($\dot{Q}$), as shown in figure 6.10c, is the total volume of blood pumped by the ventricle per minute, the product of HR and SV. The SV at rest in the standing posture averages between 60 and 80 ml of blood in most adults. Thus, at a RHR of 70 beats/min, the resting cardiac output will vary between 4.2 and 5.6 L/min. The average adult body contains about 5 L of blood, so this means that the equivalent of our total blood volume is pumped through our hearts about once every minute.

Understanding the electrical and mechanical activity of the heart provides a basis for understanding the cardiovascular system, but the heart is only one part of this system. In addition to this pump, the cardiovascular system contains an intricate network of tubes that serve as a delivery system that carries the blood to all body tissues.

(a) Calculation of stroke volume (SV), the difference between end-diastolic volume (EDV) and end-systolic volume (ESV)

$$EF = \frac{SV}{EDV} \times 100 = \frac{60\ ml}{100\ ml} = 60\%$$

(b) Calculation of ejection fraction (EDV)

$$\dot{Q} = HR \times SV$$

70 beats/min × 60 ml/beat = 4,200 ml/min = 4.2 L/min

(c) Calculation of cardiac output ($\dot{Q}$)

FIGURE 6.10 Calculations of stroke volume, ejection fraction, and cardiac output based on volumes of blood flowing into and out of the heart.

In review

➤ The electrical and mechanical events that occur in the heart during one heartbeat comprise the cardiac cycle. A Wiggers diagram depicts the intricate timing of these events.

➤ Cardiac output, the volume of blood pumped by each ventricle per minute, is the product of heart rate and stroke volume.

➤ Not all of the blood in the ventricles is ejected during systole. The ejected volume is the stroke volume, while the percentage of blood pumped with each beat is the ejection fraction.

➤ To calculate stroke volume, ejection fraction, and cardiac output:

$$SV\ (ml/beat) = EDV - ESV.$$
$$EF\ (\%) = (SV/EDV) \times 100.$$
$$\dot{Q}\ (L/min) = HR \times SV.$$

Vascular System

The vascular system contains a series of vessels that transport blood from the heart to the tissues and back: the arteries, arterioles, capillaries, venules, and veins.

Arteries are large muscular, elastic, conduit vessels for transporting blood away from the heart to the arterioles. The aorta is the largest artery, transporting blood from the left ventricle to all regions of the body as it eventually branches into smaller and smaller arteries, finally branching into arterioles. The **arterioles** are the site of greatest control of the circulation by the sympathetic nervous system, so arterioles are sometimes called resistance vessels. Arterioles are heavily innervated by the sympathetic nervous system and are the main site of control of blood flow to specific tissues.

From the arterioles, blood enters the **capillaries**, the narrowest and simplest vessels in terms of their structure, with walls only one cell thick. Virtually all exchange between the blood and the tissues occurs at the capillaries. Blood leaves the capillaries to begin the return trip to the heart in the **venules**, and the venules form larger vessels—the **veins**. The vena cava is the great vein transporting blood back to the right atrium from all regions of the body above (superior vena cava) and below (inferior vena cava) the heart.

Blood Pressure

Blood pressure is the pressure exerted by the blood on the vessel walls, and the term usually refers to arterial blood pressure. It is expressed by two numbers: the **systolic blood pressure (SBP)** and the **diastolic blood pressure (DBP)**. The higher number is the SBP; it represents the highest pressure in the artery that occurs during ventricular systole. Ventricular contraction pushes the blood through the arteries with tremendous force, and that force exerts high pressure on the arterial walls. The lower number is the DBP and represents the lowest pressure in the artery, corresponding to ventricular diastole when the ventricle is filling.

Mean arterial pressure (MAP) represents the average pressure by the blood as it travels through the arteries. Since diastole takes about twice as long as systole in a normal cardiac cycle, mean arterial pressure can be estimated from the DBP and SBP as follows:

$$\text{MAP} = 2/3 \ \text{DBP} + 1/3 \ \text{SBP}.$$

Alternately,

$$\text{MAP} = \text{DBP} + [0.333 \times (\text{SBP} - \text{DBP})].$$

(SBP – DBP) is also called pulse pressure.

To illustrate, with a normal resting blood pressure of 120 mmHg over 80 mmHg, the MAP = 80 + [0.333 × (120 – 80)] = 93 mmHg.

In focus

Systolic blood pressure is the highest pressure within the vascular system, whereas diastolic blood pressure is the lowest pressure. Mean arterial pressure is the average pressure on the vessel walls during a cardiac cycle.

General Hemodynamics

The cardiovascular system is a continuous closed-loop system. Blood flows through this closed-loop system because of the pressure gradient that exists between the arterial and venous sides of the circulation. To understand regulation of blood flow to the tissues it is necessary to understand the intricate relationship between pressure, flow, and resistance.

In order for blood to flow in a vessel there must be a pressure difference from one end of the vessel to the other end. Blood will flow from the region of the vessel with high pressure to the region of the vessel with low pressure. Alternatively, if there is no pressure difference across the vessel, there is no driving force and therefore no blood flow. In the circulatory system, the mean arterial pressure in the aorta is approximately 100 mmHg at rest, and the pressure in the right atrium is very close to 0 mmHg. Therefore, the pressure difference across the entire circulatory system is 100 mmHg – 0 mmHg = 100 mmHg.

The reason for the pressure differential from the arterial to the venous circulation is that the blood vessels themselves provide resistance or impedance to blood flow. The resistance that the vessel provides is largely dictated by the properties of the blood vessels and the blood itself. These properties include the length and radius of the blood vessel and the viscosity or thickness of the blood flowing through the vessel. Resistance to flow can be calculated as

$$\text{resistance} = \eta \times \text{L} / \text{r}^4.$$

where η is the viscosity (thickness) of the blood, L is the length of the vessel, and r is the radius of the vessel, which is raised to the fourth power. Blood flow is proportional to the pressure difference across the system and is inversely proportional to resistance. This relationship can be illustrated by the following equation:

$$\text{blood flow} = \Delta\text{pressure} / \text{resistance}.$$

Notice that blood flow can increase by either an increase in the pressure difference (Δpressure), a decrease in resistance, or a combination of the two. Altering resistance to control blood flow is much more advantageous because very small changes in

Blood Flow to the Heart: Coronary Artery Blood Flow

The mechanism of blood flow to and through the coronary arteries is quite different from that of blood flow to the rest of the body. During contraction, when blood is forced out of the left ventricle under high pressure, the aortic semilunar valve is forced open. When this valve is open, its flaps block the entrances to the coronary arteries. As the pressure in the aorta decreases, the semilunar valve closes, and these entrances are exposed so that blood can then enter the coronary arteries. This design ensures that the coronary arteries are spared the very high blood pressure created by contraction of the left ventricle, thus protecting these vessels from damage.

blood vessel radius equate to large changes in resistance. This is due to the fourth-power mathematical relationship between vascular resistance and vessel radius.

Changes in vascular resistance are largely due to changes in the radius or diameter of the vessels, as the viscosity of the blood and the length of the vessels do not change significantly under normal conditions. Therefore, regulation of blood flow to organs is accomplished by small changes in vessel radius through **vasoconstriction** and **vasodilation**. This allows the cardiovascular system to divert blood flow to the areas where it is needed most.

As mentioned earlier, most resistance to blood flow occurs in the arterioles. Figure 6.11 shows the blood pressure changes across the entire vascular system. The arterioles are responsible for ~70% to 80% of the drop in mean arterial pressure across the entire cardiovascular system. This is important because small changes in arteriole radius can greatly affect the regulation of mean arteriole pressure and the local control of blood flow. At the capillary level, changes due to systole and diastole are no longer evident and the flow is smooth (laminar) rather than turbulent.

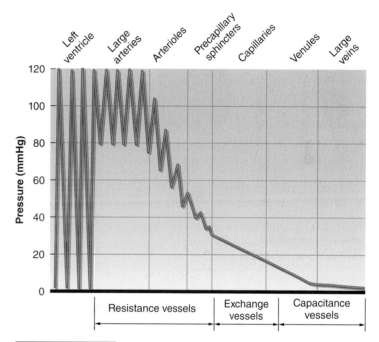

FIGURE 6.11 Pressure changes across the systemic circulation. Notice the large pressure drop that occurs across the arteriole portion of the system.

In focus

In terms of the entire cardiovascular system, cardiac output is the blood flow to the entire system; the Δpressure is the difference between aortic pressure when blood leaves the heart and venous pressure when blood returns to the heart; and resistance is the impedance to blood flow from the blood vessels. Blood flow is mainly controlled by small changes in blood vessel (arteriole) radius that greatly affect resistance.

Distribution of Blood

Distribution of blood to the various body tissues varies considerably depending on the immediate needs of a specific tissue compared with those of other areas of the body. As a general rule, the most metabolically active tissues receive the greatest blood supply. At rest under normal conditions, the liver and kidneys combine to receive almost half of the cardiac output, while resting skeletal muscles receive only about 15% to 20%.

During exercise, blood is redirected to the areas where it is needed most. During heavy endurance exercise, contracting muscles may receive up to 80% or more of the blood flow, and flow to the liver and kidneys decreases. This redistribution, along with increases in cardiac output (to be discussed in chapter 8), allows up to 25 times more blood flow to active muscles (see figure 6.12).

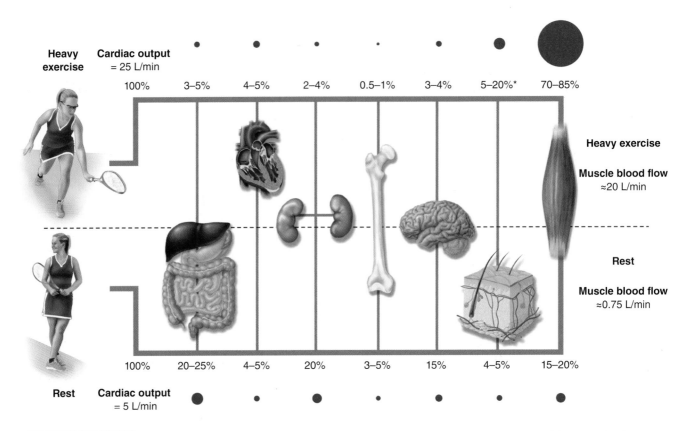

Heavy exercise
Cardiac output
= 25 L/min

100% 3–5% 4–5% 2–4% 0.5–1% 3–4% 5–20%* 70–85%

Heavy exercise

Muscle blood flow
≈20 L/min

Rest

Muscle blood flow
≈0.75 L/min

100% 20–25% 4–5% 20% 3–5% 15% 4–5% 15–20%

Rest
Cardiac output
= 5 L/min

FIGURE 6.12 Distribution of cardiac output at rest and maximal exercise. *Depends on ambient and body temperatures.

Reprinted, by permission, from P.O. Åstrand et al., 2003, *Textbook of work physiology: Physiological bases of exercise,* 4th ed. (Champaign, IL: Human Kinetics), 143.

Alternately, after one eats a big meal, the digestive system receives more of the available cardiac output than when the digestive system is empty. Along the same lines, during increasing environmental heat stress, skin blood flow increases to a greater extent as the body attempts to maintain normal temperature. The cardiovascular system responds accordingly to redistribute blood, whether it is to the exercising muscle to match metabolism, for digestion, or to facilitate thermoregulation. These changes in the distribution

of cardiac output are controlled by the sympathetic nervous system, primarily by increasing or decreasing arteriolar diameter. Arterioles have a strong muscular wall that can significantly alter vessel diameter, are highly innervated by sympathetic nerves, and have the capacity to respond to local control mechanisms.

Intrinsic Control of Blood Flow

Intrinsic control of blood distribution refers to the ability of the local tissues to dilate or constrict the arterioles that serve them and alter regional blood flow depending on the immediate needs of those tissues. With exercise and the increased metabolic demand of the exercising skeletal muscles, the arterioles undergo locally mediated vasodilation, opening up to allow more blood to enter that highly active tissue.

There are essentially three types of intrinsic control of blood flow. The strongest stimulus for the release of local vasodilating chemicals is metabolic, in particular an increased oxygen demand. As oxygen uptake by metabolically active tissues increases, available oxygen is diminished. Local arterioles vasodilate to allow

In focus

The cardiovascular system has a tremendous capacity to redistribute blood away from areas where the need is low to areas where there is increased need. Skeletal muscle normally receives about 15% of the cardiac output at rest. This can increase to 80% or more during heavy endurance exercise. Distribution of blood to various areas is controlled primarily at the level of the arterioles.

more blood to perfuse that area, delivering more oxygen. Other chemical changes that can stimulate increased blood flow are decreases in other nutrients and increases in by-products (carbon dioxide, K^+, H^+, lactic acid) or inflammatory chemicals.

Second, many dilator substances can be produced within the endothelium (inner lining) of arterioles and can initiate vasodilation in the vascular smooth muscle of those arterioles. These substances include nitric oxide (NO), prostaglandins, and endothelium-derived hyperpolarizing factor (EDHF). These endothelium-derived vasodilators are important regulators of blood flow at rest and during exercise in humans. Third, pressure changes within the vessels themselves can also cause vasodilation and vasoconstriction. This is referred to as the myogenic response. The vascular smooth muscle contracts in response to an increase in pressure across the vessel wall and relaxes in response to a decrease in pressure across the vessel wall. Additionally, acetylcholine and adenosine also have been proposed as potential vasodilators for the increase in muscle blood flow during exercise. Figure 6.13 illustrates the three types of intrinsic control of vascular tone.

Extrinsic Neural Control

The concept of intrinsic local control explains redistribution of blood within an organ or tissue mass; however, the cardiovascular system must divert blood flow to where it is needed, beginning at a site upstream of the local environment. Redistribution at the system or organ level is controlled by neural mechanisms. This is known as **extrinsic neural control** of blood flow, because the control comes from outside the specific area (extrinsic) instead of from locally inside the tissues (intrinsic).

Blood flow to all body parts is regulated largely by the sympathetic nervous system. The circular layers of smooth muscle within the artery and arteriole walls are abundantly innervated by sympathetic nerves. In most vessels, an increase in sympathetic nerve activity causes these circular smooth muscle cells to contract, constricting blood vessels and thereby decreasing blood flow.

Under normal resting conditions, sympathetic nerves transmit impulses continuously to the blood vessels (in particular, the arterioles), keeping the vessels moderately constricted to maintain adequate blood pressure. This state of tonic vasoconstriction is referred to as vasomotor tone. When sympathetic stimulation increases, further constriction of the blood vessels in a specific area decreases blood flow into that area and allows more blood to be distributed elsewhere. But if sympathetic stimulation decreases below the level needed to maintain tone, constriction of vessels in the area is lessened, so the vessels passively vasodilate,

(a) Metabolic regulation

(b) Myogenic contraction

↓ Pressure = dilation   ↑ Pressure = constriction

(c) Endothelium-mediated vasodilation

FIGURE 6.13 Intrinsic control of blood flow. Arterioles are signaled to dilate or constrict at the local level by (*a*) changes in the local concentration of oxygen or metabolic products, (*b*) by the effects of local pressure within the arterioles, and (*c*) by endothelium-derived factors.

Figure courtesy of Dr. Donna H. Korzick, Pennsylvania State University.

increasing blood flow into that area. Therefore, sympathetic stimulation will cause vasoconstriction in most vessels. Blood flow can passively be increased through a lowering of the normal tonic level of sympathetic outflow.

In focus

Blood flow can be controlled at the local tissue level (intrinsic control) by the release of locally acting metabolic dilators, endothelium-dependent vasodilators (NO, prostaglandins, EDHF), and the myogenic response to changes in pressure within the vessel. The sympathetic nervous system plays a major role in extrinsic control of blood flow, redirecting blood flow from areas of low need to areas of high need.

Distribution of Venous Blood

While flow to tissues is controlled by changes on the arterial side of the system, most of the blood *volume* normally resides in the venous side of the system. At rest, the blood volume is distributed among the vasculature as shown in figure 6.14. The venous system has a great capacity to hold blood volume because veins have little vascular smooth muscle and are very elastic and "balloon-like." Thus, the venous system provides a large reservoir of blood available to be rapidly distributed back to the heart (venous return) and from there, to the arterial circulation. This is accomplished through sympathetic stimulation of the venules and veins, which causes the vessels to constrict (venoconstriction).

Integrative Control of Blood Pressure

Blood pressure is normally maintained by reflexes within the autonomic nervous system. Specialized pressure sensors located in the aortic arch and the carotid

arteries, called **baroreceptors**, are sensitive to changes in arterial pressure. When the pressure in these large arteries changes, afferent signals are sent to the cardiovascular control centers in the brain where autonomic reflexes are initiated, and efferent signals are sent to respond to changes in blood pressure. For example, when blood pressure is elevated, the baroreceptors are stimulated by an increase in stretch. They relay this information to the cardiovascular control center in the brain. In response to the increased pressure there is a reflex increase in vagal tone, to decrease heart rate, and a decrease in sympathetic activity to both the heart and the arterioles, all of which serves to normalize blood pressure. In response to a decrease in blood pressure, less stretch is sensed by the baroreceptors, and the response is to increase heart rate by vagal withdrawal and to increase sympathetic nervous activity, thus correcting the low-pressure signal.

There are also other specialized receptors, called **chemoreceptors** and **mechanoreceptors**, that send information about the chemical environment in the muscle and the length and tension of the muscle, respectively, to the cardiovascular control centers. These receptors also modify the blood pressure response and are especially important during exercise.

Return of Blood to the Heart

Because humans spend so much time in an upright position, the cardiovascular system requires mechanical assistance to overcome the force of gravity and assist the return of venous blood from the lower parts of the body to the heart. Three basic mechanisms assist in this process:

- Valves in the veins
- The muscle pump
- The respiratory pump

The veins contain valves that allow blood to flow in only one direction, thus preventing backflow and further pooling of blood in the lower body. These venous valves also complement the action of the skeletal **muscle pump**, the rhythmic mechanical compression of the veins that occurs during the rhythmic skeletal muscle contraction accompanying many types of movement and exercise, for example during walking and running (figure 6.15). The muscle pump pushes blood volume in the veins back toward the heart. Finally, changes in pressure in the abdominal and thoracic cavities during breathing assist blood return to the heart by creating a pressure gradient between the veins and the chest cavity.

FIGURE 6.14 Blood volume distribution within the vasculature when the body is at rest.

To heart

Opened valve

Contracted
skeletal
muscles

Closed valve

Vein

FIGURE 6.15 The muscle pump. As the skeletal muscles contract, they squeeze the veins in the legs and assist in the return of blood to the heart. Valves within the veins ensure the unidirectional flow of blood back to the heart.

In review

➤ Blood is distributed throughout the body based primarily on the metabolic needs of the individual tissues. The most active tissues receive the highest blood flows.

➤ Redistribution of blood flow is controlled locally by the release of dilators from either the tissue (metabolic regulation) or the endothelium of the blood vessel (endothelium-mediated dilation). A third type of intrinsic control involves the response of the arteriole to pressure. Decreased arteriolar pressure causes vasodilation, thus increasing blood flow to the area, while increased pressure causes local constriction.

➤ Extrinsic neural control of blood flow distribution is accomplished by the sympathetic nervous system, primarily through vasoconstriction of small arteries and arterioles.

➤ Blood returns to the heart through the veins, assisted by valves within the veins, the muscle pump, and changes in respiratory pressure.

Blood

Blood serves many diverse purposes in regulating normal body function. The three functions of primary importance to exercise and sport are

- transportation,
- temperature regulation, and
- acid–base (pH) balance.

We are most familiar with blood's transporting functions, delivering oxygen and fuel substrates and carrying away metabolic by-products. In addition, blood is critical in temperature regulation during physical activity because it picks up heat from the exercising muscle and delivers it to the skin where it can be dissipated to the environment (see chapter 12). Blood also buffers the acids produced by anaerobic metabolism and maintains proper pH for metabolic processes (see chapters 2 and 7).

Blood Volume and Composition

The total volume of blood in the body varies considerably with an individual's size, body composition, and state of training. Larger blood volumes are associated with greater lean body mass and higher levels of endurance training. The blood volume of people of average body size and normal physical activity generally ranges from 5 to 6 L in men and 4 to 5 L in women.

Blood is composed of plasma and formed elements (see figure 6.16). Plasma normally constitutes about 55% to 60% of total blood volume but can decrease by 10% of its normal amount or more with intense exercise in heat, or can increase by 10% or more with endurance training or acclimation to heat. Approximately 90% of the plasma volume is water; 7% consists of plasma proteins; and the remaining 3% includes cellular nutrients, electrolytes, enzymes, hormones, antibodies, and wastes.

The formed elements, which normally constitute the other 40% to 45% of total blood volume, are the red blood cells (erythrocytes), white blood cells (leukocytes), and platelets (thrombocytes). Red blood cells constitute more than 99% of the formed-element volume; white blood cells and platelets together account for less than 1%. The percentage of the total blood volume composed of cells or formed elements is referred to as the **hematocrit**. Hematocrit varies among individuals, but a normal range is 41% to 50% in adult men and 36% to 44% in adult women.

White blood cells protect the body from infection either by directly destroying the invading agents through phagocytosis (ingestion) or by forming antibodies to destroy them. Adults have about 7,000 white blood cells per cubic millimeter of blood.

The remaining formed elements are the blood platelets. These are cell fragments that are required for blood coagulation (clotting), which prevents excessive blood loss. Exercise physiologists are most concerned with red blood cells.

In focus

The hematocrit is the ratio of the formed elements in the blood (red cells, white cells, and platelets) to the total blood volume. An average hematocrit for adult men is 42% and for adult women is 38%.

Centrifuged whole blood

— 55% plasma
  90% H_2O
  7% plasma proteins
  3% other

— 45% formed elements
  >99% red blood cells
  <1% white blood cells
    and platelets

$$\text{Hematocrit} = \frac{45\% \text{ formed elements}}{100\% \text{ total blood volume}} = 45$$

FIGURE 6.16 The composition of whole blood, illustrating the plasma volume (fluid portion) and the cellular volume (red cells, white cells, and platelets) after the blood sample has been centrifuged. A centrifuge is shown at right.

Red Blood Cells

Mature red blood cells (erythrocytes) have no nucleus, so they cannot reproduce as other cells can. They must be replaced with new cells on a reoccurring basis, a process called **hematopoiesis**. The normal life span of a red blood cell is about four months. Thus, these cells are continuously produced and destroyed at equal rates. This balance is very important, because adequate oxygen delivery to tissues depends on having a sufficient number of red blood cells to transport oxygen. Decreases in their number or function can hinder oxygen delivery and thus affect exercise performance.

Red blood cells transport oxygen, which is primarily bound to hemoglobin. **Hemoglobin** is composed of a protein (globin) and a pigment (heme). Heme contains iron, which binds oxygen. Each red blood cell contains approximately 250 million hemoglobin molecules, each able to bind four oxygen molecules—so each red blood cell can bind up to a billion molecules of oxygen! There is an average of 15 g of hemoglobin per 100 ml of whole blood. Each gram of hemoglobin can combine with 1.33 ml of oxygen, so as much as 20 ml of oxygen can be bound for each 100 ml of blood. Therefore, when arterial blood is saturated with oxygen, it has an oxygen-carrying capacity of 20 ml of oxygen per 100 ml of blood.

In focus

When we donate blood, the removal of one "unit," or nearly 500 ml, represents approximately an 8% to 10% reduction in both the total blood volume and the number of circulating red blood cells. Donors are advised to drink plenty of fluids. Because plasma is primarily water, simple fluid replacement returns plasma volume to normal within 24 to 48 h. However, it takes at least six weeks to reconstitute the red blood cells because they must go through full development before they are functional. Blood loss greatly compromises the performance of endurance athletes by reducing oxygen delivery capacity.

Blood Viscosity

Viscosity refers to the thickness of the blood. Recall from our discussion of vascular resistance that the more viscous a fluid, the more resistant it is to flow. The

In focus

During endurance training, athletes respond with both a higher red cell volume (RCV) and an expanded plasma volume (PV). Since the PV increase is higher than the RCV increase, the hematocrit in these athletes tends to be somewhat lower than that of sedentary individuals.

viscosity of blood is normally about twice that of water and increases as hematocrit increases.

Because of oxygen transport by the red blood cells, an increase in their number would be expected to maximize oxygen transport. But if an increase in red blood cell count is not accompanied by a similar increase in plasma volume, blood viscosity and vascular resistance will increase, which could result in reduced blood flow. This generally is not a problem unless the hematocrit reaches 60% or more.

Conversely, the combination of a low hematocrit with a high plasma volume, which decreases the blood's viscosity, appears to have certain benefits for the blood's transport function because the blood can flow more easily. Unfortunately, a low hematocrit frequently results from a reduced red blood cell count, as in diseases such as anemia. Under these circumstances, the blood can flow easily, but it contains fewer carriers, so oxygen transport is impeded. For optimal physical performance, a low-normal hematocrit with a normal or slightly elevated number of red blood cells is desirable. This combination facilitates oxygen transport. Many endurance athletes achieve this combination as part of their cardiovascular system's normal adaptation to training. This adaptation is discussed in chapter 11.

In review

- ➤ Blood is about 55% to 60% plasma and 40% to 45% formed elements. Red blood cells compose about 99% of the formed elements.
- ➤ Oxygen is transported primarily by binding to the hemoglobin in red blood cells.
- ➤ As blood viscosity increases, so does resistance to flow. Increasing the number of red blood cells is advantageous to aerobic performance but only up to the point (a hematocrit approaching 60%) where viscosity limits flow.

In this chapter, we reviewed the structure and function of the cardiovascular system. We learned how blood flow and blood pressure are regulated to meet the body's needs, and explored the role of the cardiovascular system in transporting and delivering oxygen and nutrients to the body's cells while clearing away metabolic wastes, including carbon dioxide. Knowing how substances are moved within the body, we now look more closely at the movement of oxygen and carbon dioxide. In the next chapter, we explore the role of the respiratory system in delivering oxygen to, and removing carbon dioxide from, the cells of the body.

Key Terms

arteries

arterioles

atrioventricular (AV) node

baroreceptor

bradycardia

capillaries

cardiac cycle

cardiac output (Q̇)

chemoreceptor

diastolic blood pressure (DBP)

ejection fraction (EF)

electrocardiogram (ECG)

electrocardiograph

end-diastolic volume (EDV)

end-systolic volume (ESV)

extrinsic neural control

hematocrit

hematopoiesis

hemoglobin

intercalated disks

mean arterial pressure (MAP)

mechanoreceptors

muscle pump

myocardium

pericardium

premature ventricular contraction (PVC)

Purkinje fibers

sinoatrial (SA) node

stroke volume (SV)

systolic blood pressure (SBP)

tachycardia

vasoconstriction

vasodilation

veins

ventricular fibrillation

ventricular tachycardia

venules

Study Questions

1. Describe the structure of the heart, the pattern of blood flow through the valves and chambers of the heart, how the heart as a muscle is supplied with blood, and what happens when the resting heart must suddenly supply an exercising body.

2. What events take place that allow the heart to contract, and how is heart rate controlled?

3. What is the difference between systole and diastole, and how do they relate to SBP and DBP?

4. What is the relationship between pressure, flow, and resistance?

5. How is blood flow to the various regions of the body controlled?

6. Describe the three important mechanisms for returning blood back to the heart when someone is exercising in an upright position.

7. Describe the primary functions of blood.

Study Guide Activities

In addition to the activities listed in the chapter opening outline on page 139, two other activities are available in the web study guide, located at

www.HumanKinetics.com/PhysiologyOfSportAndExercise

 The **KEY TERMS** activity reviews important terms, and the end-of-chapter **QUIZ** tests your understanding of the material covered in the chapter.

The Respiratory System and Its Regulation

7

In this chapter and in the web study guide

 ACTIVITY 7.1   Anatomy of the Respiratory System looks at the basic structures of the lung.

 ACTIVITY 7.2   Inspiration and Expiration explores the key events of pulmonary ventilation.

  ACTIVITY 7.3   Pulmonary Diffusion includes an offline experiment that investigates gas exchange and pulmonary diffusion.

 ACTIVITY 7.4   Arterial–Venous Oxygen Difference looks at differences in oxygen content in the blood of resting and active people.

 ACTIVITY 7.5   Regulation of Pulmonary Ventilation provides an in-depth review of the involuntary regulation of pulmonary ventilation.

By any standard, Beijing, China, is one of the most polluted cities on the planet. In preparation for the 2008 Olympic Games, nearly $17 billion was spent in attempts to temporarily improve air quality, including cloud seeding to increase the likelihood of rain showers in the region overnight. Factories were closed, traffic was halted, and construction was put on hold for the duration of the Games. Yet air pollution at the Olympics was still about two to four times higher than that of Los Angeles on an average day, exceeding levels considered safe by the World Health Organization. Several athletes opted out of events because of respiratory problems or concerns, including Ethiopian marathon record holder Haile Gebreselassie and 2004 cycling silver medalist Sergio Paulinho of Portugal. Athletes previously diagnosed with asthma were allowed to use rescue inhalers; and for the first time ever, soccer matches were interrupted to give athletes time to recover from the pollutants, smog, heat, and humidity. Athletes and spectators endured these conditions for a few weeks, and there are no reports of long-term health problems among athletes or spectators from exposure to the Beijing air. However, the residents of Beijing encounter these adverse respiratory conditions on a daily basis.

The respiratory and cardiovascular systems combine to provide an effective delivery system that carries oxygen to, and removes carbon dioxide from, all tissues of the body.

This transportation involves four separate processes:

- Pulmonary ventilation (breathing): movement of air into and out of the lungs
- Pulmonary diffusion: the exchange of oxygen and carbon dioxide between the lungs and the blood
- Transport of oxygen and carbon dioxide via the blood
- Capillary diffusion: the exchange of oxygen and carbon dioxide between the capillary blood and metabolically active tissues

The first two processes are referred to as **external respiration** because they involve moving gases from outside the body into the lungs and then the blood. Once the gases are in the blood, they must be transported to the tissues. When blood arrives at the tissues, the fourth step of respiration occurs. This gas exchange between the blood and the tissues is called **internal respiration**. Thus, external and internal respiration are linked by the circulatory system. The following sections examine all four components of respiration.

Pulmonary Ventilation

Pulmonary ventilation, or breathing, is the process by which we move air into and out of the lungs. The anatomy of the respiratory system is illustrated in figure 7.1. Air is typically drawn into the lungs through the nose, although the mouth must also be used when the demand for air exceeds the amount that can comfortably be brought in through the nose. Nasal breathing is advantageous because the air is warmed and humidified as it swirls through the bony irregular sinus surfaces (turbinates or conchae). Of equal importance, the turbinates churn the inhaled air, causing dust and other particles to contact and adhere to the nasal mucosa. This filters out all but the tiniest particles, minimizing irritation and the threat of respiratory infections. From the nose and mouth, the air travels through the pharynx, larynx, trachea, and bronchial tree.

These anatomical structures serve as the transport zone of the lungs because gas exchange does not occur in these structures. Exchange of oxygen and carbon dioxide occurs when air finally reaches the smallest respiratory units: the respiratory bronchioles and the alveoli. The respiratory bronchioles are primarily transport tubes also but are included in this region because they contain clusters of alveoli. This is known as the respiratory zone because it is the site of gas exchange in the lungs.

In focus

Breathing through the nose helps humidify and warm the air during inhalation and filters out foreign particles from the air.

The lungs are not directly attached to the ribs. Rather, they are suspended by the pleural sacs. The pleural sacs have a double wall: the parietal pleura, which lines the thoracic wall, and the visceral or pulmonary pleura, which lines the outer aspects of the lung. These pleural walls envelop the lungs and have a thin film of fluid between them that reduces friction during respiratory movements. In addition, these sacs are connected to the lungs and to the inner surface of the thoracic cage, causing the lungs to take the shape and size of the rib or thoracic cage as the chest expands and contracts.

The anatomy of the lungs, the pleural sacs, the diaphragm muscle, and the thoracic cage determines airflow into and out of the lungs, that is, inspiration and expiration.

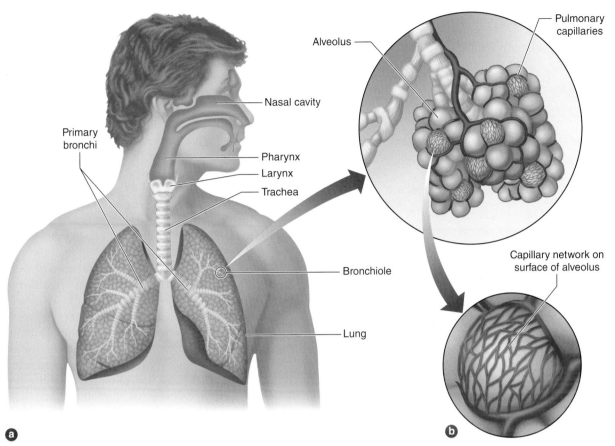

FIGURE 7.1 (*a*) The anatomy of the respiratory system, illustrating the respiratory tract (i.e., nasal cavity, pharynx, trachea, and bronchi). (*b*) The enlarged view of an alveolus shows the regions of gas exchange between the alveolus and pulmonary blood in the capillaries.

Inspiration

Inspiration is an active process involving the diaphragm and the external intercostal muscles. Figure 7.2*a* shows the resting positions of the diaphragm and the thoracic cage, or thorax. With inspiration, the ribs and sternum are moved by the external intercostal muscles. The ribs swing up and out and the sternum swings up and forward. At the same time, the diaphragm contracts, flattening down toward the abdomen.

These actions, illustrated in figure 7.2*b*, expand all three dimensions of the thoracic cage, increasing the volume inside the lungs. When the lungs are expanded they have a greater volume and the air within them has more space to fill. According to **Boyle's gas law**, which states that pressure × volume is constant (at a constant temperature), the pressure within the lungs decreases. As a result, the pressure in the lungs (intrapulmonary pressure) is less than the air pressure outside the body. Because the respiratory tract is open to the outside, air rushes into the lungs to reduce this pressure difference.

This is how air moves into the lungs during inspiration.

During forced or labored breathing, as during heavy exercise, inspiration is further assisted by the action of other muscles, such as the scaleni (anterior, middle, and posterior) and sternocleidomastoid in the neck and the pectorals in the chest. These muscles help raise the ribs even more than during regular breathing.

In focus

The pressure changes required for adequate ventilation at rest are really quite small. For example, at the standard atmospheric pressure at sea level (760 mmHg), inspiration may decrease the pressure in the lungs (intrapulmonary pressure) by only about 2 to 3 mmHg. However, during maximal respiratory effort, such as during exhaustive exercise, the intrapulmonary pressure can decrease by 80 to 100 mmHg.

Atmospheric pressure = 760 mmHg

Sternum

Diaphragm

Ribs

Intrapulmonic pressure = 760 mmHg

Intrapleural pressure = 756 mmHg

ⓐ Resting positions of the diaphragm and the thoracic cage, or thorax. Note the size of the rib cage at rest.

Atmospheric pressure = 760 mmHg

Intrapulmonic pressure = 758 mmHg

Intrapleural pressure = 754 mmHg

ⓑ The dimensions of the lungs and the thoracic cage increase during inspiration, forming a negative pressure that draws air into lungs.

Atmospheric pressure = 760 mmHg

Intrapulmonic pressure = 763 mmHg

Intrapleural pressure = 756 mmHg

ⓒ During expiration, the lung volume decreases, thereby forcing air out of the lung.

FIGURE 7.2 The process of inspiration and expiration, showing how movement of the ribs and diaphragm can increase and decrease the size of the thorax.

Expiration

At rest, **expiration** is a passive process involving relaxation of the inspiratory muscles and elastic recoil of the lung tissue. As the diaphragm relaxes, it returns to its normal upward, arched position. As the external

intercostal muscles relax, the ribs and sternum move back into their resting positions (figure 7.2*c*). While this happens, the elastic nature of the lung tissue causes it to recoil to its resting size. This increases the pressure in the lungs and causes a proportional decrease in volume in the thorax, and therefore air is forced out of the lungs.

During forced breathing, expiration becomes a more active process. The internal intercostal muscles actively pull the ribs down. This action can be assisted by the latissimus dorsi and quadratus lumborum muscles. Contracting the abdominal muscles increases the intra-abdominal pressure, forcing the abdominal viscera upward against the diaphragm and accelerating its return to the domed position. These muscles also pull the rib cage down and inward.

The changes in intra-abdominal and intrathoracic pressure that accompany forced breathing also help return venous blood back to the heart, working together with the muscle pump in the legs to assist the return of venous volume. As intra-abdominal and intrathoracic pressure increases, it is transmitted to the great veins—the pulmonary veins and superior and inferior venae cavae—that transport blood back to the heart. When the pressure decreases, the veins return to their original size and fill with blood. The changing pressures within the abdomen and thorax squeeze the blood in the veins, assisting its return through a milking action. This phenomenon is known as the **respiratory pump** and is essential in maintaining adequate venous return.

Pulmonary Volumes

The volume of air in the lungs can be measured with a technique called **spirometry**. A spirometer measures the volumes of air inspired and expired and therefore changes in lung volume. Although more sophisticated spirometers are used today, a simple spirometer contains a bell filled with air that is partially submerged in water. A tube runs from the subject's mouth under the water and emerges inside the bell, just above the water level. As the person exhales, air flows down the tube and into the bell, causing the bell to rise. The bell is attached to a pen, and its movement is recorded on a simple rotating drum (figure 7.3).

This technique is used clinically to measure lung volumes, capacities, and flow rates as an aid in diagnosing such respiratory diseases as asthma, chronic obstructive pulmonary disease (COPD), and emphysema.

The amount of air entering and leaving the lungs with each breath is known as the **tidal volume**. The **vital capacity (VC)** is the greatest amount of air that can be expired after a maximal inspiration. Even after a full expiration, some air remains in the lungs. The

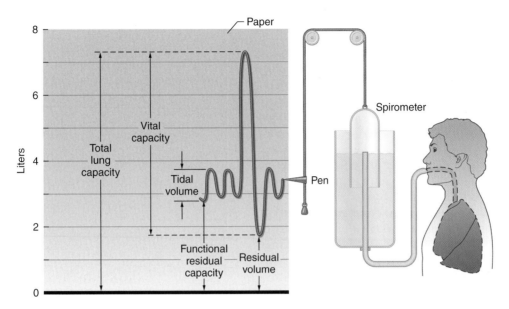

FIGURE 7.3 Lung volumes measured by spirometry.

Reprinted, by permission, from J. West, 2000, *Respiratory physiology: The essentials* (Baltimore, MD: Lippincott, Williams, and Wilkins), 14.

amount of air remaining in the lungs after a maximal expiration is the **residual volume (RV)**. The residual volume cannot be measured with spirometry. The **total lung capacity (TLC)** is the sum of the vital capacity and the residual volume.

In review

➤ Pulmonary ventilation (breathing) is the process by which air is moved into and out of the lungs. It has two phases: inspiration and expiration.

➤ Inspiration is an active process in which the diaphragm and the external intercostal muscles contract, increasing the dimensions, and thus the volume, of the thoracic cage. This decreases the pressure in the lungs, causing air to flow in.

➤ Expiration at rest is normally a passive process. The inspiratory muscles and diaphragm relax and the elastic tissue of the lungs recoils, returning the thoracic cage to its smaller, normal dimensions. This increases the pressure in the lungs and forces air out.

➤ Forced or labored inspiration and expiration are active processes and involve accessory muscle actions.

➤ Lung volumes and capacities, along with rates of airflow into and out of the lungs, are measured by spirometry.

Pulmonary Diffusion

Gas exchange in the lungs between the alveoli and the capillary blood, called **pulmonary diffusion**, serves two major functions:

• It replenishes the blood's oxygen supply, which is depleted at the tissue level as it is used for oxidative energy production.

• It removes carbon dioxide from returning systemic venous blood.

Air is brought into the lungs during pulmonary ventilation, enabling gas exchange to occur through pulmonary diffusion. Oxygen from the air diffuses from the alveoli into the blood in the pulmonary capillaries, and carbon dioxide diffuses from the blood into the alveoli in the lungs. The **alveoli** are grapelike clusters, or air sacs, at the ends of the terminal bronchioles.

Blood from the body (except for that returning from the lungs) returns through the vena cava to the right side of the heart. From the right ventricle, this blood is pumped through the pulmonary artery to the lungs, ultimately working its way into the pulmonary capillaries. These capillaries form a dense network around the alveolar sacs. These vessels are so small that the red blood cells must pass through them in single file, such that each cell is exposed to the surrounding lung tissue. This is where pulmonary diffusion occurs.

Blood Flow to the Lungs at Rest

At rest the lungs receive approximately 4 to 6 L/min of blood flow, depending on body size. Because cardiac output from the right side of the heart approximates cardiac output from the left side of the heart, blood flow to the lungs matches blood flow to the systemic circulation. However, pressure and vascular resistance in the blood vessels in the lungs are different than in the system circulation. The mean pressure in the pulmonary artery is ~15 mmHg (systolic pressure is ~25 mmHg and diastolic pressure is ~8 mmHg) compared to the mean pressure in the aorta of ~95 mmHg. The pressure in the left atrium where blood is returning to the heart from the lungs is ~5 mmHg; thus there is not a great pressure difference across the pulmonary circulation (15 – 5 mmHg). Figure 7.4 illustrates the differences in pressures between the pulmonary and systemic circulation.

Recalling the discussion of blood flow in the cardiovascular system from chapter 6, pressure = flow × resistance. Since blood flow to the lungs is equal to that of the systemic circulation, and there is a substantially lower change in pressure across the pulmonary vascular system, resistance is proportionally lower compared to that in the systemic circulation. This is reflected in differences in the anatomy of the vessels in the pulmonary versus systemic circulation: The pulmonary blood vessels are thin walled, with relatively little smooth muscle.

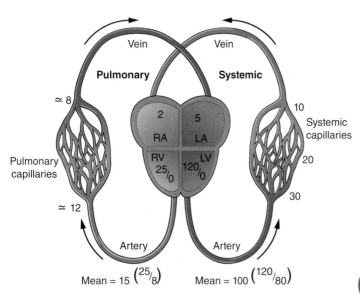

FIGURE 7.4 Comparison of pressures (mmHg) in the pulmonary and systemic circulations.

Reprinted, by permission, from J. West, 2000, *Respiratory physiology: The essentials* (Baltimore, MD: Lippincott, Williams, and Wilkins), 36.

Respiratory Membrane

Gas exchange between the air in the alveoli and the blood in the pulmonary capillaries occurs across the **respiratory membrane** (also called the alveolar-capillary membrane). This membrane, depicted in figure 7.5, is composed of

- the alveolar wall,
- the capillary wall, and
- their respective basement membranes.

The primary function of these membranous surfaces is for gas exchange. The respiratory membrane is very thin, measuring only 0.5 to 4 μm. As a result, the gases in the nearly 300 million alveoli are in close proximity to the blood circulating through the capillaries.

Partial Pressures of Gases

The air we breathe is a mixture of gases. Each exerts a pressure in proportion to its concentration in the gas mixture. The individual pressures from each gas in a mixture are referred to as **partial pressures**. According to **Dalton's law**, the total pressure of a mixture of gases equals the sum of the partial pressures of the individual gases in that mixture.

Consider the air we breathe. It is composed of 79.04% nitrogen (N_2), 20.93% oxygen (O_2), and 0.03% carbon dioxide (CO_2). These percentages remain constant regardless of altitude. At sea level, the atmospheric (or barometric) pressure is approximately 760 mmHg, which is also referred to as standard atmospheric pressure. Thus, if the total atmospheric pressure is 760 mmHg, then the partial pressure of nitrogen (PN_2) in air is 600.7 mmHg (79.04% of the total 760 mmHg pressure). Oxygen's partial pressure (PO_2) is 159.1 mmHg (20.93% of 760 mmHg), and carbon dioxide's partial pressure (PCO_2) is 0.2 mmHg (0.03% of 760 mmHg).

In the human body, gases are usually dissolved in fluids, such as blood plasma. According to **Henry's law**, gases dissolve in liquids in proportion to their partial pressures, depending also on their solubilities in the specific fluids and on the temperature. A gas's solubility in blood is a constant, and blood temperature also remains relatively constant at rest. Thus, the most critical factor for gas exchange between the alveoli and the blood is the pressure gradient between the gases in the two areas.

In focus

Dalton's law states that the total pressure of a mixture of gases equals the sum of the partial pressures of the individual gases in that mixture.

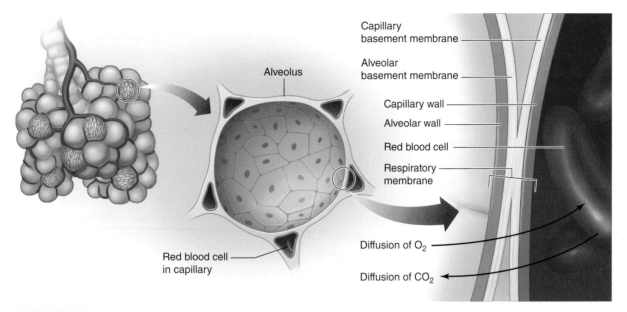

FIGURE 7.5 The anatomy of the respiratory membrane, showing the exchange of oxygen and carbon dioxide between an alveolus and pulmonary capillary blood.

Gas Exchange in the Alveoli

Differences in the partial pressures of the gases in the alveoli and the gases in the blood create a pressure gradient across the respiratory membrane. This forms the basis of gas exchange during pulmonary diffusion. If the pressures on each side of the membrane were equal, the gases would be at equilibrium and would not move. But the pressures are not equal, so gases move according to partial pressure gradients.

Oxygen Exchange

The PO_2 of air outside the body at standard atmospheric pressure is 159 mmHg. But this pressure decreases to about 105 mmHg when air is inhaled and enters the alveoli, where it is moistened and mixes with the air in the alveoli. The alveolar air is saturated with water vapor (which has its own partial pressure) and contains more carbon dioxide than the inspired air. Both the increased water vapor pressure and increased partial pressure of carbon dioxide contribute to the total pressure in the alveoli. Fresh air that ventilates the lungs is constantly mixed with the air in the alveoli while some of the alveolar gases are exhaled to the environment. As a result, alveolar gas concentrations remain relatively stable.

The blood, stripped of much of its oxygen by the metabolic needs of the tissues, typically enters the pulmonary capillaries with a PO_2 of about 40 mmHg (see figure 7.6). This is about 60 to 65 mmHg less than the PO_2 in the alveoli. In other words, the pressure gradient

for oxygen across the respiratory membrane is typically about 65 mmHg. As noted earlier, this pressure gradient drives the oxygen from the alveoli into the blood to equilibrate the pressure of the oxygen on each side of the membrane.

The PO_2 in the alveoli stays relatively constant at about 105 mmHg. As the deoxygenated blood enters the pulmonary artery, the PO_2 in the blood is only about 40 mmHg. But as the blood moves along the pulmonary capillaries, gas exchange occurs. By the time the pulmonary blood reaches the venous end of these capillaries, the PO_2 in the blood equals that in the alveoli (approximately 105 mmHg), and the blood is now considered to be saturated with oxygen at its full carrying capacity. The blood leaving the lungs through the pulmonary veins and subsequently returning to the systemic (left) side of the heart has a rich supply of oxygen to deliver to the tissues. Notice, however, that the PO_2 in the pulmonary vein is 100 mmHg, not the 105 mmHg found in the alveolar air and pulmonary capillaries. This difference is attributable to the fact that about 2% of the blood is shunted from the aorta directly to the lung to meet the oxygen needs of the lung itself. This blood has a lower PO_2 and reenters the pulmonary vein along with fully saturated blood returning to the left atrium that has just completed gas exchange. This blood mixes and thus decreases the PO_2 of the blood returning to the heart.

Diffusion through tissues is described by **Fick's law** (figure 7.7). Fick's law states that the rate of diffusion through a tissue such as the respiratory membrane is

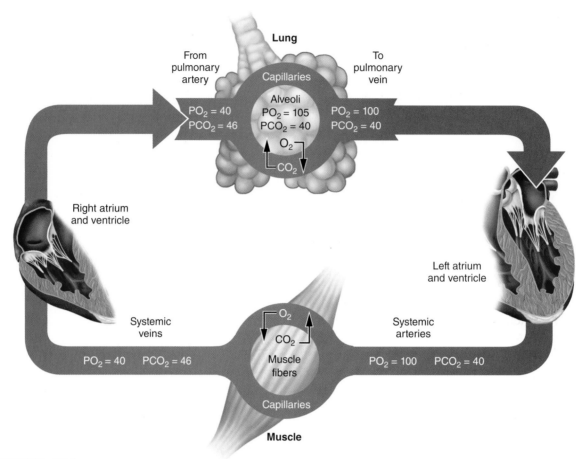

FIGURE 7.6 Partial pressure of oxygen (PO_2) and carbon dioxide (PCO_2) in blood as a result of gas exchange in the lungs and gas exchange between the capillary blood and tissues.

FIGURE 7.7 Diffusion through a sheet of tissue. The amount of gas ($\dot{V}_{gas}$) transferred is proportional to the area (A), a diffusion constant (D), and the difference in partial pressure ($P_1 - P_2$) and is inversely proportional to the thickness (T). The constant is proportional to the gas solubility (Sol) but inversely proportional to the square root of its molecular weight (MW).

Reprinted, by permission, from J. West, 2000, *Respiratory physiology: The essentials* (Baltimore, MD: Lippincott, Williams, and Wilkins), 26.

proportional to the surface area and the difference in the partial pressure of gas between the two sides of the tissue. The rate of diffusion is also inversely proportional to the thickness of the tissue in which the gas must diffuse. Additionally, the diffusion constant, which is unique to each gas, influences the rate of diffusion across the tissue. Carbon dioxide has a much lower diffusion constant than oxygen; therefore, even though there is not as great a difference between alveolar and capillary partial pressure of carbon dioxide as there is for oxygen, carbon dioxide still diffuses easily.

In focus

The greater the pressure gradient across the respiratory membrane, the more rapidly oxygen diffuses across it.

The rate at which oxygen diffuses from the alveoli into the blood is referred to as the **oxygen diffusion capacity** and is expressed as the volume of oxygen that diffuses through the membrane each minute for a pressure difference of 1 mmHg. At rest, the oxygen diffusion capacity is about 21 ml of oxygen per minute per 1 mmHg of pressure difference between the alveoli and the pulmonary capillary blood. Although the partial pressure gradient between venous blood coming into the lung and the alveolar air is about 65 mmHg (105 mmHg – 40 mmHg), the oxygen diffusion capacity is calculated on the basis of the mean pressure in the pulmonary capillary, which has a substantially higher PO_2. The gradient between the mean partial pressure of the pulmonary capillary and the alveolar air is approximately 11 mmHg, which would provide a diffusion of 231 ml of oxygen per minute through the respiratory membrane. During maximal exercise, the oxygen diffusion capacity may increase by up to three times the resting rate, because blood is returning to the lungs severely desaturated and thus there is a greater partial pressure gradient from the alveoli to the blood. In fact, rates of more than 80 ml/min have been observed among highly trained athletes.

The increase in oxygen diffusion capacity from rest to exercise is caused by a relatively inefficient, sluggish circulation through the lungs at rest, which results primarily from limited perfusion of the upper regions of the lungs attributable to gravity. If the lung is divided into three zones as depicted in figure 7.8, at rest only the bottom third (zone 3) of the lung is perfused with blood. During exercise, however, blood flow through the lungs is greater, primarily as a result of elevated blood pressure, which increases lung perfusion.

Carbon Dioxide Exchange

Carbon dioxide, like oxygen, moves along a partial pressure gradient. As shown in figure 7.6, the blood passing from the right side of the heart through the alveoli has a PCO_2 of about 46 mmHg. Air in the alveoli has a PCO_2 of about 40 mmHg. Although this results

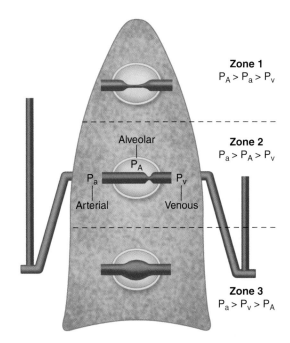

FIGURE 7.8 Explanation of the uneven distribution of blood flow in the lung.

Reprinted, by permission, from J. West, 2000, *Respiratory physiology: The essentials* (Baltimore, MD: Lippincott, Williams, and Wilkins), 44.

in a relatively small pressure gradient of only about 6 mmHg, it is more than adequate to allow for exchange of CO_2. Carbon dioxide's diffusion coefficient is 20 times greater than that of oxygen, so CO_2 can diffuse across the respiratory membrane much more rapidly.

The partial pressures of gases involved in pulmonary diffusion are summarized in table 7.1. Note that the total pressure in the venous blood is only 706 mmHg, 54 mmHg lower than the total pressure in dry air and alveolar air. This is the result of a much greater decrease in PO_2 compared with the increase in PCO_2 as the blood goes through the body's tissues.

TABLE 7.1 Partial Pressures of Respiratory Gases at Sea Level

		PARTIAL PRESSURE (mmHG)				
Gas	% in dry air	Dry air	Alveolar air	Arterial blood	Venous blood	Diffusion gradient
H_2O	0	0	47	47	47	0
O_2	20.93	159.1	105	100	40	60
CO_2	0.03	0.2	40	40	46	6
N_2	79.04	600.7	568	573	573	0
Total	**100.00**	**760**	**760**	**760**	**706[a]**	**0**

[a]See text for an explanation of the decrease in total pressure.

In review

➤ Pulmonary diffusion is the process by which gases are exchanged across the respiratory membrane in the alveoli.

➤ The amount and rate of gas exchange that occur across the membrane depend primarily on the partial pressure of each gas, although other factors are also important, as shown by Fick's law. Gases diffuse along a pressure gradient, moving from an area of higher pressure to one of lower pressure. Thus, oxygen enters the blood and carbon dioxide leaves it.

➤ Oxygen diffusion capacity increases as one moves from rest to exercise. When exercising muscles require more oxygen to be used in the metabolic processes, venous oxygen is depleted and oxygen exchange at the alveoli is facilitated.

➤ The pressure gradient for carbon dioxide exchange is less than for oxygen exchange, but carbon dioxide's diffusion coefficient is 20 times greater than that of oxygen, so carbon dioxide crosses the membrane readily without a large pressure gradient.

Transport of Oxygen and Carbon Dioxide in the Blood

We have considered how air moves into and out of the lungs via pulmonary ventilation and how gas exchange occurs via pulmonary diffusion. Next we consider how gases are transported in the blood to deliver oxygen to the tissues and to remove the carbon dioxide that the tissues produce.

Oxygen Transport

Oxygen is transported by the blood either combined with hemoglobin in the red blood cells (greater than 98%) or dissolved in the blood plasma (less than 2%). Only about 3 ml of oxygen is dissolved in each liter of plasma. Assuming a total plasma volume of 3 to 5 L, only about 9 to 15 ml of oxygen can be carried in the dissolved state. This limited amount of oxygen cannot adequately meet the needs of even resting body tissues, which generally require more than 250 ml of oxygen per minute (depending on body size). However, hemoglobin, a protein contained within each of the body's 4 to 6 billion red blood cells, allows the blood to transport nearly 70 times more oxygen than can be dissolved in plasma.

Hemoglobin Saturation

As just noted, over 98% of oxygen is transported in the blood bound to hemoglobin. Each molecule of hemoglobin can carry four molecules of oxygen. When oxygen binds to hemoglobin, it forms oxyhemoglobin; hemoglobin that is not bound to oxygen is referred to as deoxyhemoglobin. The binding of oxygen to hemoglobin depends on the PO_2 in the blood and the bonding strength, or affinity, between hemoglobin and oxygen. The curve in figure 7.9 is an oxygen–hemoglobin dissociation curve, which shows the amount of hemoglobin saturated with oxygen at different PO_2 values. The shape of the curve is extremely important for its function in the body. The relatively flat upper portion means that, at high PO_2 concentrations such as in the lungs, large drops in PO_2 result in only small changes in hemoglobin saturation. This is called the "loading" portion of the curve. A high blood PO_2 results in almost complete hemoglobin saturation, which means that the maximal amount of oxygen is bound. But as the PO_2 decreases, so does hemoglobin saturation.

The steep portion of the curve coincides with PO_2 values typically found in the tissues of the body. Here, relatively small changes in PO_2 result in large changes in saturation. This is also advantageous because this is the "unloading" portion of the curve where hemoglobin loses its oxygen to the tissues.

Many factors determine the hemoglobin saturation. If, for example, the blood becomes more acidic, the

FIGURE 7.9 Oxyhemoglobin dissociation curve.

Reprinted from S.K. Powers and E.T. Howley, 2004, *Exercise physiology: Theory and application to fitness and performance*, 5th ed. (New York: McGraw-Hill Companies), 205. With permission from The McGraw-Hill Companies.

dissociation curve shifts to the right. This indicates that more oxygen is being unloaded from the hemoglobin at the tissue level. This rightward shift of the curve (see figure 7.10a), attributable to a decline in pH, is referred to as the Bohr effect. The pH in the lungs is generally high, so hemoglobin passing through the lungs has a strong affinity for oxygen, encouraging high saturation.

a Effects of changing pH

b Effects of blood changing temperature

FIGURE 7.10 The effects of changing blood pH and blood temperature on the oxyhemoglobin dissociation curve.

Reprinted, by permission, from S.K. Powers and E.T. Howley, 2004, *Exercise physiology: Theory and application to fitness and performance*, 5th ed. (New York: McGraw-Hill Companies), 206. With permission from The McGraw-Hill Companies.

At the tissue level, especially during exercise, the pH is lower, causing oxygen to dissociate from hemoglobin, thereby supplying oxygen to the tissues. With exercise, the ability to unload oxygen to the muscles increases as the muscle pH decreases.

Blood temperature also affects oxygen dissociation. As shown in figure 7.10b, increased blood temperature shifts the dissociation curve to the right, indicating that oxygen is unloaded from hemoglobin more readily at higher temperatures. Because of this, the hemoglobin unloads more oxygen when blood circulates through the metabolically heated active muscles.

In focus

Increased temperature and hydrogen ion (H+) concentration (lowered pH) in an exercising muscle shift the oxygen dissociation curve rightward, allowing more oxygen to be unloaded to supply the active muscle. Because of the sigmoid shape of the curve, loading of hemoglobin with oxygen in the lungs is only minimally affected by the shift.

Blood Oxygen-Carrying Capacity

The oxygen-carrying capacity of blood is the maximal amount of oxygen the blood can transport. It depends primarily on the blood hemoglobin content. Each 100 ml of blood contains an average of 14 to 18 g of hemoglobin in men and 12 to 16 g in women. Each gram of hemoglobin can combine with about 1.34 ml of oxygen, so the oxygen-carrying capacity of blood is approximately 16 to 24 ml per 100 ml of blood when blood is fully saturated with oxygen. At rest, as the blood passes through the lungs, it is in contact with the alveolar air for approximately 0.75 s. This is sufficient time for hemoglobin to become 98% to 99% saturated. At high intensities of exercise, the contact time is greatly reduced, which can reduce the binding of hemoglobin to oxygen and slightly decrease the saturation, although the unique S shape of the curve guards against large drops.

People with low hemoglobin concentrations, such as those with anemia, have reduced oxygen-carrying capacities. Depending on the severity of the condition, these people might feel few effects of anemia while they are at rest because their cardiovascular system can compensate for reduced blood oxygen content by increasing cardiac output. However, during activities in which oxygen delivery can become a limitation, such as highly intense aerobic effort, reduced blood oxygen content limits performance.

Carbon Dioxide Transport

Carbon dioxide also relies on the blood for transportation. Once carbon dioxide is released from the cells, it is carried in the blood primarily in three forms:

- As bicarbonate ions resulting from the dissociation of carbonic acid
- Dissolved in plasma
- Bound to hemoglobin (called carbaminohemoglobin)

Bicarbonate Ion

The majority of carbon dioxide is carried in the form of bicarbonate ion. Bicarbonate accounts for the transport of 60% to 70% of the carbon dioxide in the blood. Carbon dioxide and water molecules combine to form carbonic acid (H_2CO_3). This reaction is catalyzed by the enzyme carbonic anhydrase, which is found in red blood cells. Carbonic acid is unstable and quickly dissociates, freeing a hydrogen ion (H^+) and forming a bicarbonate ion (HCO_3^-):

$$CO_2 + H_2O \rightarrow H_2CO_3 \rightarrow H^+ + HCO_3^-$$

The H^+ subsequently binds to hemoglobin, and this binding triggers the Bohr effect, mentioned previously, which shifts the oxygen–hemoglobin dissociation curve to the right. The bicarbonate ion diffuses out of the red blood cell and into the plasma. In order to prevent electrical imbalance from the shift of the negatively charged bicarbonate ion into the plasma, a chloride ion diffuses from the plasma into the red blood cell. This is called the chloride shift.

In focus

The majority of carbon dioxide produced by the active muscle is transported back to the lungs in the form of bicarbonate ions.

Additionally, the formation of hydrogen ions through this reaction enhances oxygen unloading at the level of the tissue. Through this mechanism, hemoglobin acts as a buffer, binding and neutralizing the H^+ and thus preventing any significant acidification of the blood. Acid–base balance is discussed in more detail in chapter 8.

When the blood enters the lungs, where the PCO_2 is lower, the H^+ and bicarbonate ions rejoin to form carbonic acid, which then dissociates into carbon dioxide and water:

$$H^+ + HCO_3^- \rightarrow H_2CO_3 \rightarrow CO_2 + H_2O.$$

The carbon dioxide that is thus re-formed can enter the alveoli and be exhaled.

Dissolved Carbon Dioxide

Part of the carbon dioxide released from the tissues is dissolved in plasma; but only a small amount, typically just 7% to 10%, is transported this way. This dissolved carbon dioxide comes out of solution where the PCO_2 is low, as in the lungs. There it diffuses from the pulmonary capillaries into the alveoli to be exhaled.

Carbaminohemoglobin

Carbon dioxide transport also can occur when the gas binds with hemoglobin, forming carbaminohemoglobin. The compound is so named because carbon dioxide binds with amino acids in the globin part of the hemoglobin molecule, rather than with the heme group as oxygen does. Because carbon dioxide binding occurs on a different part of the hemoglobin molecule than does oxygen binding, the two processes do not compete. However, carbon dioxide binding varies with the oxygenation of the hemoglobin (deoxyhemoglobin binds carbon dioxide more easily than oxyhemoglobin) and the partial pressure of CO_2. Carbon dioxide is released from hemoglobin when PCO_2 is low as it is in the lungs. Thus, carbon dioxide is readily released from the hemoglobin in the lungs, allowing it to enter the alveoli to be exhaled.

In review

➤ Oxygen is transported in the blood primarily bound to hemoglobin (as oxyhemoglobin), although a small part of it is dissolved in plasma.

➤ To better respond to increased oxygen demand, hemoglobin unloading of oxygen (desaturation) is enhanced when

  • PO_2 decreases,
  • pH decreases, or
  • temperature increases.

➤ In the arteries, hemoglobin is usually about 98% saturated with oxygen. This is a higher oxygen content than our bodies require, so the blood's oxygen-carrying capacity seldom limits performance in healthy individuals.

➤ Carbon dioxide is transported in the blood primarily as bicarbonate ion. This prevents the formation of carbonic acid, which can cause H+ to accumulate and lower the pH. Smaller amounts of carbon dioxide are either dissolved in the plasma or bound to hemoglobin.

Gas Exchange at the Muscles

We have considered how the respiratory and cardiovascular systems bring air into our lungs, exchange oxygen and carbon dioxide in the alveoli, and transport oxygen to the muscles and carbon dioxide to the lungs. We now consider the delivery of oxygen from the capillary blood to the muscle tissue.

Arterial–Venous Oxygen Difference

At rest, the oxygen content of arterial blood is about 20 ml of oxygen per 100 ml of blood. As shown in figure 7.11a, this value decreases to 15 to 16 ml of oxygen per 100 ml after the blood has passed through the capillaries into the venous system. This difference in oxygen content between arterial and venous blood is referred to as the **arterial–mixed venous oxygen difference**, or **(a-v̄)O$_2$ difference**. The term *mixed venous* (v̄) refers to the oxygen content of blood in the right atrium, which comes from all parts of the body, both active and inactive. The difference between arterial and mixed venous oxygen content reflects the 4 to 5 ml of oxygen per 100 ml of blood taken up by the tissues. The amount of oxygen taken up is proportional to its use for oxidative energy production. Thus, as the rate of oxygen use increases, the (a-v̄)O$_2$ difference also increases. It can increase to 15 to 16 ml per 100 ml of

a Muscle at rest

b Muscle during intense aerobic exercise

FIGURE 7.11 The arterial–mixed venous oxygen difference, or (a-v̄)O$_2$ difference, across the muscle.

blood during maximal levels of endurance exercise (figure 7.11*b*). However, at the level of the contracting muscle, the **(a-v)O₂ difference** during intense exercise can increase to 17 to 18 ml per 100 ml of blood. Note that there is not a bar over the *v* in this instance because we are now looking at local muscle venous blood, not mixed venous blood in the right atrium. During intense exercise, more oxygen is unloaded to the active muscles because the PO_2 in the muscles is substantially lower than in arterial blood.

In focus

The (a-v̄)O₂ difference increases from a resting value of about 4 to 5 ml per 100 ml of blood up to values of 15 to 16 ml per 100 ml of blood during intense exercise. This increase reflects an increased extraction of oxygen from arterial blood by active muscle, thus decreasing the oxygen content of the venous blood. It is important to remember that the blood returning to the right atrium is coming from all parts of the body, active and inactive. Therefore, the mixed venous oxygen content will not decrease to values much lower than 4 to 5 ml of oxygen per 100 ml of venous blood.

Oxygen Transport in the Muscle

Oxygen is transported in the muscle to the mitochondria by a molecule called **myoglobin** where it is used in oxidative metabolism. Myoglobin is similar in structure to hemoglobin, but myoglobin has a much greater affinity for oxygen than hemoglobin. This concept is illustrated in figure 7.12. At PO_2 values less than 20, the myoglobin dissociation curve is much steeper than the dissociation curve for hemoglobin. Myoglobin releases its oxygen content only under conditions in which the PO_2 is very low. Note from figure 7.12 that at a PO_2 at which venous blood is unloading oxygen, myoglobin is loading oxygen. It is estimated that the PO_2 in the mitochondria of an exercising muscle may be as low as 1 to 2 mmHg; thus myoglobin readily delivers oxygen to the mitochondria.

In focus

Myoglobin releases its oxygen only at a very low PO_2. This is compatible with the PO_2 found in exercising muscle, which may be as low as 1 to 2 mmHg.

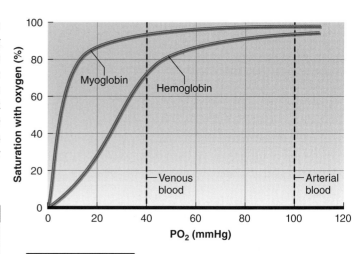

FIGURE 7.12 A comparison of the dissociation curves for myoglobin and hemoglobin.

Reprinted from S.K. Powers and E.T. Howley, 2004, *Exercise physiology: Theory and application to fitness and performance,* 5th ed. (New York: McGraw-Hill Companies), 207. With permission from The McGraw-Hill Companies.

Factors Influencing Oxygen Delivery and Uptake

The rates of oxygen delivery and uptake depend on three major variables:

- Oxygen content of blood
- Blood flow
- Local conditions (e.g., pH, temperature)

With exercise, each of these variables is adjusted to ensure increased oxygen delivery to active muscle. Under normal circumstances, hemoglobin is about 98% saturated with oxygen. Any reduction in the blood's normal oxygen-carrying capacity would hinder oxygen delivery and reduce cellular uptake of oxygen. Likewise, a reduction in the PO_2 of the arterial blood would lower the partial pressure gradient, limiting the unloading of oxygen at the tissue level. Exercise increases blood flow through the muscles. As more blood carries oxygen through the muscles, less oxygen must be removed from each 100 ml of blood (assuming the demand is unchanged). Thus, increased blood flow improves oxygen delivery.

Many local changes in the muscle during exercise affect oxygen delivery and uptake. For example, muscle activity increases muscle acidity because of lactate production. Also, muscle temperature and carbon dioxide concentration both increase because of increased metabolism. All these changes increase oxygen unloading from the hemoglobin molecule, facilitating oxygen delivery and uptake by the muscles.

Carbon Dioxide Removal

Carbon dioxide exits the cells by simple diffusion in response to the partial pressure gradient between the tissue and the capillary blood. For example, muscles generate carbon dioxide through oxidative metabolism, so the PCO_2 in muscles is relatively high compared with that in the capillary blood. Consequently, CO_2 diffuses out of the muscles and into the blood to be transported to the lungs.

In review

> - Within muscle, oxygen is transported to the mitochondria by a molecule called myoglobin. Compared to the oxyhemoglobin dissociation curve, the myoglobin-O_2 dissociation curve is much steeper at low PO_2 values.

> - The (a-$\bar{v}$)O_2 difference is the difference in the oxygen content of arterial and mixed venous blood throughout the body. This measure reflects the amount of oxygen taken up by the tissues.

> - Oxygen delivery to the tissues depends on the oxygen content of the blood, blood flow to the tissues, and local conditions (e.g., tissue temperature and PO_2).

> - Carbon dioxide exchange at the tissues is similar to oxygen exchange, except that carbon dioxide leaves the muscles, where it is formed, and enters the blood to be transported to the lungs for clearance.

Regulation of Pulmonary Ventilation

Maintaining homeostatic balance in blood PO_2, PCO_2, and pH requires a high degree of coordination between the respiratory and circulatory systems. Much of this coordination is accomplished by involuntary regulation of pulmonary ventilation. This control is not yet fully understood, although many of the intricate neural controls have been identified.

The respiratory muscles are under the direct control of motor neurons, which are in turn regulated by **respiratory centers** (inspiratory and expiratory) located within the brain stem (in the medulla oblongata and pons). These centers establish the rate and depth of breathing by sending out periodic impulses to the respiratory muscles. The cortex can override these centers if voluntary control of respiration is desired.

Additionally, input from other parts of the brain occurs under certain conditions.

The inspiratory area of the brain (dorsal respiratory group) contains cells that intrinsically fire and control the basic rhythm of ventilation. The expiratory area is quiet during normal quiet breathing (recall that expiration is a passive process at rest). However, during forceful breathing such as during exercise, the expiratory area actively sends signals to the muscles of expiration. Two other brain centers aid in the control of respiration. The apneustic area has an excitatory effect on the inspiratory center and results in prolonged firing of the inspiratory neurons. Finally, the pneumotaxic center inhibits or "switches off" inspiration, helping to regulate inspiratory volume.

The respiratory centers do not act alone in controlling breathing. Breathing also is regulated and modified by the changing chemical environment in the body. For example, sensitive areas in the brain respond to changes in carbon dioxide and H^+ levels. The central chemoreceptors in the brain are stimulated by an increase in H^+ ions in the cerebrospinal fluid. The blood–brain barrier is relatively impermeable to H^+ ions or bicarbonate. However, CO_2 readily diffuses across the blood–brain barrier and then reacts to increase H^+ ions. This, in turn, stimulates the inspiratory center, which then activates the neural circuitry to increase the rate and depth of respiration. This increase in respiration, in turn, increases the removal of carbon dioxide and H^+.

Chemoreceptors in the aortic arch (the aortic bodies) and in the bifurcation of the common carotid artery (the carotid bodies) are sensitive primarily to blood changes in PO_2 but also respond to changes in H^+ concentration and PCO_2. The carotid chemoreceptors are more sensitive to changes in H^+ concentrations and PCO_2. Overall, PCO_2 appears to be the strongest stimulus for the regulation of breathing. When carbon dioxide levels become too high, carbonic acid forms, then quickly dissociates, giving off H^+. If H^+ accumulates, the blood becomes too acidic (pH decreases). Thus, an increased PCO_2 stimulates the inspiratory center to increase respiration—not to bring in more oxygen but to rid the body of excess carbon dioxide and limit further pH changes.

In addition to the chemoreceptors, other neural mechanisms influence breathing. The pleurae, bronchioles, and alveoli in the lungs contain stretch receptors. When these areas are excessively stretched, that information is relayed to the expiratory center. The expiratory center responds by shortening the duration of an inspiration, which decreases the risk of overinflating the respiratory structures. This response is known as the Hering-Breuer reflex.

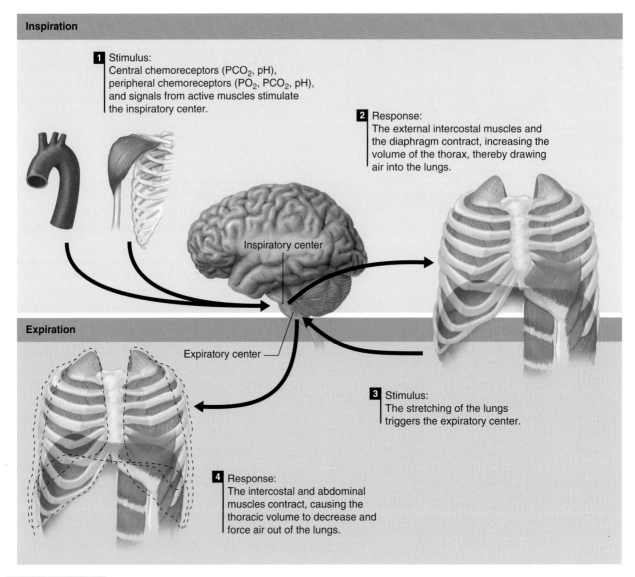

Inspiration

1 Stimulus:
Central chemoreceptors (PCO_2, pH),
peripheral chemoreceptors (PO_2, PCO_2, pH),
and signals from active muscles stimulate
the inspiratory center.

2 Response:
The external intercostal muscles and
the diaphragm contract, increasing the
volume of the thorax, thereby drawing
air into the lungs.

Inspiratory center

Expiration

Expiratory center

3 Stimulus:
The stretching of the lungs
triggers the expiratory center.

4 Response:
The intercostal and abdominal
muscles contract, causing the
thoracic volume to decrease and
force air out of the lungs.

FIGURE 7.13 An overview of the processes involved in respiratory regulation.

Many control mechanisms are involved in the regulation of breathing, as shown in figure 7.13. Such simple stimuli as emotional distress or an abrupt change in the temperature of the surroundings can affect breathing. But all these control mechanisms are essential. The goal of respiration is to maintain appropriate levels of the blood and tissue gases and to maintain proper pH for normal cellular function. Small changes in any of these, if not carefully controlled, could impair physical activity and jeopardize health.

In closing

In chapter 6, we discussed the role of the cardiovascular system during exercise. In this chapter we looked at the role played by the respiratory system. In the next chapter, we examine how the cardiovascular and respiratory systems respond to an acute bout of exercise.

Key Terms

alveoli

arterial–mixed venous oxygen difference, or (a-v̄)O$_2$ difference

arterial–venous oxygen difference, or (a-v)O$_2$ difference

Boyle's gas law

Dalton's law

expiration

external respiration

Fick's law

Henry's law

inspiration

internal respiration

myoglobin

oxygen diffusion capacity

partial pressure

pulmonary diffusion

pulmonary ventilation

residual volume (RV)

respiratory centers

respiratory membrane

respiratory pump

spirometry

tidal volume

total lung capacity (TLC)

vital capacity (VC)

Study Questions

1. Describe and differentiate between external and internal respiration.
2. Describe the mechanisms involved in inspiration and expiration.
3. What is a spirometer? Describe and define the lung volumes measured using spirometry.
4. Explain the concept of partial pressures of respiratory gases—oxygen, carbon dioxide, and nitrogen. What is the role of gas partial pressures in pulmonary diffusion?
5. Where in the lung does the exchange of gases occur with the blood? Describe the role of the respiratory membrane.
6. How are oxygen and carbon dioxide transported in the blood?
7. How is oxygen unloaded from the arterial blood to the muscle and carbon dioxide removed from the muscle into the venous blood?
8. What is meant by the arterial–mixed venous oxygen difference? How and why does this change from resting to exercise conditions?
9. Describe how pulmonary ventilation is regulated. What are the chemical stimuli that control the depth and rate of breathing? How do they control respiration during exercise?

Study Guide Activities

In addition to the activities listed in the chapter opening outline on page 163, two other activities are available in the web study guide, located at

www.HumanKinetics.com/PhysiologyOfSportAndExercise

 The **KEY TERMS** activity reviews important terms, and the end-of-chapter **QUIZ** tests your understanding of the material covered in the chapter.

Cardiorespiratory Responses to Acute Exercise

8

In this chapter and in the web study guide

 ACTIVITY 8.1 Cardiovascular Response to Exercise reviews cardiovascular changes occurring during exercise.

ACTIVITY 8.2 Cardiovascular Response Scenarios explores how cardiovascular responses contribute to real-life situations.

 ACTIVITY 8.3 Pulmonary Ventilation During Exercise investigates the response of pulmonary ventilation to exercise and the factors that affect the phases of pulmonary ventilation.

 ACTIVITY 8.4 Pulmonary Ventilation and Energy Metabolism reviews the key terms related to pulmonary ventilation and energy metabolism.

After reviewing the basic anatomy and physiology of the cardiovascular and respiratory systems, this chapter looks specifically at how these systems respond to the increased demands placed on the body during acute exercise. With exercise, oxygen demand by the active muscles increases significantly, and more nutrients are used. Metabolic processes speed up, so more waste products are created. During prolonged exercise or exercise in a hot environment, body temperature increases. In intense exercise, H^+ concentration increases in the muscles and blood, lowering their pH.

Cardiovascular Responses to Acute Exercise

Numerous interrelated cardiovascular changes occur during dynamic exercise. The primary goal of these adjustments is to increase blood flow to working muscle; however, cardiovascular control of virtually every tissue and organ in the body is also altered. To better understand the changes that occur, we must examine the function of both the heart and the peripheral circulation. In this section we examine changes in all components of the cardiovascular system from rest to acute exercise, looking specifically at the following:

- Heart rate
- Stroke volume
- Cardiac output
- Blood pressure
- Blood flow
- The blood

We then see how these changes are integrated to maintain adequate blood pressure and provide for the exercising body's needs.

Heart Rate

Heart rate (HR) is one of the simplest physiological responses to measure, and yet one of the most informative in terms of cardiovascular stress and strain.

Measuring HR involves simply taking the subject's pulse, usually at the radial or carotid artery. Heart rate is a good indicator of relative exercise intensity.

Resting Heart Rate

Resting heart rate (RHR) averages 60 to 80 beats/min in most individuals. In highly conditioned, endurance-trained athletes, resting rates as low as 28 to 40 beats/min have been reported. This is mainly due to an increase in parasympathetic (vagal) tone that accompanies endurance exercise training. Resting heart rate can also be affected by environmental factors; for example, it increases with extremes in temperature and altitude.

Just before the start of exercise, preexercise HR usually increases above normal resting values. This is called the anticipatory response. This response is mediated through release of the neurotransmitter norepinephrine from the sympathetic nervous system and the hormone epinephrine from the adrenal medulla. Vagal tone also decreases. Because preexercise HR is elevated, reliable estimates of the true RHR should be made only under conditions of total relaxation, such as early in the morning before the subject rises from a restful night's sleep.

In focus

Preexercise HR is not a reliable estimate of RHR because of the anticipatory HR response.

Heart Rate During Exercise

When exercise begins, HR increases directly in proportion to the increase in exercise intensity (figure 8.1), until near-maximal exercise is achieved. As maximal exercise intensity is approached, HR begins to plateau even as the exercise workload continues to increase. This indicates that HR is approaching a maximal value. The **maximum heart rate (HR_{max})** is the highest HR value achieved in an all-out effort to the point of volitional fatigue. Once accurately determined, HR_{max} is a highly reliable value that remains constant from day to day. However, this value changes slightly from year to year due to a normal age-related decline.

FIGURE 8.1 Changes in heart rate (HR) as a subject progressively walks, jogs, then runs on a treadmill as intensity is increased. Heart rate is plotted against exercise intensity shown as a percent of the subject's $\dot{V}O_{2max}$, at which point the rise in HR begins to plateau. The HR at this plateau is the subject's maximal HR or HR_{max}.

HR_{max} is often estimated based on age because HR_{max} shows a slight but steady decrease of about one beat per year beginning at 10 to 15 years of age. Subtracting one's age from 220 beats/min provides a reasonable approximation of one's predicted HR_{max}. However, this is only an estimate—individual values vary considerably from this average value. To illustrate, for a 40-year-old woman, HR_{max} would be estimated to be 180 beats/min (HR_{max} = 220 – 40 beats/min). However, 68% of all 40-year-olds have actual HR_{max} values between 168 and 192 beats/min (mean ± 1 standard deviation), and 95% fall between 156 and 204 beats/min (mean ± 2 standard deviations). This demonstrates the potential for error in estimating a person's HR_{max}. A similar but more accurate equation has been developed to estimate HR_{max} from age. In this equation, HR_{max} = 208 – (0.7 × age).[5]

In focus

To estimate HR_{max}:

$$HR_{max} = 220 - \text{age in years}$$
or
$$HR_{max} = 208 - (0.7 \times \text{age in years}).$$

When the exercise intensity is held constant at any submaximal workload, HR increases fairly rapidly until it reaches a plateau. This plateau is the **steady-state heart rate**, and it is the optimal HR for meeting the circulatory demands at that specific rate of work. For each subsequent increase in intensity, HR will reach a new steady-state value within 2 to 3 min. However, the more intense the exercise, the longer it takes to achieve this steady-state value.

The Fick Principle and the Fick Equation

In the 1870s, a cardiovascular physiologist by the name of Adolph Fick developed a principle critical to our understanding of the basic relationship between metabolism and cardiovascular function. In its simplest form, the Fick principle states that the oxygen consumption of a tissue is dependent on blood flow to that tissue and the amount of oxygen extracted from the blood by the tissue. This principle can be applied to the whole body or to regional circulations. Oxygen consumption is the product of blood flow and the difference in concentration of oxygen in the blood between the arterial blood supplying the tissue and the venous blood draining out of the tissue—the $(a-\bar{v})O_2$ difference. Whole-body oxygen consumption ($\dot{V}O_2$) is calculated as the product of the cardiac output ($\dot{Q}$) and $(a-\bar{v})O_2$ difference.

Fick equation:

$$\dot{V}O_2 = \dot{Q} \times (a-\bar{v})O_2 \text{ diff,}$$

which can be rewritten as

$$\dot{V}O_2 = HR \times SV \times (a-\bar{v})O_2 \text{ diff.}$$

This basic relationship is an important concept in exercise physiology and comes up frequently throughout the remainder of this book.

The concept of steady-state heart rate forms the basis for simple exercise tests that have been developed to estimate cardiorespiratory (aerobic) fitness. In one such test, individuals are placed on an exercise device, such as a cycle ergometer, and then perform exercise at two or three standardized exercise intensities. Those with better cardiorespiratory endurance capacity will have a lower steady-state HR at each exercise intensity than those who are less fit. Thus, a lower steady-state HR at a fixed exercise intensity is a valid predictor of greater cardiorespiratory fitness.

Figure 8.2 illustrates results from a submaximal graded exercise test performed on a cycle ergometer by two different individuals of the same age. Steady-state HR is measured at three or four distinct workloads, and a line of best fit is drawn through the data points. Because there is a consistent relation between exercise intensity and energy demand, steady-state HR can be plotted against the corresponding energy ($\dot{V}O_2$) required to do work on the cycle ergometer. The resultant line can be extrapolated to the age-predicted HR_{max} to estimate an individual's maximal exercise capacity. In this figure, subject A has a higher fitness level than subject B because (1) at any given submaximal intensity, his HR is lower; and (2) extrapolation to age-predicted HR_{max} yields a higher estimated maximal exercise capacity ($\dot{V}O_{2max}$).

FIGURE 8.2 The increase in heart rate with increasing power output on a cycle ergometer and oxygen uptake is linear within a wide range. The predicted maximal oxygen uptake can be extrapolated using the subject's estimated maximum heart rate as demonstrated here for two subjects with similar estimated maximum heart rates but quite different maximal workloads and $\dot{V}O_{2max}$ values.

Reprinted, by permission, from P.O. Åstrand et al., 2003, *Textbook of work physiology*, 4th ed. (Champaign, IL: Human Kinetics), 285.

Stroke Volume

Stroke volume (SV) also changes during acute exercise to allow the heart to meet the demands of exercise. At near-maximal and maximal exercise intensities, as heart rate approaches its maximum, SV is a major determinant of cardiorespiratory endurance capacity.

Stroke volume is determined by four factors:

1. The volume of venous blood returned to the heart (the heart can only pump what returns)
2. Ventricular distensibility (the capacity to enlarge the ventricle, to allow maximal filling)
3. Ventricular contractility (the inherent capacity of the ventricle to contract forcefully)
4. Aortic or pulmonary artery pressure (the pressure against which the ventricles must contract)

The first two factors influence the filling capacity of the ventricle, determining how much blood fills the ventricle and the ease with which the ventricle is filled at the available pressure. Together, these factors determine the end-diastolic volume (EDV), sometimes referred to as the **preload**. The last two characteristics influence the ventricle's ability to empty during systole, determining the force with which blood is ejected and the pressure against which it must be expelled into the arteries. The latter factor, the aortic mean pressure, which represents resistance to blood being ejected from the left ventricle (and to a less important extent, the pulmonary artery pressure resistance to flow from the right ventricle), is referred to as the **afterload**. These four factors combine to determine the SV during acute exercise.

Stroke Volume Increase With Exercise

Stroke volume increases above resting values during exercise. Most researchers agree that SV increases with increasing exercise intensity up to intensities somewhere between 40% and 60% of $\dot{V}O_{2max}$. At that point, SV typically plateaus, remaining essentially unchanged up to and including the point of exhaustion as shown in figure 8.3. However, other researchers have reported that SV continues to increase beyond 40% to 60% $\dot{V}O_{2max}$, even up through maximal exercise intensities. This is discussed in more detail in the sidebar on page 187.

When the body is in an upright position, SV can approximately double from resting to maximal values. For example, in active but untrained individuals, SV increases from about 60 to 70 ml/beat at rest to 110 to 130 ml/beat during maximal exercise. In highly trained endurance athletes, SV can increase from 80 to 110

FIGURE 8.3 Changes in stroke volume (SV) as a subject exercises on a treadmill at increasing intensities. Stroke volume is plotted as a function of percent $\dot{V}O_{2max}$. The SV increases with increasing intensity up to approximately 40% to 60% of $\dot{V}O_{2max}$, before reaching a maximum (SV_{max}).

ml/beat at rest to 160 to 200 ml/beat during maximal exercise. During supine exercise, such as recumbent cycling, SV also increases but usually by only about 20% to 40%—not nearly as much as in an upright position. Why does body position make such a difference?

When the body is in the supine position, blood does not pool in the lower extremities. Blood returns more easily to the heart in a supine posture, which means that resting SV values are higher in the supine position than in the upright position. Thus, the increase in SV with maximal exercise is not as great in the supine position as in the upright position because SV starts out higher. Interestingly, the highest SV attainable in upright exercise is only slightly greater than the resting value in the reclining position. The majority of the SV increase during low to moderate intensities of exercise in the upright position appears to be compensating for the force of gravity that causes blood to pool in the extremities.

Explanations for the Stroke Volume Increase

One explanation for the increase in SV with exercise is that the primary factor determining SV is increased preload or the extent to which the ventricle fills with blood and stretches, that is, the EDV. When the ventricle stretches more during filling, it subsequently contracts more forcefully. For example, when a larger volume of

blood enters and fills the ventricle during diastole, the ventricular walls stretch to a greater extent. To eject that greater volume of blood, the ventricle responds by contracting more forcefully. This is referred to as the **Frank-Starling mechanism**. At the level of the muscle fiber, the greater the stretch of the myocardial cells, the more actin-myosin cross-bridges formed, and greater force is developed.

Additionally, SV will increase during exercise if the ventricle's contractility (an inherent property of the ventricle) is enhanced. Contractility can increase by increasing sympathetic nerve stimulation or circulating catecholamines (epinephrine, norepinephrine), or both. An improved force of contraction can increase SV with or without an increased EDV by increasing the ejection fraction. Finally, when mean arterial blood pressure is low, SV is greater since there is less resistance to outflow into the aorta. These mechanisms all combine to determine the SV at any given intensity of dynamic exercise.

Stroke volume is much more difficult to measure than HR. Some clinically used cardiovascular diagnostic techniques have made it possible to determine exactly how SV changes with exercise. Echocardiography (using sound waves) and radionuclide techniques ("tagging" of red blood cells with radioactive tracers) have elucidated how the heart chambers respond to increasing oxygen demands during exercise. With either technique, continuous images of the heart can be taken at rest and up to near-maximal intensities of exercise.

Figure 8.4 illustrates the results of one study of normally active but untrained subjects.[3] In this study, participants were tested during both supine and upright cycle ergometry at rest and at three exercise intensities, which are depicted on the x-axis of figure 8.4.

Going from resting conditions to exercise of increasing intensity, there is an increase in left ventricular EDV (a greater filling or preload), which serves to increase SV through the Frank-Starling mechanism. There is also a decrease in the left ventricular end-systolic volume (greater emptying), indicating an increased force of contraction.

This figure shows that both the Frank-Starling mechanism and increased contractility are important in increasing SV during exercise. The Frank-Starling mechanism appears to have its greatest influence at lower exercise intensities, and improved contractile force becomes more important at higher exercise intensities.

Recall that HR also increases with exercise intensity. The plateau or small decrease in left ventricular EDV at high exercise intensities could be caused by a reduced ventricular filling time due to the high HR. One study

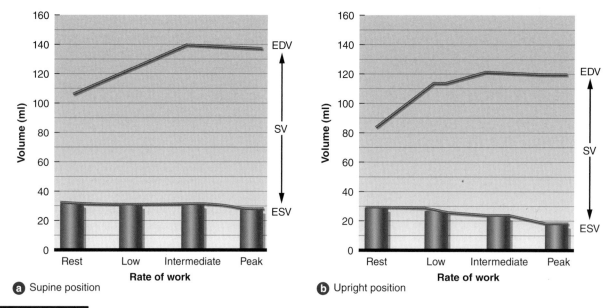

FIGURE 8.4 Changes in left ventricular end-diastolic volume (EDV), end-systolic volume (ESV), and stroke volume (SV) at rest and during low-, intermediate-, and peak-intensity exercise when the subject is in the supine and upright positions. Note that SV = EDV − ESV.

Reprinted, by permission, from L.R. Poliner et al., 1980, "Left ventricular performance in normal subjects: A comparison of the responses to exercise in the upright and supine position," *Circulation* 62: 528-534.

showed that ventricular filling time decreased from about 500 to 700 ms at rest to about 150 ms at HRs between 150 and 200 beats/min.[6] Therefore, with increasing intensities approaching $\dot{V}O_{2max}$ (and HR_{max}), the diastolic filling time could be shortened enough to limit filling. As a result, EDV might plateau or even start to decrease.

For the Frank-Starling mechanism to increase SV, left ventricular EDV must increase, necessitating an increased venous return to the heart. As discussed in chapter 5, the muscle pump and respiratory pump both aid in increasing venous return. In addition, redistribution of blood flow and volume from inactive tissues such as the splanchnic and renal circulations increases the available central blood volume.

To review, two factors that can contribute to an increase in SV with increasing intensity of exercise are increased venous return (preload) and increased ventricular contractility. The third factor that contributes to the increase in SV during exercise—a decrease in afterload—results from a decrease in total peripheral resistance. Total peripheral resistance decreases because of vasodilation of the blood vessels in exercising skeletal muscle. This decrease in afterload allows the left ventricle to expel blood against less resistance, facilitating greater emptying of the ventricle.

Cardiac Output

Since cardiac output is the product of heart rate and stroke volume ($\dot{Q} = HR \times SV$), cardiac output predictably increases with increasing exercise intensity (figure 8.5). Resting cardiac output is approximately 5.0 L/min but varies in proportion to the size of the person. Maximal cardiac output varies between less than 20 L/min in sedentary individuals to 40 or more L/min in elite endurance athletes. Maximal $\dot{Q}$ is a function of both body size and endurance training. The linear relationship between cardiac output and exercise intensity is expected because the major purpose of the increase in cardiac output is to meet the muscles' increased demand for oxygen. Like $\dot{V}O_{2max}$, when cardiac output approaches maximal exercise intensity it may reach a plateau (figure 8.5). In fact, it is likely that $\dot{V}O_{2max}$ is ultimately limited by the inability of cardiac output to increase further.

In focus

During exercise, cardiac output increases in proportion to exercise intensity to match the need for increased blood flow to exercising muscles.

Conflicting Research on Stroke Volume During Exercise

Although researchers agree that SV increases as exercise intensity increases up to approximately 40% to 60% $\dot{V}O_{2max}$, reports about what happens after that point differ. A review of studies conducted between the 1960s and the early 1990s reveals no clear pattern of SV increase beyond the 40% to 60% range. Several studies have shown a plateau in SV at approximately 50% of $\dot{V}O_{2max}$, with little or no change occurring with further increases, while other studies have shown that SV continues to increase beyond that rate.

This apparent disagreement might result from differences among studies in the mode of exercise testing or the participants' training status. Studies that show plateaus in the 40% to 60% $\dot{V}O_{2max}$ range typically have used cycle ergometers as the mode of exercise. This makes intuitive sense since blood is pooled in the legs during cycle ergometer exercise, resulting in decreased venous return of blood from the legs. Thus, the plateau in SV might be unique to cycling exercise.

Alternatively, in those studies in which SV continued to increase up to maximal exercise intensities, subjects were generally highly trained athletes. Many highly trained athletes, including highly trained cyclists tested on a cycle ergometer, can continue to increase their SV beyond 40% to 60% $\dot{V}O_{2max}$, perhaps because of adaptations caused by aerobic training. The increases in

Cardiac output and stroke volume responses to increasing exercise intensities measured in untrained subjects, trained distance runners, and elite runners.

cardiac output and SV with increasing work, as represented by increasing HR, in elite athletes, trained university distance runners, and untrained university students, are illustrated in the graphs shown here.

As one final caveat, SV is difficult to accurately measure at very high exercise intensities, so differences between studies could result from differences in the techniques used to measure cardiac output or SV as well as the accuracy of those techniques at high exercise intensities.

Adapted, by permission, from B. Zhou et al., 2001, "Stroke volume does not plateau during graded exercise in elite male distance runners," *Medicine and Science in Sports and Exercise* 33: 1849-1854.

FIGURE 8.5 The cardiac output ($\dot{Q}$) response to walking/running on a treadmill at increasing intensities plotted as a function of percent $\dot{V}O_{2max}$. Cardiac output increases in direct proportion to increasing intensity, eventually reaching a maximum ($\dot{Q}_{max}$).

The Integrated Cardiac Response to Exercise

To see how HR, SV, and $\dot{Q}$ vary under various conditions of rest and exercise, consider the following example. An individual first moves from a reclining position to a seated posture and then to standing. Next the person begins walking, then jogging, and finally breaks into a fast-paced run. How does the heart respond?

In a reclining position, HR is ~50 beats/min; it increases to about 55 beats/min during sitting and to about 60 beats/min during standing. When the body shifts from a reclining to a sitting position and then to a standing position, gravity causes blood to pool in the legs, which reduces the volume of blood returning to the heart and thus decreases SV. To compensate for the reduction in SV, HR increases in order to maintain cardiac output; that is, $\dot{Q} = HR \times SV$.

During the transition from rest to walking, HR increases from about 60 to about 90 beats/min. Heart rate increases to 140 beats/min with moderate-paced jogging and can reach 180 beats/min or more with a fast-paced run. The initial increase in HR—up to about 100 beats/min—is mediated by a withdrawal of parasympathetic (vagal) tone. Further increases in HR are mediated by increased activation of the sympathetic nervous system. Stroke volume also increases with exercise, further increasing cardiac output. These relationships are illustrated in figure 8.6.

During the initial stages of exercise in untrained individuals, increased cardiac output is caused by an increase in both HR and SV. When the level of exercise

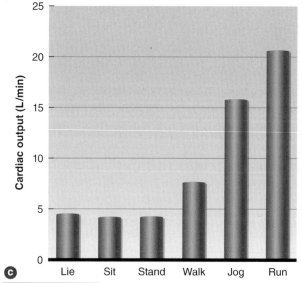

FIGURE 8.6 Changes in (*a*) heart rate, (*b*) stroke volume, and (*c*) cardiac output with changes in posture (lying supine, sitting, and standing upright) and with exercise (walking at 5 km/h [3.1 mph], jogging at 11 km/h [6.8 mph], and running at 16 km/h [9.9 mph]).

exceeds 40% to 60% of the individual's maximal exercise capacity, SV either plateaus or continues to increase at a much slower rate. Thus, further increases in cardiac output are largely the result of increases in HR. Further stroke volume increases contribute more to the rise in cardiac output at high intensities of exercise in highly trained athletes.

In review

➤ As exercise intensity increases, HR increases proportionately, approaching HR_{max} near the maximal exercise intensity.

➤ Stroke volume (the amount of blood ejected with each contraction) also increases proportionately with increasing exercise intensity but usually achieves its maximal value at about 40% to 60% of $\dot{V}O_{2max}$ in untrained individuals. Highly trained individuals can continue to increase SV, sometimes up to maximal exercise intensity.

➤ Increases in HR and SV combine to increase cardiac output. Thus, more blood is pumped during exercise, ensuring that an adequate supply of oxygen and metabolic substrates reach the exercising muscles and that the waste products of muscle metabolism are cleared away.

Blood Pressure

During endurance exercise, systolic blood pressure increases in direct proportion to the increase in exercise intensity. However, diastolic pressure does not change significantly and may even decrease. As a result of the increased systolic pressure, mean arterial blood pressure increases. A systolic pressure that starts out at 120 mmHg in a normal healthy person at rest can exceed 200 mmHg at maximal exercise. Systolic pressures of 240 to 250 mmHg have been reported in normal, healthy, highly trained athletes at maximal intensities of aerobic exercise.

Increased systolic blood pressure results from the increased cardiac output ($\dot{Q}$) that accompanies increasing rates of work. This increase in pressure helps facilitate the increase in blood flow through the vasculature. Also, blood pressure (that is, hydrostatic pressure) in large part determines how much plasma leaves the capillaries, entering the tissues and carrying needed supplies. Thus increased systolic pressure aids substrate delivery to working muscles.

After increasing initially, mean arterial pressure reaches a steady state during submaximal steady-state endurance exercise. As work intensity increases, so does systolic blood pressure. If steady-state exercise is

prolonged, the systolic pressure might start to decrease gradually, but diastolic pressure remains constant. The slight decrease in systolic blood pressure, if it occurs, is a normal response and simply reflects increased vasodilation in the active muscles, which decreases the **total peripheral resistance**, or **TPR** (since mean arterial pressure = cardiac output × total peripheral resistance).

Diastolic blood pressure changes little during submaximal dynamic exercise; however, at maximal exercise intensities, diastolic blood pressure may increase slightly. Remember that diastolic pressure reflects the pressure in the arteries when the heart is at rest (diastole). With dynamic exercise there is an overall increase in sympathetic tone to the vasculature, causing overall vasoconstriction. However, this vasoconstriction is blunted in the exercising muscles by the release of local vasodilators, a phenomenon called **sympatholysis**. Thus, because there is a balance between vasoconstriction to inactive regional circulations and vasodilation in active skeletal muscle, diastolic pressure does not change substantially. However, in some cases of cardiovascular disease, increases in diastolic pressure of 15 mmHg or more occur in response to exercise and are one of several indications for immediately stopping a diagnostic exercise test. Figure 8.7 illustrates a typical blood pressure response in a healthy subject during leg and arm cycling exercise with increasing exercise intensities.

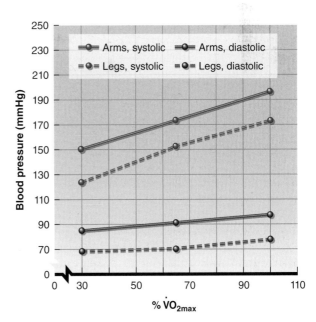

FIGURE 8.7 Systolic and diastolic blood pressure responses to leg and arm cycling as relative exercise intensity (%$\dot{V}O_{2max}$) increases.

Adapted from P.O. Åstrand et al., 1965, "Intraarterial blood pressure during exercise with different muscle groups," *Journal of Applied Physiology* 20: 253-256. Used with permission.

As seen in figure 8.7, upper body exercise causes a greater blood pressure response than leg exercise at the same absolute rate of energy expenditure. This is most likely attributable to the smaller exercising muscle mass of the upper body compared with the lower body, plus an increased energy demand needed to stabilize the upper body during arm exercise. This difference in the systolic blood pressure response to upper and lower body exercise has important implications for the heart. Myocardial oxygen uptake and myocardial blood flow are directly related to the product of HR and systolic blood pressure (SBP). This value is referred to as the **rate–pressure product (RPP)**, or double product (RPP = HR × SBP). With static or dynamic resistance exercise or upper body work, the rate–pressure product is elevated, indicating increased myocardial oxygen demand. The use of rate–pressure product as an indirect index of myocardial oxygen demand is important in clinical exercise testing.

Periodic blood pressure increases during resistance exercise, such as weightlifting, can be extreme. With high-intensity resistance training, blood pressure can briefly reach 480/350 mmHg. Very high pressures like these are more commonly seen when the exerciser performs a **Valsalva maneuver** to aid heavy lifts. This maneuver occurs when a person tries to exhale while the mouth, nose, and glottis are closed. This action causes an enormous increase in intrathoracic pressure. Much of the subsequent blood pressure increase results from the body's effort to overcome the high internal pressures created during the Valsalva maneuver.

Blood Flow

Acute increases in cardiac output and blood pressure during exercise allow for increased total blood flow to the body. These responses facilitate increased blood to areas where it is needed, primarily the exercising muscles. Additionally, sympathetic control of the cardiovascular system redistributes blood so that areas with the greatest metabolic need receive more blood than areas with low demands.

Redistribution of Blood During Exercise

Blood flow patterns change markedly in the transition from rest to exercise. Through the vasoconstrictor action of the sympathetic nervous system on local arterioles, blood flow is redirected away from areas where elevated flow is not essential to those areas that are active during exercise (refer back to figure 6.12, p. 154). Only 15% to 20% of the resting cardiac output goes to muscle, but during high-intensity exercise, the muscles may receive 80% to 85% of the cardiac output. This shift in blood flow to the muscles is accomplished primarily by reducing blood flow to the kidneys and the so-called splanchnic circulation (which includes the liver, stomach, pancreas, and intestines). Figure 8.8 illustrates a typical distribution of cardiac output throughout the body at rest and during heavy exercise. Because cardiac output increases greatly with increasing intensity of exercise, the values are shown both as the relative percentage of cardiac output and as the abso-

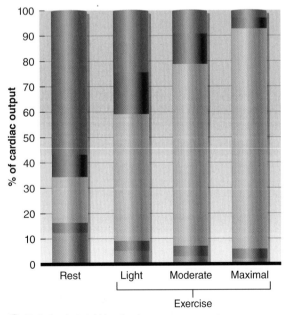

(a) Relative to total blood volume

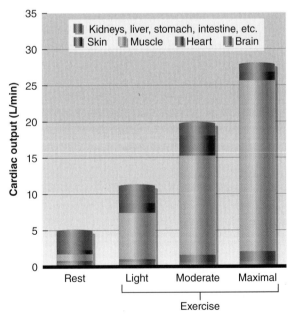

(b) Absolute values

FIGURE 8.8 The distribution of cardiac output at rest and during exercise.

Data from A.J. Vander, J.H. Sherman, & D.S. Luciano, 1985, *Human physiology: The mechanisms of body function*, 4th ed. (New York: McGraw-Hill).

lute cardiac output going to each regional circulation at rest and at three intensities of exercise.

Although several physiological mechanisms are responsible for the redistribution of blood flow during exercise, they work together in an integrated fashion. To illustrate this, consider what happens to blood flow during exercise, focusing on the primary driver of the response, namely the increased blood flow requirement of the exercising skeletal muscles.

As exercise begins, active skeletal muscles rapidly require increased oxygen delivery. This need is partially met through sympathetic stimulation of vessels in those areas to which blood flow is to be reduced (e.g., the splanchnic and renal circulations). The resulting vasoconstriction in those areas allows for more of the (increased) cardiac output to be distributed to the exercising skeletal muscles. In the skeletal muscles, sympathetic stimulation to the constrictor fibers in the arteriolar walls also increases; however, local vasodilating substances are released from the exercising muscle and overcome sympathetic vasoconstriction, producing an overall vasodilation in the muscle (sympatholysis).

Many local vasodilating substances are released in exercising skeletal muscle. As the metabolic rate of the muscle tissue increases during exercise, metabolic waste products begin to accumulate. Increased metabolism causes an increase in acidity (increased hydrogen ions and lower pH), carbon dioxide, and temperature in the muscle tissue. These are some of the local changes that trigger vasodilation of, and increasing blood flow through, the arterioles feeding local capillaries. Local vasodilation is also triggered by the low partial pressure of oxygen in the tissue or a reduction in oxygen bound to hemoglobin (increased oxygen demand), the act of muscle contraction, and possibly other vasoactive substances (including adenosine) released as a result of skeletal muscle contraction.

When exercise is performed in a hot environment, there is also an increase in blood flow to the skin to help dissipate the body heat. The sympathetic control of skin blood flow is unique in that there are sympathetic vasoconstrictor fibers (similar to skeletal muscle) and sympathetic active vasodilator fibers interacting over most of the skin surface area. During dynamic exercise, as body core temperature rises, there is initially a reduction in sympathetic vasoconstriction, causing a passive vasodilation. Once a specific body core temperature threshold is reached, skin blood flow begins to dramatically increase by activation of the sympathetic active vasodilator system. The increase in skin blood flow during exercise promotes heat loss, because

metabolic heat from deep in the body can be released only when blood moves close to the skin. This limits the rate of rise in body temperature, as discussed in more detail in chapter 12.

Cardiovascular Drift

With prolonged aerobic exercise or aerobic exercise in a hot environment at a steady-state intensity, SV gradually decreases and HR increases. Cardiac output is well maintained, but arterial blood pressure also declines. These alterations, illustrated in figure 8.9, have been referred to collectively as **cardiovascular drift**, and they are generally associated with increasing body temperature and dehydration. Cardiovascular drift is associated with a progressive increase in the fraction of cardiac output directed to the vasodilated skin to facilitate heat loss and attenuate the increase in body core temperature. With more blood in the skin for the purpose of cooling the body, less blood is available to return to the heart, thus decreasing preload. There is also a small decrease in blood volume resulting from sweating and from a generalized shift of plasma across the capillary membrane into the surrounding tissues. These factors combine to decrease ventricular filling pressure, which decreases venous return to the heart and reduces the EDV. With the reduction in EDV, SV is reduced (SV = EDV − ESV). In order to maintain cardiac output ($\dot{Q}$ = HR × SV), HR increases to compensate for the decrease in SV.

FIGURE 8.9 Circulatory responses to prolonged, moderately intense exercise in the upright posture in a thermoneutral 20 °C environment, illustrating cardiovascular drift. Values are expressed as the percentage of change from the values measured at the 10 min point of the exercise.

Adapted, by permission, from L.B. Rowell, 1986, *Human Circulation: Regulation during physical stress* (New York: Oxford University Press), 230.

Competition for Blood Supply

When the demands of exercise are added to blood flow demands for all other systems of the body, competition for a limited available cardiac output can occur. This competition for available blood flow can develop among several vascular beds, depending on the specific conditions. For example, there may be competition for available blood flow between active skeletal muscle and the gastrointestinal system following a meal. McKirnan and coworkers[2] studied the effects of feeding versus fasting on the distribution of blood flow during exercise in miniature pigs. The pigs were divided into two groups. One group fasted for 14 to 17 h before exercise. The other group ate their morning ration in two feedings: Half the ration was fed 90 to 120 min before exercise and the other half 30 to 45 min before exercise. Both groups of pigs then ran at approximately 65% of their $\dot{V}O_{2max}$.

Blood flow to the hindlimb muscles during exercise was 18% lower, and gastrointestinal blood flow was 23% higher, in the fed group than in the fasted group. Similar results in humans suggest that the redistribution of gastrointestinal blood flow to the working muscles is attenuated after a meal. As a practical application, these findings suggest that athletes should be cautious in timing their meals before competition to maximize blood flow to the active muscles during exercise.

In focus

During exercise, blood is redistributed in the body primarily to meet the demands of active tissues, particularly contracting skeletal muscle fibers.

Another example of the competition for blood flow is seen in exercise in a hot environment. In this scenario, competition for available cardiac output can occur between the skin circulation for thermoregulation and the exercising muscles. This will be discussed in more detail in chapter 12.

Blood

We have now examined how the heart and blood vessels respond to exercise. The remaining component of the cardiovascular system is the blood: the fluid that carries oxygen and nutrients to the tissues and clears away waste products of metabolism. As metabolism increases during exercise, several aspects of the blood itself become increasingly critical for optimal performance.

Oxygen Content

At rest, the blood's oxygen content varies from 20 ml of oxygen per 100 ml of arterial blood to 14 ml of oxygen per 100 ml of venous blood returning to the right atrium. The difference between these two values (20 ml – 14 ml = 6 ml) is referred to as the arterial–mixed venous oxygen difference, or $(a-\bar{v})O_2$ difference. This value represents the extent to which oxygen is extracted, or removed, from the blood as it passes through the body.

With increasing exercise intensity, the $(a-\bar{v})O_2$ difference increases progressively and can almost triple from rest to maximal exercise intensities (see figure 8.10). This increased difference really reflects a decreasing venous oxygen content, because arterial oxygen content changes little from rest up to maximal exertion. With exercise, more oxygen is required by the active muscles; therefore more oxygen is extracted from the blood. The venous oxygen content decreases, approaching zero in the active muscles. However, mixed venous blood in the right atrium of the heart rarely decreases below 4 ml of oxygen per 100 ml of blood because the blood

In review

➤ Mean arterial blood pressure increases immediately in response to exercise, and the magnitude of the increase is proportional to the intensity of exercise. During whole-body endurance exercise, this is accomplished primarily by an increase in systolic blood pressure, with minimal changes in diastolic pressure.

➤ Systolic blood pressure can exceed 200-250 mmHg at maximal exercise intensity, the result of increases in cardiac output. Upper body exercise causes a greater blood pressure response than leg exercise at the same absolute rate of energy expenditure, likely due to the smaller muscle mass involved and the need to stabilize the trunk during dynamic arm exercise.

➤ Blood flow is redistributed during exercise from inactive or low-activity tissues of the body like the liver and kidneys to meet the increased metabolic needs of exercising muscles.

➤ With prolonged aerobic exercise, or aerobic exercise in the heat, SV gradually decreases and HR increases proportionately to maintain cardiac output. This is referred to as cardiovascular drift and is associated with a progressive increase in blood flow to the vasodilated skin and losses of fluid from the vascular space.

FIGURE 8.10 Changes in the oxygen content of arterial and mixed venous blood and the (a-v̄)O₂ difference (arterial–mixed venous oxygen difference) as a function of exercise intensity.

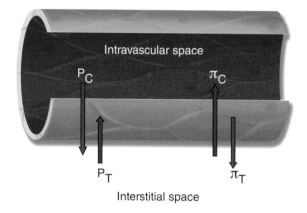

Net capillary filtration = $(P_C + \pi_T) - (P_T - \pi_C)$

FIGURE 8.11 Filtration of plasma from the micro-vasculature. Both the blood pressure (P_C) inside the blood vessel and the oncotic pressure (π_T) in the tissue cause plasma to flow from the intravascular space to the interstitial space. The pressure that the tissue (P_T) exerts on the blood vessel and the oncotic pressure of the blood (π_C) inside the blood vessel cause plasma to be reabsorbed. Net filtration of plasma can be determined by summing the outward forces ($P_C + \pi_T$) and subtracting the inward forces ($P_T - \pi_C$); net capillary filtration = $(P_C + \pi_T) - (P_T - \pi_C)$.

returning from the active tissues is mixed with blood from inactive tissues as it returns to the heart. Oxygen extraction by the inactive tissues is far lower than in the active muscles.

Plasma Volume

Upon standing, or with the onset of exercise, there is an almost immediate loss of plasma from the blood to the interstitial fluid space. The movement of fluid out of the capillaries is dictated by the pressures inside the capillaries, which include the **hydrostatic pressure** exerted by increased blood pressure and the **oncotic pressure**, the pressure exerted by the proteins in the blood, mostly albumin. The pressures that influence fluid movement outside the capillaries are the pressure provided by the surrounding tissue as well as the oncotic pressures from proteins in the interstitial fluid (figure 8.11). Osmotic pressures, those exerted by electrolytes in solution on both sides of the capillary wall, also play a role. As blood pressure increases with exercise, the hydrostatic pressure within the capillaries increases. This increase in blood pressure forces water from the intravascular compartment to the interstitial compartment. Also, as metabolic waste products build up in the active muscle, intramuscular osmotic pressure increases, which draws fluid out of the capillaries to the muscle.

Approximately a 10% to 15% reduction in plasma volume can occur with prolonged exercise, with the largest falls occurring during the first few minutes. During resistance training, the plasma volume loss is proportional to the intensity of the effort, with similar transient losses of fluid from the vascular space of 10% to 15%.

If exercise intensity or environmental conditions cause sweating, additional plasma volume losses may occur. Although the major source of fluid for sweat formation is the interstitial fluid, this fluid space will be diminished as sweating continues. This increases the oncotic (since proteins do not move with the fluid) and osmotic (since sweat has fewer electrolytes than interstitial fluid) pressures in the interstitial space, causing even more plasma to move out of the vascular compartment into the interstitial space. Intracellular fluid volume is impossible to measure directly and accurately, but research suggests that fluid is also lost from the intracellular compartment during prolonged exercise and even from the red blood cells, which may shrink in size.

A reduction in plasma volume can impair performance. For long-duration activities in which dehydration occurs and heat loss is a problem, blood flow to active tissues may be reduced to allow increasingly more blood to be diverted to the skin in an attempt to lose body heat. Note that a decrease in muscle blood flow occurs only in conditions of dehydration and only at high intensities. Severely reduced plasma volume also increases blood viscosity, which can impede blood flow and thus limit oxygen transport, especially if the hematocrit exceeds 60%.

In activities that last a few minutes or less, body fluid shifts are of little practical importance. As exercise duration increases, however, body fluid changes and temperature regulation become important for performance. For the football player, the Tour de France cyclist, or the marathon runner, these processes are crucial, not only for competition but also for survival. Deaths have occurred from dehydration and hyperthermia during, or as a result of, various sport activities. These issues are discussed in detail in chapter 12.

Hemoconcentration

When plasma volume is reduced, hemoconcentration occurs. When the fluid portion of the blood is reduced,

the cellular and protein portions represent a larger fraction of the total blood volume; that is, they become more concentrated in the blood. This hemoconcentration increases red blood cell concentration substantially—by up to 20% or 25%. Hematocrit can increase from 40% to 50%. However, the total number and volume of red blood cells do not change substantially.

The net effect, even without an increase in the total number of red blood cells, is to increase the number of red blood cells per unit of blood; that is, the cells are more concentrated. As the red blood cell concentration increases, so does the blood's per unit hemoglobin content. This substantially increases the blood's oxygen-carrying capacity, which is advantageous during exercise and provides a distinct advantage at altitude, as we will see in chapter 13.

Central Regulation of the Cardiorespiratory System During Dynamic Exercise

The cardiovascular and respiratory adjustments to dynamic exercise are profound and rapid. Within 1 s of the initiation of muscle contraction, HR dramatically increases by vagal withdrawal and respiration increases. Increases in cardiac output and blood pressure increase blood flow to the active skeletal muscle to meet its metabolic demands. What causes these extremely rapid early changes in the cardiovascular system, since they take place well before metabolic needs of working muscle occur?

Over the years there has been considerable debate over what causes the cardiovascular system to be "turned on" at the onset of exercise. One explanation is the theory of **central command**,

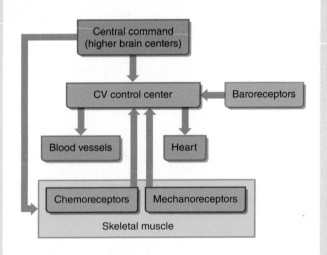

A summary of cardiovascular (CV) control during exercise.

which involves parallel "coactivation" of both the motor and the cardiovascular control centers of the brain. Activation of central command rapidly increases HR and blood pressure. In addition to central command, the cardiovascular responses to exercise are modified by mechanoreceptors, chemoreceptors, and baroreceptors. As discussed in chapter 6, baroreceptors are sensitive to stretch and send information back to the cardiovascular control centers about blood pressure. Signals from the periphery are sent back to the cardiovascular control centers through the stimulation of mechanoreceptors that are sensitive to the stretch of the skeletal muscle and through the chemoreceptors that are sensitive to an increase in metabolites in the muscle. Feedback about blood pressure and the local muscle environment helps to fine-tune and adjust the cardiovascular response. These relationships are illustrated in the figure in this sidebar.

Adapted from S.K. Powers and E.T. Howley, 2004. *Exercise physiology: Theory and application to fitness and performance* 5th ed. (New York, McGraw-Hill), 188. With permission of The McGraw-Hill Companies.

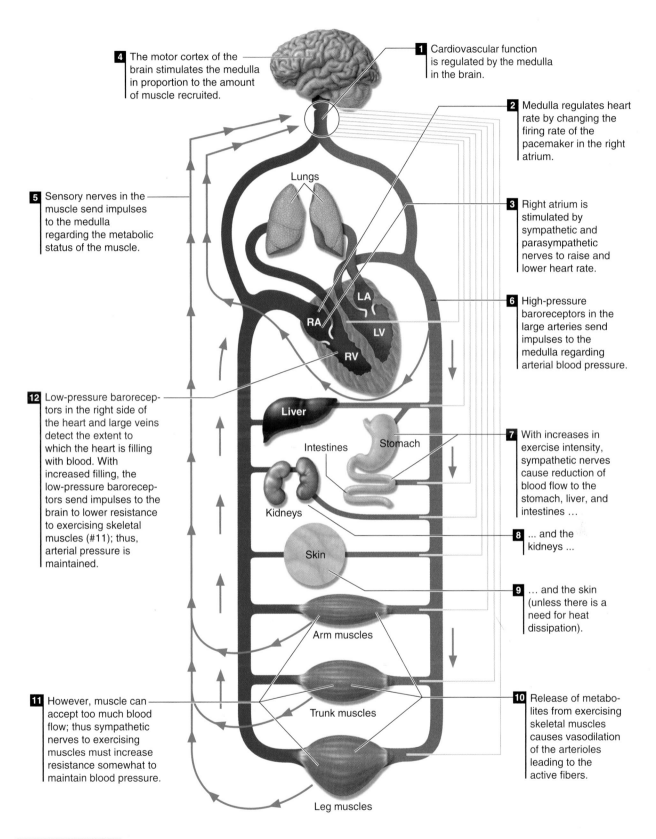

4 The motor cortex of the brain stimulates the medulla in proportion to the amount of muscle recruited.

1 Cardiovascular function is regulated by the medulla in the brain.

2 Medulla regulates heart rate by changing the firing rate of the pacemaker in the right atrium.

5 Sensory nerves in the muscle send impulses to the medulla regarding the metabolic status of the muscle.

3 Right atrium is stimulated by sympathetic and parasympathetic nerves to raise and lower heart rate.

6 High-pressure baroreceptors in the large arteries send impulses to the medulla regarding arterial blood pressure.

12 Low-pressure baroreceptors in the right side of the heart and large veins detect the extent to which the heart is filling with blood. With increased filling, the low-pressure baroreceptors send impulses to the brain to lower resistance to exercising skeletal muscles (#11); thus, arterial pressure is maintained.

7 With increases in exercise intensity, sympathetic nerves cause reduction of blood flow to the stomach, liver, and intestines …

8 … and the kidneys …

9 … and the skin (unless there is a need for heat dissipation).

11 However, muscle can accept too much blood flow; thus sympathetic nerves to exercising muscles must increase resistance somewhat to maintain blood pressure.

10 Release of metabolites from exercising skeletal muscles causes vasodilation of the arterioles leading to the active fibers.

Lungs · LA · RA · LV · RV · Liver · Intestines · Stomach · Kidneys · Skin · Arm muscles · Trunk muscles · Leg muscles

FIGURE 8.12 Integration of the cardiovascular system's response to exercise.

Adapted, by permission, from E.F. Coyle, 1991, "Cardiovascular function during exercise: Neural control factors," *Sports Science Exchange* 4(34): 1-6. Copyright 1991 by Gatorade Sports Science Institute.

Integration of the Exercise Response

As is evident from all of the changes in cardiovascular function that take place during exercise, the cardiovascular system is extremely complex but responds exquisitely to deliver oxygen to meet the demands of exercising muscle. Figure 8.12 is a simplified flow diagram that illustrates how the body integrates all these cardiovascular responses to provide for its needs during exercise. Key areas and responses are labeled and summarized to help illustrate how these complex control mechanisms are coordinated. It is important to note that although the body attempts to meet the blood flow needs of the muscle, it can do so only if blood pressure is not compromised. Maintenance of arterial blood pressure appears to be the highest priority of the cardiovascular system, irrespective of exercise, the environment, and other competing needs.

In review

➤ The changes that occur in the blood during exercise include the following:

1. The (a-v̄)O$_2$ difference increases, as venous oxygen concentration decreases, reflecting increased extraction of oxygen from the blood for use by the active tissues.

2. Plasma volume decreases. Plasma is pushed out of the capillaries by increased hydrostatic pressure as blood pressure increases, and fluid is drawn into the muscles by the increased oncotic and osmotic pressures in the muscle tissues, a by-product of metabolism. With prolonged exercise or exercise in hot environments, increasingly more plasma volume is lost through sweating.

3. Hemoconcentration occurs as plasma volume (water) decreases. Although the actual number of red blood cells stays relatively constant, the relative number of red blood cells per unit of blood increases, which increases oxygen-carrying capacity.

Respiratory Responses to Acute Exercise

Now that we have discussed the role of the cardiovascular system in delivering oxygen to the exercising muscle, we examine how the respiratory system responds to acute dynamic exercise.

Pulmonary Ventilation During Dynamic Exercise

The onset of exercise is accompanied by an immediate increase in ventilation. In fact, like the HR response, the marked increase in breathing may occur even before the onset of muscular contractions, that is, be an anticipatory response. This is shown in figure 8.13 for light, moderate, and heavy exercise. Because of its rapid onset, this initial respiratory adjustment to the demands of exercise is undoubtedly neural in nature, mediated by respiratory control centers in the brain (central command), although neural signals also come from receptors in the exercising muscle.

The more gradual second phase of the respiratory increase shown during heavy exercise in figure 8.13 is controlled primarily by changes in the chemical status of the arterial blood. As exercise progresses, increased metabolism in the muscles generates more CO$_2$ and H$^+$. Recall that these changes shift the oxyhemoglobin saturation curve leftward, enhancing oxygen unloading in the muscles, which increases the (a-v̄)O$_2$ difference. Increased CO$_2$ and H$^+$ are sensed by chemoreceptors primarily located in the brain, carotid bodies, and lungs, which in turn stimulate the inspiratory center, increasing rate and depth of respiration. Chemoreceptors in the muscles themselves might also be involved. In addition, receptors in the right ventricle of the heart send information to the inspiratory center so that increases in cardiac output can stimulate breathing

FIGURE 8.13 The ventilatory response to light, moderate, and heavy exercise. The subject exercised at each of the three intensities for 5 min. After an initial steep increase, the ventilation rate tended to plateau at a steady-state value at the light and moderate intensities but continued to increase somewhat at the heavy intensity.

during the early minutes of exercise. The influences of CO_2 and H^+ concentrations in the blood on breathing rate and pattern serve to fine-tune the neutrally mediated respiratory response to exercise in order to precisely match oxygen delivery with aerobic demands without overtaxing respiratory muscles.

In focus

Pulmonary ventilation increases during exercise in direct proportion to the metabolic needs of exercising muscle. At low exercise intensities, this is accomplished by increases in tidal volume (the amount of air moved in and out of the lungs during regular breathing). At higher intensities, the rate of respiration also increases. Maximal rates of pulmonary ventilation depend on body size. Maximal ventilation rates of approximately 100 L/min are common for smaller individuals but may exceed 200 L/min in larger individuals.

At the end of exercise, the muscles' energy demands decrease almost immediately to resting levels. But pulmonary ventilation returns to normal at a slower rate. If the rate of breathing perfectly matched the metabolic demands of the tissues, respiration would decrease to the resting level within seconds after exercise. But respiratory recovery takes several minutes, which suggests that postexercise breathing is regulated primarily by acid–base balance, the partial pressure of dissolved carbon dioxide (PCO_2), and blood temperature.

Breathing Irregularities During Exercise

Ideally, breathing during exercise is regulated in a way that maximizes aerobic performance. However, respiratory dysfunction during exercise can hinder performance.

Dyspnea

The sensation of **dyspnea** (shortness of breath) during exercise is common among individuals with poor aerobic fitness levels who attempt to exercise at intensities that significantly elevate arterial CO_2 and H^+ concentrations. As discussed in chapter 7, both stimuli send strong signals to the inspiratory center to increase the rate and depth of ventilation. Although exercise-induced dyspnea is sensed as an inability to breathe, the underlying cause is an inability to adjust breathing to blood PCO_2 and H^+.

Failure to reduce these stimuli during exercise appears to be related to poor conditioning of respiratory muscles. Despite a strong neural drive to ventilate

the lungs, the respiratory muscles fatigue easily and are unable to reestablish normal homeostasis.

Hyperventilation

The anticipation of or anxiety about exercise, as well as some respiratory disorders, can cause an increase in ventilation in excess of that needed to support exercise. Such overbreathing is termed **hyperventilation**. At rest, hyperventilation can decrease the normal PCO_2 of 40 mmHg in the alveoli and arterial blood to about 15 mmHg. As arterial CO_2 concentrations decrease, blood pH increases. These effects combine to reduce the ventilatory drive. Because the blood leaving the lungs is almost always about 98% saturated with oxygen, an increase in the alveolar PO_2 does not increase the oxygen content of the blood. Consequently, the reduced drive to breathe—along with the improved ability to hold one's breath after hyperventilating—results from carbon dioxide unloading rather than increased blood oxygen. This is sometimes referred to as "blowing off CO_2." Even when performed for only a few seconds, such deep, rapid breathing can lead to light-headedness and even loss of consciousness. This phenomenon reveals the sensitivity of the respiratory system's regulation by carbon dioxide and pH.

Valsalva Maneuver

The Valsalva maneuver is a potentially dangerous respiratory procedure that frequently accompanies certain types of exercise, in particular the lifting of heavy objects. This occurs when the individual

- closes the glottis (the opening between the vocal cords),
- increases the intra-abdominal pressure by forcibly contracting the diaphragm and the abdominal muscles, and
- increases the intrathoracic pressure by forcibly contracting the respiratory muscles.

As a result of these actions, air is trapped and pressurized in the lungs. The high intra-abdominal and intrathoracic pressures restrict venous return by collapsing the great veins. This maneuver, if held for an extended period of time, can greatly reduce the volume of blood returning to the heart, decreasing cardiac output and lowering arterial blood pressure. Although the Valsalva maneuver can be helpful in certain circumstances, this maneuver can be dangerous and should be avoided.

Ventilation and Energy Metabolism

During long periods of mild steady-state activity, ventilation matches the rate of energy metabolism, varying in proportion to the volume of oxygen consumed and the volume of carbon dioxide produced ($\dot{V}O_2$ and $\dot{V}CO_2$, respectively) by the body.

Ventilatory Equivalent for Oxygen

The ratio between the volume of air expired or ventilated ($\dot{V}_E$) and the amount of oxygen consumed by the tissues ($\dot{V}O_2$) in a given amount of time is referred to as the **ventilatory equivalent for oxygen**, or $\dot{V}_E/\dot{V}O_2$. It is typically measured in liters of air breathed per liter of oxygen consumed per minute.

At rest, the $\dot{V}_E/\dot{V}O_2$ can range from 23 to 28 L of air per liter of oxygen. This value changes very little during mild exercise, such as walking. But when exercise intensity increases to near-maximal levels, the $\dot{V}_E/\dot{V}O_2$ can be greater than 30 L of air per liter of oxygen consumed. In general, however, the $\dot{V}_E/\dot{V}O_2$ remains relatively constant over a wide range of exercise intensities, indicating that the control of breathing is properly matched to the body's demand for oxygen.

Ventilatory Threshold

As exercise intensity increases, at some point ventilation increases disproportionately to oxygen consumption. The point at which this occurs, typically between ~55% to 70% of $\dot{V}O_{2max}$, is called the **ventilatory threshold**, illustrated in figure 8.14. At approximately the same intensity as the ventilatory threshold, more lactate starts to appear in the blood. This may result from greater production of lactate or less clearance of lactate or both. This lactic acid combines with sodium bicarbonate (which buffers acid) and forms sodium lactate, water, and carbon dioxide. As we know, the increase in carbon dioxide stimulates chemoreceptors that signal the inspiratory center to increase ventilation. Thus, the ventilatory threshold reflects the respiratory response to increased carbon dioxide levels. Ventilation increases dramatically beyond the ventilatory threshold, as seen in figure 8.14.

In focus

Ventilation increases during exercise in direct proportion to the exercise intensity, up to the ventilatory threshold. Beyond this point, ventilation increases disproportionately as the body tries to clear excess CO_2.

FIGURE 8.14 Changes in pulmonary ventilation ($\dot{V}_E$) during running at increasing velocities, illustrating the concept of ventilatory threshold.

In review

➤ During exercise, ventilation shows an almost immediate increase due to increased inspiratory center stimulation. This is caused by both central command and neural feedback from muscle activity itself. This phase is followed by a plateau (during light exercise) or a much more gradual increase in respiration (during heavy exercise) that results from chemical changes in the arterial blood resulting from exercise metabolism.

➤ Altered breathing patterns and sensations associated with exercise include dyspnea, hyperventilation, and performance of the Valsalva maneuver.

➤ During mild, steady-state exercise, ventilation increases to match the rate of energy metabolism; that is, ventilation parallels oxygen uptake. The ratio of air ventilated to oxygen consumed is the ventilatory equivalent for oxygen ($\dot{V}_E/\dot{V}O_2$).

➤ The ventilatory threshold is the point at which ventilation begins to increase disproportionately to the increase in oxygen consumption. This increase in $\dot{V}_E$ reflects the need to remove excess carbon dioxide.

➤ We can estimate lactate threshold with reasonable accuracy by identifying that point at which $\dot{V}_E/\dot{V}O_2$ starts to increase while $\dot{V}_E/\dot{V}CO_2$ continues to decline.

Estimating Lactate Threshold

The disproportionate increase in ventilation without an equivalent increase in oxygen consumption led to early speculation that the ventilatory threshold might be related to the lactate threshold (that point at which blood lactate production exceeds lactate reuptake and clearance as described in chapter 5). The ventilatory threshold reflects a disproportionate increase in the volume of carbon dioxide produced per minute ($\dot{V}CO_2$) relative to the oxygen consumed. Recall from chapter 5 that the respiratory exchange ratio (RER) is the ratio of carbon dioxide production to oxygen consumption. Thus, the disproportionate increase in carbon dioxide production also causes RER to increase.

The increased $\dot{V}CO_2$ was thought to result from excess carbon dioxide being released from bicarbonate buffering of lactic acid. Wasserman

Changes in the ventilatory equivalent for carbon dioxide ($\dot{V}_E/\dot{V}CO_2$) and the ventilatory equivalent for oxygen ($\dot{V}_E/\dot{V}O_2$) during increasing intensities of exercise on a cycle ergometer. Note that the breakpoint of the estimated lactate threshold at a running velocity of 14.4 km/h (8.9 mph) is evident only in the $\dot{V}_E/\dot{V}O_2$ ratio.

and McIlroy[7] coined the term **anaerobic threshold** to describe this phenomenon because they assumed that the sudden increase in CO_2 reflected a shift toward more anaerobic metabolism. They believed that this was a good noninvasive alternative to blood sampling for detecting the onset of anaerobic metabolism. It should be noted that a number of scientists objected to their use of the term "anaerobic threshold" to describe this respiratory phenomenon.

Over the years, the anaerobic threshold concept has been refined considerably to provide a relatively accurate estimate of lactate threshold. One of the more accurate techniques for identifying this threshold involves monitoring both the ventilatory equivalent for oxygen ($\dot{V}_E/\dot{V}O_2$) and the **ventilatory equivalent for carbon dioxide ($\dot{V}_E/\dot{V}CO_2$)**, which is the ratio of the volume of air expired ($\dot{V}_E$) to the volume of carbon dioxide produced ($\dot{V}CO_2$). Using this technique, the threshold is defined as that point where there is a systematic increase in $\dot{V}_E/\dot{V}O_2$ without a concomitant increase in $\dot{V}_E/\dot{V}CO_2$. This is illustrated in the figure in this sidebar. Both the $\dot{V}_E/\dot{V}CO_2$ and $\dot{V}_E/\dot{V}O_2$ decline with increasing exercise intensity at the lower intensities. However, the $\dot{V}_E/\dot{V}O_2$ starts to increase at about 75 W while the $\dot{V}_E/\dot{V}CO_2$ continues to decline. This indicates that the increase in ventilation to remove CO_2 is disproportionate to the body's need to provide O_2. In general, this respiratory threshold technique provides a reasonably close estimate of the lactate threshold, eliminating the need for repeated blood sampling.

Respiratory Limitations to Performance

Like all tissue activity, respiration requires energy. Most of this energy is used by the respiratory muscles during pulmonary ventilation. At rest, the respiratory muscles account for only about 2% of the total oxygen uptake.

As the rate and depth of ventilation increase, so does the energy cost of respiration. The diaphragm, the intercostal muscles, and the abdominal muscles can account for up to 11% of the total oxygen consumed during heavy exercise and can receive up to 15% of the cardiac output. During recovery from dynamic exercise, sustained elevations in ventilation continue to demand increased energy, accounting for 9% to 12% of the total oxygen consumed postexercise.

Although the muscles of respiration are heavily taxed during exercise, ventilation is sufficient to prevent an increase in alveolar PCO_2 or a decline in alveolar PO_2 during activities lasting only a few minutes. Even during maximal effort, ventilation usually is not pushed to its maximal capacity to voluntarily move air in and out of the lungs. This capacity is called the **maximal voluntary ventilation** and is significantly greater than ventilation at maximal exercise. However, considerable evidence suggests that pulmonary ventilation might be a limiting factor during very high intensity (95-100% $\dot{V}O_{2max}$) exercise in highly trained subjects.

Can heavy breathing for several hours (such as during marathon running) cause glycogen depletion and fatigue of the respiratory muscles? Animal studies have shown a substantial sparing of their respiratory muscle glycogen compared with muscle glycogen in exercising muscles. Although similar data are not available for humans, our respiratory muscles are better designed for long-term activity than are the muscles in our extremities. The diaphragm, for example, has two to three times more oxidative capacity (oxidative enzymes and mitochondria) and capillary density than other skeletal muscle. Consequently, the diaphragm can obtain more energy from oxidative sources than can skeletal muscles.

Similarly, airway resistance and gas diffusion in the lungs do not limit exercise in a normal, healthy individual. The volume of air inspired can increase 20- to 40-fold with exercise—from ~5 L/min at rest up to 100 to 200 L/min with maximal exertion. Airway resistance, however, is maintained at near-resting levels by airway dilation (through an increase in the laryngeal aperture and bronchodilation). During submaximal and maximal efforts in untrained and moderately trained individuals, blood leaving the lungs remains nearly saturated with oxygen (~98%). However, with maximal exercise in some highly trained elite endurance athletes, there is too large a demand on lung gas exchange, resulting in a decline in arterial PO_2 and arterial oxygen saturation (i.e., **exercise-induced arterial hypoxemia**, or **EIAH**). Approximately 40% to 50% of elite endurance athletes experience a significant reduction in arterial oxygenation during exercise approaching exhaustion.[4] Arterial hypoxemia at maximal exercise is likely the result of a mismatch between ventilation and perfusion of the lung. Since cardiac output is extremely high in elite athletes, blood is flowing through the lungs at a high rate and thus there may not be sufficient time for that blood to become saturated with oxygen. Thus, in healthy individuals, the respiratory system is well designed to accommodate the demands of heavy breathing during short- and long-term physical effort.

However, some highly trained individuals who consume unusually large amounts of oxygen during exhaustive exercise can face respiratory limitations.

In focus

In some cases, the performance of highly trained distance runners can be limited by their respiratory systems. A mismatch may occur between ventilation and perfusion in the lungs, resulting in a decrease in arterial blood PO_2 and reduced hemoglobin saturation.

The respiratory system also can limit performance in patient populations with restricted or obstructed airways. For example, asthma causes constriction of the bronchial tubes and swelling of the mucous membranes. These effects cause considerable resistance to ventilation, resulting in a shortness of breath. Exercise is known to bring about symptoms of asthma, or to worsen those symptoms in select individuals. The mechanism or mechanisms through which exercise induces airway obstruction in individuals with so-called exercise-induced asthma remain unknown, despite extensive study.

In review

➤ Respiratory muscles can account for up to 10% of the body's total oxygen consumption and 15% of the cardiac output during heavy exercise.

➤ Pulmonary ventilation is usually not a limiting factor for performance even during maximal effort, although it can limit performance in some elite endurance athletes.

➤ The respiratory muscles are well designed to avoid fatigue during long-term activity.

➤ Airway resistance and gas diffusion usually do not limit performance in normal, healthy individuals exercising at sea level.

➤ The respiratory system can, and often does, limit performance in people with various types of restrictive or obstructive respiratory disorders.

Respiratory Regulation of Acid–Base Balance

As noted earlier, high-intensity exercise results in the production and accumulation of lactate and H^+. Although regulation of acid–base balance involves more than control of respiration, it is discussed here because the respiratory system plays such a crucial role in rapid adjustment of the body's acid–base status during and immediately after exercise.

Acids, such as lactic acid and carbonic acid, release hydrogen ions (H⁺). As noted in chapter 2, the metabolism of carbohydrate, fat, or protein produces inorganic acids that dissociate, increasing the H^+ concentration in body fluids, thus lowering the pH. To minimize the effects of free H^+, the blood and muscles contain base substances that combine with, and thus buffer or neutralize, the H^+:

$$H^+ + buffer \rightarrow H\text{-buffer}.$$

Under resting conditions, body fluids have more bases (such as bicarbonate, phosphate, and proteins) than acids, resulting in a slightly alkaline tissue pH that ranges from 7.1 in muscle to 7.4 in arterial blood. The tolerable limits for arterial blood pH extend from 6.9 to 7.5, although the extremes of this range can be tolerated only for a few minutes (see figure 8.15). An H^+ concentration above normal (low pH) is referred to as acidosis, whereas a decrease in H^+ below the normal concentration (high pH) is termed alkalosis.

The pH of intra- and extracellular body fluids is kept within a relatively narrow range by

- chemical buffers in the blood,
- pulmonary ventilation, and
- kidney function.

The three major chemical buffers in the body are bicarbonate (HCO_3^-), inorganic phosphates (P_i), and

proteins. In addition to these, hemoglobin in the red blood cells is also a major buffer. Table 8.1 illustrates the relative contributions of these buffers in handling acids in the blood. Recall that bicarbonate combines with H^+ to form carbonic acid, thereby eliminating the acidifying influence of free H^+. The carbonic acid in turn forms carbon dioxide and water in the lungs. The CO_2 is then exhaled and only water remains.

The amount of bicarbonate that combines with H^+ equals the amount of acid buffered. When lactic acid decreases the pH from 7.4 to 7.0, more than 60% of the bicarbonate initially present in the blood has been used. Even under resting conditions, the acid produced by the end products of metabolism would use up a major portion of the bicarbonate from the blood if there were no other way of removing H^+ from the body. Blood and chemical buffers are required only to transport metabolic acids from their sites of production (the muscles) to the lungs or kidneys, where they can be removed. Once H^+ is transported and removed, the buffer molecules can be reused.

In the muscle fibers and the kidney tubules, H^+ is primarily buffered by phosphates, such as phosphoric acid and sodium phosphate. Less is known about the capacity of the buffers intracellularly, although cells contain more protein and phosphates and less bicarbonate than do the extracellular fluids.

As noted earlier, any increase in free H^+ in the blood stimulates the respiratory center to increase ventilation. This facilitates the binding of H^+ to bicarbonate and the removal of carbon dioxide. The end result is a decrease in free H^+ and an increase in blood pH. Thus, both the chemical buffers and the respiratory system provide short-term means of neutralizing the acute effects of exercise acidosis. To maintain a constant buffer reserve, the accumulated H^+ is removed from the body via excretion by the kidneys and eliminated in urine. The kidneys filter H^+ from the blood along with other waste products. This provides a way to eliminate

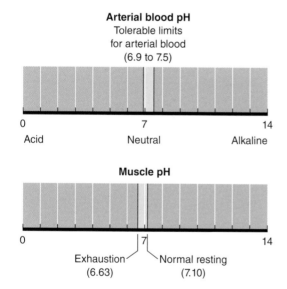

FIGURE 8.15 Tolerable limits for arterial blood pH and muscle pH at rest and at exhaustion. Note the small range of physiological tolerance for both muscle and blood pH.

TABLE 8.1 Buffering Capacity of Blood Components

Buffer	Slykes[a]	%
Bicarbonate	18.0	64
Hemoglobin	8.0	29
Proteins	1.7	6
Phosphates	0.3	1
Total	**28.0**	**100**

[a]Milliequivalents of hydrogen ions taken up by each liter of blood from pH 7.4 to 7.0.

H^+ from the body while maintaining the concentration of extracellular bicarbonate.

During sprint exercise, muscle glycolysis generates a large amount of lactate and H^+, which lowers the muscle pH from a resting level of 7.1 to less than 6.7. As shown in table 8.2, an all-out 400 m sprint decreases leg muscle pH to 6.63 and increases muscle lactate from a resting value of 1.2 mmol/kg to almost 20 mmol/kg of muscle. Such disturbances in acid–base balance can impair muscle contractility and its capacity to generate adenosine triphosphate (ATP). Lactate and H^+ accumulate in the muscle, in part because they do not freely diffuse across the skeletal muscle fiber membranes. Despite the great production of lactate and H^+ during the ~60 s required to run 400 m, these by-products diffuse throughout the body fluids and reach equilibrium after only about 5 to 10 min of recovery. Five minutes after the exercise, the runners described in table 8.2 had blood pH values of 7.10 and blood lactate concentrations of 12.3 mmol/L, compared with a resting pH of 7.40 and a resting lactate level of 1.5 mmol/L.

Reestablishing normal resting concentrations of blood and muscle lactate after such an exhaustive exercise bout is a relatively slow process, often requiring 1 to 2 h. As shown in figure 8.16, recovery of blood lactate to the resting level is facilitated by continued lower-intensity exercise, called active recovery.[1] After a series of exhaustive sprint bouts, the participants in this study either sat quietly (passive recovery) or exercised at an intensity of 50% $\dot{V}O_{2max}$. Blood lactate is removed more quickly during active recovery because the activity maintains elevated blood flow through the active muscles, which in turn enhances both lactate diffusion out of the muscles and lactate oxidation.

FIGURE 8.16 Effects of active and passive recovery on blood lactate levels after a series of exhaustive sprint bouts. Note that the blood lactate removal rate is faster when the subjects perform exercise during recovery than when they rest during recovery.

Although blood lactate remains elevated for 1 to 2 h after highly anaerobic exercise, blood and muscle H^+ concentrations return to normal within 30 to 40 min of recovery. Chemical buffering, principally by bicarbonate, and respiratory removal of excess carbon dioxide are responsible for this relatively rapid return to normal acid–base homeostasis.

TABLE 8.2 Blood and Muscle pH and Lactate Concentration 5 min After a 400 m Run

| Runner | Time (s) | MUSCLE | | BLOOD | |
		pH	Lactate (mmol/kg)	pH	Lactate (mmol/L)
1	61.0	6.68	19.7	7.12	12.6
2	57.1	6.59	20.5	7.14	13.4
3	65.0	6.59	20.2	7.02	13.1
4	58.5	6.68	18.2	7.10	10.1
Average	**60.4**	**6.64**	**19.7**	**7.10**	**12.3**

Air Pollution

Over the past 30 years there has been increasing concern about possible problems associated with exercising in polluted air. The air in many cities is contaminated with small quantities of gases and particles not naturally found in the air we breathe. When air becomes stagnant or when a temperature inversion occurs, some of these pollutants reach concentrations that can significantly impair athletic performance. The major contaminants of concern are carbon monoxide, ozone, and sulfur oxides.

Carbon monoxide (CO) is an odorless gas, derived from the burning of various fuels and also present in tobacco smoke. Carbon monoxide rapidly enters the blood; and the affinity of hemoglobin to bind CO is approximately 250 times greater than its affinity for oxygen, so hemoglobin preferentially binds the CO, displacing oxygen molecules. Blood CO concentrations are directly proportional to the CO concentrations in the inspired air. Several studies have reported linear decreases in $\dot{V}O_{2max}$ with increases in blood levels of CO; but the reduction in $\dot{V}O_{2max}$ is not statistically significant until blood CO levels exceed 4%, although performance time on the treadmill has been reduced at CO levels as low as 3%. Submaximal exercise at less than 60% of $\dot{V}O_{2max}$ does not appear to be affected until blood CO levels exceed 15%.

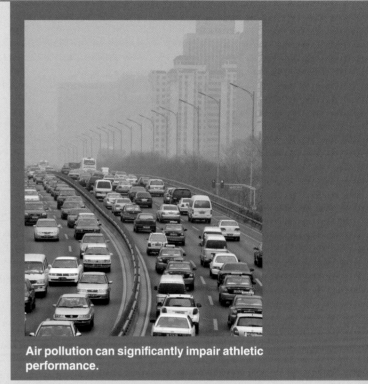

Air pollution can significantly impair athletic performance.

Ozone (O_3) is the most common photochemical oxidant and is the result of the reaction between ultraviolet light and internal combustion engine emissions. When people are exposed to high O_3 concentrations, eye irritation, chest tightness, dyspnea, coughing, and nausea are common complaints. Decrements in lung function occur with increased O_3 concentrations and prolonged exposure. $\dot{V}O_{2max}$ is significantly decreased following 2 h of intermittent exercise with exposure to 0.75 parts per million (ppm) of O_3. This decrease in $\dot{V}O_{2max}$ is likely associated with reduced oxygen transfer at the lung, resulting from reduced alveolar air exchange.

Sulfur dioxide (SO_2) generated from the burning of fossil fuels is another pollutant of concern. Research on SO_2 and exercise is limited, but atmospheric concentrations of this gas above 1.0 ppm can cause significant discomfort and decreases in aerobic exercise performance. Sulfur dioxide is primarily an upper airway and bronchial irritant.

In review

➤ Excess H^+ (decreased pH) impairs muscle contractility and ATP generation.

➤ The respiratory and renal systems play integral roles in maintaining acid–base balance. The renal system is involved in more long-term maintenance of acid–base balance through the secretion of H^+.

➤ Whenever H^+ concentration starts to increase, the inspiratory center responds by increasing the rate and depth of respiration. Removing carbon dioxide is an essential means for reducing H^+ concentrations.

➤ Carbon dioxide is transported in the blood primarily bound to bicarbonate. Once it reaches the lungs, carbon dioxide is formed again and exhaled.

➤ Whenever H^+ concentration begins to increase, whether from carbon dioxide or lactate accumulation, bicarbonate ion can buffer the H^+ to prevent acidosis.

In this chapter, we discussed the responses of the cardiovascular and respiratory systems to exercise. We also considered the limitations that these systems can impose on abilities to perform sustained aerobic exercise. The next chapter presents basic principles of exercise training, allowing us to better understand in the subsequent chapters how the body adapts to resistance training as well as aerobic and anaerobic training.

Key Terms

afterload

anaerobic threshold

cardiovascular drift

central command

dyspnea

exercise-induced arterial hypoxemia (EIAH)

Frank-Starling mechanism

hydrostatic pressure

hyperventilation

maximal voluntary ventilation

maximum heart rate (HR_{max})

oncotic pressure

preload

rate–pressure product (RPP)

resting heart rate (RHR)

steady-state heart rate

sympatholysis

total peripheral resistance (TPR)

Valsalva maneuver

ventilatory equivalent for carbon dioxide ($\dot{V}_E/\dot{V}CO_2$)

ventilatory equivalent for oxygen ($\dot{V}_E/\dot{V}O_2$)

ventilatory threshold

Study Questions

1. Describe how heart rate, stroke volume, and cardiac output respond to increasing rates of work. Illustrate how these three variables are interrelated.

2. How do we determine HR_{max}? What are alternative methods using indirect estimates? What are the major limitations to these indirect estimates?

3. Describe two important mechanisms for returning blood back to the heart during exercise in an upright position.

4. What is the Fick principle, and how does this apply to our understanding of the relationship between metabolism and cardiovascular function?

5. Define the Frank-Starling mechanism. How does this work during exercise?

6. How does blood pressure respond to exercise?

7. What are the major cardiovascular adjustments that the body makes when someone is overheated during exercise?

8. What is cardiovascular drift? Why might this be a problem with prolonged exercise?

9. Describe the primary functions of blood.

10. What changes occur in the plasma volume and red blood cells with increasing levels of exercise? With prolonged exercise in the heat?

11. How does pulmonary ventilation respond to increasing intensities of exercise?

12. Define the terms dyspnea, hyperventilation, Valsalva maneuver, and ventilatory threshold.

13. What role does the respiratory system play in acid–base balance?

14. What is the normal resting pH for arterial blood? For muscle? How are these values changed as a result of exhaustive sprint exercise?

15. What are the primary buffers in the blood? In muscles?

Study Guide Activities

In addition to the activities listed in the chapter opening outline on page 181, two other activities are available in the web study guide, located at

www.HumanKinetics.com/PhysiologyOfSportAndExercise

 The **KEY TERMS** activity reviews important terms, and the end-of-chapter **QUIZ** tests your understanding of the material covered in the chapter.

PART III

Exercise Training

Traditionally, the study of exercise physiology can be divided into how the body responds during acute bouts of exercise and how it adapts to repeated exercise sessions (i.e., training responses). In the two previous sections of the book, we examined the control and function of skeletal muscle during acute exercise (part I) and the role of the cardiovascular and respiratory systems in supporting those functions (part II). In part III we examine how these systems adapt when exposed to repeated bouts of exercise (i.e., adaptations to training). Chapter 9, "Principles of Exercise Training," lays the groundwork for the two subsequent chapters by describing the terminology and principles used by exercise physiologists. The principles presented in this chapter can be used to optimize the physiological adaptations to repeated exercise. In chapter 10, "Adaptations to Resistance Training," we consider the mechanisms through which muscular strength and muscular endurance may be improved in response to resistance training. Finally, in chapter 11, "Adaptations to Aerobic and Anaerobic Training," we discuss the changes in various systems of the body that result from performing regular physical activity involving various combinations of exercise intensity and duration. The adaptations occurring across the physiological systems that ultimately lead to improvements in exercise capacity and athletic performance are specific to the training to which those systems are exposed.

Principles of Exercise Training

9

In this chapter and in the web study guide

 ACTIVITY 9.1 Basic Training Principles reviews the basic training principles and connects them to a real-life situation.

 ACTIVITY 9.2 Forms of Resistance Training explores the characteristics of the different forms of resistance training.

 ACTIVITY 9.3 Evaluating an Aerobic Training Program provides an opportunity to evaluate a basic aerobic power training program.

Bryan Clay, an outstanding American track and field athlete, won the gold medal in the decathlon during the 2008 Olympic Games in Beijing, China, taking the title from the 2004 Olympic champion, Roman Šebrle. In 2001, Šebrle had set the world record for the decathlon that still stands, scoring over 9,000 points. Decathletes are considered by many to be the "ultimate" athletes, since they have to compete in events that test their speed, strength, power, agility, and endurance. The decathlon is a two-day event, comprising the 100 m sprint, long jump, shot put, high jump, and 400 m run on the first day, and the 110 m hurdles, discus, pole vault, javelin, and 1,500 m run on the second day. As we will see in this chapter and the following two chapters, training is very specific to the sport or event. Intense muscular power training to increase the distance one can heave a 16 lb shot put will do little if anything to improve one's 1,500 m run time. Decathletes must spend countless hours training specifically for each of their 10 events, fine-tuning their training techniques to maximize performance in each event.

Previous chapters examining the acute response to exercise were concerned with the body's immediate response to a single exercise bout. We now investigate how the body responds to repeated bouts of exercise performed over a period of time, that is, exercise training. When one performs regular exercise over a period of weeks, the body adapts physiologically. The physiological adaptations that occur when proper training principles are followed improve both exercise capacity and sport performance. With resistance training, the muscles become stronger. With aerobic training, the heart and lungs become more efficient and endurance capacity increases. With high-intensity anaerobic training, the neuromuscular, metabolic, and cardiovascular systems adapt, allowing the person to generate more adenosine triphosphate (ATP) per unit of time, thus increasing muscular endurance and speed of movement over short periods of time. These adaptations are highly specific to the type of training done. Before investigating specific adaptations to training, this chapter first looks at the basic terminology and general principles used in exercise training, then gives an overview of the elements of proper training programs.

Terminology

Before discussing the principles of exercise training, we first define key terms that will be used throughout the rest of this book.

Muscular Strength

Strength is defined as the maximal force that a muscle or muscle group can generate. Someone with a maximal capacity to bench press 100 kg (220 lb) has twice the strength of someone who can bench press 50 kg (110 lb). In this example, strength is defined as the maximal weight the individual can lift with one single effort. This is referred to as **1-repetition maximum**, or **1RM**. To determine 1RM in the weight room or fitness center, people select a weight that they know they can

lift at least one time. After a proper warm-up, they try to execute several repetitions. If they can perform more than one repetition, they add weight and try again to execute several repetitions. This continues until the person is unable to lift the weight more than a single repetition. This last weight that can be lifted only once is the 1RM for that particular exercise.

Muscle strength can be accurately measured in the research laboratory through use of specialized equipment that allows quantification of static strength and dynamic strength at various speeds and at various angles in the joint's range of motion (see figure 9.1). Gains in muscular strength involve changes in both the structure

FIGURE 9.1 An isokinetic testing and training device.

and neural control of muscle. These will be discussed in the following chapter (chapter 10).

Muscular Power

Power is defined as the rate of performing work, thus the product of force and velocity. Maximal muscular power, generally referred to simply as power, is the explosive aspect of strength, the product of strength and speed of movement.

$$power = force \times distance/time,$$
where force = strength
and distance/time = speed.

Consider an example. Two individuals can each bench press 200 kg (441 lb), moving the weight the same distance, from where the bar touches the chest to the position of full extension. But the one who can do it in half the time has twice the power of the slower individual. This is illustrated in table 9.1.

In focus

Maximal muscular power is the functional application of both strength and speed of movement. It is a key component in almost every sport.

Although absolute strength is an important component of performance, power is even more important for most competitive activities. In football, for example, an offensive lineman with a bench press 1RM of 200 kg (441 lb) may be unable to control a defensive lineman with a bench press 1RM of only 150 kg (330 lb) if the defensive lineman can move his 1RM at a much faster speed. The offensive lineman is 50 kg (110 lb) stronger, but the defensive lineman's faster speed coupled with adequate strength could give him the performance edge. Although field tests are available to estimate power, these tests are generally not very specific to power because their results are affected by factors other than power. Power can be measured, however, through use of more sophisticated electronic devices such as the one depicted in figure 9.1.

Throughout this book, the primary concern is with issues of muscular strength, with only brief mention of muscular power. Recall that power has two components: strength and speed. Speed is a more innate quality that changes little with training. Thus, improvements in power follow improvements in strength gained through traditional resistance training programs. However, high power output exercises, such as vertical jump training, have been shown to increase power for those specific movements.[1]

Muscular Endurance

Many sporting activities depend on the muscles' ability to repeatedly develop or sustain submaximal forces or to do both. This capacity to perform repeated muscle contractions, or to sustain a contraction over time, is termed **muscular endurance**. Examples of muscular endurance include performing sit-ups or push-ups, or sustaining force for an extended period of time, as when one is attempting to pin an opponent in wrestling. Although several excellent laboratory techniques are available to directly measure muscular endurance, a simple way to estimate it is to assess the maximum number of repetitions one can perform at a given percentage of 1RM. For example, a man who has a 1RM for the bench press of 100 kg (220 lb) could evaluate his muscular endurance independent of his muscular strength by measuring how many repetitions he could perform at, for example, 75% of that load (75 kg, or 165 lb). Muscular endurance is increased through gains in muscular strength and through changes in local metabolic and cardiovascular function. Metabolic and circulatory adaptations that occur with training are discussed in chapter 11.

Table 9.1 illustrates the functional differences between strength, power, and muscular endurance in three athletes. The actual values have been greatly exaggerated for the purpose of illustration. From this table we can see that although Athlete A has half the strength of Athletes B and C, he has twice the power of Athlete B and is equal in power to Athlete C. Therefore, because of his fast speed of movement, his lack of strength does not seriously limit his power output. Also, for purposes of designing training programs, the analysis of these three athletes indicates that Athlete A should focus training on developing strength, without losing speed; Athlete B should focus training on developing speed of movement, although this is unlikely to change much; and Athlete C should focus training on developing muscular endurance. These recommendations are made assuming that each athlete needs to optimize performance in each of these three areas.

Aerobic Power

Aerobic power is defined as the rate of energy release by cellular metabolic processes that depend upon the availability and involvement of oxygen. Maximal aerobic power refers to the maximal capacity for aerobic resynthesis of ATP and is synonymous with the terms aerobic capacity and maximal oxygen uptake ($\dot{V}O_{2max}$). Maximal aerobic power is limited primarily by the cardiovascular system and to a lesser extent by respiration and metabolism. The best laboratory test of aerobic power is a graded exercise test to exhaustion during which

TABLE 9.1 Strength, Power, and Muscular Endurance of Three Athletes Performing the Bench Press

Component	Athlete A	Athlete B	Athlete C
Strength[a]	100 kg	200 kg	200 kg
Power[b]	100 kg lifted 0.6 m in 0.5 s = 120 kg · m/s = 1,177 J/s or 1,177 W	200 kg lifted 0.6 m in 2.0 s = 60 kg · m/s = 588 J/s or 588 W	200 kg lifted 0.6 m in 1.0 s = 120 kg · m/s = 1,177 J/s or 1,177 W
Muscular endurance[c]	10 repetitions with 75 kg	10 repetitions with 150 kg	5 repetitions with 150 kg

[a]Strength was determined by the maximum amount of weight the athlete could bench press just once (i.e., the 1RM).

[b]Power was determined as the athlete performed the 1RM test as explosively as possible. Power was calculated as the product of force (weight lifted) times the distance lifted from the chest to full arm extension (0.6 m or about 2 ft), divided by the time it took to complete the lift.

[c]Muscular endurance was determined by the greatest number of repetitions that could be completed using 75% of the 1RM.

$\dot{V}O_2$ is measured and $\dot{V}O_{2max}$ is determined, as discussed in detail in chapter 5. A number of submaximal and maximal field tests, using walking, jogging, running, cycling, swimming, and rowing, have been developed to estimate $\dot{V}O_{2max}$ without the need to actually measure it in the laboratory.

Anaerobic Power

Anaerobic power is defined as the rate of energy release by cellular metabolic processes that function without the involvement of oxygen. Maximal anaerobic power, or anaerobic capacity, is defined as the maximal capacity of the anaerobic system (ATP-PCr system and anaerobic glycolytic system) to produce ATP. Unlike the situation with aerobic power, there is no universally accepted laboratory test to determine anaerobic power. Several tests provide estimates of maximal anaerobic power, as discussed in chapter 5, and include the maximal accumulated oxygen deficit, the critical power test, and the Wingate anaerobic test.

In review

➤ Muscular strength refers to the ability of a muscle to exert force.

➤ Muscular power is the rate of performing work, or the product of force and velocity.

➤ Muscular endurance is the capacity to sustain a static contraction or to maintain repeated muscle contractions.

➤ Maximal aerobic power, or aerobic capacity, is the maximal capacity for aerobic resynthesis of ATP.

➤ Maximal anaerobic power, or anaerobic capacity, is defined as the maximal capacity of the anaerobic system to produce ATP.

General Principles of Training

The next two chapters present in detail the specific physiological adaptations that result from resistance training, aerobic training, and anaerobic training. Several principles, however, can be applied to all forms of physical training.

Principle of Individuality

Athletes are not all created with the same ability to respond to an acute exercise bout, or the same capacity to adapt to exercise training. Heredity plays a major role in determining the body's response to a single bout of exercise, as well as chronic changes to a training program. This is the **principle of individuality**. Except for identical twins, no two people have exactly the same genetic characteristics, so individuals are unlikely to show precisely the same responses. Variations in cellular growth rates, metabolism, cardiovascular and respiratory regulation, and neural and endocrine regulation lead to tremendous individual variation. Such individual variation likely explains why some people show great improvement after participating in a given program (high responders) whereas others experience little or no change after following the same program (low responders). We discuss this phenomenon of high and low responders in more detail in chapter 11. For these reasons, any training program must take into account the specific needs and abilities of the individuals for whom it is designed. Do not expect all individuals to have exactly the same degree of improvement!

Principle of Specificity

Training adaptations are highly specific to the type of activity and to the volume and intensity of the exercise performed. To improve muscular power, for example, the shot-putter would not emphasize distance running or slow, low-intensity resistance training. He or she needs to develop explosive power. Similarly, the marathon runner would not concentrate on sprint-type interval training. This is likely the reason that athletes who train for strength and power, such as weightlifters, often have great strength but don't have highly developed aerobic endurance when compared to untrained people. According to the **principle of specificity**, exercise adaptations are specific to the mode and intensity of training, and the training program must stress the physiological systems that are critical for optimal performance in the given sport in order to achieve specific training adaptations.

Principle of Reversibility

Most athletes would agree that resistance training improves muscle strength and capacity to resist fatigue. Likewise, endurance training improves the ability to perform aerobic exercise at higher intensities and for longer periods. But if training is decreased or stopped (detraining), the physiological adaptations that caused those improvements in performance will be reversed. Any gains achieved with training will eventually be lost. The **principle of reversibility** lends scientific support to the saying, "Use it or lose it." A training program must include a maintenance plan. In chapter 14 we examine specific physiological changes that occur when the training stimulus stops.

Principle of Progressive Overload

Two important concepts, overload and progressive training, form the foundation of all training programs. According to the **principle of progressive overload**, systematically increasing the demands on the body is necessary for further improvement. For example, when undergoing a strength training program, in order to gain strength the muscles must be overloaded, which means they must be loaded beyond the point to which they are normally loaded. Progressive resistance training implies that as the muscles become stronger, either increased resistance or increased repetitions are required to stimulate further strength increases.

As an example, consider a young man who can perform only 10 repetitions of a bench press before reaching fatigue, using 50 kg (110 lb) of weight. With a week or two of resistance training, he should be able to increase to 14 or 15 repetitions with the same weight. He then adds 2.3 kg (5 lb) to the bar, and his repetitions decrease to 8 or 10. As he continues to train, the repetitions continue to increase; and within another

week or two, he is ready to add another 5 lb of weight. Thus, improvement depends on a progressive increase in the amount of weight lifted. Similarly, with anaerobic and aerobic training, training volume (intensity and duration) can be increased progressively.

Principle of Variation

The **principle of variation**, also called the **principle of periodization**, first proposed in the 1960s, has become very popular over the past 30 years in the area of resistance training. Periodization is the systematic process of changing one or more variables in the training program—mode, volume, or intensity—over time to allow for the training stimulus to remain challenging and effective.[1] Training intensity and volume of training are the most commonly manipulated aspects of training to achieve peak levels of fitness for competition. Classical periodization involves high initial training volume with low intensity; then, as training progresses, volume decreases and intensity gradually increases. Undulating periodization uses more frequent variation within a training cycle.

For sport-specific training, the volume and intensity of training are varied over a macrocycle, which is generally up to a year of training. A macrocycle is composed of two or more mesocycles that are dictated by the dates of major competitions. Each mesocycle is subdivided into periods of preparation, competition, and transition. This principle is discussed in greater detail in chapter 14.

In review

- According to the principle of individuality, each person responds uniquely to training, and training programs must be designed to allow for individual variation.

- According to the principle of specificity, to maximize benefits, training must be specifically matched to the type of activity or sport the person engages in. An athlete involved in a sport that requires tremendous strength, such as weightlifting, would not expect great strength gains from endurance running.

- According to the principle of reversibility, training benefits are lost if training is either discontinued or reduced abruptly. To avoid this, all training programs must include a maintenance program.

- According to the principle of progressive overload, as the body adapts to training at a given volume and intensity, the stress placed on the body must be increased progressively for the training stimulus to remain effective in producing further improvements.

➤ According to the principle of variation (or periodization), one or more aspects of the training program should be altered over time to maximize effectiveness of training. The systematic variation of volume and intensity is most effective for long-term progression.

Resistance Training Programs

Over the past 50 to 75 years, research has provided a substantial knowledge base concerning resistance training and its application to health and sport. The health aspects of resistance training are discussed in chapter 20. This section concerns primarily the use of resistance training for sport.

Training Needs Analysis

Fleck and Kraemer[3] suggest that a **needs analysis** is an appropriate first step in designing and prescribing a resistance training program for athletes. The needs analysis should include the following assessment:

- What major muscle groups need to be trained?
- What type of training should be used to achieve the desired outcome (improved strength, power, etc.)?
- What energy system should be stressed?
- What are the primary sites of concern for injury prevention?

Once such a needs analysis has been completed, the resistance training program can be designed and prescribed in terms of

- the exercises that will be performed;
- the order in which they will be performed;
- the number of sets for each exercise;
- the rest periods between sets and between exercises; and
- the amount of resistance, the number of repetitions, and the velocity of movement to be used.

In 2009, the American College of Sports Medicine (ACSM) revised its position stand on progressive resistance training for healthy adults.[1] Previous statements specified a minimum of one set of 8 to 12 reps for each of 8 to 10 different exercises that together involve all of the major muscle groups for all adults. The new position stand recommends resistance training models specific to desired outcomes, that is, improvements in strength, muscle hypertrophy, power, local muscular endurance, or gross motor performance.

Improving Strength, Hypertrophy, and Power

Resistance programs aimed at improving strength should involve repetitions with both concentric (CON, muscle shortening) and eccentric (ECC, muscle lengthening) actions. Isometric contractions play a beneficial, but secondary, role and may be included as well. CON strength improvement is greatest when ECC exercises are included, and ECC training has been shown to produce specific benefits for those action-specific movements. Large muscle groups should be stressed before smaller groups, multiple-joint exercises before single-joint exercises, and higher-intensity efforts before those of lower intensity. Table 9.2 provides a summary of the ACSM recommendations on loading, volume (sets and reps), velocity of movements, and frequency of training.

It is recommended that rest periods of 2 to 3 min or more be used between heavy loads for novice and intermediate lifters; for advanced lifters 1 or 2 min may suffice. Once an individual can perform the current workload at or above the desired number of reps for two consecutive training sessions, a 2% to 10% increase in load should be applied. While both machine-based exercises and free weights can be used for novice and intermediate lifters, for advanced lifters the emphasis should be placed on free weights.

When muscle hypertrophy, for example in bodybuilders, or development of muscular power is the goal, recommendations for sequencing, rest periods, and so on are the same as those for strength development. However, as shown in table 9.2, other aspects of the program differ.

Types of Resistance Training

Resistance training can use static contractions, dynamic contractions, or both. Dynamic contractions can include either or both concentric and eccentric contractions using free weights, variable resistance, isokinetic actions, and plyometrics.

Static-Contraction Resistance Training

Static-contraction resistance training, also called **isometric training**, evolved in the early 20th century but gained great popularity and support in the mid-1950s

TABLE 9.2 American College of Sports Medicine's Recommendations for Resistance Training Programs[a]

Primary goal of resistance training program	Training level	Loading	Volume	Velocity	Frequency (times per week)
Strength development	Novice	60-70% 1RM	1-3 sets, 8-12 reps	Slow, moderate	2-3
	Intermediate	70-80% 1RM	Multiple sets, 6-12 reps	Moderate	3-4
	Advanced	80-100% 1RM	Multiple sets, 1-12 reps	Unintentionally slow to fast	4-6
Muscle hypertrophy	Novice	70-85% 1RM	1-3 sets, 8-12 reps	Slow, moderate	2-3
	Intermediate	70-85% 1RM	1-3 sets, 6-12 reps	Slow, moderate	4
	Advanced	70-100% 1RM; emphasis on 70-85%	3-6 sets, 1-12 reps	Slow, moderate, fast	4-6
Development of muscle power	Novice	0-60% 1RM—lower body; 30-60% 1RM—upper body	1-3 sets, 3-6 reps	Moderate	2-3
	Intermediate	0-60% 1RM—lower body; 30-60% 1RM—upper body	1-3 sets, 3-6 reps	Fast	3-4
	Advanced	85-100% 1RM	3-6 sets, 1-6 reps, various strategies	Fast	4-5
Increased local muscular endurance	Novice	Light	1-3 sets, 10-15 reps	Slow—moderate reps Moderate—high reps	2-3
	Intermediate	Light	1-3 sets, 10-15 reps	Slow—moderate reps Moderate—high reps	3-4
	Advanced	30-80% 1RM	Various strategies, 10-25 reps or more	Slow—moderate reps Moderate—high reps	4-6

[a]These recommendations also include the type of muscle action (eccentric and concentric), single-joint versus multiple-joint exercises, order or sequence of exercises, and the rest intervals. See text for further information.

[b]Periodized—see text for explanation of periodization.

Adapted from ACSM, 2009.

as a result of research by several German scientists. These studies indicated that static resistance training caused tremendous strength gains and that those gains exceeded the gains resulting from dynamic-contraction procedures. Subsequent studies were unable to reproduce the original studies' results, yet static contractions remain an important form of training, particularly for core stabilization and enhancing grip strength.[1] Additionally, in postsurgical rehabilitation when the limb is immobilized and thus incapable of dynamic contractions, static contractions facilitate recovery and reduce muscle atrophy and strength loss.

Free Weights Versus Machines

With **free weights**, such as barbells and dumbbells, the resistance or weight lifted remains constant throughout the dynamic range of movement. If a 50 kg (110 lb) weight is lifted, it will always weigh 50 kg. In contrast, a variable-resistance contraction involves varying the resistance to try to match it to the strength curve. Figure 9.2 illustrates how strength varies throughout the range of motion in a two-arm curl. Maximal strength production by the elbow flexors occurs at approximately 100° in the range of movement. These muscles are weakest at 60° (elbows fully flexed) and at 180° (elbows fully extended). In these positions, one is able to generate only 67% and 71%, respectively, of the maximal force-producing capabilities at the optimal angle of 100°.

(In) focus

The ability of a muscle or muscle group to generate force varies throughout the full range of movement.

When one is using free weights, the range of motion is less restricted than with machines, and the resistance or weight used to train the muscle is limited by the weakest point in that range of motion. If the person in figure 9.2 had the capacity to lift only 45 kg (100 lb) at the optimal angle of 100°, then he would be able to lift only 32 kg (71 lb) at the fully extended position of 180°. Therefore, if he is starting with a barbell loaded with 32 kg, he can just barely move it from the fully extended position to start his lift. However, by the time he gets to an angle of 100° in his full range of motion, he is lifting only 70% of what he could maximally lift at that angle. Thus, free weights maximally tax the weakest points in the range of motion and provide moderate resistance at the midrange (90-140°). Individuals performing the two-arm curl tend to greatly reduce their range of motion as they start to fatigue (referred to as "cheating"). They are simply trying to stay out of the weakest portion of their range of motion. The bottom

FIGURE 9.2 The variation in strength relative to the angle of the elbow during the two-arm curl. Strength is optimized at an angle of 100°. The maximal force-development capacity of the muscle group at a given angle is given as a percentage of the capacity at the optimal angle of 100°.

line is that with free weights, the maximum weight one can lift is limited by the weakest portion in the range of motion, which means that the strongest position in the range of motion is never maximally taxed! However, free weights do offer some distinct advantages, especially for the expert lifter.

Starting in the 1970s, a number of resistance training machines or devices were introduced that used stacked weights, variable-resistance, and isokinetic techniques. Weight machines have been regarded as safer, are easy to use, and allow performance of some exercises that are difficult to do with free weights. Machines help stabilize the body, especially for novice lifters, and limit the muscle action to that desired without extraneous muscle groups firing.

On the other hand, free weights offer some advantages that resistance machines do not provide. The athlete must control the weight being lifted. An athlete must recruit more motor units—not only in the muscles being trained but also in supporting muscles—to gain control of the bar, to stabilize the weight lifted, and to maintain body balance. The athlete must both balance and stabilize the weight. In that regard, when an athlete is training for a sport such as football, the experience with free weights more closely resembles actions associated with actual sport competition. Also, because free weights do not limit the range of motion of a particular exercise, optimal training specificity can be achieved. Whereas a bicep curl on a machine may be done only in the vertical plane, an athlete using free weights can perform the curl in any plane, choosing one, for example, that reflects a sport-specific motion. And finally, data show that if significant strength gains are to be achieved over a shortened training period, free weights may provide greater strength gains than many types of weight machines.

Both machine-based resistance programs and free-weight training programs result in measurable gains in strength, hypertrophy, and power. Free-weight programs result in greater improvements in free-weight tests, and vice versa. The choice to use weight machines versus free weights depends on the experience of the lifter and the desired outcomes.

In focus

When neutral testing devices are used, strength gains from free-weight programs and machine-based programs are similar.

Eccentric Training

Another form of dynamic-contraction resistance training, called **eccentric training**, emphasizes the eccentric phase. With eccentric contractions, the muscle's ability to resist force

is considerably greater than with concentric contractions (see chapter 1). Subjecting the muscle to this greater training stimulus theoretically produces greater strength gains.

Early research was unable to show a clear advantage of eccentric training over either concentric- or static-contraction training. However, more recent well-controlled studies have shown the importance of including the eccentric phase of muscle contraction along with the concentric phase to maximize gains in strength and size. Further, eccentric contraction is important to muscle hypertrophy, as discussed in the next chapter.

Variable-Resistance Training

With a variable-resistance device, the resistance is decreased at the weakest points in the range of movement and increased at the strongest points. **Variable-resistance training** is the basis for several popular resistance training machines. The underlying theory is that the muscle can be more fully trained if it is forced to act at higher constant percentages of its capacity throughout each point in its range of movement. Figure 9.3 illustrates a variable-resistance device in which a cam alters the resistance through the range of motion.

Isokinetic Training

Isokinetic training is conducted with equipment that keeps movement speed constant. Whether one applies very light force or an all-out maximal

FIGURE 9.3 A variable-resistance training device that uses a cam to alter the resistance through the range of motion.

muscle contraction, the speed of movement does not vary. Using electronics, air, or hydraulics, the device can be preset to control the speed of movement (angular velocity) from 0°/s (static contraction) up to 300°/s or higher. An isokinetic device is illustrated in figure 9.1. Theoretically, if properly motivated, the individual can contract the muscles at maximal force at all points in the range of motion.

Plyometrics

Plyometrics, or stretch–shortening cycle exercise, became popular during the late 1970s and early 1980s primarily for improving jumping ability. Proposed to bridge the gap between speed and strength training, plyometrics uses the stretch reflex to facilitate recruitment of motor units. It also stores energy in the elastic and contractile components of muscle during the eccentric contraction (stretch) that can be recovered during the concentric contraction. As an example, to develop knee extensor muscle strength and power, a person goes from standing upright to a deep squat position (eccentric contraction) and then jumps up onto a

box (concentric contraction), landing in a squat position on top of the box. The person then jumps off the box onto the ground, landing in a squat position, and repeats the sequence with the next box (see figure 9.4).

Electrical Stimulation

One can stimulate a muscle by passing an electric current directly across it or its motor nerve. This technique, called **electrical stimulation**, has proven effective in a clinical setting to reduce the loss of strength and muscle size during periods of immobilization and to restore strength and size during rehabilitation. Electrical stimulation training also has been used experimentally in healthy subjects (including athletes). However, the gains reported are no greater than those achieved with more conventional training. Athletes have used this technique to supplement their regular training programs, but no evidence shows any additional gains in strength, power, or performance from this supplementation.

Core Stability and Strength

In recent years a significant emphasis has been placed on "core" stability and strengthening exercises. While there are varying opinions on what anatomical features constitute the "core," the general consensus is that the core is the group of trunk muscles that surround the spine and abdominal viscera and include the abdominal, gluteal, hip girdle, paraspinal, and other accessory muscles.

Initially, this type of core-specific exercise training was explored in the rehabilitation setting, specifically for the treatment of low back pain, but its benefits have also been recognized in sport performance. Greater core stability may benefit sporting performance by providing a foundation for greater force production and force transfer to the extremities. For example, having the core stabilized and engaged in the simple action of throwing a ball allows for greater biomechanical efficiency in the limb transmitting the force to throw the ball and for the activation of stabilizing muscles in the contralateral arm. The principle of core stabilization promotes proximal stability for distal mobility.

There has been little definitive research on the benefits of core stability and core strength training. One reason is that there are no standardized tests for evaluating core strength and stability. Further, the studies that have been done have been mainly with injured populations and not specific to athletic performance. However, the limited research does show that this type of training decreases the likelihood of injury, especially in the lower back and the lower extremities, during sport performance. The physiological explanation for this finding is that core stability training increases the

FIGURE 9.4 Plyometric box jumping (see the text for a detailed explanation).

sensitivity of the muscle spindles, thereby permitting a greater state of readiness for loading joints during movement[5] and protecting the body from injury.

The many different types of core stability and strengthening training include balance and instability resistance (e.g., physioball). It is thought that because the core is composed mainly of type I muscle fibers, the core musculature may respond well to multiple sets of exercises with high repetitions.[2] Yoga, Pilates, tai chi, and the physioball are commonly incorporated into athletes' training programs to promote core stability and strength. Further research is needed to determine the benefits of core training and the underlying mechanisms.

In review

➤ A needs analysis should be completed before a resistance training program is designed to tailor the program to the athlete's specific needs.

➤ Low-repetition, high-resistance training enhances strength development, whereas high-repetition, low-intensity training optimizes the development of muscular endurance.

➤ Variation (or periodization), through which various aspects of the training program are altered, is important to optimize results and to prevent overtraining or burnout.

➤ Resistance programs aimed at improving strength should involve repetitions with both concentric (muscle shortening) and eccentric (muscle lengthening) actions. Isometric contractions play a beneficial, but secondary, role and may be included as well.

➤ Large muscle groups should be stressed before smaller groups, multiple-joint exercises before single-joint exercises, and higher-intensity efforts before those of lower intensity.

➤ Rest periods of 2 to 3 min or more should be incorporated between heavy loads for novice and intermediate lifters; for advanced lifters, 1 to 2 min may suffice.

➤ While both machine-based exercises and free weights can be used for novice and intermediate lifters, for advanced lifters the emphasis should be placed on free weights.

➤ Electrical stimulation can be successfully used in rehabilitating athletes but has no additional benefits when used to supplement resistance training in healthy athletes.

➤ Exercises aimed at improving core stability benefit sport performance by providing a foundation for greater force production and force transfer to the extremities while stabilizing other parts of the body.

Anaerobic and Aerobic Power Training Programs

Anaerobic and aerobic power training programs, while quite different at the extremes (e.g., training for the 100 m dash vs. the 42.2 km [26.2 mi] full marathon), are designed along a continuum. Table 9.3 illustrates how training requirements vary in competitive running events as one goes from short sprints to long distances. With this table serving as an example that can be applied to all sports, the primary emphasis for the short sprints is on training the ATP-PCr system. For longer sprints and middle distances, the primary emphasis is on the glycolytic system; and for the longer distances, the primary emphasis is on the oxidative system. Anaerobic power is represented by the ATP-PCr and anaerobic glycolytic systems, while aerobic power is represented by the oxidative system. Note, however, that even at the extremes, more than one energy system must be trained.

Different types of training programs can be used to meet the specific training requirements of each event, such as in running and swimming, and each sport. We will first describe some of the more popular types of training programs and how they are used to improve the specific energy systems.

Interval Training

The concept of **interval training** can be traced back to at least the 1930s, when the famous German coach Woldemar Gerschler formalized a structured system of interval training. Interval training consists of repeated bouts of high- to moderate-intensity exercise interspersed with periods of rest or reduced-intensity exercise. Research has shown that athletes can perform a considerably greater volume of exercise by breaking the total exercise period into shorter, more intense bouts, with rest or active recovery intervals inserted between the intense bouts.

The vocabulary used to describe an interval training program is similar to that used in resistance training and includes the terms sets, repetitions, training time, training distance and frequency, exercise interval, and rest or active recovery interval. Interval training is frequently prescribed in these terms as illustrated in the following example for a middle-distance runner:

- Set 1: 6×400 m (436 yd) at 75 s (90 s slow jog)
- Set 2: 6×800 m (872 yd) at 180 s (200 s jog-walk)

For the first set, the athlete would run six repetitions of 400 m each, completing the exercise interval in 75 s and recovering for 90 s between exercise intervals with slow jogging. The second set consists of running six repetitions of 800 m each, completing the exercise interval in 180 s, and recovering for 200 s with walking-jogging.

TABLE 9.3 Percentage of Emphasis on the Three Metabolic Energy Systems in Training for Various Running Events

Running event	Anaerobic speed (ATP-PCr system)	Anaerobic endurance (anaerobic glycolytic system)	Aerobic endurance (oxidative system)
100 m (109 yd)	95	3	2
200 m (218 yd)	95	2	3
400 m (436 yd)	80	15	5
800 m (872 yd)	30	65	5
1,500 m (0.93 mi)	20	55	25
3,000 m (1.86 mi)	20	40	40
5,000 m (3.10 mi)	10	20	70
10,000 m (6.2 mi)	5	15	80
Marathon (42.2 km; 26.2 mi)	5	5	90

Adapted from *Exercise physiology*, F. Wilt, "Training for competitive running," edited by H.B. Falls. Copyright Elsevier 1968.

While interval training is traditionally associated with track, cross country, and swimming, it is appropriate for all sports and activities. One can adapt interval training procedures for each sport or event by first selecting the form or mode of training and then manipulating the following primary variables to fit the sport and athlete:

- Rate of the exercise interval
- Distance of the exercise interval
- Number of repetitions and sets during each training session
- Duration of the rest or active recovery interval
- Type of activity during the active recovery interval
- Frequency of training per week

Exercise Interval Intensity

One can determine the intensity of the exercise interval either by establishing a specific duration for a set distance, as illustrated in our previous example for set 1 (i.e., 75 s for 400 m), or by using a fixed percentage of the athlete's maximal heart rate (HR_{max}). Setting a specific duration is more practical, particularly for short sprints. One typically determines this by using the athlete's best time for the set distance and then adjusting the duration according to the relative intensity that the athlete wants to achieve, with 100% equal to the athlete's best time. As an example, to develop the ATP-PCr system, the intensity should be near maximal (e.g., 90-98%); to develop the anaerobic glycolytic system, it should be high (e.g., 80-95%); and to develop the aerobic system, it should be moderate to high (e.g., 75-85%). These estimated percentages are only approximations and are dependent on the athlete's genetic potential and fitness level, duration of the interval (e.g., 10 s vs. 10 min), number of repetitions and sets, and duration of the active recovery interval.

Using a fixed percentage of the athlete's HR_{max} might provide a better index of the physiological stress experienced by the athlete. Heart rate monitors are now readily available and relatively inexpensive (see figure 9.5). HR_{max} can be determined during a maximal exercise test in the laboratory as described in chapter 8 or during an all-out run on the track using the heart rate monitor. Training the ATP-PCr system will require training at very high percentages of the athlete's HR_{max} (e.g., 90-100%), as will training to develop the anaerobic glycolytic system (e.g., 85-100% HR_{max}). To develop the aerobic system, the intensity should be moderate to high (e.g., 70-90% HR_{max}).

Figure 9.6 illustrates changes in blood lactate concentration in a single runner using interval training at three different intensities corresponding to those

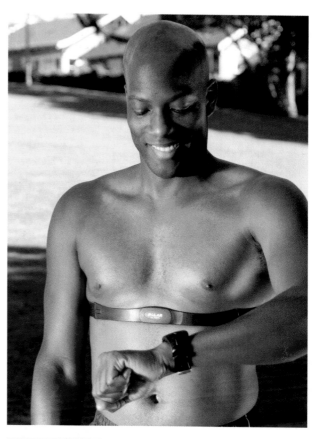

FIGURE 9.5 A runner outfitted with a heart rate monitor. The receiving unit, attached to the chest strap, picks up and transmits electrical impulses from the heart to the digital monitor and memory device worn on the wrist. After the workout, the contents of the memory device can be downloaded to a computer.

intensities needed to train the ATP-PCr system, the glycolytic system, and the oxidative system. The runner performed five repetitions at each intensity within a single set on different days, and the lactate concentrations were obtained from a blood sample taken after the last repetition of each intensity.

Distance of the Exercise Interval

The distance of the exercise interval is determined by the requirements of the event, sport, or activity. Athletes who run or sprint short distances, such as track sprinters, basketball players, and soccer players, will utilize short intervals of 30 m to 200 m (33-219 yd), although a 200 m sprinter will frequently run over distances of 300 to 400 m (328-437 yd). A 1,500 m runner may run intervals as short as 200 m to increase speed; but most of his or her training would be at distances of 400 to 1,500 m (437-1,640 yd), or even longer distances, to increase endurance and decrease fatigue or exhaustion in the race.

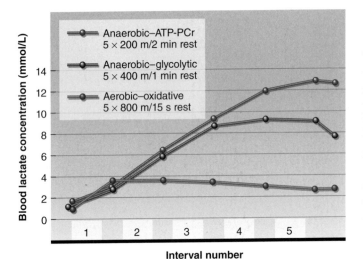

FIGURE 9.6 Blood lactate concentrations in a single runner following a single set of five repetitions of interval training at three different paces, each on different days, corresponding to the appropriate pace for training each energy system.

Number of Repetitions and Sets During Each Training Session

The number of repetitions and sets will also be largely determined by the needs of the sport, event, or activity. Generally, the shorter and more intense the interval, the greater should be the number of repetitions and sets. As the training interval is lengthened in both distance and duration, the number of repetitions and sets is correspondingly reduced.

Duration of the Rest or Active Recovery Interval

The duration of the rest or active recovery interval will depend on how rapidly the athlete recovers from the exercise interval. The extent of recovery is best determined by the reduction of the athlete's heart rate to a predetermined level during the rest or active recovery period. For younger athletes (30 years of age or younger), heart rate is generally allowed to drop to between 130 and 150 beats/min before the next exercise interval begins. For those over 30 years, since HR_{max} decreases ~1 beat/min per year, we subtract the difference between the athlete's age and 30 years from both 130 and 150. So, for a 45-year-old, we would subtract 15 beats/min to obtain the athlete's recovery range of 115 to 135 beats/min. The recovery interval between sets can be established in a similar manner, but generally the heart rate should be below 120 beats/min.

Type of Activity During the Active Recovery Interval

The type of activity performed during the active recovery interval for land-based training can vary from slow walking to rapid walking or jogging. In the pool, slow swimming using alternative strokes or the primary stroke is appropriate. In some cases, usually in the pool, total rest can be used. Generally, the more intense the exercise interval, the lighter or less intense the activity performed in the recovery interval. As the athlete becomes better conditioned, he or she will be able to increase the intensity of the exercise interval or decrease the duration of the rest interval, or both.

Frequency of Training per Week

The frequency of training will depend largely on the purpose of the interval training. A world-class sprinter or middle-distance runner typically works out five to seven days a week, although not every workout will include interval training. Swimmers use interval training almost exclusively. Team sport athletes can benefit from two to four days of interval training per week when interval training is used only as a supplement to a general conditioning program.

The coach or athlete who is interested in the specific details of how to organize and administer an interval training program should refer to the classic text by Fox and Mathews (1974).[4] These authors have provided many excellent examples of how interval training can be used for various types of sports.

Continuous Training

Continuous training involves continuous activity without rest intervals. This can vary from long, slow distance (LSD) training to high-intensity endurance training. Continuous training is structured primarily to affect the oxidative and glycolytic energy systems. High-intensity continuous activity is usually performed at intensities representing 85% to 95% of the athlete's HR_{max}. For swimmers and track and cross-country athletes, this could be above, or at or near, race pace. This pace would likely match or exceed the pace associated with the athlete's lactate threshold. Scientific evidence has clearly demonstrated that marathon runners typically race at, or very close to, their lactate threshold.

LSD training became extremely popular in the 1960s. With this form of training, introduced in the 1920s by Dr. Ernst Van Auken, a German physician and coach, the athlete typically trains at relatively low intensities, between 60% and 80% of HR_{max}, which is approximately the equivalent of 50% to 75% of VO_{2max}. Distance, rather than speed, is the main objective. Distance runners may train 15 to 30 mi (24-48 km) per day using LSD techniques, with weekly distances of 100 to 200 mi (161-322 km). The pace of the run is substantially slower than the runner's maximal pace. While less stressful to the cardiovascular and respiratory systems, extreme distances can result in overuse injuries and general breakdown of muscles and joints. Further, the serious runner needs to train at or near race pace on a regular basis to develop leg speed and strength. Thus, most runners will vary their workout from one day to the next, from week to week, and from month to month.

Long, slow distance training is probably the most popular and safest form of aerobic endurance conditioning for the nonathlete who just wants to get into shape and stay in shape for health-related purposes. More vigorous or burst types of activity generally are not encouraged in older, sedentary people. Long, slow distance is also a good training program for athletes in team sports for maintaining aerobic endurance during the season as well as the off-season.

Fartlek training, or speed play, is another form of continuous exercise that has a flavor of interval training. This form of training was developed in Sweden in the 1930s and is used primarily by distance runners. The athlete varies the pace from high speed to jogging speed at his or her discretion. This is a free form of training in which fun is the primary goal, and distance and time are not even considered. Fartlek training is normally performed in the countryside where there are a variety of hills. Many coaches have used Fartlek training to supplement either high-intensity, continuous training, or interval training, since it provides variety to the normal training routine.

Interval-Circuit Training

Introduced in the Scandinavian countries in the 1960s and 1970s, **interval-circuit training** combines interval and circuit training into one workout. The circuit may be 3,000 to 10,000 m in length, with stations every 400 to 1,600 m (437-1,750 yd). The athlete jogs, runs, or sprints the distance between stations; stops at each station to perform a strength, flexibility, or muscular endurance exercise in a manner similar to that in actual circuit training; and continues on, jogging, running, or sprinting to the next station. These courses are typically located in parks or in the country where there are many trees and hills. Such a training regimen can benefit almost any type of athlete and provide diversity to what might be an otherwise monotonous training regimen.

In review

➤ Anaerobic and aerobic power training programs are designed to train the three metabolic energy systems: ATP-PCr system, anaerobic glycolytic system, and oxidative system.

➤ Interval training consists of repeated bouts of high- to moderate-intensity exercise interspersed with periods of rest or reduced intensity exercise. For short intervals, the rate or pace of activity and the number of repetitions are usually high, and the recovery period is usually short. Just the opposite is the case for long intervals.

➤ Both the exercise rate and the recovery rate can be closely monitored with use of a heart rate monitor.

➤ Interval training is appropriate for all sports. The length and intensity of intervals can be adjusted based on the sport requirements.

➤ Continuous training has no rest intervals and can vary from LSD training to high-intensity training. Long, slow distance training is very popular for general fitness training.

➤ Fartlek training, or speed play, is an excellent activity for recovering from several days or more of intense training.

➤ Interval-circuit training combines interval training and circuit training into one workout.

In this chapter, we reviewed general principles of training and the terminology used to describe these principles. We then learned the essential ingredients of successful resistance training and anaerobic and aerobic power training programs. With this background, we can now focus on how the body adapts to these different types of training programs. In the next chapter, we will see how the body responds to resistance training.

Key Terms

1-repetition maximum (1RM)

aerobic power

anaerobic power

continuous training

eccentric training

electrical stimulation

Fartlek training

free weights

hypertrophy

interval-circuit training

interval training

isokinetic training

isometric training

LSD training

muscular endurance

needs analysis

plyometrics

power

principle of individuality

principle of periodization

principle of progressive overload

principle of reversibility

principle of specificity

principle of variation

static-contraction resistance training

strength

variable-resistance training

Study Questions

1. Define and differentiate the terms *strength*, *power*, and *muscular endurance*. How does each component relate to athletic performance?

2. Define aerobic and anaerobic power. How does each relate to athletic performance?

3. Describe and provide examples for the principles of individuality, specificity, reversibility, progressive overload, and variation.

4. What factors need to be considered when one is conducting a needs analysis for designing a resistance training program?

5. What would be the appropriate range for resistance and repetitions when one is designing a resistance training program targeted to develop strength? Muscular endurance? Muscular power? Hypertrophy?

6. Describe the various types of resistance training and explain the advantages and disadvantages of each.

7. What type of training program would likely provide the greatest improvement for sprinters? Marathon runners? Football players?

8. Describe the various forms of interval and continuous training programs and discuss the advantages and disadvantages of each. Indicate the sport or event most likely to benefit from each one.

Study Guide Activities

In addition to the activities listed in the chapter opening outline on page 209, two other activities are available in the web study guide, located at

www.HumanKinetics.com/PhysiologyOfSportAndExercise

 The **KEY TERMS** activity reviews important terms, and the end-of-chapter **QUIZ** tests your understanding of the material covered in the chapter.

Adaptations to Resistance Training

10

In this chapter and in the web study guide

 ACTIVITY 10.1 What Causes Strength Gains? explores the roles of neural adaptations and hypertrophy in strength gains.

 ACTIVITY 10.2 Muscle Soreness and Cramps investigates the causes and treatment of muscle soreness and muscle cramps.

We know that athletes who use resistance training become much stronger. For a number of years, Dr. William Gonyea and his colleagues at the University of Texas Health Sciences Center Dallas have examined how the athlete's muscle gets stronger with resistance training. But Dr. Gonyea and his colleagues have been working with a different type of athlete—cats! The cats receive food rewards for their daily workouts that entice them to work very hard "pumping iron." Like their human counterparts, these cats experience substantial increases in strength and in muscle size. This chapter discusses the results of Dr. Gonyea's studies as he and his colleagues have challenged traditional thinking on how a muscle increases its size.

With chronic exercise, many adaptations occur in the neuromuscular system. The extent of the adaptations depends on the type of training program followed: Aerobic training, such as jogging or swimming, results in little or no gain in muscular strength and power, but major neuromuscular adaptations occur with **resistance training**.

Resistance training was once considered inappropriate for athletes except those in competitive weightlifting; weight events in track and field; and—on a limited basis—football, wrestling, and boxing. Women were essentially banned from the weight room! But in the late 1960s and early 1970s, coaches and researchers discovered that strength and power training are beneficial for almost all sports and activities, and for women as well as men. Finally, in the late 1980s and early 1990s, health professionals began to recognize the importance of resistance training to overall health and fitness.

Most athletes now include strength and power training as important components of their overall training programs. Much of this attitude change is attributable to research that has proven the performance benefits of resistance training and to innovations in training techniques and equipment. Resistance training is now an important part of the exercise prescription for all those who seek the health-related benefits of exercise.

Resistance Training and Gains in Muscular Fitness

Throughout this book, we see how important muscular fitness is to athletic performance as well as general health. How do we get stronger, and how do we increase muscle power and muscle endurance? Maintaining an active lifestyle is important in maintaining muscular fitness, but resistance training programs are necessary to increase strength, power, and endurance. In this section, we briefly review the changes that result from resistance training. We focus on strength, with only a brief mention of power and muscular endurance—topics that are discussed in more detail later in this book.

The neuromuscular system is one of the most responsive systems in the body to training. Resistance training programs can produce substantial strength gains. Within three to six months, one can see from 25%

to 100% improvement, sometimes even more. These estimates of percentage gains in strength are, however, somewhat misleading. Most subjects in strength training research studies have never lifted weights or participated in any other form of resistance training. Much of their early gains in strength is the result of learning how to more effectively produce force and produce a true maximal movement, such as moving a barbell from the chest to a fully extended position in the bench press. This learning effect can account for as much as 50% of the overall gain in strength.[14]

Gains in strength appear to be similar when we compare women to men, children to adults, and elderly people to young and middle-aged adults, when these gains are expressed as a percentage of their initial strength. However, the increase in the absolute weight lifted will generally be greater in men compared to women, in adults compared to children, and in young adults compared to older adults. For example, after 20 weeks of resistance training, assume that a 12-year-old boy and a 25-year-old man each improves his bench press strength by 50%. If the man's initial bench press strength (1-repetition maximum, 1RM) was 50 kg (110 lb), he would have improved by 25 kg (55 lb) to a new 1RM of 75 kg (165 lb). If the boy's initial 1RM was 25 kg, he would have improved by 12.5 kg (28 lb) to a new 1RM of 37.5 kg (83 lb).

Muscle is very plastic, increasing in size and strength with exercise training and decreasing in size and strength when immobilized. The remainder of this chapter details what physiological adaptations occur that allow people to become stronger. We also address what causes the acute pain in the specific muscles trained during the first week or two of training.

Mechanisms of Gains in Muscle Strength

For many years, strength gains were assumed to result directly from increases in muscle size (hypertrophy). This assumption was logical because most who strength trained regularly often developed large, bulky muscles. Also, muscles associated with a limb immobilized in a cast for weeks or months start to decrease in size (**atrophy**) and lose strength almost immediately. Gains in muscle size are generally paralleled by gains in strength, and losses in muscle size correlate highly with losses in strength. Thus, it is tempting to conclude that a direct cause-and-effect relationship exists between muscle size and muscle strength. While there is an association between size and strength, muscle strength involves far more than mere muscle size.

This does not mean that muscle size is unimportant in the ultimate strength potential of the muscle. Size is extremely important, as revealed by the existing men's and women's world records for competitive weight-lifting, shown in figure 10.1. As weight classification increases (implying increased muscle mass), so does the record for the total weight lifted. However, the mechanisms associated with strength gains are very complex and are not completely understood. What, in addition to increased size of the muscle, explains strength gains with training? There is increasing evidence that the neural control of the trained muscle is also altered, allowing a greater force production from the muscle.

Neural Control of Strength Gains

An important neural component explains at least some of the strength gains that result from resistance training. Enoka has made a convincing argument that strength gains can be achieved without structural changes in muscle but not without neural adaptations.[5] Thus, strength is not solely a property of muscle. Rather, it is a property of the motor system. Motor unit recruitment, stimulation frequency, and other neural factors are also important to strength gains. They may well explain most, if not all, strength gains that occur in the absence of hypertrophy, as well as episodic superhuman feats of strength.

Synchronization and Recruitment of Additional Motor Units

Motor units are generally recruited asynchronously; they are not all called on at the same instant. They are controlled by a number of different neurons that can transmit either excitatory or inhibitory impulses (see chapter 3). Whether the muscle fibers contract or stay relaxed depends on the summation of the many impulses received by the given motor unit at any one time. The motor unit is activated and its muscle fibers contract only when the incoming excitatory impulses exceed the inhibitory impulses and the threshold is met or exceeded.

Strength gains may result from changes in the connections between motor neurons located in the spinal cord, allowing motor units to act more synchronously, facilitating contraction, and increasing the muscle's ability to generate force. There is good evidence to

FIGURE 10.1 World records for (*a*) the snatch, (*b*) the clean and jerk, and (*c*) total weight for men and women through 2010.

support increased motor unit synchronization with resistance training, but there is still controversy as to whether synchronization of motor unit activation produces a more forceful contraction. It is clear, however, that synchronization does improve the rate of force development and the capability to exert steady forces.[3]

An alternate possibility is simply that more motor units are recruited to perform the given task, independent of whether these motor units act in unison. Such improvement in recruitment patterns could result from an increase in neural drive to the α-motor neurons during maximal contraction. This increase in neural drive could also increase the frequency of discharge (rate coding) of the motor units. It is also possible that the inhibitory impulses are reduced, allowing more motor units to be activated or to be activated at a higher frequency.

Increased Rate Coding of Motor Units

The increase in neural drive of α-motor neurons could also increase the frequency of discharge, or rate coding, of their motor units. Recall from chapter 1 that as the frequency of stimulation of a given motor unit is increased, the muscle eventually reaches a state of tetanus, producing the absolute peak force or tension of the muscle fiber or motor unit (see figure 1.12, p. 45). There is limited evidence that rate coding is increased with resistance training. Rapid movement or ballistic-type training appears to be particularly effective in stimulating increases in rate coding.

Autogenic Inhibition

Inhibitory mechanisms in the neuromuscular system, such as the Golgi tendon organs, might be necessary to prevent the muscles from exerting more force than the bones and connective tissues can tolerate. This control is referred to as **autogenic inhibition**. However, under extreme situations when larger forces are sometimes produced, significant damage can occur to these structures, suggesting that the protective inhibitory mechanisms can be overridden.

The function of Golgi tendon organs was discussed in chapter 3. When the tension on a muscle's tendons and internal connective tissue structures exceeds the threshold of the embedded Golgi tendon organs, motor neurons to that muscle are inhibited; that is, autogenic inhibition occurs. Both the reticular formation in the brain stem and the cerebral cortex function to initiate and propagate inhibitory impulses.

Training can gradually reduce or counteract these inhibitory impulses, allowing the muscle to reach greater levels of strength. Thus, strength gains may be achieved by reduced neurological inhibition. This theory is attractive because it can at least partially explain superhuman feats of strength and strength gains in the absence of hypertrophy.

Autogenic inhibition may be attenuated with resistance training, allowing a greater force production from trained muscles independent of increases in muscle mass.

Other Neural Factors

In addition to increasing motor unit recruitment or decreasing neurological inhibition, other neural factors can contribute to strength gains with resistance training. One of these is referred to as coactivation of agonist and antagonist muscles (the agonist muscles are the primary movers, and the antagonist muscles act to impede the agonists). If we use forearm flexor concentric contraction as an example, the biceps is the primary agonist and the triceps is the antagonist. If both were contracting with equal force development, no movement would occur. Thus, to maximize the force generated by an agonist, it is necessary to minimize the amount of coactivation. Reduction in coactivation could explain a portion of strength gains attributed to neural factors, but its contribution likely would be small.

Changes also have been noted in the morphology of the neuromuscular junction, with both increased and decreased activity levels that might be directly related to the muscle's force-producing capacity.

Muscle Hypertrophy

How does a muscle's size increase? Two types of hypertrophy can occur: transient and chronic. **Transient hypertrophy** is the increased muscle size that develops during and immediately following a single exercise bout. This results mainly from fluid accumulation (edema) in the interstitial and intracellular spaces of the muscle that comes from the blood plasma. Transient hypertrophy, as its name implies, lasts only for a short time. The fluid returns to the blood within hours after exercise.

Chronic hypertrophy refers to the increase in muscle size that occurs with long-term resistance training. This reflects actual structural changes in the muscle that can result from an increase in the size of existing individual muscle fibers (**fiber hypertrophy**), in the number of muscle fibers (**fiber hyperplasia**), or both. Controversy surrounds the theories that attempt to explain the underlying cause of this phenomenon. Of importance, however, is the finding that the eccentric component of training is important in maximizing increases in muscle fiber cross-sectional area. A number of studies have shown greater hypertrophy and strength resulting solely from eccentric contraction training as compared to concentric contraction or combined eccentric and concentric contraction training. Further, higher-velocity eccentric training appears to result in greater hypertrophy and strength gains than slower-velocity training.[18] These greater increases appear to be related to disruptions in the sarcomere Z-lines. This disruption had originally been labeled as muscle damage but is now thought to represent fiber protein remodeling.[18] Thus, training with only concentric actions could limit muscle hypertrophy and increases in muscle strength. The following section examines the two postulated mechanisms for increasing muscle size with resistance training: fiber hypertrophy and fiber hyperplasia.

Fiber Hypertrophy

Early research suggested that the number of muscle fibers in each of a person's muscles is established by birth or shortly thereafter and that this number remains fixed throughout life. If this were true, then whole-muscle hypertrophy could result only from individual muscle fiber hypertrophy. This could be explained by

- more myofibrils,
- more actin and myosin filaments,
- more sarcoplasm,
- more connective tissue, or
- any combination of these.

As seen in the micrographs in figure 10.2, intense resistance training can significantly increase the cross-sectional area of muscle fibers. In this example, fiber hypertrophy is probably caused by increased numbers of myofibrils and actin and myosin filaments, which would provide more cross-bridges for force production during maximal contraction. Such dramatic enlargement of muscle fibers does not occur, however, in all cases of muscle hypertrophy.

Individual muscle fiber hypertrophy from resistance training appears to result from a net increase in muscle protein synthesis. The muscle's protein content is in a continual state of flux. Protein is always being synthesized and degraded. But the rates of these processes vary with the demands placed on the body. During exercise, protein synthesis decreases, while protein degradation apparently increases. This pattern reverses during the postexercise recovery period, even to the point of a net synthesis of protein. The provision of a carbohydrate and protein supplement immediately after a training bout can create a more positive nitrogen balance, facilitating protein synthesis and maximizing the skeletal muscles' adaptive response to resistance exercise.[13]

The hormone testosterone is thought to be at least partly responsible for these changes, because one of its primary functions is the promotion of muscle growth. For example, males experience a significantly greater increase in muscle growth starting at puberty, which is largely due to a 10-fold increase in testosterone production. Testosterone is a steroidal hormone with major anabolic functions. It has been well established that massive doses of anabolic steroids coupled with resistance training markedly increase muscle mass and strength (see chapter 16).

Fiber Hyperplasia

Research on animals suggests that hyperplasia may also be a factor in the hypertrophy of whole muscles. Studies on cats provide fairly clear evidence that fiber splitting occurs with extremely heavy weight training.[6]

FIGURE 10.2 Microscopic views of muscle cross sections taken from the leg muscle of a man who had not trained during the previous two years, (a) before he resumed training and (b) after he completed six months of dynamic strength training. Note the significantly larger fibers (hypertrophy) after training.

Cats were trained to move a heavy weight with a forepaw to get their food (figure 10.3). With the use of food as a powerful incentive, they learned to generate considerable force. With this intense strength training, selected muscle fibers appeared to actually split in half, and each half then increased to the size of the parent fiber. This is seen in the cross-sectional cuts through the muscle fibers shown in figure 10.4.

Subsequent studies, however, demonstrated that hypertrophy of selected muscles in chickens, rats, and mice that resulted from chronic exercise overload was attributable solely to hypertrophy of existing fibers, not hyperplasia. In these studies, each fiber in the whole muscle was actually counted. These direct fiber counts revealed no change in fiber number.

This finding led the scientists who performed the initial cat experiments to conduct an additional resistance training study with cats. This time they used actual fiber counts to determine if total muscle hypertrophy resulted from hyperplasia or fiber hypertrophy.[7] Following a resistance training program of 101 weeks, the cats were able to perform one-leg lifts of an average of 57% of their body weight, resulting in an 11% increase in muscle weight. Most important, the researchers found a 9% increase in the total number of muscle fibers, confirming that muscle fiber hyperplasia did occur.

The difference in results between the cat studies and those with rats and mice most likely is attributable to differences in the manner in which the animals were trained. The cats were trained with a pure form of resistance training: high resistance and low repetitions. The other animals were trained with more endurance-type activity: low resistance and high repetitions.

FIGURE 10.4 Muscle fiber splitting. The drawn models were extrapolated from a series of microscopic slides.

One additional animal model has been used to stimulate muscle hypertrophy associated with hyperplasia. Scientists have placed the anterior latissimus dorsi muscle of chickens in a state of chronic stretch by attaching weights to it, with the other wing serving as the normal control condition. In many of the studies that have used this model, the chronic stretch has resulted in substantial hypertrophy and hyperplasia.

Researchers are still uncertain about the roles played by hyperplasia and individual fiber hypertrophy in increasing human muscle size with resistance training. Most evidence indicates that individual fiber hypertrophy accounts for most whole-muscle hypertrophy. However, results from selected studies indicate that hyperplasia is possible in humans.

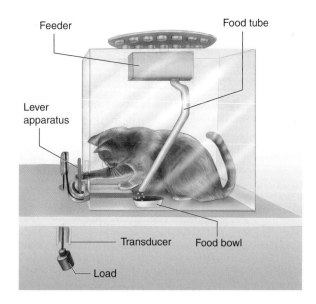

FIGURE 10.3 Heavy resistance training in cats.

In several studies of bodybuilders, swimmers, and kayakers, substantial muscle hypertrophy has been observed in trained muscles, but in the absence of individual fiber hypertrophy when compared to values in untrained control subjects. This suggests a greater number of muscle fibers in the trained muscles than in the corresponding muscles of untrained control subjects. However, other studies have shown individual fiber hypertrophy in highly trained athletes compared to untrained controls.

In focus

Muscle fiber hyperplasia has been clearly shown to occur in animal models with the use of resistance training to induce muscle hypertrophy. Only a few studies, on the other hand, suggest evidence of hyperplasia in humans.

In a cadaver study of seven previously healthy young men who had suffered sudden accidental death, the investigators compared cross sections of autopsied right and left tibialis anterior muscles (lower leg). Right-hand dominance is known to lead to greater hypertrophy of the left leg. In fact, the average cross-sectional area of the left muscle was 7.5% larger. This was associated with a 10% greater number of fibers in the left muscle. There was no difference in mean fiber size.[19]

The differences among these studies might be explained by the nature of the training load or stimulus. Training at high intensities or high resistances is thought to cause greater fiber hypertrophy, particularly of the type II (fast-twitch) fibers, than training at lower intensities or resistances.

Only one longitudinal study demonstrated the possibility of hyperplasia in men who had previous recreational resistance training experience.[15] Following 12 weeks of intensified resistance training, the muscle fiber number in the biceps brachii of several of the 12 subjects appeared to increase significantly. It appears from this study that hyperplasia can occur in humans but possibly only in certain subjects or under certain training conditions.

From the preceding information, it appears that fiber hyperplasia can occur in animals and possibly in humans. How are these new cells formed? As shown in figure 10.4, it is postulated that individual muscle fibers have the capacity to divide and split into two daughter cells, each of which can then develop into a functional muscle fiber. Importantly, satellite cells, which are the myogenic stem cells involved in skeletal muscle regeneration, are likely involved in the generation of new muscle fibers. These cells are typically activated by muscle stretching and injury; and, as we see later in this chapter, muscle injury results from intense training, particularly eccentric-action training. Muscle injury can lead to a cascade of responses, in which satellite cells become activated and proliferate, migrate to the damaged region, and fuse to existing myofibers or combine and fuse to produce new myofibers.[12] This is illustrated in figure 10.5.

Integration of Neural Activation and Fiber Hypertrophy

Research on resistance training adaptations indicates that early increases in voluntary strength, or maximal force production, are associated primarily with neural adaptations resulting in increased voluntary activation of muscle. This was clearly demonstrated in a study of both men and women who participated in an eight-week, high-intensity resistance training program, training twice per week.[20] Muscle biopsies were obtained at the beginning of the study and every two weeks during the training period. Strength, measured according to the 1RM, increased substantially over the eight weeks of training, with the greatest gains coming after the second week. Muscle biopsies, however, revealed only a small, insignificant increase in muscle fiber cross-sectional area by the end of the eight weeks of training. Thus, the strength gains were largely the result of increased neural activation.

Long-term increases in strength generally result from hypertrophy of the trained muscle. However, because it takes time to build protein through a decrease in protein degradation, an increase in protein synthesis, or both, early strength gains are typically due to changes in the pattern by which nerves activate the muscle fibers. Most research shows that neural factors contribute prominently to strength gains during the first 8 to 10 weeks of training. Hypertrophy contributes little during these initial weeks of training but progressively increases its contribution, becoming the major contributor after 10 weeks of training. However, not all studies concur with this pattern of strength development. One six-month study of strength-trained athletes showed that neural activation explained most of the strength gains during the most intensive training months and that hypertrophy was not a major factor.[11]

In focus

Early gains in strength appear to result more from changes in neural factors, but later long-term gains are largely the result of muscle hypertrophy.

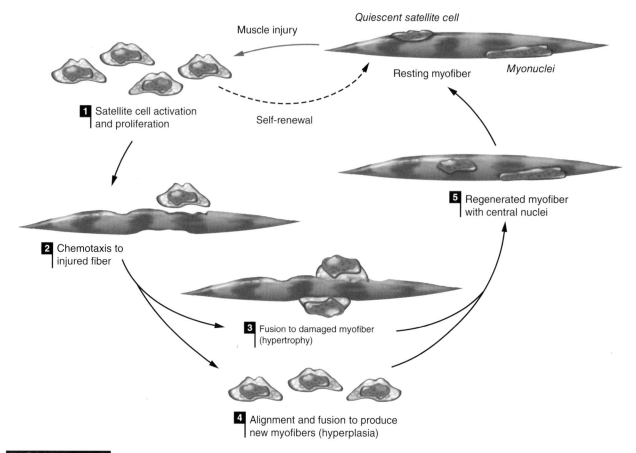

FIGURE 10.5 The satellite cell response to muscle injury.

Reprinted from T.J. Hawke and D.J. Garry, 2001, "Myogenic satellite cells: Physiology to molecular biology," *Journal of Applied Physiology* 91: 534-551. Used with permission.

Muscle Atrophy and Decreased Strength With Inactivity

When a normally active or highly trained person reduces his or her level of activity or ceases training altogether, major changes occur in both muscle structure and function. This is illustrated by the results of two types of studies: studies in which entire limbs have been immobilized and studies in which highly trained people stop training—so called detraining.

Immobilization

When a trained muscle suddenly becomes inactive through immobilization, major changes are initiated within that muscle in a matter of hours. During the first 6 h of immobilization, the rate of protein synthesis starts to decrease. This decrease likely initiates muscular atrophy, which is the wasting away or decrease in the size of muscle tissue. Atrophy results from lack of muscle use and the consequent loss of muscle protein that accompanies the inactivity. Strength decreases are most dramatic during the first week of immobilization, averaging 3% to 4% per day. This is associated with the atrophy but also with decreased neuromuscular activity of the immobilized muscle.

Immobilization appears to affect both type I and type II fibers. From various studies, researchers have observed disintegrated myofibrils, streaming Z-disks (discontinuity of Z-disks and fusion of the myofibrils), and mitochondrial damage. When muscle atrophies, the cross-sectional fiber area decreases. Several studies have shown the effect to be greater in type I fibers, including a decrease in the percentage of type I fibers, thereby increasing the relative percentage of type II fibers.

Muscles can and often do recover from immobilization when activity is resumed. The recovery period is substantially longer than the period of immobilization. See chapter 14 for more on the effects of muscle immobilization.

Cessation of Training

Similarly, significant muscle alterations can occur when people stop training. In one study, women resistance trained for 20 weeks and then stopped training for 30 to 32 weeks. Finally they retrained for six weeks.[21] The training program focused on the lower extremity, using a full squat, leg press, and leg extension. Strength increases were dramatic, as seen in figure 10.6. Compare the women's strength after their initial training period (post-20) with their strength after detraining (pre-6). This represents the strength loss they experienced with cessation of training. During the two training periods, increases in strength were accompanied by increases in the cross-sectional[21] area of all fiber types and a decrease in the percentage of type IIx fibers. Detraining had relatively little effect on fiber cross-sectional area, although the type II fiber areas tended to decrease (figure 10.7).

To prevent losses in the strength gained through resistance training, basic maintenance programs must be established once the desired goals for strength development have been achieved. Maintenance programs are designed to provide sufficient stress to the muscles to maintain existing levels of strength while allowing a reduction in intensity, duration, or frequency of training.

In one study, men and women resistance trained with knee extensions for either 10 or 18 weeks and then spent an additional 12 weeks with either no training or reduced training.[8] Knee extension strength increased 21.4% following the training period. Subjects who then stopped training lost 68% of their strength gains

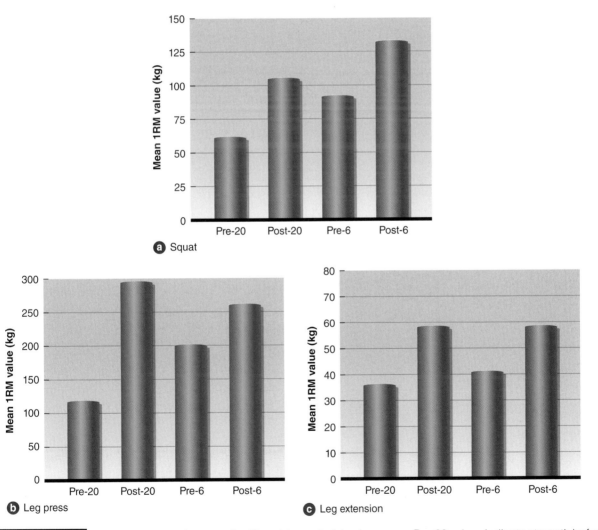

a Squat

b Leg press

c Leg extension

FIGURE 10.6 Changes in muscle strength with resistance training in women. Pre-20 values indicate strength before starting training; post-20 values indicate the changes following 20 weeks of training; pre-6 values indicate the changes following 30 to 32 weeks of detraining; and post-6 values indicate the changes following six weeks of retraining.

Adapted from R.S. Staron et al., 1991, "Strength and skeletal muscle adaptations in heavy-resistance-trained women after detraining and retraining," *Journal of Applied Physiology* 70: 631-640. Used with permission.

FIGURE 10.7 Changes in mean cross-sectional areas for the major fiber types with resistance training in women over periods of training (post-20), detraining (pre-6), and retraining (post-6). Type IIa/IIx is an intermediate fiber type. See figure 10.6 caption for more details.

during the weeks they didn't train. But subjects who reduced their training (from three days per week to two, or from two to one) did not lose strength. Thus, it appears that strength can be maintained for at least up to 12 weeks with reduced training frequency.

Fiber Type Alterations

Can muscle fibers change from one type to another through resistance training? The earliest research concluded that neither speed (anaerobic) nor endurance (aerobic) training could alter the basic fiber type, specifically from type I to type II, or from type II to type I. These early studies did show, however, that fibers begin to take on certain characteristics of the opposite fiber type if the training is of the opposite kind (e.g., type II fibers might become more oxidative with aerobic training).

Research using animal models has shown that fiber type conversion is indeed possible under conditions of cross-innervation, in which a type II motor unit is experimentally innervated by a type I motor neuron or a type I motor unit is experimentally innervated by a type II motor unit. Also, chronic, low-frequency nerve stimulation transforms type II motor units into type I motor units within a matter of weeks. Muscle fiber types in rats have changed in response to 15 weeks of high-intensity treadmill training, resulting in an increase in type I and type IIa fibers and a decrease in type IIx fibers.[9] The transition of fibers from type IIx to type IIa and from type IIa to type I was confirmed by several different histochemical techniques.

Staron and coworkers found evidence of fiber type transformation in women as a result of heavy resistance training.[22] Substantial increases in static strength and in the cross-sectional area of all fiber types were noted following a 20-week heavy resistance training program for the lower extremity. The mean percentage of type IIx fibers decreased significantly, but the mean percentage of type IIa fibers increased. The transition of type IIx fibers to type IIa fibers with resistance training has been consistently reported in a number of subsequent studies. Further, other studies demonstrate that a combination of high-intensity resistance training and short-interval speed work can lead to a conversion of type I to type IIa fibers.

In review

- Neural adaptations always accompany the strength gains that result from resistance training, but hypertrophy may or may not take place.
- Neural mechanisms leading to strength gains can include an increase in frequency of stimulation, or rate coding; recruitment of more motor units; more synchronous recruitment of motor units; and decreases in autogenic inhibition from the Golgi tendon organs.
- Transient muscle hypertrophy is the temporary enlargement of muscle resulting from edema immediately after an exercise bout.
- Chronic muscle hypertrophy occurs from repeated resistance training and reflects actual structural changes in the muscle.
- Although most muscle hypertrophy probably results from an increase in the size of individual muscle fibers (fiber hypertrophy), some evidence suggests that an increase in the number of muscle fibers (hyperplasia) may also be involved.
- Muscles atrophy (decrease in size and strength) when they become inactive, as with injury, immobilization, or cessation of training.
- Atrophy begins very quickly if training is stopped; but training can be reduced, as in a maintenance program, without resulting in atrophy or loss of strength.
- With resistance training there is a transition of type IIx to type IIa fibers.
- Evidence indicates that one fiber type can actually be converted to the other type (e.g., type I to type II, or vice versa) as a result of cross-innervation or chronic stimulation, and possibly with training.

Muscle Soreness and Cramps

Muscle soreness generally results from exhaustive or very high intensity exercise. This is particularly true when people perform a specific exercise for the first time. While muscle soreness can be felt at any time, there is generally a period of mild muscle soreness that can be felt during and immediately after exercise and then a more intense soreness felt a day or two later.

Acute Muscle Soreness

Pain felt during and immediately after exercise can result from accumulation of the end products of exercise, such as H^+, and from tissue edema, mentioned earlier, which is caused by fluid shifting from the blood plasma into the tissues. Edema is the cause of the acute muscle swelling that people feel after heavy endurance or strength training. The pain and soreness usually disappear within a few minutes to several hours after the exercise. Thus, this soreness is often referred to as **acute muscle soreness**.

Delayed-Onset Muscle Soreness

Muscle soreness felt a day or two after a heavy bout of exercise is not totally understood, yet researchers are continuing to give us greater insight into this phenomenon. Because this pain does not occur immediately, it is referred to as **delayed-onset muscle soreness (DOMS)**. DOMS is classified as a type I muscle strain and can vary from slight muscle stiffness to severe, debilitating pain that restricts movement. In the following sections, we discuss some theories that attempt to explain this form of muscle soreness.

Almost all current theories acknowledge that eccentric action is the primary initiator of DOMS. This has been clearly demonstrated in a number of studies examining the relationship of muscle soreness to eccentric, concentric, and static actions. In studies, individuals who train solely with eccentric actions experience extreme muscle soreness, whereas those who train using static and concentric actions experience little soreness. This idea has been further explored in studies in which subjects ran on a treadmill for 45 min on two separate days, one day on a level grade and the other day on a 10% downhill grade.[16, 17] No muscle soreness was associated with the level running. But the downhill running, which required extensive eccentric action, resulted in considerable soreness within 24 to 48 h, even though blood lactate concentrations, previously thought to cause muscle soreness, were much higher with level running.

In focus

Delayed-onset muscle soreness results primarily from eccentric action and is associated with actual muscle tissue disruption or damage.

In the following section we examine some of the proposed explanations for exercise-induced DOMS.

Structural Damage

The presence of increased concentrations of several specific muscle enzymes in blood after intense exercise suggests that some structural damage may occur in the muscle membranes. These enzyme concentrations in the blood increase from 2 to 10 times following bouts of heavy training. Recent studies support the idea that these changes might indicate various degrees of muscle tissue breakdown. Examination of tissue from the leg muscles of marathon runners has revealed remarkable damage to the muscle fibers after both training and marathon competition. The onset and timing of these muscle changes parallel the degree of muscle soreness experienced by the runners.

The electron micrograph in figure 10.8 shows muscle fiber damage as a result of marathon running.[10] In this

FIGURE 10.8 An electron micrograph of a muscle sample taken immediately after a marathon, showing the disruption of the cell membrane in one muscle fiber.

case, the cell membrane appears to have been totally ruptured, allowing the cell's contents to float freely between the other normal fibers. Fortunately, not all damage to muscle cells is as severe.

Figure 10.9 shows changes in the contractile filaments and Z-disks before and after a marathon race. Recall that Z-disks are the points of contact for the contractile proteins. They provide structural support for the transmission of force when the muscle fibers are activated to shorten. Figure 10.9b, after the marathon, shows moderate Z-disk streaming and major disruption of the thick and thin filaments in a parallel group of sarcomeres as a result of the force of eccentric actions or stretching of the tightened muscle fibers.

Although the effects of muscle damage on performance are not fully understood, it is generally agreed that this damage is responsible in part for the localized muscle pain, tenderness, and swelling associated with DOMS. However, blood enzyme concentrations might increase and muscle fibers might be damaged frequently during daily exercise that produces no muscle soreness. Also, remember that muscle damage appears to be a precipitating factor for muscle hypertrophy.

Inflammatory Reaction

White blood cells serve as a defense against foreign materials that enter the body and against conditions that threaten the normal function of tissues. The white blood cell count tends to increase following activities that induce muscle soreness. This observation led some investigators to suggest that soreness results from inflammatory reactions in the muscle. But the link between these reactions and muscle soreness has been difficult to establish.

In early studies, researchers attempted to use drugs to block the inflammatory reaction, but these efforts were unsuccessful in reducing either the amount of muscle soreness or the degree of inflammation. These early results did not support a link between simple inflammatory mediators and DOMS. However, more recent studies have begun to establish a link between muscle soreness and inflammation. It is now recognized that substances released from injured muscle can act as attractants, initiating the inflammatory process. Mononucleated cells in muscle are activated by the injury, providing the chemical signal to circulating inflammatory cells. Neutrophils (a type of white blood cell) invade the injury site and release cytokines (immunoregulatory substances), which then attract and activate additional inflammatory cells. Neutrophils possibly also release oxygen free radicals that can damage cell membranes. The invasion of these inflammatory cells is also associated with the incidence of pain, thought to be caused by a release of substances from the inflammatory cells

FIGURE 10.9 (*a*) An electron micrograph showing the normal arrangement of the actin and myosin filaments and Z-disk configuration in the muscle of a runner before a marathon race. (*b*) A muscle sample taken immediately after a marathon race shows moderate Z-disk streaming and major disruption of the thick and thin filaments in a parallel group of sarcomeres caused by the eccentric actions of running.

stimulating the pain-sensitive nerve endings. Macrophages (another type of cell of the immune system) then invade the damaged muscle fibers, removing debris through a process known as phagocytosis. Last, a second phase of macrophage invasion occurs, which is associated with muscle regeneration.[23]

Sequence of Events in DOMS

The general consensus among researchers is that a single theory or hypothesis cannot explain the mechanism causing DOMS. Instead researchers have proposed a sequence of events that may explain the DOMS phenomenon, including the following:

1. High tension in the contractile-elastic system of muscle results in structural damage to the muscle and its cell membrane. This is also accompanied by excessive strain of the connective tissue.

2. The cell membrane damage disturbs calcium homeostasis in the injured fiber, inhibiting cellular respiration. The resulting high calcium concentrations activate enzymes that degrade the Z-lines.

3. Within a few hours is a significant elevation in circulating neutrophils that participate in the inflammatory response.

4. The products of macrophage activity and intracellular contents (such as histamine, kinins, and K^+) accumulate outside the cells. These substances then stimulate the free nerve endings in the muscle. This process appears to be accentuated in eccentric exercise, in which large forces are distributed over relatively small cross-sectional areas of the muscle.

Recent comprehensive reviews have provided much greater insight into the cause of muscle soreness. We now are confident that muscle soreness results from injury or damage to the muscle itself, generally the muscle fiber and possibly the plasmalemma.[1, 4] This damage sets up a chain of events that includes the release of intracellular proteins and an increase in muscle protein turnover. The damage and repair process involves calcium ions, lysosomes, connective tissue, free radicals, energy sources, inflammatory reactions, and intracellular and myofibrillar proteins. But the precise cause of skeletal muscle damage and the mechanisms of repair are not well understood. As we have discussed previously, some evidence suggests that this process is an important step in muscle hypertrophy.

Up to this point, our discussion of DOMS has focused on muscle injury. Edema, or the accumulation of fluids in the muscular compartment, also can lead to DOMS. This edema is likely the result of muscle injury but could occur independently of muscle injury. An accumulation of interstitial or intracellular fluid increases the tissue fluid pressure within the muscle compartment, which in turn activates pain receptors within the muscle.

Delayed-Onset Muscle Soreness and Performance

With DOMS comes a reduction in the force-generating capacity of the affected muscles. Whether the DOMS is the result of injury to the muscle or edema independent of muscle injury, the affected muscles are not able to exert as much force when the person is asked to apply maximal force, such as in the performance of a 1RM strength test. Maximal force-generating capacity gradually returns over days or weeks. It has been proposed that the loss in strength is the result of three factors:[24]

1. The physical disruption of the muscle as illustrated in figures 10.8 and 10.9

2. Failure within the excitation–contraction coupling process

3. Loss of contractile protein

Failure in excitation–contraction coupling appears to be the most important, particularly during the first five days. This is illustrated in figure 10.10.

Muscle glycogen resynthesis also is impaired when a muscle is damaged. Resynthesis is generally normal for the first 6 to 12 h after exercise, but it slows or stops completely as the muscle undergoes repair, thus limiting the fuel-storage capacity of the injured muscle. Figure 10.11 illustrates the time sequence of the various factors associated with intense eccentric exercise, including pain, edema, plasma creatine kinase (a plasma enzyme marker of muscle fiber damage), glycogen depletion, ultrastructural damage in the muscle, and muscular weakness.

Reducing the Negative Effects of DOMS

Reducing the negative effects of DOMS is important for maximizing training gains. The eccentric component of muscle action could be minimized during early training, but this is not possible for athletes in most sports. An alternative approach is to start training at a very low intensity and progress slowly through the first few weeks. Yet another approach is to initiate the training program with a high-intensity, exhaustive training bout. Muscle soreness would be great for the first few days, but evidence suggests that subsequent training bouts would cause considerably less muscle soreness. Because the factors associated with DOMS are also potentially important in stimulating muscle hypertrophy, DOMS is most likely necessary to maximize the training response.

FIGURE 10.10 Estimated contributions of excitation–contraction (EC) coupling failure, decreased contractile protein content, and physical disruption to the decrease in strength following muscle injury.

Reprinted, by permission, from G. Warren et al., 2001, "Excitation-contraction uncoupling: Major role in contraction-induced muscle injury," *Exercise and Sport Sciences Reviews* 29(2): 82-87

In review

➤ Acute muscle soreness occurs late in an exercise bout and during the immediate recovery period.

➤ Delayed-onset muscle soreness usually peaks a day or two after the exercise bout. Eccentric action seems to be the primary instigator of this type of soreness.

➤ Proposed causes of DOMS include structural damage to muscle cells and inflammatory reactions within the muscles. The proposed sequence of events includes structural damage, impaired calcium homeostasis, inflammatory response, and increased macrophage activity.

➤ Muscle strength is reduced in muscles injured by eccentric contractions and is likely the result of physical disruption of the muscle, failure of the excitation–contraction process, and loss of contractile protein.

➤ Muscle soreness can be minimized by using lower intensity and fewer eccentric contractions early in training. However, muscle soreness may ultimately be an important part of maximizing the resistance training response.

FIGURE 10.11 The delayed response to eccentric exercise of various physiological markers. The darkness of the shaded bar corresponds to the intensity of the response at the indicated time. CK = creatine kinase.

Adapted, by permission, from W.J. Evans and J.G. Cannon, 1991, "The metabolic effects of exercise induced muscle damage," *Exercise and Sport Sciences Reviews* 19: 99-125.

Exercise-Induced Muscle Cramps

Skeletal muscle cramps are a frustrating problem in sport and physical activity and commonly occur even in highly fit athletes. Skeletal muscle cramps can come during the height of competition, immediately after competition, or at night in the deep of sleep. Muscle cramps are equally frustrating to scientists, because there are multiple and unknown causes of muscle cramping and little is known about the best treatment and prevention strategies. Although muscle cramps can be the result of rare medical conditions, most exercise-induced or exercise-associated muscle cramps are unrelated to disease or medical disorder. **Exercise-associated muscle cramps (EAMCs)** have been defined as painful, spasmodic, involuntary contractions of skeletal muscles that occur during or immediately after exercise. Nocturnal muscle cramps may or may not be associated with exercise.

It is becoming increasingly clear that there are two distinct types of exercise-associated muscle cramps.[2] The first type is associated with skeletal muscle overload and fatigue from overuse, insufficient conditioning, or both. The underlying mechanisms for this type of fatigue-associated cramp involve excitation of the muscle spindle and inhibition of the Golgi tendon organ, resulting in abnormal α-motor neuron control. This type of cramping is generally localized to the overworked muscle. Risk factors associated with this type of cramping are age, poor stretching habits, cramping history, and excessive exercise intensity and duration.

The second type of exercise-associated muscle cramp involves electrolyte deficits. This type of muscle cramp typically occurs in athletes who have been sweating extensively and have significant electrolyte disturbances, mainly sodium and chloride. These types of electrolyte disturbances can occur during or after a long race, game, or match, or as a consequence of multiple exercise bouts in which sweat sodium and chloride losses exceed intake. In the muscle, to compensate for the large electrolyte sweat and plasma volume losses, fluids shift from the interstitial compartment to the intravascular compartment. It is thought that this shift in fluid can cause the neuromuscular junctions to become hyperexcitable, leading to spontaneous discharge and initiation of action potentials in the muscles. This type of exertional heat cramp usually evolves from small localized visible muscle fasciculations to severe and debilitating muscle spasms. These cramps often begin in the legs but can become widespread.

Treatment of exercise-associated muscle cramps depends on the type of cramping. For fatigue-related cramps, treatments includes rest, passive stretching of the affected muscle or muscle groups, and holding the muscle in a stretched position until muscle activation is relieved.

Treatment of heat cramps involves the prompt ingestion of a high-salt solution (3 g in 500 ml of a sodium electrolyte beverage every 5-10 min). In addition, massage and ice application can help to calm the affected muscles and relieve pain. Fluids also should be taken in if dehydration and electrolyte loss are suspected.

To prevent EAMCs, the athlete should

- be well conditioned, to reduce the likelihood of muscle fatigue;
- regularly stretch the muscle groups prone to EAMCs;
- maintain fluid and electrolyte balance and carbohydrate stores; and
- reduce exercise intensity and duration if necessary.

In review

➤ Exercise-associated muscle cramps are attributable to either fluid or electrolyte imbalances or both.

➤ Muscle fatigue-associated cramps are related to sustained α-motor neuron activity, with increased muscle spindle activity and decreased Golgi tendon organ activity.

➤ Heat-associated cramps, which typically occur in athletes who have been sweating excessively, involve a shift in fluid from the interstitial space to the intravascular space, resulting in a hyperexcitable neuromuscular junction.

➤ Rest, passive stretching, holding the muscle in the stretched position, and fluid and electrolyte restoration can be effective in treating EAMCs. Proper conditioning, stretching, and nutrition are also possible prevention strategies.

Resistance Training for Special Populations

Until the 1970s, resistance training was widely regarded as appropriate only for young, healthy male athletes. This narrow concept led many people to overlook the benefits of resistance training when planning their own activities. In this section, we first consider sex and age, and then we summarize the importance of this form of training for all athletes, regardless of their sex, age, or sport.

Sex and Age Differences

In recent years, considerable interest has focused on training for women, children, and people who are elderly. As mentioned earlier in this chapter, the widespread use of resistance training by women, either for sport or for health-related benefits, is rather recent. Substantial knowledge has developed since the early 1970s revealing that women and men have the same ability to develop strength but that, on average, women may not be able to achieve peak values as high as those attained by men. This difference in strength is attributable primarily to muscle size differences related to sex differences in anabolic hormones. Resistance training techniques developed for and applied to men's training seem equally appropriate for women's training. Issues of strength and resistance training for women are covered in more detail in chapter 19.

In focus

In 1984, the University of Arizona was the first NCAA Division I school to hire a woman as head strength coach for both the men's and women's athletic programs. The position went to Meg Ritchie Stone, former Scottish discus thrower and shot-putter for Great Britain's Olympic team.

The wisdom of resistance training for children and adolescents has long been debated. The potential for injury, particularly growth plate injuries from the use of free weights, has caused much concern. Many people once believed that children would not benefit from resistance training, based on the assumption that the hormonal changes associated with puberty are necessary for gaining muscle strength and mass. We now know that children and adolescents can train safely with minimal risk of injury if appropriate safeguards are implemented. Furthermore, they can indeed gain both muscular strength and muscle mass (chapter 17).

Interest in resistance training procedures for elderly people has also increased. A substantial loss of fat-free body mass accompanies aging. This loss reflects mainly the loss of muscle mass, largely because most people become less active as they age. When a muscle isn't used regularly, it loses function, with predictable atrophy and loss of strength.

Can resistance training in elderly people reverse this process? People who are elderly can indeed gain strength and muscle mass in response to resistance training. This fact has important implications for both their health and the quality of their lives (chapter 18). With maintained or improved strength, they are less likely to fall. This is a significant benefit because falls are a major source of injury and debilitation for elderly people and often lead to death.

Resistance Training for Sport

Gaining strength, power, or muscular endurance simply for the sake of being stronger, being more powerful, or having greater muscular endurance is of relatively little importance to athletes unless it also improves their athletic performance. Resistance training by field-event athletes and competitive weightlifters makes intuitive sense. The need for resistance training by the gymnast, distance runner, baseball player, high jumper, or ballet dancer is less obvious.

We do not have extensive research to document the specific benefits of resistance training for every sport or for every event within a sport. But clearly each has basic strength, power, and muscular endurance requirements that must be met to achieve optimal performance. Training beyond these requirements may be unnecessary.

Training is costly in terms of time, and athletes can't afford to waste time on activities that won't result in better athletic performances. Thus, some performance measurement is imperative to evaluate any resistance training program's efficacy. To resistance train solely to become stronger, with no associated improvement in performance, is of questionable value. However, it should also be recognized that resistance training to improve muscular endurance can reduce the risk of injury for most sports, because fatigued individuals are at an increased risk of injury.

In review

➤ Resistance training can benefit almost everyone, regardless of the person's sex, age, or athletic involvement.

➤ Most athletes in most sports can benefit from resistance training if an appropriate program is designed for them. But to ensure that the program is working, performance should be assessed periodically and the training regime adjusted as needed.

In closing

In this chapter we have carefully considered the role of resistance training in increasing muscular strength and improving performance. We have examined how muscle strength is gained through both muscular and neural adaptations, what factors can lead to muscle soreness and muscle cramps, and how resistance training is of importance for both health and sport, irrespective of age or sex. In the next chapter, we turn our attention away from resistance training and begin exploring how the body adapts to aerobic and anaerobic training.

Key Terms

acute muscle soreness

atrophy

autogenic inhibition

chronic hypertrophy

delayed-onset muscle soreness (DOMS)

exercise-associated muscle cramps (EAMCs)

fiber hyperplasia

fiber hypertrophy

resistance training

transient hypertrophy

Study Questions

1. What is a reasonable expectation for percentage strength gains following a six-month resistance training program? How do these percentage gains differ by age, sex, and previous resistance training experience?
2. Discuss the different theories that have attempted to explain how muscles gain strength with training.
3. What is autogenic inhibition? How might it be important to resistance training?
4. Differentiate transient and chronic muscle hypertrophy.
5. What is fiber hyperplasia? How might it occur? How might it be related to gains in size and muscle strength with resistance training?
6. What is the physiological basis for hypertrophy?
7. What is the physiological basis for atrophy?
8. What is the physiological basis for delayed-onset muscle soreness?
9. What is the physiological basis for exercise-associated muscle cramps?

Study Guide Activities

In addition to the activities listed in the chapter opening outline on page 227, two other activities are available in the web study guide, located at

www.HumanKinetics.com/PhysiologyOfSportAndExercise

 The **KEY TERMS** activity reviews important terms, and the end-of-chapter **QUIZ** tests your understanding of the material covered in the chapter.

Adaptations to Aerobic and Anaerobic Training

11

In this chapter and in the web study guide

 ACTIVITY 11.1 Adaptations reviews the cardiovascular, respiratory, and metabolic responses to training.

 ACTIVITY 11.2 Individual Response considers the factors affecting individual response to training.

 ACTIVITY 11.3 Aerobic Training explores adaptations in response to aerobic training by applying them to real-life situations.

 ACTIVITY 11.4 Anaerobic Training explores adaptations in response to anaerobic training by applying them to real-life situations.

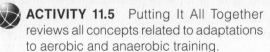 **ACTIVITY 11.5** Putting It All Together reviews all concepts related to adaptations to aerobic and anaerobic training.

On October 9, 2010, the Ironman World Championships were held in Kona, on the Big Island of Hawaii, for the 34th time. Some 1,800 triathletes swam 2.4 mi through tough ocean waves, biked 112 mi through hot lava fields, then ran 26.2 mi in temperatures reaching into the 90s. Chris McCormack completed this grueling event in 8 h, 10 min, and 37 s to win the Ford Ironman World Championship for the second time in four years. In the women's division, Miranda Carfrae earned her first Ironman title, finishing the course in 8:58:36—a rare women's sub-9 h performance. How are these athletes able to compete in this race? While there is little doubt that they are genetically gifted with a high $\dot{V}O_{2max}$, rigorous training is also required specifically to develop their cardiorespiratory endurance capacities.

During a single bout of aerobic exercise, the human body precisely adjusts its cardiovascular and respiratory function to meet the energy and oxygen demands of actively contracting muscle. When these systems are challenged repeatedly, as happens with regular exercise training, they adapt in ways that allow the body to improve $\dot{V}O_{2max}$ and overall endurance performance. **Aerobic training**, or cardiorespiratory endurance training, improves cardiac function and peripheral blood flow and enhances the capacity of the muscle fibers to generate greater amounts of adenosine triphosphate (ATP). In this chapter, we examine adaptations in cardiovascular and respiratory function in response to endurance training and how such adaptations improve an athlete's endurance capacity and performance. Additionally, we examine adaptations to anaerobic training. **Anaerobic training** improves anaerobic metabolism; short-term, high-intensity exercise capacity; tolerance for acid–base imbalances; and in some cases, muscle strength. Both aerobic and anaerobic training induce a variety of adaptations that benefit exercise and sport performance.

The effects of training on cardiovascular and respiratory, or aerobic, endurance is well known to endurance athletes like distance runners, cyclists, cross-country skiers, and swimmers but is often ignored by other types of athletes. Training programs for many nonendurance athletes often ignore the aerobic endurance component. This is understandable, because for maximum improvement in performance, training should be highly specific to the particular sport or activity in which the athlete participates, and endurance is frequently not recognized as important to nonendurance activities. The reasoning is, why waste valuable training time if the result is not improved performance?

The problem with this reasoning is that most nonendurance sports do indeed have an endurance, or aerobic, component. For example, in football, players and coaches might fail to recognize the importance of cardiorespiratory endurance as part of the total training program. From all outward appearances, football is an anaerobic, or burst-type, activity consisting of repeated bouts of high-intensity work of short duration. Seldom does a run exceed 40 to 60 yd (37-55 m), and even this is usually followed by a substantial rest interval. The need for endurance may not be readily apparent. What athletes and coaches might fail to consider is that this burst-type activity must be repeated many times during the game. With a higher aerobic endurance capacity, an athlete could maintain the quality of each burst activity throughout the game and would still be relatively "fresh" (less drop-off in performance, fewer feelings of fatigue) during the fourth quarter.

A parallel question arises concerning the importance of including resistance training as a part of the total training program for sports that do not demand high levels of strength, or high-intensity sprint training for sports that do not require speed or high anaerobic capacities. Yet athletes in almost all endurance sports are doing some resistance training to increase, or at least maintain, basic strength levels, as well as some sprint training to facilitate their ability to sustain speed when needed (e.g., sprinting to the finish line at the end of a marathon).

Chapters 9 and 14 cover the principles of training for sport performance—the "how," "when," and "how much" questions about training. The focus here is on those physiological changes that occur within the body systems when aerobic or anaerobic exercise is repeated regularly to induce a training response.

Adaptations to Aerobic Training

Improvements in endurance that accompany regular (daily, every other day, etc.) aerobic training, such as running, cycling, or swimming, result from multiple adaptations to the training stimuli. Some adaptations occur within the muscles themselves, promoting more efficient utilization of oxygen and fuel substrates. Still other important changes occur in the cardiovascular system, improving circulation to and within the muscles. Pulmonary adaptations, as will be noted later, occur to a lesser extent.

Endurance: Muscular Versus Cardiorespiratory

Endurance is a term that refers to two separate but related concepts: muscular endurance and cardiorespiratory endurance. Each makes a unique contribution to athletic performance, and each differs in its importance to different athletes.

For sprinters, endurance is the quality that allows them to sustain a high speed over the full distance of, for example, a 100 or 200 m race. This component of fitness is termed muscular endurance, the ability of a single muscle or muscle group to maintain high-intensity, repetitive, or static contractions. This type of endurance is also exemplified by a weightlifter doing multiple repetitions, a boxer, or a wrestler. The exercise or activity can be rhythmic and repetitive in nature, such as multiple repetitions of the bench press for the weightlifter and jabbing for the boxer. Or the activity can be more static, such as a sustained muscle action when a wrestler attempts to pin an opponent. In either case, the resulting fatigue is confined to a specific muscle group, and the activity's duration is usually no more than 1 or 2 min. Muscular endurance is highly related to muscular strength and to anaerobic power development.

While muscular endurance is specific to individual muscles or muscle groups, **cardiorespiratory endurance** relates to the ability to sustain prolonged, dynamic whole-body exercise using large muscle groups. Cardiorespiratory endurance is related to the development of the cardiovascular and respiratory systems' ability to maintain oxygen delivery to working muscles during prolonged exercise, as well as the muscles' ability to utilize energy aerobically (discussed in chapters 2 and 5). This is why the terms cardiorespiratory endurance and aerobic endurance are sometimes used synonymously.

In focus

Cardiorespiratory endurance, or aerobic endurance, is the ability to sustain prolonged rhythmic exercise involving relatively large muscle groups.

Evaluating Cardiorespiratory Endurance Capacity

Studying the effects of training on endurance requires an objective, repeatable means of measuring an individual's cardiorespiratory endurance capacity. In that way, the exercise scientist, coach, or athlete can monitor improvements as physiological adaptations occur during the training program.

Maximal Endurance Capacity: $\dot{V}O_{2max}$

Most exercise scientists regard $\dot{V}O_{2max}$, sometimes called maximal aerobic power or maximal aerobic capacity, as the best objective laboratory measure of cardiorespiratory endurance. Recall from chapter 5 that $\dot{V}O_{2max}$ is defined as the highest rate of oxygen consumption

attainable during maximal or exhaustive exercise. $\dot{V}O_{2max}$ as defined by the Fick equation is determined by maximal cardiac output (delivery of oxygen and blood flow to working muscles) and the maximal $(a-\bar{v})O_2$ difference (the ability of the active muscles to extract and use the oxygen). As exercise intensity increases, oxygen consumption eventually either plateaus or decreases slightly even with further increases in workload, indicating that a true maximal $\dot{V}O_2$ has been achieved.

With endurance training, more oxygen can be delivered to, and utilized by, active muscles than in an untrained state. Previously untrained subjects demonstrate average increases in $\dot{V}O_{2max}$ of 15% to 20% after a 20-week training program. These improvements allow individuals to perform endurance activities at a higher intensity, improving their performance potential. Figure 11.1 illustrates the increase in $\dot{V}O_{2max}$ after 12 months of aerobic training in a previously untrained individual. In this example, $\dot{V}O_{2max}$ increased by about 30%. Note that the $\dot{V}O_2$ "cost" of running at a certain submaximal intensity, referred to as running economy, did not change but that higher running speeds could be attained after training.

FIGURE 11.1 Changes in $\dot{V}O_{2max}$ with 12 months of endurance training. $\dot{V}O_{2max}$ increased from 44 to 57 ml · kg^{-1} · min^{-1}, a 30% increase. Peak speed during the treadmill test increased from 13 km/h (8 mph) to 16 km/h (~10 mph).

Submaximal Endurance Capacity

In addition to increasing maximal endurance capacity, endurance training also increases **submaximal endurance capacity**, which is much more difficult to evaluate.

A lower steady-state heart rate at the same submaximal exercise intensity is one physiological variable that can be used to objectively quantify the effect of training. Additionally, exercise scientists have used performance measures to quantify submaximal endurance capacity. For example, one test used to determine submaximal endurance capacity is the average peak absolute power output a person can maintain over a fixed period of time on a cycle ergometer. For running, the average peak speed or velocity a person can maintain during a set period of time would be a similar type of test. Generally, these tests last 30 min to an hour.

Submaximal endurance capacity is more closely related to actual competitive endurance performance than $\dot{V}O_{2max}$ and is likely determined by both $\dot{V}O_{2max}$ and the threshold for the onset of blood lactic acid accumulation (OBLA)—that point at which lactate begins to appear at a disproportionate rate in the blood (see chapter 5). With endurance training, submaximal endurance capacity increases.

Cardiovascular Adaptations to Training

Multiple cardiovascular adaptations occur in response to exercise training, including changes in the following:

- Heart size
- Stroke volume
- Heart rate
- Cardiac output
- Blood flow
- Blood pressure
- Blood volume

To fully understand adaptations in these variables, it is important to review how these components relate to oxygen transport.

Oxygen Transport System

Cardiorespiratory endurance is related to the cardiovascular and respiratory systems' ability to deliver sufficient oxygen to meet the needs of metabolically active tissues.

Recall from chapter 8 that the ability of the cardiovascular and respiratory systems to deliver oxygen to active tissues is defined by the **Fick equation**, which states that whole-body oxygen consumption is determined by both the delivery of oxygen via blood flow (cardiac output) and the amount of oxygen extracted by the tissues, the $(a-\bar{v})O_2$ difference. The product of cardiac output and the $(a-\bar{v})O_2$ difference determines the rate at which oxygen is being consumed:

$$\dot{V}O_2 = \text{stroke volume} \times \text{heart rate} \times (a-\bar{v})O_2 \text{ diff}$$

and

$$\dot{V}O_{2\,max} = \text{maximal stroke volume} \times \text{maximal heart rate} \times \text{maximal } (a-\bar{v})O_2 \text{ diff.}$$

Because HR_{max} either stays the same or decreases slightly with training, increases in $\dot{V}O_{2\,max}$ depend on adaptations in maximal stroke volume and maximal $(a-\bar{v})O_2$ difference.

The oxygen demand of exercising muscles increases with increasing exercise intensity. Aerobic endurance depends on the cardiorespiratory system's ability to deliver sufficient oxygen to these active tissues to meet their heightened demands for oxygen for oxidative metabolism. As maximal levels of exercise are achieved, heart size, blood flow, blood pressure, and blood volume can all potentially limit the maximal ability to transport oxygen. Endurance training elicits numerous changes in these components of the **oxygen transport system** that enable it to function more effectively.

Heart Size

As an adaptation to the increased work demand, cardiac muscle mass and ventricular volume increase with training. Cardiac muscle, like skeletal muscle, undergoes morphological adaptations as a result of chronic endurance training. At one time, **cardiac hypertrophy** induced by exercise—**"athlete's heart,"** as it was called—was viewed with concern because experts incorrectly believed that enlargement of the heart always reflected a pathological state, as sometimes occurs with severe hypertension. Training-induced cardiac hypertrophy, on the other hand, is now recognized as a normal adaptation to chronic endurance training.

The left ventricle, as discussed in chapter 6, does the most work and thus undergoes the greatest adaptation in response to endurance training. The type of ventricular adaptation depends on the type of exercise training performed. For example, during resistance training, the left ventricle must contract against increased afterload from the systemic circulation. From chapter 8 we learned that blood pressure during resistance exercise can exceed 480/350 mmHg. This presents a considerable resistance that must be overcome by the left ventricle. To overcome this high afterload, the heart muscle compensates by increasing left ventricular wall thickness, thereby increasing its contractility. Thus, the increase in its muscle mass is in direct response to repeated exposure to the increased afterload with resistance training. However, there is little change in ventricular volume.

With endurance training, left ventricular chamber size increases. This allows for increased left ventricular filling and consequently an increase in stroke volume. The increase in left ventricular dimensions is largely

Measuring Heart Size

The measurement of heart size has been of interest to cardiologists for years because a hypertrophied, or enlarged, heart is typically a pathological condition indicating the presence of cardiovascular disease. More recently, exercise scientists have been interested in heart size as it relates to the training state and performance of the athlete or exercising individual. Since the 1970s, studies of athletes and people who participate in endurance training have used echocardiography to accurately measure the size of the heart and its chambers. Echocardiography involves the technique of ultrasound, which uses high-frequency sound waves directed through the chest wall to the heart. These sound waves are emitted from a transducer placed on the chest; and once they contact the various structures of the heart, they rebound back to a sensor, which is able to capture the deflected sound waves and provide a moving picture of the heart. A trained physician or technician can visualize the size of the heart's chambers, thicknesses of its walls, and heart valve action. There are several forms of echocardiography: M-mode echocardiography, which provides a one-dimensional view of the heart; two-dimensional echocardiography; and Doppler echocardiography, which is used more often to measure blood flow through large arteries. The figures in this sidebar illustrate two-dimensional echocardiography being conducted and the resulting echocardiogram.

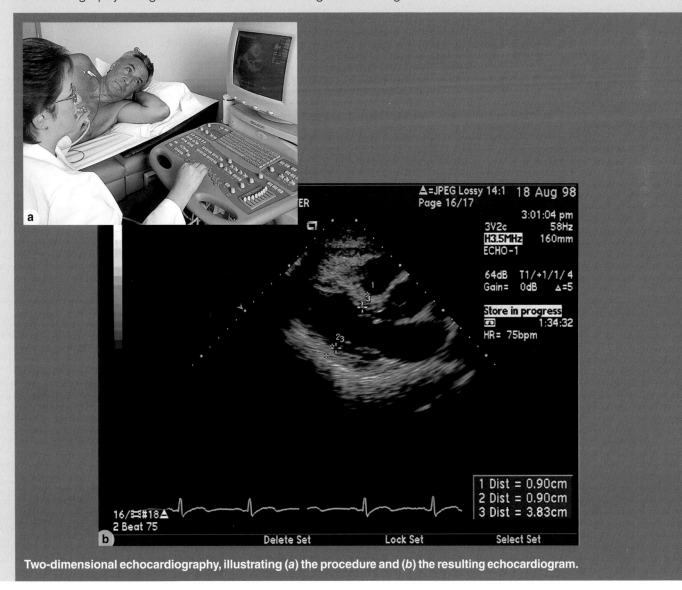

Two-dimensional echocardiography, illustrating (a) the procedure and (b) the resulting echocardiogram.

attributable to a training-induced increase in plasma volume (discussed later in this chapter) that increases left ventricular end-diastolic volume (increased pre-load). In concert with this, a decrease in heart rate at rest caused by increased parasympathetic tone, and during exercise at the same rate of work, allows a longer diastolic filling period. The increases in plasma volume and diastolic filling time increase left ventricular chamber size at the end of diastole. This effect of endurance training on the left ventricle is often called a volume loading effect.

It was originally hypothesized that this increase in left ventricular dimensions was the only change in the left ventricle caused by endurance training. Additional research has revealed that, similar to what happens in resistance training, myocardial wall thickness increases with endurance training.[11] Using magnetic resonance imaging, Milliken and colleagues[23] found that highly trained endurance athletes (competitive cross-country skiers, endurance cyclists, and long-distance runners) had greater left ventricular masses than did non-endurance-trained control subjects. Furthermore, left ventricular mass was highly correlated with $\dot{V}O_{2max}$ powering these subjects.

Fagard[11] conducted the most extensive review of the existing research literature in 1996, focusing on long-distance runners (135 athletes and 173 controls), cyclists (69 athletes and 65 controls), and strength athletes (178 athletes, including weight- and powerlifters, bodybuilders, wrestlers, throwers, and bobsledders, and 105 controls). For each group, the athletes were matched by age and body size with a group of sedentary control subjects. For each group of runners, cyclists, and strength athletes, the internal diameter of the left ventricle (LVID, an index of chamber size) and the total left ventricular mass (LVM) were greater in the athletes compared with their age- and sized-matched controls (figure 11.2). Thus, from this large cross-sectional study support the hypothesis that both left ventricular chamber size and wall thickness increase with endurance training.

Most studies of heart size changes with training have been cross-sectional, comparing trained individuals with sedentary, untrained individuals. Certainly a portion of the differences that we see in figure 11.2 can be attributed to genetics, not training. However, a number of longitudinal studies have followed individuals from an untrained state to a trained state, and others have followed individuals from a trained state to an untrained state. These studies have reported increases in heart size with training and decreases with detraining. So, training does bring about changes, but they are likely not as large as the differences shown in figure 11.2.

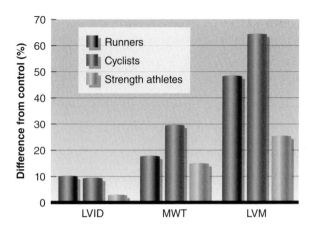

FIGURE 11.2 Percentage differences in heart size of three groups of athletes (runners, cyclists, and strength athletes) compared with their age- and size-matched sedentary controls (0%). Percentage differences are presented for left ventricular internal diameter (LVID), mean wall thickness (MWT), and left ventricular mass (LVM). Data are from Fagard (1996).

In review

➤ Cardiorespiratory endurance (also called maximal aerobic power) refers to the ability to perform prolonged, dynamic exercise using a large muscle mass.

➤ $\dot{V}O_{2max}$—the highest rate of oxygen consumption obtainable during maximal or exhaustive exercise—is the best single measure of cardiorespiratory endurance.

➤ Cardiac output, the product of heart rate and stroke volume, represents how much blood leaves the heart each minute, whereas $(a-\bar{v})O_2$ difference is a measure of how much oxygen is extracted from the blood by the tissues. According to the Fick equation, the product of these values is the rate of oxygen consumption: $\dot{V}O_2$ = stroke volume × heart rate × $(a-\bar{v})O_2$ difference.

➤ Of the chambers of the heart, the left ventricle adapts the most in response to endurance training.

➤ With endurance training, the internal dimensions of the left ventricle increase, mostly in response to an increase in ventricular filling secondary to an increase in plasma volume.

➤ Left ventricular wall thickness and mass also increase with endurance training, allowing for a greater force of contraction.

Stroke Volume

Stroke volume at rest is substantially higher after an endurance training program than it is before training. This endurance training–induced increase is also seen at a given submaximal exercise intensity and at maximal exercise. This increase is illustrated in figure 11.3, which shows the changes in stroke volume of a subject who exercised at increasing intensities up to a maximal intensity before and after a six-month endurance training program. Typical values for stroke volume at rest and during maximal exercise in untrained, trained, and highly trained athletes are listed in table 11.1. The wide range of stroke volume values for any given cell within this table is largely attributable to differences in body size. Larger people typically have larger hearts and a greater blood volume, and thus higher stroke volumes—an important point when one is comparing stroke volumes of different people.

After aerobic training, the left ventricle fills more completely during diastole. Plasma volume expands with training, which allows for more blood to enter the ventricle during diastole, increasing end-diastolic volume (EDV). The heart rate of a trained heart is also lower at rest and at the same absolute exercise intensity than that of an untrained heart, allowing more time for the increased diastolic filling. More blood entering the ventricle increases the stretch on the ventricular walls; by the Frank-Starling mechanism (see chapter 8), this results in an increased force of contraction.

The thickness of the posterior and septal walls of the left ventricle also increases slightly with endurance training. Increased ventricular muscle mass results in increased contractile force, in turn causing a lower end-systolic volume.

The decrease in end-systolic volume is facilitated by the decrease in peripheral resistance that occurs with training. Increased contractility resulting from an increase in left ventricular thickness and greater diastolic filling (Frank-Starling mechanism), coupled with the reduction in systemic peripheral resistance, increases the ejection fraction [equal to $(EDV - ESV)/EDV$] in the trained heart. More blood enters the left ventricle, and a greater percentage of what enters is forced out with each contraction, resulting in an increase in stroke volume.

FIGURE 11.3 Changes in stroke volume with endurance training during walking, jogging, and running on a treadmill at increasing velocities.

In focus

Increased left ventricular dimensions, reduced systemic peripheral resistance, and a greater blood volume account for the increases in resting, submaximal, and maximal stroke volume after an endurance training program.

Adaptations in stroke volume during endurance training are illustrated by a study in which older men trained aerobically for one year.[9] Their cardiovascular function was evaluated before and after training. The subjects performed running, treadmill, and cycle ergometer exercise for 1 h each day, four days per week. They exercised at intensities of 60% to 80% of $\dot{V}O_{2max}$, with brief bouts of exercise exceeding 90% of $\dot{V}O_{2max}$. End-diastolic volume increased at rest and throughout submaximal exercise. The ejection fraction increased, which was associated with a decreased ESV, suggesting increased contractility of the left ventricle. $\dot{V}O_{2max}$ increased by 23%, a substantial improvement in endurance.

It is clear that central stroke volume adaptations occur with endurance training, but there are also peripheral adaptations that contribute to the increase in $\dot{V}O_{2max}$, at least in middle-aged exercisers. This was demonstrated in a unique longitudinal study involving both exercise training and a bed rest deconditioning model.[22] Five 20-year-old men were tested (baseline

TABLE 11.1 Stroke Volumes at Rest (SV_rest) and During Maximal Exercise (SV_max) for Different States of Training

Subjects	SV_rest (ml/beat)	SV_max (ml/beat)
Untrained	50-70	80-110
Trained	70-90	110-150
Highly trained	90-110	150-220+

values), placed on bed rest for 20 days (deconditioning), and then trained for 60 days, starting immediately at the conclusion of bed rest. These same five men were restudied 30 years later at the age of 50; they were tested at baseline in a relatively sedentary state and after six months of endurance training. The average percentage increases in $\dot{V}O_{2max}$ were similar for the subjects at age 20 (18%) and at age 50 (14%). However, the increase in $\dot{V}O_{2max}$ at age 20 was explained by increases in both maximal cardiac output and maximal $(a-\bar{v})O_2$ difference; at age 50, the increase was explained primarily by an increase in $(a-\bar{v})O_2$ difference, while maximal cardiac output was unchanged. Maximal stroke volume was increased after training at both age 20 and age 50 but to a lesser degree at age 50 (+16 ml/beat at age 20 vs. +8 ml/beat at age 50).

In review

➤ Following endurance training, stroke volume (SV) is increased at rest and during submaximal and maximal exercise.

➤ A major factor leading to the SV increase is an increased EDV caused by an increase in plasma volume and a greater diastolic filling time secondary to a lower heart rate.

➤ Another contributing factor to increased SV is an increased left ventricular force of contraction. This is caused by hypertrophy of the cardiac muscle and increased ventricular stretch resulting from an increase in diastolic filling (increased preload), leading to greater elastic recoil (Frank-Starling mechanism).

➤ Reduced systemic vascular resistance (decreased afterload) also contributes to the increased volume of blood pumped from the left ventricle with each beat.

Heart Rate

Aerobic training has a major impact on heart rate at rest, during submaximal exercise, and during the postexercise recovery period. The effect of aerobic training on maximal heart rate is rather negligible.

Resting Heart Rate Resting heart rate decreases markedly as a result of endurance training. Some studies have shown that a sedentary individual with an initial resting heart rate of 80 beats/min can decrease resting heart rate by approximately 1 beat/min with each week of aerobic training, at least for the first few weeks. After 10 weeks of moderate endurance training, resting heart rate can decrease from 80 to 70 beats/min

or lower. On the other hand, well-controlled studies with large numbers of subjects have shown much smaller decreases in resting heart rate, that is, fewer than 5 beats/min following up to 20 weeks of aerobic training. The actual mechanisms responsible for this decrease are not entirely understood, but training appears to increase parasympathetic activity in the heart while decreasing sympathetic activity.

In focus

Highly conditioned endurance athletes often have resting heart rates lower than 40 beats/min, and some have values lower than 30 beats/min.

Recall from chapter 6 that bradycardia is a term indicating a heart rate of fewer than 60 beats/min. In untrained individuals, bradycardia can be the result of abnormal cardiac function or heart disease. Therefore, it is necessary to differentiate between training-induced bradycardia, which is a normal response to endurance training, and pathological bradycardia, which can be cause for concern.

Submaximal Heart Rate During submaximal exercise, aerobic training results in a lower heart rate at any given absolute exercise intensity. This is illustrated in figure 11.4, which shows the heart rate of an individual exercising on a treadmill before and after training. At each walking or running speed, the posttraining heart rate is lower than the heart rate before training. The

FIGURE 11.4 Endurance training–induced changes in heart rate during progressive walking, jogging, and running on a treadmill at increasing speeds.

training-induced decrease in heart rate is typically greater at higher intensities.

While maintaining a cardiac output appropriate to meet the needs of working muscle, a trained heart performs less work (lower heart rate, higher stroke volume) than an untrained heart at the same absolute workload.

Maximum Heart Rate   A person's maximal heart rate (HR_{max}) tends to be stable and typically remains relatively unchanged after endurance training. However, several studies have suggested that for people whose untrained HR_{max} values exceed 180 beats/min, HR_{max} might be slightly lower after training. Also, highly conditioned endurance athletes often have lower HR_{max} values than untrained individuals of the same age, although this is not always the case. Athletes over 60 years old sometimes have higher HR_{max} values than untrained people of the same age.

In focus

Resting heart rate is typically lower (by more than 10 beats/min) following endurance training. After endurance training, submaximal heart rate is likewise lower during exercise at the same absolute workload, generally by 10 to 20 beats/min or more. Maximal heart rate generally does not change or decrease slightly with endurance training.

Interactions Between Heart Rate and Stroke Volume   During exercise, the product of heart rate and stroke volume provides a cardiac output appropriate to the intensity of the activity being performed. At maximal or near-maximal intensities, heart rate may change to provide the optimal combination of heart rate and stroke volume to maximize cardiac output. If heart rate is too fast, diastolic filling time is reduced, and stroke volume might be compromised. For example, if HR_{max} is 180 beats/min, the heart beats three times per second. Each cardiac cycle thus lasts for only 0.33 s. Diastole is as short as 0.15 s or less. This fast heart rate allows very little time for the ventricles to fill. As a consequence, stroke volume may decrease at high heart rates when filling time is compromised.

However, if the heart rate slows, the ventricles have longer to fill. This has been proposed as one reason highly trained endurance athletes tend to have lower HR_{max} values: Their hearts have adapted to training by drastically increasing their stroke volumes, so lower HR_{max} values can provide optimal cardiac output.

Which comes first? Does increased stroke volume result in a decreased heart rate, or does a lower heart rate result in an increased stroke volume? This question

remains unanswered. In either case, the combination of increased SV and decreased HR is a more efficient way for the heart to meet the metabolic demands of the exercising body. The heart expends less energy by contracting less often but more forcefully than it would if contraction frequency were increased. Reciprocal changes in heart rate and stroke volume in response to training share a common goal: to allow the heart to pump the maximal amount of oxygenated blood at the lowest energy cost.

Heart Rate Recovery   During exercise, as discussed in chapter 6, heart rate must increase to increase cardiac output to meet the blood flow demands of active muscles. When the exercise bout is finished, heart rate does not instantly return to its resting level. Instead, it remains elevated for a while, slowly returning to its resting rate. The time it takes for heart rate to return to its resting rate is called the heart rate recovery period.

After endurance training, as shown in figure 11.5, heart rate returns to its resting level much more quickly after an exercise bout than it does before training. This is true after both submaximal and maximal exercise.

Because the heart rate recovery period is shorter after endurance training, this measurement has been proposed as an indirect index of cardiorespiratory fitness. In general, a more fit person recovers faster after a standardized rate of work than a less fit person, so this measure may have some utility in field settings when more direct measures of endurance capacity are not possible or feasible. However, factors other than training can also affect heart rate recovery time. For example, an elevated core temperature or an enhanced

FIGURE 11.5   Changes in heart rate during recovery after a 4 min, all-out bout of exercise before and after endurance training.

sympathetic nervous system response can prolong heart rate elevation.

The heart rate recovery curve is a useful tool for tracking a person's progress during a training program. But because of the potential influence of other factors, it should not be used to compare between individuals.

Cardiac Output

We have looked at the effects of training on the two components of cardiac output: stroke volume and heart rate. While stroke volume increases, heart rate generally decreases at rest and during exercise at a given absolute intensity.

Because the magnitude of these reciprocal changes is similar, cardiac output at rest and during submaximal exercise at a given exercise intensity does not change much following endurance training. In fact, cardiac output can decrease slightly. This is likely the result of an increase in the $(a-\bar{v})O_2$ difference (reflecting greater oxygen extraction by the tissues) or a decrease in the rate of oxygen consumption (reflecting an increased mechanical efficiency). Generally, cardiac output matches the oxygen consumption required for any given intensity of effort.

Maximal cardiac output, however, increases considerably in response to aerobic training, as seen in figure 11.6, and is largely responsible for the increase in $\dot{V}O_{2max}$. This increase in cardiac output must result from an increase in maximal stroke volume, because HR_{max} changes little, if any. Maximal cardiac output ranges from 14 to 20 L/min in untrained individuals and from 25 to 35 L/min in trained individuals, and can be 40 L/min or more in highly conditioned endurance athletes. These absolute values, however, are greatly influenced by body size.

In review

- Resting heart rate decreases as a result of endurance training. In a sedentary person, the decrease is typically about 1 beat/min per week during the initial weeks of training, but smaller decreases have been reported. Highly trained endurance athletes may have resting heart rates of 40 beats/min or lower.

- Heart rate during submaximal exercise is also lower, with larger decreases seen at higher exercise intensities.

- Maximal heart rate either remains unchanged or decreases slightly with training.

- Heart rate during the recovery period decreases more rapidly after training, making it an indirect but convenient way of tracking the adaptations that occur with training. However, this value is not useful for comparing fitness levels of different people.

- Cardiac output at rest and at submaximal levels of exercise remains unchanged (or may decrease slightly) after endurance training.

- Cardiac output during maximal exercise increases considerably and is largely responsible for the increase in $\dot{V}O_{2max}$. The increased maximal cardiac output is the result of the substantial increase in maximal stroke volume, made possible by training-induced changes in cardiac structure and function.

Blood Flow

Active muscles need considerably more oxygen and fuel substrates than inactive ones. To meet these increased needs, more blood must be delivered to these muscles during exercise. With endurance training, the cardiovascular system adapts to increase blood flow to exercising muscles to meet their higher demand for oxygen and metabolic substrates. Four factors account for this enhanced blood flow to muscle following training:

- Increased capillarization
- Greater recruitment of existing capillaries
- More effective blood flow redistribution from inactive regions
- Increased total blood volume

To permit increased blood flow, new capillaries develop in trained muscles. This allows the blood flow-

FIGURE 11.6 Changes in cardiac output with endurance training during walking, then jogging, and finally running on a treadmill as velocity increases.

ing into skeletal muscle from arterioles to more fully perfuse the active fibers. This increase in capillaries usually is expressed as an increase in the number of capillaries per muscle fiber, or the **capillary-to-fiber ratio**. Table 11.2 illustrates the differences in capillary-to-fiber ratios between well-trained and untrained men, both before and after exercise.[14]

In all tissues, including muscle, not all capillaries are open at any given time. In addition to new capillarization, existing capillaries in trained muscles can be recruited and open to flow, which also increases blood flow to muscle fibers. The increase in new capillaries with endurance training and increased capillary recruitment combine to increase the cross-sectional area for exchange between the vascular system and the metabolically active muscle fibers. Because endurance training also increases blood volume, shifting more blood into the capillaries will not severely compromise venous return.

A more effective redistribution of cardiac output also can increase blood flow to the active muscles. Blood flow is directed to the active musculature and shunted away from areas that do not need high flow. Blood flow can increase to the more active fibers even within a specific muscle group. Armstrong and Laughlin[1] demonstrated that endurance-trained rats could redistribute blood flow to their most active tissues during exercise better than untrained rats could. The total blood flow to the exercising hindlimbs did not differ between the trained and untrained rats. However, the trained rats distributed more of their blood to the most oxidative muscle fibers, effectively redistributing the blood flow away from the glycolytic muscle fibers. These findings are difficult to replicate in humans because of measurement challenges, as well as the fact that human skeletal muscle is a mosaic with mixed fiber types among individual muscles.

Finally, the body's total blood volume increases with endurance training, providing more blood to meet the body's many blood flow needs during endurance activity. The mechanisms responsible for this are discussed later in this chapter.

In focus

The increase in blood flow to muscle is one of the most important factors supporting increased endurance capacity and aerobic exercise performance. This increase is attributable to increased capillary density (both new capillaries and greater capillary recruitment), diversion of a larger portion of the cardiac output to the active muscles, and increased blood volume.

Blood Pressure

Resting blood pressure does not change significantly in healthy subjects in response to endurance training, but some studies have shown modest reductions after training in borderline or moderately hypertensive individuals. Reductions in both systolic and diastolic blood pressure of approximately 6 to 7 mmHg may result in hypertensive subjects. The mechanisms underlying this reduction are unknown. Following endurance training, blood pressure is reduced at a given submaximal exercise intensity; but at maximal exercise capacity, systolic blood pressure is increased and diastolic pressure is decreased.

Although resistance exercise can cause large transient increases in both systolic and diastolic blood pres-

TABLE 11.2	Muscle Fiber Capillarization in Well-Trained and Untrained Men			
Stage	Capillaries per mm²	Muscles fibers per mm²	Capillary-to-fiber ratio	Diffusion distance[a]
Well-trained				
Preexercise	640	440	1.5	20.1
Postexercise	611	414	1.6	20.3
Untrained				
Preexercise	600	557	1.1	20.3
Postexercise	599	576	1.1	20.5

Note. This table illustrates the larger size of the muscle fibers in the well-trained men in that they had fewer fibers for a given area (fibers per mm²). They also had an approximately 50% higher capillary-to-fiber ratio than the untrained men.

[a]Diffusion distance is expressed as the average half-distance between capillaries on the cross-sectional view expressed in micrometers.

Adapted from L. Hermansen and M. Wachtlova, 1971, "Capillary density of skeletal muscle in well trained and untrained men," *Journal of Applied Physiology* 30: 860-863. Used with permission.

sure during lifting of heavy weights, chronic exposure to these high pressures does not elevate resting blood pressure. Hypertension is not common in competitive weightlifters or in strength and power athletes. In fact, a few studies have even shown that resistance training may lower resting systolic blood pressure. Hagberg and coworkers[13] followed a group of borderline-hypertensive adolescents through five months of weight training. The subjects' resting systolic blood pressures decreased significantly.

Blood Volume

Endurance training increases total blood volume, and this effect is larger at higher training intensities. Furthermore, the effect occurs rapidly. This increased blood volume results primarily from an increase in plasma volume, but there is also an increase in the volume of red blood cells. The time course and mechanism for the increase of each of these components of blood are quite different.[29]

Plasma Volume The increase in plasma volume with training is thought to result from two mechanisms. The first mechanism, which has two phases, results in increases in plasma proteins, particularly albumin. Recall from chapter 8 that plasma proteins are the major driver of oncotic pressure in the vasculature. As plasma protein concentration increases, so does oncotic pressure, and fluid is reabsorbed from the interstitial fluid into the blood vessels. During an intense bout of exercise, proteins leave the vascular space and move into the interstitial space. They are then returned in greater amounts through the lymph system. It is likely that the first phase of rapid plasma volume increase is the result of the increased plasma albumin, which is noted within the first hour of recovery from the first training bout. In the second phase, protein synthesis is turned on (upregulated) by repeated exercise, and new proteins are formed. With the second mechanism, exercise increases the release of antidiuretic hormone and aldosterone, hormones that cause reabsorption of water and sodium in the kidneys, which increases blood plasma. That increased fluid is kept in the vascular space by the oncotic pressure exerted by the proteins. Nearly all of the increase in blood volume during the first two weeks of training can be explained by the increase in plasma volume.

Red Blood Cells An increase in red blood cell volume with endurance training also contributes to the overall increase in blood volume, but this is an inconsistent finding. Although the actual number of red blood cells may increase, the hematocrit—the ratio of the red blood cell volume to the total blood volume—may actually decrease. Figure 11.7 illustrates this apparent paradox. Notice that the hematocrit is reduced even though there has been a slight increase in red blood cells. A trained athlete's hematocrit can decrease to such an extent that the athlete appears to be anemic on the basis of a relatively low concentration of red cells and hemoglobin ("pseudoanemia").

The increased ratio of plasma to cells resulting from a greater increase in the fluid portion reduces the blood's viscosity, or thickness. Reduced viscosity may aid the smooth flow of blood through the blood vessels, particularly through the smaller vessels such as the capillaries. One of the physiological benefits of decreasing blood viscosity is that it enhances oxygen delivery to the active muscle mass.

Both the total amount (absolute values) of hemoglobin and the total number of red blood cells are typically elevated in highly trained athletes, although these values relative to total blood volume are below normal. This ensures that the blood has more than ample oxygen-carrying capacity. The turnover rate of red blood cells also may be higher with intense training.

In focus

The increase in blood volume following endurance training is attributable to increases in both plasma volume and red blood cell volume; both changes facilitate the delivery of oxygen to active muscles.

FIGURE 11.7 Increases in total blood volume and plasma volume occur with endurance training. Note that although the hematocrit (percentage of red blood cells) decreased from 44% to 42%, the total volume of red blood cells increased by 10%.

In review

- Blood flow to active muscle is increased by endurance training.
- Increased muscle blood flow results from four factors:
    1. Increased capillarization
    2. Greater opening of existing capillaries (capillary recruitment)
    3. More effective blood flow distribution
    4. Increased blood volume
- Resting blood pressure generally is reduced by endurance training in those with borderline or moderate hypertension but not in healthy, normotensive subjects.
- Endurance training results in a reduction in blood pressure during submaximal exercise at the same exercise intensity, but at maximal exercise intensity the systolic blood pressure is increased and diastolic blood pressure is decreased compared to pretraining values.
- Blood volume increases as a result of endurance training.
- Plasma volume is expanded through increased protein content (returned from lymph and upregulated protein synthesis). This effect is maintained and supported by fluid-conserving hormones.
- Red blood cell volume also increases, but the increase in plasma volume is typically higher.
- Increased plasma volume decreases blood viscosity, which can improve tissue perfusion and oxygen availability.

Respiratory Adaptations to Training

No matter how proficient the cardiovascular system is at supplying blood to exercising muscle, endurance would be hindered if the respiratory system were not able to bring in enough oxygen to fully oxygenate red blood cells. Respiratory system function does not usually limit performance because ventilation can be increased to a much greater extent than cardiovascular function. But, as with the cardiovascular system, the respiratory system undergoes specific adaptations to endurance training to maximize its efficiency.

Pulmonary Ventilation

After training, pulmonary ventilation is essentially unchanged at rest. Although endurance training does not change the structure or basic physiology of the lung, it does decrease ventilation during submaximal exercise by as much as 20% to 30% at a given submaximal intensity. Maximal pulmonary ventilation is substantially increased from a rate of about 100 to 120 L/min in untrained sedentary individuals to about 130 to 150 L/min or more following endurance training. Pulmonary ventilation rates typically increase to about 180 L/min in highly trained athletes and can exceed 200 L/min in very large, highly trained endurance athletes. Two factors can account for the increase in maximal pulmonary ventilation following training: increased tidal volume and increased respiratory frequency at maximal exercise.

Ventilation is not usually a limiting factor for endurance exercise performance. However, some evidence suggests that at some point in a highly trained person's adaptation, the pulmonary system's capacity for oxygen transport may not be able to meet the demands of the limbs and the cardiovascular system.[8] This results in what has been termed exercise-induced arterial hypoxemia, in which arterial oxygen saturation decreases below 96%. As discussed in chapter 7, this desaturation in highly trained elite athletes likely results from the large right heart cardiac output directed to the lung during exercise and consequently a decrease in the time the blood spends in the lung.

Pulmonary Diffusion

Pulmonary diffusion, or gas exchange occurring in the alveoli, is unaltered at rest and during submaximal exercise following training. However, it increases at maximal exercise intensity. Pulmonary blood flow (blood coming from the right side of the heart to the lungs) increases following training, particularly flow to the upper regions of the lungs when a person is sitting or standing. This increases lung perfusion. More blood is brought into the lungs for gas exchange, and at the same time ventilation increases so that more air is brought into the lungs. This means that more alveoli will be involved in pulmonary diffusion. The net result is that pulmonary diffusion increases.

Arterial–Venous Oxygen Difference

The oxygen content of arterial blood changes very little with endurance training. Even though total hemoglobin is increased, the amount of hemoglobin per unit of blood is the same or even slightly reduced. The $(a-\bar{v})O_2$ difference, however, does increase with training, particularly at maximal exercise intensity. This increase results from a lower mixed venous oxygen content, which means that the blood returning to the heart (which is a mixture of venous blood from all body parts, not just the active tissues) contains less oxygen than it would in an untrained person. This reflects both greater oxygen extraction by active tissues and a more effective distribution of blood flow to active tissues. The increased extraction results in part from an increase in oxidative capacity of active muscle fibers as described later in this chapter.

(In)focus

Although the largest part of the increase in $\dot{V}O_{2max}$ results from the increases in cardiac output and muscle blood flow, an increase in $(a-\bar{v})O_2$ difference also plays a key role. This increase in $(a-\bar{v})O_2$ difference is attributable to a more effective distribution of arterial blood away from inactive tissue to the active tissue and an increased ability of active muscle to extract oxygen.

In summary, the respiratory system is quite adept at bringing adequate oxygen into the body. For this reason, the respiratory system seldom limits endurance performance. Not surprisingly, the major training adaptations noted in the respiratory system are apparent mainly during maximal exercise, when all systems are being maximally stressed.

(In)review

➤ Unlike what happens with the cardiovascular system, endurance training has little effect on lung structure and function.

➤ To support increases in $\dot{V}O_{2max}$, there is an increase in pulmonary ventilation during maximal effort following training as both tidal volume and respiratory rate increase.

➤ Pulmonary diffusion at maximal intensity increases, especially to upper regions of the lung that are not normally perfused.

➤ The $(a-\bar{v})O_2$ difference widens with training, reflecting increased oxygen extraction by the tissues and more effective blood distribution to the active tissues.

Adaptations in Muscle

Repeated excitation and contraction of muscle fibers during endurance training stimulate changes in their structure and function. Our main interest here is in aerobic training and the changes it produces in muscle fiber type, mitochondrial function, and oxidative enzymes.

Muscle Fiber Type

As noted in chapter 1, low- to moderate-intensity aerobic activities rely extensively on type I (slow-twitch) fibers. In response to aerobic training, type I fibers become larger. More specifically, they develop a larger cross-sectional area, although the magnitude of change depends on the intensity and duration of each training bout and the length of the training program. Increases in cross-sectional area of up to 25% have been reported. Fast-twitch (type II) fibers, because they are not being recruited to the same extent during endurance exercise, generally do not increase cross-sectional area.

Most early studies showed no change in the percentage of type I versus type II fibers following aerobic training, but subtle changes were noted among the type II fiber subtypes. Type IIx fibers are recruited less often than IIa fibers, and for that reason they have a lower aerobic capacity. Long-duration exercise may eventually recruit these fibers to perform in a manner resembling IIa fibers. This can cause some IIx fibers to take on the characteristics of the more oxidative IIa fibers. Recent evidence suggests that not only is there a transition of type IIx to IIa fibers; there can also be a transition of type II to type I fibers. The magnitude of change is generally small, not more than a few percent. As an example, in the HERITAGE Family Study,[26] a 20-week program of aerobic training increased type I fibers from 43% pretraining to almost 47% posttraining and decreased type IIx fibers from 20% to 15%, with type IIa remaining essentially unchanged. These more recent studies have included larger numbers of subjects and have taken advantage of improved measurement technology; both might explain why fiber type composition changes within a muscle are now recognized.

Capillary Supply

One of the most important adaptations to aerobic training is an increase in the number of capillaries surrounding each muscle fiber. Table 11.2 illustrates that endurance-trained men have considerably more capillaries in their leg muscles than sedentary individuals.[14] With long periods of aerobic training, the number of capillaries may increase by more than 15%.[26] Having more capillaries allows for greater exchange of gases, heat, nutrients, and metabolic by-products between the blood and contracting muscle fibers. In fact, the increase in capillary density (i.e., increase in capillaries per muscle fiber) is potentially one of the most important alterations in response to training that causes the increase in $\dot{V}O_{2max}$. It is now clear that the diffusion of oxygen from the capillary to the mitochondria is a major factor limiting the maximal rate of oxygen consumption by the muscle. Increasing capillary density facilitates this diffusion, thus maintaining an environment well suited to energy production and repeated muscle contractions.

In focus

Aerobic training increases both the number of capillaries per muscle fiber and the number of capillaries for a given cross-sectional area of muscle. These changes improve blood perfusion through the muscles, enhancing the diffusion of oxygen, carbon dioxide, nutrients, and by-products of metabolism between the blood and muscle fibers.

Myoglobin Content

When oxygen enters the muscle fiber, it binds to myoglobin, a molecule similar to hemoglobin. This iron-containing molecule shuttles the oxygen molecules from the cell membrane to the mitochondria. Type I fibers contain large quantities of myoglobin, which gives these fibers their red appearance (myoglobin is a pigment that turns red when bound to oxygen). Type II fibers, on the other hand, are highly glycolytic, so they contain (and require) little myoglobin—hence their whiter appearance. More important, their limited myoglobin supply limits their oxidative capacity, resulting in poor endurance for these fibers.

Myoglobin transports oxygen and releases it to the mitochondria when oxygen becomes limited during muscle action. This oxygen reserve is used during the transition from rest to exercise, providing oxygen to the mitochondria during the lag between the beginning of exercise and the increased cardiovascular delivery of oxygen. Endurance training has been shown to increase muscle myoglobin content by 75% to 80%. This adaptation clearly supports a muscle's increased capacity for oxidative metabolism after training.

Mitochondrial Function

As noted in chapter 2, oxidative energy production takes place in the mitochondria. Not surprisingly, then, aerobic training also induces changes in mitochondrial function that improve the muscle fibers' capacity to produce ATP. The ability to use oxygen and produce ATP via oxidation depends on the number and size of the muscle mitochondria. Both increase with aerobic training.

During one study that involved endurance training in rats, the actual number of mitochondria increased approximately 15% during 27 weeks of exercise.[15] Average mitochondrial size also increased by about 35% over that training period. As with other training-

induced adaptations, the magnitude of change depends on training volume.

In focus

Skeletal muscle fiber mitochondria increase both in number and size with aerobic training, providing the muscle with an increased capacity for oxidative metabolism.

Oxidative Enzymes

Regular endurance exercise has been shown to induce major adaptations in skeletal muscle, including an increase in the number and size of the muscle fiber mitochondria as just discussed. These changes are further enhanced by an increase in mitochondrial capacity. The oxidative breakdown of fuels and the ultimate production of ATP depend on the action of **mitochondrial oxidative enzymes**, the specialized proteins that catalyze (i.e., speed up) the breakdown of nutrients to form ATP. Aerobic training increases the activity of these important enzymes.

Figure 11.8 illustrates the changes in the activity of succinate dehydrogenase (SDH), a key muscle oxidative enzyme, over seven months of progressive swim training. While the rate of increases in $\dot{V}O_{2max}$ slowed after the first two months of training, activity of this key oxidative enzyme continued to increase throughout the entire training period. This suggests that training-induced increases in $\dot{V}O_{2max}$ might be limited more by the circulatory system's ability to transport oxygen than by the muscles' oxidative potential.

The activities of muscle enzymes such as SDH and citrate synthase are dramatically influenced by aerobic training. This is seen in figure 11.9, which compares the activities of these enzymes in untrained people, moderately trained joggers, and highly trained runners.[7] Even moderate daily exercise increases the activity of these enzymes and thus the oxidative capacity of the muscle. For example, jogging or cycling for as little as 20 min per day has been shown to increase SDH activity in leg muscles by more than 25%. Training more vigorously, for example for 60 to 90 min per day, produces a two- to threefold increase in this enzyme's activity.

One metabolic consequence of mitochondrial changes induced by aerobic training is **glycogen sparing**, a slower rate of utilization of muscle glycogen and enhanced reliance on fat as a fuel source at a given exercise intensity. Enhanced glycogen sparing with endurance training most likely improves the ability to sustain a higher exercise intensity, such as maintaining a faster race pace in a 10 km run.

FIGURE 11.8 The percentage change in maximal oxygen uptake ($\dot{V}O_{2max}$) and the activity of succinate dehydrogenase (SDH), one of the muscles' key oxidative enzymes, during seven months of swim training. Interestingly, although this enzyme activity continues to increase with increasing levels of training, the swimmers' maximal oxygen uptake appears to level off after the first 8 to 10 weeks of training. This implies that mitochondrial enzyme activity is not a direct indication of whole-body endurance capacity.[8]

(a) Succinate dehydrogenase

(b) Citrate synthase

FIGURE 11.9 Leg muscle (gastrocnemius) enzyme activities of untrained (UT) subjects, moderately trained (MT) joggers, and highly trained (HT) marathon runners. Enzyme levels are shown for two of many key enzymes that participate in the oxidative production of adenosine triphosphate.

Adapted from D.L. Costill et al., 1979, "Lipid metabolism in skeletal muscle of endurance-trained males and females." *Journal of Applied Physiology* 28: 251-255 and from D.L. Costill et al., 1979, "Adaptations in skeletal muscle following strength training," *Journal of Applied Physiology* 46: 96-99. Used with permission.

In review

➤ Aerobic training selectively recruits type I muscle fibers and fewer type II fibers. Consequently, the type I fibers increase their cross-sectional area with aerobic training.

➤ After training, there appears to be a small increase in the percentage of type I fibers, as well as a transition of some type IIx to type IIa fibers.

➤ Capillary density—the number of capillaries supplying each muscle fiber—increases with training.

➤ Aerobic training increases muscle myoglobin content by as much as 75% to 80%. Myoglobin transports oxygen from cell membranes to the mitochondria.

➤ Aerobic training increases both the number and the size of muscle fiber mitochondria.

➤ Activities of many oxidative enzymes are increased with aerobic training.

➤ These changes occurring in the muscles, combined with adaptations in the oxygen transport system, enhance the capacity of oxidative metabolism and improve endurance performance.

Metabolic Adaptations to Training

Now that we have discussed training changes in both the cardiovascular and respiratory systems, as well as skeletal muscle adaptations, we are ready to examine how these integrated adaptations are reflected by changes in three important physiological variables related to metabolism:

- Lactate threshold
- Respiratory exchange ratio
- Oxygen consumption

Lactate Threshold

Lactate threshold, discussed in chapter 5, is a physiological marker that is closely associated with endurance performance—the higher the lactate threshold, the better the performance capacity. Figure 11.10*a* illustrates the difference in lactate threshold between an endurance-trained individual and an untrained individual. This figure also accurately represents the changes in lactate threshold that would occur following a 6- to 12-month program of endurance training.

In either case, in the trained state, one can exercise at a higher percentage of one's $\dot{V}O_{2max}$ before lactate begins to accumulate in the blood. In this example, the trained runner could sustain a race pace of 70% to 75% of his $\dot{V}O_{2max}$, an intensity that would result in continued lactate accumulation in the blood of the untrained runner. This translates into a much faster race pace (see figure 11.10*b*). Above the lactate threshold, the lower lactate at a given rate of work is likely attributable to a combination of reduced lactate production and increased lactate clearance. As athletes become better trained, their postexercise blood lactate concentrations are lower for the same rate of work.

FIGURE 11.10 Changes in lactate threshold (LT) with training expressed as (*a*) a percentage of maximal oxygen uptake (%$\dot{V}O_{2max}$) and (*b*) an increase in speed on the treadmill. Lactate threshold occurs at a speed of 8.4 km/h (5.2 mph) in the untrained state and at 11.6 km/h (7.2 mph) in the trained state.

Challenging Traditional Training Regimens

Traditionally, exercise physiologists have recommended one of three regimens to improve aerobic power: continuous exercise at a moderate to high intensity; long, slow (low-intensity) exercise; or interval training. However, a growing body of research suggests that **high-intensity interval training** (HIT) is a time-efficient way to induce many adaptations normally associated with traditional endurance training. Scientists at McMaster University in Canada have studied the effects of training using short bursts of very intense cycling, interspersed with up to a few minutes of rest or low-intensity cycling for recovery.[12] A common training mode employed is based on the Wingate test, a test that consists of 30 s of "all-out" cycling and generally produces mean power outputs that are two to three times higher than what subjects typically generate during a maximal oxygen uptake test.

In one study, young healthy subjects performed four to six 30 s sprints separated by 4 min of recovery, three times a week. These men showed the same beneficial changes in their heart, blood vessels, and muscles as another group who underwent a traditional training program involving up to an hour of continuous cycling, five days per week. Improvements in exercise performance—whether measured as cycling time to exhaustion at a fixed work intensity or in time trials that more closely resemble normal athletic competition—were comparable between groups, despite considerable differences in training time commitment.[12] HIT appears to stimulate some of the same molecular signaling pathways that regulate skeletal muscle remodeling in response to endurance training, including mitochondrial biogenesis and changes in the capacity for carbohydrate and fat transport and oxidation. Researchers are currently studying whether "modified" forms of HIT—which might be safer and better tolerated by older, less fit individuals or people with metabolic diseases such as type 2 diabetes—are equally effective for improving health and fitness.

(a) Maximal cytochrome oxidase activity

(b) Cycling performance

(a) Maximal activity of the mitochondrial enzyme cytochrome oxidase (COX) measured in skeletal muscle biopsy samples. This oxidative enzyme increased similarly with high-intensity interval training (HIT) and traditional moderate-intensity endurance training (END). (b) Cycling time trial performance before and after two weeks of training. Importantly, the total training time commitment was 10 h for the END group but only ~2.5 h for the HIT group. The total exercise volume was ~90% lower in the HIT group.

Adapted, by permission, from M. Gibala et al., 2006, "Short-term sprint interval versus traditional endurance training: similar initial adaptations in human skeletal muscle and exercise performance," *Journal of Physiology* 575: 901-911.

Respiratory Exchange Ratio

Recall from chapter 5 that the respiratory exchange ratio (RER) is the ratio of carbon dioxide released to oxygen consumed during metabolism. The RER reflects the composition of the mixture of substrates being used as an energy source, with a lower RER reflecting an increased reliance on fats for energy production and a higher RER reflecting a higher contribution of carbohydrates.

After training, the RER decreases at both absolute and relative submaximal exercise intensities. These changes are attributable to a greater utilization of free fatty acids instead of carbohydrate at these work rates following training.

Resting and Submaximal Oxygen Consumption

Oxygen consumption ($\dot{V}O_2$) at rest is unchanged following endurance training. While a few cross-sectional comparisons have suggested that training elevates resting $\dot{V}O_2$, the HERITAGE Family Study—with a large number of subjects and with duplicate measures of resting metabolic rate both before and after 20 weeks of training—showed no evidence of an increased resting metabolic rate after training.[32]

During submaximal exercise at a given intensity, $\dot{V}O_2$ is either unchanged or slightly reduced following training. In the HERITAGE Family Study, training reduced submaximal $\dot{V}O_2$ by 3.5% at a work rate of 50 W. There was a corresponding reduction in cardiac output at 50 W, reinforcing the strong interrelationship between $\dot{V}O_2$ and cardiac output.[31] This small decrease in $\dot{V}O_2$ during submaximal exercise, not seen in many studies, could have resulted from an increase in exercise economy (performing the same exercise intensity with less extraneous movement).

Maximal Oxygen Consumption

$\dot{V}O_{2max}$ is the best indicator of cardiorespiratory endurance capacity and increases substantially in response to endurance training. While small and very large increases have been reported, an increase of 15% to 20% is typical for a previously sedentary person who trains at 50% to 85% of his or her $\dot{V}O_{2max}$ three to five times per week, 20 to 60 min per day, for six months. For example, the $\dot{V}O_{2max}$ of a sedentary individual could reasonably increase from 35 ml · kg⁻¹ · min⁻¹ to 42 ml

· kg⁻¹ · min⁻¹ as a result of such a program. This is far below the values we see in world-class endurance athletes, whose values generally range from 70 to 94 ml · kg⁻¹ · min⁻¹. The more sedentary an individual is when starting an exercise program, the larger the increase in $\dot{V}O_{2max}$.

What Limits Aerobic Power and Endurance Performance?

A number of years ago, exercise scientists were divided on what major physiological factor or factors actually limit $\dot{V}O_{2max}$. Two contrasting theories had been proposed.

One theory held that endurance performance was limited by the lack of sufficient concentrations of oxidative enzymes in the mitochondria. Endurance training programs substantially increase these oxidative enzymes, allowing active tissue to use more of the available oxygen, resulting in a higher $\dot{V}O_{2max}$. In addition, endurance training increases both the size and number of muscle mitochondria. Thus, this theory argued, the main limitation of maximal oxygen consumption is the inability of the existing mitochondria to use the available oxygen beyond a certain rate. This theory was referred to as the utilization theory.

The second theory proposed that central and peripheral cardiovascular factors limit endurance capacity. These circulatory influences would preclude delivery of sufficient amounts of oxygen to the active tissues. According to this theory, improvement in $\dot{V}O_{2max}$ following endurance training results from increased blood volume, increased cardiac output (via stroke volume), and a better perfusion of active muscle with blood.

Evidence strongly supports the latter theory. In one study, subjects breathed a mixture of carbon monoxide (which irreversibly binds to hemoglobin, limiting hemoglobin's oxygen-carrying capacity) and air during exercise to exhaustion.[24] $\dot{V}O_{2max}$ decreased in direct proportion to the percentage of carbon monoxide breathed. The carbon monoxide molecules bonded to approximately 15% of the total hemoglobin; this percentage agreed with the percentage reduction in $\dot{V}O_{2max}$. In another study, approximately 15% to 20% of each subject's total blood volume was removed.[10] $\dot{V}O_{2max}$ decreased by approximately the same relative amount. Reinfusion of the subjects' packed red blood cells approximately four weeks later increased $\dot{V}O_{2max}$ well above baseline or control conditions. In both studies, the reduction in the oxygen-carrying capacity of the blood—via either blocking hemoglobin or removing whole blood—resulted in the delivery of less oxygen to the active tissues and a corresponding reduction in

$\dot{V}O_{2max}$. Similarly, studies have shown that breathing oxygen-enriched mixtures, in which the partial pressure of oxygen in the inspired air is substantially increased, increases endurance capacity.

These and subsequent studies indicate that the available oxygen supply is the major limiter of endurance performance. Saltin and Rowell[28] reviewed this topic and concluded that oxygen transport to the working muscles, not the available mitochondria and oxidative enzymes, limits $\dot{V}O_{2max}$. They argued that increases in $\dot{V}O_{2max}$ with training are largely attributable to increased maximal blood flow and increased muscle capillary density in the active tissues. Skeletal muscle adaptations (including increased mitochondrial content and respiratory capacity of the muscle fibers) contribute importantly to the ability to perform prolonged, high-intensity, submaximal exercise.

Table 11.3 summarizes the typical physiological changes that occur with endurance training. The changes pre- to posttraining in a previously inactive man are compared with values for a world-class male endurance runner.

In review

➤ Lactate threshold increases with endurance training, allowing performance of higher exercise intensities without significantly increasing blood lactate concentration.

➤ With endurance training, the RER decreases at submaximal work rates, indicating greater utilization of free fatty acids as an energy substrate (carbohydrate sparing).

➤ Oxygen consumption generally remains unchanged at rest and remains unaltered or decreases slightly during submaximal exercise following endurance training.

➤ $\dot{V}O_{2max}$ increases substantially following endurance training, but the extent of increase possible is genetically limited in each individual. The major limiting factor appears to be oxygen delivery to the active muscles.

Long-Term Improvement in Aerobic Power and Cardiorespiratory Endurance

Although an individual's highest attainable $\dot{V}O_{2max}$ is usually achieved within 12 to 18 months of intense endurance training, endurance *performance* can continue to improve. Improvement in endurance performance without improvement in $\dot{V}O_{2max}$ is likely

attributable to improvements in the ability to perform at increasingly higher percentages of $\dot{V}O_{2max}$ for extended periods. Consider, for example, a young male runner who starts training with an initial $\dot{V}O_{2max}$ of 52.0 ml · kg^{-1} · min^{-1}. He reaches his genetically determined peak $\dot{V}O_{2max}$ of 71.0 ml · kg^{-1} · min^{-1} after two years of intense training, after which no further increases occur, even with more frequent or more intense workouts. At this point, as shown in figure 11.11, the young runner is able to run at 75% of his $\dot{V}O_{2max}$ (0.75 × 71.0 = 53.3 ml · kg^{-1} · min^{-1}) in a 10 km (6.2 mi) race. After an additional two years of intensive training, his $\dot{V}O_{2max}$ is unchanged, but he is now able to compete at 88% of his $\dot{V}O_{2max}$ (0.88 × 71.0 = 62.5 ml · kg^{-1} · min^{-1}). Obviously, by being able to sustain an oxygen uptake of 62.5 ml · kg^{-1} · min^{-1}, he is able to run at a much faster race pace.

This ability to sustain exercise at a higher percentage of $\dot{V}O_{2max}$ is partly the result of an increase in the ability to buffer lactate, because race pace is directly related to the $\dot{V}O_2$ value at which lactate begins to accumulate.

FIGURE 11.11 Change in race pace with continued training after maximal oxygen uptake stops increasing beyond 71 ml · kg^{-1} · min^{-1}.

Factors Affecting an Individual's Response to Aerobic Training

We have discussed general trends in adaptations that occur in response to endurance training. However, we must always remember that we are talking about adaptations in individuals and that everyone does not respond in the same manner. Several factors that can affect individual response to aerobic training must be considered.

TABLE 11.3 Typical Effects of Endurance Training in a Previously Inactive Man, Contrasted With Values for a Male World-Class Endurance Athlete

Variables	Pretraining, sedentary male	Posttraining, sedentary male	World-class endurance athlete
CARDIOVASCULAR			
HR_{rest} (beats/min)	75	65	45
HR_{max} (beats/min)	185	183	174
SV_{rest} (ml/beat)	60	70	100
SV_{max} (ml/beat)	120	140	200
$\dot{Q}$ at rest (L/min)	4.5	4.5	4.5
$\dot{Q}_{max}$ (L/min)	22.2	25.6	34.8
Heart volume (ml)	750	820	1,200
Blood volume (L)	4.7	5.1	6.0
Systolic BP at rest (mmHg)	135	130	120
Systolic BP_{max} (mmHg)	200	210	220
Diastolic BP at rest (mmHg)	78	76	65
Diastolic BP_{max} (mmHg)	82	80	65
RESPIRATORY			
$\dot{V}_E$ at rest (L/min)	7	6	6
$\dot{V}_{Emax}$ (L/min)	110	135	195
TV at rest (L)	0.5	0.5	0.5
TV_{max} (L)	2.75	3.0	3.9
VC (L)	5.8	6.0	6.2
RV (L)	1.4	1.2	1.2
METABOLIC			
$(a-\bar{v})O_2$ diff at rest (ml/100 ml)	6.0	6.0	6.0
$(a-\bar{v})O_2$ diff max (ml/100 ml)	14.5	15.0	16.0
$\dot{V}O_2$ at rest (ml · kg^{-1} · min^{-1})	3.5	3.5	3.5
$\dot{V}O_{2max}$ (ml · kg^{-1} · min^{-1})	40.7	49.9	81.9
Blood lactate at rest (mmol/L)	1.0	1.0	1.0
Blood lactate max (mmol/L)	7.5	8.5	9.0
BODY COMPOSITION			
Weight (kg)	79	77	68
Fat weight (kg)	12.6	9.6	5.1
Fat-free weight (kg)	66.4	67.4	62.9
Fat (%)	16.0	12.5	7.5

Note. HR = heart rate; SV = stroke volume; $\dot{Q}$ = cardiac output; BP = blood pressure; $\dot{V}_E$ = ventilation; TV = tidal volume; VC = vital capacity; RV = residual volume; $(a-\bar{v})O_2$ diff = arterial–mixed venous oxygen difference; $\dot{V}O_2$ = oxygen consumption.

Training Status and $\dot{V}O_{2max}$

The higher the initial state of conditioning, the smaller the relative improvement for the same volume of training. For example, if two people, one sedentary and the other partially trained, undergo the same endurance training program, the sedentary person will show the greatest relative (%) improvement.

In fully mature athletes, the highest attainable $\dot{V}O_{2max}$ is reached within 8 to 18 months of intense endurance training, indicating that each athlete has a finite maximal attainable level of oxygen consumption. This finite range is genetically determined but may potentially be influenced by training in early childhood during the development of the cardiovascular system.

Heredity

The ability to increase maximal oxygen consumption levels is genetically limited. This does not mean that each individual has a preprogrammed $\dot{V}O_{2max}$ that cannot be exceeded. Rather, a range of $\dot{V}O_{2max}$ values seems to be predetermined by an individual's genetic makeup, with that individual's highest attainable $\dot{V}O_{2max}$ somewhere in that range. Each individual is born into a predetermined genetic window, and that individual can shift up or down within that window with exercise training or detraining, respectively.

Research on the genetic basis of $\dot{V}O_{2max}$ began in the late 1960s and early 1970s.[18] Recent research has shown that identical (monozygous) twins have similar $\dot{V}O_{2max}$ values, whereas the variability for dizygous (fraternal) twins is much greater (shown in figure 11.12).[5] Each symbol represents a pair of brothers. Brother A's $\dot{V}O_{2max}$ value is indicated by the symbol's position on the x-axis, and brother B's $\dot{V}O_{2max}$ value is on the y-axis. Similarity in the siblings' $\dot{V}O_{2max}$ values is noted by comparing the x and y coordinates of the symbol (i.e., how close it falls to the diagonal line $x = y$ on the graph). Similar results were found for endurance capacity, determined by the maximal amount of work performed in an all-out, 90 min ride on a cycle ergometer.

Bouchard and colleagues[4] concluded that heredity accounts for between 25% and 50% of the variance in $\dot{V}O_{2max}$ values. This means that of all factors influencing $\dot{V}O_{2max}$, heredity alone is responsible for one-quarter to one-half of the total influence. World-class athletes who have stopped endurance training continue for many years to have high $\dot{V}O_{2max}$ values in their sedentary, deconditioned state. Their $\dot{V}O_{2max}$ values may decrease from 85 to 65 ml · kg⁻¹ · min⁻¹, but this "deconditioned" value is still very high compared with the general population.

Heredity also potentially explains the fact that some people have relatively high $\dot{V}O_{2max}$ values yet have no

FIGURE 11.12 Comparisons of $\dot{V}O_{2max}$ in twin (monozygous and dizygous) and nontwin brothers.

Adapted, by permission, from C. Bouchard et al., 1986, "Aerobic performance in brothers, dizygotic and monozygotic twins," *Medicine and Science in Sports and Exercise* 18: 639-646.

history of endurance training. In a study that compared untrained men who had $\dot{V}O_{2max}$ values below 49 ml · kg⁻¹ · min⁻¹ with untrained men who had $\dot{V}O_{2max}$ values above 62.5 ml · kg⁻¹ · min⁻¹, those with high values were distinguished by having higher blood volumes, which contributed to higher stroke volumes and cardiac outputs at maximal intensities. The higher blood volumes in the high $\dot{V}O_{2max}$ group were most likely genetically determined.[20]

(In)focus

Heredity is a major determinant of aerobic power, accounting for as much as 25% to 50% of the variation in $\dot{V}O_{2max}$ among individuals.

Thus, both genetic and environmental factors influence $\dot{V}O_{2max}$ values. The genetic factors probably establish the boundaries for the athlete, but endurance training can push $\dot{V}O_{2max}$ to the upper limit of these boundaries. Dr. Per-Olof Åstrand, one of the most highly recognized exercise physiologists during the second half of the 20th century, stated on numerous occasions that the best way to become a champion Olympic athlete is to be selective when choosing one's parents!

Sex

Healthy untrained girls and women have significantly lower $\dot{V}O_{2max}$ values (20-25% lower) than healthy untrained boys and men. Highly conditioned female endurance athletes have values much closer to those of highly trained male endurance athletes (i.e., only about 10% lower). This is discussed in greater detail in chapter 19. Representative ranges of $\dot{V}O_{2max}$ values for athletes and nonathletes are presented in table 11.4 by age, sex, and sport.

TABLE 11.4 **Maximal Oxygen Uptake Values (ml · kg^{-1} · min^{-1}) for Nonathletes and Athletes**

Group or sport	Age	Males	Females
Nonathletes	10-19	47-56	38-46
	20-29	43-52	33-42
	30-39	39-48	30-38
	40-49	36-44	26-35
	50-59	34-41	24-33
	60-69	31-38	22-30
	70-79	28-35	20-27
Baseball and softball	18-32	48-56	52-57
Basketball	18-30	40-60	43-60
Bicycling	18-26	62-74	47-57
Canoeing	22-28	55-67	48-52
Football	20-36	42-60	–
Gymnastics	18-22	52-58	36-50
Ice hockey	10-30	50-63	–
Jockey	20-40	50-60	–
Orienteering	20-60	47-53	46-60
Racquetball	20-35	55-62	50-60
Rowing	20-35	60-72	58-65
Skiing, alpine	18-30	57-68	50-55
Skiing, Nordic	20-28	65-94	60-75
Ski jumping	18-24	58-63	–
Soccer	22-28	54-64	50-60
Speed skating	18-24	56-73	44-55
Swimming	10-25	50-70	40-60
Track and field, discus	22-30	42-55	*
Track and field, running	18-39	60-85	50-75
	40-75	40-60	35-60
Track and field, shot put	22-30	40-46	*
Volleyball	18-22	–	40-56
Weightlifting	20-30	38-52	*
Wrestling	20-30	52-65	–

*Data not available.

High Responders and Low Responders

For years, researchers have found wide variations in the amount of improvement in $\dot{V}O_{2max}$ with endurance training. Studies have demonstrated individual improvements in $\dot{V}O_{2max}$ ranging from 0% to 50% or more, even in similarly fit subjects completing exactly the same training program.

In the past, exercise physiologists have assumed that these variations result from differing degrees of compliance with the training program. People who comply with the program should, and do, have the highest percentage of improvement, and poor compliers should show little or no improvement. However, given the same training stimulus and full compliance with the program, substantial variations still occur in the percent improvement in $\dot{V}O_{2max}$ for different people.

It is now evident that the response to a training program is also genetically determined.[2] This is illustrated in figure 11.13. Ten pairs of identical twins completed a 20-week endurance training program; the improvements in $\dot{V}O_{2max}$, expressed as percentages, are plotted for each twin pair—Twin A on the x-axis and Twin B on the y-axis.[25] Notice the similarity in response for each twin pair. Yet across twin pairs, improvement in $\dot{V}O_{2max}$ varied from 0% to 40%. These results, and those from other studies, indicate that there will be **high responders** (showing large improvement) and **low responders** (showing little or no improvement) among groups of people who participate in identical training programs.

Results from the HERITAGE Family Study also support a strong genetic component in the magnitude of increase in $\dot{V}O_{2max}$ with endurance training. Families, including the biological mother and father and three or more of their children, trained three days a week for 20 weeks, initially exercising at a heart rate equal to 55% of their $\dot{V}O_{2max}$ for 35 min per day and progressing to a heart rate equal to 75% of their $\dot{V}O_{2max}$ for 50 min per day by the end of the 14th week, which they maintained for the last six weeks.[3] The average increase in $\dot{V}O_{2max}$ was about 17% but varied from 0% to more than 50%. Figure 11.14 illustrates the improvement in $\dot{V}O_{2max}$ for each subject in each family. Maximal heritability was estimated at 47%. Note that subjects who are high responders tend to be clustered in the same families, as are those who are low responders.

It is clear that this is a genetic phenomenon, not a result of compliance or noncompliance. One must consider this important point when conducting training studies and designing training programs. Individual differences must always be accounted for.

In focus

Individual differences cause substantial variation in subjects' responses to a given training program. Genetics accounts for much of this variation in response.

In review

➤ Although improvements in $\dot{V}O_{2max}$ eventually plateau, endurance performance can continue to improve for years with continued training.

➤ An individual's genetic makeup predetermines a range for his or her $\dot{V}O_{2max}$ and accounts for 25% to 50% of the variance in $\dot{V}O_{2max}$ values. Heredity also largely explains individual variations in response to identical training programs.

➤ Highly conditioned female endurance athletes have $\dot{V}O_{2max}$ values only about 10% lower than those of highly conditioned male endurance athletes.

Cardiorespiratory Endurance in Nonendurance Sports

Many people regard cardiorespiratory endurance as the most important component of physical fitness. It is an

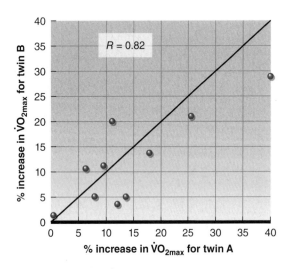

FIGURE 11.13 Variations in the percentage increase in $\dot{V}O_{2max}$ for identical twins undergoing the same 20-week training program.

Reprinted, by permission, from D. Prud'homme et al., 1984, "Sensitivity of maximal aerobic power to training is genotype-dependent," *Medicine and Science in Sports and Exercise* 16(5): 489-493.

FIGURE 11.14 Variations in the improvement in V̇O₂ₘₐₓ following 20 weeks of endurance training by families. Values represent the changes in V̇O₂ₘₐₓ in ml/min, with an average increase of 393 ml/min. Data for each family are enclosed within a bar, and each family member's value is represented as a dot within the bar.

Adapted from C. Bouchard et al., 1999, "Familial aggregation of V̇O₂ₘₐₓ response to exercise training. Results from HERITAGE Family Study," *Journal of Applied Physiology* 87: 1003-1008. Used with permission.

athlete's major defense against fatigue. Low endurance capacity leads to fatigue, even in activities that are not aerobic in nature. For any athlete, regardless of the sport or activity, fatigue represents a major deterrent to optimal performance. Even minor fatigue can hinder the athlete's total performance:

- Muscular strength is decreased.
- Reaction and movement times are prolonged.
- Agility and neuromuscular coordination are reduced.
- Whole-body movement speed is slowed.
- Concentration and alertness are reduced.

The decline in concentration and alertness associated with fatigue is particularly important. The athlete can become careless and more prone to serious injury, especially in contact sports. Even though these decrements in performance might be small, they can be just enough to cause an athlete to miss the critical free throw in basketball, the strike zone in baseball, or the 20 ft (6 m) putt in golf.

All athletes can benefit from improving their cardiorespiratory endurance. Even golfers, whose sport demands little in the way of aerobic endurance, can benefit. Improved endurance can allow golfers to complete a round of golf with less fatigue and to better withstand long periods of walking and standing.

For the sedentary, middle-aged adult, numerous health factors indicate that cardiovascular endurance should be the primary emphasis of training. Training for health and fitness is discussed at length in part VII of this book.

The extent of endurance training needed varies considerably from one sport to the next and from one athlete to the next. It depends on the athlete's current endurance capacity and the endurance demands of the chosen activity. The marathon runner uses endurance training almost exclusively, with limited attention to strength, flexibility, and speed. The baseball player, however, places very limited demands on endurance capacity, so endurance conditioning is not as highly emphasized. Nevertheless, baseball players could gain substantially from endurance running, even if only at a moderate pace for 5 km (3.1 mi) per day, three days a week. As a benefit of training, baseball players would have little or no leg trouble (a frequent complaint), and they would be able to complete a doubleheader with little or no fatigue.

Adequate cardiovascular conditioning must be the foundation of any athlete's general conditioning program. Many athletes in nonendurance activities never incorporate even moderate endurance training into their training programs. Those who have done so are well aware of their improved physical condition and its impact on their athletic performance.

In focus

All athletes can benefit from maximizing their cardiorespiratory endurance.

Adaptations to Anaerobic Training

In muscular activities that require near-maximal force production for relatively short periods of time, such as sprinting, much of the energy needs are met by the ATP-phosphocreatine (PCr) system and the anaerobic breakdown of muscle glycogen (glycolysis). The following sections focus on the trainability of these two systems.

Changes in Anaerobic Power and Anaerobic Capacity

Exercise scientists have had difficulty agreeing on an appropriate laboratory or field test to measure anaerobic power. Unlike the situation with aerobic power, for which $\dot{V}O_{2max}$ is generally agreed to be the gold standard measurement, no single test adequately measures

Individual Differences in Response to Training: The HERITAGE Family Study

It has been clearly established that individuals differ considerably in their responses to a specific intervention, such as a drug or diet. The same is true for the response of $\dot{V}O_{2max}$ to aerobic training, with studies showing improvements ranging from 0% to 50% or more in response to exactly the same training program.

The HERITAGE Family Study was funded by the National Institutes of Health in 1992 to study the possible genetic control of these variations in $\dot{V}O_{2max}$ in response to aerobic training and the associated variations in the major risk factors for cardiovascular disease and type 2 diabetes. Funding continued through the year 2004. Families were recruited for this study, including the biological mother and biological father and three or more of their children. Some families were included that did not meet these criteria.

Dr. Claude Bouchard.

Led by Dr. Claude Bouchard, executive director of the Pennington Biomedical Research Center, Louisiana State University, and a pioneer in genetics research in the exercise sciences, a team of researchers from several universities (Arizona State University, Indiana University, Laval University in Canada, the University of Minnesota, and the University of Texas at Austin) recruited a total of 742 sedentary subjects from families of both white and black descent who completed the study. Another university, Washington University, was responsible for data quality control and analyses. The subjects completed a comprehensive battery of tests both before and after a 20-week program of aerobic training, with each training session supervised by an exercise physiologist. The test battery included physiological measures associated with aerobic fitness and clinical medicine markers associated with risk for cardiovascular disease and diabetes. The average increase in $\dot{V}O_{2max}$ expressed per kilogram of body weight was 18%, but the range of increases varied from 0% to 53%. The increase in $\dot{V}O_{2max}$ was influenced by genetics (maximal heritability = 47%; see figure 11.14) but was influenced very little by age, sex, and race. It is important to recognize from these data that each individual responds differently to the exact same exercise stimulus. We cannot expect the same improvement in all people! The mere fact that a person has a low response to training does not mean the person did not follow the training program. For more information on the HERITAGE Family Study, go to its website (www.pbrc.edu/heritage/home.htm).

anaerobic power. Most research has been conducted through use of three different tests of either anaerobic power, anaerobic capacity, or both: the Wingate anaerobic test, the critical power test, and the maximal accumulated oxygen deficit test (see chapter 5). Of these three, the Wingate test has been the most widely used.

With the Wingate anaerobic test, the subject pedals a cycle ergometer at maximal speed for 30 s against a high braking force. The braking force is determined by the person's weight, sex, age, and level of training. Power output can be determined instantaneously throughout the 30 s test but is generally averaged over 3 to 5 s intervals. The peak power output is the highest mechanical power achieved at any stage in the test; it is generally achieved during the first 5 to 10 s and is considered an index of anaerobic power. The mean power output is computed as the average power output over the total 30 s period, and one obtains total work simply by multiplying the mean power output by 30 s. Mean power output and total work have both been used as indexes of anaerobic capacity.

With anaerobic training, such as sprint training on the track or on a cycle ergometer, there are increases in both peak anaerobic power and anaerobic capacity. However, results have varied widely across studies, from those that showed only minimal increases to those showing increases of up to 25%.

Adaptations in Muscle With Anaerobic Training

With anaerobic training, which includes sprint training and resistance training, there are changes in skeletal muscle that specifically reflect muscle fiber recruitment for these types of activities. As discussed in chapter 1, at higher intensities, type II muscle fibers are recruited to a greater extent, but not exclusively, because type I fibers continue to be recruited. Overall, sprint and resistance activities use the type II muscle fibers significantly more than do aerobic activities. Consequently, both type IIa and type IIx muscle fibers undergo an increase in their cross-sectional areas. The cross-sectional area of type I fibers also is increased but usually to a lesser extent. Furthermore, with sprint training there appears to be a reduction in the percentage of type I fibers and an increase in the percentage of type II fibers, with the greatest change in type IIa fibers. In two of these studies, in which subjects performed 15 s to 30 s all-out sprints, the type I percentage decreased from 57% to 48% and type IIa increased from 32% to 38%.[16, 17] This shift of type I to type II fibers is not typically seen with resistance training.

Adaptations in the Energy Systems

Just as aerobic training produces changes in the aerobic energy system, anaerobic training alters the ATP-PCr and anaerobic glycolytic energy systems. These changes are not as obvious or predictable as those that result from endurance training, but they do improve performance in anaerobic activities.

Adaptations in the ATP-PCr System

Activities that emphasize maximal muscle force production, such as sprinting and weightlifting events, rely most heavily on the ATP-PCr system for energy. Maximal effort lasting less than about 6 s places the greatest demands on the breakdown and resynthesis of ATP and PCr. Costill and coworkers reported their findings from a study of resistance training and its effects on the ATP-PCr system.[6] Their participants trained by performing maximal knee extensions. One leg was trained using 6 s maximal work bouts that were repeated 10 times. This type of training preferentially stressed the ATP-PCr energy system. The other leg was trained with repeated 30 s maximal bouts, which instead preferentially stressed the glycolytic system.

The two forms of training produced the same muscular strength gains (about 14%) and the same resistance to fatigue. As seen in figure 11.15, the activities of the anaerobic muscle enzymes creatine kinase and myokinase increased as a result of the 30 s training bouts but were almost unchanged in the leg trained with repeated 6 s maximal efforts. This finding leads us to conclude that maximal sprint bouts (6 s) might improve muscular strength but contribute little to the mechanisms responsible for ATP and PCr breakdown.

FIGURE 11.15 Changes in creatine kinase (CK) and muscle myokinase (MK) activities as a result of 6 s and 30 s bouts of maximal anaerobic training.

Summary of Cardiovascular Adaptation to Chronic Endurance Exercise

Physiologists commonly establish models to help explain how various physiological factors or variables work together to affect a specific outcome or component of performance. Dr. Donna H. Korzick, an exercise physiologist at Pennsylvania State University, has created a unifying figure to model the factors that contribute to the cardiovascular adaptation to chronic endurance training (see the figure in this sidebar).

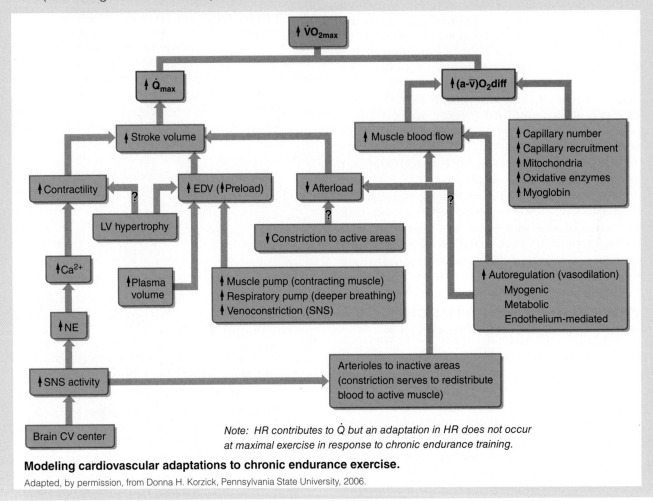

Note: HR contributes to $\dot{Q}$ but an adaptation in HR does not occur at maximal exercise in response to chronic endurance training.

Modeling cardiovascular adaptations to chronic endurance exercise.

Adapted, by permission, from Donna H. Korzick, Pennsylvania State University, 2006.

Data have been published, however, that show improvements in ATP-PCr enzyme activities with training bouts lasting only 5 s.

Regardless of the conflicting results, these studies suggest that the major value of training bouts that last only a few seconds (sprints) is the development of muscular strength. Such strength gains enable the individual to perform a given task with less effort, which reduces the risk of fatigue. Whether these changes allow the muscle to perform more anaerobic work remains unanswered, although a 60 s sprint-fatigue test suggests that short sprint-type anaerobic training does not enhance anaerobic endurance.[6]

Adaptations in the Glycolytic System

Anaerobic training (30 s bouts) increases the activities of several key glycolytic enzymes. The most frequently studied glycolytic enzymes are phosphorylase, phosphofructokinase (PFK), and lactate dehydrogenase (LDH). The activities of these three enzymes increased 10% to 25% with repeated 30 s training bouts but changed little with short (6 s) bouts that stress primarily the ATP-PCr system.[6] In a more recent study, 30 s maximal all-out sprints significantly increased hexokinase (56%) and PFK (49%) but not total phosphorylase activity or LDH.[19]

Because both PFK and phosphorylase are essential to the anaerobic yield of ATP, such training might enhance glycolytic capacity and allow the muscle to develop greater tension for a longer period of time. However, as seen in figure 11.16, this conclusion is not supported by results of a 60 s sprint performance test, in which the subjects performed maximal knee extension and flexion. Power output and the rate of fatigue (shown by a decrease in power production) were affected to the same degree after sprint training with either 6 or 30 s training bouts. Thus, we must conclude that performance gains with these forms of training result from improvements in strength rather than improvements in the anaerobic yield of ATP.

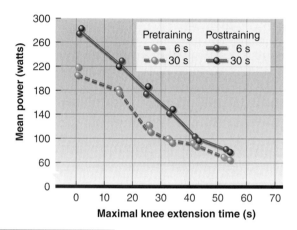

FIGURE 11.16 Performance in a 60 s sprint bout after training with 6 s and 30 s anaerobic bouts. Subjects are the same as in figure 11.15.

In focus

Anaerobic training increases the ATP-PCr and glycolytic enzymes but has no effect on the oxidative enzymes. Conversely, aerobic training increases the oxidative enzymes but has little effect on the ATP-PCr or glycolytic enzymes. This fact reinforces a recurring theme: Physiological alterations that result from training are highly specific to the type of training.

In review

➤ Anaerobic training bouts improve both anaerobic power and anaerobic capacity.

➤ The performance improvement noted with sprint-type anaerobic training appears to result more from strength gains than from improvements in the functioning of the anaerobic energy systems.

➤ Anaerobic training increases the ATP-PCr and glycolytic enzymes but has no effect on the oxidative enzymes.

Specificity of Training and Cross-Training

Physiological adaptations in response to physical training are highly specific to the nature of the training activity. Furthermore, the more specific the training program is to a given sport or activity, the greater the improvement in performance in that sport or activity. The concept of **specificity of training** is very important for all physiological adaptations.

This concept is also important in testing of athletes. As an example, to accurately measure endurance improvements, athletes should be tested while engaged in an activity similar to the sport or activity in which they usually participate. Consider one study of highly trained rowers, cyclists, and cross-country skiers. Their $\dot{V}O_{2max}$ was measured while they performed two types of work: uphill running on a treadmill and maximal performance of their sport-specific activity.[30] A key finding, shown in figure 11.17, was that the $\dot{V}O_{2max}$ attained by all the athletes during their sport-specific activity was as high as or higher than the values obtained on the treadmill. For many of these athletes, $\dot{V}O_{2max}$ was substantially higher during their sport-specific activity.

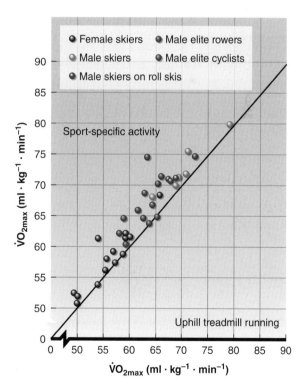

FIGURE 11.17 $\dot{V}O_{2max}$ values measured during uphill treadmill running versus sport-specific activities in selected groups of athletes.

Adapted from S.B. Strømme, F. Ingjer, and H.D. Meen, 1977, "Assessment of maximal aerobic power in specifically trained athletes," *Journal of Applied Physiology* 42: 833-837. Used with permission.

A highly creative design for studying the concept of specificity of training involves one-legged exercise training, with the untrained opposite leg used as the control. In one study, subjects were placed in three groups: a group that sprint trained one leg and endurance trained the other, a group that sprint trained one leg and did not train the other, and a group that endurance trained one leg and did not train the other.[27] Improvement in $\dot{V}O_{2max}$ and lowered heart rate and blood lactate response at submaximal work rates were found only when exercise was performed with the endurance-trained leg.

Much of the training response occurs in the specific muscles that have been trained, possibly even in individual motor units in a specific muscle. This observation applies to metabolic as well as cardiorespiratory responses to training. Table 11.5 shows the activities of selected muscle enzymes from the three energy systems for untrained, anaerobically trained, and aerobically trained men. The table shows that aerobically trained muscles have significantly lower glycolytic enzyme activities. Thus, they might have less capacity for anaerobic metabolism or might rely less on energy from glycolysis. More research is needed to explain the implications of the muscular changes accompanying both anaerobic and aerobic training, but this table clearly illustrates the high degree of specificity to a given training stimulus.

In focus

Close attention must be given to selecting an optimal training program. The program must be carefully matched with the athlete's individual needs to maximize the physiological adaptations to training, thereby optimizing performance.

Cross-training refers to training for more than one sport at the same time or training several different fitness components (such as endurance, strength, and flexibility) at one time. The athlete who trains by swimming, running, and cycling in preparation for competing in a triathlon is an example of the former, and the athlete involved in heavy resistance training and high-intensity cardiorespiratory training at the same time is an example of the latter.

For the athlete training for cardiorespiratory endurance and strength at the same time, the studies conducted to date indicate that gains in strength, power,

TABLE 11.5 Selected Muscle Enzyme Activities (mmol · g^{-1} · min^{-1}) for Untrained, Anaerobically Trained, and Aerobically Trained Men

	Untrained	Anaerobically trained	Aerobically trained
AEROBIC ENZYMES			
Oxidative system			
Succinate dehydrogenase	8.1	8.0	20.8[a]
Malate dehydrogenase	45.5	46.0	65.5[a]
Carnitine palmityl transferase	1.5	1.5	2.3[a]
ANAEROBIC ENZYMES			
ATP-PCr system			
Creatine kinase	609.0	702.0[a]	589.0
Myokinase	309.0	350.0[a]	297.0
Glycolytic system			
Phosphorylase	5.3	5.8	3.7[a]
Phosphofructokinase	19.9	29.2[a]	18.9
Lactate dehydrogenase	766.0	811.0	621.0

[a]Significant difference from the untrained value.

and endurance can result. However, the gains in muscular strength and power are less when strength training is combined with endurance training than when strength training alone is done. The opposite does not appear to be true: Improvement of aerobic power with endurance training does not appear to be attenuated by inclusion of a resistance training program. In fact, short-term endurance can be increased with resistance training. Although earlier studies supported the conclusion that concurrent strength and endurance training limits gains in strength and power, one well-controlled study did not show this. McCarthy and colleagues[21] reported similar gains in strength, muscle hypertrophy, and neural activation in a group of previously untrained subjects who underwent concurrent high-intensity strength training and cycle endurance training compared with a group who performed only high-intensity strength training.

In review

➤ For athletes to maximize cardiorespiratory gains from training, the training should be specific to the type of activity that an athlete usually performs.

➤ Resistance training in combination with endurance training does not appear to restrict improvement in aerobic power and may increase short-term endurance, but it can limit improvement in strength and power when compared with gains from resistance training alone.

In this chapter, we examined how the cardiovascular, respiratory, and metabolic systems adapt to aerobic and anaerobic training. We concentrated on how these adaptations can improve both aerobic and anaerobic performance. This chapter concludes our review of how body systems respond to both acute and chronic exercise. Now that we have completed our examination of how the body responds to internal changes, we can turn our attention to the external world. In the next part of the book, we focus on the body's adaptations to varying environmental conditions, beginning in the next chapter by considering how external temperature can affect performance.

Key Terms

aerobic training

anaerobic training

athlete's heart

capillary-to-fiber ratio

cardiac hypertrophy

cardiorespiratory endurance

cross-training

Fick equation

glycogen sparing

high-intensity interval training

high responders

low responders

mitochondrial oxidative enzymes

oxygen transport system

specificity of training

submaximal endurance capacity

Study Questions

1. Differentiate between muscular endurance and cardiovascular endurance.

2. What is maximal oxygen uptake ($\dot{V}O_{2max}$)? How is it defined physiologically, and what determines its limits?

3. Of what importance is $\dot{V}O_{2max}$ to endurance performance? Why does the competitor with the highest $\dot{V}O_{2max}$ not always win?

4. Describe the changes in the oxygen transport system that occur with endurance training.

5. What is possibly the most important adaptation that the body makes in response to endurance training, which allows for an increase in both $\dot{V}O_{2max}$ and performance? Through what mechanisms do these changes occur?

6. What metabolic adaptations occur in response to endurance training?

7. Explain the two theories that have been proposed to account for improvements in $\dot{V}O_{2max}$ with endurance training. Which of these has the greatest validity today? Why?

8. How important is genetic potential in a developing young athlete?

9. What adaptations have been shown to occur in muscle fibers with anaerobic training?

10. Discuss specificity of anaerobic training with respect to enzyme changes in muscle.

11. Why is cross-training beneficial to endurance athletes? How does it benefit sprint and power athletes?

Study Guide Activities

In addition to the activities listed in the chapter opening outline on page 247, two other activities are available in the web study guide, located at

www.HumanKinetics.com/PhysiologyOfSportAndExercise

 The **KEY TERMS** activity reviews important terms, and the end-of-chapter **QUIZ** tests your understanding of the material covered in the chapter.

PART IV

Environmental Influences on Performance

In previous sections, we discussed how the various body systems coordinate their activities to allow us to perform physical activity. We also saw how these systems adapt when exposed to the stress of various types of training. In part IV, we turn our attention to how the body responds and adapts when challenged to exercise under extreme environmental conditions. In chapter 12, "Exercise in Hot and Cold Environments," we examine the mechanisms by which the body regulates its internal temperature both at rest and during exercise. Then we consider how the body responds and adapts to exercise in the heat and cold, along with the health risks associated with physical activity in hot and cold environments. In chapter 13, "Exercise at Altitude," we discuss the unique challenges that the body faces when performing physical activity under conditions of reduced atmospheric pressure (altitude) and how the body adapts to extended periods at altitude. We then discuss the best way to prepare for competing at altitude and whether altitude training might help people perform better at sea level. Finally, health risks associated with ascent to high altitude are discussed.

Exercise in Hot and Cold Environments

12

In this chapter and in the web study guide

ACTIVITY 12.1   Control of Heat Exchange explores the role of the hypothalamus in controlling body temperature.

ACTIVITY 12.2   Exercise in the Heat reviews changes in physiological responses due to exercising in the heat.

ACTIVITY 12.3   Heat-Related Disorders investigates three athlete scenarios covering the signs, symptoms, and treatment of heat cramps, heat exhaustion, and heatstroke.

ACTIVITY 12.4   Exercise in the Cold reviews changes in physiological responses due to exercising in the cold.

ACTIVITY 12.5   Cold-Related Health Risks investigates two recreational scenarios covering the signs, symptoms, and treatment of health risks related to exercising in the cold.

Korey Stringer, a 335 lb (152 kg) right tackle for the Minnesota Vikings professional football team, was one of the National Football League's top offensive linemen. On the first day of training camp in 2001, Korey failed to complete the full team workout, leaving the practice field with symptoms of severe heat strain, including nausea, dizziness, and vomiting. On the second day of camp, July 31, 2001, Stringer collapsed on the practice field during an intense practice at the Vikings' training facility in Mankato, Minnesota. He had been training on a hot, cloudless day wearing a full football uniform and helmet. Having lost consciousness, he was taken to an air-conditioned trailer and eventually was taken by ambulance to a local hospital, where a core body temperature of 108.8 °F was recorded. He died approximately 13 h later—a victim of exertional heatstroke. Athletes are susceptible to heatstroke, particularly athletes who start training for their sport during the hottest months of the year. In addition to a lack of acclimation to the heat, the padding and uniforms worn by football players limit their ability to lose heat, and this strain on the body may be exacerbated if the athlete does not drink sufficient fluids.

Source: Sports Illustrated; July 29, 2002; 97 (4): 56-60.

The stresses of physical exertion often are complicated by environmental thermal conditions. Performing in extreme heat or cold places a heavy burden on the mechanisms that regulate body temperature. Although these mechanisms are amazingly effective in regulating body temperature under normal conditions, mechanisms of **thermoregulation** can be inadequate when we are subjected to extreme heat or cold. Fortunately, our bodies are able to adapt to such environmental stresses with continued exposure over time, a process known as **acclimation** (which refers to a short-term adaptation, e.g., days to weeks) or **acclimatization** (the proper term when we are referring to natural adaptations gained over long periods of time, e.g., months to years).

In the following discussion, we focus on the physiological responses to acute and chronic exercise in both hot and cold environments. Specific health risks are associated with exercise in both temperature extremes, so we also discuss the prevention of temperature-related illness and injuries during exercise.

Body Temperature Regulation

Humans are homeothermic, which means that internal body temperature is physiologically regulated to keep it nearly constant even when environmental temperature changes. Although a person's temperature varies from day to day, and even from hour to hour, these fluctuations are usually no more than about 1.0 °C (1.8 °F). Only during prolonged heavy exercise, fever due to illness, or extreme conditions of heat or cold do body temperatures deviate from the normal baseline range of 36.1 to 37.8 °C (97.0-100.0 °F). Body temperature reflects a careful balance between heat production and heat loss. Whenever this balance is disturbed, body temperature changes.

In focus

In physiology, temperatures are expressed as degrees Centigrade. To convert from °F to °C and vice versa, use the following transformations:

- To go from °F to °C: Subtract 32°, then divide by 1.8.
- To go from °C to °F: Multiply by 1.8, then add 32°.

Metabolic Heat Production

Only a small part (usually less than 25%) of the energy (adenosine triphosphate, ATP) the body produces is used for physiological functions such as muscle contraction; the rest is converted to heat. All active tissues produce metabolic heat (M) that must be intricately offset by heat loss to the environment to maintain the internal temperature of the body. If the body's heat production exceeds its heat loss, as it often does during moderate- to heavy-intensity aerobic activity, the body stores the excess heat and internal temperature increases. People's ability to maintain a constant internal temperature depends on their ability to balance the metabolic heat they produce and the heat they gain from the environment with the heat the body loses. This balance is depicted in figure 12.1.

Transfer of Body Heat to and From the Environment

Let's examine the mechanisms by which heat is transferred between a person and his or her surroundings. For the body to transfer heat to the environment, the heat produced in the body must first move from deep in the body (the core) to the skin (the shell), where it has access to the outside environment. Heat is primarily moved from the core to the skin by the blood. Only

FIGURE 12.1 To maintain a steady-state core temperature, the body must balance the heat gained from metabolism and from external environmental factors with the heat lost through the avenues of radiation, conduction, convection, and evaporation.

when heat reaches the skin can it be transferred to the environment by any of four mechanisms: conduction, convection, radiation, and evaporation. These are illustrated in figure 12.2.

Conduction and Convection

Heat **conduction** (K) involves the transfer of heat from one solid material to another through direct molecular contact. As an example, heat can be lost from the body when the skin is in contact with a cold object, as when one sits on cold metal bleachers watching a soccer match on a chilly day. Conversely, if a hot object is pressed against the skin, heat from the object will be conducted to the skin and heat will be gained by the body. If the contact is prolonged, heat from the skin surface can be transferred to the blood as it flows through the skin and transferred to the core, raising internal (core) temperature. During exercise, conduction is usually negligible as a source of heat exchange because the body surface area in contact with solid objects (for example, soles of the feet on hot playing fields) is small. Therefore, many environmental physiologists treat conductive heat exchange as negligible in their calculations of heat balance and exchange.

Convection (C), on the other hand, involves transferring heat by the motion of a gas or a liquid across the heated surface. When the body is still and there is little air movement, a thin unstirred "boundary" layer of air surrounds the body. However, the air around us is usually in constant motion, especially so during exercise as we move either the whole body or body segments (e.g., the arms pumping as we run) through the air. As air moves around us, passing over the skin,

FIGURE 12.2

The removal of heat from the skin. Heat is delivered to the body surface via the arterial blood and to a lesser extent by conduction through the subcutaneous tissue. When the temperature of the skin is greater than that of the environment, heat is removed by conduction (if the skin is in contact with an object), convection, radiation, and sweat evaporation; when the environmental temperature exceeds skin temperature, heat can be removed only by evaporation.

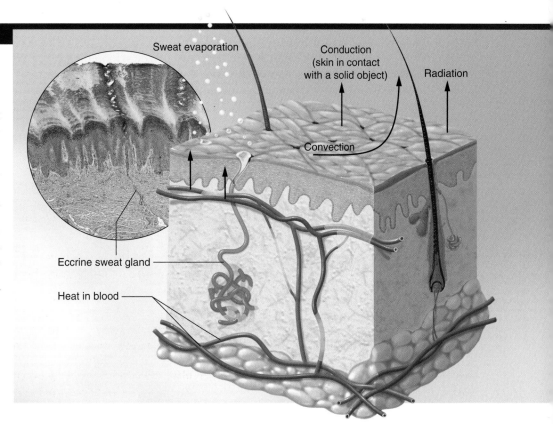

heat is exchanged with the air molecules. The greater the movement of the air (or liquid, such as water), the greater the rate of heat exchange by convection. Thus, in an environment in which air temperature is cooler than the skin temperature, convection permits the transfer of heat from the skin to the air (heat loss); however, if air temperature is higher than skin temperature, heat is gained by the body through convection. We often consider these processes to be mechanisms of heat loss, forgetting that when the environmental temperature exceeds skin temperature, the gradient works in the opposite direction.

Convection is important on a daily basis, since it constantly removes the metabolic heat we generate at rest and during activities of daily living, as long as the air temperature is lower than the skin temperature. However, if a person is submerged in cold water, the amount of heat dissipated from the body to the water by convection can be nearly 26 times greater than when the person is exposed to a similar cold air temperature.

Radiation

At rest, **radiation** (R) and convection are the primary methods for eliminating the body's excess heat. At normal room temperature (typically 21-25 °C, or ~70-77 °F), the nude body loses about 60% of its excess heat by radiation. The heat is given off in the form of infrared rays, which are a type of electromagnetic wave. Figure 12.3 shows two infrared thermograms of an individual.

The skin constantly radiates heat in all directions to objects around it, such as clothing, furniture, and walls, but it also can receive radiant heat from surrounding objects that are warmer. If the temperature of the surrounding objects is greater than that of the skin, the body will experience a net heat gain via radiation. A

tremendous amount of radiant heat is received from exposure to the sun.

Taken together, conduction, convection, and radiation are considered avenues of **dry heat exchange**. Resistance to dry heat exchange is called **insulation**, a concept known to everyone as it relates to clothing and home heating and cooling. The ideal insulator is a layer of still air (remember that moving air causes convective heat loss), which we achieve by trapping layers of air in fibers (down, fiberglass, etc.). Adding insulation in this way minimizes unwanted heat loss in cold environments. However, during exercise we want to dissipate heat to the environment, which is best accomplished by wearing thin, light-colored clothing (to limit radiant heat absorption) that allows for maximally exposed skin surface area. The effect of clothing on sweat evaporation is discussed later.

Evaporation

By far, **evaporation** (E) is the primary avenue for heat dissipation during exercise. As a fluid evaporates and turns into its gaseous form, heat is lost. Evaporation accounts for about 80% of the total heat loss when one is physically active and is therefore an extremely important avenue for heat loss. Even at rest, evaporation accounts for 10% to 20% of body heat loss, since some evaporation occurs without our awareness (termed *insensible water loss*).

As body core temperature increases, once a threshold core temperature is reached sweat production increases dramatically. As sweat reaches the skin, it is converted from a liquid to a vapor, and heat is lost from the skin in the process, the latent heat of vaporization. Thus, sweat evaporation becomes increasingly important as body temperature increases.

a 0.0500°C 27.250°C   37.450°C

b 0.0500°C 27.250°C   37.450°C

FIGURE 12.3 Thermograms of the body showing the variations in radiant (infrared) heat leaving the (*a*) front and (*b*) rear surfaces of the body both before (left) and after (right) running outside at a temperature of 30 °C and 75% humidity. The color scale at the bottom of each photo provides the temperature variation for the changes in color.

Evaporation of 1 L of sweat in an hour results in the loss of 680 W (2,428 kJ) of heat. An important concept to remember is that sweat must evaporate for any heat loss to occur. Some sweat drips off the body or stays on the skin or in the clothing, especially if the air is humid. This unevaporated sweat contributes nothing to body cooling and simply represents a wasteful loss of body water.

Analogous to insulation, which limits dry heat exchange, clothing also adds resistance to sweat evaporation. While some cooling of the skin does occur as sweat evaporates from wet clothing surfaces, the cooling power is less than for evaporation directly from skin to air. Clothing that fits loosely and comprises fabrics that promote wicking or free movement of water vapor molecules through the fabric enhances evaporative cooling.

In focus

Sweat must evaporate to provide cooling. Sweat that drips off the skin provides little or no cooling.

Figure 12.4 shows the complex interaction between the mechanisms of body heat balance (production and loss) and environmental conditions.[2] Using the symbols defined in the previous paragraphs, we can represent the state of heat balance by a simple equation:

$$M - W \pm R \pm C \pm K - E = 0,$$

where W represents any useful work being performed as a result of muscle contraction. Notice that while R, C, and K can be either positive (heat gain) or negative (heat loss), E can only be negative. When $M - W \pm R \pm C \pm K - E > 0$, heat is stored in the body and core temperature rises.

Humidity and Heat Loss

The water vapor pressure of the air (the pressure exerted by water vapor molecules suspended in the air) plays a major role in evaporative heat loss. Relative humidity is a more commonly used term that relates the water vapor pressure of the air to that of fully saturated air (100% humidity). When humidity is high, the air already contains many water molecules. This decreases

FIGURE 12.4 The complex interaction between the body's mechanisms for heat balance and environmental conditions.

Cooling Capacity of Sweat Evaporation

When air temperature is close to skin temperature, the only available means of cooling is evaporation. To put the tremendous cooling power of sweat evaporation into perspective, a 70 kg (154 lb) marathon runner who runs a 2:30 marathon will produce about 1,000 W of metabolic heat. If every drop of sweat he produced evaporated, he would need to sweat at a rate of about 1.5 L per hour. Because inevitably some sweat drips off, a better estimated sweating rate necessary to maintain core temperature would be 2 L per hour. Without replacing fluids, however, our hypothetical runner would lose about 5 L (5.3 qt) of water, or over 7% of his body weight.

its capacity to accept more water because the vapor pressure gradient between the skin and the air is decreased. Thus, high humidity limits sweat evaporation and heat loss, while low humidity offers an ideal opportunity for sweat evaporation and heat loss. But this efficient cooling mechanism can also pose a problem. If sweating is prolonged without adequate fluid replacement, dehydration can occur.

Thermoregulatory Control

We live our entire lives within a very small, fiercely protected range of internal body temperatures. If sweating and evaporation were unlimited, we could withstand extreme ambient heat (e.g., even oven temperatures >200 °C for short periods!) if we were protected from contact with hot surfaces. On the other hand, the temperature limits for living cells range from about 0 °C (where ice crystals form) to about 45 °C (where intracellular proteins start to unravel), and humans can tolerate *internal* temperatures below 35 °C or above 41 °C for only very brief periods of time. To maintain internal temperature within these limits, we have developed very effective and, in some instances specialized, physiological responses to heat and cold. These responses involve the finely controlled coordination of several body systems.

Internal body temperature at rest is regulated at approximately 37 °C (98.6 °F). During exercise, the body is often unable to dissipate heat as rapidly as it is produced. In rare circumstances, a person can reach internal temperatures exceeding 40 °C (104 °F), with a temperature above 42 °C (107.6 °F) in active muscles. The muscles' energy systems become more chemically efficient with a small increase in muscle temperature, but internal body temperatures above 40 °C can adversely affect the nervous system and reduce further efforts to unload excess heat. How does the body regulate its internal temperature? The hypothalamus plays a central role (see figure 12.5).

The Preoptic-Anterior Hypothalamus: The Body's Thermostat

A simple way to envision the mechanisms that control internal body temperature is to compare them to the thermostat that controls the air temperature in a home, although the body's mechanisms function in a more complex manner and generally with greater precision than a home heating and cooling system. Sensory receptors called **thermoreceptors** detect changes in temperature and relay this information to the body's thermostat, located in a region of the brain called the **preoptic-anterior hypothalamus (POAH)**. In response, the hypothalamus activates mechanisms that regulate the heating or cooling of the body. Like a home thermostat, the hypothalamus has a predetermined temperature, or set point, that it tries to maintain. This is the normal body temperature. The smallest deviation from this set point signals this thermoregulatory center to readjust the body temperature.

Thermoreceptors are located throughout the body but especially in the skin and central nervous system. The peripheral receptors located in the skin monitor the skin temperature, which varies with changes in the temperature around a person. They provide information not only to the POAH but also to the cerebral cortex, which allows one to consciously perceive temperature and voluntarily control one's exposure to heat or cold. Because the skin temperature changes long before core temperature, these receptors serve as an early warning system for impending thermal challenges.

Central receptors are located in the hypothalamus, other brain regions, and the spinal cord and monitor the temperature of the blood as it circulates through these sensitive areas. These central receptors are responsive to blood temperature changes as small as 0.01 °C (0.018 °F) and to the rate of change as well. Because of this exquisite sensitivity, very small changes in the temperature of the blood passing through the hypothalamus quickly trigger reflexes that help one conserve or eliminate body heat as needed.

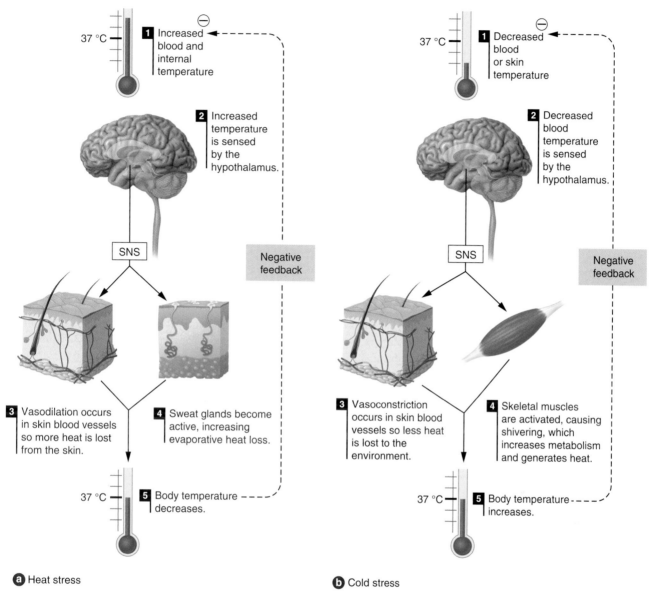

a Heat stress   **b** Cold stress

FIGURE 12.5 A simplified overview of the role of the hypothalamus in controlling body temperature.

Environmental challenges to thermal homeostasis present challenges in parallel to many other bodily control systems, such as those that control blood pressure, body fluid and electrolyte balance, and circadian rhythms. In testimony to the intricate design of the human brain, these other hypothalamus-based controllers are located in close proximity to the POAH, with neural connections that finely coordinate control across all of these systems.

Thermoregulatory Effectors

When the POAH senses temperatures above or below normal, signals are sent through the sympathetic nervous system to four sets of effectors:

- **Skin arterioles.** When skin or core temperature changes, the POAH sends signals via the sympathetic nervous system (SNS) to the smooth muscle in the walls of the arterioles that supply the skin, causing them to dilate or constrict. This either increases or decreases skin blood flow. Skin vasoconstriction results primarily from SNS release of the neurotransmitter norepinephrine, although other neurotransmitters are involved as well, and facilitates heat conservation by minimizing dry heat exchange. Skin vasodilation in response to heat stress is a more complex and less well understood process. Increased skin blood flow aids in heat dissipation to the environment through conduction, convection, and radiation (and indirectly, evaporation as skin temperature goes up). Fine-tuning of skin blood flow is the mechanism by which minute-to-minute adjustments are made in heat balance and exchange. These adjustments are rapid and occur with no real energy cost to the body.

- **Eccrine sweat glands.** When either the skin or core temperature is elevated sufficiently, the POAH also sends impulses through the SNS to the **eccrine sweat glands**, resulting in active secretion of sweat onto the skin surface. The primary neurotransmitter involved is acetylcholine; thus we refer to sweat gland activation as sympathetic cholinergic stimulation. Like skin arterioles, sweat glands are about 10 times more responsive to increases in core temperature than to similar increases in skin temperature. The evaporation of this moisture, as discussed earlier, removes heat from the skin surface.

- **Skeletal muscle.** Skeletal muscle is called into action when a person needs to generate more body heat. In a cold environment, thermore-ceptors in the skin or core sense cold and send signals to the hypothalamus. In response to this integrated neural input, the hypothalamus activates the brain centers that control muscle tone. These centers stimulate shivering, which is a rapid, involuntary cycle of contraction and relaxation of skeletal muscles. This increased muscle activity is ideal for generating heat to either maintain or increase the body temperature, because no useful work results from the shivering, only heat production.

- **Endocrine glands.** The effects of several hormones cause the cells to increase their metabolic rates. Increased metabolism affects heat balance because it increases heat production. Cooling the body stimulates the release of thyroxine from the thyroid gland. Thyroxine can elevate the metabolic rate throughout the body by more than 100%. Also, recall that epinephrine and norepinephrine (the catecholamines) mimic and enhance the activity of the SNS. Thus, they directly affect the metabolic rate of virtually all body cells.

In review

- ➤ Humans are homeothermic, meaning that internal body temperature is regulated through physiological mechanisms, usually keeping it in the resting range of 36.1 to 37.8 °C (97.0-100.0 °F) despite changes in environmental temperatures.

- ➤ Body heat is transferred by conduction, convection, radiation, and evaporation. At rest, most heat is lost via radiation and convection, but during exercise, evaporation becomes the most important avenue of heat loss.

- ➤ At any given air temperature, higher humidity (i.e., a higher water vapor pressure of the surrounding air) decreases evaporative heat loss.

- ➤ The preoptic-anterior area of the hypothalamus (POAH) is the primary thermoregulatory center. It acts as a thermostat, monitoring temperature and accelerating heat loss or heat production as needed.

- ➤ Two sets of thermoreceptors provide temperature information to the POAH. Peripheral receptors in the skin relay information about the temperature of the skin and the environment around it. Central receptors in the hypothalamus, other brain areas, and the spinal cord transmit information about the internal body temperature. Central thermoreceptors are far more sensitive to temperature change than peripheral receptors. The main role of the peripheral receptors is anticipatory, allowing for early adjustments to be made.

- ➤ Effectors stimulated by the hypothalamus through the sympathetic nervous system can alter the body temperature. Increased skeletal muscle activity (voluntary or involuntary as in the case of shivering) increases the temperature by increasing metabolic heat production. Increased sweat gland activity decreases the temperature by increasing evaporative heat loss. Smooth muscle in the skin arterioles can cause these vessels to dilate to direct blood to the skin for heat transfer or to constrict to retain heat deep in the body. Metabolic heat production can also be stimulated by the actions of hormones such as thyroxine and the catecholamines.

Physiological Responses to Exercise in the Heat

Heat production is beneficial during exercise in a cold environment because it helps maintain normal body temperature. However, even when exercise is performed in a cool environment, the metabolic heat load places a considerable burden on the mechanisms that control body temperature. In this section, we examine some physiological changes that occur in response to exercise while the body is exposed to heat stress and the impact that these changes can have on performance. For this discussion, heat stress means any environmental condition that increases body temperature and jeopardizes homeostasis.

Cardiovascular Function

As we learned in chapter 8, exercise increases the demands on the cardiovascular system. When the need to regulate body temperature is added during exercise in the heat, the burden placed on the cardiovascular system is enhanced. During exercise in hot conditions, the circulatory system has to continue to transport blood not only to working muscle but also to the skin, where the tremendous heat generated by the muscles can be transferred to the environment. To meet this dual demand during exercise in the heat, two changes occur. First, cardiac output increases further (above that associated with a similar exercise intensity in cool conditions) by increasing both heart rate and contractility. Second, blood flow is shunted away from nonessential areas like the gut, liver, and kidneys and to the skin.

Consider what happens during prolonged running on a hot day. The aerobic exercise increases both metabolic heat production and the demand for blood flow and oxygen delivery to the working muscles. This excess heat can be dissipated only if blood flow increases to the skin.

In response to the elevated core temperature (and to a lesser extent, the higher skin temperature), the SNS signals sent from the POAH to the skin arterioles cause these blood vessels to dilate, delivering more metabolic heat to the body surface. Sympathetic nervous system signals also go to the heart to increase heart rate and cause the left ventricle to pump more forcefully. However, the ability to increase stroke volume is limited as blood pools in the periphery and less returns to the left atrium. To maintain cardiac output under such circumstances, the heart rate gradually creeps upward to help compensate for the decrease in stroke volume.

In focus

Exercising in hot environments sets up a competition between the active muscles and the skin for a limited blood supply. The muscles need blood and the oxygen it delivers to sustain activity; the skin needs blood to facilitate heat loss to keep the body cool.

This phenomenon, known as cardiovascular drift, was discussed in chapter 8. Because blood volume stays constant or even decreases (as fluid is lost in sweat), another phase of cardiovascular adjustment occurs simultaneously. Sympathetic nerve signals to the kidneys, liver, and intestines cause vasoconstriction of those regional circulations, which allows more of the available cardiac output to reach the skin without compromising muscle blood flow.

What Limits Exercise in the Heat?

Seldom are records set in endurance events, such as distance running, when the environmental heat stress is great. The factors that cause early fatigue when heat stress is superimposed on prolonged exercise have been the topic of some debate, and several theories have been proposed. None of these theories captures every situation, but taken together they demonstrate the multiple control systems at work during thermoregulation.

At some point, the cardiovascular system can no longer compensate for the increasing demands of continuing endurance exercise and efficiently regulating the body's heat. Consequently, any factor that tends to overload the cardiovascular system or to interfere with heat dissipation can drastically impair performance, increase the risk of overheating, or both. Exercise in the heat becomes limited when heart rate approaches maximum, especially in untrained or non-heat-acclimated exercisers, as illustrated in figure 12.6. Interestingly, working muscle blood flow is well maintained even at very high core temperatures unless significant dehydration occurs.

Another theory that helps explain limitations to exercise in the heat—especially in well-trained, acclimated athletes—is the **critical temperature theory**. This theory proposes that, regardless of the rate at which core temperature (and thus brain temperature) increases, the brain will send signals to stop exercise when some critical temperature is reached, usually between 40 and 41 °C (104 and 105.8 °F).

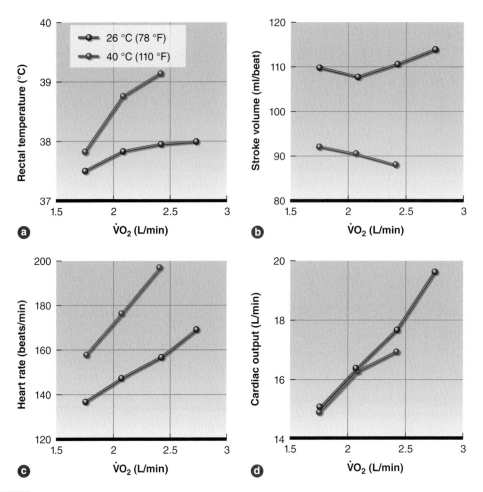

FIGURE 12.6 (a) Rectal temperature and (b-d) cardiovascular responses to graded exercise in thermoneutral (26 °C, 78 °F; open circles) and hot (43 °C, 110 °F; filled circles) environments. Note that, in addition to directional changes in these variables caused by heat stress, the maximal intensity is decreased when exercise is performed in a hot environment.

Based on Rowell, 1974.

Body Fluid Balance: Sweating

On hot summer days, it is not uncommon for the temperature of the environment to exceed both the skin and deep body temperatures. As mentioned earlier, this makes evaporation far more important for heat loss, because radiation, convection, and conduction now become avenues of heat gain from the environment. Increased dependence on evaporation means an increased demand for sweating.

The eccrine sweat glands are controlled by stimulation of the POAH. Elevated blood temperature causes this region of the hypothalamus to transmit impulses through the sympathetic nerve fibers to the millions of eccrine sweat glands distributed over the body's surface. The sweat glands are fairly simple tubular structures extending through the dermis and epidermis, opening onto the skin, as illustrated in figure 12.7.

A second type of sweat gland, the apocrine gland, is clustered in particular regions of the body including the face, axilla, and genital regions. These are the glands associated with "nervous perspiration," and they do not contribute significantly to heat loss by evaporation. The eccrine sweat glands, on the other hand, play a purely thermoregulatory role. They are located over most of skin surface, with ~2 to 5 million covering the whole body. They are most densely distributed on the palms of the hands, the soles of the feet, and the forehead. The lowest densities are found on the forearms, lower legs, and thighs. When sweating begins, there are large regional variations in sweating rate. During exercise, the highest local sweating rates are typically measured on the middle and lower back and on the forehead, while the lowest sweating rates are observed on the hands and feet.

Sweat is formed in the coiled secretory portion of the sweat gland and at this stage has an electrolyte

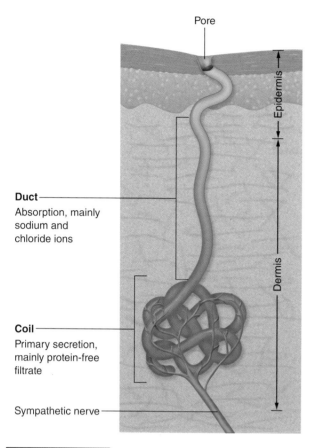

Duct
Absorption, mainly
sodium and
chloride ions

Coil
Primary secretion,
mainly protein-free
filtrate

Sympathetic nerve

Pore

Epidermis

Dermis

FIGURE 12.7 Anatomy of an eccrine sweat gland that is innervated by a sympathetic cholinergic nerve.

composition similar to that of the blood, since plasma is the source of sweat formation. As this filtrate of plasma passes through the uncoiled duct of the gland, sodium and chloride are reabsorbed back into the surrounding tissues and then into the blood. As a result, the final sweat that is extruded onto the skin surface through the sweat gland pores is hypotonic to (has less electrolytes than) plasma. During light sweating, the filtrate sweat travels slowly enough through the duct that there is time for reabsorption of sodium and chloride. Thus,

the sweat that forms during light sweating contains very little of these electrolytes by the time it reaches the skin. However, when the sweating rate increases during exercise, the filtrate moves more quickly through the tubules, allowing less time for reabsorption, and the sodium and chloride content of sweat can be considerably higher.

As seen in table 12.1, the electrolyte concentration of trained and untrained subjects' sweat is significantly different. With training and repeated heat exposure (acclimation), more sodium is reabsorbed and the sweat is more dilute, in part because the sweat glands become more sensitive to the hormone aldosterone. Unfortunately, the sweat glands apparently do not have a similar mechanism for conserving other electrolytes. Potassium, calcium, and magnesium are not reabsorbed by the sweat glands and are therefore found in similar concentrations in both sweat and plasma. In addition to heat acclimation and aerobic training, genetics is a major determinant of both sweating rate and sweat sodium losses.

While performing heavy exercise in hot conditions, the body can lose more than 1 L of sweat per hour per square meter of body surface. This means that during intense effort on a hot and humid day (high level of heat stress), an average-sized female athlete (50-75 kg, or 110-165 lb) might lose 1.6 to 2.0 L of sweat, or about 2.5% to 3.2% of body weight, each hour. A person can lose a critical amount of body water in only a few hours of exercise in these conditions.

A high rate of sweating maintained for a prolonged time ultimately reduces blood volume. This limits the volume of blood returning to the heart, increasing heart rate and eventually decreasing cardiac output, which in turn reduces performance potential, particularly for endurance activities. In long-distance runners, sweat losses can approach 6% to 10% of body weight. Such severe dehydration can limit subsequent sweating and make the individual susceptible to heat-related illnesses. Chapter 15 provides a detailed discussion of dehydration and the value of fluid replacement.

TABLE 12.1 Example of Sodium, Chloride, and Potassium Concentrations in the Sweat of Trained and Untrained Subjects During Exercise

Subjects	Sweat Na⁺ (mmol/L)	Sweat Cl⁻ (mmol/L)	Sweat K⁺ (mmol/L)
Untrained men	90	60	4
Trained men	35	30	4
Untrained women	105	98	4
Trained women	62	47	4

Individual sweat electrolyte concentrations are highly variable, but training and heat acclimation decrease sodium losses in the sweat.

Data from the Human Performance Laboratory, Ball State University.

Loss of both electrolytes and water in the sweat triggers the release of both aldosterone and antidiuretic hormone (ADH), also known as **vasopressin** or **arginine vasopressin**. Recall that aldosterone is responsible for maintaining appropriate sodium concentrations in the blood and that ADH plays a key role in maintaining fluid balance (chapter 4). Aldosterone is released from the adrenal cortex in response to stimuli such as decreased blood sodium content, reduced blood volume, or reduced blood pressure. During acute exercise in the heat and during repeated days of exercise in the heat, this hormone limits sodium excretion from the kidneys. More sodium is retained by the body, which in turn promotes water retention. This allows the body to retain water and sodium in preparation for additional exposures to the heat and subsequent sweat losses.

Similarly, exercise and body water loss stimulate the posterior pituitary gland to release ADH. This hormone stimulates water reabsorption from the kidneys, which further promotes fluid retention in the body. Thus, the body attempts to compensate for loss of electrolytes and water during periods of heat stress and heavy sweating by reducing their loss in urine. Also, recall that blood flow to the kidneys is substantially reduced during exercise in the heat, which likewise aids in fluid retention.

Health Risks During Exercise in the Heat

Despite the body's defenses against overheating, excessive heat production by active muscles, heat gained from the environment, and conditions that prevent the dissipation of excess body heat may elevate the internal body temperature to levels that impair normal cellular functions. Under such conditions, excessive heat gains pose a risk to one's health, as we highlighted in the opening anecdote of this chapter. Air temperature alone is not an accurate index of the total physiological stress imposed on the body in a hot environment. Six variables must be taken into account:

- Metabolic heat production
- Air temperature
- Ambient water vapor pressure (humidity)
- Air velocity
- Radiant heat sources
- Clothing

All of these factors influence the degree of heat stress that a person experiences. The contribution of each factor to the total heat stress under various environmental conditions can actually be predicted mathematically using advanced heat balance equations.

An individual exercising on a bright, sunny day with an air temperature of 23 °C (73.4 °F) and no measurable wind experiences considerably more heat stress than someone exercising in the same air temperature but under cloud cover and with a slight breeze. At temperatures above skin temperature, which is normally 32 to 33 °C (92 °F), radiation, conduction, and convection substantially add to the body's heat load rather than acting as avenues for heat loss. How, then, can we estimate the amount of heat stress to which an individual may be exposed?

Measuring Heat Stress

It has become common to hear about the "heat index" on local weather channels. The heat index, a complex equation involving air temperature and relative humidity, is a measure of how hot it *feels*, that is, how we perceive the heat. However, the heat index does not do a good job of reflecting the physiological stress on humans, so its use is limited in exercise physiology. Through the years, efforts have been made to quantify atmospheric variables into a single index that would reflect the physiological heat stress on an individual. In the 1970s, **wet-bulb globe temperature (WBGT)** was devised to simultaneously account for conduction, convection, evaporation, and radiation (see figure 12.8). It is based on three different thermometer readings and provides a single temperature reading to estimate the cooling capacity of the surrounding environment.

The dry-bulb temperature (T_{db}) is the actual air temperature one would measure with a typical thermometer. A second thermometer has a bulb that is kept moist by a wetted cotton "sock" dipped in distilled water. As water evaporates from this wet bulb, its temperature (T_{wb}) will be lower than the T_{db}, reflecting the effect of sweat evaporating from the skin. The difference between the wet- and dry-bulb temperatures indicates the environment's capacity for cooling by evaporation. In still air with 100% relative humidity,

these two bulb temperatures are the same because evaporation is impossible. Lower ambient water vapor pressure and increased air movement promote evaporation, increasing the difference between these two bulb temperatures. The third thermometer, placed inside a black globe, typically shows a temperature higher than T_{db} as the globe, painted a flat black, maximally absorbs radiant heat. Thus, its temperature (T_g) is a good indicator of the environment's radiant heat load.

The temperatures from these three thermometers can be combined into the following equation to estimate the overall atmospheric challenge to body temperature in outdoor environments:

$$WBGT = 0.1\,T_{db} + 0.7\,T_{wb} + 0.2\,T_g.$$

The fact that the coefficient for T_{wb} is the largest reflects the importance of sweat evaporation in the physiology of heat exchange. Also note that WBGT reflects only the environment's impact on heat stress and is most effectively used along with a measure or estimate of metabolic heat production. Clothing further influences heat stress.

WBGT, as an index of **thermal stress**, is now routinely used by coaches, team physicians, and athletic trainers to anticipate the health risks associated with athletic practices and competitions in thermally stressful environments.

FIGURE 12.8 A wet-bulb globe temperature stand showing the three separate thermometers for air (dry bulb) temperature, wet-bulb temperature reflecting the cooling effect of evaporation, and globe temperature, which measures additional effects of radiant heat.

Labels on figure:
Globe thermometer for globe temperature
Wet-bulb thermometer for evaporative water loss
Dry-bulb thermometer for air temperature

In focus

To calculate wet-bulb globe temperature:
Outdoor WBGT = $0.1\,T_{db} + 0.7\,T_{wb} + 0.2\,T_g.$
Indoor WBGT = $0.7\,T_{wb} + 0.3\,T_g.$

Heat-Related Disorders

Exposure to the combination of external heat stress and metabolically generated heat can lead to three heat-related disorders (see figure 12.9): heat cramps, heat exhaustion, and heatstroke.

Heat Cramps

Heat cramps, the least serious of the three heat disorders, are characterized by severe and painful cramping of large skeletal muscles. They involve primarily the muscles that are most heavily used during exercise, and such instances of athletes "locking up" are different from cramps everyone has experienced in small muscles. Heat cramps are brought on by sodium losses and dehydration that accompany high rates of sweating, and thus are most common in heavy sweaters who lose a lot of sodium in their sweat. (A common

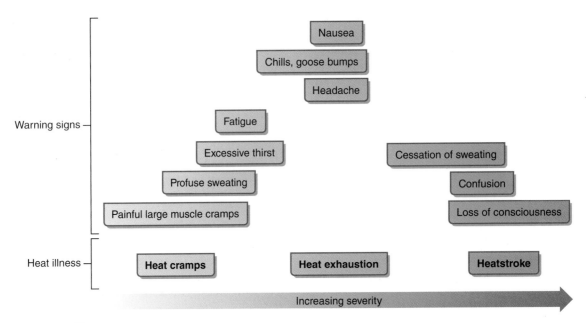

FIGURE 12.9 Schematic diagram of heat illnesses of increasing severity, showing some of the typical warning signs that accompany these illnesses. Not all of these warning signs are present in every case of heat illness, nor do they always occur in this, or any predictable, order. Any of the three heat illnesses shown here can occur suddenly without prior symptoms; that is, there is not always a progression from heat exhaustion to heatstroke.

Adapted by permission of All Sport, Inc.

misconception is that potassium is involved in cramping and that eating potassium-rich foods like bananas will prevent heat cramps.) Heat cramps can be prevented or minimized in susceptible athletes by proper hydration practices involving liberal salt intake with foods and in beverages consumed during exercise. Treatment for these cramps involves moving the stricken individual to a cooler location and administering a saline solution, either orally or intravenously if necessary.

Heat Exhaustion

Heat exhaustion typically is accompanied by such symptoms as extreme fatigue, dizziness, nausea, vomiting, fainting, and a weak, rapid pulse. It is caused by the cardiovascular system's inability to adequately meet the body's needs as it becomes severely dehydrated. Recall that during exercise in heat, active muscles and skin compete for a share of a limited, and decreasing, blood volume. Heat exhaustion may result when these simultaneous demands cannot be met, and it typically occurs when blood volume decreases as a result of excessive fluid loss from profuse sweating. A second form of heat exhaustion, from sodium depletion, is rare in athletes. Therefore, heat exhaustion can be thought of as a syndrome of dehydration and should be treated as such.

With heat exhaustion, the thermoregulatory mechanisms are functioning but cannot dissipate heat quickly enough because insufficient blood volume is available to allow adequate blood flow to the skin. Although the condition often occurs during moderate to heavy exercise in the heat, it is not necessarily accompanied by extremely high core temperatures. Some people who collapse from heat exhaustion have core temperatures well below 39 °C (102.2 °F). People who are unfit or not acclimated to the heat are more susceptible than others to heat exhaustion.

Treatment for victims of heat exhaustion involves rest in a cooler environment with their feet elevated to facilitate return of blood to the heart. If the person is conscious, administration of salt water is usually recommended. If the person is unconscious, medically supervised intravenous administration of saline solution is recommended.

Heatstroke

As we saw in our opening story, **heatstroke** is a life-threatening heat disorder that requires immediate medical attention. Heatstroke is caused by failure of the body's thermoregulatory mechanisms. It is characterized by

- an increase in internal body temperature to a value exceeding 40 °C (104 °F); and

- confusion, disorientation, or unconsciousness.

The final element—altered mental status—is the key to recognizing impending heatstroke because neural tissues in the brain are particularly sensitive to extreme heat. In heatstroke, cessation of active sweating may also occur, but sweat may remain on the skin. The notion that heatstroke is always accompanied by dry, red skin is outdated, and this sign should never be used to distinguish heatstroke from heat exhaustion.

In addition to a severely elevated core temperature, altered mental status or cognitive function is a hallmark sign of heatstroke.

If heatstroke is left untreated, core temperature will continue to rise, progressing to coma and ultimately death. Appropriate treatment always involves cooling the body as rapidly as possible. In the field, this can best be accomplished by immersing the victim—as much of the body as possible excluding the head—in a bath of cold water or ice water. While cold water immersion provides the fastest cooling rates, in cases where cold or ice water are not available, temperate water immersion is the next best option. Where immersion is not feasible, the combined effect of wrapping the entire body in cold, wet sheets and fanning vigorously may be used. Cooling methods that place ice bags on small areas, like the armpits, neck, and groin, are not effec-tive in rapidly lowering core temperature because of the small surface area covered.

For the athlete, heatstroke is not just a problem associated with extreme conditions. Studies have reported rectal temperatures above 40.5 °C (104.9 °F) in marathon runners who successfully completed races conducted in temperate and even in cool conditions.

Preventing Hyperthermia

We can do little about the prevailing weather conditions. In threatening conditions, athletes must either move the exercise session to a less stressful environment (for example, moving practice indoors) or decrease their effort (and thus their metabolic heat production) in order to reduce their risk of overheating. Athletes, coaches, and sport organizers should all be able to recognize the symptoms of heat illness.

To prevent heat disorders, several simple precautions should be taken. Competition and practice should not be held outdoors when the WBGT is more than 28 °C (82.4 °F) unless special precautions are taken. Scheduling practices and events either in the early morning or in the late evening avoids the severe heat stress of midday. Fluids should be readily available, and drink breaks should be scheduled every 15 to 30 min, with a goal of matching fluid intake to sweat loss. Because individual sweating rates and sweat sodium losses vary tremendously and cannot easily be predicted across individual athletes, athletes should customize their fluid

Guidelines for Practicing and Competing Under Conditions of Heat Stress

1. Athletic events (distance races, tennis matches, sport team practices, etc.) should be sched-uled to avoid the hottest times of the day. As a general rule, if the WBGT is above 28 °C (82-83 °F), consider canceling, moving indoors, decreasing intensity of practice, or otherwise altering the event.

2. An adequate supply of palatable fluid must be available. Athletes should be educated and encouraged to prevent excessive (>2%) weight loss—that is, to replace their sweat losses to prevent dehydration but not overdrink to the point where they gain weight during the event.

3. Because individual sweating rates and sweat sodium losses vary tremendously, athletes should customize their fluid intake based on their individual sweating rate. Sweating rate can be estimated by measuring body weight before and after exercise. Fluids containing electrolytes and carbohydrates can provide benefits over water alone.

4. Athletes should be aware of signs and symptoms of heat illness. Cold-water immersion is the most efficient method for cooling hyperthermic athletes in the field.

5. Organizers of events and medical personnel should have the right to cancel or terminate events and to stop individual athletes who exhibit clear signs of heat exhaustion or heatstroke.

intake based on their individual sweating rate. This is best accomplished by having athletes weigh themselves before and after exercise sessions and learn to estimate their approximate fluid needs.

Clothing is another important consideration. Obviously, the more clothing worn, the less body area exposed to the environment to allow for direct heat loss. The foolish practice of exercising in a rubberized suit to promote weight loss is an excellent illustration of how a dangerous microenvironment (the isolated environment inside the suit) can be created in which temperature and humidity are sufficiently high to block virtually all heat loss from the body. This can rapidly lead to heat exhaustion or heatstroke. Football uniforms are another example of clothing that impedes heat loss. Coaches and athletic trainers should avoid practice sessions in full uniforms whenever possible, especially early in the season when temperatures tend to be hottest and players tend to be less fit and not well acclimated.

Distance athletes should wear as little clothing as possible when heat stress is a potential limitation to thermoregulation. They should tend toward underdressing rather than overdressing, because the metabolic heat generated will soon make extra clothing an unnecessary burden. Clothing should be loosely woven to allow sweat to be absorbed and wicked away from the skin and should be light colored to reflect heat back to the environment. Hats should be worn during exercise in bright sunlight or when cloud cover is limited.

It is also important to maintain adequate hydration, since the body loses considerable water through sweating. This is discussed in considerable detail in chapter 15. Briefly, drinking fluid both before and during exercise can greatly reduce the negative effects of exercising in the heat. Adequate fluid intake will attenuate the increase in core body temperature and heart rate normally seen when a person exercises in the heat and will allow exercise to be continued longer. This is illustrated in figure 12.10.

FIGURE 12.10 Effects of fluid intake on core (rectal) body temperature during a 2 h run. Subjects were given fluid during one trial, and on a separate day they completed a second trial without fluid. Note that fluid intake did not have much influence until about 45 min, after which time body heat storage was reduced compared with that in the no-fluid trial.

Adapted, by permission, from D.L. Costill, 1970, "Fluid ingestion during distance running," *Archives of Environmental Health* 1: 520-525.

In review

➤ Heat stress involves more than just the air temperature. Perhaps the most widely used measurement of the combined physiological effects of heat stress is the WBGT, which measures air temperature and accounts for the heat exchange potential through convection, evaporation, and radiation in a specific environment. Exercise intensity and clothing must be considered separately along with the WBGT.

➤ Heat cramps are caused by the loss of fluids and salt (sodium) that results from excessive sweating in susceptible athletes. High dietary sodium intakes and proper hydration strategies can prevent heat cramps.

➤ Heat exhaustion results from the inability of the cardiovascular system to adequately meet the blood flow needs of the active muscles and the skin. It is often brought on by dehydration caused by excessive loss of fluids and electrolytes, which results in reduced blood volume. Although it is not in itself life threatening, heat exhaustion can deteriorate to heatstroke if untreated.

➤ Heatstroke is caused by failure of the body's thermoregulatory mechanisms. If untreated, core temperature continues to rise quickly and can be fatal.

Acclimation to Exercise in the Heat

How can athletes prepare for prolonged activity in the heat? Does repeated exercise in the heat make us better able to tolerate thermal stress? Many studies have addressed these questions and have concluded that repeated exercise in the heat causes a series of relatively rapid adaptations that enable us to perform better, and more safely, in hot conditions. When these physiological changes occur over short periods of time, like days to weeks, or if they are artificially induced as in a climatic chamber, those adaptations are termed *heat acclimation*. A similar but much more gradual set of adaptations occurs in people who adapt to hot conditions by living in hot environments for months to years. This is known as *acclimatization* (note that the word "climate" is part of this latter term).

Effects of Heat Acclimation

Repeated bouts of prolonged, low-intensity exercise in the heat cause a relatively rapid improvement in the ability to maintain cardiovascular function and eliminate excess body heat, which reduces physiological strain. This process, termed **heat acclimation**, involves changes in plasma volume, cardiovascular function, sweating, and skin blood flow that allow for subsequent exercise bouts in the heat to be performed

with a lower core temperature and heart rate response (figure 12.11). Because the body's heat loss capacity at a given rate of work is enhanced by acclimation, core temperature during exercise increases less than before acclimation (figure 12.11*a*), and heart rate increases less in response to standardized submaximal exercise after heat acclimation (figure 12.11*b*). In addition, after heat acclimation, more work can be done before adverse symptoms occur or a maximal tolerable core temperature or heart rate is reached.

This series of positive adaptations typically takes a period of 9 to 14 days of exercise in the heat to fully occur, as shown in figure 12.12. Well-trained individuals need fewer exposures than untrained individuals to fully acclimate. A critical physiological adjustment that occurs over the first one to three days of acclimation is the expansion of plasma volume. The exact mechanism by which plasma volume expands after these initial exercise-heat exposures is not universally agreed upon. The process likely involves (1) proteins being forced out of the circulation as muscles contract, (2) these same proteins then being returned to the blood through the lymph, and (3) fluid moving into the blood because of the oncotic pressure exerted by the increased protein content. However, this change is temporary, and blood volume usually returns to original levels within 10 days. This early expansion of blood volume is important because it supports stroke volume, allowing the body to maintain cardiac output while additional physiological adjustments are made.

FIGURE 12.11 Typical (*a*) rectal temperature and (*b*) heart rate responses during an acute bout of exercise at the same intensity before and after heat acclimation. Note that, in addition to lower physiological strain, exercise time is usually longer after acclimation.

Data from King et al., 1985.

a

b

c

FIGURE 12.12 Changes in (a) rectal temperature, (b) heart rate, and (c) hourly sweating rate in a group of men exercising for 100 min per day for 12 consecutive days in a hot environment.

As shown in figure 12.12, end-exercise heart rate and core temperature decrease early in the acclimation process, while the increase in sweating rate during exercise in the heat occurs somewhat later. An additional adaptation is a more even distribution of sweat over the body, with increased sweating on the most exposed body areas such as the arms and legs, areas that are most effective at dissipating body heat. At the beginning of exercise, sweating also starts earlier in an acclimated person, which improves heat tolerance; and the sweat that is produced becomes more dilute, conserving sodium. This latter effect occurs in part because the eccrine sweat glands become more sensitive to the effects of circulating aldosterone.

Achieving Heat Acclimation

Heat acclimation requires more than merely resting in a hot environment. The benefits of acclimation, as well as the rate at which we acclimate, depend on

- the environmental conditions during each exercise session,
- the duration of exercise-heat exposure, and
- the rate of internal heat production (exercise intensity).

An athlete must exercise in a hot environment to attain full acclimation that sustains exercise in the heat. Simply sitting in a hot environment, such as a sauna or steam room, for long periods each day will not fully or adequately prepare the individual for physical exertion in the heat, at least not to the same extent as will exercising in the heat.

In focus

One can adapt to heat (undergo heat acclimation) by low- to moderate-intensity exercise in the heat for 1 h or more each day for 9 to 14 days. Cardiovascular changes generally occur first, starting with—and supported by—plasma volume expansion during the first one to three days. The major changes in the sweating mechanisms generally occur later, between days 3 and 10.

How can the athlete maximize heat acclimation? Because body temperature is elevated and sweating occurs, athletes gain partial heat tolerance simply by training, even in a cooler environment. Therefore, athletes are "preacclimated" to heat and need fewer exercise-heat exposures to fully acclimate. To gain maximal benefits, athletes who train in environments cooler than those in which they will compete must achieve heat acclimation before the contest or event. Heat acclimation will improve their performance and reduce the associated physiological stress and risk of heat injury.

In review

> ➤ Repeated exercise exposure to heat stress gradually improves one's ability to support cardiovascular function and lose excess heat during subsequent bouts of exercise-heat stress. This process is called heat acclimation.

> ➤ With heat acclimation, people start to sweat earlier and sweating rate increases, particularly on areas that are well exposed and are the most efficient at promoting heat loss. This reduces skin temperature, which increases the thermal gradient from the skin to the environment and promotes heat loss.

> ➤ Exercise core temperature and heart rate during are reduced with heat acclimation. Plasma volume increases early in the process, contributing to an increase in stroke volume that supports the delivery of blood to both active muscles and skin.

> ➤ Full heat acclimation requires exercise in a hot environment, not merely exposure to heat.

> ➤ The rate of heat acclimation depends on training status, the conditions to which one is exposed during each session, the duration of the exposure, and the rate of internal heat production.

Exercise in the Cold

Humans can be thought of as tropical animals. Most of our adjustments to heat stress are physiological, whereas many adjustments to cold environments involve behavior, like putting on more clothing or seeking shelter. Increased year-round participation in sport has sparked new interest in, and concerns about, exercise in the cold. In addition, certain occupations and military endeavors require people to work in cold conditions—conditions that often limit performance. For these reasons, the physiological responses and health risks associated with cold stress are important issues in exercise science. We define cold stress here as any environmental condition causing a loss of body heat that threatens homeostasis. In the following discussion we focus on the two major cold environments: air and water.

The hypothalamus has a temperature "set point" of about 37 °C (98.6 °F), but daily fluctuations in the body temperature can be as much as 1 °C. A decrease in either skin or blood temperature provides feedback to the thermoregulatory center (POAH) to activate mechanisms that conserve body heat and increase heat production. The primary means by which our bodies avoid excessive heat loss (in the order in which they are invoked) are peripheral vasoconstriction, nonshivering thermogenesis, and shivering. Because these mechanisms or effectors of heat production and conservation are often inadequate, we also must rely on behavioral responses such as huddling behavior (decreasing exposed body surface area) and putting on more clothing to help insulate our deep body tissues from the environment.

In focus

When exercising in the cold, people should not overdress. Overdressing can cause the body to become hot and initiate sweating. As the sweat soaks through the clothing, evaporation removes the heat, and heat is lost at an even faster rate.

Peripheral vasoconstriction occurs as a result of sympathetic stimulation of the smooth muscle layers of the arterioles in the skin. This stimulation causes the vascular smooth muscle to contract, which constricts the arterioles, reduces the blood flow to the shell of the body, and minimizes heat loss. Even at thermoneutral temperatures, there is tonic (constant baseline) skin vasoconstriction, and continuous adjustment of skin vascular tone occurs at all times to offset small heat imbalances in the body. When changing skin blood flow alone is not adequate to prevent heat loss, **nonshivering thermogenesis**—stimulation of metabolism by the SNS—is increased. Increasing the metabolic rate increases heat production. The next line of defense of body temperature during cold stress is **shivering**, a rapid, involuntary cycle of contraction and relaxation of skeletal muscles, which can cause a four- to fivefold increase in the body's rate of heat production. The overall adjustments in blood flow and metabolism that serve to maintain body core temperature are shown in figure 12.13.

Habituation and Acclimation to Cold

Whether humans truly acclimate to cold environments in the physiological sense—and if so, how this occurs—is far less clear than the process of heat acclimation. When one looks at studies of people who undergo repeated daily cold exposures, the results seem controversial.

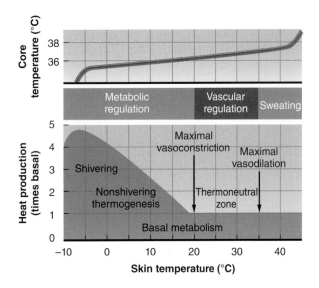

FIGURE 12.13 Thermoregulatory mechanisms by which humans strive to maintain a relatively constant body core temperature. In the thermoneutral zone, minor adjustments to skin blood flow minimize heat loss or gain. When maximal vasoconstriction is not sufficient to maintain core temperature, metabolic regulation, first in the form of nonshivering thermogenesis and then by shivering, serves to increase metabolic heat production.

However, Dr. Andrew Young of the U.S. Army Research Institute for Environmental Medicine and others have proposed a scheme to explain the development of different patterns of cold adaptation that are observed in humans.[4] People who are regularly exposed to repeated cold environments under which significant body heat loss does not occur typically undergo **cold habituation**, in which skin vasoconstrictor and shivering responses are blunted, and core temperature falls to a greater extent than before the chronic cold exposures. This pattern of adaptation often occurs when small areas of skin—often the hands and face—are exposed repeatedly to cold air.

However, when heat loss is more severe or occurs at a faster rate, total-body heat loss may occur. In cases where increased metabolic heat production alone can sufficiently minimize heat loss, enhanced nonshivering and shivering thermogenesis develop (**metabolic acclimation**). Yet a third distinct pattern of cold adaptation, called **insulative acclimation**, tends to occur in situations in which increased metabolism is unable to maintain core temperature. In insulative acclimation, enhanced skin vasoconstriction occurs, which increases peripheral insulation and minimizes heat loss.

Other Factors Affecting Body Heat Loss

The mechanisms of conduction, convection, radiation, and evaporation, which usually perform effectively in dissipating metabolically produced heat during exercise in warm conditions, can dissipate heat to the environment faster than the body produces it in extremely cold environments.

Pinpointing the exact conditions that permit excessive body heat loss and eventual **hypothermia** (low body core temperature) is difficult. Thermal balance depends on a wide variety of factors that affect the balance between body heat production and heat loss. Generally speaking, the larger the thermal gradient between the skin and the cold environment, the greater the heat loss. However, a number of anatomical and environmental factors can influence the rate of heat loss.

Body Size and Composition

Insulating the body against the cold is the most obvious form of protection against hypothermia. Recall that insulation is defined as resistance to dry heat exchange through radiation, convection, and conduction. Both inactive peripheral muscles and subcutaneous fat are excellent insulators. Skinfold measurements of subcutaneous fat thickness are a good indicator of an individual's tolerance for cold exposure. The thermal conductivity of fat (its capacity for transferring heat) is relatively low, so fat impedes heat transfer from the deep tissues to the body surface. People who have more fat mass conserve heat more efficiently than smaller, leaner individuals in the cold.

The rate of heat loss also is affected by the ratio of body surface area to body mass. Larger individuals have a small surface area-to-mass ratio, which makes them less susceptible to hypothermia. As shown in table 12.2, small children have a large surface area-to-mass

TABLE 12.2 Body Weight, Height, Surface Area, and Surface Area/Mass Ratios for an Average-Sized Adult and Child

Person	Weight (kg)	Height (m)	Surface area (m²)	Area/mass ratio (m²/kg)
Adult	85	1.83	2.07	0.024
Child	25	1.00	0.79	0.032

ratio compared with adults, leading to proportionately greater heat loss. This makes it more difficult for them to maintain normal body temperature in the cold.

Women tend to have more body fat than men, but true sex differences in cold tolerance are minimal. Some studies have shown that the added subcutaneous fat in women may give them an advantage during cold-water immersion, but when men and women of similar body fat mass and size are compared, no real difference is noted in body temperature regulation during exposure to the cold. As people age, they often tend to lose overall muscle mass, making them more susceptible to hypothermia.

The body's insulating shell consists of two regions: the superficial skin and subcutaneous fat and the underlying muscle. Increased skin vasoconstriction, increased subcutaneous fat thickness, and increased inactive muscle mass, especially in the limbs, can all increase total body insulation.

Windchill

As with heat, the air temperature alone does not provide a valid index of the heat loss experienced by the individual. Air movement, or wind, increases convective heat loss and therefore increases the rate of cooling. **Windchill** is an index based on the cooling effect of wind and is an often misunderstood and misused concept. Windchill is typically presented in charts of windchill equivalent temperatures showing various combinations of air temperature and wind speed that result in the same cooling power as that seen with no wind (figure 12.14). It is important to remember that windchill is not the temperature of the wind or the air (windchill does *not* change air temperature). True windchill refers to the cooling power of the environment. As windchill increases, so does the risk of freezing of tissues (figure 12.14).

Heat Loss in Cold Water

More research has been conducted on cold-water exposure than exposure to cold air. Whereas radiation and sweat evaporation are the primary mechanisms for heat

Air temperature (°C)

Wind speed (km/h)	−10	−15	−20	−25	−30	−35	−40	−45	−50
5	−13	−19	−24	−30	−36	−41	−47	−53	−58
10	−15	−21	−27	−33	−39	−45	−51	−57	−63
15	−17	−23	−29	−35	−41	−48	−54	−60	−66
20	−18	−24	−30	−37	−43	−49	−56	−62	−68
25	−19	−25	−32	−38	−44	−51	−57	−64	−70
30	−20	−26	−33	−39	−46	−52	−59	−65	−72
35	−20	−27	−33	−40	−47	−53	−60	−66	−73
40	−21	−27	−34	−41	−48	−54	−61	−68	−74
45	−21	−28	−35	−42	−48	−55	−62	−69	−75
50	−22	−29	−35	−42	−49	−56	−63	−69	−76
55	−22	−29	−36	−43	−50	−57	−63	−70	−77
60	−23	−30	−36	−43	−50	−57	−64	−71	−78
65	−23	−30	−37	−44	−51	−58	−65	−72	−79
70	−23	−30	−37	−44	−51	−58	−65	−72	−80
75	−24	−31	−38	−45	−52	−59	−66	−73	−80
80	−24	−31	−38	−45	−52	−60	−67	−74	−81

Very low — Freezing is possible, but unlikely
Likely — Freezing is likely > 30 min
High — Freezing risk < 30 min
Severe — Freezing risk < 10 min
Extreme — Freezing risk < 3 min

FIGURE 12.14 Windchill equivalent temperature chart showing various combinations of temperature and wind speed that result in the same cooling power as that seen with no wind. For example, a wind speed of 20 km/h at −10 °C would result in the same heat loss as −30 °C with no wind. Also shown in the figure is the risk of tissues freezing as windchill—the cooling power of the environment—increases.

loss in air, convection allows the greatest heat transfer during immersion in water (recall that convection involves heat loss to moving fluids, which include both liquids and gases). As mentioned earlier, water has a thermal conductivity about 26 times greater than air. This means that heat loss by convection is 26 times faster in cold water than in cold air. When all heat-transfer mechanisms are considered (radiation, conduction, convection, and evaporation), the body generally loses heat four times faster in water than it does in air of the same temperature.

Humans generally maintain a constant internal temperature when they remain inactive in water at temperatures down to about 32 °C (89.6 °F). But when the water temperature decreases further, they may become hypothermic. Because of the large loss of heat from a body immersed in cold water, prolonged exposure or unusually cold water can lead to extreme hypothermia and death. Individuals immersed in water at 15 °C (59 °F) experience a decrease in rectal temperature of about 2.1 °C (3.8 °F) per hour. In 1995, four U.S. Army rangers died of hypothermia after exposure to 11 °C (52 °F) swamp water in Florida, tragically publicizing the fact that hypothermia can occur when water temperature is well above freezing.

If the water temperature were lowered to 4 °C (39.2 °F), rectal temperature would decrease at a rate of 3.2 °C (5.8 °F) per hour. The rate of heat loss is further accelerated if the cold water is moving around the individual because heat loss by convection increases. As a result, survival time in cold water under these conditions is quite brief. Victims can become weak and lose consciousness within minutes.

If the metabolic rate is low, as when the person is at rest, then even moderately cool water can cause hypothermia. But exercise in water increases the metabolic rate and offsets some of the heat loss. For example, although heat loss increases when one is swimming at high speeds (because of convection), the swimmer's accelerated rate of metabolic heat production more than compensates for the greater heat transfer. For competition and training, water temperatures between 23.9 and 27.8 °C (75-82 °F) are appropriate.

Physiological Responses to Exercise in the Cold

We have seen how the body adapts to maintain its internal temperature when exposed to a cold environment. Now consider what happens when the demands of physical performance are added to those of thermoregulation in the cold. How does the body respond to exercise in cold environmental conditions?

Muscle Function

Cooling a muscle causes it to contract with less force. The nervous system responds to muscle cooling by altering the normal muscle fiber recruitment patterns for force development, which may decrease the efficiency of the muscle's actions. Both muscle shortening velocity and power decrease significantly when muscle temperature is lowered. Luckily, large deep muscles seldom experience such low temperatures because they are protected from heat loss by a continuous supply of warm blood flow.

If clothing insulation and exercise metabolism are sufficient to maintain an athlete's body temperature in the cold, aerobic exercise performance may be unimpaired. However, as fatigue sets in and intensity decreases, so does metabolic heat production. Long-distance running, swimming, and skiing in the cold can expose the participant to such conditions. At the beginning of these activities, the athlete can exercise at an intensity that generates sufficient metabolic heat to maintain core temperature. However, late in the activity, when the energy reserves have diminished, exercise intensity declines, and this reduces metabolic heat production. The resulting decrease in core temperature causes the individual to become even more fatigued and less capable of generating heat. Under these conditions, the athlete may be confronted with a potentially dangerous situation.

Cold conditions affect muscle function in another way. As small muscles in the periphery like the fingers become cold, muscle function can be severely affected. This results in a loss of manual dexterity and limits the ability to perform fine motor skills like writing and manual labor tasks.

Metabolic Responses

Prolonged exercise increases the mobilization and oxidation of free fatty acids (FFAs) as a fuel source. The primary stimulus for this increased lipid metabolism is the release of catecholamines (epinephrine and norepinephrine). Exposure to cold markedly increases epinephrine and norepinephrine secretion, but FFA levels increase substantially less than during prolonged exercise in warmer conditions. Cold exposure triggers vasoconstriction in the vessels supplying not only the skin but fatty subcutaneous tissues as well. The subcutaneous fat is a major storage site for lipids (as adipose tissue), so this vasoconstriction reduces the blood flow to an important area from which the FFAs would be mobilized. Thus, FFA levels do not increase as much as the elevated levels of epinephrine and norepinephrine would predict.

Blood glucose plays an important role in both cold tolerance and exercise endurance. For example, hypoglycemia (low blood sugar) suppresses shivering. The reasons for these changes are unknown. Fortunately, blood glucose concentrations are maintained reasonably well during cold exposure. Muscle glycogen, on the other hand, is used at a somewhat higher rate in the cold than in warmer conditions. However, studies on exercise metabolism in the cold are limited, and our knowledge regarding hormonal regulation of metabolism in the cold is too limited to support any definitive conclusions.

In review

> Peripheral vasoconstriction decreases the transfer of heat from the skin to the air, thus decreasing heat loss to the environment. This is the body's first line of defense in the cold.

> Nonshivering thermogenesis increases metabolic heat production through activation of the SNS and by the action of hormones. Shivering thermogenesis increases metabolic heat production further to help maintain or increase body temperature.

> There are three distinct patterns of adaptation to repeated cold exposure: cold habituation, metabolic acclimation, and insulative acclimation.

> Body size is an important consideration for heat loss. Both a higher surface area-to-mass ratio and a lower peripheral muscle mass or subcutaneous fat increase the loss of body heat to the environment.

> Wind increases heat loss by convection. The cooling power of the environment, known as windchill, is typically expressed as equivalent temperatures.

> Immersion in cold water tremendously increases heat loss through convection. In some cases, exercise may generate enough metabolic heat to offset some of this loss.

> When muscle is cooled, it is less able to produce force, and fatigue can occur more rapidly.

> During prolonged exercise in the cold, as fatigue causes exercise intensity to decline, metabolic heat production decreases and exercisers may become susceptible to hypothermia.

> Exercise triggers release of catecholamines, which increase the mobilization and use of FFAs for fuel. But in the cold, vasoconstriction impairs circulation to peripheral fat stores, so this process is attenuated.

Health Risks During Exercise in the Cold

If humans had retained the ability of lower animals, such as reptiles, to tolerate low body temperatures, we could survive extreme hypothermia. Unfortunately, the evolution of thermoregulation in humans has been accompanied by a diminished ability of tissues to function effectively outside a narrow range of temperatures. This section deals briefly with those health risks associated with cold stress. The American College of Sports Medicine published a comprehensive position stand, "Prevention of Cold Injuries During Exercise," in 2006 that addresses these topics in much greater detail.[1]

Hypothermia

Individuals immersed in near-freezing water will die within a few minutes when their rectal temperature decreases from a normal level of 37 °C (98.6 °F) to 24 or 25 °C (75.2 or 77 °F). Cases of accidental hypothermia, as well as data obtained from surgical patients who are intentionally made hypothermic, reveal that the lethal lower limit of body temperature is usually between 23 and 25 °C (73.4-77 °F), although patients have recovered after having rectal temperatures below 18 °C (64.4 °F).

Once core temperature falls below about 34.5 °C (94.1 °F), the hypothalamus begins to lose its ability to regulate body temperature. This ability is completely lost when the internal temperature decreases to about 29.5 °C (85.1 °F). This loss of function is associated with a slowing of metabolic reactions. For each 10 °C (18 °F) drop in cellular temperature, the metabolism of the cell decreases by half. As a result, low core temperatures can cause drowsiness, lethargy, and even coma.

Cardiorespiratory Effects

The hazards of excessive cold exposure include potential injury to both peripheral tissues and the cardiovascular and respiratory systems. Death from hypothermia has resulted from cardiac arrest while respiration was still functional. Cooling primarily influences the sinoatrial node, the heart's pacemaker, leading to a substantial decrease in heart rate and, ultimately, cardiac arrest.

People have questioned whether rapid, deep breathing of cold air can damage or freeze the respiratory tract. In fact, the cold air that passes into the mouth and trachea is rapidly warmed, even when the temperature of inhaled air is less than –25 °C (–13 °F).[3] Even at this temperature, when a person is at rest and breathing primarily through his or her nose, the air is warmed to about 15 °C (59 °F) by the time it has traveled about 5 cm (2 in.) into the nasal passage. As shown in figure

12.15, extremely cold air entering the nose is quite warm by the time it reaches the back of the nasal passage, thereby posing no threat of damage to the throat, trachea, or lungs. Mouth breathing, which often occurs during exercise, may result in cold irritation to the mouth, pharynx, trachea, and even bronchi when the air temperature is below –12 °C (10 °F). Excessive cold exposure also affects respiratory function by decreasing respiratory rate and volume.

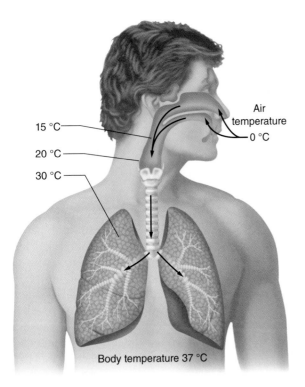

FIGURE 12.15 The warming of inspired air as it moves through the respiratory tract.

15 °C
20 °C
30 °C
Air temperature
0 °C
Body temperature 37 °C

Treatment for Hypothermia

Mild hypothermia can be treated by removing the affected person from the cold, providing dry clothing and blankets for insulation, and providing warm beverages. Moderate to severe cases of hypothermia require gentle handling to avoid initiating a cardiac arrhythmia. This requires slowly rewarming the victim. Severe cases of hypothermia require hospital facilities and medical care. Recommendations to prevent cold exposure–related injuries were outlined by the American College of Sports Medicine in its 2006 position stand on cold illness during exercise.[1]

Frostbite

Exposed skin can freeze when its temperature is lowered just a few degrees below the freezing point (0 °C,

32 °F). Because of the warming influence of circulation and metabolic heat production, the environmental air temperature (including windchill; see figure 12.14) required to freeze exposed fingers, nose, and ears is about –29 °C (–20 °F). Recall from our earlier discussion that peripheral vasoconstriction helps the body retain heat. Unfortunately, during exposure to extreme cold, the circulation in the skin can decrease to the point that the tissue dies from lack of oxygen and nutrients. This is commonly called **frostbite**. If not treated early, frostbite injuries can be serious, leading to gangrene and loss of tissue. Frostbitten parts should be left untreated until they can be thawed, preferably in a hospital, without risk of refreezing.

Exercise-Induced Asthma

While not strictly a cold-related illness, exercise-induced asthma is a common problem that affects as many as 50% of winter sport athletes. The main cause of this syndrome is the drying of the airways due to the combination of the high respiration rate associated with exercise and the extremely dry air as temperature drops. The resultant airway narrowing often leaves athletes gasping for breath. Luckily, there are preventive medications available such as β-agonists plus inhalers that can quickly deliver corticosteroids and bronchodilating medications to relieve symptoms.

In review

➤ The hypothalamus begins to lose its ability to regulate body temperature when core temperature drops below about 34.5 °C (94.1 °F).

➤ Hypothermia critically affects the heart's sinoatrial node, decreasing the heart rate, which in turn reduces cardiac output.

➤ Breathing cold air does not freeze the respiratory passages or the lungs because the inspired air is progressively warmed as it moves through the respiratory tract.

➤ Exposure to extreme cold decreases respiratory rate and volume.

➤ Frostbite occurs as a consequence of the body's attempt to prevent heat loss by skin vasoconstriction. If vasoconstriction is prolonged, the skin cools rapidly and the reduced blood flow, combined with the lack of oxygen and nutrients, ultimately can cause cutaneous tissue to die.

➤ Because cold air is inherently dry, many athletes experience the symptoms of exercise-induced asthma during high-intensity exercise in cold environments.

In closing

In this chapter, we began our examination of how the external environment affects the body's ability to perform physical work. We looked at the effects of extreme heat and cold stress and the body's responses to these conditions. We considered the health risks associated with these temperature extremes and how the body adapts to these conditions through acclimation. In the next chapter, we examine additional environmental extremes associated with exercise at altitude.

Key Terms

acclimation, or heat acclimation

acclimatization

arginine vasopressin

cold habituation

conduction

convection

critical temperature theory

dry heat exchange

eccrine sweat glands

evaporation

frostbite

heat cramps

heat exhaustion

heatstroke

hypothermia

insulation

insulative acclimation

metabolic acclimation

nonshivering thermogenesis

peripheral vasoconstriction

preoptic-anterior hypothalamus (POAH)

radiation

shivering

thermal stress

thermoreceptors (also called thermoceptors)

thermoregulation

vasopressin

wet-bulb globe temperature (WBGT)

windchill

Study Questions

1. What are the four major avenues for loss of body heat?
2. Which of these four pathways is most important for controlling body temperature at rest? During exercise?
3. What happens to the body temperature during exercise, and why?
4. Why is water vapor pressure in the air an important factor when one is performing in the heat? Why are wind and cloud cover important?
5. What factors may limit the ability to continue to exercise in hot environments?
6. What is the purpose of the wet-bulb globe temperature (WBGT)? What does it measure?
7. Differentiate between heat cramps, heat exhaustion, and heatstroke.
8. What physiological adaptations occur that allow a person to acclimate to exercise in the heat?
9. How does the body minimize excessive heat loss during cold exposure?
10. What are the three patterns of cold adaptation and when might each be expected to occur?
11. What factors should be considered to provide maximal protection when people are exercising in the cold?

Study Guide Activities

In addition to the activities listed in the chapter opening outline on page 283, two other activities are available in the web study guide, located at

www.HumanKinetics.com/PhysiologyOfSportAndExercise

The **KEY TERMS** activity reviews important terms, and the end-of-chapter **QUIZ** tests your understanding of the material covered in the chapter.

Exercise at Altitude

13

In this chapter and in the web study guide

 ACTIVITY 13.1 Exercise at Altitude—The Environment reviews characteristics of hypobaric (high-altitude) environments.

 ACTIVITY 13.2 Physiological Responses to Acute Altitude Exposure explores how physiological responses to altitude change.

 ACTIVITY 13.3 Training and Performance at Altitude considers the effects of less extreme altitudes on the performance of three athletes.

 ACTIVITY 13.4 Health Risks of Hypobaric Environments investigates the health risks of acute exposure to altitude.

Athletic competitions held at high-altitude venues have traditionally yielded poorer performances. As a result, there were many complaints when it was announced that the 1968 Olympic Games would be held in Mexico City, at an altitude of 2,240 m (7,350 ft) above sea level. Ethiopian Mamo Wolde won the marathon, but his time of 2:20:26 was slower than those of previous Olympic winners. Australian distance runner Ron Clarke was the favorite and world-record holder in the 10,000 m. With two laps to go in Mexico City, Clarke had positioned himself for a final burst to the finish; however, with 500 m to go, he began to stagger, dropped to sixth place, then collapsed unconscious after finishing. But at least two athletes who participated in those games were glad they had performed in the rarified air of Mexico City. Bob Beamon broke the previous world record in the long jump by almost 0.6 m (2 ft), and Lee Evans beat the world record in the 400 m run by nearly 0.24 s. These records stood for nearly 20 years, leading some sport scientists to suggest that the low air density that accompanies the high-altitude conditions of Mexico City, which adversely affected athletes in longer, aerobic events, likely contributed to the stellar performances in short-duration, explosive events.

Our previous discussions of the physiological responses to exercise have all been based on the conditions that exist at or near sea level, where the **barometric** (air) **pressure** (P_b) averages about 760 mmHg. Recall from chapter 7 that barometric pressure is a measure of the total pressure that all of the gases composing the atmosphere exert on the body (and everything else). Regardless of the P_b, oxygen molecules always make up 20.93% of the air. The **partial pressure of oxygen (PO_2)** is that portion of P_b exerted only by the oxygen molecules in the air. At sea level, PO_2 is therefore 0.2093 times 760 mmHg, or 159 mmHg. Partial pressure is an important concept in understanding altitude physiology, because it is primarily the low PO_2 at altitude that limits exercise performance. Although the human body tolerates small fluctuations in PO_2, large variations pose problems. This is evident when mountain climbers ascend to high altitudes where significantly reduced PO_2 can substantially impair physical performance and can even jeopardize life.

The reduced barometric pressure at altitude is referred to as a **hypobaric** environment or simply hypobaria (low atmospheric pressure). The lower atmospheric pressure also means a lower PO_2 in the inspired air, which limits pulmonary diffusion of oxygen from the lungs and oxygen transport to the tissues. The low PO_2 in the air is termed **hypoxia** (low oxygen), while the resulting low PO_2 in the blood is called **hypoxemia**.

In this chapter, we examine the unique characteristics of hypobaric, hypoxic environments and how these conditions alter physiological responses at rest and during physical activity, training, and sport performance. We cover these changes upon acute ascent to altitude, the ways in which these responses change as humans acclimate to high altitude, and specialized training strategies used by athletes to improve performance at altitude. We also examine several specific health risks associated with hypoxic environments.

Environmental Conditions at Altitude

Clinical problems associated with altitude were reported as early as 400 BC. However, most of the early concerns about ascent to high altitudes focused on the cold conditions at altitude rather than the limitations imposed by low air pressure. The initial landmark discoveries that led to our current understanding of the reduced P_b and PO_2 at altitude can be credited primarily to four scientists spanning the 17th through the 19th centuries. Torricelli (ca. 1644) developed the mercury barometer, an instrument that permitted the accurate measurement of atmospheric pressure. Only a few years later (1648), Pascal demonstrated a reduction in barometric pressure at high altitudes. Nearly 130 years after that (1777), Lavoisier described oxygen and the other gases that contribute to the total barometric pressure.[14] Finally, in 1801, John Dalton set forth the principle (called Dalton's law of partial pressures) that states that the total pressure exerted by a mixture of gases is equal to the sum of the partial pressures of those individual gases.

The deleterious effects of high altitude on humans that are caused by low PO_2 (hypoxia) were recognized in the late 1800s. More recently, a team of scientists led by the late John Sutton performed an intricate series of laboratory studies in the hypobaric chamber at the U.S. Army Institute of Environmental Medicine. These experiments, known collectively as *Operation Everest II*, have significantly added to our understanding of exercise at altitude.[13]

Based on the effects of altitude on performance, the following definitions are useful[1]:

- Near sea level [below 500 m (1,640 ft)]: no effects of altitude on well-being or exercise performance.

- Low altitude [500-2,000 m (1,640-6,560 ft)]: no effects on well-being, but performance may be diminished, especially in athletes performing above 1,500 m (4,920 ft). These performance decrements may be overcome with acclimation.

- Moderate altitude [2,000-3,000 m (6,560-9,840 ft)]: effects on well-being in unacclimated individuals and decreased maximal aerobic capacity and performance likely. Optimal performance may or may not be restored with acclimation.

- High altitude [3,000-5,500 m (9,840-18,000 ft)]: adverse health effects (including acute mountain sickness, discussed later in this chapter) in a large percentage of individuals and significant performance decrements even after full acclimation.

- Extreme altitude [above 5,500 m (~18,000 ft)]: severe hypoxic effects. The highest permanent human settlements are at 5,200-5,800 m (17,000-19,000 ft).

For our discussion, the term *altitude* refers to elevations above 1,500 m (4,920 ft), because few negative physiological effects on exercise or sport performance are seen below that altitude.

While the major impact of altitude on exercise physiology is attributable to the low PO_2 that ultimately limits oxygen availability to the tissues, the atmosphere at altitude also differs in other ways from sea-level conditions.

Atmospheric Pressure at Altitude

Air has weight. The barometric pressure at any place on earth is related to the weight of the air in the atmosphere above that point. At sea level, for example, the air extending to the outermost reaches of the earth's atmosphere (approximately 38.6 km, or 24 mi) exerts a pressure equal to 760 mmHg. At the summit of Mount Everest, the highest point on earth (8,848 m, or 29,028 ft), the pressure exerted by the air above is only about 250 mmHg. These and related altitude differences are depicted in figure 13.1.

The barometric pressure on earth does not remain constant. Rather, it varies somewhat with changes in climatic conditions, time of year, and the specific site

Altitude (ft) (m)	0 (sea level) 0	5,202 1,610	7,251 2,210	14,108 4,300	29,028 8,048
Barometric pressure P_b (mmHg)	760	631	585	430	253
% O_2 in the air	20.93	20.93	20.93	20.93	20.93
Partial pressure of oxygen PO_2 (mmHg) in the air	159	132	122	90	53
Typical temperature (°C) (°F)	15 59	9 47	2 36	−11 12	−43 −46

FIGURE 13.1 Differences in atmospheric conditions at sea level as altitude increases and barometric pressure falls accordingly. Note that the partial pressure of oxygen in the air decreases from 159 mmHg at sea level to only 53 mmHg at the summit of Mount Everest.

at which the measurement is taken. On Mount Everest, for example, the mean barometric pressure varies from 243 mmHg in January to nearly 255 mmHg in June and July. These minor variations are of little interest at sea level except to meteorologists because of their effect on weather patterns but are of considerable physiological importance for a climber attempting to ascend Mount Everest without supplemental oxygen.

Although barometric pressure varies, the percentages of gases in the air that we breathe remain unchanged from sea level to high altitude. At any elevation, the air always contains 20.93% oxygen, 0.03% carbon dioxide, and 79.04% nitrogen. Only the partial pressures of these gases change. As shown in figure 13.1, the pressure that oxygen molecules in the air exert at various altitudes drops proportionally with decreases in the barometric pressure. The consequent changes in PO_2 have significant effects on the partial pressure of oxygen that reaches the lungs, as well as the gradients between the alveoli of the lungs and the blood (where oxygen is loaded) and between the blood and the tissues (where oxygen is unloaded). These effects are discussed in detail later in this chapter.

Air Temperature and Humidity at Altitude

Clearly, the low PO_2 at altitude has the greatest impact on exercise physiology. However, other environmental factors contribute to the ability to perform exercise as well. For example, air temperature decreases at a rate of about 1 °C (1.8 °F) for every 150 m (about 490 ft) of ascent. The average temperature near the summit of Mount Everest is estimated to be about −40 °C (−40 °F), whereas at sea level the temperature would be about 15 °C (59 °F). The combination of low temperatures, low ambient water vapor pressure, and high winds at altitude poses a serious risk of cold-related disorders, such as hypothermia and windchill injuries.

Water vapor has its own partial pressure, also known as the water vapor pressure (P_{H2O}). Because of the cold temperatures at altitude, the water vapor pressure in the air is extremely low. Cold air holds very little water. Thus, even if air is fully saturated with water (100% relative humidity), the actual vapor pressure of water contained in the air is low. The extremely low P_{H2O} at high altitude promotes evaporation of moisture from the skin (or clothing) surface, because of the high gradient between skin and air, and can lead quickly to dehydration. In addition, a large volume of water is lost through respiratory evaporation due to a combination of a large vapor pressure gradient between warmed air leaving the mouth and nose and the dry air in the environment plus an increased respiration rate (discussed later) experienced at altitude.

In focus

Air temperature typically decreases as altitude increases. This decrease in temperature is accompanied by a decrease in the water vapor pressure of the air. This drier air can lead to dehydration through increased insensible water loss, increased respiratory water loss, and increased sweat evaporation.

Solar Radiation at Altitude

The intensity of solar radiation increases at high altitude for two reasons. First, at high altitudes, light travels through less of the atmosphere before reaching the earth. For this reason, less of the sun's radiation, especially ultraviolet rays, is absorbed by the atmosphere at higher altitudes. Second, because atmospheric water normally absorbs a substantial amount of the sun's radiation, the low water vapor in the air at altitude also increases radiant exposure. Solar radiation may be further amplified by reflective light from snow, which is usually found at higher elevations.

In review

➤ Altitude presents a hypobaric environment (one in which the atmospheric barometric pressure is reduced). Altitudes of 1,500 m (4,921 ft) or higher have a notable physiological impact on exercise performance.

➤ Although the percentages of the gases in the air that we breathe remain constant regardless of altitude, the partial pressure of each of these gases decreases with the decreased barometric pressure at higher altitudes.

➤ The low partial pressure of oxygen (PO_2) in the air at altitude is the environmental condition with the most profound physiological impact. Because the PO_2 in the lungs is low, PO_2 gradients between the alveoli of the lungs and the blood (where oxygen is loaded) and between the blood and the tissues (where oxygen is unloaded) are decreased.

➤ Air temperature typically decreases as altitude increases. Cold air can hold little water, so the air at altitude is dry. These two factors combine to increase susceptibility to cold-related disorders and dehydration.

➤ Because the atmosphere is thinner and drier at altitude, solar radiation is more intense at higher elevations. This effect is magnified when the ground is snow covered.

Physiological Responses to Acute Altitude Exposure

This section deals with how the human body responds to acute altitude exposure, emphasizing responses that can affect exercise and sport performance. The main concerns are respiratory, cardiovascular, and metabolic responses. Most physiological studies have been performed on healthy, fit young men; unfortunately, few studies on the effects of altitude have included women, children, or the elderly—populations whose responses to the conditions of altitude might differ from those described here.

Respiratory Responses to Altitude

Adequate oxygen supply to exercising muscles is essential to physical performance and, as seen in chapter 8, depends on an adequate supply of oxygen being brought into the body, moved from the lungs to the blood, transported to the muscles, and adequately taken up into the muscles. A limitation in any of these steps can impair performance.

Pulmonary Ventilation

The sequence of steps leading to the transport of oxygen to working muscle begins with pulmonary ventilation, the active movement of gas molecules into the alveoli of the lungs (breathing). Ventilation increases within seconds of exposure to high altitude, both at rest and during exercise, because chemoreceptors in the aortic arch and carotid arteries are stimulated by the low PO_2 and signals are sent to the brain to increase breathing. The increased ventilation is primarily associated with an increased tidal volume and an even greater increase in respiratory rate. Over the next several hours and days, ventilation remains elevated to a level proportional to the altitude.

In focus

Ventilation increases noticeably and almost immediately upon exposure to hypoxia because the decreased PO_2 stimulates peripheral chemoreceptors. The increased rate and depth of breathing help offset even larger decreases in PO_2 in the body.

Increased ventilation acts much the same as hyperventilation at sea level. The amount of carbon dioxide in the alveoli is reduced. Carbon dioxide follows the pressure gradient, so more diffuses out of the blood, where its pressure is relatively high, and into the lungs to be exhaled. This "blowing off" of CO_2 causes blood PCO_2 to fall and blood pH to increase, a condition known as **respiratory alkalosis**. This alkalosis has two effects. First, it causes the oxyhemoglobin saturation curve to shift to the left (discussed in the next section). Second, it helps keep the rise in ventilation caused by the hypoxic (low PO_2) drive from increasing even further. At a given submaximal exercise intensity, ventilation is higher at altitude than at sea level, but maximal exercise ventilation is similar.

In an effort to offset respiratory alkalosis, the kidneys excrete more bicarbonate ion, the ions that buffer the carbonic acid formed from carbon dioxide. Thus, a reduction in bicarbonate ion concentration reduces the blood's buffering capacity. More acid remains in the blood, and the alkalosis is minimized.

Pulmonary Diffusion

Under resting conditions, pulmonary diffusion (diffusion of O_2 from the alveoli to the arterial blood) does not limit the exchange of gases between the alveoli and the blood. If gas exchange were limited or impaired at altitude, less oxygen would enter the blood, so the arterial PO_2 would be much lower than the alveolar PO_2. Instead, these two values are almost equal (figure 13.2). Therefore, the low arterial blood PO_2, or hypoxemia, is a direct reflection of the low alveolar PO_2 and not a limitation of oxygen diffusion from the alveoli to the arterial blood.

Oxygen Transport

As shown in figure 13.2, the inspired PO_2 at sea level is 159 mmHg; however, it decreases to about 104 in the alveoli primarily because of the addition of water vapor molecules ($PH_2O = 47$ mmHg at 37 °C). When the alveolar PO_2 drops at altitude, fewer binding sites on the hemoglobin in the blood perfusing the lungs become saturated with O_2. As depicted in figure 13.3, the oxygen-binding (or oxyhemoglobin dissociation) curve for hemoglobin has a distinct S shape. At sea level, when alveolar PO_2 is about 104 mmHg, 96% to 97% of hemoglobin has O_2 bound to it. When PO_2 in the lungs is decreased to 46 mmHg at 4,300 m (14,108 ft), only about 80% of hemoglobin sites are saturated with O_2. If the oxygen-loading portion of the curve were not relatively flat, far less O_2 would be taken up by the blood as it passes through the lungs, and binding would be extremely limited at altitude. Therefore, while arterial blood is still desaturated at altitude, the inherent shape of the oxyhemoglobin dissociation curve serves to minimize this problem.

FIGURE 13.2 Comparison of the partial pressure of oxygen (PO_2) in the inspired air and in body tissues at sea level and at 4,300 m (14,108 ft) of altitude, the altitude of Pikes Peak, Colorado. As inspired PO_2 decreases, so does alveolar PO_2. Arterial PO_2 is similar to that in the lungs, but the gradient for diffusion of O_2 into tissues, including muscle, is greatly reduced.

A second adaptation occurs very early in altitude exposure that also aids in preventing the fall in arterial oxygen content. As mentioned earlier, a respiratory alkalosis accompanies the increased ventilation caused by acute altitude exposure. This increase in blood pH actually shifts the oxyhemoglobin dissociation curve to the left, as shown in figure 13.3. The result is that, rather than 80% binding of oxygen to hemoglobin, 89% of hemoglobin is saturated with O_2. Because of this shift, more oxygen binds to hemoglobin in the lungs and more oxygen is unloaded to the tissues at higher altitudes, where PO_2 is lower in both tissues.

Gas Exchange at the Muscles

Figure 13.2 illustrates that arterial PO_2 at sea level is about 100 mmHg, and the PO_2 in body tissues is consistently about 40 mmHg at rest; so the difference, or the pressure gradient, between the arterial PO_2 and the tissue PO_2 at sea level is about 60 mmHg. However, when one moves to an elevation of 4,300 m (14,108 ft), arterial PO_2 decreases to about 42 mmHg and the tissue PO_2 drops to 27 mmHg. Thus, the pressure gradient decreases from 60 mmHg at sea level to only 15 mmHg at the higher altitude. This represents a 75% reduction in the diffusion gradient! Because the diffusion gradient is responsible for driving the oxygen from the hemoglobin in the blood into the tissues, this change in

FIGURE 13.3 The S-shaped oxyhemoglobin dissociation curve at sea level (red line). When alveolar PO_2 is about 104 mmHg, 96% to 97% of hemoglobin is saturated with O_2. The respiratory alkalosis with acute altitude exposure shifts the oxyhemoglobin dissociation curve leftward (blue line), partially compensating for the desaturation resulting from a drop in PO_2.

arterial PO_2 at altitude is a much greater consideration for exercise performance than the small reduction in hemoglobin saturation that occurs in the lungs.

Cardiovascular Responses to Altitude

As the respiratory system becomes increasingly limiting at altitude, the cardiovascular system likewise undergoes substantial changes to attempt to compensate for the decrease in arterial PO_2 that accompanies hypoxia.

Blood Volume

Within the first few hours of arrival at altitude, plasma volume begins to progressively decrease, and the decline reaches a plateau by the end of the first few weeks. This decrease in plasma volume is the result of both respiratory water loss and increased urine production. This combination of respiratory water loss and the excretion of fluid can reduce total plasma volume by up to 25%. Initially, the result of the plasma loss is an increase in the hematocrit, the percentage of the blood volume composed of red blood cells (containing hemoglobin). This adaptation—more red blood cells for a given blood flow—allows more oxygen to be delivered to the muscles for a given cardiac output. Over a period of weeks at altitude, this diminished plasma volume eventually returns to normal if adequate fluids are ingested.

Continued exposure to high altitude triggers the release of erythropoietin (EPO) from the kidneys, the hormone responsible for stimulating erythrocyte (red blood cell) production. This increases the total number of red blood cells and creates a greater total blood volume, which allows the person to partially compensate for the lower PO_2 experienced at altitude. However, this compensation is slow, taking weeks to months to fully restore red cell mass.

(In) focus

The decrease in plasma volume at altitude is the body's short-term response to improve oxygen delivery by increasing hemocrit. An actual increase in the total number of red blood cells better compensates for the lower PO_2 at altitude, but this adaptation can take weeks to months.

Cardiac Output

The preceding discussion clearly illustrates that the amount of oxygen delivered to the muscles by a given volume of blood is limited at altitude because of the reduced arterial PO_2. A logical means to compensate for this is to increase the volume of blood delivered to the active muscles. At rest and during submaximal exercise, this is accomplished by increasing cardiac output.

Since cardiac output is the product of stroke volume and heart rate, increasing either or both of these variables will increase cardiac output. Upon ascent to altitude, the sympathetic nervous system is stimulated, releasing norepinephrine and epinephrine, the main hormones that alter cardiac function. The increase in norepinephrine in particular persists for several days of acute altitude exposure.

When submaximal exercise is performed during the first few hours at altitude, stroke volume is decreased compared to sea-level values (attributable to the reduced plasma volume). Fortunately, heart rate is increased disproportionately to not only compensate for the decrease in stroke volume but to actually slightly increase cardiac output. However, this extra cardiac workload is not an efficient way to ensure sufficient oxygen delivery to the body's active tissues for prolonged periods. Consequently, after a few days at altitude, the muscles begin extracting more oxygen from the blood (increasing the arterial–venous oxygen difference), which reduces the demand for increased cardiac output, since $\dot{V}O_2 = \dot{Q} \times (a\text{-}\bar{v})O_2$ diff. The increase in heart rate and cardiac output peaks after about 6 to 10 days at high altitude, after which cardiac output and heart rate during a given exercise bout start to decrease.

At maximal or exhaustive work levels at higher altitudes, both maximal stroke volume and maximal heart rate are decreased, as is cardiac output. The decrease in stroke volume is directly related to the decrease in plasma volume. Maximal heart rate may be somewhat lower at high altitude as a consequence of a decrease in the response to sympathetic nervous system activity, possibly attributable to a reduction in β-receptors (receptors in the heart that respond to sympathetic nerve activation, thus increasing the heart rate). With a decreased diffusion gradient to move oxygen from the blood into the muscles coupled with this reduction in maximal cardiac output, it is apparent why both $\dot{V}O_{2max}$ and submaximal aerobic performance are hindered at altitude. In summary, hypobaric conditions significantly limit oxygen delivery to the muscles, reducing the capacity to perform high-intensity or prolonged aerobic activities.

Metabolic Responses to Altitude

Ascent to altitude increases the basal metabolic rate, possibly due to increases in both thyroxin and catecholamine concentrations. This increased metabolism must be balanced by an increased food intake to prevent body weight from decreasing, a common occurrence during the first few days at altitude because appetite declines as well. In individuals who maintain their body weight at altitude, there is an increased reliance on carbohydrate for fuel, both at rest and during submaximal exercise. Because glucose yields more energy than fats or proteins per liter of oxygen, this adaptation is beneficial.

Table 13.1 summarizes the acute responses to altitude at rest and during submaximal exercise. Given the hypoxic conditions at altitude, and because any fixed amount of work at altitude represents a higher percentage of $\dot{V}O_{2max}$, we would expect anaerobic metabolism to be increased. If this occurs, we would expect lactic acid production to increase at any given work rate above the lactate threshold. This is in fact the case upon arrival at altitude. However, with longer exposure to altitude, the lactate concentration in the muscles and venous blood at a given intensity of exercise (including maximal exertion) is lower, despite the fact that muscle $\dot{V}O_2$ does not change with adaptation to altitude. To date, there is no universally accepted explanation for this so-called lactate paradox.[3]

Nutritional Needs at Altitude

In addition to the alterations in physiological systems and processes already described, several other considerations are important to note with ascent to altitude. At altitude, the body has a natural tendency to lose fluids through the skin (insensible water loss), the respiratory system, and the kidneys. This water loss is exaggerated with exercise as sweat evaporation increases from the wetted skin to the relatively dry air. These avenues of fluid loss dramatically increase the risk of dehydration, and one should pay careful attention to staying hydrated. A rule of thumb at altitude is to consume at least 3 to 5 L of fluid per day; however, this must be tailored to individual needs. It may seem counterproductive to increase fluid intake when the decrease in plasma volume is taking place to help "pack" red blood cells. However, dehydration can negatively alter the body water balance among fluid compartments, so staying well hydrated and allowing the natural decrease in plasma volume to occur is sound advice.

Appetite decreases at altitude, and decreased food intake often accompanies that decline. Decreased energy consumption coupled with increased metabolic rates can lead to daily energy deficits of up to 500 kcal/day, resulting in weight loss over time. Consuming adequate calories to support exercise and recreational activities is important, and climbers should be taught to eat more calories than their appetite suggests.

Finally, successful acclimation and acclimatization to high altitude depend on adequate iron stores in the body. Iron deficiency may prevent the increase in red

---

TABLE 13.1 **Effects of Acute Hypoxia (Initial 48 h) on Physiological Responses at Rest and During Submaximal Exercise**

System	Acute hypoxic effect at rest	Acute hypoxic effect at a given submaximal exercise intensity
Respiratory and oxygen transport	Immediate increase in ventilation (increased frequency > increased tidal volume) Decreased 2,3-DPG concentration Leftward shift in the oxyhemoglobin dissociation curve Stimulation of peripheral chemoreceptors Respiratory alkalosis	Increased ventilation
Cardiovascular	Decreased plasma volume Increased heart rate Decreased stroke volume Increased cardiac output Increased blood pressure	Increased heart rate Decreased stroke volume (due to decreased plasma volume) Increased cardiac output Increased $\dot{V}O_2$
Metabolic	Increased basal metabolic rate Decreased $(a-\bar{v})O_2$ difference	Greater utilization of carbohydrates for energy Increased lactate production initially, then lower Decreased blood pH
Renal	Diuresis Excretion of bicarbonate ions Increased release of erythropoietin	

blood cell production that occurs progressively over the first four weeks or so at altitude. Consumption of iron-rich foods and perhaps even iron supplements is recommended before and during altitude exposure.

In review

➤ Altitude causes hypobaric hypoxia, resulting in a decreased partial pressure of oxygen in the inspired air, in the alveoli, in the blood, and at the tissue level.

➤ With acute exposure to altitude, a series of adaptations occur in an attempt to minimize the drop in oxygen delivery to the tissues. Pulmonary ventilation increases and pulmonary diffusion is reasonably well maintained, but oxygen transport is slightly impaired because hemoglobin saturation at altitude is reduced.

➤ The diffusion gradient that allows oxygen exchange between the blood and active tissue is substantially reduced at moderate and high altitudes; thus, oxygen uptake by muscle is impaired.

➤ A decrease in plasma volume initially increases red blood cell concentration, allowing more oxygen to be transported per unit of blood, partially compensating for this impaired oxygen binding to hemoglobin.

➤ Upon initial ascent to altitude, cardiac output increases during submaximal work to compensate for the decreased oxygen content per liter of blood. It does so by increasing heart rate, because stroke volume falls with the fall in plasma volume.

➤ During maximal work at altitude, stroke volume and heart rate are both lower, which reduces cardiac output. This reduced cardiac output, combined with the decreased pressure gradient, severely impairs oxygen delivery to tissues.

➤ Ascent to altitude increases metabolic rate by increasing sympathetic nervous system activity. There is an increased reliance on carbohydrate for fuel, both at rest and during submaximal exercise.

➤ The exaggerated fluid loss and general loss of appetite at altitude increase the risk of dehydration.

➤ Decreased energy intake coupled with the increased energy expenditure of activity at altitude can lead to daily energy deficits and weight loss.

Exercise and Sport Performance at Altitude

The difficulty of demanding physical exertion at high altitude has been described by many climbers. In 1925, E.G. Norton[9] gave the following account of climbing without supplemental oxygen at 8,600 m (28,208 ft): "Our pace was wretched. My ambition was to do 20 consecutive paces uphill without a pause to rest and pant, elbow on bent knee, yet I never remember achieving it—13 was nearer the mark." In this section we briefly consider how exercise and sport performance is affected by altitude.

Maximal Oxygen Uptake and Endurance Activity

Maximal oxygen uptake decreases as altitude increases (see figure 13.4). $\dot{V}O_{2max}$ decreases little until the atmospheric PO_2 drops below 131 mmHg. This

FIGURE 13.4 Changes in maximal oxygen uptake ($\dot{V}O_{2max}$) with decrements in barometric pressure (P_b) and partial pressure of oxygen (PO_2). Values for $\dot{V}O_{2max}$ are recorded as percentages of $\dot{V}O_{2max}$ attained at sea level (P_b = 760 mmHg). Note that the decline in $\dot{V}O_{2max}$ begins at about 1,500 m and is fairly linear. At the altitudes of Mexico City (2,240 m), Leadville, Colorado (3,180 m), and Nuñoa, Peru (4,000 m), one's $\dot{V}O_{2max}$ would be significantly below one's capacity at sea level or in Denver (1,600 m).[5]

Data from E.R. Buskirk et al., 1967, "Maximal performance at altitude and on return from altitude in conditioned runners," *Journal of Applied Physiology* 23: 259-266.

decline generally begins at an altitude of about 1,500 m (approximately 5,000 ft)—about the elevation of Denver, Colorado, and Albuquerque, New Mexico. At altitudes between 1,500 m and 5,000 m (16,400 ft), the decreased $\dot{V}O_{2max}$ is due primarily to the reduced arterial PO_2; at higher elevations, a decreased maximal cardiac output further limits $\dot{V}O_{2max}$. $\dot{V}O_{2max}$ decreases approximately 8% to 11% for every 1,000 m increase (or 3% for every 1,000 ft increase) in altitude above 1,500 m. The rate of decline may become even steeper at very high altitudes (figure 13.5). When men and women are matched for their initial aerobic fitness level, there are no sex differences in the rate of decline in $\dot{V}O_{2max}$.

As shown in figure 13.5, men climbing Mount Everest in a 1981 expedition experienced a change in $\dot{V}O_{2max}$ from about 62 ml · kg^{-1} · min^{-1} at sea level to only about 15 ml · kg^{-1} · min^{-1} near the mountain's peak. Because resting oxygen requirements are about 3.5 ml · kg^{-1} · min^{-1}, without supplemental oxygen these men had little capacity for physical effort at this elevation. A study by Pugh and coworkers,[10] also shown in figure 13.5, showed that men with $\dot{V}O_{2max}$ values of 50 ml · kg^{-1} · min^{-1} at sea level would be unable to exercise, or even to move, near the peak of Mount Everest because their $\dot{V}O_{2max}$ values at that altitude would decrease to 5 ml · kg^{-1} · min^{-1}. Thus, most normal people with sea-level $\dot{V}O_{2max}$ values below 50 ml · kg^{-1} · min^{-1} would not be able to survive without supplemental oxygen at the summit of Mount Everest because their $\dot{V}O_{2max}$ values at such an altitude would be too low to sustain their body tissues. Enough oxygen would be consumed to barely meet their resting requirements.

Obviously, activities of long duration that place considerable demands on oxygen transport and uptake by the tissues are those that are most severely affected by the hypoxic conditions at altitude. At the summit of Mount Everest, $\dot{V}O_{2max}$ is reduced to 10% to 25% of its sea-level value. This severely limits the body's exercise capacity. Because $\dot{V}O_{2max}$ is reduced by a certain percentage, individuals with larger aerobic capacities can perform a standard work task with less perceived effort and with less cardiovascular and respiratory stress at altitude than those with a lower $\dot{V}O_{2max}$. This may explain how Messner and Habeler were able to reach the summit of Everest without supplemental oxygen in 1978—they obviously possessed high sea-level $\dot{V}O_{2max}$ values.

In focus

Endurance athletes with a high $\dot{V}O_{2max}$ at sea level have a competitive advantage at altitude if everything else is equal. As $\dot{V}O_{2max}$ declines on arrival at altitude, competition at any given pace will be performed at a lower percentage of $\dot{V}O_{2max}$.

Anaerobic Sprinting, Jumping, and Throwing Activities

Whereas endurance events are impaired at altitude, anaerobic sprint activities that last less than a minute (such as 100 m to 400 m track sprints) are generally not impaired by moderate altitude and can sometimes be improved. Such activities place minimal demands on the oxygen transport system and aerobic metabolism. Instead, most of the energy is provided through the adenosine triphosphate (ATP), phosphocreatine, and glycolytic systems.

In addition, the thinner air at altitude provides less aerodynamic resistance to athletes' movements. At the 1968 Olympic Games, for example, the thinner air of Mexico City clearly aided the performances of certain athletes, as noted at the beginning of this chapter. At Mexico City, world or Olympic records were set or tied in the men's 100 m, 200 m, 400 m, 800 m, long jump, and triple jump events and in the women's 100 m, 200 m, 400 m, 800 m, 4 × 100 relay, and long jump events. Because similar results occurred in swimming events up to 800 m, some exercise scientists have questioned the role of lower air density in improved sprint performance. Interestingly, while shot put performance was not affected at the altitude of Mexico City, discus

FIGURE 13.5 $\dot{V}O_{2max}$ relative to the partial pressure of oxygen (PO_2) of the inspired air for two expeditions to Mount Everest.

Adapted from J.B. West et al., 1983, "Maximal exercise at extreme altitudes on Mount Everest," *Journal of Applied Physiology* 55: 688-698. Used by permission.

performance declined because there is less "lift" at low barometric pressures.

In review

➤ Prolonged endurance performance suffers the most at high altitude because oxidative energy production is limited.

➤ Maximal oxygen consumption decreases in proportion to the decrease in atmospheric pressure, beginning to decline at about 1,500 m (4,921 ft).

➤ Anaerobic sprint activities that last 2 min or less are generally not impaired at moderate altitude. In some instances, sprint performance may be improved because the thinner air at altitude provides less resistance to movement.

Acclimation: Chronic Exposure to Altitude

When people are exposed to altitude over days, weeks, and months, their bodies gradually adjust to the lower oxygen partial pressure in the air. But, however well they acclimate to the conditions at high altitude, they never fully compensate for the hypoxia. Even endurance-trained athletes who live at altitude for years never attain the level of performance or the $\dot{V}O_{2max}$ values that they might achieve at sea level. In this regard, acclimation to altitude is similar to heat acclimation discussed in chapter 12. Heat acclimation improves performance and attenuates physiological strain during exercise in the heat compared to that experienced during the first few days; however, performance is still poorer than in cooler environments.

The following sections cover some of the physiological adaptations that occur with prolonged altitude exposure. These include changes at the pulmonary, cardiovascular, and muscle tissue (cellular) level. In general, these adaptations take longer to fully develop (several weeks to several months) than those associated with heat acclimation (typically one to two weeks). Generally, about three weeks are needed for full acclimation to even moderate altitude. For each additional 600 m (1,970 ft) altitude increase, another week is needed on average. All of these beneficial effects are lost within a month of return to sea level. Many of these adjustments in resting and maximal exercise variables are shown in figure 13.6.

Pulmonary Adaptations

One of the most important adaptations to altitude is an increase in pulmonary ventilation, both at rest and during exercise. Within three or four days at 4,000 m (13,123 ft), the increased resting ventilation rate levels off at a value about 40% higher than at sea level. Submaximal exercise ventilatory rate also plateaus at about 50% higher but over a longer time frame. Increases in ventilation during exercise remain elevated at altitude and are more pronounced at higher exercise intensities.

Rest

Maximal exercise

FIGURE 13.6 Physiological variables measured at sea level, after two or three days at altitude, and after weeks and months at altitude (3,00-3,500 m or 9,843-11,483 ft). Both resting (left) and maximal exercise (right) variables are shown.

Drawn from data presented in Bartsch and Saltin, 2008.

Blood Adaptations

During the first two weeks at altitude, the number of circulating erythrocytes increases. The lack of oxygen at altitude stimulates the renal release of erythropoietin or EPO. Within the first 3 h after the athlete arrives at a high elevation, the blood's EPO concentration increases; it then continues to increase for two or three days. Although blood EPO concentrations return to baseline levels in about a month, the **polycythemia** (increased red blood cells) may be evident for three months or more. After a person lives at 4,000 m (13,123 ft) for about six months, his or her total blood volume (composed mainly of the red cell volume and the plasma volume) increases by about 10%, not only as a result of the altitude-induced stimulation of erythrocyte production but also because of plasma volume expansion (discussed later).[10]

The percentage of total blood volume composed of erythrocytes is referred to as the hematocrit. Residents in the central Andes of Peru (4,540 m, or 14,895 ft) have an average hematocrit of 60% to 65% (properly termed an acclimatization rather than an acclimation response, see chapter 12). This is considerably higher than the average hematocrit of sea-level residents, which is only 45% to 48%. However, during six weeks of exposure to the Peruvian altitude, sea-level residents have shown remarkable increases in their hematocrit levels, up to an average of 59%.

As the volume of erythrocytes increases, so does the blood's hemoglobin content (and concentration after an initial decline, see figure 13.6). As noted in figure 13.7, blood hemoglobin concentration tends to increase proportionately with increases in the elevation at which people reside. The data presented are for men. For women, however, the limited available data show a similar trend but with a lower concentration than for men at a given altitude. These adaptations improve oxygen-carrying capacity of a fixed volume of blood.

The reduction in plasma volume during acute altitude exposure reduces total blood volume, thus reducing submaximal and maximal cardiac output. But with acclimatization, as plasma volume increases over several weeks at altitude and as red blood cells continue to increase, maximal cardiac output increases. However, it does not return to sea-level values as shown in figure 13.6. Thus, overall oxygen delivery capacity is increased with acclimatization but not to the extent needed to achieve sea-level $\dot{V}O_{2max}$ values.

There is some debate about whether acclimation alters oxygen transport in the blood by changing the shape and position of the oxyhemoglobin dissociation curve (figure 13.3). The concentration of 2,3-diphosphoglycerate (2,3-DPG) increases in red blood cells, which shifts the curve to the right. This would favor the unloading of oxygen at the tissues (because more oxygen would be unloading from hemoglobin at any given low arterial PO_2), but this effect opposes the loading benefit of the respiratory alkalosis, a leftward shift. The net effect of both mechanisms is variable.

Muscle Adaptations

Although few attempts have been made to study muscle changes that occur during exposure to altitude, sufficient muscle biopsy data exist to indicate that muscles undergo significant structural and metabolic changes during ascent to altitude. In a study of climbers experiencing four to six weeks of chronic hypoxia on expeditions, muscle fiber cross-sectional area decreased, thus decreasing total muscle area. Capillary density in the muscles increased, which allowed more blood and oxygen to be delivered to the muscle fibers. Muscles' inability to meet exercise demands at high altitude might be related to a decrease in their mass and their ability to generate ATP.

The cause of the decreased muscle cross-sectional area within the first days and weeks at altitude is not fully understood. As previously discussed, prolonged exposure to high altitude frequently causes a loss of appetite and a noticeable weight loss. During a 1992 expedition to climb Mount McKinley, six men experienced an average weight loss of 6 kg, or 13 lb (D.L. Costill et al., unpublished data). Although part of this loss represents a general decrease in body weight and extracellular water, all of the men experienced a noticeable decrease in muscle mass. It seems logical to assume that much of this decrease in muscle mass is associated with loss of appetite and a wasting of muscle

FIGURE 13.7 Hemoglobin (Hb) concentrations of men living at, and acclimatized to, various altitudes.

protein. Perhaps future studies on nutrition and body composition in mountain climbers will provide a more thorough explanation of the incapacitating influences of high altitude on muscle structure and function.

Several weeks at altitudes above 2,500 m (8,202 ft) reduce the metabolic potential of muscle, although this may not occur at lower elevations. Both mitochondrial function and glycolytic enzyme activities of the leg muscles (vastus lateralis and gastrocnemius) are significantly reduced after four weeks at altitude. This suggests that, in addition to receiving less oxygen, muscles lose some of their capacity to perform oxidative phosphorylation and generate ATP. Unfortunately, no muscle biopsy data have been obtained from long-term residents at high altitudes to determine whether those individuals experience any muscular adaptations as a consequence of lifelong residence at these elevations.

Cardiovascular Adaptations

Studies conducted on endurance-trained runners in the late 1960s indicated that the decrement in $\dot{V}O_{2max}$ when they first reached high altitude improved little for the duration of their exposure to hypoxia. Aerobic capacity remained unchanged for up to two months at altitude.[5] Although the runners who previously had been exposed to altitude were more tolerant of hypoxia, their $\dot{V}O_{2max}$ values and running performance were not significantly improved with acclimation. Because of the many adaptations that occur during acclimation to altitude, this lack of improvement in aerobic capacity and performance was unexpected. Perhaps these trained subjects had already attained maximal training adaptations and were unable to further adapt in response to altitude exposure. Or perhaps the reduced PO_2 of altitude made it more difficult for them to train at the same intensity and volume as at sea level.

Altitude: Optimizing Training and Performance

We have considered the major changes that occur as the human body becomes acclimated to altitude and how these adaptations affect performance at altitude. But is there any advantage to training at altitude to improve performance at sea level? Are there advantages to training at altitude versus training at sea level when one must compete at altitude? And what about the relatively new concept of "live high, train low" to optimize performance?

Does Altitude Training Improve Sea-Level Performance?

Athletes have hypothesized for decades that training under hypoxic conditions, for example in an altitude chamber by simply breathing low-oxygen gas mixtures, can improve sea-level endurance performance. Since many of the beneficial changes associated with altitude acclimation are similar to those conferred by aerobic training, can combining the two be even more beneficial? Can altitude training improve sea-level performance?

A strong theoretical argument can be made for altitude training. First, altitude training evokes substantial tissue hypoxia (reduced oxygen supply). This is thought to be essential for initiating the conditioning response.

Second, the altitude-induced increase in red blood cell mass and hemoglobin content improves oxygen delivery on return to sea level. Although evidence suggests that these latter changes are transient, lasting

In review

➤ Hypoxic conditions stimulate the renal release of erythropoeitin (EPO), which increases erythrocyte (red blood cell) production in bone marrow. More red blood cells means more hemoglobin. Although plasma volume decreases initially, which helps concentrate the hemoglobin, it eventually returns to normal. Normal plasma volume plus additional red blood cells increases total blood volume. All these changes increase the blood's oxygen-carrying capacity.

➤ Total muscle mass decreases after a few weeks at altitude, as does total body weight. Part of this decrease is from dehydration and appetite suppression. However, there is also protein breakdown in the muscles.

➤ Other muscle adaptations include decreased fiber area, increased capillary supply, and decreased oxidative enzyme activities.

➤ While work capacity improves with altitude acclimation, the decrease in $\dot{V}O_{2max}$ with initial exposure to altitude does not improve much over several weeks of exposure and typically never returns to sea-level values.

Live High, Train Low

Researchers at the Institute for Exercise and Environmental Medicine in Dallas, Texas, conducted a series of studies in the mid-1990s to investigate altitude training for enhancing endurance performance. When living and training at altitude, athletes are faced with the problem that the intensity of training is reduced because aerobic capacity and cardiorespiratory function are reduced at altitude. Thus, although athletes gain certain physiological benefits from being at altitude, they lose training adaptations associated with higher intensities of training. The researchers investigated the potential of having subjects *live* at moderate altitude but *train* at low altitude, where training intensity is not compromised. In one study,[8] researchers divided 39 competitive runners into three equal groups: One (the high–low group) lived at moderate altitude (2,500 m, or 8,202 ft) and trained at low altitude (1,250 m, or 4,100 ft); one group (high–high) lived and trained at moderate altitude (2,500 m); and one group (low–low) lived and trained at low altitude (150 m, or 490 ft). Using a 5,000

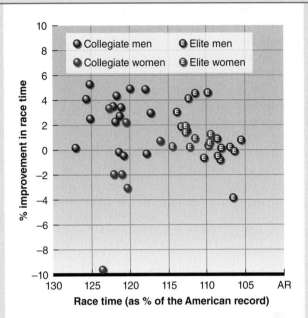

Improvement in race time (%) in elite male and female runners[11] and collegiate male and female runners[8] following four weeks of living at altitude but training at 1,250 m (4,100 ft). See the text for details.

m time trial as the primary performance outcome measure, the researchers found that the high–low group was the only group to significantly improve their running performance, even though both the high–low and high–high groups increased their $\dot{V}O_{2max}$ values by 5% in direct proportion to their increase in red cell mass. Thus, there appear to be performance benefits from living at moderate altitude but going to lower elevations to maximize training intensity.

This was tested more recently by the same scientists working with a group of 14 elite male and 8 elite female runners, with all but two ranked in the U.S. top 50 for their event. These athletes lived at 2,500 m (8,202 ft) and trained at 1,250 m (4,100 ft) over a period of 27 days. Testing was conducted at sea level both the week before and the week after the 27 days of living at altitude. Sea-level 3,000 m time trial performance increased by 1.1% and $\dot{V}O_{2max}$ increased by 3.2% as a result of this intervention.[11] The figure in this sidebar illustrates the difference in race time performance for both studies, with values expressed as the percentage change before and after altitude exposure. These differences are plotted by the pre-altitude race time, expressed as a percentage of the existing U.S. record in the event at the time of the time trial.

only several days, in theory this still should provide an advantage for the athlete. Furthermore, studies from the 1960s and 1970s appeared to show that training at altitude did indeed enhance performance at sea level. Unfortunately, those studies did not test control groups who trained and competed at sea level, so it was impossible to tell whether the improved performance of the altitude-trained athletes was due to the training or the altitude.

More recent studies have shown that there is no additional benefit of living and training at altitude for increasing sea-level $\dot{V}O_{2max}$ or improving sea-level aerobic performance. In addition, living at sea level and training in a hypobaric chamber to simulate altitude do not appear to provide any advantage over the same volume of sea-level training. In the few studies in which altitude training was found to influence postaltitude sea-level performance, the subjects were not well trained before going to altitude. This makes it difficult to determine how much of their postaltitude improvement was attributable solely to training, independent of altitude.

Studying athletes at altitude poses additional problems because they are often unable to train at the same volume and intensity of effort as when at sea level. This was demonstrated in a group of elite female cyclists who performed self-selected maximum power outputs during high-intensity interval training. They completed trials under the following conditions: breathing atmospheric air (normoxia) and breathing a hypoxic gas mixture simulating 2,100 m (6,888 ft). The athletes' sustained (10 min) and short-term (15 s) power outputs at maximal intensity were reduced under hypoxic conditions.[4] Training at even higher elevations, where acclimatization effects would be even more beneficial, causes even greater disruptions in training.

In addition, living and training at moderate to high altitude often causes athletes to dehydrate and to lose blood volume and muscle mass. These and other side effects tend to diminish the athletes' fitness and their motivation and tolerance for intense training. As a result, studies are difficult to interpret, but the value of altitude training for optimal sea-level performance has not been validated.

In focus

Athletes have used altitude training in an attempt to improve sea-level endurance performance; however, the existing research on endurance athletes does not support its effectiveness.

Is there a better way to use altitude to prepare endurance athletes for competition at sea level while not interfering with training intensity or duration? If athletes lived at high altitudes but trained at sea level or low altitudes, could this combination provide physiological benefits without compromising training? This concept is discussed in the sidebar "Live High, Train Low" on page 323.

Optimizing Performance at Altitude

What can athletes who normally train at sea level but must compete at altitude do to prepare most effectively for competition? Although not all combinations have been attempted and research thus far is not conclusive, it appears that athletes have two viable options. One option is to compete as soon as possible after arriving at altitude, and certainly within 24 h of arrival. This does

not provide the beneficial effects of acclimation, but the altitude exposure is brief enough that the classic symptoms of altitude sickness are not yet totally manifest. After the first 24 h, the athlete's physical condition often worsens because of the untoward effects of acute altitude exposure, such as dehydration, headache, and sleep disturbances.

Another option is to train at higher altitudes for a minimum of two weeks before competing. But not even two weeks is sufficient for total acclimation. Total altitude acclimation would require a minimum of three to six weeks, and usually even longer. As previously mentioned, several weeks of intense aerobic training at sea level to increase the athletes' $\dot{V}O_{2max}$ will allow them to compete at altitude at a lower relative intensity (% $\dot{V}O_{2max}$) than if they had not trained aerobically.

In focus

In 2006, the eventual Super Bowl champion Pittsburgh Steelers had a road play-off game in Denver against the Broncos one week after beating Indianapolis. The team had a choice to make—travel to Denver early in the week to adapt to the altitude or wait until the last minute and fly into the Mile High City. After consultation with exercise physiologists and medical personnel, they chose the latter and went on to win 34-17, with few players experiencing altitude-related problems.

Extended training for optimal performance at altitude requires an elevation between 1,500 m (4,921 ft), which is considered the lowest level at which an effect will be noticed, and 3,000 m (9,840 ft), which is the highest level for efficient conditioning. Work capacity is reduced during the initial days at altitude. For this reason, when first reaching higher altitudes, athletes should reduce workout intensity to between 60% and 70% of sea-level intensity, gradually working up to full intensity within 10 to 14 days.

Artificial "Altitude" Training

The largest and most important adaptations to altitude are physiological changes caused by the hypoxia experienced there, so we might anticipate that people could achieve similar adaptations simply by breathing gases with a low PO_2. But no evidence supports the idea that brief periods (1-2 h per day) of breathing hypoxic gases or hypobaric mixtures induce even a partial adaptation similar to that observed at altitude. On the other hand, alternating periods (lasting between 5 and 14 days) of training at 2,300 m (7,546 ft) and at sea level adequately stimulated altitude acclimation in a group of elite middle-distance runners.[6] Staying at sea level for up to 11 days did not interfere with the usual adjustments to altitude as long as training was maintained.

The favorable results of the studies on "living high and training low" have stimulated considerable interest in how this concept might be applied without having to send athletes to altitude to live. One approach has been to develop a hypoxic apartment where athletes sleep and live. The gas mixture inside the apartment is adjusted so that nitrogen represents a higher percentage of the inspired air, reducing the percentage of oxygen in the inspired air as well as its partial pressure. Pioneered by Finnish sport scientists, these apartments can simulate altitudes between 2,000 m (6,560 ft) and 3,000 m (9,840 ft) as the nitrogen and oxygen percentages in the inspired air are adjusted to reduce the partial pressure of oxygen to those levels associated with 2,000 to 3,000 m altitude. Hypoxic sleeping devices or tents have also been proposed.

Unfortunately, at this time, few carefully controlled scientific data exist to confirm whether these apartments or sleeping devices actually improve performance and physiological function. One recent meta-analysis (a statistical approach that combines data from multiple studies to draw conclusions) reported that natural "live high, train low" approaches provide the best results for enhancing performance in elite athletes, while some nonelite exercisers seem to benefit from the artificial approaches.[2] However those authors cautioned that improvements seen in these subelite athletes could have been due to a placebo effect. Ethical issues have also been raised about the use of such devices.

In review

➤ Most studies show that training at altitude leads to no significant improvement in sea-level performance. Living at high altitudes and training at low altitudes currently appears to be the best alternative.

➤ Athletes who must perform at altitude should do so as early after arrival as possible, certainly within 24 h of arrival, before the detrimental side effects that occur at altitude become too great.

➤ Alternatively, athletes who must perform at altitude could train at an altitude between 1,500 m (4,921 ft) and 3,000 m (9,840 ft) for a minimum of two weeks (more is better) prior to performing.

➤ There is no evidence that brief periods (1-2 h per day) of breathing hypoxic gases or hypobaric gas mixtures induce even a partial adaptation similar to that observed at altitude.

Health Risks of Acute Exposure to Altitude

A large proportion of people who ascend to moderate and high altitudes experience symptoms of **acute altitude (mountain) sickness**. This disorder is characterized by symptoms such as headache, nausea, vomiting, dyspnea (difficult breathing), and insomnia. These symptoms can begin anywhere from 6 to 48 h after arrival at high altitude and are most severe on days 2 and 3. Although not life threatening, acute altitude sickness can be incapacitating for several days or longer. In some cases, the condition can worsen. The victim can develop the more lethal altitude-related illnesses of high-altitude pulmonary edema or high-altitude cerebral edema.

Acute Altitude (Mountain) Sickness

The incidence of acute altitude sickness varies with the altitude, the rate of ascent, and the individual's experience and susceptibility. Several studies have been conducted to determine the incidence of acute altitude sickness in groups of novice hikers (tourists) and more experienced climbers. Results vary widely, ranging from a frequency of less than 1% to almost 60% at altitudes of 3,000 to 5,500 m (9,840-18,045 ft) (figure 13.8). Forster,[7] however, reported that 80% of those who ascended to the top of Mauna Kea (4,205 m, or 13,796 ft) on the island of Hawaii experienced some symptoms of acute altitude sickness. Another study showed that at elevations of 2,500 to 3,500 m (8,202-11,483 ft), altitudes commonly experienced by recreational skiers and hikers, the incidence of acute altitude sickness was about 7% for men and 22% for women, but the reason for this sex difference is unclear.[12]

Although the precise underlying cause of acute altitude sickness is not fully understood, it appears that people who experience the greatest distress also have a low ventilatory response to hypoxia. This inadequate ventilation allows PO_2 to decrease further and carbon dioxide to accumulate in the tissues, and these two factors may induce most of the symptoms associated with altitude sickness.

Headache is the most common symptom associated with ascent to high altitude. Headache is rarely experienced below 2,500 m (~8,000 ft), but ascent to 3,600 m (~12,000 ft) results in headache in the majority of people. The headache at altitude, which many sufferers describe as continuous and throbbing, is typically worse in the morning and after exercise. Alcohol consumption worsens the symptoms. The precise mechanism is unsubstantiated, but hypoxia causes dilation of the cerebral blood vessels, so stretching of pain receptors in these structures is a likely cause.

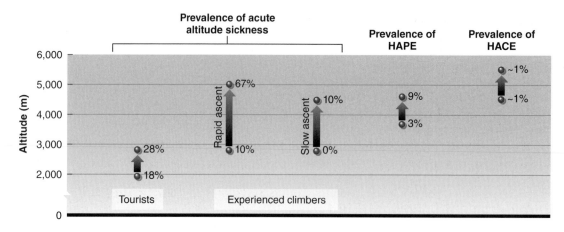

FIGURE 13.8 Reported prevalence of acute altitude sickness, high-altitude pulmonary edema (HAPE), and high-altitude cerebral edema (HACE) as a function of altitude, experience, and, in the case of acute altitude sickness, rate of ascent.

Adapted from compiled data presented in Bartsch and Saltin, 2008.

Another consequence of acute altitude sickness is an inability to sleep even if the individual is markedly fatigued. Studies have shown that the inability to achieve satisfying sleep at altitude is associated with an interruption in the sleep stages. Additionally, some people suffer from a pattern of interrupted breathing, called **Cheyne-Stokes breathing**, which prevents them from falling asleep and remaining asleep. Cheyne-Stokes breathing is characterized by alternating periods of rapid breathing and slow, shallow breathing, usually including intermittent periods in which breathing completely stops. The incidence of this irregular breathing pattern increases with altitude, occurring in 24% of people at 2,440 m (8,005 ft), 40% of people at 4,270 m (14,009 ft), and almost everyone at altitudes above 6,300 m (20,669 ft).[15]

How can athletes avoid acute altitude sickness? Even athletes who are highly endurance trained before altitude exposure seem to have little protection against the effects of hypoxia; and it is difficult to determine which athletes may be susceptible to these symptoms unless this is suggested by a prior history of acute altitude sickness.

People can usually prevent acute altitude sickness by gradually ascending to altitude, spending periods of a few days at lower elevations. A gradual ascent of no more than 300 m (984 ft) per day at elevations above 3,000 m (9,840 ft) has been suggested to minimize the risks of altitude sickness. Of the drugs that have been used to reduce the symptoms of those who develop acute altitude sickness, acetazolamide started the day before ascent is the only established preventive

measure. Acetazolamide is sometimes combined with a steroid such as dexamethazone. Both drugs must be used with medical supervision. Of course, the definitive treatment for severe acute mountain sickness is a retreat to lower altitude, but high-flow oxygen and the use of hyperbaric rescue bags are also effective in severe cases.

High-Altitude Pulmonary Edema

Unlike acute mountain sickness, **high-altitude pulmonary edema (HAPE)**, which is the accumulation of fluids in the lungs, is life threatening. The cause of HAPE is likely related to the pulmonary vasoconstriction resulting from hypoxia, causing blood clots to form in the lungs. Remaining tissue becomes overperfused, and fluid and protein leak out of the capillaries. This seems to occur most frequently in unacclimatized people who rapidly ascend to altitudes above 2,500 m (8,202 ft). The disorder occurs in otherwise healthy people and has been reported more often in children and young adults. The fluid accumulation interferes with air movement into and out of the lungs, leading to shortness of breath, a persistent cough, chest tightness, and excessive fatigue. Disruption of normal breathing impairs blood oxygenation and, if severe enough, cyanosis (a bluish tint) of the lips and fingernails, mental confusion, and loss of consciousness may occur. High-altitude pulmonary edema is treated via administration of supplemental oxygen and movement of the victim to a lower altitude.

High-Altitude Cerebral Edema

Rare cases of **high-altitude cerebral edema (HACE)**, which is fluid accumulation in the cranial cavity, have been reported. The condition is often a subsequent complication of HAPE. This neurological condition is characterized by mental confusion, lethargy, and ataxia (difficulty walking), progressing to unconsciousness and death. Most cases have been reported at altitudes greater than 4,300 m (14,108 ft). Similar to that for HAPE, the cause of HACE involves hypoxia-induced leakage of fluids from cerebral capillaries, causing edema and a resultant pressure buildup in the confined intracranial space. Treatment involves administration of supplemental oxygen, a hyperbaric bag, and prompt descent to a lower altitude. If descent is delayed, permanent impairment may ensue.

In review

➤ Acute altitude sickness, often called acute mountain sickness, typically causes symptoms such as headaches, nausea, dyspnea, and insomnia. Symptoms usually appear 6 to 48 h after arrival at altitude.

➤ The exact cause of acute altitude sickness is not known, but many researchers suspect that the symptoms may result from the combination of hypoxia and carbon dioxide accumulation in the tissues.

➤ Acute altitude sickness usually can be avoided by a slow, gradual ascent to altitude, climbing no more than 300 m (984 ft) per day at elevations above 3,000 m (9,840 ft).

➤ High-altitude pulmonary and cerebral edema (HAPE and HACE)—which involve accumulation of fluid in the lungs and cranial cavity, respectively—are life-threatening conditions. Both are treated by oxygen administration, hyperbaric bags, and descent.

Activities are seldom conducted under ideal environmental conditions. Heat, cold, humidity, and altitude—alone or in combination—present unique problems that are superimposed on the physiological demands of exercise. This chapter and the preceding chapter summarized the nature of these common environmental stresses and how we can cope with them.

Much of our discussion thus far has dealt with how physiological variables and environmental stress can hinder our performance. In the next part of the book, we examine various ways to optimize performance. We begin by looking at the importance of the amount of training, considering what happens when we train either too much or too little.

Key Terms

acute altitude (mountain) sickness

barometric pressure

Cheyne-Stokes breathing

high-altitude cerebral edema (HACE)

high-altitude pulmonary edema (HAPE)

hypobaric

hypoxemia

hypoxia

partial pressure of oxygen (PO_2)

polycythemia

respiratory alkalosis

Study Questions

1. Describe the conditions at altitude that could limit the ability to perform physical activity.

2. What types of exercise are detrimentally influenced by exposure to high altitude and why?

3. When someone ascends to an altitude of over 1,500 m, describe the physiological adjustments that occur within the first 24 h.

4. Differentiate the physiological adjustments that accompany acclimation to altitude over a period of days, weeks, and months.

5. Would an endurance athlete who trained at altitude be able to perform better during subsequent sea-level performance? Why or why not?

6. Describe the theoretical advantage of living high and training low.

7. What are the best strategies for preparing athletes for high-altitude competition?

8. What are the health risks associated with acute exposure to high altitude and how can they be minimized?

Study Guide Activities

In addition to the activities listed in the chapter opening outline on page 309, two other activities are available in the web study guide, located at

www.HumanKinetics.com/PhysiologyOfSportAndExercise

 The **KEY TERMS** activity reviews important terms, and the end-of-chapter **QUIZ** tests your understanding of the material covered in the chapter.

PART V

Optimizing Performance in Sport

Previous chapters explain how the body responds to an acute bout of exercise, how it adapts to chronic training, and how it adjusts to environmental extremes. We can now apply that knowledge to optimizing performance in exercise and sport. In part V, we focus on how athletes can best prepare for competition from a physiological standpoint. In chapter 14, "Training for Sport," we discuss how to optimize athletes' training regimen and explore how too much or too little training can impair performance. In chapter 15, "Body Composition and Nutrition for Sport," we address the issues of assessing body composition, relating body composition to sport performance, and sports that have weight standards. We then evaluate athletes' dietary needs and consider how nutritional supplementation and dietary manipulation might improve performance. In chapter 16, "Ergogenic Aids and Sport," we discuss various pharmacological, hormonal, and physiological agents that have been proposed to improve performance. We examine the potential benefits, proven effects, and health risks associated with their use.

Training for Sport

14

In this chapter and in the web study guide

ACTIVITY 14.1 Overtraining Syndrome considers the effects of overtraining on two swimmers and explores symptoms of sympathetic and parasympathetic overtraining.

ACTIVITY 14.2 Tapering for Peak Performance investigates the effects of tapering on the performance of two distance runners.

ACTIVITY 14.3 Detraining explores how physiological responses to detraining change.

Throughout his college career, Eric had trained at swimming 4 h each day, covering as much as 13.7 km (8.5 mi) per day. Despite this effort, his performance time for the 200 yd (183 m) butterfly event had not improved since his freshman year. With a best performance of 2 min 15 s for the event, he was seldom given a chance to compete because several teammates could perform the event in less than 2 min 5 s. During Eric's senior year, his coach made a major change in the team's training plan. The swimmers trained only 2 h per day and swam an average of 4.5 to 4.8 km (2.8-3.0 mi) per day. Further, they swam each interval at a faster pace and had a longer rest period between intervals. Suddenly Eric's performance began to improve. After three months, his time dropped to 2 min 10 s, still not good enough to make him a major contender. But as a reward for Eric's improvement, the coach chose him to swim the 200 yd butterfly event at the conference championship meet, which was preceded by three weeks of reduced training (tapering) of only 1.6 km (1 mi) per day. Subsequently, with less volume of training than in previous years and well rested after the taper, Eric was able to make the finals of the event at the championship meet. His preliminary time was 2 min 1 s. In the finals, he improved even further, posting a third-place finish with a time of 1 min 57.7 s, an impressive performance for a swimmer who performed better with lower-volume but higher-quality training.

Repeated days and weeks of training can be considered a positive "stress" because the adaptations caused by training improve the capacity for energy production, oxygen delivery, muscle contraction, and other mechanisms that enhance exercise performance. The major changes associated with training occur within the first 6 to 10 weeks. The magnitude of these adaptations depends on the volume and intensity of exercise performed during training, which has led many coaches and athletes to believe erroneously that the athlete who trains the longest and hardest will be the best performer. However, the quantity and the quality of training are two separate things. Too often, training sessions are judged by the total volume (e.g., the distance ran, cycled, or swum) performed in each training session, leading coaches to design training programs that are not optimal for improving performance and often impose unrealistic demands on the athlete.

The rate at which an individual adapts to training is genetically limited. Too much training can reduce the athlete's optimal potential for improvement and in some cases can cause a breakdown in the adaptation process, eventually reducing performance. When training is taken to extremes, serious illness or injury can occur.

In focus

Every athlete's rate of adaptation to training is genetically limited. Each individual responds differently to the same training stress, so what might be excessive training for one person could be well below the capacity of another for optimal adaptations. Therefore, it is important to recognize individual differences when designing and carrying out training programs.

Although the volume of work performed in training is an important stimulus for physiological improvements in performance, a proper balance should be established between volume and intensity. Training can be overdone, leading to chronic fatigue, illness, overuse injury, overtraining syndrome, and performance decrements. In contrast, proper rest and achieving the proper balance between training volume and intensity can, and will, enhance performance. Much effort has been directed toward determining the appropriate volume and intensity required to achieve optimal adaptation. Exercise physiologists have tested numerous training regimens to determine both the minimal and maximal stimuli needed for cardiovascular and muscular improvements. The next section examines those factors that can affect the response to a given training program, developing a model for optimizing the training stimulus.

Optimizing Training: A Model

All well-designed training programs incorporate the principle of progressive overload. As discussed in chapter 9, this principle states that to continue to provide the benefits of training, the training stimulus must be progressively increased as the body adapts to the current stimulus. The only way to continue to improve with training is to progressively increase the training stimulus. However, when this concept is carried too far, training may become excessive, pushing the body beyond its ability to adapt, producing no additional improvement in conditioning or performance and leading to performance decrements. Conversely, if the volume or the intensity of training is too low, the resulting physiological changes will be hindered and optimal performance will not be achieved. Thus, the coach and athlete face the challenge of determining the optimal training stimulus for each particular athlete, recognizing that what works for one athlete might not work for another.

Figure 14.1 provides a model demonstrating the continuum of training stages that a competitive athlete might go through during a full year. This model is based on the principle of periodization, which was described in chapter 9 and is illustrated further in figure 14.2. In this model, **undertraining** represents the type of training an athlete would undertake between competitive seasons or during active rest. Generally, physiological adaptations will be minor, and there will be no improvement in performance during this stage. **Acute overload** represents what might be considered an "average" training load, whereby the athlete is stressing the body to the extent necessary to improve both physiological function and performance. **Overreaching** is a relatively new term that refers to a brief period of heavy overload without adequate recovery, thus exceeding the athlete's adaptive capacity. There will be a brief performance decrement, from several days to several weeks, but eventually performance will improve. Finally, **overtraining** refers to that point at which an athlete experiences physiological maladaptations and chronic performance decrements. This generally leads to the **overtraining syndrome.**[1] **Excessive training**, not shown in the model, refers to training that is well above what is needed for peak performance but does not strictly meet the criteria for either overreaching or overtraining.

Overreaching

Overreaching is a systematic attempt to intentionally overstress the body for a short period of training. Done correctly, this allows the body to adapt to the increased training stimulus, that is, beyond the level of adaptation attained during "normal" overload. As with overtraining, there is a brief decrement in performance lasting several days to several weeks, followed by increased physiological function and increased performance. Obviously, this is the critical phase of training—on the edge, leading either to improved physiological function and performance or to overtraining if one goes too far. With overreaching, the period of full recovery from training takes several days to several weeks; however, with overtraining, recovery can take many months or, in some cases, years. The key to overreaching is to push the athlete hard enough to accomplish the desired positive physiological and performance improvements but to avoid going into the stage of overtraining. This is not an easy task to accomplish!

Excessive Training

With excessive training, either the volume or the intensity of training, or both, are increased to extreme levels. A "more is better" philosophy often drives the training schedule. For many years, athletes were undertrained. As coaches and athletes became bolder and started to push the envelope by increasing both training volume and intensity, they found that athletes responded well, and world records began to tumble. However, one can take this philosophy only so far. At a certain point, performance begins to either plateau or decline. Let's take a look at some examples.

FIGURE 14.1 Model of the continuum of training stages.

Adapted, by permission, from L.E. Armstrong and J.L. VanHeest, 2002, "The unknown mechanism of the overtraining syndrome," *Sports Medicine* 32(1): 185-209.

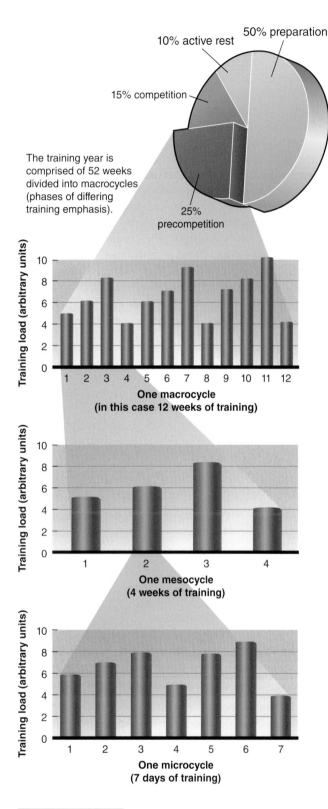

FIGURE 14.2 The structure of a periodized training program. This model varies the training load over time to achieve acute overload and some overreaching, while avoiding overtraining.

Adapted, by permission, from R.W. Fry, A.R. Morton, and D. Keast, 1991, "Overtraining in athletes: An update," *Sports Medicine* 12: 32-65.

Most of the research on excessive training has been conducted on swimmers, but the principles also apply to most other forms of training. Research shows that swim training 3 to 4 h per day, five or six days each week, provides no greater benefits than training only 1 to 1.5 h per day.[6] In fact, such excessive training has been shown to significantly decrease muscular strength and sprint swimming performance.

Few studies have compared the physical conditioning and performance benefits of single versus multiple daily training sessions. Studies conducted to date reveal no scientific evidence that multiple daily training sessions enhance fitness and performance more than a single daily session. This is illustrated by the data in figure 14.3, which show the responses of two groups of swimmers who trained once per day (group 1) or twice per day (group 2) for weeks 5 through 10 of a 25-week training program.[5] All swimmers began the program following the same training regimen: one session per day. But from the beginning of the 5th week through the end of the 10th week, group 2 trained twice per day. After six weeks on the different regimens, both groups returned to the once-daily program. All the swimmers' heart rates and blood lactate values decreased dramatically when training began, and no significant differences were seen in the two groups' results in response to the change in training volume. The swimmers who trained twice per day showed no additional improvements over those who trained only once per day. In fact, their blood lactate concentrations (figure 14.3a) and heart rates (figure 14.3b) appeared to be slightly higher for the same fixed-pace swim.

To determine the influence of long-term excessive training, performance improvements of swimmers who trained twice daily for a total distance of more than 10,000 m (10,936 yd) per day (the LS, or long-swim, group) were compared with improvements of those who swam approximately half that distance in a single session each day (the SS, or short-swim, group).[6] Changes in performance time for the 100 yd (91 m) front crawl were examined over a four-year period for both groups. The LS swimmers and SS swimmers experienced an identical average improvement of 0.8% per year. Similar findings also were observed for competitors in other events, such as the 200, 500, and 1,650 yd (183, 457, and 1,509 m) front crawl.

The concept of training specificity (see chapter 9) implies that several hours of daily training will not provide the adaptations needed for athletes who participate in events of short duration. Most competitive swimming events last less than 2 min. How can training for 3 to 4 h per day at speeds that are markedly slower than competitive pace prepare the swimmer for the maximal efforts of competition? Such a large training volume

FIGURE 14.3 Changes in swimmers' (*a*) blood lactate concentrations and (*b*) heart rates during a standardized 366 m (400 yd) swim over 25 weeks of training. From the beginning of the 5th through the end of the 10th week, one group trained once per day (group 1) whereas the other group trained twice each day (group 2).

prepares the athlete to tolerate a high volume of training but likely does little to benefit actual performance.

The need for long daily workouts (high volume) is now being seriously questioned by researchers. For certain sports, it appears that training volume could be reduced significantly, possibly by as much as one-half in some sports, without reducing the benefits, and with less risk of overtraining athletes to the point of decreased performance. The principle of training specificity suggests that low-intensity, high-volume training does not improve sprint-type performance.

Training intensity is also an important factor and refers to both the relative force of muscle action (i.e., resistance training) and the relative stress placed on the metabolic and cardiovascular systems (i.e., anaerobic and aerobic training). There is a strong interaction between training intensity and training volume: As intensity is reduced, training volume must be increased to achieve adaptation. Training at very high intensities requires substantially less training volume, but the adaptations that occur will be significantly different from those achieved with low-intensity, high-volume training. This concept applies to all three types of training, that is, resistance, anaerobic, and aerobic.

High-intensity, low-volume training can be tolerated only for brief periods. While this type of training does increase muscular strength in resistance training and total body speed and anaerobic capacity in high-intensity interval training, it provides little or no improvement in aerobic capacity. Conversely, low-intensity, high-volume training stresses the oxygen transport and oxidative metabolic systems, causing greater gains in aerobic capacity, but has little or no effect on muscular strength, anaerobic capacity, or total body speed.

Attempts to perform large amounts of high-intensity training can have negative effects on adaptation. The energy needs of high-intensity exercise place greater demands on the glycolytic system, rapidly depleting muscle glycogen. If such training is attempted too often, for example daily, the muscles can become chronically depleted of their energy reserves, and the person might demonstrate signs of chronic fatigue or overtraining, as we discuss later in this chapter.

In review

➤ Optimal training involves following a model that incorporates the principles of periodization, because the body needs to systematically go through stages of undertraining, acute overload, and overreaching to maximize performance.

➤ Excessive training refers to training that is done with an unnecessarily high volume, intensity, or both. It provides little or no additional improvements in conditioning or performance and can lead to decreased performance and health problems.

➤ Training volume can be increased through increase in the duration or frequency of training bouts or both. Many studies have shown no significant differences in improvement between athletes who train with typical training volumes and those who train with twice the volume (training

> conducted for twice the duration or twice a day instead of once a day).
>
> ➤ Training intensity determines the specific adaptations that occur in response to the training stimulus. As training intensity increases, training volume must be reduced, and vice versa.

Overtraining

With overly intense training, athletes may experience an unexplained decline in performance and physiological function that extends over weeks, months, or years. This condition is termed overtraining, and the precise cause or causes of the resulting performance decrement are not fully understood. Research has pointed to both psychological and physiological causes. Furthermore, overtraining can occur with each of three major forms of training—resistance, anaerobic, and aerobic training—so it is likely that the cause(s) and symptoms will vary by the type of training.

All athletes experience some degree of fatigue during repeated days and weeks of training, so not all fatigue-producing situations can be classified as overtraining (as we noted previously with overreaching). Fatigue that follows one or more exhaustive training sessions is typically relieved by a few days of reduced training or rest and a proper calorie- and carbohydrate-rich diet. Overtraining, on the other hand, is characterized by a sudden decline in performance and physiological function that cannot be remedied by a few days of reduced training, rest, or dietary intervention.

Effects of Overtraining: The Overtraining Syndrome

Most of the symptoms that result from overtraining, collectively referred to as the overtraining syndrome, are subjective and identifiable only after the individual's performance and physiological function have suffered. Unfortunately, these symptoms can be highly individualized, making it very difficult for athletes, trainers, and coaches to recognize that performance decrements have indeed been caused by overtraining. A decline in physical performance with continued training is usually the first indication of the overtraining syndrome (see figure 14.4). The athlete senses a loss of muscular strength, coordination, and exercise capacity and generally feels fatigued. Other primary signs and symptoms of the overtraining syndrome include[1]

- change in appetite;
- body weight loss;
- sleep disturbances;
- irritability, restlessness, excitability, anxiousness;
- loss of motivation and vigor;
- lack of mental concentration;
- feelings of depression; and
- lack of appreciation for things—including exercise—that normally are enjoyable.

a Acute overload and overreaching

b Overtraining

FIGURE 14.4 Typical pattern of the expected improvement in performance with acute overload and overreaching in contrast to the pattern seen with overtraining.

Adapted, by permission, from M.L. O'Toole, 1998, Overreaching and overtraining in endurance athletes. In *Overtraining in sport,* edited by R.B. Kreider, A.C. Fry, and M.L. O'Toole (Champaign, IL: Human Kinetics), 10, 13.

The underlying causes of overtraining syndrome are often a complex combination of emotional and physiological factors. Hans Selye[19] noted that a person's stress tolerance can break down as often from a sudden increase in anxiety as from an increase in physical distress. The emotional demands of competition, the desire to win, the fear of failure, unrealistically high goals, and others' expectations can be sources of intolerable emotional stress. Because of this, overtraining is typically accompanied by a loss of competitive desire and a loss of enthusiasm for training. Furthermore, Armstrong and VanHeest[1] make the important observation that the overtraining syndrome and clinical depression involve remarkably similar signs and symptoms, brain structures, neurotransmitters, endocrine pathways, and immune responses, suggesting that they have similar etiologies.

The physiological factors responsible for the detrimental effects of overtraining are likewise not fully understood. However, research suggests that overtraining is associated with alterations in the nervous, endocrine, and immune systems. Although a cause-and-effect relationship between these changes and the symptoms of overtraining has not been clearly established, these symptoms can help determine whether an individual is overtrained. In the following discussion, we focus on some of the observed changes associated with overtraining and on potential causes of the overtraining syndrome.

Autonomic Nervous System Responses to Overtraining

Some studies suggest that overtraining is associated with abnormal responses of the autonomic nervous system. Physiological symptoms accompanying the decline in performance often reflect changes in those organs or systems that are controlled by either the sympathetic or parasympathetic branches of the autonomic nervous system (see chapter 3). Sympathetic nervous system abnormalities due to overtraining can lead to

- increased resting heart rate,
- increased blood pressure,
- loss of appetite,
- decreased body mass,
- sleep disturbances,
- emotional instability, and
- elevated basal metabolic rate.

This form of overtraining occurs predominantly among athletes who emphasize very high intensity resistance training methods.

Other studies suggest that the parasympathetic nervous system might be dominant in some cases of overtraining, usually in endurance athletes. In these cases, the performance decrements markedly differ from those associated with sympathetic overtraining. Signs of parasympathetic overtraining, assumed to be the result of high training volume overload, include

- early onset of fatigue,
- decreased resting heart rate,
- rapid heart rate recovery after exercise, and
- decreased resting blood pressure.

Thus, it appears that athletes in different sports or events will likely exhibit unique signs and symptoms of the overtraining syndrome that are related to their training regimens. In fact, some authorities have named these forms of overtraining "intensity related" and "volume related," recognizing that specific training stressors result in unique signs and symptoms when applied excessively.[13]

Some of the symptoms associated with autonomic nervous system overtraining are also seen in people who are not overtrained. For this reason, the presence of these symptoms cannot be used to confirm overtraining. Of the two conditions, symptoms of sympathetic overtraining are the most frequently observed.

Although there is not strong scientific evidence to support the autonomic nervous system overtraining theory, the autonomic nervous system definitely is affected by overtraining.

Hormonal Responses to Overtraining

Measurements of various blood hormone concentrations during periods of overreaching suggest that marked disturbances in endocrine function accompany excessive stress. As shown in figure 14.5, when swimmers increase their training 1.5- to 2-fold, blood concentrations of thyroxine and testosterone usually decrease and blood concentrations of cortisol increase. The ratio of testosterone to cortisol is thought to regulate anabolic processes in recovery, so a change in this ratio is considered an important indicator, and perhaps a cause, of the overtraining syndrome. Decreased testosterone coupled with increased cortisol might lead to more protein catabolism than anabolism in the cells. Other research, however, suggests that although cortisol concentrations increase with overreaching and the early stages of overtraining, both resting and exercise cortisol concentrations eventually decrease in the overtraining syndrome. Further, most overtraining studies have been conducted on aerobically trained endurance athletes. Fewer studies exist on anaerobically trained and resistance-trained athletes. Using the terminology introduced in the last section, intensity-related overtraining (anaerobic and resistance training) does not appear to alter resting hormonal concentrations.[13]

Overtrained athletes often have higher blood concentrations of urea because urea is produced by the breakdown of protein, that is, protein catabolism. This protein loss is the mechanism that is thought to be responsible for the weight loss seen in many overtrained athletes.

Elevated resting blood concentrations of epinephrine and norepinephrine have also been reported during periods of intensified aerobic training. These two hormones elevate heart rate and blood pressure. This has led some exercise physiologists to suggest that the blood concentrations of these catecholamines could be measured to confirm overtraining. However, other studies have found no change in these catecholamines during intensified training, and some have even found decreased resting concentrations.

Acute overload training and overreaching can produce most of the same hormonal changes reported in overtrained athletes. For this reason, measuring these and other hormones does not provide valid confirmation of overtraining. Athletes whose hormone concentrations appear to be abnormal may simply be experiencing the normal effects of training. Further, the time interval between the last training bout and the resting blood sample is very important. Some potential markers remain elevated for more than 24 h and might not reflect a true resting state. These hormonal changes simply might reflect the stress of training rather than a breakdown in the adaptive process. Consequently, experts generally agree that no blood marker conclusively defines the overtraining syndrome.

Armstrong and VanHeest[1] proposed that the various stressors associated with the overtraining syndrome act primarily through the hypothalamus. They postulated that these stressors activate the following two predominant hormonal axes involved in the body's response to stressors:

- The sympathetic–adrenal medullary axis (SAM), involving the sympathetic branch of the autonomic nervous system
- The hypothalamic-pituitary-adrenocortical axis (HPA)

This is illustrated in figure 14.6a. Figure 14.6b illustrates the brain and immune system interactions with these two axes. These two figures are quite complex and go well beyond the scope of an introductory-level exercise physiology text. However, a cursory study of the interactions depicted in these figures will give one

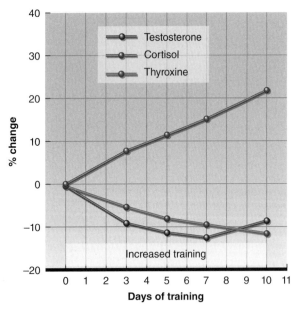

FIGURE 14.5 Changes in resting blood concentrations of testosterone, cortisol, and thyroxine over a period of intensified training. During the 10-day period shown here, swimmers increased their training from about 4,000 m to 8,000 m (4,374-8,749 yd) per day. These data show that resting cortisol concentrations increased in response to the added stress, whereas testosterone and thyroxine showed an unwanted decline during this period.

an appreciation of the complexity of this syndrome. Importantly, note that the stressors have their initial effect on the brain (hypothalamus). Thus, it is highly likely that brain neurotransmitters play an important role in the overtraining syndrome. Serotonin is a major neurotransmitter that is suspected to play a significant role in the overtraining syndrome. Unfortunately, plasma concentrations of this important neurotransmitter do not accurately reflect concentrations in the brain. Advances in technology should provide the necessary tools to help us better understand what is going on inside the brain.

A major role for cytokines in the overtraining syndrome recently has been proposed,[21] providing support for the Armstrong and VanHeest model in figure 14.6b. Elevated circulating cytokines result from infection as well as from skeletal muscle, bone, and joint trauma associated with overtraining. They appear to be a normal part of the body's inflammatory response to infection and injury. It is theorized that excessive musculoskeletal stress, coupled with insufficient rest and recovery, sets up a cascade of events whereby a local acute inflammatory response evolves into chronic inflammation and eventually into systemic inflammation. Systemic inflammation activates circulating monocytes, which can then synthesize large quantities of cytokines. Cytokines then act on most of the brain and body functions in a manner consistent with symptoms expressed in the overtraining syndrome.[21]

In focus

Overtraining syndrome appears to be associated with systemic inflammation and increased synthesis of cytokines. These changes are associated with depressed immune function and may place the athlete at an increased risk for infection and disease.

Immunity and Overtraining

The immune system provides a line of defense against invading bacteria, parasites, viruses, and tumor cells. This system depends on the actions of specialized cells (such as lymphocytes, granulocytes, and macrophages) and antibodies. These primarily eliminate or neutralize foreign invaders that might cause illness (pathogens). Unfortunately, one of the most serious consequences of overtraining is the negative effect it has on the body's immune system. In fact, from the model proposed in figure 14.6, compromised **immune function** is potentially a major factor in the initiation of the overtraining syndrome.

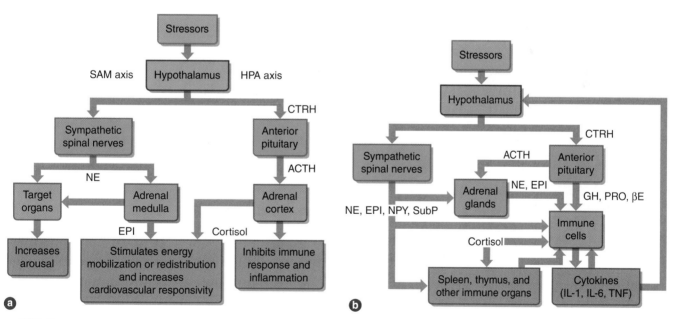

FIGURE 14.6 (a) The role of the hypothalamus and the sympathetic–adrenal medullary (SAM) and hypothalamic-pituitary-adrenocortical (HPA) axes as possible mediators of the overtraining syndrome. (b) The brain–immune system interactions with this model, with the cytokines playing a potentially major role in mediating overtraining. Symbols: ACTH = adrenocorticotropin; EPI = adrenaline (epinephrine); CTRH = corticotrophin releasing hormone; GH = growth hormone; IL-1 = interleukin-1; IL-6 = interleukin-6; NE = noradrenaline; SubP = substance P; PRO = prolactin; TNF = tumor necrosis factor; NPY = neuropeptide Y; βE = β-endorphin.

Adapted, by permission, from L.E. Armstrong and J.L. VanHeest, 2002, "The unknown mechanism of the overtraining syndrome," *Sports Medicine* 32: 185-209.

Many studies have shown that excessive training suppresses normal immune function, increasing the overtrained athlete's susceptibility to infections. This is illustrated in figure 14.7. Studies also show that short bouts of intense exercise can temporarily impair the immune response, and successive days of heavy training can amplify this suppression. Several investigators have reported an increased incidence of illness following a single, exhaustive exercise bout, such as running a full 42 km (26.2 mi) marathon. This immune suppression is characterized by abnormally low concentrations of both lymphocytes and antibodies. Invading organisms or substances are more likely to cause illness when these concentrations are low. Also, intense exercise during illness might decrease one's ability to fight off the infection and increase the risk of even greater complications.[17]

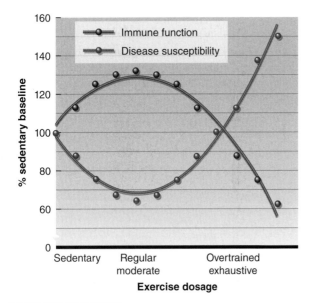

FIGURE 14.7 The inverted J-shaped model of the relationship between amount of exercise and immune function. This model suggests that moderate exercise may lower the risk of infection or disease, whereas overtraining may increase the risk.

Data from D.C. Nieman, 1997, "Immune response to heavy exertion," *Journal of Applied Physiology* 82: 1385-1394.

Predicting the Overtraining Syndrome

We must remember that the underlying cause or causes of the overtraining syndrome are not fully known, although it is likely that physical or psychological (or emotional) overload, or a combination of the two, might trigger this condition. Trying not to exceed an athlete's stress tolerance by regulating the amount of

physiological and psychological stress experienced during training is difficult. Most coaches and athletes use intuition to determine training volume and intensity, but few can accurately assess the true impact of a workout on the athlete. No preliminary symptoms warn athletes that they are on the verge of becoming overtrained. By the time coaches realize that they have pushed an athlete too hard, it is often too late. The damage done by repeated days of excessive training or stress can be repaired only by days, and in some cases weeks or months, of reduced training or complete rest.

Numerous investigators have tried to identify markers of the overtraining syndrome in its early stages by using assorted physiological and psychological measurements. A list of potential markers is provided in table 14.1. Unfortunately, none has proven totally effective. It is often difficult to determine whether the measurements obtained are related to overtraining or whether they simply reflect normal responses to overload or overreaching training.

One good method to identify the overtraining syndrome is to monitor the athlete's heart rate during a standardized workout, such as a fixed-paced run or swim, using a digital heart rate monitor (figure 9.5 on p. 221). The data presented in figure 14.8 illustrate a runner's heart rate response during a 1 mi (1.6 km) run performed at a fixed pace of 6 min/mi (3.7 min/km), or 10 mph (16 km/h). This response was monitored when the runner was untrained (UT), after the runner

FIGURE 14.8 A runner's heart rate responses during a fixed-pace treadmill run at 10 mph (16 km/h) performed before training (UT), after training (T), and when the runner showed symptoms of overtraining (OT).

TABLE 14.1 Potential Markers of Overreaching (OR), Overtraining (OT), and Overtraining Syndrome (OTS)

Physiological and psychological marker	Response	OR	OT	OTS
HR_{rest} and HR_{max}	Decreased		X	X
HR_{submax} and $\dot{V}O_{2submax}$	Increased	X		X
$\dot{V}O_{2max}$	Decreased			X
Anaerobic metabolism	Impaired		X	
Basal metabolic rate	Increased			X
RER_{submax} and RER_{max}	Decreased		X	X
Nitrogen balance	Negative			X
Nerve excitability	Increased			X
Sympathetic nervous response	Increased			X
Psychological mood states	Altered	X		
Risk of infection	Increased	X		
Hematocrit and hemoglobin	Decreased		X	
Leukocytes and immunophenotypes	Decreased		X	
Serum iron and ferritin	Decreased		X	
Serum electrolyte levels	Decreased			X
Serum glucose and free fatty acids	Decreased		X	
Plasma lactate concentration, submax, max	Decreased		X	X
Ammonia	Increased		X	X
Serum testosterone and cortisol	Decreased	X		
ACTH, growth hormone, prolactin	Decreased			X
Catecholamines, rest, night	Decreased			X
Creatine kinase	Increased			X

HR = heart rate; RER = respiratory exchange ratio; ACTH = adrenocorticotropic hormone.

Adapted from Armstrong and VanHeest, 2002.

had trained (TR), and during a period when the runner demonstrated symptoms of overtraining syndrome (OT). This figure shows that heart rate was higher when the runner was in the overtrained state than when the runner was responding well to training. Similar findings have been reported for swimmers.[5] Such a test provides a simple and objective way to monitor training and can possibly provide an early warning sign of overtraining.

In focus

Possibly the best predictor of the overtraining syndrome is the heart rate response to a standardized bout of exercise at a given submaximal intensity. Performance decrements are also good indicators but often appear later in the syndrome.

Overtraining, Chronic Fatigue, and Fibromyalgia Syndromes

Chronic fatigue syndrome is very similar to the overtraining syndrome.[20] There is likely considerable overlap between the two. Furthermore, there is considerable overlap between chronic fatigue syndrome and **fibromyalgia**. Chronic fatigue and fibromyalgia, however, also occur in nonathletes and in individuals who are not physically active. Other than that, there are many similarities in symptoms across the three syndromes. These similarities can include chronic fatigue at rest and during exercise, psychological distress, immune system dysfunction, hormonal dysfunction, HPA axis dysfunction, and neurotransmitter dysfunction. Furthermore, all three syndromes are difficult to diagnose, and generally the specific cause or causes remain unknown.

Reducing the Risk of and Recovering From the Overtraining Syndrome

Recovery from the overtraining syndrome is possible with a marked reduction in training intensity or complete rest. Although most coaches recommend a few days of easy training, overtrained athletes require considerably more time for full recovery. This might necessitate the total cessation of training for a period of weeks or months. In some cases, counseling might be needed to help the athletes cope with other stress in their lives that might contribute to this condition.

The best way to minimize the risk of overtraining is to follow periodization training procedures, alternating easy, moderate, and hard periods of training as discussed in chapter 9. Although individual tolerance varies tremendously, even the strongest athletes have periods when they are susceptible to the overtraining syndrome. As a rule, one or two days of intense training should be followed by an equal number of easy training days. Likewise, a week or two of hard training should be followed by a week of reduced effort with little or no emphasis on anaerobic exercise.

Endurance athletes (such as swimmers, cyclists, and runners) should pay particular attention to their caloric and carbohydrate intakes. Repeated days of hard training gradually reduce muscle glycogen. Unless these athletes consume extra carbohydrate during these periods, muscle and liver glycogen reserves can be depleted. As a consequence, the most heavily recruited muscle fibers are not able to generate the energy needed for exercise.

In review

➤ Overtraining stresses the body beyond its capacity to adapt, decreasing physiological capacity and performance.

➤ The symptoms of overtraining syndrome are subjective and vary from individual to individual; many also accompany regular training, making prevention or diagnosis of the overtraining syndrome difficult.

➤ Possible explanations for the overtraining syndrome include changes in function of the autonomic nervous system, altered endocrine responses, suppressed immune function, and altered brain neurotransmitters.

➤ Many potential signs and symptoms of overtraining have been proposed to diagnose overtraining in its earliest stages. However, at this time, the heart rate response to a fixed-pace exercise bout appears to be the easiest and most accurate technique.

➤ Overtraining syndrome is treated by a marked reduction in training intensity or complete rest, for weeks or months. Prevention can best be accomplished through use of periodization training procedures that vary training intensity and volume.

➤ For endurance athletes, it is important to ensure adequate calorie and carbohydrate intake to meet energy needs.

Tapering for Peak Performance

Peak performance requires maximal physical and psychological tolerance for the stress of the activity. But periods of intense training can reduce muscular strength, decreasing athletes' performance capacity. For this reason, to compete at their peak, many athletes reduce their training intensity and volume before a major competition to give their bodies a break from the rigors of intense training, a practice commonly referred to as **tapering**. The **taper period**, during which intensity and volume are reduced, provides adequate time for healing of tissue damage caused by intense training and for the body's energy reserves to be fully replenished. Tapering periods range from 4 to 28 days or longer,[16] depending on the sport, the event, and the athlete's needs. Tapering is not appropriate for all sports, particularly those in which competition occurs once a week or more frequently. Still, rested athletes generally perform better.

The most notable change during the taper period is a marked increase in muscular strength, which explains at least part of the performance improvement that occurs. It is difficult to determine whether strength improvements result from changes in the muscles' contractile mechanisms or improved muscle fiber recruitment. However, examination of individual muscle fibers taken from swimmers' arms before and after 10 days of intensified training showed that the type II (fast-twitch) fibers exhibited a significant reduction in their maximal shortening velocity.[8] This change has been attributed to changes in the fibers' myosin molecules. In these cases, the myosin in the type II fibers became more like that in the type I fibers. We assume from these data that such

Exertional Rhabdomyolysis

Rhabdomyolysis is an acute disease, which can be fatal, that is distinguished by the breakdown of skeletal muscle fibers. Exertional rhabdomyolysis refers to muscle fiber breakdown that occurs in response to unaccustomed, strenuous exercise. Normally, it is not dangerous and the symptoms are tolerable, similar to delayed-onset muscle soreness (see chapter 10). However, exertional rhabdomyolysis becomes clinically relevant when the muscle damage is severe. This usually occurs when proteins leak out of the damaged muscle and precipitate in the kidneys, causing acute renal failure and even death.

The most common signs and symptoms of rhabdomyolysis include

- severe muscle aching throughout the entire body,
- muscle weakness, and
- dark or cola-colored urine.

Clinically relevant exertional rhabdomyolysis (CRER) represents a relatively small percentage of all rhabdomyolysis cases. Clinically relevant exertional rhabdomyolysis is usually reported in case studies, since it is such a rare disease. The case studies that have been reported to date describe incidents that occurred during military training and during normal fitness training. Two cases reported in 2003 involved healthy adults who were encouraged to perform excessive strenuous exercise in a health club setting. Both developed symptoms and required hospitalization to prevent renal failure.[22] Symptoms include severe muscle pain and weakness, muscle swelling, and dark brown urine indicating the presence of myoglobin. In severe cases, patients may develop fever, leukocytosis, renal failure, and electrolyte abnormalities. Clinically relevant exertional rhabdomyolysis can be precipitated by high-intensity exercise, particularly excessive eccentric exercise, and is exacerbated by exercising in the heat or at altitude.[14] Particular caution must be taken when training unfit people not to push them beyond their comfort level. It is best to start at low intensities and volumes and gradually increase the training stress over weeks or months.

Rhabdomyolysis can be caused by intense exertion in combination with some drugs. Some people who take statin medications to lower their cholesterol may have severe muscle pain and be prone to rhabdomyolysis. Rhabdomyolysis can also be caused by the use of alcohol, heroin, or cocaine. Dehydration exacerbates the problem.

changes in the muscle fibers cause the power loss that swimmers and runners experience during prolonged periods of intense training. We can also assume that the recovery of strength and power that occurs with tapering is linked to modifications of the muscles' contractile mechanisms. Tapering also allows time for the muscle to repair any damage incurred during intense training and for the energy reserves (i.e., muscle and liver glycogen) to be restored.

In focus

In some sports, tapering for competition is crucial to optimal performance. Reduced training volume and intensity, coupled with quality rest, are needed to allow muscle to repair itself and to restore its energy reserves for competition.

Although tapering is widely practiced in a variety of sports, many coaches fear that reduced training for such a long period before a major competition will decrease conditioning and impair performance. But numerous studies clearly show that this fear is unwarranted. Developing optimal $\dot{V}O_{2max}$ initially requires a considerable amount of training, but once it has been developed, much less training is needed to maintain it at its highest level. In fact, the training level of $\dot{V}O_{2max}$ can be maintained even when training frequency is reduced by two-thirds.[10]

Runners and swimmers who reduce their training by about 60% for 15 to 21 days show no losses in $\dot{V}O_{2max}$ or endurance performance.[4, 11] One study showed that swimmers' blood lactate concentrations after a standard swim were lower after a taper period than before. More important, the swimmers experienced a 3% improvement in performance as a result of the reduced training and demonstrated an 18% to 25% increase in arm strength and power.[4]

In a study of distance runners, those runners who went through a seven-day taper decreased their running time in a 5 km time trial by 3% compared to no improvement in those who did not taper. Submaximal oxygen uptake during running at 80% $\dot{V}O_{2max}$ was decreased by 6% in those who tapered, indicating a greater economy of effort. Blood lactate concentrations at 80% of $\dot{V}O_{2max}$ were unchanged, as were $\dot{V}O_{2max}$ and leg extension peak force.[12]

Unfortunately, little information is available to demonstrate the influence of tapering on performance in team sports and in long-duration endurance events such as cycling and marathon running. Before guidelines can be offered for athletes in these sports, research is needed to demonstrate that similar benefits can be generated by such periods of reduced training.

In review

➤ Many athletes decrease their training intensity and volume before a competition to increase strength, power, and performance capacity. This practice is called tapering.

➤ The optimal duration of the taper is between 4 and 28 days, or longer, and is dependent on the sport or event and the athlete's needs.

➤ Muscular strength increases significantly during the tapering period.

➤ Tapering allows time for the muscle to repair any damage incurred during intense training and for the energy reserves (i.e., muscle and liver glycogen) to be restored.

➤ Less training is needed to maintain previous gains than was originally needed to attain them, so tapering does not decrease conditioning.

➤ Aerobic performance may improve by an average of about 3% with proper tapering.

Detraining

Tapering, by reducing the training stimulus, can facilitate performance. What happens to highly conditioned athletes who have fine-tuned their performance abilities to a peak level but then come to the end of their competitive season? Many athletes welcome the opportunity to completely relax, purposely avoiding any strenuous physical activity. But how does physical inactivity affect highly trained athletes from a physiological standpoint?

Detraining is defined as the partial or complete loss of training-induced adaptations in response to either the cessation of training or a substantial decrement in the training load—in contrast to tapering, which is a gradual reduction of the peak training load over only a few days to a few weeks. Some of our knowledge about detraining comes from clinical research with patients who have been forced into inactivity because of injury or surgery. Most athletes fear that all they have gained through hard training will be lost during even a brief period of inactivity. But recent studies reveal that a few days of rest or reduced training will not impair and might even enhance performance, similar to what we see with tapering. Yet at some point, training reduction or complete inactivity will decrease physiological function and performance.

Muscular Strength and Power

When a broken limb is immobilized in a rigid cast, changes begin almost immediately in both the bone and the muscles. Within only a few days, the cast that was applied tightly around the injured limb is loose. After several weeks, a large space separates the cast and the limb. Skeletal muscles undergo a substantial decrease in size, a process known as atrophy, when they remain inactive. Not surprisingly, muscle atrophy is accompanied by considerable loss of muscular strength and power. Total inactivity leads to rapid losses, but even prolonged periods of reduced activity lead to gradual losses that eventually can become quite significant.

Similarly, research confirms that muscular strength and power both are reduced once athletes stop train-

Detraining in Space

As astronauts orbit the earth, they are in an environment where the gravitational forces are considerably less than those on earth, that is, microgravity. While astronauts experience a sense of weightlessness in orbit, the gravitational forces of earth (i.e., 1 g) do not reach 0 g. During an extended stay in microgravity, astronauts undergo physiological changes that are nearly identical to those of detraining. But what could be perceived as maladaptations might in fact be necessary adaptations to microgravity. Let's briefly review the changes that take place when astronauts leave the 1 g environment on earth to spend weeks or months in space.

Muscle mass and strength decline in microgravity, particularly in the postural muscles, that is, those muscles that maintain the body upright countering the force of gravity. The cross-sectional area of both type I and type II muscle fibers is decreased as well. The extent of decline depends on the muscle group, the duration of the flight, and the type and extent of the in-flight exercise program. Microgravity also affects bone, with average bone mineral losses of about 4% in the weight-bearing bones; but the magnitude of this loss depends on the length of exposure to microgravity.

The cardiovascular system also undergoes major adaptations to microgravity. When the body is in microgravity, there is a reduction in hydrostatic pressure so that blood no longer pools in the lower extremities as it does at 1 g. As a consequence, more blood returns to the heart, which leads to transient increases in stroke volume. Reductions in plasma volume occur over time but are likely due to reduced fluid intake rather than increased production of urine (diuresis) by the kidneys. Transcapillary fluid shifts between the microcirculation and surrounding tissues can also account for some of the reduction in plasma volume, most likely upper body capillary filtration; for example, blood relocates into the facial tissues, creating a puffy facial appearance. Red cell mass also decreases, so total blood volume is reduced as well.[24] The reduced blood volume serves astronauts well while they remain in microgravity. But it presents a serious problem on their return to a 1 g environment, where the body is again subjected to the hydrostatic pressure effect. Astronauts have experienced postural (orthostatic) hypotension and fainting during their first few hours back in a normal 1 g environment because their blood volume was insufficient to meet all their circulatory needs.

Maximal aerobic power ($\dot{V}O_{2max}$) is generally reduced immediately postflight, likely due to the reduction in plasma volume and leg strength during flight. However, data are limited on directly measured $\dot{V}O_{2max}$ in astronauts preflight, during flight, and postflight. Head-down tilt (−6°) bed rest, used as an earth-based model of spaceflight, shows consistent reductions in $\dot{V}O_{2max}$ that are associated with reductions in total blood volume, plasma volume, and, consequently, maximal stroke volume. The head-down tilt model has been shown to provide $\dot{V}O_{2max}$ (or $\dot{V}O_{2peak}$) data comparable to actual pre- to postflight changes.[23]

Importantly, having an understanding of the general decline in physiological function that occurs during spaceflight has led the scientific and medical community to the realization that in-flight exercise programs are essential to preserve the long-term health of astronauts. Research is now under way to design the most appropriate programs and exercise equipment to meet this objective.

ing. The rate and magnitude of loss vary with the level of training. Highly skilled, accomplished weightlifters appear to have a rather rapid decline in strength within a few weeks of discontinuing intense training.[9] With previously untrained people, strength gains can be maintained from several weeks up to over seven months. In a study of young (20-30 years) and older (65-75 years) men and women who trained for nine weeks, the increase in strength (1-repetition maximum) averaged 34% for the younger subjects and 28% for the older subjects, with no differences between men and women. After 12 weeks of detraining, none of the four groups had significant losses in strength from the end of the nine-week training program values. After 31 weeks of detraining, there was only an 8% loss in the younger subjects and a 13% decrease in the older subjects.[15]

A study with collegiate swimmers revealed that even with up to four weeks of inactivity, terminating training did not affect arm or shoulder strength.[2] No strength changes were seen in these swimmers, whether they spent four weeks at complete rest or whether they reduced their training frequency to one or three ses-

sions per week. But swimming power was reduced by 8% to about 14% during the four weeks of reduced activity, whether the swimmers underwent complete rest or merely reduced training frequency. Although muscular strength might not have diminished during the four weeks of rest or reduced training, the swimmers might have lost their ability to apply force during swimming, possibly attributable to a loss of skill.

The physiological mechanisms responsible for the loss of muscular strength as a consequence of either immobilization or inactivity are not clearly understood. Muscle atrophy causes a noticeable decrease in muscle mass and water content, which could partly account for a loss in the development of maximal muscle fiber tension. Changes occur in the rates of protein synthesis and degradation as well as in specific fiber type characteristics. When muscles aren't used, the frequency of their neurological stimulation is reduced, and normal fiber recruitment is disrupted. Thus, part of the strength loss associated with detraining could result from an inability to activate some muscle fibers.

The retention of muscular strength, power, and size is extremely important for the injured athlete. The athlete can save much time and effort during rehabilitation by performing even a low level of exercise with the injured limb, starting in the first few days of recovery. Simple isometric contractions are very effective for rehabilitation because their intensity can be graded and they do not require joint movement. Any program of rehabilitation, however, must be designed in cooperation with the supervising physician and physical therapist.

Muscular Endurance

Muscular endurance decreases after only two weeks of inactivity. At this time, not enough evidence is available to determine whether this performance decrement results from changes in the muscle or from changes in cardiovascular capacity. In this section, we examine muscle changes that are known to accompany detraining and that could decrease muscular endurance.

The localized muscle adaptations that occur during periods of inactivity are well documented, but the exact role that these changes play in the loss of muscular endurance is unclear. We know from postsurgery cases that after a week or two of cast immobilization, the activities of oxidative enzymes such as succinate dehydrogenase (SDH) and cytochrome oxidase decrease by 40% to 60%. Data collected from swimmers, shown in figure 14.9, indicate that the muscles' oxidative potential decreases much more rapidly than the subjects' maximal oxygen uptake with detraining. Reduced oxidative enzyme activity would be expected to impair muscular endurance, and this most likely relates to submaximal endurance capacity rather than to maximal aerobic capacity, or $\dot{V}O_{2max}$.

In contrast, when athletes stop training, the muscles' glycolytic enzymes, such as phosphorylase and phosphofructokinase, change little, if at all, for at least four weeks. In fact, Coyle and colleagues[7] observed no change in glycolytic enzyme activities with up to 84 days of detraining compared with a nearly 60% decrease in the activities of various oxidative enzymes. This might at least partly explain why performance times in sprint events are unaffected by a month or more of inactivity, but the ability to perform longer endurance events may decrease significantly with as little as two weeks of detraining.

One notable change in muscle during detraining is a change in its glycogen content. Endurance-trained

FIGURE 14.10 Changes in glycogen content of the deltoid muscle in competitive swimmers during four weeks of detraining. Muscle glycogen returned almost to the untrained concentration at the end of this period.

muscle tends to increase glycogen storage. But four weeks of detraining has been shown to decrease muscle glycogen by 40%.[3] Figure 14.10 illustrates the decrease in muscle glycogen accompanying four weeks of detraining in competitive collegiate swimmers and in untrained subjects (serving as a time control). The untrained people showed no change in muscle glycogen content after four weeks of inactivity, but the swimmers' values, which were high initially, decreased until they were about equal to those of the untrained people. This indicates that the trained swimmers' improved capacity for muscle glycogen storage was reversed during detraining.

Measurements of blood lactate and pH after a standard work bout have been used to assess the physiological changes that accompany training and detraining. For example, a group of collegiate swimmers were required to perform a standard-paced, 200 yd (183 m) swim at 90% of their seasonal best following five months of training and then to repeat this test at the same absolute pace once a week for the following four weeks of detraining. The results are shown in table 14.2. Blood lactate concentrations, taken immediately after this standard swim, increased from week to week during a month of inactivity. At the end of the fourth week of detraining, the swimmers' acid–base balance was significantly disturbed. This was reflected by a significant increase in blood lactate concentrations and a

FIGURE 14.9 Percentage decreases in $\dot{V}O_{2max}$, muscle succinate dehydrogenase (SDH) activity, and cytochrome oxidase activity during six weeks of detraining. These findings suggest that muscles experience a decline in metabolic potential, although tests of $\dot{V}O_{2max}$ show little change over this period of detraining.

TABLE 14.2 Blood Lactate, pH, and Bicarbonate (HCO$_3^-$) Values in Collegiate Swimmers Undergoing Detraining

Measurement	WEEKS OF DETRAINING			
	0[a]	1[b]	2	4
Lactate (mmol/L)	4.2	6.3	6.8	9.7[c]
pH	7.26	7.24	7.24	7.18[c]
HCO$_3^-$ (mmol/L)	21.1	19.5[c]	16.1[c]	16.3[c]
Swim time (s)	130.6	130.1	130.5	130.0

Note. Measurements were taken immediately after a fixed-pace swim.

[a]The values at week 0 represent the measurements taken at the end of five months of training.

[b]The values for weeks 1, 2, and 4 are the results obtained after one, two, and four weeks of detraining, respectively.

[c]Significant difference from the value at the end of training.

significant decrease in the concentrations of bicarbonate (a buffer).

Speed, Agility, and Flexibility

Training produces less improvement in speed and agility than it does in strength, power, muscular endurance, flexibility, and cardiorespiratory endurance. Consequently, losses of speed and agility that occur with inactivity are relatively small, and peak speed and agility can be maintained with a limited amount of training. But this does not imply that the track sprinter can get by with training only a few days a week. Success in actual competition relies on factors other than basic speed and agility, such as correct form, skill, and the ability to generate a strong finishing sprint. Many hours of practice are required to tune performance to its optimal level, but most of this time is spent developing performance qualities other than speed and agility.

Flexibility, on the other hand, is lost rather quickly during inactivity. Stretching exercises should be incorporated into both in-season and off-season training programs. Reduced flexibility has been proposed to increase athletes' susceptibility to serious injury.

Cardiorespiratory Endurance

The heart, like other muscles in the body, is strengthened by endurance training. Inactivity, on the other hand, can substantially decondition the heart and the cardiovascular system. The most dramatic example of this is seen in a classic study conducted on subjects undergoing long periods of total bed rest; they weren't allowed to leave their beds, and physical activity was kept to an absolute minimum.[18] Cardiovascular and metabolic function were assessed at a constant submaximal rate of work and at maximal rates of work both before and after the 20-day period of bed rest. The cardiovascular effects that accompanied bed rest included

- a considerable increase in submaximal heart rate,
- a 25% decrease in submaximal stroke volume,
- a 25% reduction in maximal cardiac output, and
- a 27% decrease in maximal oxygen consumption.

The reductions in cardiac output and $\dot{V}O_{2max}$ appear to result from reduced stroke volume; this appears to be largely attributable to a decreased plasma volume, with reductions in heart volume and ventricular contractility playing a smaller role.

It is interesting that the two most highly conditioned subjects in this study (the two who had the highest $\dot{V}O_{2max}$ values) experienced greater decrements in $\dot{V}O_{2max}$ than the three less fit people, as shown in figure 14.11. Furthermore, the untrained subjects regained their initial conditioning levels (before bed rest) in the first 10 days of reconditioning, but the well-trained subjects needed about 40 days for full recovery. This suggests that highly trained individuals cannot afford long periods with little or no endurance training. The athlete who totally abstains from endurance training at the completion of the season will experience great difficulty getting back into top aerobic fitness when the new season begins.

Inactivity can significantly reduce $\dot{V}O_{2max}$. How much activity is needed to prevent such considerable losses of

FIGURE 14.11 Changes in $\dot{V}O_{2max}$ with 20 days of bed rest for five individual subjects. Note that the subjects who were least fit (lowest $\dot{V}O_{2max}$ values) at the start of bed rest showed smaller decrements with inactivity and greater gains when they trained after bed rest. Highly fit individuals, on the other hand, were far more affected by the period of inactivity.

Adapted, by permission, from B. Saltin et al., 1968, "Response to submaximal and maximal exercise after bed rest and training," *Circulation* 38(7): 75.

physical conditioning? Although a decrease in training frequency and duration reduces aerobic capacity, the losses are significant only when frequency and duration are reduced by two-thirds of the regular training load.

However, training intensity apparently plays a more crucial role in maintaining aerobic power during periods of reduced training. Training at 70% $\dot{V}O_{2max}$ appears to be necessary to maintain maximal aerobic capacity.[10]

In focus

The body rapidly loses many of the benefits of training if training is discontinued. Some minimal level of training is necessary to prevent these losses. Research indicates that at least three training sessions per week at an intensity of at least 70% $\dot{V}O_{2max}$ are needed to maintain aerobic conditioning.

In review

➤ Detraining is defined as the partial or complete loss of training-induced adaptations in response to either the cessation of training or a substantial decrement in the training load.

➤ Detraining causes muscle atrophy, which is accompanied by losses of muscular strength and endurance. However, muscles require only minimal stimulation to retain these qualities during periods of reduced activity.

➤ Muscular endurance decreases after only two weeks of inactivity. Possible explanations for this are
  1. decreased oxidative enzyme activity,
  2. decreased muscle glycogen storage, or
  3. disturbance of acid–base balance.

➤ Detraining-induced losses in speed and agility are small, but flexibility seems to be lost quickly.

➤ With detraining, losses of cardiorespiratory endurance are much greater than losses of muscular strength, power, and endurance over the same time period.

➤ To maintain cardiorespiratory endurance, training must be conducted at least three times per week, and training intensity should be at least 70% of $\dot{V}O_{2max}$.

In closing

In this chapter we have examined how the quantity of training can affect performance. We saw that too much training, in the form of either excessive training or overtraining, can actually impair performance. Then we looked at the effects of too little training—detraining—as a result of either inactivity or immobilization after an injury. Finally, we saw that with detraining, many of the gains achieved during regular training are quickly lost, especially cardiovascular endurance.

Now that we have dispelled the myth that more training always means better performance, in what other ways can athletes try to optimize their performance? In the next chapter, we turn our attention to optimal body composition and nutrition for the serious athlete.

Key Terms

acute overload

chronic fatigue syndrome

detraining

excessive training

fibromyalgia

immune function

overreaching

overtraining

overtraining syndrome

rhabdomyolysis

tapering

taper period

undertraining

Study Questions

1. Describe the model used to optimize training. Define the terms undertraining, acute overload, overreaching, and overtraining.

2. What is excessive training? How does it relate to the model for optimizing training?

3. Define and describe the overtraining syndrome. What are the general symptoms of the overtraining syndrome? How do these differ between sympathetic and parasympathetic overtraining?

4. How might the hypothalamus be involved in the overtraining syndrome? What role might cytokines play?

5. Describe the relationship between physical activity and immune function and disease susceptibility.

6. What appears to be the best predictor of the overtraining syndrome?

7. How do we treat the overtraining syndrome?

8. What physiological changes occur during the taper period that can be credited with improvements in performance?

9. What alterations occur in strength, power, and muscular endurance with physical detraining?

10. What alterations occur in speed, agility, and flexibility with physical detraining?

11. What changes occur in cardiovascular function as one becomes deconditioned?

12. What similarities do we see between spaceflight and detraining? Why does the body make these adaptations during spaceflight?

Study Guide Activities

In addition to the activities listed in the chapter opening outline on page 333, two other activities are available in the web study guide, located at

www.HumanKinetics.com/PhysiologyOfSportAndExercise

 The **KEY TERMS** activity reviews important terms, and the end-of-chapter **QUIZ** tests your understanding of the material covered in the chapter.

Body Composition and Nutrition for Sport

15

In this chapter and in the web study guide

ACTIVITY 15.1   Assessing Body Composition looks at several methods, or tests, for assessing body composition.

ACTIVITY 15.2   Body Composition and Sport Performance reviews how body composition affects sport performance.

ACTIVITY 15.3   Weight Standards reviews some appropriate uses of weight standards.

ACTIVITY 15.4   Nutrients explores the primary functions of key nutrients.

ACTIVITY 15.5   Water and Electrolyte Balance explores how the body gains and loses water at rest and during exercise.

ACTIVITY 15.6   The Athlete's Diet is a case study that examines the effectiveness of an athlete's diet.

A former Major League Baseball player made minimum salary during his first few years in the majors. Early preseason polls projected his team to finish the season at the bottom of its division, but the team ended up in the World Series. This player became one of the best at his position in the National League during that season, and once the World Series was over, he asked management for a substantial salary increase ($75,000 in the mid-1970s). Management agreed to his salary demands contingent on his loss of 25 lb (11 kg)! The player refused to lose the weight, so the parties were deadlocked.

The team physician suggested sending the player to a major university for an accurate body composition assessment, and both parties agreed. A hydrostatic weighing was performed, and the results showed that the player had less than 6% body fat, with a total of only 11 lb (5 kg) of fat! Because 3% to 4% body fat is necessary for survival, this player had only 4 to 5 lb (about 2 kg) of fat to lose, and that loss wasn't advised because he was already at the lower range of acceptable body fat levels recommended for athletes. Management was satisfied; this player received his salary increase, and he did not have to lose weight.

This athlete's weight was well above the weight range recommended for his height, and he had a peculiar gait commonly referred to as a waddle. The combination of being overweight by the standard height–weight charts and having a waddle led management to demand the 25 lb (11 kg) weight loss. Had the athlete agreed to management's demand, he would have likely destroyed his career as a professional athlete. How many athletes have been faced with a similar situation? How many gave in?

Coaches and athletes today are acutely aware of the importance of achieving and maintaining optimal body weight for peak performance in sport. Appropriate size, build, and body composition are critical to success in almost all athletic endeavors. Compare the specific performance requirements of the 152 cm, 45 kg (5 ft, 100 lb) Olympic gymnast and those of the 206 cm, 147 kg (6 ft 9 in., 325 lb) defensive lineman in professional football. Although size and body build can be altered only slightly, body composition can change substantially with dieting and exercise. Resistance training can substantially increase muscle mass, and sound dieting combined with vigorous exercise can significantly decrease body fat. Such changes can be of major importance in achieving optimal athletic performance.

Peak performance also requires a careful dietary balance of the essential nutrients. The U.S. government has established standards for optimal nutrient intake that are termed Dietary Reference Intakes (DRIs). The DRIs provide estimates of the range of intakes of various food substances needed to maintain good health.

The nutritional needs of very active athletes can exceed the DRIs considerably. Individual caloric needs are quite variable, depending on the athlete's size, sex, and sport choice. Cyclists competing in the Tour de France and Norwegian cross-country skiers during training have been reported to expend up to 9,000 kcal per day. One ultra-endurance runner expended an average of 10,750 kcal per day over a 5.2-day 600 mi (966 km) race![20] Also, some competitive sports require adherence to rigid weight standards. Athletes who participate in these sports must closely monitor their weight and thus their caloric intake. Too often, this leads to nutritional abuses, drug use, dehydration, and serious health risks. In addition, the dietary tactics used by some athletes to achieve excessive weight loss are of increasing concern because of the potential association with eating disorders, such as anorexia nervosa and bulimia nervosa.

Body Composition in Sport

Body composition refers to the body's chemical composition. Figure 15.1 illustrates three models of body composition. The first two divide the body into its

Chemical model   **Anatomical model**   **Two-compartment model**

 Three models of body composition (see text for description).

Adapted, by permission, from J.H. Wilmore, 1992, Body weight and body composition. In *Eating, body weight, and performance in athletes: Disorders of modern society*, edited by K.D. Brownell, J. Rodin, and J.H. Wilmore (Philadelphia, PA: Lippincott, Williams, and Wilkins), 77-93.

various chemical or anatomical components; the last one simplifies body composition into two components, the fat mass and the fat-free mass, which is the model used in this book. **Fat mass** is often discussed in terms of **relative body fat**, which is the percentage of the total body mass that is composed of fat. **Fat-free mass** simply refers to all body tissue that is not fat.

In focus

The fat-free mass (FFM) is composed of all of the body's nonfat tissue, including bone, muscle, organs, and connective tissue.

Assessing Body Composition

Assessment of body composition provides additional information beyond the basic measures of height and weight to both the coach and the athlete. As an example, if the center fielder of a Major League Baseball team is 188 cm (6 ft 2 in.) tall and weighs 91 kg (200 lb), is he at his ideal playing weight? Knowing that 4.5 kg (10 lb) of a total weight of 91 kg (200 lb) is fat weight and that the remaining 86.5 kg (190 lb) is fat-free weight provides considerably more insight than knowing height and weight alone. In this example, only 5% of his body weight is fat, which is about as low as any athlete should go, as discussed in our chapter opening. Armed with this knowledge, both athlete and coach realize that this athlete's body composition is ideal. They should not be concerned with weight loss, even though standard height–weight charts indicate that the athlete is overweight. However, another baseball player of the same height and weight who has 23 kg (50 lb) of fat would be 25% fat. This would constitute a serious weight problem for an elite athlete: He would be overfat.

In most sports, the higher the percentage of body fat, the poorer the performance. An accurate assessment of the athlete's body composition provides valuable insight into the weight that allows optimal performance. But how do we determine an athlete's body composition?

Densitometry

Densitometry involves measuring the density of the athlete's body. Density (D) is defined as mass (M) divided by volume (V):

$$D_{body} = M_{body} / V_{body}.$$

The mass of the body is the athlete's scale weight. Body volume can be obtained by several different tech-

In focus

Although total body size and weight are important for most athletes, an athlete's body composition is generally of greater concern. Standard height–weight tables do not provide accurate estimates of what an athlete should weigh because they do not take into account the composition of the weight. An athlete can be overweight according to these tables yet have very little body fat.

niques, but the most common is **hydrostatic weighing**, also called underwater weighing, in which the athlete is weighed while totally immersed in water. The difference between the athlete's scale weight and underwater weight, when corrected for the density of water, equals the body's volume. This volume must be further corrected to account for air trapped in the body. The amount of air trapped in the gastrointestinal tract is small, difficult to measure, and usually ignored. The gas trapped in the lungs, however, must be measured because its volume is generally large.

Figure 15.2 shows the hydrostatic weighing technique. The density of the fat-free mass is higher than the density of water, while that of fat is lower than the density of water. Observe a swimming pool full of people of varying body types. Those who have an abundance of body fat have a low total body density and can easily float, while those who are very lean have a higher total body density and tend to sink. This is an oversimplification but hopefully helpful for understanding the concept.

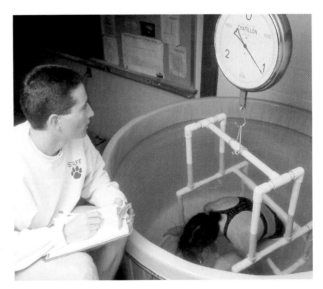

FIGURE 15.2 Hydrostatic weighing (the underwater weighing technique) to determine the density of the body.

Densitometry has long been the technique of choice for assessing body composition. New techniques typically are compared against densitometry to determine their accuracy. However, densitometry has its limitations. If body weight, underwater weight, and lung volume during underwater weighing are measured correctly, the resulting **body density** value is accurate. Densitometry's major weakness is in the conversion of body density to an estimate of relative body fat. Accurate estimates of the individual densities of fat mass and fat-free mass are required when the two-component model of body composition is used. The equation most often used to convert body density to an estimate of relative or percentage body fat is the standard equation of Siri:

$$\% \text{ body fat} = (495 \, / \, D_{body}) - 450.$$

This equation assumes that the densities of the fat mass and the fat-free mass are relatively constant in all people. Indeed, the density of fat at different sites is very consistent in the same individual and relatively consistent between people. The value generally used is 0.9007 g/cm. But determining the density of the fat-free mass (D_{FFM}), which the equation of Siri assumes is 1.100, is more problematic. To determine this density, we must make two assumptions:

1. The density of each tissue constituting the fat-free mass is known and remains constant.

2. Each tissue type represents a constant proportion of the fat-free mass (e.g., we assume that bone always represents 17% of the fat-free mass).

Exceptions to either of these assumptions cause error when we convert body density to relative body fat. Unfortunately, the density of the fat-free mass does vary between people.

In focus

Densitometry is the most accurate commonly used method for body composition assessment. Small inaccuracies inherent in densitometry largely reflect the variation in the density of the fat-free mass from one individual to another.

Other Laboratory Techniques

Many other laboratory techniques are available for assessing body composition. These include radiography, computed tomography (CT), magnetic resonance imaging (MRI), hydrometry (for measuring total body water), total body electrical conductivity, and neutron activation. Most of these techniques are complex and require expensive equipment. None of them is likely to be used for the general assessment of athletic populations, so we will not discuss them further in this chapter; however, we do discuss CT in chapter 22. Two other techniques hold considerable promise—dual-energy X-ray absorptiometry and air displacement.

Dual-energy X-ray absorptiometry (DEXA) evolved from the earlier single- and dual-photon absorptiometry techniques used between 1963 and 1984. The earlier techniques were used to estimate regional bone mineral content and bone mineral density, primarily in the spine, pelvis, and femur. The newer DEXA technique (see figure 15.3) allows the quantification of not only bone but also soft tissue composition. Furthermore, it

a

b

FIGURE 15.3 The dual-energy X-ray absorptiometry (DEXA) machine used to estimate bone density and bone mineral content as well as total body composition (fat mass and fat-free mass): (a) the machine, (b) a regional scan of the body.

is not limited to regional estimates but can provide total body estimates. Research to date suggests that DEXA provides precise and reliable estimates of body composition. The advantages of DEXA over the underwater weighing technique include the ability to estimate bone density and bone mineral content in addition to fat and fat-free mass. Furthermore, it is a passive technique whereby the athlete simply lies on a table during the scan, as opposed to having to be submerged underwater multiple times. The disadvantage is the cost of the equipment and technical support.

Air plethysmography is a densitometric technique. Volume is determined by air displacement rather than by water immersion. This technique, developed in the early 1900s, was used largely in research laboratories up until the 1990s, when a commercial model became widely available (see figure 15.4). The principle of

FIGURE 15.4 The Bod Pod air plethysmography device uses the air displacement technique to estimate total body volume.

operation is rather simple. It involves a closed chamber of room air at atmospheric pressure, which has a known volume. The individual to be tested opens the chamber door, enters the chamber, sits in a fixed position, and then closes the chamber door, forming an airtight seal. The new volume of the air in the chamber is determined, which is then subtracted from the total volume of the chamber to provide an estimate of the person's volume.

Although this is a relatively simple technique for the subject, it requires considerable accuracy in controlling for changes in temperature, gas composition, and the subject's breathing while in the chamber. Studies have confirmed the accuracy of this technique under most conditions. It appears to provide a relatively precise measurement of body volume. Just as with the underwater weighing technique, one can obtain relatively accurate measurements of total body volume and thus obtain accurate estimates of total body density. However, one still must use the subject's body density in an equation to estimate relative body fat, recognizing the uncertainties of the D_{FFM} for that subject.

Field Techniques

Several field techniques are also available for assessing body composition. These techniques are more accessible than laboratory techniques because the equipment is less costly and cumbersome; they are techniques that therefore can be used more easily by the coach, the trainer, or even the athlete, outside the laboratory.

Skinfold Fat Thickness The most widely applied field technique involves measuring the **skinfold fat thickness** (see figure 15.5) at one or more sites and using the values obtained to estimate body composition. It generally is recommended that the sum of the measurements from three or more skinfold sites be used in a quadratic, curvilinear equation to estimate body density. A curvilinear equation more accurately describes the relationship between the sum of skinfold measurements and body density. Linear equations underestimate the density of lean people, which causes overestimation of body fat. Just the opposite happens for obese people: Body density is overestimated, and body fat is underestimated. Skinfold fat thickness measurements that use quadratic equations provide reasonably accurate estimates of total body fat or relative fat.

Bioelectric Impedance **Bioelectric impedance** is a simple procedure introduced in the 1980s that takes just a few minutes to perform. Four electrodes are attached to the body, at the ankle, the foot, the wrist, and the back of the hand, as shown in figure 15.6. An undetectable current is passed through the distal electrodes (hand and foot). The proximal electrodes (wrist and

FIGURE 15.5 Measuring skinfold fat thickness at the triceps skinfold site.

ankle) receive the current flow. Electrical conduction through the tissues between the electrodes depends on the water and electrolyte distribution in that tissue. Fat-free mass contains almost all the body water and the conducting electrolytes, so conductivity is much greater in the fat-free mass than in the fat mass. The fat mass has a much greater impedance, meaning that it is much more difficult for the current to flow through the fat mass. Thus, the amount of current flow through the tissues reflects the relative amount of fat contained in those tissues.

With the bioelectric impedance technique, measurements of the impedance, the conductivity, or both are transformed into estimates of relative body fat. Estimates of relative body fat based on bioelectric impedance highly correlate with body fat measurements obtained through hydrostatic weighing. However, the relative body fat in lean athletic populations tends to be overestimated with bioelectric impedance because of the nature of the equations used. Furthermore, hydration alters the bioimpedance and therefore must be tightly controlled. More sport-specific equations are being developed; and a newer technique, multi-frequency bioelectric impedance spectroscopy, could possibly improve the accuracy of measurement in these lean athletic populations.

Body Composition and Sport Performance

Many athletes in certain sports (e.g., football and basketball) believe that they must be big to be good in their sport because size traditionally has been associated with performance quality: The bigger the athlete, the better the performance. But big does not always mean better. In certain other sports, smaller and lighter are considered better for performance (e.g., gymnastics, figure skating, and diving). Yet this can be taken to extremes, compromising the athlete's health and performance. In the following sections, we consider how performance can be affected by body composition.

FIGURE 15.6 The bioelectric impedance technique for assessing relative body fat.

Fat-Free Mass

Rather than be concerned with total body size or weight, most athletes should be concerned specifically with fat-free mass. Maximizing fat-free mass is desirable for athletes involved in activities that require strength, power, and muscular endurance. But increased fat-free mass is likely to be undesirable for the endurance athlete, such as a distance runner, who must move his or her total body mass horizontally for extended periods. A higher fat-free mass is an additional load that must be carried and might impair the athlete's performance. This might also be true for the high jumper, long jumper, triple jumper, and pole-vaulter, who must maximize their vertical or horizontal distances or both. Additional weight, even though it is active fat-free mass, could decrease rather than facilitate performance in these events.

Techniques eventually will be available to estimate not only athletes' fat mass and fat-free mass but also the potential for increasing their fat-free mass. Such techniques would allow athletes to design training programs that would develop their fat-free mass to this projected maximum while maintaining their fat mass at relatively low levels. Combining resistance training with the ingestion of carbohydrate, or carbohydrate and protein, during recovery from resistance training appears to be effective for increasing the fat-free mass.[17, 30] This routine appears to stimulate the release of the anabolic hormones.

Relative Body Fat

Relative body fat is a major concern of athletes. Adding more fat to the body just to increase the athlete's weight and overall size is generally detrimental to performance. Many studies have shown that the higher the percentage of body fat, above optimal values, the poorer the person's performance. This is true of all activities in which the body weight must be moved through space, such as running and jumping. It is less important for more stationary activities, such as archery and shooting. In general, leaner athletes perform better.

Endurance athletes try to minimize their fat stores because excess weight has been proven to impair their performance. Both absolute fat and relative body fat can profoundly influence running performance in highly trained distance runners. Less fat generally leads to better performance.

Heavyweight weightlifters might be exceptions to the general rule that less fat is better. These athletes add large amounts of fat weight just before competition under the premise that the additional weight will lower their center of gravity and give them a greater mechanical advantage in lifting. The sumo wrestler is another notable exception to the theory that overall size is not the major determinant of athletic success. In this sport, the larger individual has a decided advantage; but even so, the wrestler with the higher fat-free mass should have the best overall success. Performance in swimming also seems to be an exception to this general rule. Body fat might provide some advantage to the swimmer because it improves buoyancy, which can reduce body drag in the water and reduce the metabolic cost of staying on the surface.

In review

➤ Knowing a person's body composition is more valuable for predicting performance potential than merely knowing height and weight.

➤ Densitometry is one of the best methods for assessing body composition and has long been considered the most accurate, although it does carry certain risks of error. It involves calculating the density of the athlete's body by dividing body mass by body volume, which is typically determined via hydrostatic weighing or air displacement. Body composition can be calculated, although there is some margin of error.

➤ Dual-energy X-ray absorptiometry, originally developed for estimating bone density and bone mineral content, is now capable of providing accurate estimates of not only total body composition—fat mass and fat-free mass—but also segmental body composition and bone mass.

➤ Field techniques for assessing body composition include measuring skinfold fat thickness and bioelectric impedance. These techniques are less costly and more accessible to the athlete and the coach than are laboratory techniques.

➤ Maximizing fat-free mass is desirable for athletes in sports that require strength, power, and muscular endurance but could be a hindrance to endurance athletes, who must be able to move their total body mass for extended periods, and jumpers, who must move their body mass vertically or horizontally for distance.

➤ The degree of fatness has more influence on performance than does total body weight. In general, the greater the relative body fat, the poorer the performance. Possible exceptions include heavyweight weightlifters, sumo wrestlers, and swimmers.

Weight Standards

Weight standards have been used in several sports for many years. More recently, their use has become more widespread, and most sports have now adopted weight standards intended to ensure that athletes have the optimal body size and composition for maximal performance. Unfortunately, this is not always the result.

The elite athlete has long been esteemed as representing the most desirable physical and physiological characteristics for performance in a sport or activity. Theoretically, the elite athlete's genetic foundation and years of intense training have combined to provide the ultimate athletic profile for the given sport. These elite athletes set the standards toward which others aspire. However, this can be misleading as we see in figure 15.7.[28] The figure displays the percentage body fat values of elite female track and field athletes. If we look just at the distance runners, many of the best were below 12% body fat. The two top distance runners had only about 6% fat. One of these had won six consecutive international cross-country championships, and the other held what was then the best time in the world for the marathon. From these results, we could be tempted to suggest that any female distance runner should have between 6% and 12% relative body fat if she has world-class aspirations. However, one of the best distance runners in the United States at that time, who was within two years of taking over the top spot, had a relative body fat of 17%. Furthermore, one of the women in this study had a relative body fat of 37%, and she set the best time in the world for the 50 mi (80 km) run within six months of her evaluation! More than likely, neither of these women would have gained an advantage if she had been forced to decrease her weight to achieve 12% body fat or lower.

Inappropriate Use of Weight Standards

Weight standards have been seriously abused. Coaches have seen that athletes' performances generally improve as body weight decreases. This has led some coaches to adopt the philosophy that if small weight losses improve performance a little, then major weight losses should improve it even more. Not only coaches are guilty of making this assumption: Athletes and their parents also are drawn into this way of thinking. As an example, a university athlete, considered one of the best in the United States in her sport, had dieted and exercised down to such a low body weight that her relative body fat was less than 5%. If anyone who joined the team appeared to be leaner, she would work even harder to reduce her weight and fat content. This woman's athletic performance began to deteriorate, and she started to develop injuries that never seemed to heal. She was eventually diagnosed with anorexia nervosa (chapter 19) and underwent professional treatment. But her career as an elite athlete was over.

In focus

Many sports enforce weight standards with the goal of ensuring that the athletes are of optimal body size for participation. Unfortunately, athletes often turn to questionable, ineffective, or even dangerous methods of weight loss to reach their weight goal.

Risks With Severe Weight Loss

Many schools, districts, or state- and national-level organizations organize sports (e.g., wrestling) on the basis of size, with weight as the predominant factor. Athletes in these sports often attempt to achieve the lowest possible weight to gain an advantage over opponents. In so doing, many have jeopardized their health. In the following sections, we examine a few of the consequences of severe weight loss in athletes, both male and female.

Dehydration Fasting or very low calorie diets lead to large amounts of weight loss, primarily through dehydration. As we discuss later in this chapter, for every gram of carbohydrate stored, there is an obligatory gain of 2.6 g of water. When carbohydrates are

FIGURE 15.7 Relative body fat in elite female track and field athletes (see text for explanation).

Data from Wilmore et al., 1977.

used for energy, that water is lost. Thus, with fasting and very low calorie diets, carbohydrate stores are substantially depleted during the first few days. This results in a significant loss of weight attributable to the loss of body water.

Furthermore, athletes trying to make weight might exercise in rubberized sweat suits, sit in steam and sauna baths, chew on towels to lose saliva, and minimize their fluid intake. Such severe water losses compromise kidney and cardiovascular function and are potentially dangerous. Losses of as little as 2% of the athlete's weight through dehydration can impair performance, even impair skilled performance in short-duration high-intensity sports like tennis, soccer, and basketball.[12]

Chronic Fatigue   Pushing body weight too low can have major repercussions. When weight drops below a certain optimal level, the athlete is likely to experience performance decrements and increased incidence of illness and injury. The performance decrements can be attributable to many factors, including chronic fatigue that often accompanies major weight losses. The causes of this fatigue have not been established, but there are several likely possibilities.

The symptoms of an athlete who is chronically underweight (below optimal competitive weight) mimic those seen with overtraining (chapter 14). Both neural and hormonal components are involved in the phenomenon of overtraining. In some cases, the sympathetic nervous system appears to be inhibited, and

the parasympathetic system dominates. In addition, the hypothalamus does not function normally, and immune function is likely impaired. These alterations lead to a cascade of symptoms that include chronic fatigue.

This chronic fatigue also could be attributed to substrate depletion. Energy for almost all athletic activities is derived predominantly from carbohydrate. Carbohydrate also represents the smallest source of stored energy. The combined carbohydrate storage in muscle, liver, and extracellular fluid accounts for approximately 2,500 kcal of stored energy. When athletes are training hard and are not eating an adequate diet (when they are deficient either in total calories or in total carbohydrate calories), their carbohydrate energy stores become depleted. Most important to the athlete, liver and muscle glycogen levels decrease, which in turn reduces blood glucose levels. The combined effect of these decreases can be chronic fatigue and considerable declines in performance. In addition, under these conditions the body also uses its protein stores as an energy substrate for exercise. This can, over time, gradually deplete muscle protein.

During the 1990s, we started to see athletes diagnosed with chronic fatigue syndrome. This syndrome might or might not be related to what we have termed chronic fatigue. At this time, very little is known about chronic fatigue syndrome, although it does appear to involve immune system dysfunction (see chapter 14). Patients have incapacitating fatigue, and the symptoms may vary in severity over time but generally last for months or years. The symptoms include prolonged, debilitating fatigue; sore throat; muscle tenderness or pain (myalgia); and cognitive dysfunction.

Eating Disorders The constant attention given to achieving and maintaining a prescribed weight goal, particularly if the weight goal is inappropriate, can lead to disordered eating. A high proportion of athletes, predominantly females, have disordered eating. This term can simply refer to restricting food intake to levels that are well below energy expenditure, but disordered eating can also involve pathological behaviors to control body weight, such as self-induced vomiting and laxative abuse. Disordered eating can lead to clinical eating disorders, such as anorexia nervosa or bulimia nervosa. These disorders have become prevalent among female athletes. Each has strict criteria for diagnosis that set it apart from disordered eating in general.

More than 90% of individuals with eating disorders are young girls and women. Among athletes, those in appearance and lean physique sports (such as gymnastics, figure skating, diving, and dancing) and in endurance sports (such as running and swimming) seem at greatest risk.[25] On certain teams, particularly in lean physique and endurance sports, the prevalence of eating disorders might approach or even exceed 50% at the elite or world-class level.[26] Athletes and coaches must realize the potential link between weight standards and eating disorders. The issue of eating disorders and disordered eating in athletes is a major focus of chapter 19.

A female athlete who is prone to disordered eating is vulnerable to several medical problems that are interrelated, including low energy availability, menstrual dysfunction, and bone mineral loss.[1] This group of medical problems is now referred to as the **Female Athlete Triad**. This topic is discussed in much more detail in chapter 19.

In focus

Disordered eating appears to be more prevalent among female athletes, particularly those engaged in lean physique sports like cross-country running, figure skating, gymnastics, and ballet, than in the general population.[1, 25] It is important to incorporate discussion of nutritional requirements to meet the energy needs associated with exercise training in an effort to maximize performance and minimize the risk of disordered eating.

Menstrual Dysfunction Menstrual dysfunction, or abnormal menstrual patterns and cycles, is often observed in female athletes.[10] A high prevalence of oligomenorrhea (irregular and inconsistent cycle lengths of 36-90 days), amenorrhea (cessation of menstrual cycles for three months), and delayed menarche (first period) has been associated with sports that emphasize low body weight or low body fat.[11] The combination of caloric restriction and a vegetarian diet is also common among female athletes.[1]

The cause of these menstrual disturbances is the failure of the athlete to consume an adequate number of calories to meet the energy expenditure needs associated with exercise training.[1] As a result, the athlete is in a state of energy deficiency or low energy availability. Menstrual dysfunction is the body's adaptation to an energy deficit, in which energy is shunted away from growth and reproduction to support more essential processes like thermoregulation, immune function, and cell maintenance.[27] This is discussed more fully in chapter 19.

A strong link exists between anorexia nervosa and menstrual dysfunction. In fact, amenorrhea is one of the strict criteria necessary for the diagnosis of anorexia nervosa in girls and women. A similar relationship has not yet been established with bulimia (see chapter 19), but an increasing number of athletes are found to be both bulimic and amenorrheic.

Bone Mineral Disorders Bone mineral disorders are recognized as a potentially serious consequence of menstrual dysfunction. The link between the two was first reported in 1984. Now a number of scientists are researching the relationship between athletic amenorrhea and low bone mineral content or density. Past studies suggested that bone density increases with the resumption of normal menses (menstruation), but more recent observations suggest that the amount of bone that is regained might be limited and that bone density might remain well below normal even with reestablishment of normal menstrual function. The long-term consequences of a chronically low bone density in athletic populations with respect to fracture risk have not yet been established; but certainly in anorexic women, the fracture rate is seven times that observed in normal women.[21]

Establishing Appropriate Weight Standards

The potential for abuse of weight standards is clearly established. If standards are not set appropriately, athletes could be pushed well below optimal body weight. Thus, it is critically important to properly set weight standards.

Body weight standards should be based on an athlete's body composition. Thus, establishing weight standards should translate into establishing standards of relative body fat for each sport and, where appropriate, for each event within a sport. With this in mind, what is the recommended relative body fat for an elite athlete in any given sport? For each sport, an optimal range of values for relative body fat should be established, outside of which the athlete's performance is likely impaired. And because fat distribution shows definite sex differences, the weight standards should be sex specific. Representative ranges for men and women in various sports are presented in table 15.1. In most cases, these values represent the elite athletes in those sports.

The recommended values might not be appropriate for all athletes who engage in a specific activity. The existing techniques for measuring body composition include inherent errors, as we discussed earlier. Even with the better laboratory techniques, measurement of body density can introduce a 1% to 3% error, and an even greater error is associated with converting that density to relative body fat. In addition, we must understand the concept of individual variability. Not every female distance runner will have her best performance at 12% body fat or lower. While some will improve performance with these low values, others won't be able to get down to such low relative fat values, or they will find that their performance starts to decline before they reach the suggested values. For these reasons, a range of values should be set for males and females in specific activities, recognizing individual variability, methodological error, and sex differences.

Achieving Optimal Weight

Many athletes discover that they are considerably above their assigned playing weight with only a few weeks remaining before they report to training camp. Consider a 25-year-old professional basketball player who realizes that his weight is 9 kg (20 lb) above his playing weight during the previous season. He must lose this excess weight by the start of the preseason training camp, a mere four weeks away. Failure to do so will cost him a fine of $1,000 per day for each pound

TABLE 15.1 Ranges of Relative Body Fat Values for Male and Female Athletes in Various Sports

Group or sport	Men (% FAT)	Women (% FAT)	Group or sport	Men (% FAT)	Women (% FAT)
Baseball or softball	8-14	12-18	Rugby	6-16	*
Basketball	6-12	10-16	Skating	5-12	8-16
Bodybuilding	5-8	6-12	Skiing (alpine and Nordic)	7-15	10-18
Canoeing or kayaking	6-12	10-16	Ski jumping	7-15	10-18
Cycling	5-11	8-15	Soccer	6-14	10-18
Fencing	8-12	10-16	Swimming	6-12	10-18
Football	5-25	–	Synchronized swimming	–	10-18
Golf	10-16	12-20	Tennis	6-14	10-20
Gymnastics	5-12	8-16	Track and field, field events	8-18	12-20
Horse racing (jockey)	6-12	10-16	Track and field, running events	5-12	8-15
Ice or field hockey	8-16	12-18	Triathlon	5-12	8-15
Orienteering	5-12	8-16	Volleyball	7-15	10-18
Pentathlon	*	8-15	Weightlifting	5-12	10-18
Racquetball	6-14	10-18	Wrestling	5-16	–
Rowing	6-14	8-16			

*Data not available.

over his assigned weight. Exercise alone is of little value because he would need 9 to 12 months to lose this much weight through that means. Will this athlete accomplish his goal?

Avoiding Fasting and Crash Dieting

Our basketball player must lose 2.3 kg (5 lb) per week for the next four weeks, so he decides to embark on a crash diet, selecting whatever diet is in vogue, knowing that a person can lose 2.7 to 3.6 kg (6-8 lb) per week with such diets. He is not unique. Many athletes find themselves overweight and out of shape because of overeating and reduced activity during the off-season, and they typically wait until the last few weeks before their reporting dates to attack the problem. In our example, the football player might be able to shed 9 kg (20 lb) in four weeks with his crash diet. But much of this weight loss would be from body water and very little from stored fat. Several studies have shown that substantial weight losses occur with very low calorie diets (500 kcal per day or less) over the first several weeks, but that of the weight lost, more than 60% comes from the body's fat-free tissue and less than 40% from fat depots.

While much of the football player's weight is lost from water stores, a substantial amount of protein is lost as well. Also, most crash diets are based on a major reduction in carbohydrate intake. This reduced intake is insufficient to supply the body's needs for carbohydrate, and as a result the body's carbohydrate stores become depleted. Because water storage accompanies carbohydrate storage, the water stores are also reduced as the carbohydrate stores diminish, as discussed earlier in this chapter.

In addition, the body relies more heavily on free fatty acids for energy because its carbohydrate stores are depleted. As a result, ketone bodies, a by-product of fatty acid metabolism, accumulate in the blood, causing a condition known as ketosis. This condition further increases water loss. Much of this water loss occurs during the first week of the diet. Athletes that have attempted this ill-advised shortcut to rapid weight loss will lose substantial weight, but, because much of the weight loss is from the fat-free mass, their performance has been severely compromised.

Optimal Weight Loss: Decreasing Fat Mass and Increasing Fat-Free Mass

The sensible approach to reducing body fat stores is to combine moderate dietary restriction with increased exercise.

When athletes exceed the upper end of the weight range for their sport, they should work toward achieving the upper-end goal weight slowly, losing no more than 0.5 to 1 kg (less than 2.2 lb) per week. Losing more weight than that per week leads to losses in fat-free mass, which is usually not the desired outcome. When the upper limit of the range is reached, further weight loss should be undertaken only with close supervision of the coach, athletic trainer, or team physician. This weight loss should be achieved at an even slower rate—less than 0.5 kg (1.1 lb) per week—to ensure that performance is not negatively affected. The rate of this loss should be reduced still more, or the weight loss program terminated, if performance is affected or if medical symptoms are noted.

Decreasing caloric intake by 200 to 500 kcal per day will allow weight losses of about 0.5 kg (1.1 lb) per week, particularly if combined with a sound exercise program. This is a realistic goal, and such losses add up to a substantial weight loss over time. When trying to reduce, athletes should consume their total daily calories over at least three meals per day. Many athletes make the mistake of eating only one or two meals per day, skipping breakfast, lunch, or both, and then consuming a large dinner. Research in animals has shown that, given the same number of total calories, the animals that eat their daily food ration in one or two meals gain more weight than those that nibble their ration throughout the day. Human research is less clear.

In focus

Athletes who are above their ideal performance weight should lose weight gradually, not more than 1 kg (about 2 lb) per week, to preserve their fat-free mass. They should accomplish this by integrating a good diet containing 200 to 500 kcal less than their daily energy expenditure with a reasonable increase in resistance and endurance activities.

The purpose of weight loss programs is to lose body fat, not fat-free mass. Therefore the combination of diet and exercise is the preferred approach. Combining increased activity with caloric reduction prevents any significant loss of fat-free mass. In fact, body composition can be significantly altered with physical training. Chronic exercise can increase fat-free mass and decrease fat mass. The magnitude of these changes varies with the type of exercise used for training. Resistance training promotes gains in fat-free mass, and both resistance training and endurance training promote loss of fat mass. To lose weight, athletes should combine a moderate resistance and endurance training program with modest caloric restriction.

As a final point, a balanced diet is, of course, essential to ensure that the athlete receives all necessary vitamins and minerals. Vitamin supplementation might or might not be necessary: Results of research thus far are in conflict. But if the nutritional adequacy of the diet is at all questionable, a simple multivitamin that meets the person's needs is suggested.

In review

➤ Severe weight loss in athletes can cause potential health problems, such as dehydration, chronic fatigue, disordered eating, menstrual dysfunction, and bone mineral disorders.

➤ The chronic fatigue symptoms that often accompany severe weight loss mimic those of overtraining. This fatigue also can be caused by substrate depletion.

➤ Body weight standards should be based on body composition and should emphasize relative body fat rather than total body mass.

➤ For each sport, a range of values should be established, recognizing the importance of individual variation, methodological error, and sex differences.

➤ When severe (very low Calorie) diets are followed, much of the weight loss that occurs is from water, not fat.

➤ Most severe diets limit carbohydrate intake, depleting carbohydrate stores. Water is lost along with the carbohydrates, exacerbating the problem of dehydration. Also, the increased reliance on free fatty acids can lead to ketosis, which further increases water loss.

➤ The combination of diet and exercise is the preferred approach to optimal weight loss.

➤ Athletes should lose no more than about 1 kg (about 2 lb) per week until reaching the upper end of the desired weight range. After that, weight loss should be less than 0.5 kg (1 lb) per week until goal weight is reached. More rapid weight losses result in loss of fat-free mass. People can accomplish weight loss at the recommended rate by reducing dietary intake 200 to 500 kcal per day, especially in combination with a sound exercise program.

➤ For fat loss, moderate resistance and endurance training is most effective. Resistance training promotes gains in fat-free mass.

Nutrition and Sport

Having established weight and body composition standards, we now turn to the nutritional aspects of preparing the athlete for optimal performance. As we will see in this section, it is important to maintain a diet that will provide general health benefits, maintain an appropriate weight and body composition, and maximize athletic performance.

A person's diet should contain a relative balance of carbohydrate, fat, and protein. Of the total calories consumed, the recommended balance for most people is

- carbohydrate—55% to 60%;
- fat—no more than 35% (less than 10% saturated); and
- protein—10% to 15%.

Interestingly, this recommended percentage distribution of total calories consumed appears to be optimal for both athletic performance and health. A similar distribution of caloric intake is recommended for the prevention of cardiovascular disease, diabetes, obesity, and cancer as discussed in greater detail later in this chapter. Although all foods ultimately can be broken down to carbohydrate, fat, or protein, these nutrients are not all that the body needs, as we see in the next section.

Classification of Nutrients

Energy from ingested foods is essential to our ability to sustain physical activity, but we rely on foods for much more than energy. Food can be categorized into six classes of nutrients, each with specific functions in the body:

- Carbohydrate
- Fat (lipid)
- Protein
- Vitamins
- Minerals
- Water

The following discussion examines the physiological importance to the athlete of each class of nutrients.

Carbohydrate

A carbohydrate (CHO) is classified as either a monosaccharide, disaccharide, or polysaccharide. Monosaccharides are the simple one-unit sugars, such as glucose, fructose, and galactose, that cannot be reduced to a simpler form. Disaccharides (such as sucrose, maltose, and lactose) are composed of two monosaccharides. For example, sucrose (table sugar) consists of glucose and fructose. Oligosaccharides are short chains of 3 to 10 monosaccharides linked together. Polysaccharides are composed of large chains of linked monosaccharides. Glycogen is the polysaccharide found in animals, including man, and is stored in muscle and liver. Starch and fiber are the two plant polysaccharides and are commonly referred to as complex carbohydrates. Simple carbohydrates are carbohydrates derived from processed foods or foods high in sugar. All carbohydrates must be broken down to monosaccharides before the body can use them.

Carbohydrate serves many functions in the body:

- It is a major energy source, particularly during high-intensity exercise.
- Its presence regulates fat and protein metabolism.
- The nervous system relies exclusively on carbohydrate for energy.
- Muscle and liver glycogen are synthesized from carbohydrate.

Major sources of carbohydrate include grains, fruits, vegetables, milk, and concentrated sweets. Refined sugar, syrup, and cornstarch are nearly pure carbohydrates. Many concentrated sweets such as candy, honey, jellies, molasses, and soft drinks contain few if any other nutrients.

Carbohydrate Consumption and Glycogen Storage
The body stores excess carbohydrate, primarily in the muscles and liver, as glycogen. Because of this, carbohydrate consumption directly influences muscle glycogen storage and the ability to train and compete in endurance events. As shown in figure 15.8, athletes who trained intensely over three consecutive days and ate a low-carbohydrate diet (40% of total calories) experienced a day-to-day decrease in muscle glycogen.[8] When these same athletes consumed a high-carbohydrate diet (70% of total calories), their muscle glycogen levels recovered almost completely within the 22 h between training bouts. In addition, athletes perceive training as easier when their muscle glycogen is maintained throughout a workout.

Early studies demonstrated that when men eat a diet containing a normal amount of carbohydrate (about 55% of total calories ingested), their muscles store about 100 mmol of glycogen per kilogram of muscle. One study showed that diets containing less than 15% carbohydrate led to storage of only 53 mmol/kg, but carbohydrate-rich diets (60-70% CHO) led to storage of 205 mmol/kg. When subjects exercised to exhaustion at 75% of their maximal oxygen uptake, their exercise times were proportional to the amount of muscle glycogen stored before the test, as shown in figure 15.9.

Most studies have shown that glycogen storage replacement is not determined simply by carbohydrate intake. Exercise with an eccentric (muscle lengthening) component, such as running and weightlifting, can induce some muscle damage and impair glycogen resynthesis. In these situations, muscle glycogen levels can appear quite normal during the first 6 to 12 h after exercise, but glycogen resynthesis slows or stops completely as muscle repair begins.

FIGURE 15.8 The influence of dietary carbohydrate (CHO) on muscle glycogen stores during repeated days of training. Note that when a low-CHO diet was consumed, muscle glycogen gradually declined over the three days of study, whereas the CHO-rich diet was able to return the glycogen to near normal each day.

D.L. Costill and J.M. Miller, "Nutrition for endurance sport: Carbohydrate and fluid balance," 1980, *International Journal of Sports Medicine*, 1: 2-14. Reprinted by permission.

FIGURE 15.9 The relation between preexercise muscle glycogen content and exercise time to exhaustion. The exercise time to exhaustion and muscle glycogen were nearly four times greater when the subjects ate a carbohydrate-rich diet than when the diet was composed mostly of fat and protein.

The precise cause for this response is unknown, but conditions in the muscle could inhibit muscle glucose uptake and glycogen storage. For example, within 12 to 24 h after intense eccentric exercise, damaged muscle fibers are infiltrated with inflammatory cells (leukocytes, macrophages) that remove cellular debris resulting from damage to the cells' membranes (see chapters 10 and 14). This repair process can require a significant amount of the blood glucose, reducing the amount of glucose available for resynthesizing muscle glycogen. In addition, some evidence suggests that eccentrically exercised muscle is less sensitive to insulin, which would limit muscle fiber uptake of glucose. Perhaps future studies will more fully explain why eccentric-type activities delay glycogen storage. But for now, we can only observe that glycogen recovery from various forms of exercise can differ and that this should be considered for optimal diet, training, and competition.

When athletes eat only as much food as hunger dictates, they often fail to consume enough carbohydrate to compensate for the amount used during training or competition. This imbalance between glycogen use and carbohydrate intake might explain in part why some athletes become chronically fatigued and need 48 h or more to restore normal muscle glycogen levels. Athletes who train exhaustively on successive days require a diet rich in carbohydrate to reduce the heavy, tired feeling associated with muscle glycogen depletion.

In focus

Carbohydrate is the primary fuel source for most athletes and should constitute at least 50% of their total caloric intake. For endurance athletes, carbohydrate intake as a percentage of total caloric intake might need to be higher: 55% to 65%. However, most important is the total number of grams of carbohydrate ingested. It appears that athletes need from 5 to 13 g/kg of body weight per day in order to maintain glycogen stores. This wide range is necessary to account for the training intensity and total daily energy expenditure, sex, and environmental conditions. For example, during periods of moderate-intensity training, 5 to 7 g/kg per day should be adequate. However, with long-duration, high- and extremely high-intensity training, intake should be increased to 7 to 10 g/kg per day and 10 to 13 g/ kg per day, respectively.[20]

The Glycemic Index It has long been known that the rapid increase in blood sugar levels (hyperglycemia) with the intake of carbohydrate usually is associated with simple carbohydrates, such as glucose, sucrose, fructose, and high-fructose corn syrup. However, this is not always the case. Scientists have discovered that the glycemic response (i.e., increase in blood sugar) to carbohydrate intake varies considerably for both simple and complex carbohydrates. This has led to the use of what has been termed the glycemic index of foods (GI). The ingestion of glucose or white bread leads to a rapid increase in blood sugar. Their response is used as a standard and has been arbitrarily assigned a GI of 100. The glycemic response for all other foods is referenced against the response for glucose or white bread, using 50 g of both the test food and glucose or white bread as the standard. The GI is calculated as follows: GI = 100 ü (blood glucose response over 2 h to 50 g of test food/blood glucose response over 2 h to 50 g of glucose or white bread). Three categories of GI have been established:[20]

- High glycemic index foods (GI >70) such as sport drinks, jelly beans, baked potato, french fries, popcorn, cornflakes, Corn Chex, and pretzels

- Moderate glycemic index foods (GI 56-70) such as pastry, pita bread, boiled white rice, bananas, Coca-Cola, and regular ice cream

- Low glycemic index foods (GI ≤55) such as white boiled spaghetti, kidney and baked beans, milk, grapefruit, apples, pears, peanuts, M&M's, and yogurt

Food items were classified according to the 2002 International Table of Glycemic Index and Glycemic Load Values.[14]

While the GI is a useful tool for rating foods, it is not without controversy. First, the GI for a given food can vary considerably between individuals as well as between mean values for research studies with large numbers of subjects. Second, some complex carbohydrates have high GIs. Third, adding small amounts of fat to a high-GI carbohydrate can greatly reduce the GI of that food. Finally, GI values differ substantially depending on whether glucose or white bread is used as the reference food, with white bread producing substantially higher values.[14, 20] An additional index has been proposed that might be of importance during exercise. The glycemic load (GL) considers both the GI and the amount of carbohydrate (CHO) in a single serving, and is calculated as follows: GL = (GI ü CHO, g)/100.

With this as background, we can now consider the implications of the GI for sport nutrition. Before exercise, low-GI foods would be preferred to reduce the likelihood of hyperinsulinemia. However, high-GI

foods should be an advantage during exercise by helping maintain blood glucose levels. This should also be the case during recovery from intense and prolonged exercise, as the higher blood sugar level should increase muscle and liver glycogen storage.

Carbohydrate Intake and Performance   As noted earlier, muscle glycogen provides a major source of energy during exercise. Muscle glycogen depletion has been shown to be a major cause of fatigue and ultimate exhaustion in high-intensity exercise of short duration or in moderate-intensity exercise lasting more than an hour. This is clearly illustrated in figure 15.10, which shows the marked depletion of muscle glycogen at very high intensities (150% and 120% of $\dot{V}O_{2max}$) for durations of less than 30 min and at lower intensities (83%, 64%, and 31% of $\dot{V}O_{2max}$) for durations of an hour to 3 h. The original data for this figure were from Gollnick, Piehl, and Saltin.[16] Scientists speculated that loading the muscle with extra glycogen before starting exercise should enhance performance.

Studies in the 1960s demonstrated that men who ate a carbohydrate-rich diet for three days stored nearly twice their normal amounts of muscle glycogen.[4] When they were asked to exercise to exhaustion at 75% of $\dot{V}O_{2max}$, their exercise times significantly increased (see figure 15.9). This practice, called **glycogen loading** or

FIGURE 15.10   The influence of exercise intensity (31%, 64%, 83%, 120%, and 150% of $\dot{V}O_{2max}$) on the reduction in muscle glycogen stores. At relatively high intensities, the rate of muscle glycogen use is extremely high compared to that at the moderate and lower intensities.

Adapted, by permission, from A. Jeukendrup and M. Gleeson, 2004, *Sport nutrition: An introduction to energy production and performance* (Champaign, IL: Human Kinetics). Original data from Gollnick, Piehl, and Saltin.

carbohydrate loading, is widely used by distance runners, cyclists, and other athletes who must perform exercise for several hours. We discuss this practice in greater detail later in this chapter.

Blood glucose levels become low (hypoglycemia) during exhaustive high-intensity, long-duration exercise, and this might contribute to fatigue. Numerous studies have shown that subjects' performances improve when they are given carbohydrate feedings during exercise lasting 1 to 4 h. Comparisons of subjects receiving carbohydrate feedings and those receiving placebos generally reveal no performance differences during the early stages of exercise; but during the final stages, performance is greatly improved with carbohydrate feedings.

Although all of the precise mechanisms by which carbohydrate feedings improve performance are not fully understood, maintaining blood glucose near normal concentrations allows the muscles to obtain more energy from this source. Carbohydrate feedings during exercise generally do not spare muscle glycogen use. Instead, they may preserve liver glycogen or even promote glycogen synthesis during exercise, enabling the exercising muscles to rely more on blood glucose for energy late in the exercise. Carbohydrate feedings might also enhance central nervous system function, reducing the perception of effort. Endurance performance (more than 1 h) can be enhanced when carbohydrate is consumed within 5 min before the exercise begins, more than 2 h before exercise (such as during the precompetition meal), and at frequent intervals during the activity.

An athlete should use caution when ingesting carbohydrate foods during the period from 15 to 45 min before exercise because this could cause hypoglycemia shortly after the exercise begins, which could lead to early exhaustion by depriving the muscle of its primary energy sources. Carbohydrate ingested during that period stimulates insulin secretion, elevating insulin when the activity begins. In response to the elevated insulin level, glucose uptake by the muscles reaches an abnormally high rate, leading to hypoglycemia and early fatigue (figure 15.11). Not everyone experiences this reaction, but sufficient evidence indicates that high-GI carbohydrates (those that cause a large increase in blood insulin) should be avoided or moderated in the period from 15 to 45 min before exercise.

Why don't carbohydrate feedings during exercise produce the same hypoglycemic effects observed with preexercise feedings? Sugar feedings during exercise result in smaller increases in both blood glucose and insulin, lessening the threat of an overreaction that leads to a sudden decrease in blood glucose. This

FIGURE 15.11 The effects of preexercise carbohydrate (sugar) feeding on blood glucose levels during exercise. Note the decrease in blood glucose to hypoglycemic levels with sugar feeding 45 min before exercise. Also, subjects during the sugar-feeding trial were unable to complete the full 90 min at 70% of $\dot{V}O_{2max}$, achieving only 75 min.

Adapted from D.L. Costill et al., 1977, "Effects of elevated plasma FFA and insulin on muscle glycogen usage during exercise," *Journal of Applied Physiology* 43(4): 695-699. Used with permission.

finer control of blood glucose during exercise might be caused by increased muscle fiber permeability that decreases the need for insulin, or insulin-binding sites may be altered during muscular activity. Regardless of the cause, carbohydrate intake during exercise appears to supplement the carbohydrate supply needed for muscular activity.

Finally, it is important to consume carbohydrates immediately after high-intensity and long-duration exercise during which carbohydrate stores have been reduced or depleted. Rates of glycogen resynthesis are very high during the first 2 h of recovery, and progressively decrease thereafter. In a study by Ivy and colleagues,[18] cyclists exercised continuously for 70 min on a cycle ergometer on two separate occasions, a week apart, at moderate to high work rates to deplete the active muscles' glycogen stores. During one trial, a 25% carbohydrate solution was ingested immediately after exercise, while in the other trial the solution was ingested after 2 h of recovery. Glycogen storage rates were three times higher during the first 2 h in the trial in which the solution was provided immediately after exercise compared to the trial in which the solution was not provided until after 2 h of recovery. The storage rates were the same for the two trials during the second 2 h (see figure 15.12). More recently, it has been shown that adding protein to the carbohydrate supplement enhances the replenishment of muscle glycogen stores during the recovery period. Adding protein to the carbohydrate supplement maximizes glycogen synthesis with less frequent supplementation and less carbohydrate.[20] Further, it also appears to stimulate muscle tissue repair.

The importance of maximizing carbohydrate storage in the liver and muscles prior to exercise, and of providing carbohydrate during and immediately following exercise, has led food and nutrition companies to develop products to meet these needs, as discussed at the end of this chapter.

FIGURE 15.12 Replenishment of muscle glycogen stores after 70 min of muscle glycogen-depleting exercise using two different regimens of carbohydrate replacement. In the trial in which the carbohydrate solution was provided immediately following exercise (left), muscle glycogen storage was three times higher during the first 2 h of recovery compared to the other trial in which the carbohydrate solution was not given until after 2 h of recovery (right). There was no difference in muscle glycogen storage during the next 2 h.

Adapted from J.L. Ivy et al., 1988, "Muscle glycogen synthesis after exercise: Effect of time of carbohydrate ingestion," *Journal of Applied Physiology* 64: 1480-1485. Used with permission.

Fat

Fat, also termed lipid, is a class of organic compounds with limited water solubility. It exists in the body in many forms, such as triglycerides, free fatty acids (FFAs), phospholipids, and sterols. The body stores most of its fat as triglycerides, composed of three molecules of fatty acids and one molecule of glycerol. Triglycerides are our most concentrated source of energy.

Dietary fat, especially cholesterol and triglycerides, plays a major role in cardiovascular disease (chapter 21); and excessive fat intake also has been linked to other diseases such as cancer, diabetes, and obesity. But despite the negative publicity, fat serves many vital functions in the body:

- It is an essential component of cell membranes and nerve fibers.
- It is a primary energy source, providing up to 70% of our total energy in the resting state.
- It supports and cushions vital organs.
- All steroid hormones in the body are produced from cholesterol.
- Fat-soluble vitamins gain entry into, are stored in, and are transported through the body via fat.
- Body heat loss is minimized by the insulating effect of subcutaneous fat.

The most basic unit of fat is the fatty acid, which is the part used for energy production. Fatty acids occur in two forms: saturated and unsaturated. Unsaturated fats contain one (monounsaturated) or more (polyunsaturated) double bonds between carbon atoms, and each double bond takes the place of two hydrogen atoms. A saturated fatty acid possesses no double bonds, so it has the maximum amount of hydrogen bound to the carbons. Excessive saturated fat consumption is a risk factor for numerous diseases.

Fats derived from animal sources generally contain more saturated fatty acids than fats derived from plants. Fats that are more highly saturated tend to be solids at room temperature, whereas less saturated fats tend to be liquid. The tropical oils are notable exceptions: Palm, palm kernel, and coconut oil are plant-derived fats that are liquid at room temperature but are very high in saturated fat. And although many vegetable oils are low in saturated fats, they are often used in foods as hydrogenated shortening. The process of hydrogenation adds hydrogen to the fat, increasing its saturation.

Fat Consumption Fat can enhance food's palatability by absorbing and retaining flavors and by affecting the food's texture. For this reason, it is quite common in our diets. The dietary fat intake for both men and women was as high as 45% of total calories consumed in 1965 but decreased to about 33% in 2000. Most likely this decrease is attributable to the recent media attention on the health risks of dietary fat. Most nutritionists recommend that fat consumption should not exceed 35% of total calories consumed. The Dietary Guidelines for Americans 2005, produced by the U.S. Departments of Health and Human Services and Agriculture, advise us to limit saturated fats to less than 10% of total caloric intake, cholesterol to less than 300 mg per day, and trans fatty acids to as low a level as possible.

Fat Intake and Performance   For the athlete, fat is especially important as an energy source. Muscle and liver glycogen stores in the body are limited, so the use of fat (or FFA) for energy production can delay exhaustion. Clearly, any change that allows the body to use more fat would be an advantage, particularly for endurance performance. In fact, one adaptation that occurs in response to endurance training is an increased ability to use fat as an energy source. Unfortunately, merely eating fat does not stimulate the muscles to burn fat. Instead, eating fatty foods tends only to elevate plasma triglycerides, which then must be broken down before the FFA can be used for energy production. To increase the use of fat, the FFA levels in the blood, not the triglyceride levels, must be increased.

Highly trained athletes can adapt to a high-fat diet. However, is this beneficial to overall performance? We have just seen how glycogen loading can improve endurance performance. Does "fat loading" have similar benefits? While several studies have demonstrated limited benefit of a high-fat versus a high-carbohydrate diet, most studies have shown either no benefit or decreased performance. The body adapts to a high-fat diet by increasing the supply of fat and the capacity for fat oxidation in the muscle, thus increasing fat oxidation during exercise. But this usually occurs at the expense of a reduction in muscle glycogen stores, thus negating any beneficial effects. Drawing conclusions from the studies conducted so far is complicated by the fact that the type of fat (medium-chain vs. long-chain triglycerides) and the duration of the high-fat diet intervention (less than a week vs. several weeks or more) have varied widely in these studies.

Protein

Protein is a class of nitrogen-containing compounds formed by amino acids. Protein serves numerous functions in our bodies:

- It is the major structural component of the cell.
- It is used for growth, repair, and maintenance of body tissues.
- Hemoglobin, enzymes, and many hormones are produced from protein.
- It is one of the three primary buffers in the control of acid–base balance.
- Proteins in the plasma help maintain normal blood osmotic pressure.
- Antibodies for disease protection are formed from protein.
- Energy can be produced from protein.

Twenty amino acids have been identified as necessary for human growth and metabolism (table 15.2). Of these, 11 (for children) or 12 (for adults) are termed **nonessential amino acids**, meaning that our bodies synthesize them, so we don't rely on dietary intake for their supply. The remaining eight or nine are termed **essential amino acids** because we cannot synthesize them; thus, they are an essential part of our daily diets. Absence of one of these essential amino acids from the diet precludes formation of any proteins that contain that amino acid, and thus any tissue requiring those proteins cannot be maintained.

A dietary protein source that contains all the essential amino acids is called a complete protein. Meat, fish, poultry, eggs, and milk are examples. The proteins in vegetables and grains are called incomplete proteins because they do not supply all the essential amino acids. This concept is important for people on vegetarian diets (discussed later in this chapter). However, combining several incomplete protein sources at one meal should resolve this problem.

Protein Consumption   Protein accounts for approximately 15% of the total calories consumed per day in the United States. The RDAs for protein are 0.95 g/kg of body weight per day for 4- to 13-year-olds, 0.85 g/kg per day for 14- to 18-year-olds, and 0.80 g/kg per day for adults. Men typically require more protein than women

TABLE 15.2   Essential and Nonessential Amino Acids

Essential	Nonessential
Isoleucine	Alanine
Leucine	Arginine
Lysine	Asparagine
Methlonine	Aspartic acid
Phenylalanine	Cysteine
Threonine	Glutamic acid
Tryptophan	Glutamine
Valine	Glycine
Histidine (children)[a]	Proline
	Serine
	Tyrosine
	Histidine (adult)[a]

[a]Histidine is not synthesized in infants and young children, so it is an essential amino acid for children but not for adults.

because men generally weigh more and have greater muscle mass. However, men generally eat more food per day to sustain their weight and muscle mass. Thus, an allowance of 0.8 g/kg of body weight is considered appropriate for both men and women.

Protein Intake and Performance   Should athletes who are training for muscular strength and endurance and aerobic endurance increase their protein intake? Amino acids are the body's building blocks, so protein is essential for the growth and development of body tissues. For many years, protein supplementation was believed essential for athletes. In fact, muscle was once thought to consume itself as fuel for its own actions, so protein supplementation was considered necessary to prevent muscle wasting. Over the years, nutritionists and physiologists argued against the need for supplementing proteins for optimal sport performance. It was generally believed that the RDA of 0.8 g of protein per kilogram of body weight each day would adequately meet the demands of hard training.

More recently, studies using metabolic-tracer and nitrogen-balance technologies have shown that the overall protein and specific amino acid requirements are higher for individuals in training than for normally active people. The role of protein differs for endurance- and strength-trained athletes. It appears that strength-training individuals need up to 2.1 times the RDA, or about 1.6 to 1.7 g of protein per kilogram of body weight per day, whereas athletes engaging in endurance training need 1.2 to 1.4 g of protein per kilogram of body weight per day.[2] While endurance exercise places greater demand on protein as an auxiliary fuel, strength training requires additional amino acids as the building blocks for muscle development. Of course there could be exceptions to this for athletes who are just starting a new, rigorous training program or who are doing very intense, long-duration workouts.

Is it necessary to supplement athletes' diets to optimize their intake of protein? Because most athletes consume a large number of calories each day, it is possible to obtain the additional protein by consuming as little as

In review

> Carbohydrates are sugars and starches. They exist in the body as monosaccharides, disaccharides, oligosaccharides, and polysaccharides. All carbohydrates must be broken down into monosaccharides before the body can use them as a fuel.

> Insufficient intake of carbohydrate during periods of intense training can lead to depletion of glycogen stores. Conversely, glycogen loading by consumption of a diet rich in carbohydrate offers major benefits to performance.

> Endurance performance can be enhanced when carbohydrates are consumed up to an hour before exercise, within 5 min of starting exercise, and during exercise. Exercisers can replenish carbohydrate stores rapidly by ingesting carbohydrate during the first 2 h of recovery. This can be facilitated by the addition of protein to the carbohydrate supplement.

> Fats, or lipids, exist in the body as triglycerides, FFAs, phospholipids, and sterols. They are stored primarily as triglycerides, which are the body's most concentrated energy source. A triglyceride molecule can be broken down into one glycerol and three fatty acid molecules. Only the FFAs are used by the body for energy production.

> Although fat is a major energy source, the use of high-fat diets to enhance endurance performance by sparing glycogen has generally been unsuccessful.

> The smallest unit of protein is an amino acid. All proteins must be broken down to amino acids before the body can use them. Only nonessential amino acids can be synthesized in the body; the essential amino acids must be attained through diet.

> Protein is not a primary energy source in our bodies, but it can be used for energy production during endurance exercise.

> The current RDA for protein (0.8 g/kg per day) may be too low for athletes involved in intense resistance training (1.6-1.7 g/kg per day) or for endurance athletes (1.2-1.6 g/kg per day). During the initial days of training or during periods of very intense training the requirement might be higher. However, extremely high protein diets offer no additional benefits and could offer a health risk to normal kidney function.

> Protein supplementation during recovery from resistance training can stimulate muscle protein synthesis.

10% of total calories as protein. Despite the belief that if a little extra protein is good, then diets extremely high in protein or specific amino acids must be better, there is no scientific evidence that protein intakes exceeding 1.7 g/kg per day provide an additional advantage. In fact, some health risks might be associated with excessive protein intake because it places greater demands on the kidneys to excrete the unused amino acids. A diet containing 10% or at most 15% of calories from protein should be adequate for most athletes unless their total energy intake is deficient. For example, a 100 kg (220 lb) bodybuilder with a 4,500 kcal per day intake containing 15% protein would consume 675 kcal of protein, or about 165 g per day. Thus, the bodybuilder's total protein intake would be 1.65 g/kg per day (165 g/100 kg).

Earlier, we saw that adding protein to carbohydrate solutions enhanced the synthesis of glycogen during recovery following intense aerobic exercise. Supplementing protein intake following a bout of resistance training also seems to have a beneficial effect. Studies have recently shown that the elevation of plasma amino acids during recovery stimulates muscle protein synthesis.[17, 30]

Vitamins

Vitamins are a group of unrelated organic compounds that perform specific functions to promote growth and maintain health. We need them in relatively small quantities, but without them we could not use the other nutrients we ingest. Vitamins act primarily as catalysts or cofactors in chemical reactions. They are essential for energy release, tissue building, and metabolic regulation. Vitamins can be classified into one of two major categories: fat soluble or water soluble. The fat-soluble vitamins, A, D, E, and K, are absorbed from the digestive tract bound to lipids (fats). These vitamins are stored in the body, so excessive intake can cause toxic accumulations. The B-complex vitamins, biotin, pantothenic acid, folate, and vitamin C are water soluble. They are absorbed from the digestive tract along with water. Any excess of these vitamins is excreted, mostly in the urine, but vitamin toxicity has been reported with some of these. Table 15.3 lists the various vitamins and their RDA values, or AI values when the RDA values are not available.

Most vitamins have some function important to the athlete:

- Vitamin A is crucial for normal growth and development because it plays an integral role in bone development.

- Vitamin D is essential for intestinal absorption of calcium and phosphorus and thus for bone development and strength. By regulating calcium absorption, this vitamin also has a key role in neuromuscular function.

- Vitamin K is an intermediate in the electron transport chain, making it important for oxidative phosphorylation.

Of all the vitamins, though, only the B-complex vitamins and vitamins C and E have been extensively investigated for their potential to facilitate athletic performance. In the following sections, we briefly consider these vitamins.

B-Complex Vitamins The B-complex vitamins were once thought to be a single vitamin. Now more than a dozen B-complex vitamins have been identified. These vitamins' essential roles in cellular metabolism cannot be overemphasized. Among their diverse functions, they serve as cofactors in various enzyme systems involved in the oxidation of food and the production of energy. Consider just a few examples. Vitamin B_1 (thiamin) is needed for the conversion of pyruvic acid to acetyl coenzyme A. Vitamin B_2 (riboflavin) becomes flavin adenine dinucleotide (FAD), which acts as a hydrogen acceptor during oxidation. Vitamin B_3 (niacin) is a component of nicotinamide adenine dinucleotide phosphate (NADP), a coenzyme in glycolysis. Vitamin B_{12} has a role in amino acid metabolism and is also needed for the production of red blood cells, which transport oxygen to the cells for oxidation. The B-complex vitamins have such a close interrelationship that a deficiency in one can impair utilization of the others. Symptoms of deficiencies vary with the vitamins involved.

Several studies have shown that supplementation of one or more of the B-complex vitamins facilitates performance. However, most researchers agree that this is true only if the individual being studied suffers a preexisting B-complex deficiency. Creating a deficiency in one or more of the B-complex vitamins usually impairs performance, but this is reversed when the deficiency is corrected with supplementation. No compelling evidence supports supplementation when there is no deficiency.

Vitamin C Vitamin C (ascorbic acid) is common in our foods, but deficiencies can occur in people who smoke, use oral contraceptives, have surgery, or run a fever. This vitamin is important for the formation and maintenance of collagen, a crucial protein found in

TABLE 15.3   RDAs or AIs for Vitamins and Minerals

	Dose	9-13 YEARS Male	9-13 YEARS Female	14-18 YEARS Male	14-18 YEARS Female	19-50 YEARS Male	19-50 YEARS Female	51-70 YEARS Male	51-70 YEARS Female
VITAMINS									
A (retinol)	µg/day	600	600	900	700	900	700	900	700
B$_1$ (thiamine)	mg/day	0.09	0.09	1.2	1.0	1.2	1.1	1.2	1.2
B$_2$ (riboflavin)	mg/day	0.9	0.09	1.3	1.0	1.3	1.1	1.3	1.1
B$_3$ (niacin)	mg/day	12	12	16	14	16	14	16	14
B$_6$	mg/day	1.0	1.0	1.3	1.2	1.3	1.3	1.7	1.5
B$_{12}$	µg/day	1.8	1.8	2.4	2.4	2.4	2.4	2.4	2.4
C	mg/day	45	45	75	65	90	75	90	75
D	µg/day	5[a]	5[a]	5[a]	5[a]	5[a]	5[a]	10[a]	10[a]
E	mg/day	11	11	15	15	15	15	15	15
Biotin (H)	µg/day	20[a]	20[a]	25[a]	25[a]	30[a]	30[a]	30[a]	30[a]
K	µg/day	60[a]	60[a]	75[a]	75[a]	120[a]	90[a]	120[a]	90[a]
Folate	µg/day	300	300	400	400	400	400	400	400
Pantothenic acid	mg/day	4[a]	4[a]	5[a]	5[a]	5[a]	5[a]	5[a]	5[a]
MINERALS									
Calcium	mg/day	1,300[a]	1,300[a]	1,300[a]	1,300[a]	1,000[a]	1,000[a]	1,200[a]	1,200[a]
Chloride	g/day	2.3[a]	2.3[a]	2.3[a]	2.3[a]	2.3[a]	2.3[a]	2.0[a]	2.0[a]
Chromium	µg/day	25[a]	21[a]	35[a]	24[a]	35[a]	25[a]	30[a]	20[a]
Copper	µg/day	700	700	890	890	900	900	900	900
Fluoride	mg/day	2[a]	2[a]	3[a]	3[a]	4[a]	3[a]	4[a]	3[a]
Iodine	µg/day	120	120	150	150	150	150	50	150
Iron	mg/day	8	8	11	15	8	18	8	8
Magnesium	mg/day	240	240	410	360	410[b]	315[b]	420	320
Manganese	mg/day	1.9[a]	1.6[a]	2.2[a]	1.6[a]	2.3[a]	1.8[a]	2.3[a]	1.8[a]
Molybdenum	µg/day	34	34	43	43	45	45	45	45
Phosphorus	mg/day	1,250	1,250	1,250	1,250	700	700	700	700
Potassium	g/day	4.5[a]	4.5[a]	4.7[a]	4.7[a]	4.7[a]	4.7[a]	4.7[a]	4.7[a]
Selenium	µg/day	40	40	55	55	55	55	55	55
Sodium	g/day	1.5[a]	1.5[a]	1.5[a]	1.5[a]	1.5[a]	1.5[a]	1.3[a]	1.3[a]
Zinc	mg/day	8	8	11	9	11	8	11	8

[a]AI (RDA is not available).

[b]Men: Age 19-30 years = 400 and age 31-50 years = 420; women: Age 19-30 years = 310 and age 31-50 years = 320.

Full reports can be obtained at the following U.S. government website: www.nal.usda.gov/fnic.

Note. Values are also available for infants and small children, and for pregnancy and lactation.

Food and Nutrition Board of the National Academy of Sciences, and Health Canada: 1997-2005.

connective tissue, so it is essential for healthy bones, ligaments, and blood vessels. Vitamin C also functions in

- the metabolism of amino acids;
- synthesis of some hormones, such as the catecholamines (epinephrine and norepinephrine) and the anti-inflammatory corticoids; and
- promotion of iron absorption from the intestines.

Many people also believe that vitamin C assists healing, combats fever and infection, and prevents or cures the common cold. Although evidence to date is inconclusive, the role of vitamin C in the fight against disease is an area of major interest.

Vitamin C supplementation to enhance performance has produced equivocal findings in the research conducted to date. However, those who have reviewed this area generally agree that, even with the increased requirements of training, vitamin C supplementation does not improve performance when no deficiency exists. As noted in the sidebar, it has been suggested that vitamins, including vitamin C, also may function as antioxidants to combat the cellular damage created by the metabolically generated free radicals.

Vitamin E Vitamin E is stored in muscle and fat. This vitamin's functions are not well established, although it is known to enhance the activity of vitamins A and C by preventing their oxidation. Indeed, the most important role of vitamin E is its action as an antioxidant. It disarms free radicals (highly reactive molecules) that could otherwise severely damage cells and disrupt metabolic processes. Exercise has been shown to cause DNA damage to the cell, whereas supplementing the intake of vitamin E reduces DNA damage induced by exercise. However, investigators found no benefit of 30 days of vitamin E supplementation with respect to the muscle damage that resulted from 240 maximal isokinetic eccentric knee flexion–extension actions (24 sets of 10 repetitions each) when compared with a placebo control condition.[5]

Vitamin E has received considerable media attention over the years as a potential miracle vitamin that might prevent or alleviate a number of medical conditions, such as rheumatic fever, muscular dystrophy, coronary artery disease, sterility, menstrual disorders, and spontaneous abortions. It also has been suggested that vitamin E supplements may prevent lung damage from many of the pollutants that we inhale. Such claims generally lack supporting scientific evidence.

A Major Transition: From the RDAs to the DRIs

In the early 1940s, the Food and Nutrition Board of the National Academy of Sciences established the United States Recommended Daily Allowance (RDA) guidelines for all food nutrients. The latest edition of the RDAs in their original form was released in 1989. The RDAs provide estimates of safe and adequate daily dietary intakes and estimated minimum requirements for selected vitamins and minerals. A major revision of the RDAs was initiated in the early 1990s. The RDAs have been replaced by new recommendations called Dietary Reference Intakes. The DRIs reflect a joint effort between the United States and Canada to provide dietary intake recommendations grouped by nutrient function and classification.

The new DRIs were released in a series of reports starting in 1997 and continuing through 2005. They include four different reference values:

- Estimated Average Requirement (EAR)—the intake value estimated to meet the requirement for 50% of healthy individuals in an age- and sex-specific group.
- Recommended Dietary Allowance (RDA)—the intake value sufficient to meet the nutrient requirements of nearly all (97-98%) individuals in a specific group.
- Tolerable Upper Intake Level (UL)—the highest level of daily nutrient intake that is unlikely to pose no risk of adverse health effects to almost all individuals in a specific group.
- Adequate Intake (AI)—a recommended intake value based on observed or experimentally determined approximations or estimates of nutrient intake by healthy individuals in a specific group that are assumed to be adequate. This is used when an RDA cannot be determined.

For further information on DRIs and specific recommendations for each of the nutrient classifications by sex and age, refer to the Food and Nutrition Information Center, U.S. Department of Agriculture, at www.nal.usda.gov/fnic.

Many athletes have consumed supplementary doses of vitamin E since it has been postulated to benefit performance through its relationship with oxygen use and energy supply. However, research reviews generally conclude that vitamin E supplementation does not improve athletic performance.

Minerals

A number of inorganic substances known as minerals are essential for normal cellular functions. Minerals account for approximately 4% of body weight. Some are present in high concentrations in the skeleton and teeth, but minerals are also found throughout the body, in and around every cell, dissolved in the body's fluids. They can be present either as ions or combined with various organic compounds. Mineral compounds that can dissociate into ions in the body are called **electrolytes**.

By definition, **macrominerals** are those of which the body needs more than 100 mg per day. **Microminerals**, or **trace elements**, are those needed in smaller amounts. Table 15.3 lists the essential minerals and their RDAs or AIs.

Mineral intake is less likely to be supplemented by athletes than vitamin intake, possibly because far fewer claims have been made about the performance-enhancing qualities of specific minerals. Of the minerals, calcium and iron have been most frequently investigated.

Calcium   Calcium is the most abundant mineral in the body, constituting approximately 40% of the total mineral content. Calcium is well known for its importance in building and maintaining healthy bones, and that is where most of it is stored. But it is also essential for nerve impulse transmission. Calcium plays major roles in enzyme activation and regulation of cell mem-brane permeability, both important for metabolism. And this mineral is also essential for normal muscle function: Recall from chapter 1 that calcium is stored in the sarcoplasmic reticulum of muscles and released when the muscle fibers are stimulated. It is required for formation of the actin-myosin cross-bridges that cause the fibers to contract.

Sufficient calcium intake is critical to our health. If we do not consume enough calcium, it will be removed from its storage sites in the body, especially the bones. This condition is called osteopenia. It weakens the bones and can lead to osteoporosis, a common problem in postmenopausal women and aging men and women. Unfortunately, few studies have been conducted on calcium supplementation, and their results suggest that supplementation is of no value in the presence of an adequate (RDA) dietary intake of calcium.

Phosphorus   Phosphorus is closely linked to calcium. It constitutes approximately 22% of the body's total mineral content. About 80% of this phosphorus is combined with calcium (calcium phosphate), providing strength and rigidity to the bones. Phosphorus is an essential part of metabolism, cell membrane structure, and the buffering system to maintain constant blood pH. Phosphorus plays a major role in bioenergetics: It is an essential component of adenosine triphosphate. There is no evidence to suggest the need for supplementation in athletes.

Iron   Iron—a micromineral—is present in the body in relatively small amounts (35-50 mg/kg of body weight). It plays a crucial role in oxygen transportation: Iron is required for the formation of both hemoglobin and myoglobin. Hemoglobin, located in the red blood cells, binds with oxygen in the lungs and then transports it to the body tissues via the blood. Myoglobin, found

Free Radicals and Antioxidants

Most of the oxygen consumed during aerobic exercise is used in the mitochondria for oxidative phosphorylation and is reduced to water. However, a small number of univalently produced oxygen intermediates, termed **free radicals**, may leak out of the electron transport chain. Laboratory studies have shown that free radical generation increases after acute exercise, and this has been theorized to coincide with oxidative tissue damage. Because these free radicals are highly reactive, they are theorized to modulate muscle function and accelerate the fatigue process. Fortunately, under normal conditions, muscle fibers are equipped with antioxidant enzymes that serve as an efficient defense system to prevent the damaging accumulation of free radicals. In addition, the dietary intake of antioxidants, such as vitamin E and β-carotene, also directly traps free radicals, preventing them from interfering with cellular function. Some researchers suggest that these dietary supplements may help block the negative effects of exercise-induced free radical release. Consequently, the importance of antioxidant vitamins has become a popular topic of discussion and study in the fields of nutrition and cellular biology.

in muscle, combines with oxygen and stores it until needed.

Iron deficiency is prevalent throughout the world. By some estimates, as much as 25% of the world's population is iron deficient. In the United States, approximately 20% of women and 3% of men are iron deficient, as are 50% of pregnant women. The major problem associated with this condition is iron-deficiency anemia, in which hemoglobin levels are reduced, decreasing the blood's oxygen-carrying capacity. This causes fatigue, headaches, and other symptoms. Iron deficiency is a more common problem in women than in men because both menstruation and pregnancy cause iron losses that must be replenished. This problem is compounded by the fact that women generally consume less food, and thus less iron, than men.

Iron has received much attention in the research literature. Women are considered anemic only when their hemoglobin concentration is below 10 g per 100 ml of blood. For men, the value is 12 g per 100 ml of blood. Studies generally suggest that 22% to 25% of female athletes and 10% of male athletes are iron deficient. But these numbers may be conservative. These studies also indicate that hemoglobin is not the only marker of anemia or necessarily the best. Plasma ferritin concentrations provide a good marker of the body's iron stores. Values below 20 to 30 µg/L indicate low body iron stores.

When iron supplements are given to those who are iron deficient (i.e., with low plasma ferritin levels), performance measures, particularly aerobic capacity, typically improve. However, supplementation of iron in those who are not deficient appears to have little or no benefit. In fact, iron supplements can be a health risk, because excess iron is toxic for the liver, and ferritin levels higher than 200 µg/L are associated with an increased risk for coronary artery disease.

Sodium, Potassium, and Chloride Sodium, potassium, and chloride are the important electrolytes that are distributed throughout all body fluids and tissues. Sodium and chloride are found primarily in the fluid outside of the cells and in the blood plasma, but potassium is located mainly inside the cells. This selective distribution of these three minerals establishes the separation of electrical charge across neuron and muscle cell membranes. Thus, these minerals enable neural impulses to control muscle activity (see chapter 3). In addition, they are responsible for maintaining the body's water balance and distribution, normal osmotic equilibrium, acid–base balance (pH), and normal cardiac rhythm.

Western diets are replete with sodium, so dietary deficiency is unlikely. However, minerals are lost with sweating, so any condition that causes excessive sweating, such as extreme exertion or exercise in a hot environment, can deplete these minerals. When discussing mineral imbalances, we often focus on deficiencies. However, many of these minerals also have negative effects when taken in excess. In fact, excess potassium can cause heart failure! Individual needs vary, but megadoses are never advisable.

To conclude this section on vitamins and minerals, we can say that while physical activity increases vitamin and mineral requirements, this is generally countered by an increase in food intake. For athletes who eat balanced meals in response to their bodies' increased caloric needs, it is highly likely that all vitamin and mineral needs will be met and supplementation will have no performance benefits. However, for those who are intentionally consuming a low-energy or unbalanced diet, supplementation may be necessary to maintain performance. If there is any question about the adequacy of an athlete's diet, a low-dose multivitamin and mineral supplement may be appropriate. Also, the new DRIs have a category for upper limits for most micronutrients that can be used as guidelines for excesses.

In review

- Vitamins perform numerous functions in the body and are essential for normal growth and development. Many are involved in metabolic processes, such as those leading to energy production.

- Vitamins A, D, E, and K are fat soluble. These can accumulate to toxic levels in the body. B-complex vitamins, biotin, pantothenic acid, folate, and vitamin C are water soluble. Excesses of these are excreted, so toxicity is rarely a problem. Several of the B-complex vitamins are involved in the processes of energy production.

- Macrominerals are minerals required in amounts of more than 100 mg per day. Microminerals (trace elements) are those we require in smaller amounts.

- Minerals are required for numerous physiological processes, such as muscle contraction, oxygen transport, fluid balance, and bioenergetics. Minerals can dissociate into ions, which can participate in numerous chemical reactions. Minerals that can dissociate into ions are called electrolytes.

- Vitamins and minerals do not appear to have any special performance-enhancing value. Taking them in amounts greater than the RDA will not improve performance and may have unwanted effects.

Water

Seldom is water thought of as a nutrient because it has no caloric value. Yet its importance in maintaining life is second only to oxygen's. Water constitutes about 60% of a typical young man's and 50% of a typical young woman's total body weight; but this varies with body composition, since the fat-free mass has a much higher water content (~73% water) than the fat mass (~10% water). It has been estimated that we can survive losses of up to 40% of our body weight in fat, carbohydrate, and protein. But a water loss of only 9% to 12% of body weight can be fatal.

In focus

Athletes commonly lose between 1% to 6% of their body water during intense, prolonged exercise. However, a water loss in excess of 9% of a person's total body weight can lead to death.

Approximately two-thirds of the water in our bodies is contained in our cells and is referred to as **intracellular fluid**. The remainder is outside the cells, referred to as the **extracellular fluid**. Extracellular fluid includes the interstitial fluid surrounding the cells, the blood plasma, lymph, and other body fluids.

Water plays several critical roles in exercise. Among its most important functions, water provides transportation between and delivery to the body's various tissues, regulates body temperature, and maintains blood pressure for proper cardiovascular function. In the next sections, we more closely examine the role of water in exercise and performance.

Water and Electrolyte Balance

For optimal performance, the body's water and electrolyte contents should remain relatively constant. Unfortunately, this doesn't always happen during exercise. In the next sections, we examine water content and electrolyte balance at rest, how exercise affects these, and the impact on performance when water or electrolyte balance is disturbed.

Water Balance at Rest

Under normal resting conditions, the body's water content is relatively constant: Water intake equals water output. About 60% of our daily water intake is obtained from the fluids we drink and about 30% is from the foods we consume. The remaining 10% is produced in our cells during metabolism (recall from chapter 2

that water is a by-product of oxidative phosphorylation). Metabolic water production varies from 150 to 250 ml per day, depending on the rate of energy expenditure: Higher metabolic rates produce more water. The total daily water intake from all sources averages about 33 ml per kilogram of body weight per day. For a 70 kg (154 lb) person, average intake is 2.3 L per day. Water output, or water loss, occurs from four sources:

- Evaporation from the skin
- Evaporation from the respiratory tract
- Excretion from the kidneys
- Excretion from the large intestine

Human skin is permeable to water. Water diffuses to the skin's surface, where it evaporates into the environment. In addition, the gases we breathe are constantly being humidified by water as they pass through the respiratory tract. These two types of water loss (from the skin and respiration) occur without our sensing them. Thus, they are termed insensible water losses. Under cool, resting conditions, these losses account for about 30% of daily water loss.

The majority of our daily water loss—60% at rest—occurs from our kidneys, which excrete water and waste products as urine. Under resting conditions, the kidneys excrete about 50 to 60 ml of water per hour. Another 5% of the water is lost by sweating (although this is often considered along with insensible water loss), and the remaining 5% is excreted from the large intestine in the feces. The sources of water gain and water loss at rest are depicted in figure 15.13.

Water Balance During Exercise

Water loss accelerates during exercise, as seen in table 15.4. The ability to lose the heat generated during exercise depends primarily on the formation and evaporation of sweat. As body temperature increases, sweating increases in an effort to prevent overheating (see chapter 12). But at the same time, more water is produced during exercise because of increased oxidative metabolism. Unfortunately, the amount produced even during the most intense effort has only a small impact on the **dehydration**, or water loss, that results from heavy sweating.

In general, the amount of sweat produced during exercise is determined by

- environmental temperature, radiant heat load, humidity, and air velocity;
- body size; and
- metabolic rate.

Water gain at rest	Water loss at rest	Water loss during exercise
Fluid intake (60%) + Food intake (30%) + Metabolic water production (10%)	Insensible water loss from skin and respiration (30%) + Sweat loss (5%) + Urine (60%) + Fecal loss (5%)	Sweating (90%) + Insensible water loss (10%)

FIGURE 15.13 Sources and percentages of body water gains and losses at rest and during exercise.

TABLE 15.4 Typical Values of Water Loss From the Body at Rest in a Cool Environment and During Prolonged Exhaustive Exercise

Source of loss	RESTING		PROLONGED EXERCISE	
	ml/h	% total	ml/h	% total
Skin (insensible loss)	15	15	15	1
Respiration (insensible loss)	15	15	100	7
Sweating	4	5	1,200	91
Urine	58	60	10	1
Feces	4	5	-	0
Total	**96**	**100**	**1,325**	**100**

These factors influence the body's heat storage and temperature regulation. Heat is transferred from warmer areas to cooler ones, so heat loss from the body is impaired by high environmental temperatures, radiation, high humidity, and still air. Body size, specifically the ratio between surface area and mass, is important because large individuals generally expend more energy to do a given task, so they typically have higher metabolic rates and produce more heat. But they also have more surface area (skin), which allows more sweat formation and evaporation. As exercise intensity increases, so does the metabolic rate. This increases body heat production, which in turn increases sweating. To conserve water during exercise, blood flow to the kidneys decreases in an attempt to prevent dehydration; but like the increase in metabolic water production, this too may be insufficient. During high-intensity exercise under environmental heat stress, sweating can cause losses of as much as 2 to 3 L of water per hour. (Chapter 12 contains additional information about body water losses during exercise in warm environments.)

In focus

During an event such as the marathon, sweating may reduce body water content by 6% or more. In cold, dry environments or at altitude, water loss from respiration contributes to the overall loss of body water as well.

Dehydration and Exercise Performance

Even minimal changes in the body's water content can impair endurance performance. Without adequate fluid replacement, an athlete's exercise tolerance shows a pronounced decrease during long-term activity because of water loss through sweating. The impact of dehydration on the cardiovascular and thermoregulatory systems is quite predictable. Fluid loss decreases plasma volume. This decreases blood pressure, which in turn reduces blood flow to the muscles and skin. In an effort to overcome this, heart rate increases. Because less blood reaches the skin, heat dissipation is hindered, and the body retains more heat. Thus, when a person is dehydrated by 2% of body weight or more, both heart rate and body temperature are elevated during exercise above values observed when normally hydrated.

As one might expect, these physiological changes will decrease exercise performance. Figure 15.14 illustrates the effects of an approximate 2% decrease in body weight attributable to dehydration from the use of a diuretic on distance runners' performance in 1,500 m, 5,000 m, and 10,000 m time trials on an outdoor track.[3] The dehydration condition resulted in plasma volume decreases between 10% and 12%. Although the average $\dot{V}O_{2max}$ did not differ between the normally hydrated and dehydrated trials, mean running velocity decreased by 3% in the 1,500 m run and by more than 6% in the 5,000 and 10,000 m runs. The greater the duration of the performance, the greater is the expected decline in performance for the same degree of dehydration. These trials were conducted in relatively cool weather. The higher the temperature, humidity, and radiation, the greater the expected decrement in performance for the same degree of dehydration. The decrement in performance would be progressively greater with greater levels of dehydration.

The effect of dehydration on performance in muscular strength, muscular endurance, and anaerobic types of activities is not as clear. Decrements have been seen in some studies, whereas other studies have shown no change in performance. In one of the best-controlled studies, researchers at Penn State University reported that 2% dehydration resulted in significant deterioration of basketball skills in 12- to 15-year-old boys who were skilled basketball players.[12]

Wrestlers and other weight-category athletes commonly dehydrate to get a weight advantage during the weigh-in for a competition. Most rehydrate after the weigh-in before the competition and experience only small decrements in performance. A summary of the effects of dehydration on exercise performance is shown in table 15.5.

(In)review

➤ Water balance depends on electrolyte balance, and vice versa.

➤ At rest, water intake equals water output. Water intake includes water ingested from foods and fluids and produced as a metabolic by-product. The majority of water output at rest occurs from the kidneys, but water also is lost from the skin, from respiratory tract, and in feces.

➤ During exercise, metabolic water production increases as metabolic rate increases.

➤ Water loss during exercise increases because as heat in the body increases, more water is lost with increased sweating. Sweat becomes the primary avenue for water loss during exercise. In fact, the kidneys decrease urine production in an effort to prevent dehydration.

➤ When dehydration reaches 2% of body weight, aerobic endurance performance, and even skilled performance in sports such as shooting free throws in basketball, is notably impaired. Heart rate and body temperature increase in response to dehydration.

Electrolyte Balance During Exercise

Normal body function depends on a balance between water and electrolytes. We have discussed the effects of water loss on performance. Now we turn our attention to the effects of the other component of this delicate balance: electrolytes. When large amounts of water are lost from the body, as during exercise, the balance between water and electrolytes can be disrupted quickly. In the next sections, we examine the effects of exercise on electrolyte balance. Our focus is on the two

FIGURE 15.14 The decline in running velocity (meters per minute) with dehydration of about 2% of body weight for 1,500 m, 5,000 m, and 10,000 m time trials compared with velocity in the normally hydrated condition.

Reprinted, by permission, from L.E. Armstrong, D.L. Costill, and W.J. Fink, 1985, "Influence of diuretic-induced dehydration on competitive running performance," *Medicine and Science in Sports and Exercise,* 17: 456-461.

TABLE 15.5 Alterations in Physiological Function and Performance During Dehydration of 2% or Greater

Variable	Dehydration
CARDIOVASCULAR	
Blood volume/Plasma volume	Decreased
Cardiac output	Decreased
Stroke volume	Decreased
Heart rate	Increased
METABOLIC	
Aerobic capacity ($\dot{V}O_{2max}$)	No change or decreased
Anaerobic power (Wingate test)	No change or decreased
Anaerobic capacity (Wingate test)	No change or decreased
Blood lactate, peak value	Decreased
Buffer capacity of the blood	Decreased
Lactate threshold, velocity	Decreased
Muscle and liver glycogen	Decreased
Blood glucose during exercise	Possibly decreased
Protein degradation with exercise	Possibly decreased
THERMOREGULATION AND FLUID BALANCE	
Electrolytes, muscle and blood	Decreased
Exercise core temperature	Increased
Sweat rate	Decreased, delayed onset
Skin blood flow	Decreased
PERFORMANCE	
Muscular strength	No change or decreased
Muscle endurance	No change or decreased
Muscular power	Unknown
Speed of movement	No change or decreased
Run time to exhaustion	Decreased
Total work performed	Decreased
Attention and focus	Decreased
Some skill aspects of performance	Decreased

Note. Data for this table were derived from the following reviews: M. Fogelholm, 1994, "Effects of bodyweight reduction on sports performance," *Sports Medicine* 18: 249-267; C.A. Horswill, 1994, Physiology and nutrition for wrestling, in D.R. Lamb, H.G. Knutten, & R. Murray (Eds.), *Physiology and nutrition for competitive sport* (Vol. 7, pp. 131-174); H.L. Keller, S.E. Tolly, & P.S. Freedson, 1994, "Weight loss in adolescent wrestlers," *Pediatric Exercise Science* 6: 211-224; and R. Opplinger, H. Case, C. Horswill, G. Landry, & A. Shelter, 1996, "Weight loss in wrestlers: An American College of Sports Medicine position stand," *Medicine and Science in Sports and Exercise* 28: ix-xii.

major routes for electrolyte loss: sweating and urine production.

Electrolyte Loss in Sweat Human sweat is a filtrate of blood plasma, so it contains many substances found there, including sodium (Na^+), chloride (Cl^-), potassium (K^+), magnesium (Mg^{2+}), and calcium (Ca^{2+}). Although sweat tastes salty, it contains far fewer minerals than the plasma and other body fluids. In fact, sweat is 99% water.

Sodium and chloride are the predominant ions in sweat and blood. As indicated in table 15.6, the concentrations of sodium and chloride in sweat are about one-third those found in plasma and five times those found in muscle. Each of these three fluids' **osmolarity**, which is the ratio of solutes (such as electrolytes) to fluid, is also shown. Sweat's electrolyte concentration can vary considerably between individuals. It is strongly influenced by genetics, the rate of sweating, the state of training, and the state of heat acclimatization.

At the elevated rates of sweating reported during endurance events, sweat contains large amounts of sodium and chloride but little potassium, calcium, and magnesium. Based on estimates of the athlete's total body electrolyte content, such losses would lower the body's sodium and chloride content by only about 5% to 7%. Total body levels of potassium and magnesium, two ions principally confined to the insides of cells, would decrease by about 1%. These losses probably have no measurable effect on an athlete's performance.

As electrolytes are lost in sweat, the remaining ions are redistributed among the body tissues. Consider potassium. It diffuses from active muscle fibers as they contract, entering the extracellular fluid. The increase this causes in extracellular potassium levels does not equal the amount of potassium that is released from active muscles, because potassium is taken up by inactive muscles and other tissues while the active muscles are losing it. During recovery, intracellular potassium levels normalize quickly. Some researchers suggest that these muscle potassium disturbances during exercise might contribute to fatigue by altering the membrane potentials of neurons and muscle fibers, making it more difficult to transmit impulses.

Electrolyte Loss in Urine In addition to clearing wastes from the blood and regulating water levels, the kidneys also regulate the body's electrolyte content. Urine production is the other major source of electrolyte loss. At rest, electrolytes are excreted in the urine as necessary to maintain homeostatic levels, and this is the primary route for electrolyte loss. But as the body's water loss increases during exercise, urine production rate decreases considerably in an effort to conserve water. Consequently, with very little urine being produced, electrolyte loss by this avenue is minimized.

The kidneys play another role in electrolyte management. If, for example, a person eats 250 mEq of salt (NaCl), the kidneys will normally excrete 250 mEq of these electrolytes to keep the body NaCl content constant. Heavy sweating and dehydration, however, trigger the release of the hormone aldosterone from the adrenal gland. This hormone stimulates renal reabsorption of sodium. Consequently, the body retains more sodium than usual during the hours and days after a prolonged exercise bout. This elevates the body's sodium content and increases the osmolarity of the extracellular fluids.

This increased sodium content triggers thirst, compelling the person to consume more water, which is then retained in the extracellular compartment. The increased water consumption reestablishes normal osmolality in the extracellular fluids but leaves these fluids expanded, which dilutes the other substances present there. This expansion of the extracellular fluids has no negative effects and is temporary. In fact, this is one of the major mechanisms for the increase in plasma volume that occurs with training and with acclimatization to exercise in the heat. Fluid levels return to normal within 48 to 72 h after exercise, providing there are no subsequent exercise bouts.

TABLE 15.6 **Electrolyte Concentrations and Osmolarity in Sweat, Plasma, and Muscle of Men Following 2 h of Exercise in the Heat**

| Site | ELECTROLYTES (mEq/L) | | | | Osmolarity (mOsm/L) |
	Na⁺	Cl⁻	K⁺	Mg²⁺	
Sweat	40-60	30-50	4-6	1.5-5	80-185
Plasma	140	101	4	1.5	295
Muscle	9	6	162	31	295

Note. mEq/L = milliequivalents per liter (thousandths of 1 g of solute per 1 L of solvent).

In review

- The loss of large volumes of sweat can disrupt electrolyte balance, although electrolytes are rather dilute in sweat.
- Electrolyte loss during exercise occurs primarily with water loss from sweating. Sodium and chloride are the most abundant electrolytes in sweat.
- There is substantially variability among individuals in both sweating rate and the electrolyte composition of the sweat. This makes it virtually impossible to create a "one size fits all" plan for replacing fluids and electrolytes.
- At rest, excessive electrolytes are excreted in the urine by the kidneys. But urine production declines substantially during exercise, so little electrolyte loss occurs by this route.
- Dehydration induces the hormones ADH and aldosterone to promote renal retention of water and sodium. The increased sodium and decreased fluid volume trigger thirst (see figures 4.6 [p. 106] and 4.7 [p. 107]).

Replacement of Body Fluid Losses

The body loses more water than electrolytes during heavy sweating. This raises the osmotic pressure in the body fluids because the electrolytes become more concentrated. The need to replace body water is greater than the need for electrolytes because only by replenishing water content can the electrolytes return to normal concentrations. But how does the body know when this is necessary?

Thirst When people feel thirsty, they drink. The thirst sensation is regulated largely by the osmoreceptors in the hypothalamus. Sensory signals of thirst are invoked when the plasma's osmolality is increased above a threshold value. A second set of signals arises from the low-pressure baroreceptors when low blood volume is sensed. However, relative to osmolar control of thirst, a large blood volume loss is necessary for this backup control system to be activated. Unfortunately, the body's **thirst mechanism** doesn't precisely gauge its state of dehydration. It does not sense thirst until well after dehydration begins. Even when dehydrated, people might desire fluids only at intermittent intervals.

The control of thirst is not fully understood. When permitted to drink water as their thirst dictates, people can require 24 to 48 h to completely replace water lost through heavy sweating. In contrast, dogs and burros can drink up to 10% of their total body weight within the first few minutes after exercise or heat exposure, replacing all lost water. Because of our sluggish drive to replace body water and to prevent chronic dehydration, we are advised to drink more fluid than our thirst dictates. Because of the increased water loss during exercise, it is imperative that athletes' water intake be sufficient to meet their bodies' needs, and it is essential that they rehydrate during and after an exercise bout.

Benefits of Fluids During Exercise Drinking fluids during prolonged exercise, especially in hot weather, has obvious benefits. Water intake will minimize dehydration, increases in body temperature, cardiovascular stress, and declines in performance. As seen in figure 15.15, when subjects became dehydrated during several hours of treadmill running in the heat (40 °C, or 104 °F) without fluid replacement, their heart rates increased steadily throughout the exercise.[4] When they were deprived of fluids, the subjects became exhausted and couldn't complete the 6 h exercise. Ingesting either water or a saline solution in amounts equal to weight loss prevented dehydration and kept subjects' heart rates lower. Even warm fluids (near body temperature) provide some protection against overheating, but cold fluids enhance body cooling because some of the deep body heat is used to warm cold drinks to body temperature.

FIGURE 15.15 Effects of 6 h of treadmill running in the heat on heart rate. Subjects received no fluid, a saline solution, or water. Subjects deprived of fluids became exhausted and could not complete the 6 h exercise.

Data from S.I. Barr, D.L. Costill, and W.J. Fink, 1991, "Fluid replacement during prolonged exercise: Effects of water, saline or no fluid," *Medicine and Science in Sports and Exercise*, 23, 811-817.

Hyponatremia

Fluid replacement is beneficial, but too much of a good thing could potentially be bad. In the 1980s, the first cases of hyponatremia were reported in endurance athletes. **Hyponatremia** is clinically defined as a serum sodium concentration below the normal range of 135 to 145 mmol/L. Symptoms of hyponatremia generally appear when serum sodium levels drop below 130 mmol/L. Early signs and symptoms include bloating, puffiness, nausea, vomiting, and headache. As the severity increases, due to increasing cerebral edema (swelling of the brain), the symptoms include confusion, disorientation, agitation, seizures, pulmonary edema, coma, and death. How likely is hyponatremia to occur?

The processes that regulate fluid volumes and electrolyte concentrations are highly effective, so consuming enough water to dilute plasma electrolytes is difficult under normal circumstances. Marathoners who lose 3 to 5 L of sweat and drink 2 to 3 L of water maintain normal plasma concentrations of sodium, chloride, and potassium. And distance runners who run 25 to 40 km (15.5-24.9 mi) per day in warm weather and do not salt their food don't develop electrolyte deficiencies.

Some research has suggested that during ultramarathon running (more than 42 km, or 26.2 mi), athletes can experience hyponatremia. A case study of two runners who collapsed after an ultramarathon race (160 km, or 100 mi) in 1983 revealed that their blood sodium concentrations had decreased from a normal value of 140 mmol/L to values of 123 and 118 mmol/L.[15] One of the runners experienced a grand mal seizure; the other became disoriented and confused. Examining the runners' fluid intakes and estimating their sodium intakes during the run suggested that they had diluted their sodium contents by consuming too much fluid that contained too little sodium.

The ideal resolution to prevent hyponatremia would be to replace water at the exact rate at which it is being lost or to add sodium to the ingested fluid. The problem with the latter approach is that most sport drinks contain no more than 25 mmol/L of sodium and are apparently too weak to prevent sodium dilution alone, but very strong concentrations cannot be tolerated. Exercise hyponatremia appears to be the result of a fluid overload due to overconsumption, underreplacement of sodium losses, or both. Only a small number of cases have been reported. Thus, it is probably inappropriate to form conclusions from this information to design a fluid replacement regimen for people who must exercise for long periods in the heat.

In review

➤ The need to replace lost body water is greater than the need to replace lost electrolytes, because sweat is very dilute.

➤ The thirst mechanism does not exactly match the body's hydration state, so often more fluid must be consumed than thirst dictates to maintain body weight.

➤ Fluid intake during prolonged exercise minimizes dehydration and optimizes cardiovascular and thermoregulatory function.

➤ In some rare cases, drinking too much fluid with too little sodium has led to hyponatremia (a low plasma concentration of sodium), which can cause confusion, disorientation, and even seizures, coma, and death.

The Athlete's Diet

Athletes place considerable demands on their bodies every day they train and compete. Their bodies must be as finely tuned as possible. This, by necessity, must include optimal nutrition. Too often, athletes spend considerable time and effort perfecting skills and attaining top physical condition while ignoring proper nutrition and sleep. Performance deterioration often can be traced to poor nutrition.

The previous sections of this chapter have provided guidelines for each of the nutrients and, where appropriate, how these would vary according to the athletes' training requirements. Most athletes need guidance in selecting those foods that will help them meet these requirements. As mentioned at the beginning of the chapter, the Food and Nutrition Information Center, U.S. Department of Agriculture website (www.nal.usda.gov/fnic) is an excellent source of information for the coach, trainer, and athlete that will assist them in personalizing diets to meet each athlete's specific nutritional needs.

There are, however, special situations in which additional information is needed. We will now look at vegetarian diets, the precompetition meal, and muscle glycogen replacement and loading.

Vegetarian Diet

In an effort to eat a healthy diet and increase their carbohydrate intake, many athletes have adopted vegetarianism. Vegans are strict vegetarians who eat only food from plant sources. Lactovegetarians also

Recommendations for Fluid Replacement Before, During, and After Exercise

Dehydration is a potential problem for athletes who train and compete for long periods of time, as well as in hot and humid environments. To achieve adequate hydration levels, the American College of Sports Medicine, the American Dietetic Association, and the Dietitians of Canada published guidelines to achieve adequate fluid intake before, during, and following exercise. Key recommendations include the following:

- Two hours before exercise, the athlete should consume 400 to 600 ml (14 to 22 oz) of fluid to provide hydration and allow time for excretion of excess ingested water.
- During exercise, the athlete should drink enough fluid to keep fluid losses to less than 2% of body weight. Weight gain from overdrinking should be avoided.
- After exercise, the athlete should consume adequate fluids to fully replace sweat losses that occurred during exercise.
- Sport drinks containing carbohydrate concentrations of 4% to 8%, and sodium in concentrations between 0.5 and 0.7 g/L, are recommended during intense exercise events lasting longer than 1 h.
- Including sodium in drinks or eating high-sodium foods during the recovery period can help the rehydration process.[2]

consume dairy products. Ovovegetarians add eggs to their vegetable diets, and lacto-ovovegetarians eat plant foods, dairy products, and eggs.

Can athletes perform well on a vegetarian diet? Athletes who are strict vegans must be very careful in selecting the plant foods they eat to provide a good balance of the essential amino acids, sufficient calories, and adequate sources of vitamin A, riboflavin, vitamin B_{12}, vitamin D, calcium, zinc, and iron. Adequate iron intake is of particular concern in female vegetarian athletes because of the lower bioavailability of iron in plant-based diets and because of women's greater risk for anemia and low iron stores. Some professional athletes have noted significant deterioration in athletic performance after switching to strict vegetarian diets. The problem usually is traced to unwise selection of foods. Including milk and eggs in the diet decreases the risk of nutritional deficiencies. Anyone contemplating switching to a vegetarian diet should either read authoritative material on the subject written by qualified nutritionists or consult a registered dietitian or sport nutritionist.

Precompetition Meal

For years, many athletes have eaten the traditional steak dinner several hours before competition. This practice might have originated from the early belief that muscle consumes itself to fuel its own activity and that steak would provide the necessary protein to counteract this loss. But we now know that steak is probably the worst food an athlete could eat before competing. Steak contains a relatively high percentage of fat, which requires several hours for full digestion. During competition, this would cause the digestive system to compete with the muscles for the available blood supply. Also, nervous tension is typically high before a big competition, so even the choicest steak cannot truly be enjoyed at this time. The steak would be more satisfying and less likely to disturb performance if the athlete were to eat it either the night before or after the competition. But if steak is out, what should the athlete eat before competing?

Although the meal ingested a few hours before competition might contribute little to muscle glycogen stores, it can ensure a normal blood glucose level and prevent hunger. This meal should contain only about 200 to 500 kcal and consist mostly of carbohydrate foods that are easily digested. Foods such as cereal, milk, juice, and toast are digested rather quickly and won't leave the athlete feeling full during competition. In general, this meal should be consumed at least 2 h before competition. The rates at which food is digested and nutrients are absorbed into the body are quite individual, so timing the precompetition meal might depend on prior experience. In one study of endurance cyclists, a prolonged cycling exercise trial to exhaustion at 70% of the subject's $\dot{V}O_{2max}$ was performed under two different conditions, with 14 days between trials: 100 g of carbohydrate breakfast fed 3 h before exercise (fed) and no feeding before exercise (fasted). Subjects tested under the fed condition exercised 136 min before reaching exhaustion compared with 109 min in the fasted trial, indicating the importance of the precompetition meal.[22]

A liquid precompetition meal might be less likely to result in nervous indigestion, nausea, vomiting, and abdominal cramps. Such feedings are commercially available and generally have been found useful both before and between events. Finding time for athletes to eat is often difficult when they must perform in multiple preliminary and final events. Under these circumstances, a liquid feeding that is low in fat and high in carbohydrate might be the only solution.

Muscle Glycogen Replacement and Loading

Earlier in this chapter, we established that different diets can markedly influence muscle glycogen stores and that endurance performance depends largely on these stores. The theory is that the greater the amount of glycogen stored, the better the potential endurance performance because fatigue will be delayed. Thus, an athlete's goal is to begin an exercise bout or competition with as much stored glycogen as possible.

On the basis of muscle biopsy studies conducted in the mid-1960s, Åstrand[4] proposed a plan to help runners store the maximum amount of glycogen. This process is known as glycogen or carbohydrate loading. According to Åstrand's regimen, athletes should prepare for an aerobic endurance competition by completing an exhaustive training bout seven days before the event. For the next three days, they should eat fat and protein almost exclusively to deprive the muscles of carbohydrate, which increases the activity of glycogen synthase, an enzyme responsible for glycogen synthesis and storage. Athletes should then eat a carbohydrate-rich diet for the remaining three days before the event. Because glycogen synthase activity is increased, increased carbohydrate intake results in greater muscle glycogen storage. Training intensity and volume during this six-day period should be markedly reduced to prevent additional muscle glycogen depletion, thus maximizing liver and muscle glycogen reserves. Originally, an additional intense training bout was performed four days prior to competition.

This regimen has been shown to elevate muscle glycogen stores to twice the normal level, but it is somewhat impractical for most highly trained competi-

tors. During the three days of low carbohydrate intake, athletes generally find training difficult. They are also often irritable and unable to perform mental tasks, and they typically show signs of low blood sugar, such as muscle weakness and disorientation. In addition, the exhaustive depletion bouts of exercise performed seven days before the competition have little training value and can impair glycogen storage rather than enhance it. This depletion exercise also exposes athletes to possible injury or overtraining.

Considering these limitations, many now propose that the depletion exercise and the low carbohydrate aspects of Åstrand's regimen be eliminated. Instead, the athlete should simply reduce training intensity a week before competition and eat a normal, mixed diet containing 55% of the calories from carbohydrate until three days before the competition. For these days, training should be reduced to a daily warm-up of 10 to 15 min of activity and accompanied by a carbohydrate-rich diet. Following this plan, as seen in figure 15.16, glycogen will be elevated to nearly 200 mmol/kg of

FIGURE 15.16 Two regimens for muscle glycogen loading. In one regimen, the subjects were depleted of muscle glycogen (day 0) and then ate a low-carbohydrate (CHO) diet for three days. They then switched to a CHO-rich diet, which caused muscle glycogen to increase to about 200 mmol/kg. In the other dietary regimen, the subjects ate a normal, mixed diet and reduced their training volume for the first three days; they then changed to a high-CHO diet and further reduction in training volume for three days, which also resulted in muscle glycogen of about 200 mmol/kg.

Data from P.O. Åstrand, 1979, Nutrition and physical performance. In *Nutrition and the world food problem*, edited by M. Rechcigl (Basel, Switzerland: S. Karger); and W.M. Sherman, et al., 1981.

muscle, the same level attained with Åstrand's depletion regimen, and the athlete will be better rested for competition.

It is possible to increase carbohydrate stores rapidly after even a very short near-maximal-intensity bout of exercise. In a study of seven endurance athletes, scientists found that cycling for 150 s at 130% of $\dot{V}O_{2peak}$ followed by 30 s of all-out cycling and 24 h of high carbohydrate intake was sufficient to nearly double muscle glycogen stores in just one day.[13]

In focus

A diet rich in carbohydrates is critical to the success of endurance athletes. Furthermore, carbohydrate loading is a very effective technique for increasing both muscle and liver glycogen stores.

Diet is also important in preparing the liver for the demands of endurance exercise. Liver glycogen stores decrease rapidly when a person is deprived of carbohydrates for only 24 h, even when at rest. With only 1 h of strenuous exercise, liver glycogen decreases by 55%. Thus, hard training combined with a low-carbohydrate diet can empty the liver glycogen stores. A single carbohydrate meal, however, quickly restores liver glycogen to normal. Clearly, a carbohydrate-rich diet in the days preceding competition will maximize the liver glycogen reserve and minimize the risk of hypoglycemia during the event.

Water is stored in the body at a rate of about 2.6 g of water with each gram of glycogen. Consequently, an increase or decrease in muscle and liver glycogen generally produces a change in body weight of 0.5 to 1.4 kg (1-3 lb). Some scientists have proposed monitoring changes in muscle and liver glycogen stores via recording the athlete's early morning weight immediately after rising—after emptying the bladder but before eating breakfast. A sudden decrease in weight might reflect a failure to replace glycogen, a deficit in body water, or both.

Athletes who must train or compete in exhaustive events on successive days should replace muscle and liver glycogen stores as rapidly as possible. Although liver glycogen can be depleted totally after 2 h of exercise at 70% $\dot{V}O_{2max}$, it is replenished within a few hours when a carbohydrate-rich meal is consumed. Muscle glycogen resynthesis, on the other hand, is a slower process, taking several days to return to normal after an exhaustive exercise bout such as the marathon (see figure 15.17). Studies in the late 1980s revealed that muscle glycogen resynthesis was most rapid when individuals were fed at least 50 g (about 0.7 g/kg body

FIGURE 15.17 Muscle glycogen resynthesis is a slow process, requiring several days to restore normal muscle glycogen storage following exhaustive exercise. Note that when muscle glycogen decreases with hard exercise (race), muscle glycogen synthase is markedly elevated. This triggers the muscle to store glycogen when carbohydrates are eaten, returning glycogen synthase to the baseline level.

weight) of glucose every 2 h after the exercise.[19] Feeding subjects more than this amount did not appear to accelerate the replacement of muscle glycogen. During the first 2 h after exercise, the rate of muscle glycogen resynthesis is much higher than later in recovery, as discussed earlier in this chapter. Thus, an athlete recovering from an exhaustive endurance event should ingest sufficient carbohydrate as soon after exercise as is practical. Adding protein and amino acids to the carbohydrate ingested during the recovery period enhances muscle glycogen synthesis above that achieved with carbohydrate alone.

In review

➤ Some athletes have adopted vegetarian diets and appear to perform well. However, careful consideration must be given to protein sources and to consuming adequate levels of iron, zinc, calcium, and several vitamins.

➤ The precompetition meal should be eaten no less than 2 h before competition, and it should be low in fat, high in carbohydrate, and easily digestible. A liquid precompetition meal low in fat and high in carbohydrate has advantages.

➤ Carbohydrate loading increases muscle glycogen content, which, in turn, increases endurance performance.

➤ After endurance competition or training, it is important to consume substantial carbohydrate to replace the glycogen used during activity. Replacing glycogen during the first few hours after training or competing is optimal because glycogen synthase levels are at their peak.

Sport Drinks

We mentioned earlier that ingesting carbohydrate before, during, and following exercise benefits performance by ensuring adequate fuel for energy production during exercise and for replenishing glycogen stores following exercise. While selecting a wise diet can provide for most of the athlete's nutritional needs, nutritional supplements can also be of great value. In addition, adequate fluid intake is necessary for preexercise hydration, hydration during exercise, and rehydration following exercise. Sport drinks are uniquely designed to meet both the energy and fluid needs of the athlete. Performance benefits from these drinks have been clearly documented, not only in endurance activities, but in burst activities as well (e.g., soccer and basketball).[7,12]

Composition of Sport Drinks

Sport drinks differ from one another in a number of ways besides in taste. Of major concern, however, is the rate at which energy and water are delivered. Energy delivery is primarily determined by the concentration of the carbohydrates in the drink, and fluid replacement is influenced by the sodium concentration of the drink.

Energy Delivery—the Carbohydrate Concentration A major concern is how rapidly the drink leaves the stomach, or the rate of **gastric emptying**. In general, carbohydrate solutions empty more slowly from the stomach than either water or a weak sodium chloride (salt) solution. Research suggests that a solution's caloric content, a reflection of its concentration, might be a major determinant of how quickly it empties from the stomach and is absorbed from the intestine. Since carbohydrate solutions remain in the stomach longer than either water or weak solutions, increasing the glucose concentration of a sport drink significantly reduces the gastric emptying rate. For example, 400 ml (14 oz) of a weak glucose solution (139 mmol/L) is almost completely emptied from the stomach in 20 min, but emptying a similar volume of a strong glucose solution (834 mmol/L) can require nearly 2 h.[9] However, when even a small amount of a strong glucose drink leaves the stomach, it can contain more sugar than a larger amount of a weaker solution simply because of its higher concentration. But, if an athlete is trying to

The Zone Diet

Many low-carbohydrate diets aimed at weight loss have been proposed to the general public over the past 20 years or so. In the mid-1990s, many athletes were attracted to a new diet proposed to enhance athletic performance, touted in a popular book written by Dr. Barry Sears, *The Zone*.[23] The Zone diet argues against the high-carbohydrate diet typically advocated for the athlete and the general population. The Zone diet centers on the premise that people should take in 1.8 to 2.2 g of protein per kilogram of fat-free mass. The diet approximates a 40% carbohydrate, 30% fat, and 30% protein proportion of total calories consumed. However, for athletes, a much higher percentage of calories from fat is recommended.[24] Supposedly, this low-carbohydrate diet promotes a more favorable insulin-to-glucagon ratio, ultimately improving oxygen delivery to exercising muscle.[6]

Although many anecdotal stories support the performance-enhancing qualities of the Zone diet, its efficacy has yet to be clearly established by well-designed research studies. In fact, substantial data in the sport nutrition literature strongly argue against such a diet. The diet promotes an unnecessarily high intake of protein and a relatively low intake of carbohydrate. Furthermore, if the diet is pushed to the extreme, the percentage of total calories from fat increases. So, until controlled research studies support the claims made for this diet, the athlete should follow the dietary recommendations that have been proposed in this chapter, recommendations that have the support of many studies over a number of years.[2]

prevent dehydration, this would deliver less water and thus be counterproductive.

Most sport drinks on the market contain about 6 to 8 g of carbohydrate per 100 ml (3.5 oz) of fluid (6% to 8%). The carbohydrate source is generally glucose, glucose polymers, or a combination of glucose and glucose polymers, although fructose or sucrose has also been used.[20] Research studies have confirmed enhanced endurance performance with use of solutions in this range of concentration and with these sources of carbohydrates when compared to water.[2] Carbohydrate solutions above ~6% slow gastric emptying and limit the immediate availability of fluid. However, they can provide a greater amount of carbohydrate in a given period of time to meet the increased energy needs.[2, 20]

Rehydration With Sport Drinks—the Sodium Concentration
Just adding fluid to the body during exercise lessens the risk of serious dehydration. But research indicates that adding glucose and sodium to sport drinks, aside from supplying an energy source, stimulates both water and sodium absorption. Sodium increases both thirst and palatability of the drink. Recall that when sodium is retained, this causes more water to be retained. For rehydration purposes, both during and following exercise, the sodium concentration should range between 20 mmol/L and 60 mmol/L.[20] There is an important loss of sodium from the body with sweating. With high rates of sweating and large volumes of water intake, this can lead to critical reductions in the sodium concentration of the blood and possibly lead to hyponatremia, as discussed earlier in this chapter.

What Works Best?

Athletes will not drink solutions that taste bad. Unfortunately, we all have different taste preferences. To further confound the issue, what tastes good before and after a long, hot bout of exercise will not necessarily taste good during the event. Studies of taste preferences of runners and cyclists during 60 min of exercise showed that most chose a drink with a light flavor and no strong aftertaste. But, will athletes drink more if given a sport drink as compared to water? In one study, runners ran on a treadmill for 90 min and then recovered while seated for an additional 90 min. Both exercise and recovery conditions were controlled in an environmental chamber at a temperature of 32 °C (86 °F), 50% humidity. Three trials were conducted, two with two different sport drinks (6% and 8% carbohydrate) and one with water. Subjects were encouraged to drink throughout each trial. The volume consumed during exercise was similar for all three drinks; but during recovery, the runners drank about 55% more of each of the two sport drinks than water.[29]

In focus

Sport drinks have benefits in addition to those provided by plain water. The addition of carbohydrate to sport drinks provides an important energy source, and the addition of sodium and optimization of taste likely will result in greater fluid consumption, thus delaying dehydration.

In review

➤ Sport drinks have been shown to reduce the risk of dehydration and provide an important source of energy. They also can improve the performance of the athlete in both endurance and "sudden burst" activities such as soccer and basketball.

➤ The carbohydrate concentration of a sport drink generally should not exceed 6% to 8% to maximize both CHO and fluid intake.[2]

➤ The inclusion of sodium in a sport drink facilitates the intake and storage of water.

➤ Taste is an important factor when one is considering a sport drink. Most athletes prefer a light flavor without a strong aftertaste. Each athlete should select the drink that tastes best, providing the nutritional ingredients are the same.

In closing

In this chapter we examined the body composition and nutritional needs of the athlete, considering the importance of optimizing body composition for peak performance and eating wisely to enhance athletic performance. We discovered the importance of each of the six nutrient categories and how they can be adjusted to meet the athlete's training and performance needs. We looked at the precompetition meal, how to effectively replenish and load muscle glycogen stores, and the effectiveness of commercial sport drinks. Now that we have a better understanding of the importance of an appropriate weight and a balanced diet, we turn our attention to another aspect of the athlete's quest for success. In the next chapter, we evaluate those substances that have been proposed to enhance athletic performance—ergogenic aids.

Key Terms

air plethysmography

bioelectric impedance

body composition

body density

carbohydrate loading

dehydration

densitometry

dual-energy X-ray absorptiometry (DEXA)

electrolyte

essential amino acids

extracellular fluid

fat

fat-free mass

fat mass

female athlete triad

free radicals

gastric emptying

glycogen loading

hydrostatic weighing

hyponatremia

intracellular fluid

macrominerals

microminerals

nonessential amino acids

osmolarity

protein

relative body fat

skinfold fat thickness

thirst mechanism

trace elements

vitamin

Study Questions

1. Differentiate between body size and body composition.
2. What tissues of the body constitute the fat-free mass?
3. What is densitometry? How is it used to assess the body composition of the athlete? What is the major weakness of densitometry with respect to its accuracy?
4. What are several field techniques for estimating body composition? What are their strengths and weaknesses?
5. What is the relationship of relative leanness and fatness to performance in sport?
6. What guidelines should be used to determine the athlete's goal weight?
7. What are the six categories of nutrients?
8. What role does dietary carbohydrate play in endurance performance? How about fat? Protein?
9. What is an appropriate protein allowance for a normally active adult man? For a woman?
10. Discuss the value of using protein supplements to enhance performance in strength and endurance events.
11. Should the athlete supplement vitamins and minerals?
12. How does dehydration affect exercise performance? What effect does dehydration have on exercise heart rate and body temperature?
13. Describe the recommended precompetition meal.
14. Describe the method used to maximize muscle glycogen storage (glycogen loading).
15. Discuss the value of consuming carbohydrate during and after endurance exercise. What are the potential benefits of sport drinks?

Study Guide Activities

In addition to the activities listed in the chapter opening outline on page 355, two other activities are available in the web study guide, located at

www.HumanKinetics.com/PhysiologyOfSportAndExercise

 The **KEY TERMS** activity reviews important terms, and the end-of-chapter **QUIZ** tests your understanding of the material covered in the chapter.

Ergogenic Aids and Sport

16

In this chapter and in the web study guide

ACTIVITY 16.1   Drugs in Sports explores the proven effects and risks associated with sympathomimetic amines, β-blockers, caffeine, and diuretics.

ACTIVITY 16.2   Hormonal Agents investigates the benefits and risks of using hormonal agents to improve sport performance for two athletes.

ACTIVITY 16.3   Nutritional Agents reviews how nutritional agents work and whether they have proven ergogenic effects.

In May 2006, just weeks before the start of the Tour de France, Spanish police raided the clinic of a Madrid physician. They discovered a number of performance-enhancing substances and drugs, including erythropoietin (EPO), frozen packets of blood, growth hormone, and anabolic steroids. A total of 58 national (Spain) and international elite cyclists were implicated, including 13 slated to compete in the 2006 Tour de France. Two were among the top cyclists in the world, but none of the 13 was allowed to compete. Erythropoietin and blood doping increase the number of red blood cells in the body, increasing the blood's oxygen-carrying capacity. This increase has been shown to increase

The practice of using illegal substances in the hope of improving performance occurs among athletes of many different sports.

both $\dot{V}O_{2max}$ and aerobic endurance performance. Growth hormone and anabolic steroids purportedly increase muscle mass, increase strength, and reduce body fat, enhancing performance in activities requiring muscular strength and endurance. All four have been banned for use in sport by both national and international sport governing bodies. Regrettably, the eventual winner of the 2006 Tour de France, Floyd Landis, tested positive for anabolic steroids during stage 17 of the race. Suspended from professional competition through early 2009, Landis continued to maintain his innocence until May 2010, when, after almost four years of disputing the allegations against him, he admitted to doping. Unfortunately, use of illegal substances to improve performance is not confined to competitive cyclists. Almost all, if not all, sports have to deal with this problem.

Athletes often are willing to try anything to improve their performance. Some believe that special nutritional supplements can be the deciding factor. Others might use physiological agents such as oxygen or blood doping. Still others might try certain drugs or hormones.

Substances or phenomena (e.g., hypnosis) that improve an athlete's performance are referred to as ergogenic aids. The variety of potential ergogenic aids is immense, and the effects of many **ergogenic** (work producing) substances are shrouded in myth. Most athletes have received tips about ergogenic aids from a friend or coach and assume that the information is accurate, but this is not always the case. Some athletes experiment with substances hoping for even a slight performance improvement regardless of possible harmful consequences. A concern only with maximizing performance coupled with a lack of knowledge about ergogenic substances can lead an athlete to make unwise decisions.

The list of possible ergogenic aids is long, but the number that actually possess ergogenic properties is much shorter. In fact, some allegedly ergogenic substances or phenomena actually can impair performance. These are usually drugs, and Eichner has termed them **ergolytic** (work decreasing) drugs.[17] Ironically and sometimes tragically, several ergolytic agents have been promoted as ergogenic aids.

In focus

An ergogenic aid is any substance or phenomenon that enhances performance. An ergolytic agent is one that has a detrimental effect on performance. Some substances generally thought to be ergogenic are actually ergolytic.

Many athletes indiscriminately take nutritional supplements and ingest drugs and other substances in the belief that they will improve their performance. In one study of 53 Division I university coaches and trainers, 94% provided their athletes with nutritional supplements, despite the fact that the NCAA encourages nutritional education and the use of food rather than supplements.[37] This might seem totally harmless; but as we see later in this chapter, a high percentage of nutritional supplements are contaminated, and some have been found to contain banned substances. Anecdotal stories suggest that anywhere from 20% to 90% of athletes in certain sports are using, or have used, anabolic steroids. Scientific studies, however, suggest a much lower estimate of 6%.[6] Anabolic steroid use has even been reported in the general high school population in the United States, varying from 4% to 11% in boys and up to 3% in girls.[13]

Table 16.1 provides a selected listing of substances and agents proposed to have ergogenic properties

TABLE 16.1 — Proposed Ergogenic Aids and Mechanisms Through Which They Might Work

Agent	Influence heart, blood, circulation, and aerobic endurance	Increase oxygen delivery	Supply fuel for muscle and general muscle function	Act on muscle mass and strength	Result in weight loss or weight gain	Counteract or delay onset or sensation of fatigue	Counteract central nervous system inhibition	Aid in relaxation and stress reduction
PHARMACOLOGICAL								
Amphetamines	✓					✓	✓	
β-Blockers	✓							✓
Caffeine	✓		✓			✓		
Diuretics	✓				✓			
HORMONES								
Anabolic steroids				✓	✓			
Human growth hormone			✓	✓	✓			
PHYSIOLOGICAL								
Bicarbonate loading						✓		
Blood doping	✓	✓				✓		
Erythropoietin	✓	✓				✓		
Oxygen	✓	✓				✓		
Phosphate loading	✓	✓				✓		
NUTRITIONAL								
Amino acids	✓		✓	✓	✓	✓		
Creatine			✓	✓	✓	✓		
L-Carnitine			✓			✓		

that will be discussed in this chapter. This table also lists mechanisms of action by which these ergogenic aids have been proposed to work. These have been studied in sufficient depth to establish their efficacy. Many other substances have been proposed but not adequately researched.

This chapter focuses on pharmacological agents, hormones, physiological agents, and nutritional agents. More general nutritional practices are addressed in chapter 15. Psychological phenomena and mechanical factors are beyond the scope of this book but are reviewed in depth in Williams' book *Ergogenic Aids in Sport*.[44]

Researching Ergogenic Aids

Assume that a professional athlete consumes a particular substance several hours before game time and then has a successful performance. The athlete likely will attribute the success to this substance, even though there is no proof that ingesting the substance will ensure other athletes similar success.

Anyone can claim that a certain substance is ergogenic—and many substances have been so labeled strictly because of speculation—but before a substance can be legitimately classified as ergogenic, it must be proven to enhance performance. Scientific studies in this area are essential to differentiate between a true ergogenic response and a pseudoergogenic response, in which performance improves simply because the athlete expects improvement.

Placebo Effect

As we discussed in the introductory chapter, the phenomenon by which one's expectations of a substance determine the body's response to it is known as the **placebo effect**. This effect can seriously complicate the study of ergogenic qualities because researchers must be able to distinguish between the placebo effect and true responses to the substance being tested.

The placebo effect was clearly demonstrated in one of the earliest studies of anabolic steroids.[4] Fifteen male athletes who had been involved in heavy weightlifting for the previous two years volunteered for a weight training experiment using anabolic steroids. They were told that those who made the greatest strength gains over a preliminary four-month weight training period would be selected for the second phase of the study, in which they would receive anabolic steroids.

Following the initial period, 8 of these 15 subjects were randomly selected to enter the treatment phase. Only six of these subjects passed all medical screening tests and were allowed to continue to the treatment phase. This phase consisted of a four-week period in which the subjects were told that they would receive 10 mg per day of Dianabol (an anabolic steroid), when in fact they received a **placebo**—an inactive substance typically provided in a form identical to the genuine drug.

Strength data were collected over the last seven weeks of the four-month pretreatment training period and over all four weeks of the treatment (placebo) period (see figure 16.1). Even though the subjects were experienced weightlifters, they continued to gain impressive amounts of strength during the pretreatment training period. However, strength gains while subjects were taking the placebo were substantially greater than during the pretreatment period! The group improved an average of 11 kg (24 lb) during the seven-week pretreatment period but improved 45 kg (~100 lb) during the four-week treatment (placebo) period. This represents an average gain in strength of 1.6 kg (3.5 lb) per week during the pretreatment training period and 11.3 kg (25 lb) per week during the placebo period—a more than seven times greater increase in the rate of strength gain during the placebo (supposed steroid) period over the pretreatment training period. Furthermore, placebos are inexpensive, risk free, and legal for use in sport.

FIGURE 16.1 The placebo effect on muscular strength gains. The increase in total strength and strength in each of four maximum lifts over the last seven weeks of an intense four-month pretreatment training period is compared with strength increases during a subsequent four-week treatment period, in which the subjects took placebos that they thought were anabolic steroids and continued intense resistance training.

Data from Ariel and Saville, 1972.

One of the authors of this textbook (Jack H. Wilmore) repeatedly witnessed the placebo effect while conducting a large series of studies investigating the effects of β-blocking drugs on the ability to perform single bouts of exercise or to train aerobically. The Human Subjects Committee, a committee mandated by the federal government to oversee all research conducted with human subjects in the United States, requires that all human subjects receive a full disclosure of the risks associated with any experimental intervention so that they can provide informed consent before participating. Before the start of each study, a cardiologist presented a comprehensive background on β-blocking drugs to each subject, including the drugs' significance in treating various cardiovascular diseases and potential side effects associated with their use. It was amazing to note that over the course of six years of study, the most serious side effects almost always appeared in the subjects when taking the placebo.

Witnessing an ergogenic effect does not necessarily prove that a substance is truly ergogenic. All studies of potential ergogenic substances must include a placebo group so that researchers can compare actual responses resulting from the test substance with those resulting from a placebo. In many studies, a double-blind experimental design is used, in which neither the subject nor the experimenter knows who is getting the proposed ergogenic aid and who is getting the placebo. This is done to eliminate "experimenter bias," whereby the experimenter's beliefs might affect the outcome of the study. With this design, the substances are coded and only an independent person not associated with the project has access to the codes. See the introductory chapter for more information on the proper control of experiments.

In focus

Although the placebo effect has a psychological origin, the body's physiological response to a placebo is quite real. This clearly illustrates how effective our mental state can be in altering our physiological status.

Limitations of Research

To satisfy the scientific community, scientists often rely on laboratory techniques to evaluate the efficacy of any potential ergogenic aid. Often, however, scientific studies cannot provide absolutely clear answers to the questions under study. With elite athletes, success is defined in fractions of a second or in millimeters. Laboratory tests are often unable to detect such subtle differences in performance.

Scientists can be greatly limited by the accuracy of their equipment or techniques. All research methods have some margin of error. If the results fall within that margin of error, the researcher cannot be certain that the result is an effect of the substance being tested. The results might reflect limitations of the research methodology. Unfortunately, because of measurement error, individual differences, and the day-to-day variability of subjects' responses, a potential ergogenic aid must exert a major effect before scientific tests can prove that it is ergogenic.

The testing situation can also limit accuracy. Performance in a laboratory is considerably different from performance in the usual athletic environment, so laboratory results won't always accurately reflect natural athletic results. Yet an advantage of laboratory testing is that the environment can be carefully controlled. This is not always possible in field studies conducted in the athlete's usual environment, where several uncontrollable variables—such as temperature, humidity, wind, and distractions—can affect the results. Thorough

testing of a potential ergogenic aid should include both field and laboratory studies. A strong conclusion about the potential ergogenic or ergolytic properties can be made only when many studies show consistent or similar effects.

Realizing that science is limited in its ability to unequivocally determine the efficacy of a substance, we can now examine some proposed ergogenic aids. We consider substances in four classes:

- Pharmacological agents
- Hormonal agents
- Physiological agents
- Nutritional agents

Pharmacological Agents

Numerous **pharmacological agents**, or drugs, have been suggested as having ergogenic properties. The International Olympic Committee (IOC), the United States Olympic Committee (USOC), the International Amateur Athletic Federation (IAAF), and the National Collegiate Athletic Association (NCAA) all publish extensive lists of banned substances, most of which are pharmacological agents. The IOC and the USOC use the standards established by the World Anti-Doping Agency (WADA). In the United States, these standards are administered through the United States Anti-Doping Agency (USADA). The list of banned substances is updated annually.

Each athlete, coach, athletic trainer, and team physician must know which drugs are prescribed for and taken by the athlete, and they must check these drugs periodically against the listing of banned substances because the list changes frequently. The USOC has a drug education hotline that provides up-to-date information (1-800-233-0393), and WADA (www.wada-ama.org/en/) and the USADA (www.usantidoping.org/) have websites for the same purpose.

In some cases an athlete may receive a **therapeutic use exemption** for a banned substance if medical circumstances warrant (for example, an asthmatic athlete who requires a bronchodilator). These exemptions must be approved through the governing body for that sport. Athletes have been disqualified and have had to relinquish medals, ribbons, awards, and prizes after testing positive for a banned substance in a legitimate medication that had not been approved prior to competition.

We review here only drugs for which a research base has been established. Many other drugs have been touted as ergogenic, but controlled studies have yet to be conducted to determine if they are effective. The drugs we discuss are

- sympathomimetic amines,
- β-blockers,
- caffeine,
- diuretics, and
- recreationally used drugs.

Sympathomimetic Amines

Amphetamines and related compounds are central nervous system (CNS) stimulants. They are also considered sympathomimetic amines, which means that their activity mimics that of the sympathetic nervous system. For many years they have been used as appetite suppressants in medically supervised weight loss programs.

During World War II, army troops used amphetamines to combat fatigue and to improve endurance. They are now used to treat attention deficit and hyperactivity disorder (ADHD). Also known as "speed," they soon found their way into the athletic arena, where they were considered stimulants with possible ergogenic properties. More recently, two other sympathomimetic amines have been proposed as ergogenic aids—**ephedrine** and **pseudoephedrine**. Ephedrine is derived from ephedra herbs (also known as ma huang) and is used as a decongestant and as a bronchodilator in the treatment of asthma. Pseudoephedrine is used in over-the-counter medications primarily as a decongestant and in the illicit manufacturing of methamphetamine. In the following discussion of sympathomimetic amines, we focus most of our attention on amphetamines, since they have been studied most extensively.

The World Anti-Doping Code

Performance-enhancing substances have been used in sport for more than a century. In 1968, the International Olympic Committee (IOC) began drug testing athletes during the Summer and Winter Games. As concerns over the use of performance-enhancing drugs grew through the 1970s and 1980s, an IOC-led initiative established the World Anti-Doping Association (WADA) in 1999. This independent international agency is composed of, and funded equally by, the Olympic Movement and public authorities.

WADA's actions are guided by the World Anti-Doping Code, first adopted in 2003. The Code is the core document that provides the framework for harmonized anti-doping policies, rules, and regulations within sport organizations and among public authorities. More than 600 athletic governing organizations have now adopted the Code.

An important component of the Code is the Prohibited Substances List, which is updated annually. Substances or practices are considered for inclusion if they meet two of the following three criteria:

1. There is evidence that the substance or practice has the potential to enhance sport performance.
2. There is evidence that use of the substance or practice has the potential to harm the athlete.
3. The substance or practice violates the spirit of sport.

Trafficking in any performance-enhancing substance or practice is a violation of the Code.

Compliance with the Code is monitored through a testing program administered in conjunction with the sport's governing body. On an annual basis, over 100,000 drug tests are conducted worldwide at a cost of $30 million. The program is rigorous, multi-tiered, and tightly regulated for quality. Athletes must report their whereabouts at a specified time of day for randomized testing. Those who test positive for a banned substance or practice may be subject to a variety of sanctions ranging from increased monitoring up to a lifetime ban from their sport.

The Code is based on the principle of strict liability; that is, athletes are responsible for any substance in their system, even if they were unaware of it. Athletes who have medical conditions that require the use of a banned substance may apply for a therapeutic use exemption. Athletes who believe that a positive test or sanction is incorrect or unfair may appeal the result through their governing body and the Court of Arbitration for Sport.

For more information about WADA or the Code, see http://www.wada-ama.org.

Proposed Ergogenic Benefits

Athletes have found amphetamines readily available even though they are prescription drugs. Amphetamines are used by athletes for many reasons other than weight loss. Psychologically, the drugs are thought to increase concentration and mental alertness. Their stimulating effect decreases mental fatigue. Athletes anticipate more energy and motivation and often feel more competitive when using amphetamines. The drugs also produce a state of euphoria, which is part of their attraction as so-called recreational drugs. Often athletes who use amphetamines report a sense of indestructibility, which they feel spurs them to higher performance levels. Others rely on the sympathetic stimulation to raise metabolic rate and promote fat loss.

In terms of actual performance, amphetamines are thought to help athletes run faster, throw farther, jump higher, and delay the onset of total fatigue or exhaustion. Athletes who use these drugs expect virtually every aspect of performance to be enhanced. Similar claims and expectations have been associated with the use of ephedrine and pseudoephedrine.

Proven Effects

Generally, for any physiological, psychological, or performance variable that has been investigated, some studies show that amphetamines have no effect, others demonstrate an ergogenic effect, and still others indicate an ergolytic effect. As potent CNS stimulants, amphetamines do increase the state of arousal, which leads to a sense of increased energy, self-confidence, and faster decision making. Athletes who take amphetamines experience a decreased sense of fatigue; increased heart rate, systolic and diastolic blood pressure, and blood flow to skeletal muscles; and elevation of blood glucose and free fatty acids.

Do these effects aid physical performance? Although studies are not in total agreement, the more recent studies, which have used better experimental designs and controls, show that amphetamines can enhance components of athletic performance, including

- weight loss;
- reaction time, acceleration, and speed;
- strength, power, and muscular endurance;
- possibly aerobic endurance but not $\dot{V}O_{2max}$;
- higher maximum heart rate and peak lactate concentrations at exhaustion;
- better focus; and
- fine motor coordination.

The results are not as clear for ephedrine and pseudoephedrine. While several studies have shown small improvements in markers of athletic performance with use of these substances, the general conclusion is that performance benefits are inconsistent and probably insignificant for speed, strength, power, and endurance.[1, 13]

Risks of Using Sympathomimetic Amines

Deaths have been attributed to excessive amphetamine and ephedrine use. Because heart rate and blood pressure are increased, users place greater stress on their cardiovascular systems. These drugs can trigger cardiac arrhythmias in some susceptible individuals. Also, rather than delaying the onset of fatigue, amphetamines likely delay the sensation of fatigue, enabling the athletes to push dangerously beyond normal limits to the point of circulatory failure. Deaths have occurred when athletes have pushed themselves far beyond the normal point of exhaustion. As just one example, during spring training in 2003, Baltimore Orioles' pitching prospect Steve Bechler collapsed during a workout and died less than 24 h later of complications from heatstroke. He had been taking an over-the-counter supplement containing ephedrine that was linked to his heatstroke by the medical examiner at autopsy.

Amphetamines can be psychologically addictive because of the euphoria and energized feelings they cause. But the drugs also can be physically addictive if taken regularly, and a person's tolerance to them builds with continued use, requiring increasingly larger doses over time to obtain the same effects. Amphetamines also can be toxic. Extreme nervousness, acute anxiety, aggressive behavior, and insomnia are frequently mentioned side effects of regular use. Ephedrine has side effects similar to those of amphetamines and is associated with a high incidence of cardiovascular events and heat illness.

In focus

Amphetamines can improve performance in certain sports or activities; but, in addition to being illegal, these drugs carry risks that far outweigh their benefits. Amphetamines can be addictive, and they may mask important afferent and efferent signals that are designed to prevent injury. Ephedrine and pseudoephedrine generally do not improve performance, and ephedrine has been implicated in multiple deaths and untoward events in athletes, including heatstroke and cardiac problems.

β-Blockers

The sympathetic nervous system influences bodily functions through adrenergic nerves: those that use norepinephrine as their neurotransmitter. Neural impulses traveling through these nerves trigger the release of norepinephrine, which crosses the synapses and binds to adrenergic receptors at the target cells. These adrenergic receptors are classified into two groups: α-adrenergic receptors and β-adrenergic receptors.

β-Adrenergic blockers, or **β-blockers**, are a class of drugs that block the β-adrenergic receptors, preventing binding of the neurotransmitter norepinephrine. Both nonspecific and specific (i.e., cardioselective) forms of β-blockers exist in many different formulations. They reduce the effects of stimulation by the sympathetic nervous system. β-Blockers generally are prescribed for the treatment of hypertension, angina pectoris, and certain cardiac arrhythmias. They also are prescribed as a preventive treatment for migraine headaches, to reduce the symptoms of anxiety and stage fright, and for initial recovery from heart attacks.

Proposed Ergogenic Benefits

β-Blocker use in sport has been limited mostly to sports in which anxiety and tremor could impair performance. For similar reasons β-blockers have been abused by musicians seeking to reduce tremor or anxiety during a performance. When a person stands on a force platform (a highly sophisticated device that measures mechanical forces), measurable body movement is detected each time the heart beats. This movement is sufficient to affect

a shooter's aim. Accuracy in shooting sports improves if the rifle or pistol can be shot or the arrow released between heartbeats. β-Blockers can slow a shooter's heart rate, allowing more time to stabilize the aim before shooting or releasing before the next heartbeat. They have also allegedly been used by golfers to steady their stroke, particularly when putting.

Proven Effects

β-Blockers decrease the effects of sympathetic nervous system activity. This is well illustrated by the marked reduction in maximum heart rate with β-blocker administration. It is not unusual for a 20-year-old male athlete with a normal maximum heart rate of 190 beats/min to have a maximum heart rate of only 130 beats/min when taking β-blocking drugs. Resting and submaximal heart rates also are reduced by these drugs. Theoretically, the greater time between heartbeats allows greater hand stability. Because of this, the WADA and the NCAA have banned the use of β-blockers for these sports.

(In)focus

β-Blockers can enhance performance in sports such as golf and shooting events and therefore have been banned.

Risks of β-Blocker Use

Most risks from β-blockers are associated with prolonged use, not isolated incidents of use as in athletics. By blocking the relaxation effect on smooth muscle, β-blockers can induce bronchospasm in people with asthma. They can cause cardiac failure in people who have underlying problems with cardiac function. In people with bradycardia, these drugs can lead to heart block. The decreased blood pressure they cause can result in light-headedness. Some people with type 2 diabetes can become hypoglycemic when using β-blockers because insulin secretion is no longer limited. These drugs, through their various effects, can cause pronounced fatigue, which can inhibit athletic performance and decrease motivation. For athletes who must take β-blockers for a medical condition such as hypertension or arrythmias, β-1 selective blockers are usually preferred because they have fewer negative effects on performance.

Caffeine

Caffeine, one of the most widely consumed drugs in the world, is found in coffee, tea, cocoa, soft drinks, and so-called energy drinks. This drug is also common in several over-the-counter medications, often even in simple aspirin compounds. Caffeine is a CNS stimulant acting on adenosine receptors in the brain; its sympathomimetic effects are similar to those noted previously for amphetamines, although weaker.

Proposed Ergogenic Benefits

As with sympathomimetic amines, caffeine generally is touted as improving alertness, concentration, reaction time, and energy level. Athletes taking the drug often feel stronger and more competitive. They believe that they can perform longer before the onset of fatigue and that if they are fatigued beforehand, the fatigue is reduced. Caffeine is known to have metabolic effects on adipose tissue and skeletal muscle as well as on the CNS, and it has been proposed to increase the mobilization and use of free fatty acids, thus sparing muscle glycogen and prolonging endurance activity.

Proven Effects

Because of its effects on the CNS, the general effects of caffeine include

- increased mental alertness,
- increased concentration,
- elevated mood,
- decreased fatigue and delayed onset,
- decreased reaction time (i.e., faster response),
- enhanced catecholamine release,
- increased free fatty acid mobilization, and
- increased use of muscle triglycerides.

In terms of ergogenic properties, caffeine initially was studied for potential effects that could benefit endurance activities. The first studies, conducted by Costill, Ivy, and their colleagues,[14, 28] demonstrated marked improvements in endurance performance when competitive cyclists ingested a caffeinated beverage compared with a placebo beverage. Caffeine increased endurance times in fixed-pace work bouts and decreased times in fixed-distance races.

Although several subsequent studies were unable to replicate these results, more recent studies have demonstrated substantial ergogenic effects of caffeine ingestion for aerobic endurance performance.[24, 32] It was initially postulated that this improvement was the result of an increased mobilization of free fatty acids, sparing muscle glycogen for later use. But the actual mechanisms by which caffeine improves endurance performance appear to be more complex, since glycogen sparing does not always occur. A growing number of studies now demonstrate that caffeine has its effect directly on the CNS.[39] It is now well documented that caffeine lowers the perception of effort at a given rate of work, potentially allowing the athlete to perform at a higher intensity with the same perceived effort.

Caffeine has also been shown to improve performance in sprint and strength types of activities and in high-intensity team sports. Unfortunately, fewer studies have investigated this area, but caffeine might facilitate calcium exchange at the sarcoplasmic reticulum and increase the activity of the sodium-potassium pump, better maintaining the muscle membrane potential.[24]

Risks of Caffeine Use

In people who are not accustomed to using caffeine, who are sensitive to it, or who consume high doses, caffeine can produce nervousness, restlessness, insomnia, headache, gastrointestinal problems, and tremors. Caffeine also acts as a diuretic, potentially increasing an athlete's risk for dehydration and heat-related illness while performing in hot environments. It can disrupt normal sleep patterns, contributing to fatigue. Caffeine is also physically addictive; abrupt discontinuation of caffeine intake can result in severe headache, fatigue, irritability, and gastrointestinal distress. At one time caffeine was on the WADA list of banned drugs. In 2004 it was removed from the list, but its use is still monitored. Whether or not caffeine should be banned in sport is a debate that continues today.

In focus

Caffeine can enhance performance in endurance sports and may even be of benefit in activities of much shorter duration (e.g., 1 to 6 min). However, some athletes may experience a negative reaction to caffeine, in which case caffeine would be considered ergolytic.

Diuretics

Diuretics affect the kidneys, increasing urine production. Used appropriately, they reduce blood volume and total body water. They are generally prescribed to control hypertension and reduce edema (water retention) associated with congestive heart failure or other conditions.

Proposed Ergogenic Benefits

Diuretics generally are used as ergogenic aids for weight control. For decades, diuretics have been used by jockeys, wrestlers, and gymnasts to keep their weight down. More recently, they have been used by anorexics and bulimics for weight loss (rather than fat loss).

Some athletes who are taking banned drugs also have turned to diuretics, but not to enhance their performance. Because diuretics increase fluid loss, these athletes hope that the extra fluid in their urine will dilute the concentration of banned drugs, thus decreasing the likelihood that the banned substances will be detected during drug testing. This practice and other means of altering the urine in an effort to escape drug detection are called masking.

Proven Effects

Diuretics lead to significant temporary weight loss, but no evidence suggests any other potential ergogenic effects. In fact, several side effects make diuretics ergolytic. The fluid loss results primarily from losses in extracellular fluid, including plasma. For athletes, particularly those who depend on moderate to high levels of aerobic endurance, this reduction in plasma volume reduces maximal cardiac output, which in turn reduces oxygen delivery and aerobic capacity, impairing performance.

Risks of Diuretic Use

In addition to reducing plasma volume, diuretics may hinder thermoregulation. As internal body heat increases, blood must be diverted to the skin so that the heat can be lost to the environment. However, when blood plasma volume is diminished, as with diuretic use, blood remains in the central regions of the body to maintain cardiac filling pressure and adequate blood supply and blood pressure to the vital organs. Thus, less blood is available to be shunted to the skin, and heat loss may be impaired.

Electrolyte imbalance also can occur. Many diuretics cause fluid loss by ensuring electrolyte loss. A diuretic called furosemide inhibits sodium reabsorption in the kidneys, thus allowing more of it to be excreted in the urine. Because fluid follows the sodium, more fluid also will be excreted. Electrolyte imbalances can occur with losses of either sodium or potassium. These imbalances can cause fatigue and muscle cramping. More serious imbalances can lead to exhaustion, cardiac arrhythmias, and even cardiac arrest. The topic of hyponatremia was discussed in chapter 15. Some athletes' deaths have been attributed to electrolyte imbalances caused by diuretic use.

In review

- ➤ Amphetamines are CNS stimulants that increase mental alertness, elevate mood, decrease the sense of fatigue, and produce euphoria.
- ➤ Studies indicate that amphetamines can enhance concentration, reaction time, acceleration, speed, strength, maximum heart rate, peak lactate responses during exhaustive exercise, and time to exhaustion.
- ➤ Amphetamines elevate both heart rate and blood pressure and can trigger cardiac arrhythmias. Excessive use of these drugs has been blamed for some athlete deaths, and the drugs can be both psychologically and physically addictive.
- ➤ Ephedrine and pseudoephedrine have characteristics similar to amphetamines but are not nearly as effective as ergogenic aids. Ephedrine has been demonstrated time and time again to have serious side effects.
- ➤ β-Blockers block β-adrenergic receptors, limiting the binding of catecholamines.
- ➤ β-Blockers slow the resting heart rate, which is a distinct advantage for shooters who try to release the arrow or squeeze the trigger between heartbeats to minimize the slight tremor associated with each beat, and could be an advantage for golfers when chipping and putting.
- ➤ β-Blockers can cause heart block, hypotension, bronchospasm, pronounced fatigue, and decreased motivation. Selective β-blockers have fewer side effects than nonselective blockers.
- ➤ Caffeine, one of the most widely consumed drugs in the world, has both central and peripheral effects. It is a CNS stimulant, and its effects are similar to those of amphetamines but weaker.
- ➤ Caffeine increases mental alertness and concentration, elevates mood, decreases fatigue and delays its onset, increases catecholamine release and mobilization of free fatty acids, and is purported to increase muscle use of free fatty acids to spare glycogen.
- ➤ Caffeine can cause nervousness, restlessness, insomnia, and tremors.
- ➤ Diuretics affect the kidneys, increasing urine production. They often are used by athletes for temporary weight reduction and also by those trying to mask the use of other drugs during drug testing.
- ➤ Weight loss is the only proven ergogenic effect of diuretics, but this weight loss is primarily from the extracellular fluid compartment, including blood plasma. This may lead to dehydration, increased cardiac strain, and electrolyte imbalances.

Recreationally Used Drugs

A class of drugs referred to as "recreational drugs" has been widely used by athletes for both recreation and their potential ergogenic properties. These include alcohol, cocaine, marijuana, and nicotine. None of these have been shown to have ergogenic properties and most are ergolytic. Similarly, cocktails that combine alcohol with caffeine-containing energy drinks are ergolytic. Despite their popularity, their negative effects on performance are well documented.

Hormonal Agents

The use of **hormonal agents** as ergogenic aids in competitive athletics began in the late 1940s or early 1950s. Anabolic steroids were the hormones most frequently used by athletes between the 1950s and the 1980s. During the last half of the 1980s, a new potential ergogenic aid emerged with the introduction of synthetic human growth hormone.

Although numerous scientific studies have been conducted on anabolic steroids and sport, much less is known about the effects of human growth hormone on sport performance. Both anabolic steroids and human growth hormone are banned for all sports, and the medical risks associated with their use are high.

Anabolic Steroids

Androgenic-anabolic steroids, commonly referred to simply as **anabolic steroids**, are nearly identical to the male sex hormones. The anabolic (building) properties of these steroid hormones accelerate growth by increasing the rate of bone maturation and the development of muscle mass. For years, anabolic steroids have been given to youngsters with delayed growth patterns to normalize their growth curves. Literally dozens of steroids have been synthesized, with fine-tuning of the natural chemical composition of these hormones to reduce their androgenic (masculinizing) properties, to alter their route of administration (both oral and injectable steroids are available), to change the rate at which they are metabolized or eliminated, and to increase their anabolic effects on muscle.

Proposed Ergogenic Benefits

Steroid administration is known to increase fat-free mass and strength and to reduce fat mass. Consequently, an athlete who depends on muscle size, body size, or strength might be tempted to take steroids. Early claims that aerobic capacity improves with anabolic steroid use caught the attention of endurance athletes. Anabolic steroids also have been postulated to facilitate recovery from exhaustive training bouts, allowing athletes to train hard on subsequent days. This potential benefit has stirred the interest of athletes from almost all sports.

The potential for anabolic steroid use among athletes is very high, and this continues to be a major problem in sports such as football, baseball, and track. Actual drug testing results indicate a low percentage of users, but athletes who take steroids have become very good at "beating the system" when they know that they will be tested. They have used masking agents and even the urine of friends who have not used steroids. New designer steroids introduced in the early 2000s were engineered to help the athlete avoid detection. Randomized drug testing (both "in competition" and out of season), coupled with improved analytical techniques, has increased the likelihood that steroid users will be caught. However, improved education about the health risks of steroids is the best long-term strategy to reduce steroid abuse.

Proven Effects

Results of early investigations were almost evenly divided. Many of these studies showed no significant change in body size or physical performance attributable to taking steroids, yet many of the other early studies and all recent studies found steroids to have considerable positive influence on increasing muscle mass and strength. A clear dose–response relationship between steroids and either lean body mass, muscle mass, or strength has now been demonstrated, making steroids one of the most convincing ergogenic agents.

One basic problem with almost all research conducted in this area to date is the inability to observe in the research laboratory the effects of the drug dosages being used in the athletic world. Athletes are estimated to be taking 5 to 20 times the recommended maximum daily dosage or more.[25] Few studies exist in which dosages in this range were used. However, some researchers have been able to observe athletes both when the athletes are taking high doses of steroids and when they are off the drug. Research has demonstrated the following effects of steroids on performance.

Muscle Mass and Strength   In one of the first studies involving athletes who were taking steroids on their own, the effects of relatively high doses were observed in seven male weightlifters.[27] Two treatment periods, each lasting six weeks, were separated by a six-week interval without treatment. Half the subjects received a placebo during the first treatment period and the steroid during the second treatment period. The other half received the medications in reverse order: steroid first, then placebo. When the data from all subjects were analyzed, results showed that while on the steroid, the weightlifters had significant increases in

- body mass and fat-free mass,
- total body potassium and total body nitrogen (markers of fat-free mass),
- muscle size, and
- leg strength.

These increases did not occur during the placebo period. Results of this study are summarized in figure 16.2.

In a second study, Forbes[21] observed body composition changes in a professional bodybuilder and a competitive weightlifter. Both were on self-prescribed high doses of steroids. The bodybuilder had been on the high dose for 140 days and the weightlifter for 125 days. Fat-free body mass increased an average of 19.2 kg (42.3 lb), and fat mass decreased almost 10 kg (22 lb). Forbes plotted the results of a number of studies that used different dosages (figure 16.3). He observed that minimal increases of 1 to 2 kg (2.2-4.4 lb) in fat-free body mass occur with low doses of anabolic steroids. But with high doses, fat-free body mass increases markedly. His results suggest a threshold level for steroid doses, with only regular high doses resulting in substantial increases in fat-free body mass. Similarly, brief increases in testosterone, such as those that might occur after strength training, do not appear to have a major effect on body composition.

A third study looked at supraphysiological doses of testosterone on muscle size and strength in men who were nonathletes but were experienced with weightlifting.[7] Forty men completed the study and were assigned to one of the following groups: placebo with no exercise, placebo with exercise, testosterone with no exercise, and testosterone with exercise. The men received either 600 mg of testosterone enanthate or placebo intramuscularly each week for 10 weeks. The exercise groups strength trained three days per week for 10 weeks. Body composition was measured by underwater weighing, triceps and quadriceps muscle size by magnetic resonance imaging, and upper and lower body strength by the 1-repetition maximum technique. The testosterone and exercise group showed the largest increases in fat-free mass, triceps and quadriceps muscle area, and strength, whereas the placebo and no-exercise group remained unchanged (figure 16.4). The placebo

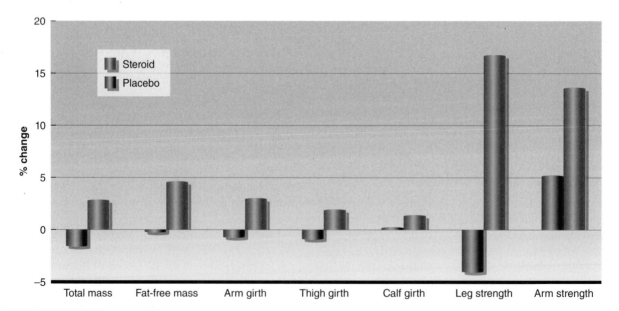

FIGURE 16.2   Percentage changes in body size, body composition, and strength when athletes used anabolic steroids or a placebo.

Adapted from Hervey et al., 1981.

FIGURE 16.3 The relationship between the total dose of steroid (mg/day) and the change in fat-free mass in kilograms. The symbols represent different anabolic steroid drugs. Steroid dose plotted logarithmically.

From an article published in *Metabolism,* vol. 34, G.B. Forbes, "The effect of anabolic steroids on lean body mass: The dose response curve," pp. 271-573, Copyright 1985, with permission from Elsevier.

and exercise group increased strength, quadriceps area, and fat-free mass; and the testosterone and no-exercise group increased squat strength and quadriceps and triceps muscle areas. This is one of the best-designed studies conducted on steroids and resistance exercise because it used placebo and no-exercise groups.

The increase in muscle mass is generally associated with increases in the cross-sectional areas of both type I and type II muscle fibers and an increase in myonuclear number. These increases are dose dependent and likely the result of increased muscle protein synthesis.[20]

Cardiorespiratory Endurance
Several early studies reported increases in $\dot{V}O_{2max}$ with the use of anabolic steroids. These results were consistent with the known effects of steroid administration on increasing red blood cell production and total blood volume. However, in these studies, $\dot{V}O_{2max}$ was estimated indirectly. In later, better-controlled studies, $\dot{V}O_{2max}$ was measured directly, and anabolic steroids produced no benefit. However, none of the studies investigating anabolic steroid use and improvement in aerobic capacity involved trained endurance athletes.

Recovery From Training
The theory that anabolic steroids facilitate recovery from high-intensity training is attractive. A major concern in training elite athletes today is how to reduce the negative physiological and

FIGURE 16.4 Changes in fat-free mass and quadriceps and triceps muscle areas from magnetic resonance imaging, and changes in squat and bench press strength over 10 weeks of placebo or testosterone, with or without exercise training.

Reprinted, by permission, from S. Bhasin et al., 1996, "The effects of supraphysiologic doses of testosterone on muscle size and strength in normal men," *New England Journal of Medicine* 335: 1-7. Copyright © 1996 Massachusetts Medical Society. All rights reserved.

psychological effects associated with high-intensity training, enabling the athlete to continue to train at peak levels day after day. At this time, only limited data support the possibility that steroid use facilitates recovery from exercise. Tamaki and colleagues[40] reported less muscle fiber damage after a single exhaustive bout of weightlifting in a group of rats that received a single injection of the long-acting androgenic-anabolic steroid nandrolone decanoate compared to a control group that had received a placebo. They also found that the rats in the steroid group had an increased rate of protein synthesis during recovery when compared to the control group.

In focus

Anabolic steroid use increases muscle mass and strength and reduces body fat, which can improve performance in strength-type sports or activities. Aerobic endurance appears to be unaffected by steroid use. Steroid use is illegal in sport and is banned by all governing bodies. Furthermore, the health risks can be considerable.

Risks of Anabolic Steroid Use

Although use of anabolic steroids can be beneficial for certain types of athletic performance, several major issues must be addressed. It is neither moral nor ethical for athletes to use drugs to improve their chances in competition. Most athletes feel that it is wrong for their competitors to artificially improve their performance. Yet many of these same athletes feel compelled to use steroids in an effort to compete with the other athletes in their sport or event who are chronic steroid users. Fair competition is not possible if an individual is the only athlete in a particular competition who has remained steroid free. This is one of the guiding principles of the World Anti-Doping Code.

The medical risks associated with steroid use are large, especially with the massive doses often used by athletes. Steroid use by people who are not physically mature can lead to early closure of the epiphyses of the long bones, so final stature can be reduced. Use of anabolic steroids suppresses the secretion of gonadotropic hormones, which control the development and function of the gonads (testes and ovaries). In males, decreased gonadotropin secretion can cause testicular atrophy, decreased secretion of testosterone, a reduced sperm count, and impotence long after steroid administration ends. Excess testosterone can lead to higher estrogen production as well, causing enlargement of the male breasts. In females, gonadotropins are required for ovulation and secretion of estrogens, so a decrease in these hormones disrupts those processes and menstruation. In addition, these hormonal disturbances in females can lead to masculinization: breast regression, enlargement of the clitoris, deepening of the voice, and growth of facial hair.

Prostate gland enlargement in males is another possible side effect of steroid use, elevating the risk for prostate cancer. Many steroids are metabolized by the liver. The tendency for steroids to accumulate can produce a form of chemical hepatitis brought on by steroid use and can lead to liver tumors.

Abnormal cardiac hypertrophy (enlarged heart), cardiomyopathy (diseased heart muscle), myocardial infarction (heart attack), thrombosis, arrhythmia, and hypertension have been reported in chronic steroid users. It is suspected that steroid use was at least partially responsible for the medical condition of one former offensive lineman in the National Football League who was on the waiting list for a heart transplant. Scientists have found markedly depressed high-density lipoprotein cholesterol (HDL-C) levels—reductions of 30% or more—in athletes on even moderate steroid doses. High-density lipoprotein cholesterol has anti-atherogenic properties, meaning that it prevents the development of atherosclerosis. Low levels of HDL-C are associated with a high risk for both coronary artery disease and heart attack (see chapter 21). Furthermore, low-density lipoprotein cholesterol (LDL-C), which has atherogenic properties, appears to be increased with steroid use.

Substantial personality changes have occurred with steroid use. The most notable change is a marked increase in aggressive behavior, or "roid rage." Some teens have become extremely violent and have attributed these drastic mood changes to steroid use. Evidence also suggests that drug dependency can result from steroid use.

It is important to note that not everyone using anabolic steroids is an athlete. In fact, it appears that the majority of those taking steroids are nonathletes who use steroids for "cosmetic" purposes. Further, many self-inject steroids, and of those a significant number share needles, increasing the risk of contracting infections such as hepatitis and HIV/AIDS.

Neither scientists nor physicians know the potential long-term effects of chronic steroid use. One study of male mice that received four different anabolic steroids of the kinds and doses taken by athletes showed that their life span was markedly decreased.[10] A number of birth defects have been reported by former East German athletes who were given steroids during their athletic career, but the cause and incidence of these abnormalities remain unknown. It is critical to develop a large database of former steroid users who can be followed over their life span. We have to remember that most diseases begin many years before symptoms develop. It is possible that the more serious health risks of steroid use won't be apparent for 20 or 30 years.

Recent reviews provide more detail on the potential ergogenic effects and health risks associated with ana-

"Andro"—Fuel for Power Hitters?

During the 1998 Major League Baseball season, Mark McGwire blasted 70 home runs to eclipse the previous home run record of 61 set by Roger Maris in 1961. During the season, McGwire admitted to using androstenedione ("Andro"), a prohormone, which is a precursor to testosterone. Andro was marketed to increase testosterone levels and, subsequently, skeletal muscle mass. Sales of Andro increased markedly in response to McGwire's revelation. Does Andro work? King and his colleagues at Iowa State University were the first to investigate the combined effects of Andro and resistance training, in twenty 19- to 29-year-old men.[30] Ten subjects were assigned to the Andro and resistance training group and 10 to the placebo and resistance training group. Andro had no effect on serum testosterone levels, and the gains in strength and muscle mass were the same for the Andro and placebo groups. However, there was an unexpected finding: Andro increased serum estradiol and estrone concentrations, female hormones that promote development of secondary female sex characteristics! Other studies have substantiated these findings in young, middle-aged, and older men.[9, 33] In 2004, Andro became the subject of congressional investigation that resulted in changes to several laws and regulations; after its popularity as a dietary supplement in the 1990s, possession of Andro is now a federal crime.

Dehydroepiandrosterone (DHEA) is another steroid hormone, as is its conjugate sulfate (DHEAS). Bloodborne DHEA can be converted to androstenedione or androstenediol, which can then be converted to testosterone. It has been proposed that DHEA increases muscle mass. Brown and colleagues investigated the potential benefit of DHEA in increasing both serum testosterone and strength in 19 young men participating in an eight-week resistance training program, 9 receiving DHEA and 10 receiving the placebo.[11] As with the Andro study, there were no changes in serum testosterone concentrations as a result of taking DHEA, and the changes in strength and muscle mass were the same for the DHEA and placebo groups. Others have obtained similar results in young and middle-aged men and women.[33] However, in a study of elderly women and men (65-78 years), 10 months of DHEA supplementation did increase strength and thigh muscle volume when combined with strength training during the last four months of the study.[43] At this time, it appears that DHEA supplementation does not affect muscle size or strength in young and middle-aged men and women.

Major League Baseball players must have anticipated the results of these studies on Andro and DHEA. The cover of the June 3, 2002, issue of *Sports Illustrated* announced, "Special Report—Steroids in Baseball: Confessions of an MVP." From this issue, it appeared that a substantial number of professional baseball players had skipped over Andro and DHEA and had gone straight to anabolic steroids. Although the actual percentage of players using steroids is unknown, some players estimate that it was as great as 50% at its peak, prior to mandated drug testing. Curt Schilling, former Arizona Diamondback pitcher and Co-Most Valuable Player of the 2001 World Series, was quoted as saying, "Guys look like Mr. Potato Head, with six or seven body parts that just don't look right."

bolic steroid use.[13, 20, 25, 29, 34, 48] Most governing bodies of sports likely to be affected by anabolic steroid use have developed educational materials for their athletes in hopes of preventing steroid use. Also, national governing bodies for most sports have instituted aggressive year-round testing programs in which athletes are tested randomly for steroid use.

Human Growth Hormone

For years, the medical treatment for hypopituitary dwarfism has been administration of **human growth hormone (hGH)**, a hormone secreted by the anterior pituitary gland. Before 1985, this hormone was obtained from cadaver pituitary extracts, and the supply was limited. Since the introduction of genetically engineered hGH in the mid-1980s, availability is no longer an issue, although the cost is still high.

During the 1980s, realizing this hormone's numerous functions, athletes started investigating hGH as a possible substitute for or complement to their use of anabolic steroids. As drug testing for anabolic steroids became more sophisticated, athletes were looking for an alternative for which there was no drug test at the time. Growth hormone appeared to be a drug of choice for athletes who wanted to increase their strength and muscle mass.

Proposed Ergogenic Benefits

Growth hormone (GH) has six functions of interest to athletes:

- Stimulation of protein and nucleic acid synthesis in skeletal muscle
- Stimulation of bone growth (elongation) if bones are not yet fused (important to young athletes)
- Stimulation of insulin-like growth factor (IGF-1) synthesis
- Increase in lipolysis, leading to an increase in free fatty acids and an overall decrease in body fat
- Increase in blood glucose levels
- Enhancement of healing after musculoskeletal injuries

Athletes have turned to this hormone thinking that it would increase muscle development and thus increase fat-free mass. Often GH is used with anabolic steroids to maximize the anabolic effects.

Proven Effects

Administration of GH to older men (>60 years) has been shown to increase fat-free mass, decrease fat mass, and increase bone density.[36] However, in studies of young men and experienced weightlifters, there appear to be few if any significant benefits.[47] More consistently, a reduction in fat mass has been reported, suggesting that GH is more useful as a "cutting" (fat reducing) agent than an anabolic.

In one study, young men were randomly assigned to either a GH or a placebo group. After 12 weeks of resistance training, the two groups had similar changes in muscle size, muscle strength, and rate of muscle protein synthesis in the quadriceps muscle. In a second study, experienced weightlifters were placed on GH treatment for 14 days while continuing to weight train. The investigators found that GH did not alter either the rate of muscle protein synthesis or the rate of whole-body protein breakdown, the two factors that would promote an increase in muscle mass. When body composition changes are noted with GH, the increase in mass and fat-free mass is associated with increased water retention.

In focus

It appears that human growth hormone has no anabolic properties in young, healthy athletes. However, major health risks are associated with the use of growth hormone.

Some athletes also take other drugs and certain amino acid supplements to stimulate GH release from the pituitary. To date, little evidence suggests that this practice is effective.

In review

➤ Anabolic steroids are more appropriately termed androgenic-anabolic steroids, because in their natural state they include both androgenic (masculinizing) and anabolic (building) properties. Synthetic steroids have been designed to maximize the anabolic effects while minimizing androgenic effects.

➤ Anabolic steroids increase muscle mass, strength, and muscular endurance.

➤ Anabolic steroids can increase muscle mass and strength, but the effect is dose dependent. They also reduce body fat. They do not increase aerobic endurance, and their ability to facilitate recovery from exhaustive exercise is questionable.

➤ Andro and DHEA, precursors of testosterone, have been proposed to have ergogenic properties—increased muscle mass and strength—but most research studies have not supported these claims.

➤ Potential risks associated with use of anabolic steroids include personality changes, "roid rage," testicular atrophy in men, reduced sperm count in men, breast enlargement in men, breast regression in women, prostate gland enlargement in men, masculinization in women, menstrual cycle disruption in women, liver damage, and cardiovascular diseases.

➤ Growth hormone has not been studied extensively for its potential ergogenic effects. The limited research available supports its ability to increase fat-free mass and decrease fat mass in older men, but GH appears to have little or no effect on increasing muscle mass and strength in younger people. Most of the increase in mass and fat-free mass is associated with increased water retention.

➤ Risks associated with GH use include acromegaly, hypertrophy of internal organs, muscle and joint weakness, diabetes, hypertension, and heart disease.

Risks of Growth Hormone Use

As with steroids, potential medical risks are associated with GH use. Acromegaly can result from taking GH after the bones have fused. This disorder results in bone thickening, which causes broadening of the hands, feet, and face; skin thickening; and soft tissue growth. Internal organs typically enlarge. Ultimately, the victim suffers muscle and joint weakness and often heart disease. Cardiomyopathy is the most common cause of death with GH use. Glucose intolerance, diabetes, and hypertension also can result from GH use.

Physiological Agents

Many **physiological agents** have been proposed as ergogenic aids. These agents are supposed to improve physiological responses during exercise. An athlete typically adds to a substance that occurs naturally in the body to try to improve performance. The reasoning is that if natural levels of a substance are beneficial to performance, higher levels should be even better. Several physiological agents have been proven effective but generally only under very specific conditions or for certain events or sports.

Athletes may assume that because these substances are normally found in the body, they must be safe. This assumption can be fatal.

We will look at only a few of the major physiological agents now being used as ergogenic aids:

- Blood doping
- Erythropoietin
- Oxygen supplementation
- Bicarbonate loading
- Phosphate loading

Blood Doping

Although altering blood composition in any way can be considered blood doping, the term has taken on a more specific meaning. **Blood doping** refers to any means by which a person's total volume of red blood cells is increased. This often is accomplished by transfusion of red blood cells, previously donated by either the recipient (autologous transfusions) or someone else with the same blood type (homologous transfusions). Blood doping also includes the use of erythropoietin, but this is discussed separately in the next section.

Proposed Ergogenic Benefits

The premise underlying blood doping is simple. Because most oxygen is carried in blood bound to hemoglobin, it seems logical that increasing the number of red blood cells available to ferry the oxygen to the tissues could benefit performance. If this happens, aerobic endurance and thus performance could be substantially improved.

Proven Effects

Ekblom and coworkers[19] created quite a stir in the sport world in the early 1970s. In a landmark study, they withdrew between 800 and 1,200 ml of blood from their subjects, refrigerated the blood, then reinfused the red blood cells into the subjects about four weeks later. Results showed a considerable improvement in $\dot{V}O_{2max}$ (9%) and treadmill performance time (23%) after reinfusion. Over the next few years, several studies confirmed these original findings, but several others failed to demonstrate an ergogenic effect.

Thus, the research literature was divided on the effectiveness of blood doping until a major breakthrough occurred in 1980 as a result of a study by Buick and colleagues.[12] Eleven highly trained distance runners were tested at different times during the study: (1) before blood withdrawal, (2) following blood withdrawal after allowing adequate time to reestablish normal red blood cell levels but before reinfusion of the removed blood, (3) following a sham reinfusion of 50 ml of saline (a placebo), (4) following reinfusion of 900 ml of the subject's own blood that originally had been withdrawn and preserved by freezing, and (5) after the elevated red blood cell levels had returned to normal.

As shown in figure 16.5, the researchers found a substantial increase in $\dot{V}O_{2max}$ and treadmill running time to exhaustion after the reinfusion of the red blood cells and no change after the sham reinfusion. This increase in $\dot{V}O_{2max}$ persisted for up to 16 weeks, but the improvement in treadmill time was lost within the first seven days.

Maximizing the Benefits   Why was the Buick study a major breakthrough? Gledhill[22] helped explain the controversy arising from the early studies. Many early studies that showed no improvement with blood doping had reinfused only small volumes of red blood cells, and the reinfusion was conducted within three to four weeks of the blood withdrawal. First, it appears 900 ml or more of whole blood must be reinfused to have an effect. Increases in $\dot{V}O_{2max}$ and performance are not as great when smaller volumes are used. In fact, some studies using smaller volumes failed to show any differences.

Second, it appears that it is necessary to wait for at least five to six weeks and possibly as long as 10 weeks before reinfusion. This is based on the time it takes to reestablish the blood's original, prewithdrawal hematocrit.

Finally, researchers who conducted early studies refrigerated the withdrawn blood. When blood is

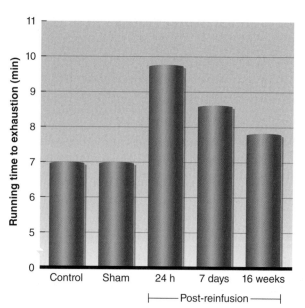

FIGURE 16.5 Changes in $\dot{V}O_{2max}$ and running time to exhaustion following reinfusion of red blood cells.

Adapted from F.J. Buick et al., 1980, "Effect of induced erythocythemia on aerobic work capacity," *Journal of Applied Physiology* 48: 636-642. Used with permission.

refrigerated, approximately 40% of the red blood cells are destroyed or lost; maximal storage time for blood under refrigeration is approximately five weeks. Later studies used frozen storage. Freezing allows an almost unlimited storage time, and only about 15% of the red blood cells are lost. Gledhill[22] concluded that blood doping significantly improves $\dot{V}O_{2max}$ and endurance performance when done under optimal conditions:

- A minimum of 900 ml of blood reinfusion
- A five- to six-week minimum interval between withdrawal and reinfusion
- Blood storage by freezing

He also showed that these improvements are the direct result of the blood's increased hemoglobin content, not an increased cardiac output caused by an expanded plasma volume.

Blood Doping and Endurance Performance Does an increase in $\dot{V}O_{2max}$ and treadmill time as a result of blood doping translate into improved endurance performance? Several studies have addressed this issue. Researchers in one study observed 5 mi (8 km) treadmill run times in a group of 12 experienced distance runners.[45] Their times were checked before and after saline (placebo) infusion and before and after blood infusion; the 5 mi (8 km) run times on the treadmill were significantly faster following blood infusion, but this difference became clear only over the second half of the trial. The blood infusion trials were 33 s (3.7%)

faster over the last 2.5 mi (4 km), and 51 s (2.7%) faster over the full 5 mi (8 km), than the placebo trials.

A second study looked at 3 mi (4.8 km) run times in a group of six trained distance runners and reported an average decrease of 23.7 s following blood doping when compared to the runners' blind control trials.[23]

Subsequent studies confirmed improvements in distance running and cross-country skiing performance with blood doping.[38] Figure 16.6 illustrates the improvement in run time with blood doping for distances of up to 11 km (6.8 mi).

Risks of Blood Doping

Although this procedure is relatively safe in the hands of competent physicians, it has inherent dangers. Adding more red blood cells to the cardiovascular system can overload it, causing the blood to become too viscous, which could lead to clotting and possibly heart failure. In an effort to control blood doping, some sport governing bodies, such as professional cycling, will not allow athletes to compete if their hematocrit is too high. With autologous blood transfusions, in which the recipient receives his or her own blood, mislabeling of the blood could occur. With homologous blood transfusions, in which blood is received from a matched donor, several other complications can occur. The reinfused blood could be mismatched. An allergic reaction could be triggered. The athlete may experience chills, fever, and nausea. The athlete also risks infection from hepatitis or HIV pathogens.[41] The potential risks of blood doping,

FIGURE 16.6 Improvements in running times for distances of up to 11 km (6.8 mi) following reinfusion of red blood cells from two units of blood preserved by freezing. Values on the y-axis reflect the reduction in time to run a specific distance on the x-axis. As an example, for a 10 km (6.2 mi) race, one could expect to run 60 s faster after reinfusion.

Adapted, by permission, from L.L Spriet, 1991, Blood doping and oxygen transport. In *Ergogenics—Enhancement of performance in exercise and sport*, edited by D.R. Lamb and M.H. Williams (Dubuque, IA: Brown & Benchmark), 213-242. Copyright 1991 Cooper Publishing Group, Carmel, IN.

even without consideration of the legal, moral, and ethical issues involved, outweigh any potential benefits.

Erythropoietin

As mentioned in the last section, erythropoietin falls into the class of blood doping, but because the mechanism is somewhat different, we examine it separately. Erythropoietin (EPO) is a naturally occurring hormone produced by the kidneys. It stimulates red blood cell production in bone marrow. In fact, this hormone is responsible for the elevated red blood cell production seen during training at altitude; training in the presence of a lower partial pressure of oxygen stimulates EPO release.

Human EPO can now be cloned through genetic engineering, so it is widely available. Several forms are available. They vary in the duration of their effect, ranging from a few days to more than a week. All are easily detectable in a blood or urine sample.

Proposed Ergogenic Benefits

EPO increases the hematocrit substantially when administered to renal failure patients, who tend to be anemic as a result of that disease. Thus, theoretically, if administered to athletes, human EPO would have the same effects as the reinfusion of red blood cells. The goal of its use is to increase the red blood cell volume, thus increasing the blood's oxygen-carrying capacity.

Proven Effects

Erythropoietin's ability to increase oxygen-carrying capacity was demonstrated in 1991 when the first study was conducted on the effects of subcutaneous injections of low doses of human EPO on maximal treadmill time and $\dot{V}O_{2max}$.[18] The study involved moderately trained to well-trained subjects. Six weeks after EPO administration, the following effects were observed:

- Both hemoglobin concentration and hematocrit increased 10%.
- $\dot{V}O_{2max}$ increased 6% to 8%.
- Time to exhaustion on the treadmill increased 13% to 17%.

Seven of the 15 subjects had been through a previous study of red blood cell reinfusion conducted four months earlier. Increases in $\dot{V}O_{2max}$ and treadmill time were almost identical in the two studies, and these improvements were attributed directly to the increase in hemoglobin.

In a second study of 20 well-trained male endurance athletes, 10 received EPO injections three times weekly for 30 days or until their hematocrit reached 50% (the safety limit imposed by some sport governing bodies), and 10 received saline injections (placebo).[8] The hematocrit in the EPO group increased from 42.7% to 50.8%, and $\dot{V}O_{2max}$ increased from 63.6 to 68.1 ml · kg^{-1} · min^{-1}. There were no changes in the placebo group. It should be noted that EPO injections stopped once the hematocrit reached 50%. Hematocrit values following EPO injections can greatly exceed this value.

Risks of Erythropoietin Use

Serious consequences can arise from EPO use. Up to 18 deaths among competitive cyclists, reported between 1987 and 1990, were alleged to be linked to EPO use, but this has not been confirmed.[3]

The outcome of EPO use is less predictable than that of red blood cell reinfusion. Once the hormone has been put into the body, no one can predict how much red blood cell production will occur. This places the athlete at great risk of substantial increases in blood viscosity. Associated problems include thrombosis (blood clot), myocardial infarction (heart attack), congestive heart failure, hypertension, stroke, and pulmonary embolism.

In focus

Blood doping and EPO can enhance aerobic capacity and the performance of aerobic sports or activities. This occurs through an increase in the oxygen-carrying capacity of blood, primarily attributable to increased red blood cells. Both procedures involve extreme risk.

Oxygen Supplementation

Watch any professional football game on television—the star running back breaks loose for a 35 yd (32 m) touchdown run, struggles back to the bench, grabs a face mask, and starts breathing 100% oxygen to facilitate his recovery. How much does he gain by using this oxygen supplementation instead of just breathing ordinary air?

Proposed Ergogenic Benefits

Obviously, the idea behind taking in more oxygen is to increase the oxygen content of the blood, as with blood doping. The aim of blood doping is to do this by increasing the oxygen-carrying capacity of the blood; with **oxygen supplementation** the idea is to achieve this by dissolving more oxygen in blood and tissues. Compared to the amount of oxygen bound to hemoglobin, however, the amount of oxygen dissolved in blood is relatively low. Breathing 100% oxygen at sea level would increase the total oxygen content of blood about 10%. By increasing this available oxygen, athletes hope to compete at higher intensities and fend off fatigue for longer periods. This technique also has been suggested as a means to speed recovery between exercise bouts.

Proven Effects

Initial attempts to investigate the ergogenic properties of pure oxygen began in the early 1900s, but it was not until the 1932 Olympic Games that oxygen was considered a potential ergogenic aid for athletic performance. That year, Japanese swimmers won impressive victories, and many attributed their success to breathing pure oxygen before competing. However, it is unclear whether their success was attributable to their use of oxygen or to their athletic abilities.

Of historical note, one of the first studies to observe the effects of breathing oxygen on performance was conducted by Sir Roger Bannister, a physician-scientist who is world renowned for his research in neurological disorders.[5] As an athlete, Dr. Bannister was the first person in the world to break the 4 min mile barrier.

Oxygen can be administered immediately before competition, during competition, during recovery from competition, or at any combination of these times.

Oxygen breathing before exercise has a limited effect on performance of that exercise bout. The total amount of work or the rate of work (exercise intensity) can be increased by breathing of oxygen if the bout is of short duration and occurs within a few seconds after the athlete breathes the oxygen. During these short bouts, submaximal work can be performed at a lower heart rate. However, no improvement occurs unless the exercise follows within seconds of breathing oxygen.

For exercise bouts exceeding 2 min, or when more than 2 min elapses between oxygen breathing and actual performance, oxygen supplementation's influence is greatly diminished. This simply reflects the limits of the body's oxygen-storage potential. Extra oxygen dissipates rapidly; little, if any, is stored.

When oxygen is administered during exercise, definite performance improvements occur. The total amount of work performed and the rate of work performed increase substantially. Likewise, submaximal work is performed with greater metabolic efficiency. Peak blood lactate levels are depressed following exhaustive exercise that the subject performs while breathing oxygen, even though considerably more work can be performed.

Studies have been unable to demonstrate any clear advantage to oxygen breathing during the recovery period. Recovery does not seem to be facilitated, nor does subsequent performance improve. In a study with professional soccer players running on a treadmill, researchers found no improvements in recovery or subsequent performance on the second exhaustive bout as a result of oxygen breathing.[46]

From a practical standpoint, oxygen administration before exercise would have little value because of the relatively short time during which oxygen stores remain elevated. The nature of most sports doesn't allow an athlete to go immediately from oxygen breathing into competition. Regardless of the ergogenic effects of oxygen intake during performance, administration during exercise has limited value for obvious reasons: Other than in high-altitude mountaineering, during which sports or events could someone carry a cylinder of oxygen?

In focus

Oxygen supplementation can increase aerobic performance but only if administered during exercise, which is not practical in sport. It has no ergogenic effect during recovery.

The recovery period seems the only practical time to administer oxygen, but this would be worthwhile only if oxygen administration were known to speed the recovery process, allowing the athlete to reenter the contest more fully recovered. However, such an effect has not been substantiated by research.

Risks of Oxygen Supplementation

At this time, no known risks are associated with oxygen supplementation. More research must be conducted to determine its safety. However, oxygen cylinders are under high pressure and oxygen promotes combustion, so oxygen equipment should never be near any heat source or flame, nor should it be allowed near anyone who is smoking.

In review

- Blood doping refers to an artificial increase in the total volume of red blood cells. It has been proposed to improve endurance performance by increasing the blood's oxygen-carrying capacity.
- Studies have shown major increases in maximal oxygen uptake, time to exhaustion, and actual performance in cross-country skiing, cycling, and distance running as a result of blood doping.
- Serious risks associated with blood doping include blood clotting; heart failure; and, if blood from another donor is used by accident or intentionally, transfusion reactions and transmission of hepatitis and HIV.
- Erythropoietin is the naturally occurring hormone that stimulates red blood cell production. It increases the number of red blood cells, and therefore the blood's oxygen-carrying capacity.
- Studies have clearly demonstrated increased maximal oxygen consumption and increased exercise time to exhaustion after administration of EPO.
- Because we cannot predict the magnitude of the response to EPO administration, it can be dangerous. The hormone can lead to death if red blood cells are overproduced, increasing the blood's viscosity. Known risks include thrombosis, myocardial infarction, congestive heart failure, hypertension, stroke, and pulmonary embolism.
- Oxygen administration during exercise improves performance but is too cumbersome to be practical. Administration before or immediately after exercise has not been proven ergogenically effective. No serious risks are associated with oxygen breathing.

Bicarbonate Loading

Recall from chapter 7 that bicarbonates are an important part of the buffering system necessary to maintain the acid–base balance of body fluids. Scientists naturally began to investigate whether performance in highly anaerobic events, in which large amounts of lactic acid are formed, could be improved by enhancing the body's buffering capacity through elevation of the blood's bicarbonate concentrations, a process called **bicarbonate loading**.

Proposed Ergogenic Benefits

By ingesting agents that increase the bicarbonate concentrations in the blood plasma, such as sodium bicarbonate (baking soda), subjects can increase blood pH, making the blood more alkaline. It was proposed that increasing plasma bicarbonate levels would provide additional buffering capacity, allowing higher concentrations of lactate in the blood. Theoretically, this could delay the onset of fatigue in short-term, all-out anaerobic work, such as all-out sprinting.

Proven Effects

Oral intake of sodium bicarbonate elevates plasma bicarbonate concentrations. However, this has little effect on intracellular concentrations of bicarbonate in muscle. Therefore, the potential benefits of bicarbonate ingestion were thought to be limited to anaerobic bouts of exercise lasting longer than 2 min, because bouts less than 2 min would be too brief to allow many hydrogen ions (H^+, from the lactic acid) to diffuse out of the muscle fibers into the extracellular fluid where they could be buffered.

In 1990, however, Roth and Brooks[35] described a cell membrane lactate transporter that operates in response to the pH gradient. Increasing the extracellular buffering capacity by ingesting bicarbonate increases the extracellular pH, which in turn increases transport of lactate from the muscle fiber via this membrane transporter to the blood plasma and other extracellular fluids. This should improve anaerobic performances even for events briefer than 2 min.

Although the theory proposing bicarbonate ingestion as an ergogenic aid for anaerobic performance is sound, the research literature is, again, conflicting. However, Linderman and Fahey,[31] in their review of the literature, found several important patterns in the research that might explain these conflicts. They concluded that bicarbonate ingestion had little or no effect on performances of less than 1 min or of more than 7 min but that for performances between 1 and 7 min, the ergogenic effects were evident. Furthermore, they found that the dose was important. Most studies that used a dose of 300 mg/kg of body mass showed a benefit, whereas most studies of lower dosage showed little or no benefit. Thus, it appears that bicarbonate ingestion of 300 mg/kg of body mass can enhance the performance of all-out, maximal anaerobic activities of 1 to 7 min duration.

An example of a study supporting these conclusions is illustrated in figure 16.7. In this study, blood bicarbonate concentrations were artificially elevated by bicarbonate ingestion before and during five sprint-cycling bouts, each lasting 1 min (see figure 16.7a).[15] Performance on the final trial improved by 42%! This elevation in blood bicarbonate levels reduced the concentration of free H^+ both during and after exercise (see figure 16.7b), thereby elevating blood pH. The authors concluded that in addition to improving buffering capacity, the extra bicarbonate appeared to speed the removal of H^+ ions from the muscle fibers, thereby reducing the decrease in intracellular pH. This conclusion essentially predicted the presence of a lactate transporter in the muscle cell membrane that Roth and Brooks[35] reported six years later.

Risks of Bicarbonate Loading

Although sodium bicarbonate has long been used as a remedy for indigestion, many authors studying bicarbonate loading have reported severe gastrointestinal discomfort in some of their subjects, including diarrhea, cramps, and bloating, from high doses of bicarbonate. These symptoms can be prevented if one ingests as much water as desired and divides the total bicarbonate dosage of at least 300 mg/kg of body mass into five equal parts over a 1 to 2 h period.[31] Also, several studies have shown sodium citrate to have similar effects on buffering capacity and performance without gastrointestinal discomfort.

Phosphate Loading

Since the early 1900s, scientists have been interested in the possibility of increasing the dietary consumption of phosphorus to improve cardiovascular and metabolic function during exercise. Several of the early studies suggested that **phosphate loading**, which involves ingestion of sodium phosphate as a dietary supplement, is an effective ergogenic aid.

Proposed Ergogenic Benefits

Phosphate loading has been proposed to have numerous potential benefits during exercise. These include elevation of extracellular and intracellular phosphate levels, which would increase the availability of phosphate for oxidative phosphorylation and phosphocreatine synthesis, thus improving energy production. Phosphate loading is also thought to enhance 2,3-diphosphoglycerate (2,3-DPG) synthesis in the red blood cells. This would shift the oxygen–hemoglobin dissociation curve to the right, permitting greater oxygen unloading in the active muscles. By reducing the affinity of hemoglobin for oxygen, higher 2,3-DPG

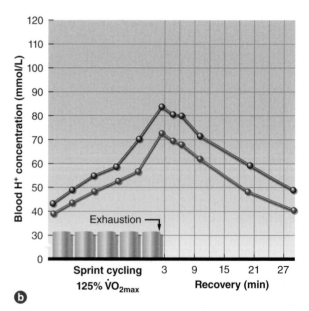

FIGURE 16.7 Concentrations of (a) blood bicarbonate (HCO_3^-) and (b) blood hydrogen ion (H^+) before, during, and after five sprint-cycling bouts with and without ingestion of sodium bicarbonate ($NaHCO_3$). The fifth sprint was performed to exhaustion. The elevated blood HCO_3^- concentrations caused attenuation of the elevation in blood H^+, a smaller decline in blood pH, a 42% increase in performance to exhaustion during the fifth sprint, and faster recovery after the sprints.

Adapted, by permission, from D.L. Costill et al., 1984, "Acid–base balance during repeated bouts of exercise: Influence of HCO_3," *International Journal of Sports Medicine* 5: 228-231.

facilitates the release of oxygen from the red blood cells. Thus, phosphate loading has been proposed to improve the cardiovascular response to exercise, improve the body's buffering capacity, and consequently improve endurance capacity and performance.

Proven Effects

Only a few studies have been conducted to determine the ergogenic benefits of phosphate loading. Unfortunately, the results are divided. Several studies showed significant improvements in $\dot{V}O_{2max}$ and time to exhaustion. However, several others showed no effects. There appear to be some potential benefits to phosphate loading, but additional research is needed to confirm this.

Risks of Phosphate Loading

At this time, no known risks are associated with phosphate loading. However, because insufficient research has been conducted to date, more studies are needed to determine its safety.

In review

- Bicarbonate is an important component of the blood's buffering system, needed to maintain normal pH by neutralizing excess acid.

- Bicarbonate loading is proposed to increase the blood's alkalinity, thus increasing the buffering capacity so that more H^+ can be neutralized. This delays the onset of fatigue.

- Bicarbonate ingestion of 300 mg/kg of body weight can delay fatigue and increase performance in all-out bouts of exercise lasting more than 1 min but less than 7 min.

- Bicarbonate loading can cause gastrointestinal distress, including cramping, bloating, and diarrhea.

- Ingestion of sodium phosphate has been postulated to improve general cardiovascular and metabolic functioning. During exercise, phosphate loading has been proposed to elevate phosphate levels throughout the body, which would increase the potential for oxidative phosphorylation and phosphocreatine synthesis, enhance oxygen release to the cells, improve cardiovascular response to exercise, improve the body's buffering capacity, and improve endurance capacity.

- Little research at this time supports the use of phosphate loading as an ergogenic aid. Existing research is conflicting, and the risks of phosphate loading are largely unknown.

Nutritional Agents

Although basic concepts of nutrition and the specific performance-enhancing properties of carbohydrates, fats, proteins, vitamins, and minerals are discussed in chapter 15, many **nutritional agents** have been proposed to have specific ergogenic properties. It is appropriate to discuss several of these in this chapter because they have received so much publicity and hype from both manufacturers and users. Most of these nutritional agents have not been adequately researched, however, so our discussion of each is brief.

Amino Acids

Specific **amino acids**, or groups of amino acids, have been proposed to have special ergogenic properties. **L-tryptophan**, an essential amino acid, has been proposed to increase aerobic endurance performance through its effects on the CNS, acting as an analgesic and delaying fatigue. L-tryptophan is the first precursor of serotonin, a potent CNS neurotransmitter. Although an initial study of L-tryptophan supplementation indicated dramatic increases in endurance performance, subsequent studies have been unable to confirm these results, showing no improvement in endurance performance.

Branched-chain amino acids (BCAA)—leucine, isoleucine, and valine—have been postulated to work in combination with L-tryptophan to delay fatigue, primarily through CNS mechanisms. There is convincing evidence that exercise-induced increases in the plasma free tryptophan/BCAA ratio are associated with increased brain serotonin and the onset of fatigue during prolonged exercise.[16] Theoretically, increasing the BCAA would reduce the ratio and delay the onset of fatigue. One study dealt with the time to exhaustion on a cycle ergometer at 70% to 75% of $\dot{V}O_{2max}$ under conditions that either increased tryptophan levels, increased BCAA levels, or reduced BCAA levels, all substantially altering the tryptophan/BCAA ratio.[42] Exercise time to exhaustion was not different among treatments (see figure 16.8). This study and others call into question the efficacy of supplementing either tryptophan or BCAA to improve endurance performance.[16]

Others have postulated that supplementation of specific amino acids increases serum GH release from the anterior pituitary gland. This also has not been clearly substantiated by research. There is some evidence that supplementation with a metabolite of leucine (β-hydroxy-β-methylbutyrate, or HMB) may increase fat-free mass and strength. It acts by decreasing the breakdown of protein that occurs with resistance training. In a recent review of the effect of HMB on fat-free mass and strength in nonathletes, only half of the

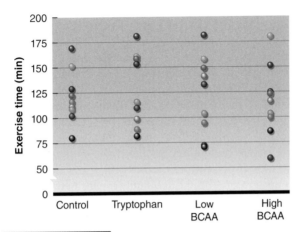

FIGURE 16.8 Time to exhaustion for individual subjects (each represented by a different color) on a cycle ergometer at 70% to 75% of $\dot{V}O_{2max}$. Each subject was tested under control conditions and after each of three treatments that increased tryptophan, reduced branched-chain amino acid (BCAA), or increased BCAA. Exercise time to exhaustion was not significantly different between control and treatment conditions.

Adapted from G. van Hall et al., 1995, "Ingestion of branched-chain amino acids and tryptophan during sustained exercise in man: Failure to affect performance," *Journal of Physiology* 486: 789-794. Used with permission.

studies showed a positive effect. There appears to be little risk associated with HMB supplementation; and, in fact, it has been reported to decrease total cholesterol, LDL-C, and systolic blood pressure.[33]

L-Carnitine

Long-chain fatty acids are the major source of energy in the body, and fatty acid oxidation provides energy both at rest and during exercise. **L-carnitine** is important in fatty acid metabolism because it assists in the transfer of fatty acids from the cytosol (the fluid portion of the cytoplasm, exclusive of organelles) across the inner mitochondrial membrane for β-oxidation. This membrane is normally impermeable to long-chain fatty acids, so the availability of L-carnitine may be a limiting factor for the rate of fatty acid oxidation.

It has been theorized that by increasing the availability of L-carnitine, athletes might facilitate the oxidation of lipids. By relying more on fat as an energy source, one could possibly spare glycogen and increase aerobic endurance capacity. The studies of L-carnitine are mixed: Several show indirect evidence of increased fat oxidation with L-carnitine supplementation, but most show no effect on fat oxidation when using either indirect or direct estimates. Studies have shown that L-carnitine supplementation does not increase muscle storage of carnitine, enhance fatty acid oxidation, spare glycogen, or postpone fatigue during exercise; nor has it been shown unequivocally to improve athletes' performance.[26]

Creatine

The use of **creatine** as an ergogenic aid has been widespread among athletes, ranging from the recreational to the professional. The primary use of creatine is based on its role in skeletal muscle, where approximately two-thirds of its total is in the form of phosphocreatine (PCr). Increasing the creatine content of skeletal muscle through creatine supplementation is theorized to increase muscle PCr levels, thus enhancing the adenosine triphosphate (ATP)-PCr energy system by better maintaining muscle ATP levels. This, in turn, theoretically would enhance peak power production during intense exercise and possibly facilitate recovery from high-intensity exercise. Creatine also serves as a buffer, helping to regulate acid–base balance, and is involved in the oxidative metabolic pathways.

Because of the popularity of creatine supplementation in the 1990s and the wide-ranging claims of its ergogenic properties, the American College of Sports Medicine (ACSM) published a consensus statement, "The Physiological and Health Effects of Oral Creatine Supplementation," in 2000.[2] A group of eminent exercise and sport scientists reviewed the existing research literature on creatine and performance at that time to develop a consensus on the actual ergogenic benefits of creatine. From their review, they concluded the following:

- Creatine supplementation can increase muscle phosphocreatine content but not in all individuals.

- The combination of creatine with a large amount of carbohydrate might increase muscle uptake of creatine.

- Exercise performance in short periods of intense, high power output activity can be enhanced, particularly with repeated bouts, consistent with the role of PCr in this type of activity.

- Maximal isometric strength, the rate of maximal force production, and maximal aerobic capacity are not enhanced by creatine supplementation.

- Creatine supplementation leads to weight gain within the first few days, likely attributable to water accumulation with creatine uptake in the muscle.

- In combination with resistance training, creatine supplementation is associated with greater gains in strength, possibly associated with the increased ability to train at higher intensities.

- The high expectations for performance enhancement exceed the true ergogenic benefits.

Subsequent scientific reviews have been published since the release of this ACSM Consensus Statement

Beware—Contamination of Nutritional Supplements

Many, if not most, athletes are taking one or more types of nutritional supplements. Most, if not all, assume that they are ingesting a substance that exactly matches the ingredients listed on the product container. The sport nutrition industry has become so big that there are now specialty stores and Internet websites selling sport nutrition products. Unfortunately, the regulations governing the purity of these products are very permissive, and many of the claims for these products have not been substantiated by scientific research studies. This lack of regulation by the U.S. Food and Drug Administration (FDA) has led to a serious problem of supplement contamination. Starting in the year 2000, researchers began to investigate the purity of many of these supplements. Their findings are sobering. In some cases products have not contained the substances listed on the label in measurable amounts, and in other cases there has been up to 150% of the listed dose. Many common supplements have been contaminated with prohibited substances that could lead to positive doping results and banning of the athlete from competition. Contaminants have included anabolic steroids, ephedrine, and caffeine. Numerous studies have now substantiated the extent and critical nature of this problem. For example, in one study conducted at the IOC-accredited laboratory in Cologne, Germany, researchers analyzed 634 nonhormonal nutritional supplements obtained from 13 countries and from 215 different suppliers. Of the 634 samples, 94 samples (14.8%) were found to contain hormones or prohormones that were not declared on the product label, and 23 samples contained compounds related to nandrolone and testosterone. The bottom line: Athletes are responsible for what they ingest; those who use supplements are taking an extremely high risk!

Information compiled by Dr. Ron J. Maughan, Loughborough University. For further information and references, refer to Maughan, R.J. (2004). Contamination of dietary supplements and positive drug tests in sport. *Journal of Sports Sciences*, 23: 883-889.

and are generally in agreement. Creatine is one of the few supplements that, in combination with resistance exercise, effectively increase both fat-free mass and strength.[33] With respect to improving athletic performance, studies are mixed. This is likely due to two factors: the physiological demands of the sport or event and the individual variability in response to the supplement. Performance enhancement is more likely to occur in sports involving brief periods of high-intensity exercise. Concerning individual variability, we discussed in chapter 9 the principle of individuality—the fact that there are high responders and low responders to any given intervention. In studies involving only a few subjects (e.g., <10), it is possible that there are more high than low responders represented in the study sample or vice versa. Finally, it is possible that creatine supplementation might enhance muscle growth by stimulating protein synthesis.

So, it appears that there is potential for ergogenic benefits from creatine supplementation. Furthermore, there appears to be little risk in supplementing creatine, particularly at the smaller doses, provided that hydration is adequate. Some weight gain, caused by water retention, is likely, which may not be desirable for some athletes.

In review

➤ A substantial risk is associated with using nutritional supplements due to potential contamination of the ingredients.

➤ Although amino acid supplementation, particularly L-tryptophan and BCAA, has been proposed to have ergogenic properties, little evidence supports this proposal; HMB, however, does appear to have ergogenic benefits.

➤ Although L-carnitine is important in fatty acid metabolism, most studies show that supplementation does not increase muscle storage of carnitine, enhance fatty acid oxidation, spare glycogen, or delay fatigue during exercise.

➤ Creatine supplementation has been shown to increase muscle creatine levels and increase performance in sports involving brief periods of high-intensity exercise.

In focus

Creatine supplementation appears to have ergogenic benefits, particularly for increasing the creatine content of skeletal muscle and for improving performance in intense, short-duration maximal exercise bouts of between 30 and 150 s.

In closing

In this chapter, we reviewed some common substances and procedures thought to have ergogenic properties. All athletes must recognize the legal, ethical, moral, and medical consequences of using any ergogenic agent. The list of banned substances continues to increase. Athletes who use banned substances risk disqualification from a particular competition, and they can be banned from competition in their sport for a year or more. In their quest for the perfect performance, athletes can easily get caught up in the hype surrounding various substances and the purported benefits they might bestow. Unfortunately, too many athletes are blinded by ambition and do not consider the consequences of their actions until their careers have been jeopardized or their health seriously impaired.

We have discussed pharmacological, hormonal, physiological, and some specific nutritional ergogenic aids. In the next part of the book, we shift our focus away from athletes in general to the unique characteristics of younger, older, and female athletes within the broader categories of growth and development, aging, and sex differences in exercise performance. We begin in chapter 17 by examining special considerations in the child and adolescent.

Key Terms

amino acids

amphetamines

anabolic steroids

β-blockers

bicarbonate loading

blood doping

branched-chain amino acids (BCAA)

caffeine

creatine

diuretics

ephedrine

ergogenic

ergolytic

hormonal agents

human growth hormone (hGH)

L-carnitine

L-tryptophan

nutritional agents

oxygen supplementation

pharmacological agents

phosphate loading

physiological agents

placebo

placebo effect

pseudoephedrine

therapeutic use exemption

Study Questions

1. What is the meaning of the term *ergogenic aid*? What is an ergolytic effect?
2. Why is it important to include control groups and placebos in studies of the ergogenic properties of any substance or phenomenon?
3. What is presently known about the use of amphetamines in athletic competition? What are the potential risks of using amphetamines?
4. Under what circumstances might β-blockers be ergogenic aids?
5. How might caffeine improve athletic performance?
6. Are diuretics ergogenic? What are some risks associated with their use?
7. What is known about alcohol, nicotine, cocaine, and marijuana as ergogenic aids?
8. What are the effects of anabolic steroid use on athletic performance? What are some of the medical risks of steroid use?
9. What is known about human growth hormone as a potential ergogenic aid? What are the risks associated with its use?
10. What is blood doping? Does blood doping improve athletic performance?
11. By what mechanism is erythropoietin theorized to benefit performance?
12. How beneficial is breathing oxygen before the start of competition, during competition, and during the recovery from competition?
13. What are the potential ergogenic properties of bicarbonate, phosphate, HMB, and creatine?

Study Guide Activities

In addition to the activities listed in the chapter opening outline on page 395, two other activities are available in the web study guide, located at

www.HumanKinetics.com/PhysiologyOfSportAndExercise

 The **KEY TERMS** activity reviews important terms, and the end-of-chapter **QUIZ** tests your understanding of the material covered in the chapter.

PART VI

Age and Sex Considerations in Sport and Exercise

From the previous parts of this book, we have gained a good understanding of the general principles of exercise and sport physiology. Historically, much of the basic and applied exercise physiology literature has focused on responses of young men. Now we turn our attention to how these principles are applied to children and adolescents, older individuals, and women. In chapter 17, "Children and Adolescents in Sport and Exercise," we examine the processes of human growth and development and how different developmental stages affect a child's physiological capacity and performance. We also consider how these stages of growth and development might alter our strategies for training young athletes for competition. In chapter 18, "Aging in Sport and Exercise," we discuss how exercise capacity and sport performance change as people age beyond the middle years, focusing on which of these changes are attributable to primary aging and what aspects of change might be attributable to an increasingly sedentary lifestyle and increased incidence of chronic diseases. We discuss the important role that training can play in minimizing the loss of performance capacity and physical conditioning that accompanies aging. In chapter 19, "Sex Differences in Sport and Exercise," we examine differences between women's and men's responses to acute exercise and exercise training and the extent to which these differences are biologically determined. We also focus on physiological and clinical issues specific to female athletes, including menstrual function, pregnancy, osteoporosis, and the high prevalence of eating disorders in female athletes.

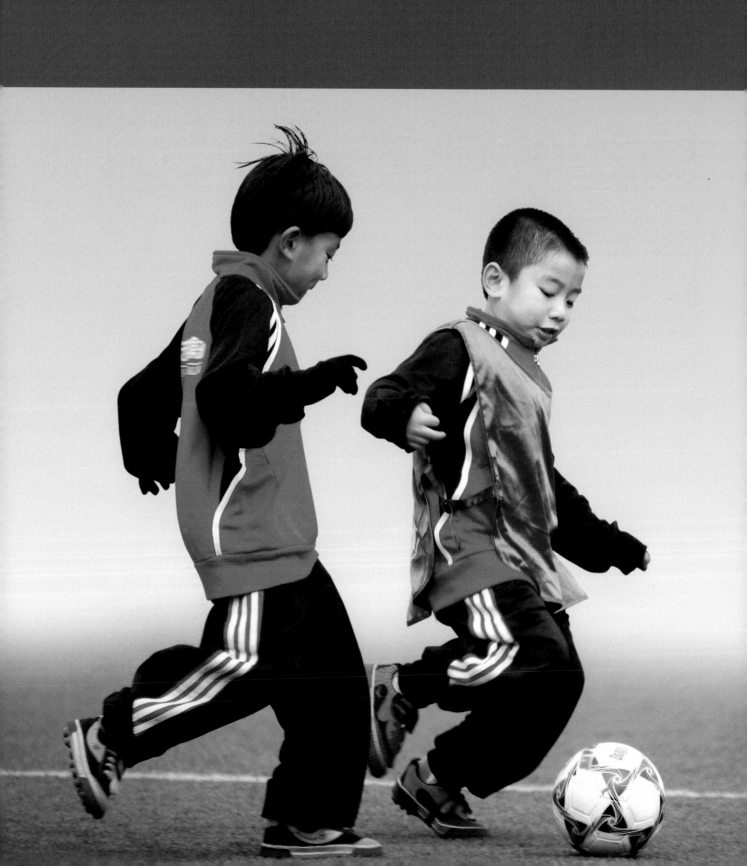

Children and Adolescents in Sport and Exercise

17

In this chapter and in the web study guide

 ACTIVITY 17.1 Terminology Examples explores real-life situations that exemplify some of the key terms used in the chapter.

 ACTIVITY 17.2 Tissue Growth and Development describes the major phases of tissue development in children and adolescents.

ACTIVITY 17.3 Response to Exercise and Training considers how physiological differences in children and adolescents affect their responses to resistance, aerobic, and anaerobic training.

The fastest man and the fastest woman in the world both come from the same small country of only 2.8 million people. Both Usain Bolt and Shelly-Ann Fraser won track and field gold medals in the 100 m race in Beijing in 2008. How can a tiny island known for sun, beaches, and reggae music produce so many track champions? While theories are plentiful, one thing that separates Jamaican athletes is their early childhood interest in track and field, nurtured by a culture that promotes, supports, and rewards childhood exercise and sport.

At a time when sustaining a program of regular physical activity in U.S. schools and those in other countries continues to face severe challenges, such is not the case in Jamaica. The Jamaican education system, through a rigorous physical education syllabus and dedicated physical education teachers, has paved the way for the island's Olympic track tradition. Exercise—and in particular, running—is engrained in the culture and widely promoted among the children of the island.

Competitions encourage children to stay active, exercise regularly, and test their athletic skills against their schoolmates. As early as age 3, while children are in preschool, they begin training and preparing for one of every school's most eagerly anticipated occasions, Sports Day. Sports Days are held at virtually every school and continued throughout high school and college. Boys and girls begin to enter nationally sponsored races when they are as young as 5; and by the time they are teenagers, top sprinters are competing before huge crowds at the National Stadium at the Inter-Secondary Schools Sports Association (ISSA) Boys' and Girls' Athletic Championship, or simply "Champs," held during the first week of April every year.

Not every child will become the next Asafa Powell or Usain Bolt, but we can learn several lessons from the Jamaican experience: Regular physical activity is good for all children, whether they grow to become world-class athletes or simply healthy and fit adults.

Throughout the previous chapters of this book, we have examined the body's physiological responses to acute bouts of exercise and its adaptations to training and the environment. But the focus has been on the adult. For many years, it was assumed that children and adolescents responded and adapted identically to adults, but few scientists actually had studied children and adolescents. There is now a cadre of researchers—pediatric exercise physiologists—who focus on the exercise responses and adaptations of the child and adolescent. It is especially important to understand how children and adolescents respond to exercise because physical activity is vital in battling the childhood obesity epidemic and in teaching children to develop lifelong healthy habits. We now have a better understanding and appreciation of both the differences and similarities between adults and children and adolescents, which we discuss in this chapter.

Growth, Development, and Maturation

Growth, development, and maturation are terms used to describe changes that occur in the body starting at conception and continuing through adulthood. **Growth** refers to an increase in the size of the body or any of its parts. **Development** refers to differentiation of cells along specialized lines of function (e.g., organ systems), so it reflects the functional changes that occur with growth. Finally, **maturation** refers to the process of taking on adult form and becoming fully

functional, and it is defined by the system or function being considered. For example, skeletal maturity refers to having a fully developed skeletal system in which all bones have completed normal growth and ossification, whereas sexual maturity refers to having a fully functional reproductive system. The state of a child's or adolescent's maturity can be defined by

- chronological age,
- skeletal age, and
- stage of sexual maturation.

Throughout this chapter we refer to the child and the adolescent. The period of life from birth to the start of adulthood is generally divided into three phases: infancy, childhood, and adolescence. **Infancy** is defined as the first year of life. **Childhood** spans the period of time between the end of infancy (the first birthday) and the beginning of adolescence and is usually divided into early childhood (preschool) and middle childhood (elementary school). The period of **adolescence** is more difficult to define in chronological years, because it varies in both its onset and its termination. Its onset generally is defined as the onset of **puberty**, when secondary sex characteristics develop and sexual reproduction becomes possible, and its termination as the completion of growth and development processes, such as attaining adult height. For most girls, adolescence ranges from 8 to 19 years and for most boys from 10 to 22 years.[20]

Given the increasing popularity of youth sport and an emphasis on increasing children's physical fitness to combat childhood obesity, we must understand the impact of exercise on the physiological aspects of growth and development. Children and adolescents must not be regarded as mere miniature versions of adults. The growth and development of their bones, muscles, nerves, and organs largely dictate their physiological and performance capacities. As children's size increases, so do almost all of their functional capacities. This is true of motor ability, strength, cardiovascular and respiratory function, and aerobic and anaerobic capacity. In the following sections, we examine age-related changes in a child's physical abilities.

To understand the physical capabilities of children and the potential impact that sport activity can have on young athletes, we must first consider the physical state of their bodies. In this section, we examine growth and development of selected body tissues.

Height and Weight

Specialists in the field of growth and development have spent considerable time analyzing the changes in height and weight that accompany growth. These two variables are most useful when we examine the rates of change. Change in height is assessed in terms of centimeters per year and change in weight in terms of kilograms per year. Figure 17.1 shows that height increases rapidly during the first two years of life. In fact, the child reaches about 50% of adult height by age 2. After this, height increases at a progressively slower rate throughout childhood; thus, there is a decline in the rate of its change. Just before puberty, the rate of change in height increases markedly, followed by an

exponential decrease in rate until full height is attained at a mean or average age of about 16 years in girls and 18 years in boys. Some boys do not reach their full height until their early 20s. The peak rate of growth in height occurs at approximately 12 years in girls and 14 years in boys. The peak rate of growth in body weight occurs at approximately 12.5 years in girls and 14.5 years in boys—slightly later than for height.

Girls mature physiologically about two years earlier than boys do.

Bone

Bones, joints, cartilage, and ligaments form the body's structural support. Bones provide points of attachment for the muscles, protect delicate tissues, and act as reservoirs for calcium and phosphorus; and some are involved in blood cell formation. Early in fetal development, the bones begin to develop from cartilage. Some flat bones, such as those of the skull, develop from fibrous membranes, but the vast majority of bones instead develop from hyaline cartilage. During fetal development, as well as during the initial 14 to 22 years of life, membranes and cartilage are transformed into bone through the process of **ossification**, or bone formation. The line of cartilage in our bones is also known as the growth plate. The average ages at which the growth plate closes and ossification is complete differ widely, but bones typically begin to fuse in the preteens, and all are fused by the early 20s. On average, girls achieve full bone maturity several years before boys. This is due to the role of different hormones, including estrogen, in signaling the growth plate to close.

The structure of mature long bones is complex. Bone is a living tissue that requires essential nutrients, so it receives a rich blood supply. Bone consists of cells distributed throughout a matrix or lattice-type arrangement, and it is dense and hard because of deposits of lime salts, mainly calcium phosphate and calcium carbonate. For this reason, calcium is an essential nutrient, particularly during periods of bone growth and in the later years of life when bone tends to become brittle because of a loss of bone mineral content associated with aging. Bones serve as a calcium reservoir. When blood calcium concentration is high, excess calcium can be deposited in the bones for storage; and when calcium concentrations are too low, bone is resorbed, or broken down, to release calcium into the blood.

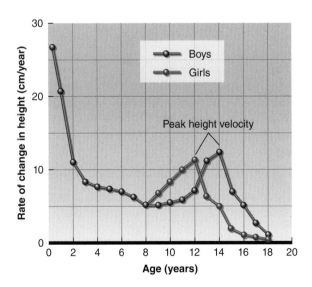

FIGURE 17.1 Changes with age in the rate of increase in height (cm/year).

When injury occurs or when extra stress is placed on a bone, more calcium is deposited. In general, bone health is evaluated by examining **bone mineral density (BMD)** as well as blood markers of bone formation and resorption. During childhood and throughout adolescence, BMD increases significantly, generally peaking sometime in the second decade and falling thereafter throughout the life span. This concept is illustrated for women across the life span in figure 17.2. Therefore, adolescence is a prime window for increasing BMD with proper nutrition and physically stressing bone through weight-bearing exercise.[21]

A recent longitudinal study incorporating jumping exercises (box jumping) for 8- to 9-year-old prepubescent boy and girls showed that simple short-term high-impact exercise confers long-term benefits. The boys and girls who participated in box jumping exercises during their school physical education class saw an increase in their BMD after seven months, and this benefit was sustained for four years after the intervention. The increase was over and above what is observed with normal growth and development. Furthermore, if the benefits of this type of exercise are sustained until BMD plateaus in early adulthood, this type of exercise could have substantial effects on reducing fracture risks later in life when BMD decreases.[12]

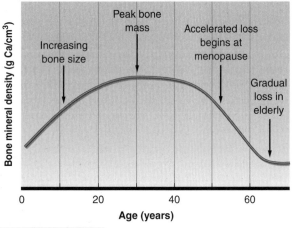

FIGURE 17.2 Changes in bone mineral density throughout the female life span. The decline after age 50 is less precipitous for men.

In focus

Exercise, along with an adequate diet, is essential for proper bone growth. Exercise affects bone width, density, and strength.

In review

> Growth in height is very rapid during the first two years of life, with a child reaching 50% of adult stature by age 2. After that, the rate is slower throughout childhood until a marked increase occurs near puberty.

> The peak rate of height growth occurs at age 12 in girls and age 14 in boys. Full height is typically attained at age 16 in girls and age 18 in boys.

> Growth in weight follows the same trend as height. The peak rate of weight increase occurs at age 12.5 in girls and age 14.5 in boys.

> Bone mineral density increases significantly throughout childhood and adolescence, peaking in early adulthood. High-impact load-bearing exercise can substantially increase BMD.

Muscle

From birth through adolescence, the body's muscle mass steadily increases, along with weight. In males, the skeletal muscle mass increases from 25% of total body weight at birth to about 40% to 45% or more in young men (20-30 years). Much of this gain occurs when the muscle development rate peaks at puberty. This peak corresponds to a sudden, almost 10-fold increase in testosterone production. Girls don't experience such rapid acceleration of muscle growth at puberty; but their muscle mass does continue to increase, although more slowly than boys', to about 30% to 35% of their total body weight as young adults. This rate difference is largely attributed to hormonal differences at puberty (see chapter 19). These percentage values for both men and women decrease with aging due to loss of muscle mass and gains in fat mass.

Increases in muscle mass with age appear to result primarily from hypertrophy (increase in size) of existing fibers, with little or no hyperplasia (increase in fiber number). Fiber hypertrophy results from increases in

In focus

The increase in muscle mass with growth and development is accomplished primarily by hypertrophy of individual muscle fibers through increases in their myofilaments and myofibrils. Muscle length increases through the addition of sarcomeres and by increases in the length of existing sarcomeres.

the myofilaments and myofibrils. Increases in muscle length as young bones elongate result from increases in the number of sarcomeres (which are added at the junction of the muscle and the tendon) and from increases in the length of existing sarcomeres. Muscle mass peaks in females at age 16 to 20 years and in males at 18 to 25 years, unless it is increased further through exercise, diet, or both.

Fat

Fat cells form and fat deposition starts in these cells early in fetal development, and this process continues indefinitely thereafter. Each fat cell can increase in size at any age from birth to death. The amount of fat that accumulates with growth and aging depends on

- diet,
- exercise habits, and
- heredity.

Heredity is unchangeable, but both diet and exercise can be altered to either increase or decrease fat stores.

At birth, 10% to 12% of total body weight is fat. At **physical maturity**, the fat content reaches approximately 15% of total body weight for males and approximately 25% for females. This sex difference, like that seen in muscle growth, is primarily attributable to hormonal differences. When girls reach puberty, their estrogen concentrations and tissue exposure increase, promoting the deposition of body fat. Figure 17.3 illustrates the changes in percent body fat, fat mass, and fat-free mass for both males and females from ages 8 to 20 years.[20] It is important to realize that both fat mass and fat-free mass increase during this time period, so an increase in absolute fat does not necessarily mean an increase in relative fat.

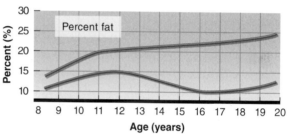

FIGURE 17.3 Changes in percent fat, fat mass, and fat-free mass for females and males from birth to 20 years of age.

Reprinted, by permission, from R.M. Malina, C. Bouchard, and O. Bar-Or, 2004, *Growth, maturation, and physical activity*, 2nd ed. (Champaign, IL: Human Kinetics), 114.

In focus

Fat storage occurs through both increases in the size of existing fat cells and increases in the number of fat cells. It appears that as existing fat cells become full, they signal the need for the development of new fat cells.

Nervous System

As children grow, they develop better balance, agility, and coordination as their nervous systems develop. Myelination of the nerve fibers must be completed before fast reactions and skilled movement can occur because conduction of an impulse along a nerve fiber

is considerably slower if myelination is absent or incomplete (chapter 3). **Myelination** of the cerebral cortex occurs most rapidly during childhood but continues well beyond puberty. Although practicing an activity or skill can improve performance to a certain extent, the full development of that activity or skill depends on full maturation (and myelination) of the nervous system. The development of strength is also likely influenced by myelination.

In review

➤ Muscle mass increases steadily along with weight gain from birth through adolescence.

➤ In boys, the rate of muscle mass increase peaks at puberty, when testosterone production increases dramatically. Girls do not experience this sharp increase in muscle mass.

➤ Muscle mass increases in boys and girls result primarily from fiber hypertrophy with little or no hyperplasia.

➤ Muscle mass peaks in girls between ages 16 and 20 and in boys between ages 18 and 25, although it can be further increased through diet and exercise.

➤ Fat cells can increase in size and number throughout life.

➤ The amount of fat accumulation depends on diet, exercise habits, and heredity.

➤ At physical maturity, the body's fat content averages 15% in young men and 25% in young women. The differences are caused primarily by higher testosterone levels in males and higher estrogen levels in females.

➤ Balance, agility, and coordination improve as children's nervous systems develop.

➤ Myelination of nerve fibers must be complete before fast reactions and skilled movements are fully developed because myelination speeds the transmission of electrical impulses.

Physiological Responses to Acute Exercise

The function of almost all physiological systems improves until full maturity is reached or shortly before. After that, physiological function plateaus for a period of time before starting to decline with advancing age. In this section, we focus on some of the changes in children and adolescents that accompany growth and development, including the following:

- Strength
- Cardiovascular and respiratory function
- Metabolic function, including aerobic capacity, running economy, anaerobic capacity, and substrate utilization

Strength

Strength improves as muscle mass increases with age. Peak strength usually is attained by age 20 in women and between ages 20 and 30 in men. The hormonal changes that accompany puberty lead to marked increases in strength in pubescent males because of the increased muscle mass noted before. The extent of development and the performance capacity of muscle

also depend on the relative maturation of the nervous system. High levels of strength, power, and skill are impossible if the child has not reached neural maturity. Myelination of many motor nerves is incomplete until sexual maturity, so the neural control of muscle function is limited before that time.

Figure 17.4 illustrates changes in leg strength in a group of boys from the Medford Boys' Growth Study.[5] The boys were followed longitudinally from age 7 to 18. The rate of strength gain (slope of line) increased noticeably around age 12, the typical age for onset of puberty. Similar longitudinal data for girls over this same age span are not available. Cross-sectional data, however, indicate that girls experience a more gradual and linear increase in absolute strength and do not exhibit any significant change in strength relative to body weight after puberty,[11] as shown in figure 17.5.

FIGURE 17.4 Gains with age in leg strength of young boys followed longitudinally over 12 years. Note the increase in the slope of the curve from 12 to 16 years of age.

Data from Clarke, 1971.

Cardiovascular and Respiratory Function

Cardiovascular function undergoes considerable change as children grow and age. Because of significant increases in aerobic power over the course of growth and development, we need to consider these changes during both submaximal and maximal exercise.

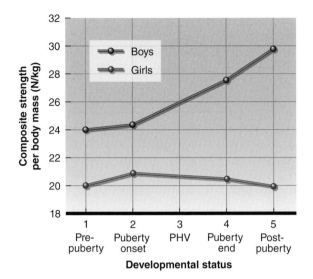

FIGURE 17.5 Changes in strength with developmental status in boys and girls. Strength is expressed as a composite static strength score from several strength testing sites, and the data are expressed per kilogram of body mass to account for differences in size between boys and girls. PHV = peak height velocity.

Reprinted from K. Froberg and O. Lammert, 1996, "Development of muscle strength during childhood." In *The child and adolescent athlete* (London: Blackwell Publishing Company) 28. Copyright 1996 by Blackwell Publishing. Reproduced with permission of Blackwell Publishing Ltd.

Rest and Submaximal Exercise

Blood pressure at rest and during submaximal exercise is lower in children than in adults but progressively increases to adult values during the late teen years. Blood pressure is directly related to body size. Larger people generally have bigger hearts and higher blood pressures, so size is at least partially responsible for children's lower blood pressure values. In addition, blood flow to active muscles during exercise in children can be greater for a given volume of muscle than in adults because children have less peripheral resistance. Thus, in children, for a given submaximal workload, blood pressure is lower and the muscle is relatively overperfused.

Recall that cardiac output is the product of heart rate and stroke volume. A child's smaller heart size and total blood volume result in a lower stroke volume, both at rest and during exercise, than in an adult. In an attempt to compensate for the lower stroke volume to maintain cardiac output, the child's heart rate response to a given rate of submaximal work (such as on a cycle ergometer), where the absolute oxygen requirement is the same, is higher than an adult's. As the child ages, heart size and blood volume increase along with body size. Consequently, stroke volume also increases as body

size increases, for the same absolute rate of work, and heart rate decreases.

However, a child's higher submaximal heart rate cannot completely compensate for the lower stroke volume. Because of this, the child's cardiac output is also somewhat lower than the adult's for a given absolute rate of work or a given oxygen consumption. To maintain adequate oxygen uptake during these submaximal levels of work, the child's arterial–mixed venous oxygen difference, or $(a-\bar{v})O_2$ difference, increases to further compensate for the lower stroke volume. The increase in $(a-\bar{v})O_2$ difference is most likely attributable to increased blood flow to the active muscles—a greater percentage of the cardiac output goes to the active muscles.[35] These submaximal relationships are illustrated in figure 17.6, in which the responses of a 12-year-old boy are compared with those of a fully mature man.

Maximal Exercise

Maximum heart rate (HR_{max}) is higher in children than in adults but decreases linearly as children age. Children under age 10 frequently have maximum heart rates exceeding 210 beats/min, whereas the average 20-year-old has a maximum heart rate of approximately 195 beats/min. With further aging (25-30 years and older), however, results of cross-sectional studies suggest that maximum heart rate decreases by slightly less than 1 beat/min per year. Longitudinal studies, on the other hand, suggest that maximum heart rate decreases only 0.5 beats/min per year. This decrease in maximum heart rate throughout the life span is caused by a decrease in sensitivity of the cardiac β-adrenergic receptors.

During maximal exercise, as also seen with submaximal exercise, the child's smaller heart and blood volume limit the maximal stroke volume that he or she can achieve. Again, the high HR_{max} cannot fully compensate for this, leaving the child with a lower maximal cardiac output than the adult. This limits the child's performance at high absolute workloads (e.g., pedaling at 100 W on a cycle ergometer or trying to achieve the same absolute $\dot{V}O_2$) because the child's capacity for oxygen delivery is less than an adult's. However, for high relative workloads in which the child is responsible for moving only his or her body mass (e.g., running on a treadmill at the same speed with no grade), this lower maximal cardiac output is not as serious a limitation. In running, for example, a 25 kg (55 lb) child requires (in proportion to body size) considerably less oxygen than a 90 kg (198 lb) man would require, yet the rate of oxygen consumption when scaled for body size is about the same for both.

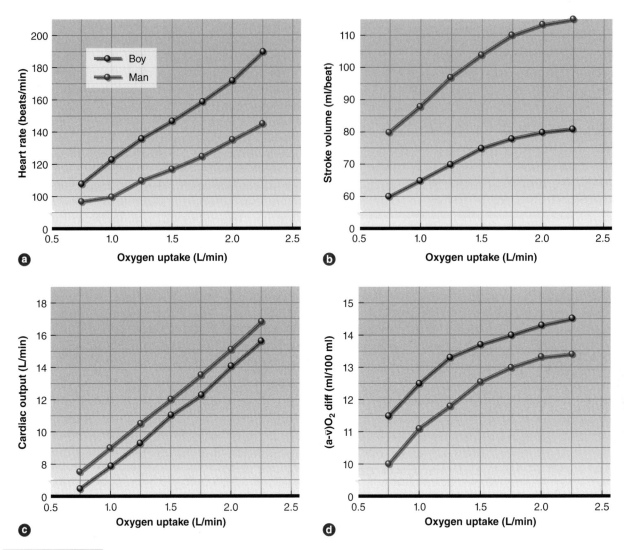

FIGURE 17.6 Submaximal (*a*) heart rate, (*b*) stroke volume, (*c*) cardiac output, and (*d*) arterial–venous oxygen difference, or (a-v̄)O$_2$ difference, in a 12-year-old boy and a fully mature man at the same rates of oxygen uptake.

In focus

Heart size is directly related to body size, so children have smaller hearts than adults. As a result of this and a smaller blood volume, children have a smaller stroke volume capacity. A child's higher maximum heart rate can only partially compensate for this lower stroke volume, and thus maximal cardiac output is lower than that of an equally trained adult.

Lung Function

Lung function changes markedly with growth. All lung volumes increase until growth is complete. Peak flow rates follow the same pattern. The changes in these volumes and flow rates are matched by the changes in the highest ventilation that can be achieved during exhaustive exercise, which is referred to as maximal expiratory ventilation ($\dot{V}_{Emax}$), or maximal minute ventilation. $\dot{V}_{Emax}$ increases with age until physical maturity and then decreases with aging. For example, cross-sectional data show that $\dot{V}_{Emax}$ averages about 40 L/min for 4- to 6-year-old boys and increases to 110 to 140 L/min at full maturity. Girls follow the same general pattern, but their absolute values are considerably lower postpuberty, primarily because of their smaller body size. These changes are associated with the growth of the pulmonary system, which parallels the general growth patterns of children. As body size increases with growth and development, so do lung size and function.

Metabolic Function

Metabolic function and substrate utilization at rest and during exercise also change as the child and adolescent grow larger, as we would expect from the changes that we have just reviewed in muscle mass and strength and cardiorespiratory function.

Aerobic Capacity

The purpose of the basic cardiovascular and respiratory adaptations that occur in response to varying levels of exercise (rates of work) is to accommodate the exercising muscles' need for oxygen. Thus, increases in cardiovascular and respiratory function that accompany growth suggest that aerobic capacity ($\dot{V}O_{2max}$) similarly increases. In 1938, Robinson[24] demonstrated this phenomenon in a cross-sectional sample of boys and men ranging in age from 6 to 91 years. He found that $\dot{V}O_{2max}$ peaks between ages 17 and 21 and then decreases linearly with age. Other studies subsequently confirmed these observations. Studies of girls and women have shown essentially the same trend, although in females the decrease begins at a much younger age, generally age 12 to 15 (refer to chapter 19), probably attributable to an earlier assumption of a more sedentary lifestyle. The changes in $\dot{V}O_{2max}$ with age, expressed in liters per minute, are illustrated in figure 17.7a.

Expressing $\dot{V}O_{2max}$ relative to body weight (ml · kg⁻¹ · min⁻¹) provides a considerably different picture, as shown in figure 17.7b. Values change little in boys from age 6 to young adulthood. For girls, little change occurs from age 6 to 13; but after age 13, aerobic capacity shows a gradual decrease. Although these observations are of

a Absolute values

b Relative to body weight

FIGURE 17.7 Changes in maximal oxygen uptake with age in children and adolescents.

general interest, they might not accurately reflect the development of the cardiorespiratory system as children grow and their physical activity levels change. Several concerns have been raised about the validity of using $\dot{V}O_{2max}$ relative to body weight to account for changes in the cardiorespiratory and metabolic systems during periods of growth and development. Rather, the sex differences that begin to emerge around puberty may reflect sex differences in increasing body mass and changing body composition. Because girls tend to increase fat mass with estrogen exposure at puberty, their $\dot{V}O_2$ relative to total body weight ($\dot{V}O_2$ per kilogram) decreases, but this may not be significant when normalized for fat-free mass.

Arguments against using body weight to scale $\dot{V}O_{2max}$ for differences in size include the following. First, although $\dot{V}O_{2max}$ values expressed relative to body weight remain relatively stable or decline with age, endurance performance steadily improves. The average 14-year-old boy can run the mile (1.6 km) almost twice as fast as the average 5-year-old boy, yet the two boys' $\dot{V}O_{2max}$ values expressed relative to body weight are similar.[27] Second, although the increases in $\dot{V}O_{2max}$ that accompany endurance training in children are relatively small compared with those in adults, the performance increases in these children are relatively large. Therefore, body weight is likely not the most appropriate way to scale $\dot{V}O_{2max}$ values for differences in body size in children and adolescents. The relationships between $\dot{V}O_{2max}$, body dimensions, and system functions during growth are extraordinarily complex.[6, 28] This is discussed in greater detail later in the chapter.

In focus

Aerobic capacity ($\dot{V}O_{2max}$), when expressed in liters per minute, is lower in children than in adults at similar levels of training. This is attributable primarily to the child's lower maximal cardiac output. When $\dot{V}O_{2max}$ values are expressed to normalize for the differences in body size between children and adults, there is little or no difference in aerobic capacity.

Running Economy

How do growth-related changes in aerobic capacity affect a child's performance? For any activity that requires a fixed rate of work, such as cycling on an ergometer, the child's lower $\dot{V}O_{2max}$ limits endurance performance. But as noted earlier, for activities in which body weight is the major resistance to movement, such as distance running, children should not be at a disadvantage, because their $\dot{V}O_{2max}$ values expressed relative to body weight are already at or near adult values.

Yet children cannot maintain a running pace as fast as adults can because of basic differences in economy of effort. At a given speed on a treadmill, a child will have a substantially higher submaximal oxygen consumption when expressed relative to body weight than an adult. As children age, their legs lengthen, their muscles become stronger, and their running skills improve. Running economy increases, and this improves their distance-running pace, even if the children are not training and if their $\dot{V}O_{2max}$ values don't increase.[7, 16] Rowland argues that increased stride frequency as children and adolescents grow is the most important factor in explaining these changes in running economy.[29] It is also possible that scaling oxygen consumption to body weight is inappropriate during growth and development, as discussed in the previous section.[25]

In review

➤ As pulmonary and cardiovascular function improve with continued development, so does aerobic capacity.

➤ $\dot{V}O_{2max}$, expressed in liters per minute, peaks between ages 17 and 21 years in males and between 12 and 15 years in females, after which it plateaus for several years and then steadily decreases.

➤ When $\dot{V}O_{2max}$ is expressed relative to body weight, it plateaus in males from ages 6 to 25 years before it begins to decline. In females, the decline in $\dot{V}O_{2max}$ is small from ages 6 to 12 years but becomes more substantial starting at about age 13. However, expressing $\dot{V}O_{2max}$ relative to body weight might not provide an accurate estimate of aerobic capacity. Such $\dot{V}O_{2max}$ values do not reflect the significant gains in endurance performance capacity that are noted with both maturation and training.

➤ The child's lower $\dot{V}O_{2max}$ value (L/min) limits endurance performance unless body weight is the major resistance to movement, as in distance running.

➤ When expressed relative to body weight, a child's $\dot{V}O_{2max}$ is similar to an adult's, yet in activities such as distance running, a child's performance is far inferior to adult performance.

➤ Running economy is lower in children compared with adults when $\dot{V}O_2$ is expressed relative to body weight. One factor has been identified to explain this difference: the difference between children and adults in stride frequency for the same fixed-pace run.

Anaerobic Capacity

Children have a limited ability to perform anaerobic-type activities due to a lower glycolytic capacity. This is demonstrated in several ways. First, muscle glycogen content in children is about 50% to 60% that of adults.[8] Children do not achieve adolescent or adult concentrations of lactate in either muscle or blood for maximal and supramaximal rates of exercise.[3] This suggests that children have a lower glycolytic capacity. The lower lactate concentrations might reflect a lower concentration of phosphofructokinase, the key rate-limiting enzyme of anaerobic glycolysis, and significantly lower (~3.5-fold) lactate dehydrogenase activity.[14] Lower blood lactate concentrations in children after exhaustive exercise may reflect their lower relative muscle mass, higher lactate clearance, a greater reliance on aerobic metabolism, or some combination of these.[23] In terms of the other anaerobic metabolic pathways, children's resting stores of adenosine triphosphate (ATP) and phosphocreatine (PCr) are similar to those of adults, so activities of less than 10 to 15 s should not be compromised. Thus, only activities that tax the anaerobic glycolytic system—those from 15 s up to 2 min in duration—will be limited.

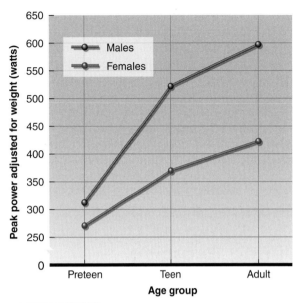

FIGURE 17.8 Optimal peak power output (anaerobic power) statistically adjusted for body mass in preteenagers (9-10 years old), teenagers (14-15 years old), and adults (mean age of 21 years). These values represent anaerobic power independent of body size.

Data from Santos et al., 2002.

In focus

Anaerobic capacity is lower in children than adults, which may reflect children's lower muscle glycogen content, lower concentration of the key rate-limiting enzyme phosphofructokinase, or decreased activity of lactate dehydrogenase.

Anaerobic mean and peak power output, as determined by the Wingate anaerobic power test (a 30 s, all-out maximal effort on a cycle ergometer), is also lower in children than in adults. Figure 17.8 illustrates the results of a similar cycle ergometer anaerobic power test.[32] In this figure, peak power is statistically adjusted for body mass to account for differences in body size when we compare values for preteenagers, teenagers, and adults. This figure demonstrates the very low peak power outputs for preteenagers (9-10 years of age) compared with both teenagers (14-15 years of age) and adults (mean age of 21 years). Teenagers were much closer to the values for adults than the preteenagers.

Bar-Or[1] summarized the development of both the aerobic and anaerobic characteristics of boys and girls from ages 9 through 16, using 18 years of age as the criterion for 100% of the adult value. The changes with age are shown in figure 17.9. Aerobic power is represented by the child's $\dot{V}O_{2max}$, whereas anaerobic power is represented by the child's performance on

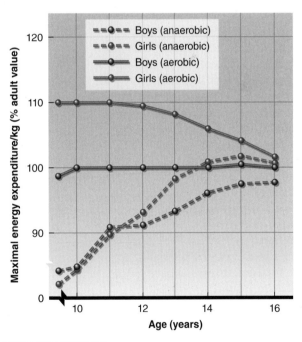

FIGURE 17.9 Development of aerobic and anaerobic characteristics in boys and girls ages 9 to 16 years. Values are expressed as a percentage of adult values (values at age 18).

Adapted, by permission, from O. Bar-Or, 1983, *Pediatric sports medicine for the practitioner: From physiologic principles to clinical applications* (New York: Springer-Verlag).

the Margaria step-running test (a field test). Maximal energy expenditure per kilogram represents the maximal energy-generating capacities of the aerobic and anaerobic systems, scaled to body weight to account for body size differences with growth. Notice that aerobic fitness remains constant for the boys but declines for the girls from 12 to 16 years. Nine- to 12-year-old girls have a higher aerobic capacity than the 18-year-old reference adult value; thus, their values are 110% of the adult value. For both boys and girls, anaerobic capacity increases from 9 through 15 years of age.

Endocrine Responses and Substrate Utilization During Exercise

As discussed in previous chapters, physical activity causes the release of several key metabolic regulatory hormones to mobilize carbohydrates and fats to be used as fuels. Many of the hormones that regulate metabolism during exercise can also influence growth and development. For example, exercise is a potent stimulus for the growth hormone (GH) and the insulin-like growth factor axis. High-intensity exercise in children can cause dramatic peaks in GH and influence the normal daily cycling of this hormone. It was once hypothesized that increased GH caused by intense exercise contributed to increased growth in adolescence. While this hypothesis has not been confirmed, it is known that children and adolescents have a different metabolic hormone profile during exercise.

In general, the pediatric-focused studies suggest that the insulin response to exercise differs with pubertal stage and sex[23] and that children have a higher stress response to exercise. This

results in differences in blood glucose control. At the start of exercise, children have a relative hypoglycemia. The reasons for this are unclear; but in addition to lower muscle glycogen content, it is thought that children have an immature capacity for hepatic glycogenolysis. It is therefore not surprising that children rely more heavily on fat oxidation for fuel during exercise. However, exogenous glucose oxidation appears to be relatively high, potentially due to reduced endogenous glucose production. This fuel utilization profile changes throughout puberty such that adolescents have a decreased relative rate of fat oxidation, more like that in adults. The change in substrate utilization during exercise may have an impact on body composition throughout development, and from a practical point of view it may also affect nutritional needs for peak performance in children.

Scaling Physiological Data to Account for Size Differences

Throughout this chapter and previous chapters, we have discussed the need to express physiological data relative to the size of the individual. In chapter 5, when we introduced the concept of $\dot{V}O_{2max}$, we mentioned that values normally are expressed relative to body mass, by dividing the absolute $\dot{V}O_{2max}$ (expressed in L/min) by body weight (ml · kg^{-1} · min^{-1}). Many scientists believe that dividing by body mass alone does not adequately account for size differences. This becomes a major issue when we compare children's values with those of adults, or men's with women's, as we see in chapter 19.

Strong cases have been made for scaling $\dot{V}O_2$, cardiac output, stroke volume, and other size-related physiological variables relative to body surface area, measured in square meters, or relative to weight, expressed to the 0.67 power or 0.75 power (wt$^{0.67}$ or wt$^{0.75}$). For years, cardiologists have expressed heart volumes relative to body surface area. Recent research suggests that using body surface area (ml · m^{-2} · min^{-1}) or wt$^{0.75}$ (ml · kg$^{-0.75}$ · min^{-1}) provides the best means by which to express the data to reduce the effect of body size.[25] One study followed young boys longitudinally from 12 to 20 years of age; one group remained untrained but active, and the other group trained.[33] There was little or no increase with run training in $\dot{V}O_{2max}$ expressed in ml · kg^{-1} · min^{-1}, whereas submaximal $\dot{V}O_2$ expressed in the same manner decreased with age, suggesting no change in aerobic capacity but an improvement in running economy. When these same data were expressed in ml · kg$^{-0.75}$ · min^{-1}, the boys who were training showed increased aerobic capacity with increased training and age but no change in running economy. This latter finding intuitively makes more sense, suggesting the use of wt$^{0.75}$ as the best way of expressing the data.

In review

- Children's ability to perform anaerobic activities is limited. A child has a lower glycolytic capacity, possibly because of a limited amount of phosphofructokinase or lactate dehydrogenase.
- Children have lower lactate concentrations in both blood and muscle at maximal and supra-maximal rates of work.
- Anaerobic mean and peak power outputs are lower in children than in adults, even when scaled for body mass.
- Children have a different insulin and stress hormone response and rely on greater fat oxidation for fuel during exercise.

Physiological Adaptations to Exercise Training

We have seen that, indeed, children are not just miniature adults. The child is physiologically distinct from the adult and must be considered differently. But how should these differences affect individualized training programs for children? Training can improve the strength, aerobic capacity, and anaerobic capacity of children. Generally, children adapt well to the same type of training routine used by adults. But training programs for children and adolescents should be designed specifically for each age-group, keeping in mind the developmental factors associated with that age. In this section, we look at training-induced changes in each of the following:

- Body composition
- Strength
- Aerobic capacity
- Anaerobic capacity

Then, where appropriate, we discuss proper training procedures to optimize performance gains and reduce the risk of injury.

Body Composition

The child and adolescent respond to physical training similarly to adults with respect to changes in body weight and composition. With both resistance and aerobic training, both boys and girls will decrease body weight and fat mass and increase fat-free mass, although the increase in fat-free mass is attenuated in the child compared with the adolescent and adult. As we have already observed, there is also evidence of significant bone growth as a result of high-impact weight-bearing exercise training,[12] above that seen with normal growth.

Strength

For many years, the use of resistance training to increase muscular strength and endurance in prepubescent and adolescent boys and girls was highly controversial. Boys and girls were discouraged from using free weights for fear that they might injure themselves and prematurely stop the growth process. Furthermore, many scientists speculated that resistance training would have little or no effect on the muscles of prepubescent boys because their circulating androgen concentrations were still low. It is now more widely accepted that certain types of resistance training are safe and provide significant benefits for youth and adolescents.[34] Kraemer and Fleck[15] concluded that the risk of injury with resistance training in youth is very low. In fact, resistance training might offer some protection against injury, for example by strengthening the muscles that cross a joint. Still, a conservative approach is recommended in prescribing resistance exercise for children, particularly preadolescents.

A number of studies conducted on both children and adolescents have clearly demonstrated that resistance training is very effective in increasing strength. The increase is largely dependent on the volume and intensity of training. Further, the percentage increases for children and adolescents are similar to those for young adults.[9]

The mechanisms allowing strength changes in children are similar to those for adults, with one minor exception: Prepubescent strength gains are accomplished largely without any changes in muscle size and likely involve improvements in neural mechanisms, including improved motor skill coordination, increased motor unit activation, and other undetermined neurological adaptations.[22] Strength gains in the adolescent result primarily from both neural adaptations and increases in muscle size and specific tension.

For actual training programs, resistance training for children should be prescribed in much the same way as for adults with a special emphasis on teaching proper lifting technique. Specific guidelines have been established by a number of professional organizations, including the American Orthopaedic Society for Sports Medicine, the American Academy of Pediatrics, the American College of Sports Medicine, the National Athletic Trainers' Association, the National Strength and Conditioning Association, the President's Council on Physical Fitness and Sports, the U.S. Olympic Committee, and the Society of Pediatric Orthopaedics. Basic guidelines have been established for the progression of resistance exercise in children, which are presented in table 17.1.[15] Further information on resistance training program designs for children is available.[10, 15]

TABLE 17.1 Basic Guidelines for Resistance Exercise Progression in Children

Age	Considerations
7 years or younger	Introduce child to basic exercises using little or no weight; develop the concept of a training session; teach exercise technique; progress from body weight calisthenics, partner exercises, and lightly resisted exercises; keep volume low.
8-10 years	Gradually increase the number of exercises; practice exercise technique in all lifts; start gradual progressive loading of exercises; keep exercises simple; gradually increase training volume; carefully monitor tolerance of the exercise stress.
11-13 years	Teach all basic exercise techniques; continue progressive loading of each exercise; emphasize exercise techniques; introduce more advanced exercises with little or no resistance. Progress to more advanced youth programs in resistance exercise; add sport-specific components; emphasize exercise techniques; increase volume.
14-15 years	Progress to more advanced youth programs in resistance exercise; add sport-specific components; emphasize exercise techniques; increase volume.
16 years or older	Move child to entry-level adult programs after all background knowledge has been mastered and a basic level of training experience has been gained.

Note. If a child of any age begins a program with no previous experience, start the child at the level for the previous age category and move him or her to more advanced levels as exercise toleration, skill, amount of training time, and understanding permit.

Reprinted, by permission, from W.J. Kraemer and S.J. Fleck, 2005, *Strength training for young athletes*, 2nd ed. (Champaign, IL: Human Kinetics), 5.

In focus

Prepubescent children can improve their strength with resistance training. These strength gains are attributable largely to neurological factors, with little or no change in the size of the muscle.

Aerobic Capacity

Do prepubescent boys and girls benefit from aerobic training to improve their cardiorespiratory systems? This also has been a highly controversial area because several early studies indicated that training prepubescent children did not change their $\dot{V}O_{2max}$ values.[26] Interestingly, even without significant increases in $\dot{V}O_{2max}$, the running performance of the children studied did improve substantially.[30] They could run a fixed distance faster following the training program. This improvement likely involved improved running economy. Other studies have shown small increases in aerobic capacity with training in prepubescent children, but these increases are less than would be expected for

adolescents or adults—about 5% to 15% in children compared with about 15% to 25% in adolescents and adults.

More substantial changes in $\dot{V}O_{2max}$ appear to occur once children have reached puberty, although the reason for this is unknown. Because stroke volume appears to be the major limitation to aerobic performance in this age-group, it is quite possible that further increases in aerobic capacity depend on heart growth. Also, as discussed earlier in this chapter, scaling of these variables is an issue. The study[33] presented in the sidebar on page 437 clearly establishes this as a key factor.

Anaerobic Capacity

Anaerobic training appears to improve children's anaerobic capacity. Following training, children have

- increased resting levels of PCr, ATP, and glycogen;
- increased phosphofructokinase activity; and
- increased maximal blood lactate levels.[1, 8]

Addressing Childhood Obesity

There is presently an epidemic of obesity in the United States, Canada, much of Europe, and other westernized countries as discussed in chapter 22. This is true not only in adults but in children and adolescents as well.

Many factors contribute to childhood obesity, including genetics, nutrition, increased screen time (television and video games), and decreased physical activity. As with adults, obese and overweight children have an increased risk for metabolic syndrome, dyslipidemia, hypertension, and type 2 diabetes.[13]

A significant amount of research is being conducted to find the causes of and appropriate treatments for childhood obesity. Observations in twin, sibling, and family studies suggest that children are more likely to be overweight if relatives are overweight and that heritability may play a role in as many as 25% to 85% of cases. However, these genetic factors are not the sole cause of the worldwide epidemic of childhood obesity. It is more likely that most of the westernized world's population carries genes that have evolved to cope with food scarcity. This, in combination with excess portion sizes of calorie-rich food and low energy expenditure, has led to an epidemic with significant world health and financial ramifications.[17]

The central treatment for childhood obesity and the cluster of diseases associated with obesity involves diet modification, increasing physical activity, and promoting healthy lifestyles. Treatment programs that involve nutritional interventions in combination with exercise have a much higher success rate than diet modification alone. However, increasing the time that children spend being physically active has its challenges. Children and adolescents spend most of their waking hours in school, yet hours spent in physical education classes have steadily declined. *Healthy People 2010* recommended increasing the amount of daily physical education; but a 2000 school health policy survey showed that only 8% of American elementary schools, 6.4% of middle schools, and 5.8% of high schools with physical education requirements provided daily physical education classes for all grades.[4] Exercise physiologists can play a significant role in battling this epidemic by showing children the benefits of exercise and making exercise a fun lifelong habit.

Ventilatory threshold, a noninvasive marker of lactate threshold, also has been reported to increase with endurance training in 10- to 14-year-old boys.[18]

When one is designing aerobic and anaerobic training programs for children and adolescents, it appears that standard training principles for adults can be applied. Children and adolescents have not been well studied, but what we do know suggests that they can be trained in a manner similar to that for adults. Again, because children and adolescents are not adults, it is prudent to be conservative to reduce the risk of injury, overtraining, and loss of interest in sport. The approach outlined earlier for resistance training is a good model to use for aerobic and anaerobic training. This is also an appropriate time in life to focus on learning a variety of motor skills by having children explore a number of activities and sports.

In review

➤ Body composition changes with training in children and adolescents are similar to those seen in adults—loss of total body weight and fat mass and increase in fat-free mass.

➤ The risk of injury from resistance training in young athletes is relatively low, and the programs they should follow are much like those for adults.

➤ Strength gains achieved from resistance training in preadolescents result primarily from improved motor skill coordination, increased motor unit activation, and other neurological adaptations. Unlike adults, preadolescents who resistance train experience little change in muscle size. Mechanisms of strength gains in adolescents are similar to those for adults.

➤ Aerobic training in preadolescents does not alter $\dot{V}O_{2max}$ as much as would be expected for the training stimulus, possibly because $\dot{V}O_{2max}$ depends on heart size. But endurance performance improves with aerobic training. Adolescents are similar to adults in their improvement.

➤ A child's anaerobic capacity increases with anaerobic training.

Motor Ability and Sport Performance

As shown in figure 17.10, the motor ability of boys and girls generally increases with age for the first 17 years, although girls tend to plateau at about the age of puberty for most items tested. These improvements result primarily from development of the neuromuscular and endocrine systems and secondarily from the increased activity. Although these data are older, they come from the last time this type of large-scale assessment was conducted.

The plateau observed in the girls at puberty is likely explained by three factors. First, as mentioned earlier, the increase in estrogen concentrations at puberty, or in the estrogen/testosterone ratio, leads to increased fat deposition. Performance tends to decrease as fat increases. Second, girls have less muscle mass. Finally, and probably of greater importance, around puberty many girls assume a much more sedentary lifestyle than boys. This is largely a matter of social conditioning, as boys are encouraged to be more active and athletic than girls. As girls become less active, their motor abilities tend to plateau. This trend appears to be changing because of changing social attitudes and more opportunities for sport and activity now available for girls (see chapter 19).

Sport performance in children and adolescents improves with growth and maturation, as can be seen for age-group records in sports such as swimming and track and field. Figure 17.11 illustrates the improvement in American records for various age-groups.

The figure gives values for the 100 m and 400 m swim and the 100 m and 1,500 m run. These events were selected because they represent a predominantly anaerobic event in swimming and running (100 m swim and run) and a predominantly aerobic activity (400 m swim and 1,500 m run). Both anaerobic and aerobic performances improve progressively with increasing age-groups, with the exception of the 1,500 m run for 17- and 18-year-old girls. Similar age-group records for weightlifting do not appear to be available, because weightlifting competition is organized by weight in broad classifications such as 16 and under, 17 to 20 years of age, and then adult classifications. On the basis of normal strength gains with growth and development, it is assumed that weightlifting records would increase markedly from late childhood through adolescence, particularly in boys.

Special Issues

During the period of growth and development from childhood through adolescence, these special issues need to be addressed:

- Thermal stress
- Growth and maturation with training

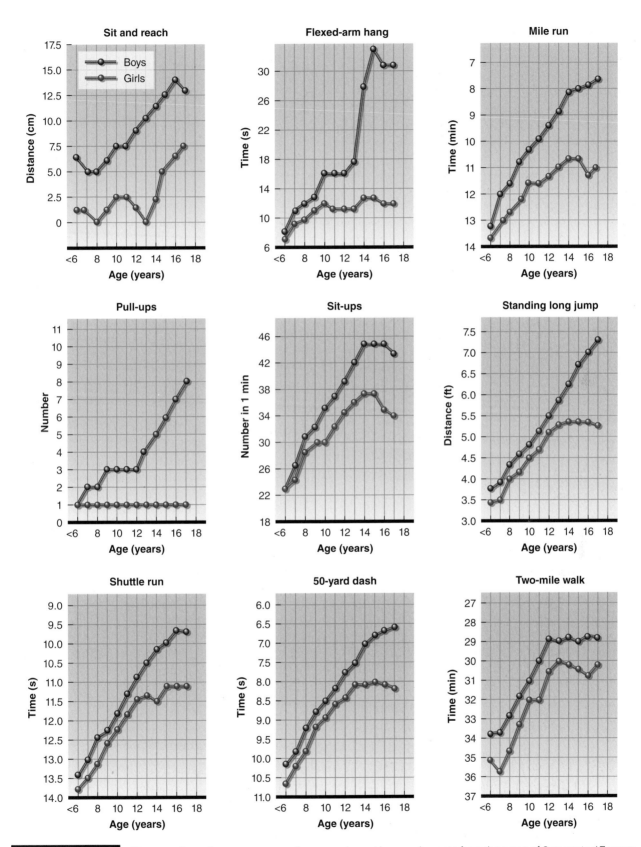

FIGURE 17.10 Changes in performance norms for several exercises and events from the ages of 6 years to 17 years.

Data from the President's Council on Physical Fitness and Sports, 1985.

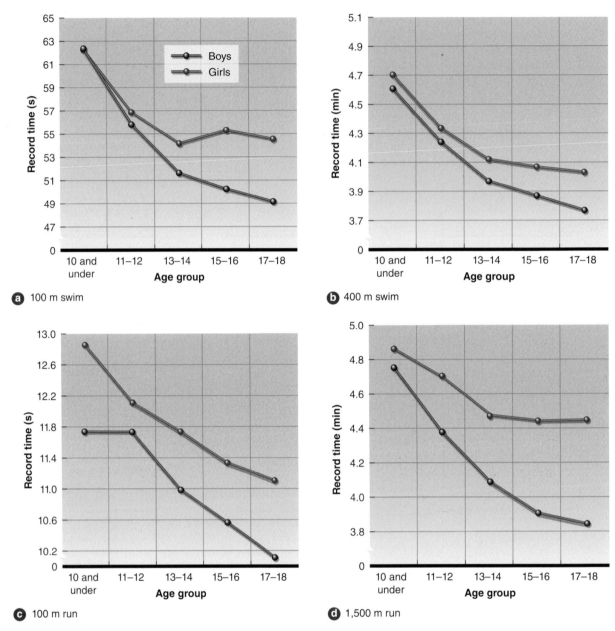

a 100 m swim

b 400 m swim

c 100 m run

d 1,500 m run

FIGURE 17.11 United States national record performances for boys and girls from 10 years of age and under up through 17 to 18 years of age in swimming and running events. Records were obtained from USA Track & Field (as of March 2011; www.usatf.org) and United States Swimming (as of March 2011; www.usaswimming.org).

Thermal Stress

Laboratory experiments suggest that children are more susceptible to heat- and cold-induced illness or injury than adults. But the number of reported cases of thermal illness or injury have not supported this theory.[31] A major concern is the child's apparently lower capacity during exercise in the heat to dissipate heat through evaporation. Children appear to rely much more on convection and radiation, which are enhanced through greater peripheral vasodilation.[2] Compared with adults,

children have a greater ratio of body surface area to mass, meaning that they have more skin surface area from which to gain or lose heat for each kilogram of body weight. Unless the environment is hot, this is an advantage, because children are better able to lose heat through radiation, convection, and conduction. However, once the environmental temperature exceeds the skin's temperature, children more readily gain heat from the environment, which is a distinct disadvantage. A child's lower capacity for evaporative heat loss is largely the result of a lower sweating rate.

Individual sweat glands in children form sweat more slowly and are less sensitive to increases in the body's core temperature than those in adults. Although young boys can acclimatize to exercise in the heat, their rate of acclimatization is slower than that of adults. Acclimatization data are not available for girls.

Only a few studies have focused on children exercising in the cold. From the limited information available, children appear to have greater conductive heat loss than adults because of a larger ratio of surface area to mass. This should be expected to place them at higher risk for hypothermia and to necessitate more clothing layers during exercise in cold temperatures.

Few studies have been conducted on children in relation to either heat or cold stress, and conclusions from existing studies sometimes have been contradictory. More research is needed in this area to determine the risks faced by children who exercise in the heat and cold. In the meantime, a conservative approach is advisable. Children may be at an increased risk of heat- and cold-related injuries compared with adults.[2]

Growth and Maturation With Training

Many people have wondered what effect physical training has on growth and maturation. Does hard physical training slow down or accelerate normal growth and development? In a comprehensive review of this area, Malina made some interesting and relevant observations.[19] Regular training has no apparent effect on growth in height. It does, however, affect weight and body composition, as discussed earlier in this chapter.

As for maturation, the age at which peak height velocity occurs generally is not affected by regular training, nor is the rate of skeletal maturation. But the data concerning the influence of regular training on indexes of sexual maturation are not as clear. Although some data suggest that menarche (the initial onset of menstruation) is delayed in highly trained girls, these data are confounded by a number of factors that generally have not been properly controlled in each study's analysis. Menarche is discussed in chapter 19.

In review

➤ Laboratory studies suggest that children may be more susceptible to injury or illness from thermal stress because they have a greater ratio of body surface area to mass when compared to adults. However, the number of reported cases does not support this.

➤ Children are limited in their evaporative heat loss compared to adults because children sweat less (less sweat is produced by each active sweat gland).

➤ Young boys acclimatize to heat more slowly than adults do. Data are not available for girls.

➤ Children appear to have greater conductive heat loss than adults, which may place children at greater risk for hypothermia in cold environments.

➤ Until more is known about children's susceptibility to thermal stress, a conservative approach should be used for children who exercise in temperature extremes.

➤ Physical training appears to have little or no negative effect on normal growth and development. Its effects on markers of sexual maturation are less clear.

In closing

In this chapter, we have discussed children and young athletes. We have seen how children gain more control of movements as their body systems grow and develop. We have seen how their developing systems can sometimes limit performance capacities and how training can improve children's performances.

We have seen that, in general, the ability to perform increases as children approach physical maturity. But as people move beyond the point of physical maturity, their physiological functioning begins to decline. Having considered the developmental process, we are now ready to consider the aging process. How is performance affected as we move beyond our physiological prime? This is our focus in the next chapter as we turn our attention to aging and the older athlete.

Key Terms

adolescence

bone mineral density

childhood

development

growth

infancy

maturation

myelination

ossification

physical maturity

puberty

Study Questions

1. Explain the concepts of growth, development, and maturation. How do they differ?
2. At what ages do height and weight reach their peak rate of growth in males and in females?
3. What typical changes occur in fat cells with growth and development?
4. How does pulmonary function change with growth?
5. What changes occur in heart rate and stroke volume for a fixed rate of work as a child grows? What factors explain these changes? What changes in these two variables occur with aerobic training?
6. What changes occur in cardiac output for a fixed rate of work as a child grows? What factors explain these changes? What changes occur with aerobic training?
7. What changes occur in maximal heart rate as a child grows?
8. What physiological variables account for the $\dot{V}O_{2max}$ increase from age 6 to 20?
9. What advice would you give to children if they wanted to improve their strength? Can they improve strength, and if so, how does this occur?
10. What happens to aerobic capacity as a prepubescent child trains aerobically?
11. What happens to anaerobic capacity as a prepubescent child trains anaerobically?
12. How do children differ from adults with respect to thermoregulation?
13. How do physical activity and regular training affect the growth and maturation processes?

Study Guide Activities

In addition to the activities listed in the chapter opening outline on page 425, two other activities are available in the web study guide, located at

www.HumanKinetics.com/PhysiologyOfSportAndExercise

 The **KEY TERMS** activity reviews important terms, and the end-of-chapter **QUIZ** tests your understanding of the material covered in the chapter.

Aging in Sport and Exercise

18

In this chapter and in the web study guide

 ACTIVITY 18.1 Physiological Changes With Age explores the physiological parameters that change with age.

 ACTIVITY 18.2 Exercise Prescription for Older Adults is a case study that examines the appropriate goals for the strength training programs of four older adults.

 ACTIVITY 18.3 Cardiorespiratory Changes provides an opportunity to estimate one's maximum heart rate and maximal oxygen uptake over the next several years, and to consider how to maintain one's aerobic capacity.

Few athletes continue to compete against younger opponents at a national level into middle and old age. One exception was Clarence DeMar, who won his seventh Boston Marathon at age 42, placed 7th at age 50, and was 78th in a field of 153 runners at age 65. In all, he ran more than 1,000 distance races, including more than 100 marathons between 1909 and 1957, a period when exercising or engaging in competition as an older adult was not popular. His performances at the Boston Marathon alone spanned 48 years, from age 20 to 68. DeMar's last race in 1957, at age 68, was 15 km (9.3 mi), which he ran despite advanced intestinal cancer and a colostomy. His best time for the Boston Marathon was 2:29:42 at age 36. Thereafter, his time gradually slowed to 3:58:37 at age 66.

The number of men and women over age 50 years who exercise regularly or participate in competitive sport has increased dramatically over the past 30 years. According to current population forecasts, the number of elderly people will increase worldwide from 6.9% of the population in 2000 to a projected 19.3% by 2050. In parallel with this overall increase in older adults, the number of middle-aged and older athletes is expected to increase as well.[35] Many of these older competitors, often termed masters or senior athletes, engage in competition for fun, general recreation, and fitness, while others train with the same enthusiasm and intensity as Olympians. Opportunities are now available for older athletes to compete in activities ranging from marathon running to powerlifting. The success achieved and the performance records set by many older athletes are phenomenal. However, although these older athletes exhibit strength and endurance capacities that are far greater than those of untrained people their age, even the most highly trained older athlete experiences a decline in performance after the fourth or fifth decade of life.

In modern societies, the level of voluntary physical activity begins to decline soon after people reach physical maturity. Technology has made virtually every aspect of life less physically demanding. Voluntary participation in strenuous physical activity on a regular basis is an unusual pattern of behavior that is not observed in most aging laboratory animals. Studies have shown that humans and other animals tend to decrease their physical activity as they grow older. As shown in figure 18.1, rats that were allowed to eat ad libitum ran an average of more than 4,000 m (4,374 yd) per week in the early months of life but covered less than 1,000 m (1,094 yd) per week during their final months.

Thus, older men and women who choose to participate in competitive sports or to train exhaustively do not follow natural human or animal behavior patterns. Why do some older individuals choose to remain physically active when the natural tendency is to become sedentary? The psychological factors that motivate these older athletes to compete are not clearly defined, but their goals probably do not differ substantially from those of their younger counterparts.

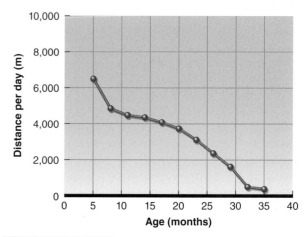

FIGURE 18.1 Voluntary running activity in rats throughout life.

Adapted from J.O. Holloszy, 1997, "Mortality rate and longevity of food-restricted exercising male rats: A reevaluation," *Journal of Applied Physiology* 82: 399-403. Used with permission.

Considering the importance of exercise for maintaining muscular and cardiorespiratory fitness, it is not surprising that inactivity can lead to deterioration of one's capacity for strenuous effort. Because of this, it is difficult to distinguish between the effects of aging by itself (known as primary aging) and those of the reduced activity and comorbidities that often accompanies aging. Researchers in aging most commonly use cross-sectional designs, but these have some important limitations compared to longitudinal study designs. For example, historical changes in medical care, diet and exercise, and other lifestyle variables may affect age cohorts differently. Selective mortality, that is, the fact that the subject population consists of the survivors of a cohort that has already experienced some degree of mortality, is an issue as well. Finally, when all but apparently healthy older individuals are excluded from exercise studies, it is difficult to apply research findings to the larger aged population with underlying disease, medication use, or both. It is important to understand the impact of primary aging alone on physiological function, but the interpretation and applicability of the results are influenced by the study design, as well as the specific population being tested.

Height, Weight, and Body Composition

As we age, we tend to lose height and gain weight, as illustrated in figure 18.2.[33] The reduction in height generally starts at about 35 to 40 years of age and is primarily attributable to compression of the intervertebral disks and poor posture early in aging. At about age 40 to 50 years in women, and 50 to 60 years in men, osteopenia and osteoporosis become a factor. **Osteopenia** is a reduction in bone mineral density below normal that occurs before **osteoporosis**, which is a severe loss of bone mass with deterioration of the microarchitecture of bone, leading to increased risk of bone fracture (see chapter 19). Genetic factors and poor diet and exercise habits throughout the life span contribute to the development of osteoporosis in both men and women, while decreased estrogen concentrations after menopause appear to be responsible for the greater rate of bone loss in women. During the adult life span a gain in weight typically occurs between age 25 and 45 and is attributable to both a decrease in physical activity levels and excess caloric intake. Beyond the age of 45, weight stabilizes for about 10 to 15 years and then decreases as the body loses bone calcium and muscle mass. Many people over 65 to 70 years of age tend to lose their appetite and thus don't consume sufficient calories to maintain body weight. An active lifestyle, however, tends to help stimulate appetite so that caloric intake more closely approximates caloric expenditure, thereby maintaining weight and preventing frailty in old age.

Beginning at about 20 years of age, humans tend to gain fat with aging. This is largely attributable to three factors: diet, physical inactivity, and reduced ability to mobilize fat stores. However, the body fat content of physically active older people, including older athletes, is significantly lower than that of age-matched sedentary people. In addition, with primary aging there tends to be a shift in the location where body fat is stored, from the periphery toward the center of the body around the organs. This centralized adiposity is associated with cardiovascular and metabolic diseases. Although physical activity cannot fully counteract the age-related gain in fat mass, in active men and women there is less of a shift of the fat stores with aging, which is more advantageous for reducing the risk of cardiovascular and metabolic disease.[24]

(In) focus

With aging, body fat content increases and is redistributed from the periphery to the center of the body while fat-free mass decreases. To a great extent these changes can be attributed to the reduction in general activity levels that often occurs with aging.

Fat-free mass decreases progressively in both men and women beginning at about the age of 40. This results primarily from decreased muscle and bone mass, with muscle having the greatest effect because it constitutes about 50% of the fat-free mass. **Sarcopenia** is the term used to describe the loss of muscle mass associated

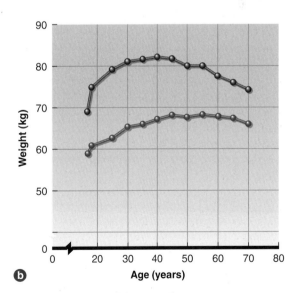

FIGURE 18.2 Changes in (*a*) height and (*b*) weight in men and women up to 70 years of age.

Reprinted, by permission, from W.W. Spirduso, 1985, *Physical dimensions of aging* (Champaign, IL: Human Kinetics), 59.

with the aging process. Figure 18.3 illustrates the changes in muscle mass with aging in a cross-sectional study of 468 men and women, aged 18 to 88 years.[19] There is almost no decline in muscle mass until about age 40, at which time the rate of decline increases, with a greater decline in men than women. Obviously, a reduction in physical activity is a major cause of this decline in muscle mass with aging, but there are other factors. While there is some controversy as to the mechanisms underlying the age-related decline in muscle mass, it is thought that the rate of muscle protein synthesis is reduced while the rate of muscle protein breakdown is unchanged or accelerated with aging, leading to negative nitrogen balance and net loss of muscle. The rate of muscle protein synthesis in 60- to 80-year-olds is about 30% lower than in a 20-year-old. This reduction in muscle protein synthesis rate in older people is likely associated with declines in growth hormone, insulin-like growth factor-1,[13] and cell signaling. Longitudinal data suggest that loss of fat-free mass and gain of fat mass offset each other. As a result, percent fat is increasing while total body mass remains relatively stable.

There is also a significant decrease in bone mineral content, starting at about age 30 to 35 in women and at age 45 to 50 in men. Throughout the life cycle, bone is constantly being formed by osteoblasts and resorbed by

In focus

The sarcopenia observed with aging is thought to be the result of decreased muscle protein synthesis in older people, accompanied by an unchanged or accelerated rate of muscle protein breakdown.

osteoclasts. Early in life, resorption occurs at a slower rate than synthesis, and bone mass increases. With aging, resorption exceeds synthesis, resulting in a net loss of bone. The loss of both muscle and bone mass is at least partially attributable to decreased physical activity, especially a lack of weight-bearing exercise. Since bone mineral accounts for less than 4% of total body mass in young adults, the contribution of osteopenia to the loss of total fat-free mass is small compared to that of sarcopenia.

These differences in weight, relative (%) body fat, fat mass, and fat-free mass with aging are illustrated in figure 18.4.[23] These data are from a study of young (18-31 years) and older (58-72 years) men and women who either were sedentary or were endurance-trained athletes. Body weight, relative body fat, and fat mass were higher in the older sedentary groups, whereas fat-free mass was lower. Similar trends, except for body weight, were noted for the endurance-trained athletes. However, the young and older endurance-trained athletes had much lower total body weight, relative body fat, and fat mass values and similar fat-free mass values compared to their sedentary age-matched counterparts.

With training, older men and women can reduce weight, percent body fat, and fat mass. Furthermore, they can increase their fat-free mass; but as in younger individuals, this is more likely with resistance training than with aerobic training. Men appear to experience greater changes in body composition than women, but the reasons for this have not been clearly established.

The most significant changes in body composition result from a combination of diet and exercise, with a modest reduction in caloric intake (500-1,000 kcal/day) being the preferred approach. A more substantial reduction in caloric intake (>1,000 kcal/day) is likely to result in a loss of fat-free mass as well as fat mass. This is not desirable, as a loss in fat-free mass is associated with a reduction in resting metabolic rate, thus decreasing the rate of weight and fat loss. Exercise that increases fat-free mass will likely increase resting metabolic rate, which would increase the rate of weight loss. It appears that older adults experience changes in body composition due to exercise training that are similar to those for younger adults.

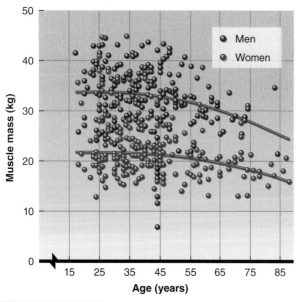

FIGURE 18.3 Changes in muscle mass with aging in 468 men and women 18 to 88 years of age. The rate of decline is greater in men than in women and is steeper after about 45 years of age.

Adapted from I. Janssen et al., 2000, "Skeletal muscle mass and distribution in 468 men and women aged 18-88 yr.," *Journal of Applied Physiology* 89: 81-88. Used with permission.

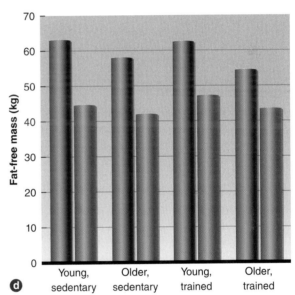

FIGURE 18.4 Differences in (*a*) total body weight, (*b*) relative body fat, (*c*) fat mass, and (*d*) fat-free mass in young and older men and women, sedentary and endurance trained.

Adapted, by permission, from W.M. Kohrt et al., 1992, "Body composition of healthy sedentary and trained, young and older men and women," *Medicine and Science in Sports and Exercise* 24: 832-837.

In review

➤ Body weight tends to increase with aging, whereas height decreases.

➤ Body fat increases with age, primarily because of increased caloric intake, decreased physical activity, and a reduced ability to mobilize fat.

➤ Beyond age 45, fat-free mass decreases, primarily because of decreased muscle and bone mass, both resulting at least partly from decreased activity.

➤ Exercise training can help attenuate these changes in body composition, even in individuals as old as 80 to 90 years of age.

Physiological Responses to Acute Exercise

As we age, muscular and cardiovascular endurance and muscular strength tend to decrease, with the extent of decrease dependent on physical activity and genetics. As activity decreases, which appears to be a natural phenomenon in both animals and humans, these reductions in physiological function are much more substantial.

Strength and Neuromuscular Function

The amount of strength needed to meet the daily demands of living (activities of daily living) remains unchanged throughout life. However, a person's maximal strength, generally well above the daily demands early in adulthood, decreases steadily with aging. Eventually, strength may decline to the point where simple activities become challenging. For example, the ability to stand up from a sitting position in a chair starts to be compromised at age 50, and before age 80 this task becomes impossible for some people (see figure 18.5a). As a further example, opening the lid on a jar that has a set resistance is a task that can easily be accomplished by almost all men and women below the age of 60. After age 60, the failure rate for this task increases significantly.

Figure 18.5b describes leg strength changes with aging in men. Knee extension strength in normally active men and women starts to decrease by age 40. But training the knee extensor muscles with resistance exercises enables older men to perform better at age 60 than most normally active men half that age. The reduction in strength with aging was highly correlated with the reduction in the cross-sectional area of the involved muscles. The reductions in strength with aging appear to be modality specific, in that losses in isokinetic strength are greatest at high angular velocities and losses in concentric strength are greater than losses in eccentric strength.

Age-related losses of muscle strength result primarily from the substantial loss of muscle mass that accompanies aging or decreased physical activity (or both), as discussed earlier in this chapter. Figure 18.6 shows a computed tomographic (CT) scan of the upper arms of three 57-year-old men of similar body weight (about 78-80 kg, or 172-176 lb). Note that the untrained subject has substantially less muscle and more fat than the others. The swim-trained subject has less fat and a markedly larger triceps muscle than the untrained subject; but his biceps muscle, a muscle seldom used during swimming, is not much different. However, both of these muscles are larger in the resistance-trained subject. The differences between these three men are likely attributable to a combination of genetics and their volume and type of training.

Aging has a marked effect on muscle mass and strength, but what about muscle fiber type? There are

FIGURE 18.5 (a) The ability to stand from a sitting position is compromised at age 50, and by age 80 this task becomes impossible for some people. (b) Changes in peak knee extension strength in untrained and trained men at various ages. Note that older men (e.g., 60-80 years) who strength train can have knee extension strength equal to or greater than that of individuals who are only a third their age.

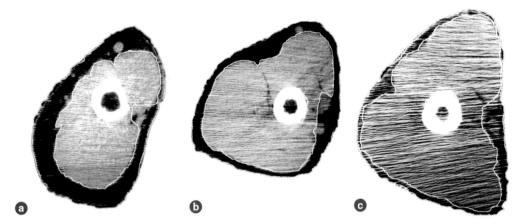

FIGURE 18.6 Computed tomography scans of the upper arms of three 57-year-old men of similar body weights. The scans show bone (dark center surrounded by white ring), muscle (striated gray area), and subcutaneous fat (dark perimeter). Note the difference in the muscle areas of (a) the untrained man, (b) the swim-trained man, and (c) the strength-trained man.

conflicting results about the effects of aging on type I and type II fibers. Cross-sectional studies of the entire vastus lateralis (quadriceps) muscle in 15- to 83-year-old subjects postmortem have suggested that fiber type remains unchanged throughout life.[20] However, results from longitudinal studies conducted over a 20-year period indicate that the amount or intensity of activity, or perhaps both, might play an important role in fiber type distribution with aging.[36, 37] Muscle biopsy samples from the gastrocnemius (calf) muscles of a group of previously elite distance runners, obtained in 1970 through 1974 and again in 1992, demonstrated that the runners who had decreased their activity (fitness trained) or become sedentary (untrained) had a significantly greater proportion of type I fibers than when they were 18 to 22 years younger (figure 18.7). Those who remained highly trained had no change. Although some of the elite runners who still competed in distance running (highly trained) showed a small increase in the percentage of type I fibers, on average these highly trained runners showed no change in their calf muscle fiber composition over the 18- to 22-year span of this study.

It has been suggested that the apparent increase in type I fibers may be attributable to an actual decrease in the number of type II fibers, resulting in a greater relative proportion of type I fibers. Although the precise cause of type II fiber loss is unclear, it has been suggested that the number of type II motor neurons decreases during aging, which eliminates innervation of these muscle fibers. This may result from death of the motor neurons in the spinal cord, causing the fibers innervated by these motor neurons to gradually atrophy. Type I motor neurons, however, may develop axonal sprouts and reinnervate some of the muscle fibers from the dead type II motor neurons. This

FIGURE 18.7 Changes in muscle fiber type composition of the gastrocnemius in elite distance runners who remained highly trained, stayed fitness trained, or became untrained during the 18 to 22 years between tests. Note that the runners who continued to compete showed almost no change in percentage of type I fibers, whereas the less fit and untrained individuals experienced an increase in the percentage of type I fibers.

increases the size of the remaining motor units, with more muscle fibers per motor neuron.

Numerous investigations have shown a decrease in both the number and the size of muscle fibers with aging. One study reported a loss of approximately 10% of the total number of muscle fibers per decade after age 50.[25] This explains in part the muscle atrophy that occurs as we get older. Additionally, it appears that the size of both type I and type II fibers decreases with aging. Endurance training (such as distance running) has little impact on the decline in muscle mass with aging. Strength training, on the other hand, reduces muscle atrophy in aging adults and can, in fact, cause older individuals to increase their muscle cross-sectional area.[25]

In focus

Muscular strength is reduced with aging. This is the result of decreases in both physical activity and muscle mass, the latter largely the result of a reduction in protein synthesis with aging and the loss of type II motor units. Whereas endurance training does little to prevent the age-associated loss in muscle mass, strength training can maintain or increase the muscle fiber cross-sectional area in older men and women.

Aging is accompanied by substantial changes in the nervous system's capacity to process information and to activate muscles. Specifically, aging affects the ability to detect a stimulus and process the information to produce a response. Simple and complex movements are slowed with aging, although people who remain physically active are only slightly slower than younger, active individuals. Motor unit activation is lower in aged adults. For example, one study showed that older men (~80 years) had lower firing rates and longer twitch contraction durations, whereas younger men (~20 years) had relatively higher firing rates and shorter contraction times.[2] However, others have shown that older individuals retain their ability to maximally recruit skeletal muscle, suggesting that reduced strength is due to local muscle rather than neural factors.[6]

Neuromuscular changes during aging are at least partially responsible for decreased strength and endurance, but active participation in exercise and sport tends to lessen the impact of aging on performance. This doesn't mean that regular physical activity can arrest biological aging, but an active lifestyle can markedly reduce many of the decrements in physical work capacity.

Saltin[30] noted that despite the loss of muscle mass in active aging men, the structural and biochemical properties of the remaining muscle mass are well maintained. The number of capillaries per unit area is similar in young and old endurance runners. Oxidative enzyme activities in the muscles of endurance-trained older athletes are only 10% to 15% lower than in endurance-trained young athletes. Thus, oxidative capacity of skeletal muscle of endurance-trained older runners is only slightly less than that of young elite runners, which suggests that aging has little effect on skeletal muscle's adaptability to endurance training.

In review

➤ Maximal strength decreases steadily with aging.

➤ Age-related losses of strength result primarily from a substantial loss of muscle mass.

➤ In general, normally active people experience a shift toward a higher percentage of type I muscle fibers as they age, possibly attributable to a reduction in type II fibers.

➤ The total number of muscle fibers and the fiber cross-sectional area decrease with age, but resistance training appears to lessen the decline in fiber area.

➤ Aging slows the nervous system's ability to respond to a stimulus, process the information, and produce a muscular contraction.

➤ Exercise training cannot arrest the process of biological aging, but it can lessen the impact of aging on performance.

Cardiovascular and Respiratory Function

To a large extent, changes in endurance performance that accompany aging can be attributed to decrements in both central and peripheral cardiovascular function. Changes in respiratory function likely play a lesser role. In this section, we examine aging effects on both the cardiovascular and respiratory systems.

Cardiovascular Function

Similar to muscle function, cardiovascular function declines as we age. One of the most notable changes that accompanies aging is a decrease in maximal heart rate (HR_{max}). Whereas children's values usually range between 195 and 215 beats/min, the average 60-year-old has an HR_{max} of approximately 166 beats/min. HR_{max} is estimated to decrease slightly less than 1 beat/min per year as we age. Traditionally, the average HR_{max} for any age has been estimated from the equation $HR_{max} = 220 - age$. However, Tanaka and colleagues developed a more accurate equation:[34]

$$HR_{max} = [208 - (0.7 \times age)].$$

This equation seems to be appropriate for all people and is not influenced by sex or activity level. The old equation tended to overestimate the HR_{max} of children and young adults and to underestimate the HR_{max} of older adults. When the old equation (HR_{max} = 220 – age) is used, individual values can deviate by ±20 beats/min or more from the predicted value. For example, the old equation predicts that an average 60-year-old would have an HR_{max} of 160 beats/min; however, this individual's actual HR_{max} might be as low as 140 beats/min or as high as 180 beats/min. While the equation proposed by Tanaka improves the prediction of the average individual's heart rate, there is still substantial interindividual variability. We will see in chapter 20 that overestimates and underestimates make a big difference when HR_{max} is used for exercise prescription.

In focus

A more accurate equation to estimate HR_{max} is HR_{max} = [208 − (0.7 × age)]. However, prediction equations estimate only the average value for people of a given age.

The reduction in HR_{max} with aging appears to be similar for sedentary and well-trained adults. At age 50, for example, normally active men have the same HR_{max} as former and still-active distance runners of the same age. This reduction in HR_{max} might be attributable to morphological and electrophysiological alterations in the cardiac conduction system, specifically in the sinoatrial (SA) node and in the bundle of His, which could slow cardiac conduction. Downregulation of the β_1 adrenergic receptors in the heart also decreases the heart's sensitivity to catecholamine stimulation.

Maximal stroke volume (SV_{max}) is modestly reduced (~10-20% reduction) in highly trained older adults. The responses to catecholamine stimulation and myocardial contractility are reduced, and recent evidence provided with more sophisticated Doppler imaging techniques indicates that the heart does not fully retain the Frank-Starling mechanism. This is likely due to left ventricular and arterial stiffening.[32] Daily exercise training started early in life may help to minimize this impairment, although this effect is limited and highly variable when training is started later in life. The decrease in maximal cardiac output with aging in highly trained men and women is primarily attributable to decreased heart rate and to a lesser extent a decrease in stroke volume. Studies of endurance runners have shown that the lower $\dot{V}O_{2max}$ values observed in older athletes result from a reduction in maximal cardiac output, despite the fact that heart volumes of older athletes are similar to those of young athletes, confirming that a decreased maximal heart rate is the primary cause of reduced $\dot{V}O_{2max}$. In untrained men and women, a number of studies have demonstrated a clear decrease in maximal stroke volume with aging.

Peripheral blood flow, such as to the legs, decreases with aging, even though capillary density in the muscles is unchanged. Studies reveal a 10% to 15% reduction in leg blood flow to the exercising muscles in middle-aged athletes at any given work rate when compared with well-trained young athletes (figure 18.8).[30] This attenuation in blood flow is due to a number of peripheral factors including blunted functional sympatholysis (i.e., a greater sympathetic outflow to the exercising muscle) and a reduction in local vasodilators.[31] But the reduced blood flow to the legs of these middle-aged and older endurance runners during submaximal exercise was apparently compensated for by a greater arterial–mixed venous oxygen difference, or (a-v̄)O_2 difference (more oxygen is extracted by the muscles). As a result, although the blood flow is lower, oxygen uptake by the exercising muscles is similar at a given submaximal work intensity in the older age-group. This was confirmed in a study of endurance-trained men, which compared subjects aged 22 to 30 years with subjects 55 to 68 years. Leg blood flow, vascular conductance, and femoral venous oxygen saturation were each 20% to 30% lower in the older men at each submaximal work rate, while leg (a-v̄)O_2 difference was higher in older subjects.[28]

It is hard to determine the extent to which the age-related changes in stroke volume, cardiac output, and peripheral blood flow result from the aging process alone, and the extent to which they are attributable to the **cardiovascular deconditioning** that accompanies

FIGURE 18.8 Leg blood flow during cycling exercise in young and middle-aged orienteers.

Adapted, by permission, from B. Saltin, 1986, The aging endurance athlete. In *Sports medicine for the mature athlete,* edited by J.R. Sutton and R.M. Brock (Indianapolis: Benchmark Press). Copyright 1986 Cooper Publishing Group, Carmel, IN.

Habitual Exercise and Vascular Aging

Regular aerobic exercise is associated with a decreased risk of cardiovascular disease in middle-aged and older adults. In general, sedentary aged humans have a decrease in their cardiac and arterial compliance caused by stiffening of the heart and the large elastic arteries. Additionally, the aged blood vessels also exhibit a change in local control of blood flow including a dysfunction in the ability of the endothelium to release and respond to vasodilators such as nitric oxide and prostaglandins, termed **endothelial dysfunction**. This change contributes to the inability of blood vessels to vasodilate and the reduction in peripheral muscle blood flow during exercise.

With habitual exercise training in older men and women, there is less arterial stiffening and endothelial dysfunction. Part of the reason for this preservation in vascular function with habitual exercise training is a preservation or restoration in vasodilator signaling including an increase in nitric oxide bioavailability. Furthermore, the mechanisms by which regular aerobic exercise favorably influences central and peripheral vascular function do not merely result from changes in other cardiovascular disease risk factors.[31] Rather, it appears that regular exercise can directly improve blood vessel function in populations with one or more risk factors.[31] Research is ongoing to determine the appropriate dose of exercise (time, duration, and intensity) needed to see these positive cardiovascular benefits in a number of healthy and clinical populations.

reduced activity. Studies suggest that both are involved, but the relative contribution of each is unknown. However, even the older athlete generally trains at a lower volume and intensity than the 20-year old athlete. Aging alone might decrease cardiovascular function and endurance less than the deconditioning that accompanies inactivity, decreased activity, or decreased intensity of training. These declines in cardiovascular function with aging are largely responsible for the declines observed in $\dot{V}O_{2max}$, as discussed later in this chapter.

Respiratory Function

Lung function changes considerably with aging in sedentary people. Both vital capacity (VC) and **forced expiratory volume in 1 s (FEV$_{1.0}$)** decrease linearly with age, starting at age 20 to 30. Whereas these decrease, residual volume (RV) increases, and the total lung capacity (TLC) remains essentially unchanged. As a result, the ratio of the residual volume to total lung capacity (RV/TLC) increases, meaning that less air can be exchanged. In our early 20s, RV accounts for 18% to 22% of the TLC, but this increases to 30% or more as we reach age 50. Smoking appears to accelerate this increase.

These changes are matched by changes in maximal ventilatory capacity during exhaustive exercise. **Maximal expiratory ventilation ($\dot{V}_{Emax}$)** increases during growth until people achieve physical maturity, and then it decreases with age. $\dot{V}_{Emax}$ values average about 40 L/min for 4- to 6-year-old boys, increase to 110 to 140 L/min for fully mature men, and then decrease to 70 to

90 L/min for 60- to 70-year-old men. Girls and women follow the same general pattern, although their absolute values are considerably lower at each age, primarily because of smaller body size. Recall from chapter 7 that TLC is directly proportional to height, which is why men often have higher values than women.

The changes in pulmonary function as adults get older result from several factors. The most important of these is a loss of elasticity of the lung tissue and chest wall as we age, which increases the work involved in breathing. The resulting stiffening of the chest wall appears to be responsible for most of the reduction in lung function. But despite all these changes, the lungs still hold a remarkable reserve and maintain an adequate diffusion capacity to permit maximal exertion, and do not appear to limit exercise capacity.

Endurance-trained older athletes have only slightly decreased pulmonary ventilation capacities. More importantly, decreased aerobic capacity among these older athletes cannot be attributed to changes in pulmonary ventilation. Also, during strenuous exercise, both normally active older people and athletes can maintain near-maximal arterial oxygen saturation. Thus, changes neither in the lungs nor in the blood's oxygen-carrying capacity appear to be responsible for the observed decrease in $\dot{V}O_{2max}$ reported in aging athletes. Rather, the primary limitation is apparently linked with oxygen transport to the muscles, that is, cardiovascular changes. As discussed earlier in this chapter, aging decreases maximum heart rate, which lowers maximal cardiac output and blood flow to the exercising muscles. Submaximal $(a\text{-}\bar{v})O_2$ difference is maintained in older exercisers, suggesting that O_2 extraction is well preserved with aging.

In review

➤ Much of the decline in endurance performance associated with aging can be attributed to decreased cardiovascular function.

➤ Maximum heart rate decreases about 1 beat/min per year as we age. The average HR_{max} for a given age can be estimated by the following equation: $HR_{max} = [208 - (0.7 \times age)]$.

➤ Maximal stroke volume is only slightly reduced in older athletes, but their cardiac output decreases with age primarily because of the reduction in HR_{max}. In untrained people, maximal stroke volume decreases due to left ventricular and arterial stiffening with aging.

➤ Peripheral blood flow also decreases with age; in trained older athletes, however, this is offset by an increased $(a\text{-}\bar{v})O_2$ difference.

➤ Habitual exercise can partially reverse or prevent many of the detrimental vascular changes that occur with aging, including reducing arterial stiffness and improving endothelial function.

➤ It is unclear how much of the decrease in cardiovascular function with aging is attributable to aging alone and how much is attributable to deconditioning because of decreased activity. However, many studies indicate that these changes are attenuated in older athletes who continue to train, which indicates that inactivity plays a substantial role.

➤ Both vital capacity and forced expiratory volume decrease linearly with age. Residual volume increases, so total lung capacity remains unchanged. This increases the RV/TLC ratio, meaning that less air can be exchanged in the lung with each breath.

➤ Maximal expiratory ventilation also decreases with age.

➤ Pulmonary changes that accompany aging are primarily caused by a loss of elasticity in the lung tissue and the chest wall. However, aging athletes have only slightly decreased pulmonary ventilation capacity.

Aerobic and Anaerobic Function

In investigating the effect of aging on aerobic and anaerobic function during exercise, the focus here is on two key variables—$\dot{V}O_{2max}$ and lactate threshold.

$\dot{V}O_{2max}$

To determine how $\dot{V}O_{2max}$ changes with aging, there are several important issues to consider. First, one must decide how to express the $\dot{V}O_{2max}$ values—in liters per minute (L/min) or in liters per minute per kilogram of body weight to adjust for size (ml $\cdot$ kg^{-1} $\cdot$ min^{-1})? In some cases, $\dot{V}O_{2max}$ expressed in liters per minute does not decrease much over a 10- to 20-year period, but when the same subjects' values are expressed relative to body weight, there is a relatively large decrease. This apparent discrepancy is simply attributable to the subjects' weight gain during the 10 to 20 years between the initial test and the final test. For non-weight-bearing exercise, such as cycling, using liters per minute is usually most appropriate. For weight-bearing activities such as running, it is usually more appropriate to express the values per unit of body weight (ml $\cdot$ kg^{-1} $\cdot$ min^{-1}).

A second issue relates to whether change values in variables with aging should be expressed as an absolute change (L/min or ml $\cdot$ kg^{-1} $\cdot$ min^{-1}) or as a percentage change, where

% change = [(final value – initial value)/initial value] × 100.

This might seem like a minor point, but it is not. As an example, a 30-year-old man has an initial $\dot{V}O_{2max}$ of 50 ml $\cdot$ kg^{-1} $\cdot$ min^{-1}, and at the age of 50 years his $\dot{V}O_{2max}$ has decreased to 40 ml $\cdot$ kg^{-1} $\cdot$ min^{-1}. A 60-year-old man has an initial $\dot{V}O_{2max}$ of 35 ml $\cdot$ kg^{-1} $\cdot$ min^{-1}, and at the age of 80 years his $\dot{V}O_{2max}$ has decreased to 25 ml $\cdot$ kg^{-1} $\cdot$ min^{-1}. In this example, both men have decreased their $\dot{V}O_{2max}$ by 10 ml $\cdot$ kg^{-1} $\cdot$ min^{-1} over a 20-year period, showing a decline of 0.5 ml $\cdot$ kg^{-1} $\cdot$ min^{-1} per year. However, the younger man has had a decrease of 20% (10/50 = 0.20, or 20%) over 20 years, or 1% per year, whereas the older man has had a decrease of 29% (10/35 = 0.29), or 1.4% per year. Although the two men have identical decreases in $\dot{V}O_{2max}$ when expressed in ml $\cdot$ kg^{-1} $\cdot$ min^{-1}, the older man has a substantially greater decrease when expressed as a percentage decrease. Many studies report both the absolute (ml $\cdot$ kg^{-1} $\cdot$ min^{-1}) and relative (%) decrease. Keeping this in mind, let's consider changes in $\dot{V}O_{2max}$ with aging, looking first at changes in normally active people and then at changes in highly trained endurance athletes.

In focus

The decrease in $\dot{V}O_{2max}$ with aging and inactivity is largely explained by decreased $\dot{Q}_{max}$ due to decreased HR_{max}. Maximal stroke volume is modestly reduced (SV_{max}), and maximal $(a-\bar{v})O_2$ difference typically does not change much with aging. The decrease in HR_{max} is attributable largely to decreases in the heart's intrinsic rate but also could be caused by decreases in sympathetic nervous system activity and alterations in the cardiac conduction system. The decrease in $\dot{V}O_{2max}$ with aging is primarily a function of reduced blood flow to the active muscles, which is associated with the reduction in maximal cardiac output.

Normally Active People   The first studies of aging and physical fitness were performed by Sid Robinson[29] in the late 1930s. He demonstrated that $\dot{V}O_{2max}$ in normally active men declined steadily from age 25 to age 75 (table 18.1). His cross-sectional data show that aerobic capacity declines an average of $0.44 \text{ ml} \cdot \text{kg}^{-1} \cdot \text{min}^{-1}$ per year up to age 75, which is about 1% per year or 10% per decade. For women between the ages of 25 and 60 years, the decline is also close to 1% per year.[1] A review of 11 cross-sectional studies on men, most under age 70, showed that the average rate of decrease in $\dot{V}O_{2max}$ was $0.41 \text{ ml} \cdot \text{kg}^{-1} \cdot \text{min}^{-1}$ per year.[1] In this same review, analysis of six cross-sectional studies of women resulted in an average rate of decline of $0.30 \text{ ml} \cdot \text{kg}^{-1} \cdot \text{min}^{-1}$ per year. This review did not publish the average rate of decline expressed as a percentage of the subjects' initial $\dot{V}O_{2max}$ values. In the mid-1990s, a large cross-sectional study of changes in $\dot{V}O_{2max}$ with aging was conducted at the NASA/Johnson Space Center in Houston, Texas. This study included 1,499 men and 409 women, all of whom were healthy and had performed a maximal treadmill test to exhaustion during which $\dot{V}O_2$ was directly measured.[17, 18] The authors reported a decline in $\dot{V}O_{2max}$ of $0.46 \text{ ml} \cdot \text{kg}^{-1} \cdot \text{min}^{-1}$ per year in men (1.2% per year) and $0.54 \text{ ml} \cdot \text{kg}^{-1} \cdot \text{min}^{-1}$ per year in women (1.7% per year).

Unfortunately, few longitudinal studies have been conducted in this area. Studies that have reexamined normally active men at various stages of their lives reveal a wide range of values for the decline in aerobic capacity. At least part of these variations can be attributed to the subjects' different activity levels and ages at the beginning of the studies. Nevertheless, the rate of decline in $\dot{V}O_{2max}$ generally is agreed to be approxi-

TABLE 18.1   Changes in $\dot{V}O_{2max}$ Among Normally Active Men

Age (years)	$\dot{V}O_{2max}$ (ml · kg⁻¹ · min⁻¹)	% change from 25 years
25	47.7	--
35	43.1	−10
45	39.5	−17
52	38.4	−20
63	34.5	−28
75	25.5	−47

Data from Robinson, 1938.

mately 10% per decade or 1% per year ($-0.4 \text{ ml} \cdot \text{kg}^{-1} \cdot \text{min}^{-1}$ per year) in relatively sedentary men. The results are similar for women, although fewer subjects have been studied.

In focus

$\dot{V}O_{2max}$ decreases by about 10% per decade with aging, starting in the midteens for women and in the mid-20s for men. This decrease is largely associated with a decrease in cardiorespiratory function.

Older Athletes   One of the most notable long-term studies of distance runners and aging was conducted by D.B. Dill and his colleagues from the Harvard Fatigue Laboratory.[4] Don Lash, world-record holder for the 2 mi run (8 min 58 s) in 1936, was among the athletes studied by the Harvard group. Although few of the former runners continued to train after leaving college, Lash was still running about 45 min per day at age 49. Despite this activity, his $\dot{V}O_{2max}$ had declined from $81.4 \text{ ml} \cdot \text{kg}^{-1} \cdot \text{min}^{-1}$ at age 24 to $54.4 \text{ ml} \cdot \text{kg}^{-1} \cdot \text{min}^{-1}$ at age 49, a 33% decline. Runners who did not continue to train during middle age showed much larger declines. On the average, their aerobic capacities declined by about 43% from age 23 to age 50 (from 70 to $40 \text{ ml} \cdot \text{kg}^{-1} \cdot \text{min}^{-1}$). These data suggest that prior training offers little advantage to endurance capacity in later life unless a person continues to engage in some form of vigorous activity. However, due to their high initial values, these individuals have a large functional reserve, and this large decrease in aerobic capacity has little effect on their ability to carry out activities of daily living. In addition, there are large individual differences in the rate of decline in $\dot{V}O_{2max}$ with aging, and genetics is a major contributor.

More recent longitudinal studies of older male runners and rowers have shown a decline in aerobic capacity and cardiovascular function and changes in muscle fiber composition with aging. These athletes were studied for 20 to 28 years, during which time some continued to train for competition whereas others became quite sedentary. Those athletes who continued high-volume and -intensity training experienced a 5% to 6% decline in $\dot{V}O_{2max}$ per decade. On the other hand, elite runners who stopped training experienced nearly a 15% decline in aerobic capacity per decade (1.5% per year), the combined effect of deconditioning and aging.

Fewer studies have been published on women, but the results show the same trends. In one study of 86 male and 49 female masters endurance runners, the authors observed both cross-sectional and longitudinal (approximately 8.5 years) changes in $\dot{V}O_{2max}$ with age.[14] Their results are illustrated in figure 18.9. The average rate of decline, as indicated by the cross-sectional data regression line, was 0.47 ml · kg^{-1} · min^{-1} per year in men (0.8% per year) and 0.44 ml · kg^{-1} · min^{-1} in women (0.9% per year). However, this figure shows that the longitudinal changes are greater than the cross-sectional changes, particularly for the older ages. In a cross-sectional study of sedentary women ($n = 2,256$), active women ($n = 1,717$), and endurance-trained women ($n = 911$) aged 18 to 89 years, $\dot{V}O_{2max}$ declined by 0.35 ml · kg^{-1} · min^{-1} per year in the sedentary women (1.2% per year), 0.44 ml · kg^{-1} · min^{-1} per year in the active women (1.1% per year), and 0.62 ml · kg^{-1} · min^{-1} per year in the endurance-trained women (1.2%).[7]

In the early 2000s, a 25-year follow-up study reexamined highly competitive, older male distance runners.[37,38] These men were initially tested at 18 to 25 years of age. During the interval between testing sessions, the runners trained at about the same relative intensity as they had when they were younger. As a consequence, their $\dot{V}O_{2max}$ values (L/min) declined only 3.6% over the 25-year period,[38] as shown in table 18.2. Although their maximal oxygen uptake decreased from 69.0 to 64.3 ml · kg^{-1} · min^{-1}, this is a decrease of only 0.19 ml · kg^{-1} · min^{-1} per year or 0.3% per year, and most of that change was attributable to a 2.1 kg (4.6 lb) increase in body weight.

This rate of decrease in these older runners' $\dot{V}O_{2max}$ values is significantly less than that of either sedentary people or those who fitness train at levels and intensities below those of these older runners. One of these runners performed a 4 min 11 s mile and a 2:29 marathon in 1992 at the age of 46! Both of these performances were significantly faster than his best in 1966. Similar findings have been reported for other athletes who continue to train with the same relative intensity and volume as they did in college.

a Men

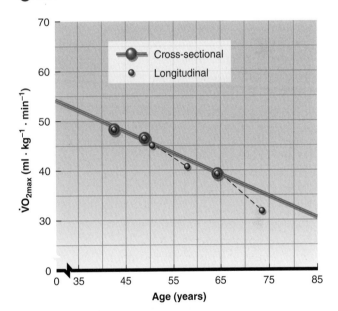

b Women

FIGURE 18.9 Cross-sectional and longitudinal declines in $\dot{V}O_{2max}$ with age in a group of 86 male and 49 female masters endurance athletes.

Adapted from Hawkins et al., 2001.

Are these performances exceptions to the natural rules of aging? Can other athletes reduce the effects of aging on their endurance by continuing to train intensely? Much depends on the training adaptability of the individual athlete, a factor that might be determined as much by heredity as by training regimen. However, advances in nutrition and training practices have contributed to masters athletes' improvements in peak performance.

TABLE 18.2 Changes in Mean Aerobic Capacity and Maximal Heart Rate With Aging in a Group of 10 Highly Trained Masters Distance Runners

Age (years)	Weight (kg)	$\dot{V}O_{2max}$		HR_{max} (beats/min)
		(L/min)	$(ml \cdot kg^{-1} \cdot min^{-1})$	
21.3	63.9	4.41	69.0	189
46.3	66.0	4.25	64.3	180

In focus

It is often difficult to differentiate between the results of biological aging and physical inactivity. A natural deterioration in physiological function occurs with aging, but this is compounded by the fact that most people also become more sedentary as they age.

The effects of aging and training on $\dot{V}O_{2max}$ in men are summarized in figure 18.10. Although the number of studies on women is much smaller, a similar trend would be expected. Note that although intense training reduces the normal aging-related decline in $\dot{V}O_{2max}$, aerobic capacity still declines. Thus, it appears that highly intense training has a slowing effect on the rate of loss in aerobic capacity during the early and middle years of adult life (e.g., 30-50 years) but less effect after 50 years of age.

In summary, $\dot{V}O_{2max}$ declines with age, and the rate of decline is approximately 1% per year. Many factors influence this rate of decline, including the following:

- Genetics
- General activity level
- Intensity of training
- Volume of training
- Increased body weight and body fat mass, decreased fat-free mass
- Age range, with older individuals experiencing greater declines

There is not universal agreement on which of these factors are most important. An integrated view of the concepts of the physiological mechanisms contributing to the reductions in endurance performance with aging is presented in figure 18.11.

Lactate Threshold

In young endurance-trained adults, the lactate threshold predicts exercise performance in distance events ranging from 2 mi to the marathon. Few studies have addressed the changes in lactate threshold, or anaerobic threshold derived from ventilatory variables, with aging. Lactate threshold was determined in a cross-sectional study of a group of masters endurance runners, 40 to 70+ years of age (111 men and 57 women).[39] Lactate threshold expressed as a percentage of $\dot{V}O_{2max}$ (LT-% $\dot{V}O_{2max}$) provides the best marker relative to endurance running performance in individuals with similar $\dot{V}O_{2max}$ values. Interestingly, LT-% $\dot{V}O_{2max}$ did not differ between men and women, but it did increase with age. More recent longitudinal studies with masters athletes reported that the change in lactate threshold over a six-year follow-up was not predictive of running performance when it was expressed as a percent of $\dot{V}O_{2max}$.[26] Another study reported similar results in 152 untrained men and 146 untrained women.[27] However, in both studies $\dot{V}O_{2max}$ was lower in the older groups, which helps explain the increase in LT-% $\dot{V}O_{2max}$. When the LT at the absolute $\dot{V}O_2$ is compared between age-groups, LT declined with age.

In review

➤ Aerobic capacity generally decreases by about 10% per decade or 1% per year in relatively sedentary men and women.

➤ Similar results have been obtained for highly trained endurance athletes, although there is a much wider variation in the results of different studies. Also, if athletes begin with a higher $\dot{V}O_{2max}$ and $\dot{V}O_{2max}$ declines at the same rate, it will remain higher than that of sedentary people of the same age.

➤ Studies of older athletes and less active people of the same age-group indicate that the decrease in $\dot{V}O_{2max}$ is not strictly a function of age. Athletes who continue to train have significantly smaller decreases in $\dot{V}O_{2max}$ as they age, particularly if they train at a high intensity.

➤ Lactate threshold, expressed as a percentage of $\dot{V}O_{2max}$, increases with aging, but it decreases when expressed in relation to the absolute $\dot{V}O_2$ at which it occurs.

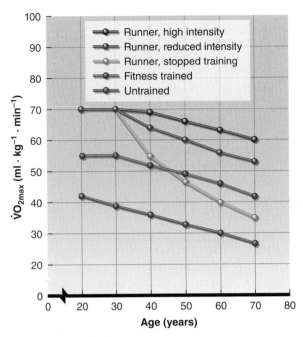

FIGURE 18.10 Changes in $\dot{V}O_{2max}$ with age for trained and untrained men.

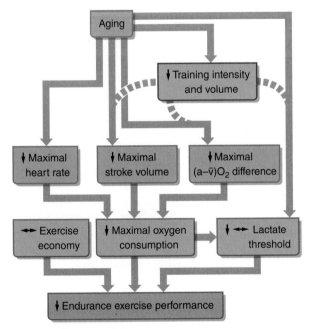

FIGURE 18.11 Factors and physiological mechanisms contributing to the reductions in endurance performance with advancing age in healthy humans. Primary aging directly contributes to the reduction in the cardiovascular determinants of $\dot{V}O_2$; however, there tends to also be a decrease in training volume and intensity with aging (dotted lines), so the relative contributions of each to reduced endurance performance are difficult to determine.

Reprinted, by permission, from H. Tanaka and D.R. Seals, 2008, "Endurance exercise performance in Masters athletes: Age-associated changes and underlying physiological mechanisms," *Journal of Physiology* 586(1): 55-63.

Physiological Adaptations to Exercise Training

Despite the decrements in body composition and exercise performance associated with aging, well-trained middle-aged and older athletes are capable of exceptional performances. Furthermore, those who train for general fitness appear to experience gains in muscular strength and endurance similar to those of young adults. In fact, four to six months of aerobic training or two to three months of resistance training in previously sedentary older individuals can restore $\dot{V}O_{2max}$ and maximal strength levels to those seen in individuals 20 years younger.

Strength

As with most other physiological functions, the loss of strength with aging is likely the result of a combination of the natural aging process and reduced physical activity that produces a decline in muscle mass and function. Although it is difficult to compare the adaptations to resistance training of younger and older people, aging appears to neither impair the ability to improve muscle strength nor prevent muscle hypertrophy. For example, when older men (ages 60-72) resistance trained for 12 weeks at 80% of their 1-repetition maximum (1RM) for extension and flexion of both knees, their extension strength increased by 107% and flexion strength increased by 227%.[8] These improvements were attributed to muscle hypertrophy as determined from midthigh CT scans. Biopsies of the vastus lateralis muscle (in the quadriceps) revealed that the cross-sectional area of type I fibers increased by 34% and that of type II fibers by 28%. The greater increase in strength compared to size was due to relatively low baseline strength values and likely neural adaptations in these previously sedentary men. In another study of older untrained men (aged 64 years), a 16-week resistance training program resulted in major increases in strength (50% for leg extension strength, 72% for leg press strength, and 83% for half-squat strength) and an increase in the mean cross-sectional area of all major muscle fiber types (46% for type I, 34% for type IIx, and 52% for type IIb).[10, 15]

In a study of older women (average age 64 years), 21 weeks of resistance training resulted in a 37% increase in leg extensor maximal force development; a 29% increase in leg extension 1RM; an increase in the cross-sectional area of the extensor muscles; and a 22% to 36% increase in type I, type IIx, and type IIb muscle fiber areas.[12]

Another study investigated the changes in leg strength, chair rise time, and muscle fiber type and fiber composition in 60- to 75-year-old men and women who performed heavy resistance and power training using squats twice a week for 24 weeks.[11] The 1RM for strength increased by 26% in women and 35% in men, whereas the time to perform three stand-ups in quick succession from a 40 cm (~16 in.) chair (chair rise time) decreased by 24% in women and 25% in men. This study illustrates the improvements in functional performance that are possible due to resistance training.

In general, older individuals can experience a significant benefit from resistance training. Older resistance-trained athletes tend to have higher muscle mass, are generally leaner, and are ~30% to 50% stronger than their sedentary peers.[40] Furthermore, compared to age-matched aerobic exercise trained subjects resistance trained athletes have more total muscle mass, have higher bone mineral densities, and maintain higher muscle strength and power. Although not to the same extent as older athletes, older sedentary individuals also experience significant strength gains from resistance training, which greatly improves their abilities to perform activities of daily living and helps to prevent falling.[40]

Aerobic and Anaerobic Capacity

Recent studies have shown that improvements in $\dot{V}O_{2max}$ with training are similar for younger (ages 21-25) and older (ages 60-71) men and women.[22, 27] Although the pretraining $\dot{V}O_{2max}$ values were, on the average, lower for the older subjects, the absolute increases of 5.5 to 6.0 ml · kg^{-1} · min^{-1} were similar in the two groups. Additionally, older men and women experienced similar increases in $\dot{V}O_{2max}$, averaging 21% for men and 19% for women, when they trained for 9 to 12 months by walking, running, or both walking and running about 4 mi (6 km) per day. Previously sedentary older subjects appear to have a peak in cardiovascular adaptation after three to six months of moderate training.[32] Together, this research indicates that endurance training produces similar gains in aerobic capacity in healthy people throughout the age range of 20 to 70 years, and this adaptation is independent of age, sex, and initial fitness level. However, this does not mean that endurance training can enable older athletes to achieve the performance standards established by younger athletes.

The precise mechanisms that trigger the body's adaptations to training at any age are not fully understood, so we do not know if improvements from training are achieved in the same way throughout life. For example, much of the improvement in $\dot{V}O_{2max}$ seen in younger subjects is associated with an increase in maximal cardiac output. But older subjects show significantly greater gains in muscle oxidative enzyme activities, which suggests that peripheral factors in older subjects' muscles might play a greater role in aerobic adaptations to training than in younger subjects.

Very little is known about the trainability of anaerobic capacity in older people. We saw earlier in this chapter that lactate threshold, expressed as a percentage of a person's $\dot{V}O_{2max}$, increases with aging and is not associated with endurance running performance.

In young and middle-aged adults, LT-% $\dot{V}O_{2max}$ is the best predictor of endurance performance—running, cycling, swimming, and cross-country skiing. As we have already seen, different rates of aging of the oxygen transport and lactic acid buffering systems are a likely explanation for this difference between older adults and young and middle-aged adults. A related point is that when one is comparing the lactate threshold of individuals with different $\dot{V}O_{2max}$ values, it is likely more appropriate to consider the absolute $\dot{V}O_2$ value to which that threshold corresponds when attempting to explain endurance performance.

In focus

The ability to adapt to training was once thought to decrease greatly with aging. However, we now know that when older subjects train at relatively high intensities they have the ability to increase their endurance capacity and strength with training.

In review

➤ Older adults appear to get the same benefits—maintenance of body weight, decreased percent body fat and fat mass, and increased fat-free mass—from exercise training as younger and middle-aged adults.

➤ It appears that aging does not impair a person's ability to increase muscle strength or muscle hypertrophy. Individual muscle fibers of older individuals also have the ability to increase in size.

➤ Endurance exercise training produces similar absolute gains in healthy people, regardless of their age, sex, or initial level of fitness. However, percent improvement is larger in those with low initial baseline levels.

➤ With endurance training, an increased $\dot{V}O_{2max}$ in older exercisers results mostly from improvement in the muscles' oxidative enzyme activities (peripheral adaptation), whereas improvement in younger people is largely attributable to increased maximal cardiac output (central adaptation).

Sport Performance

World and national records in running, swimming, cycling, and weightlifting suggest that people are in their physical prime during their 20s or early 30s. If we use a cross-sectional approach, comparing these records with national and world records for older athletes in these events allows us to examine the effects aging has on the best performers. Unfortunately, we have little longitudinal information about the effects of aging on performance because few studies have enabled us to follow physical performance in selected individuals over the span of their athletic careers. However, we can look historically at performance times in certain athletic events to gain insight into the influence of physiological function on performance with aging. In the following sections, we consider how aging affects certain types of sport performance.

Running Performance

In 1954, Roger Bannister, a 21-year-old medical student, stunned the sporting world when he became the first person to run the mile (1.61 km) in less than 4 min (3 min 59.4 s). The current record for the mile is 3:43:13, set by Hicham El Guerrouj of Morocco in 1999, which is more than 16 s faster than Bannister's record—a gap that would have placed Bannister more than 100 m (109 yd) behind today's record holder. In 1954, it would have seemed inconceivable that a sub-4-min mile could have been accomplished by someone over the age of 30 years. The oldest individual to record a sub-4-min mile was Eamonn Coghlan, who was 41 years old when he ran it indoors in 3:58:13. The oldest individual to record a sub-5-min mile was 65 years old.

Although older runners have achieved some exceptional records, running performance in general declines with age, and the rate of this decline appears to be independent of distance. Longitudinal studies of elite distance runners indicate that despite a high level of training, performance in events from the mile (1.61 km) to the marathon (42 km) declines at a rate of about 1.0% per year from the age of 27 to 47 years.[37, 38] It is interesting to note that world records for both 100 m and 10 km runs also decrease by about 1% per year from age 25 to age 60,[3] as shown in figure 18.12. Beyond age 60, however, the records for men slow by nearly 2% per year. A sprint-running test of 560 women between ages 30 and 70 revealed a steady decrease in maximal running velocity of 8.5% per decade (0.85% per year).[27] The patterns of change are about the same in sprint- and endurance-running performances.

Swimming Performance

A retrospective study of freestyle performances at the U.S. Masters swimming championships between 1991 and 1995 revealed that both men's and women's performances in the 1,500 m declined steadily from age 35

(a)

(b)

FIGURE 18.12 Change with age in men's and women's world records for (*a*) 100 m and (*b*) 10 km runs. Note that these running records slow at a much faster rate after the age of 50 to 60 years.

to about 70 years, after which swimming times slowed at a faster rate.[33] However, the rate and magnitude of the declines in both 50 m and 1,500 m performances with age were found to be greater for women than for men.

Cycling Performance

As with other strength and endurance sports, record-setting cycling performances are generally achieved in the age range of 25 to 35. Male and female cyclists' records for 40 km (24.9 mi) races decrease at about the same rate with age, an average of 20 s (approximately 0.6%) per year. The U.S. national cycling records for 20 km (12.4 mi) show a similar pattern for both men and women. For this distance, speed decreases by about 12 s (approximately 0.7%) per year from age 20 to nearly age 65.

Weightlifting

In general, maximal muscle strength is achieved between the ages of 25 and 35. Beyond that age range, as shown in figure 18.13, men's records for the sum of three power lifts decline at a steady rate of about 12.1 kg (26.7 lb), approximately 1.8%, per year. Of course, as with other measurements of human performance, individual strength performances vary considerably. Some individuals, for example, exhibit greater strength at age 60 than people half their age.

Most athletic performances decline at a steady rate during middle and advanced ages. These decreases

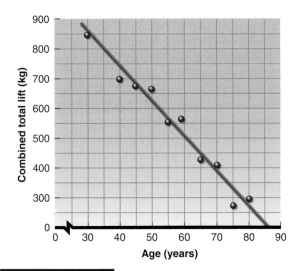

FIGURE 18.13 Changes in U.S. National Masters powerlifting records with age among male weightlifters. The values reported are combined totals for the squat, bench press, and deadlift.

result from decrements in both muscular and cardiovascular endurance and strength as discussed earlier in this chapter.

In focus

As we age, peak performances in both endurance and strength events decrease by about 1% to 2% per year, starting between ages 25 and 35.

> Records in running, swimming, cycling, and weightlifting indicate that we are in our physical and physiological prime during our 20s and early 30s.

> In all of these sports, performance generally declines with aging beyond age 30 to 35.

> Most athletic performances decline steadily during middle and older age, primarily because of decrements in endurance and strength.

Special Issues

As we age, we need to consider several special issues that can directly affect us when exercising or performing various sport activities. Here we look briefly at environmental stresses and then consider the issues of longevity, injury, and risk of death resulting from exercise and sport.

Environmental Stress

Because a variety of physiological control processes become less effective with aging, we can logically assume that older people are less tolerant of environmental stress than their younger counterparts. As we have seen elsewhere in this chapter, it is difficult to determine the separate effects of aging and physical fitness. In the fol-lowing discussion, we compare the responses of younger and older adults to exercise in the heat and comment on cold stress and altitude exposure in older athletes.

Exposure to Heat

Exposure to heat stress presents a problem for older people. A preponderance of deaths during environmental heat waves occurs in people over the age of 70. The rate of metabolic heat production is related to the absolute exercise intensity, while heat loss mechanisms are related to the relative exercise intensity, so matching young and older subjects for $\dot{V}O_{2max}$ is important. When subjects are matched for body composition and $\dot{V}O_{2max}$, there is no difference in core temperature during exercise in the heat. However, when older subjects with a normal $\dot{V}O_{2max}$ for their age are compared to young subjects, they have a higher core temperature (see figure 18.14).[21]

These results indicate that physical training affects certain thermoregulatory responses. Sweat gland density does not appear to decline with aging, but sweat gland output does; and research indicates that sweating rate is much more closely related to $\dot{V}O_{2max}$ than it is to age. As we saw in chapter 12, an increase in skin blood flow is necessary for the transfer of heat from the core to the shell for dissipation via evaporation of sweat. Even when young and older subjects are matched for $\dot{V}O_{2max}$, skin blood flow is lower in the older subjects, but higher in older highly fit compared to older normally

(a) Normal $\dot{V}O_{2max}$ for age

(b) Subjects matched for $\dot{V}O_{2max}$

FIGURE 18.14 Changes in body core temperature in response to young (red line) and older (blue line) subjects exercising in a warm environment. When subjects in each age-group have normal fitness levels for their age (*a*), core temperature rises more sharply in the older individuals. However, when the two age-groups are selected to have the same $\dot{V}O_{2max}$, the difference in core temperature disappears (*b*), suggesting that $\dot{V}O_{2max}$ is more important than chronological age in determining this response.

Adapted, by permission, from W.L. Kenney, 1997, "Thermoregulation at rest and during exercise in healthy older adults," *Exercise and Sport Sciences Reviews* 25: 41-77.

fit subjects, indicating that regular aerobic training can improve heat dissipation. Also, exercise in the heat requires significant blood flow to both the skin and exercising muscle, which is accomplished through increased cardiac output and reduced blood flow to renal and splanchnic regions. This redistribution of blood flow is less effective in older individuals; but similar to the situation with skin blood flow, improving aerobic fitness can improve this response. One study demonstrated that four weeks of endurance training in older men, which improved $\dot{V}O_{2max}$ by ~25%, improved regional redistribution, so that renal and splanchnic blood flow decreased ~200 ml more than before training, which reduced cardiovascular strain.

(In) focus

Aging reduces our ability to adapt to exercise in the heat. This is largely due to reduced aerobic capacity. Increasing aerobic capacity can increase skin blood flow, increase sweating rate by increasing sweat gland output, and improve redistribution of blood to the skin and exercising muscles.

Exposure to Cold and to Altitude

In contrast to heat exposure, exercise in the cold typically poses less of a health risk. Due to declining aerobic fitness and loss of muscle mass, older individuals do have a reduced ability to generate metabolic heat. Also, the ability of the cutaneous vasculature to constrict is impaired with aging, which may increase heat loss. As a result of these changes, older individuals fail to appropriately maintain core temperature during cold stress. This is the case even when elderly subjects are exposed to what we would consider a mild cold stress.[5] However, people can easily offset these decrements by wearing clothing that is appropriate for the environmental conditions and activity level. By performing so-called behavioral thermoregulation, older athletes can offset the decrements in physiological thermoregulation and continue to exercise safely in cold environments.

During exposure to high altitude, there is little reason to expect older athletes to respond differently than their younger counterparts. Unfortunately, data are lacking regarding aging and the rate and magnitude of acclimatization to altitude. Likewise, it is unclear whether aging per se increases the incidence of any of the altitude illnesses. We can expect the performance of an older athlete at altitude to be similar to that of a younger athlete of comparable physical fitness.

Longevity and Risk of Injury and Death

Regular physical activity is an important contributor to good health. Does training throughout adulthood affect longevity? Because the aging rate in rats is more rapid than in humans, they have been used as subjects in studies conducted to determine the influence of chronic exercise (training) on **longevity** (the length of one's life). A study by Goodrick[9] demonstrated that rats that exercised freely lived about 15% longer than sedentary rats. But an investigation at Washington University in St. Louis showed no significant increase in the life span of rats that voluntarily ran on an exercise wheel.[16] More of the active rats lived to old age, but on the average, they still died at the same age as their sedentary counterparts. It is interesting that rats that had restricted food intake and maintained a lower body weight lived 10% longer than the freely eating, sedentary rats. Although exercise training is a key component to energy balance, the only known way to increase longevity is through caloric restriction.[31]

Of course, we cannot directly apply these findings to humans, but these results raise some interesting questions that might be relevant to our health and longevity. Although it is true that an endurance exercise program can reduce a number of the risk factors associated with cardiovascular disease, only limited information supports the contention that people will live longer if they exercise regularly. Data collected from the alumni at Harvard University and the University of Pennsylvania and from participants at the Aerobic Center in Dallas suggest that there is a decrease in mortality rate and a small increase in longevity (about two years) among people who remain physically active throughout life. Therefore, regular physical activity can increase the active life span, that is, the number of years living independently and free of disability. This is sometimes referred to as compression of mortality or rectangularization of the survival curve. Perhaps future longitudinal studies will shed more light on the relationship between lifelong exercise and longevity.

What about the risk of injury and death from exercise as we get older? Studies show that as people get older they are at a greater risk for injuries involving tendons, cartilage, and bone. The most common orthopedic injuries include rotator cuff tears, quadriceps tendon ruptures, Achilles tendon ruptures, degenerative meniscus tears, focal articular cartilage defects and injuries, and stress fractures. Furthermore, when injuries occur, the healing process is usually prolonged and complete recovery can take up to a full year.[27] On the other hand, increasing the strength and endurance of elderly people reduces their risk of falls and related injury, so the benefits of maintaining a regular exercise program with aging outweigh the potential risks.

The risk of death during exercise appears to be no higher in the older athlete compared with the younger and middle-aged athlete. However, the risk in older people who are not regularly active appears to be increased, possibly due to undiagnosed or subclinical disease processes.[30] Importantly, an active lifestyle does, in fact, reduce the risk of death from many chronic diseases, something that we discuss in chapters 20 through 22.

In review

- The impaired ability of older individuals to tolerate exercise in the heat is due to reduced $\dot{V}O_{2max}$ and impaired cardiovascular adaptations rather than a direct effect of aging on thermoregulatory control or sweating.
- Regular exercise training can increase skin blood flow and sweating rate and improve the redistribution of cardiac output in older individuals as well as young men and women.
- Older people generally have an impaired ability to tolerate cold, but they can compensate by wearing appropriate clothing.
- Adaptation to altitude appears to be independent of age.
- An active lifestyle appears to be associated with a small increase in longevity. Just as important, an active lifestyle leads to a higher quality of life!
- There is an increased risk of injury from exercise as people age, and injuries tend to be slower to heal.
- The risk of death during exercise is not increased in those who are regularly active but is increased in those who seldom exercise.

In this chapter we examined the effects of aging on physical performance. We evaluated changes in cardiorespiratory endurance and strength with age. We considered the effect of aging on body composition, which we know can affect performance. And yet, in the course of our discussion, it became clear that much of the change that occurs with aging is to a great extent attributable to the inactivity that often accompanies aging. When older people participate in training, most of the changes associated with aging are lessened and the resulting degree of change is similar to that seen in young and middle-aged adults. Thus, we have dispelled many of the myths about the capacity for physical activity of older people.

In the next chapter, we turn our attention to females, who as a group are often considered less capable of physical activity than males. We consider the physiology of girls and women, the impact of this physiology on athletic ability, how performances of female athletes compare with those of male athletes, and special issues associated with being female.

Key Terms

cardiovascular deconditioning

endothelial dysfunction

forced expiratory volume in 1 second ($FEV_{1.0}$)

longevity

maximal expiratory ventilation ($\dot{V}_{Emax}$)

osteopenia

osteoporosis

peripheral blood flow

sarcopenia

Study Questions

1. What changes occur in height, weight, and body composition with aging? What accounts for these changes? How do these changes affect maximal oxygen uptake?

2. What changes occur in muscle with aging? How do they affect strength and athletic performance?

3. Describe the changes in HR_{max} with age. How does training alter this relationship?

4. How does aging affect maximal stroke volume and maximal cardiac output? What mechanisms can potentially explain these changes?

5. How does the respiratory system change with aging? What happens to vital capacity, $FEV_{1.0}$, residual volume, total lung capacity, and the ratio RV/TLC?

6. $\dot{V}O_{2max}$ declines with age across the entire population. Describe the physiological mechanisms that account for this decline. How do trained older individuals maintain a relatively high $\dot{V}O_{2max}$?

7. How does aging and habitual exercise affect blood vessel function?

8. How does age affect anaerobic function?

9. Differentiate between biological aging and physical inactivity.

10. What influence do aging and training have on body composition?

11. Describe the trainability of the older individual for both strength and aerobic endurance.

12. Describe the changes in strength and endurance performance records with aging.

13. What concerns should we have about older people exercising in hot and cold environments or at altitude?

14. Describe the risk of injury and death associated with exercise in the elderly.

15. What is the effect of exercise on longevity?

Study Guide Activities

In addition to the activities listed in the chapter opening outline on page 447, two other activities are available in the web study guide, located at

www.HumanKinetics.com/PhysiologyOfSportAndExercise

 The **KEY TERMS** activity reviews important terms, and the end-of-chapter **QUIZ** tests your understanding of the material covered in the chapter.

Sex Differences in Sport and Exercise

19

In this chapter and in the web study guide

ACTIVITY 19.1 Sex-Related Responses to Exercise and Training considers how women and men respond to acute exercise and long-term training.

ACTIVITY 19.2 Menstrual Dysfunction describes the characteristics of delayed menarche, primary amenorrhea, secondary amenorrhea, and oligomenorrhea.

ACTIVITY 19.3 Eating Disorders and Disordered Eating explores the characteristics of anorexia nervosa and bulimia nervosa and the differences between eating disorders and disordered eating.

Girls and women were prohibited from running any race longer than 800 m until the 1960s. Before 1972 they were banned from official participation in most mainstream marathons, including the Boston Marathon. Both of these restrictions resulted from the misconception that women were somehow physiologically unsuited for endurance activity. Yet at the 1984 Los Angeles Olympic Games, American runner Joan Benoit won the gold medal in the first-ever Olympic marathon for women with a time of 2:24:52. Her time would have won 11 of the previous 20 men's Olympic marathons!

Another myth that is falling by the wayside is that pregnant women should not exercise. The following story appeared in the *Austin American Statesman* on June 17, 1995. "When Sue Olsen lines up today for Grandma's Marathon in St. Paul, Minn., along the North Shore of Lake Superior, she will be precisely 16 days short of the due date for her first child. 'I get a lot of advice both ways,' Olsen, 38, said with a laugh. 'There are people who think I'm crazy and shouldn't be out there, and there are a lot of supportive people.' . . . Olsen's husband will drive parallel to the course with a cellular phone in his car, prepared to whisk her to the hospital if the situation arises." Although most would question the wisdom of this, the point is that Olsen was able to complete the marathon in 4 h. The following weekend she completed a 24 h race, delivering a baby boy the next day. In 2005, Olsen's son, John "Miles" Olsen, completed 32.7 mi in a 24 h run, just a few weeks before his 10th birthday. His mother completed 119.6 mi (Doug Grow, *Star Tribune*, Minneapolis, MN, June 7, 2005).

In the recent past, young girls typically were discouraged from participating in vigorous physical activity, while young boys climbed trees, raced against each other, and participated in various sports. The underlying notion was that boys were meant to be active and athletic but girls were weaker and less well suited to physical activity and competition. Physical education classes furthered this idea by having girls exercise differently than boys—by running shorter distances and performing modified push-ups. Less physical activity was expected of girls; and as they progressed through school, most girls could not compete on an equal basis with boys of their own age, even if given the opportunity. In athletics, girls and women were not allowed to run in long-distance races, and basketball was limited to half-court, with each team having only offensive or defensive players.

Now more athletic activities and programming are accessible to girls and women than in the past, and the results have been amazing. Their athletic accomplishments parallel those of boys and men, with performance differences of 15% or less for most sports and events. This is illustrated in table 19.1, where men's and women's world records (2006) are compared for representative events in both track and field and swimming. Do these performance differences represent true biological differences, or are there other factors we need to consider? The focus of this chapter is to determine the extent to which biological differences between females and males affect performance capacity.

TABLE 19.1 Selected Men's and Women's World Records Through 2010

Event	Men	Women	Difference
TRACK AND FIELD			
100 m	9.58 s	10.49 s	9%
1,500 m	3:26.00 min:s	3:50.46 min:s	12%
10,000 m	26:17.53 min:s	29:31.78 min:s	12%
High jump	2.45 m	2.09 m	15%
Long jump	8.95 m	7.52 m	16%
SWIMMING (FREESTYLE)			
100 m	46.91 s	52.07 s	11%
400 m	3:40.07 min:s	3:59.15 min:s	9%
1,500 m	14:34.56 min:s	15:42.54 min:s	8%

Body Size and Composition

Body size and composition are similar in boys and girls during early childhood. During late childhood, as we saw in chapter 17, girls begin to accumulate more fat than boys, and starting in early adolescence boys begin to increase their fat-free mass (FFM) at a much higher rate than girls (see figure 17.3 on p. 429).

In focus

Major differences in body size and composition between girls and boys do not start to appear until late childhood and early adolescence.

These body composition differences between the sexes occur primarily because of endocrine changes with development. Before puberty, the anterior pituitary gland secretes very small amounts of the gonadotropic hormones: follicle-stimulating hormone (FSH) and luteinizing hormone (LH). These hormones stimulate the gonads (ovaries and testes). During puberty, however, the anterior pituitary begins to secrete significantly greater amounts of both of these hormones. In females, when sufficient quantities of FSH and LH are secreted, the ovaries develop and estrogen secretion begins. In males, these same hormones trigger development of the testes and, in turn, testosterone secretion. Figure 19.1 illustrates these changes in estrogen (estradiol—the most potent form of estrogen) and testosterone from the beginning of puberty (S1) to the end of puberty (S5). **Testosterone** increases bone formation, which leads to larger bones, and increased protein synthesis, which in turn leads to increased muscle mass. As a result, adolescent males are larger and more muscular than females, and these characteristics continue into adulthood. At full maturity, men not only have a greater muscle mass, but the distribution of the muscle mass differs from that of women. Men carry a higher percentage of their muscle mass in the upper body compared with women (42.9% vs. 39.7%).[22] Testosterone also stimulates erythropoietin production by the kidneys, which leads to increased red blood cell production, as discussed later in this chapter.

Estrogen also has a significant influence on body growth by broadening the pelvis, stimulating breast development, and increasing fat deposition, particularly in the thighs and hips. This increase in fat deposition in the thighs and hips is the result of increased lipoprotein lipase activity in these areas. This enzyme

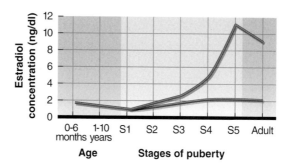

FIGURE 19.1 Changes in blood concentrations of testosterone and estrogen (estradiol) from birth to adulthood. The S1 through S5 represent stages of puberty based on secondary sex characteristics, with S1 representing the initial stages of puberty and S5 the final stages.

Reprinted, by permission, from R.M. Malina, C. Bouchard, and O. Bar-Or, 2004, *Growth, maturation, and physical activity*, 2nd ed. (Champaign, IL: Human Kinetics), 414.

is considered the gatekeeper for storing fat in adipose tissue. **Lipoprotein lipase** is produced in the fat cells (adipocytes) but is bound to the walls of the capillaries, where it exerts its influence on the chylomicrons, which are the major transporters of triglycerides in the blood. When lipoprotein lipase activity in any area of the body is high, chylomicrons are trapped and their triglycerides are hydrolyzed and transported into the adipocytes in that area for storage.

Estrogen also increases the growth rate of bone, allowing the final bone length to be reached within two to four years following the onset of puberty. As a result, females grow very rapidly for the first few years following puberty and then cease to grow. Males have a much longer growth phase, allowing them to attain a greater height. Because of these differences, compared with fully mature males, fully mature females are on average about

- 13 cm (5 in.) shorter,
- 14 to 18 kg (30-40 lb) lighter in total weight,
- 3 to 6 kg (7-13 lb) heavier in fat mass, and
- 6% to 10% higher in relative body fat.

In review

➤ Until puberty, girls and boys do not differ significantly in most measurements of body size and composition.

➤ At puberty, because of the influences of estrogen and testosterone, body composition begins to change markedly.

➤ Testosterone increases bone formation and protein synthesis, leading to a larger FFM. It also stimulates the production of erythropoietin, which increases red blood cell production.

➤ Estrogen causes increased fat deposition in females, particularly in the hips and thighs, and an increased rate of bone growth, such that bones in females reach their final length earlier than in males.

Physiological Responses to Acute Exercise

When females and males are exposed to an acute bout of exercise, whether an all-out run to exhaustion on the treadmill or a single attempt to lift the maximal weight possible, responses differ between the sexes. Differences between children and adolescent boys and girls were discussed in chapter 17. Here we focus on these differences in adults in the areas of strength and of cardiovascular, respiratory, and metabolic responses to exercise.

Strength

In terms of strength, women have traditionally been regarded as the weaker sex. In fact, women are approximately 40% to 60% weaker than men in upper body strength but only 25% to 30% weaker in lower body strength. However, because of the considerable size difference between the average man and the average woman, it is more appropriate to express strength either relative to body weight (absolute strength ÷ body weight) or relative to FFM, as a reflection of muscle mass (absolute strength ÷ FFM). When lower body strength is expressed relative to body weight, women are still 5% to 15% weaker than men, but when it is expressed relative to FFM, this difference disappears. This suggests that the innate qualities of muscle and its mechanisms of motor control are similar for women and men, a fact that was confirmed by computed tomography (CT) scans of the upper arms and thighs of physical education majors of both sexes and of male bodybuilders.[32] Computed tomography scans allow the actual estimation of muscle mass. Although both groups of men had much greater absolute levels of strength than the women, no differences between the groups were found when strength was expressed per unit of muscle cross-sectional area for both knee extensor muscles (figure 19.2a) and elbow flexor muscles (figure 19.2b).

In focus

For the same amount of muscle, there are no differences in strength between the sexes, although women have smaller muscle fiber cross-sectional areas than men and typically less muscle mass than men.

Although differences in upper body strength are reduced somewhat when expressed relative to total body weight and FFM in untrained women, substantial differences remain. There are at least two possible explanations for this. Women have a higher percentage of their muscle mass in the lower body when compared with men.[22] In addition and probably related to this muscle mass distribution, women use the muscle mass of their lower bodies much more than they use their upper body muscle mass, particularly when compared with use patterns in men. Some average-sized women

Fat Deposition: Why the Hips and Thighs?

Many women are constantly fighting fat deposition on the thighs and hips, but they are usually fighting a losing battle. Lipoprotein lipase activity is very high and lipolytic activity (fat breakdown) is low in the hips and thighs of women compared with women's other fat storage areas and compared with the hips and thighs of men. This results in a rapid storage of fat in women's thighs and hips, and the decreased lipolytic activity makes it difficult for women to lose fat from these areas. During the last trimester of pregnancy and throughout lactation, the activity of lipoprotein lipase decreases and lipolytic activity increases dramatically, which suggests that fat is stored in the hips and thighs for reproductive purposes.

a   **b**

FIGURE 19.2 No sex differences between men and women in strength [(a) maximal knee extension torque or (b) maximal elbow flexion torque] are seen when strength is expressed per unit of muscle cross-sectional area.

Reprinted, by permission, from P. Schantz et al., 1983, "Muscle fibre type distribution, muscle cross-sectional area and maximal voluntary strength in humans," *Acta Physiologica Scandinavica* 117: 219-226.

have remarkable strength, exceeding that of an average man. This indicates the importance of neuromuscular recruitment and synchronization of motor unit firing in the ultimate determination of strength (chapter 3).

Muscle biopsies have become more common among female athletes, allowing fiber type comparisons with male athletes in the same sport or event. From these data, men and women have similar fiber type distributions, as shown in figure 19.3, although men in one study reached greater extremes (greater than 90% slow-twitch [type I] or greater than 90% fast-twitch [type II]). As shown in figure 19.3, when the vastus lateralis muscle was biopsied, male distance and sprint runners varied between approximately 15% and 85% type I fiber type distribution, compared with female distance and sprint runners, whose distributions were between approximately 25% and 75%.[31] However, different results were found in two studies, one of elite female[6] and the other of elite male distance runners.[17] In these elite runners, the extremes for percentages of type I fibers were similar (41-96% for women and 50-98% for men), even though the mean values were different: Women had a mean value of 69% type I fibers compared with 79% for the men. The women had much smaller fiber areas for both type I and type II fibers (mean values of less than 4,500 μm² in women

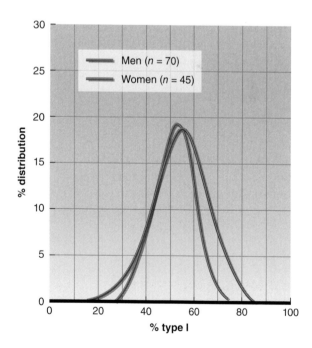

FIGURE 19.3 Distribution of type I fibers in the vastus lateralis muscle in male and female runners.

Adapted, by permission, from B. Saltin et al., 1977, "Fiber types and metabolic potentials of skeletal muscles in sedentary man and endurance runners," *Annals of the New York Academy of Sciences* 301: 3-29.

and greater than 8,000 μm² in men). Despite smaller fiber size in women, capillarization appears to be similar between men and women.

Research indicates that women have a greater resistance to fatigue compared with men. Fatigue usually is sometimes tested by having subjects maintain a constant force at a given percentage of their maximal voluntary static action for as long as possible. As an example, women would be able to maintain a constant force output at 50% of their maximal static action for a longer period of time than men at the same 50% of their maximal static action. Men, being stronger, will have to apply a greater absolute amount of force to achieve the same 50% relative force. The reason for this greater resistance to fatigue is not yet known but could be related to the amount of muscle mass recruited and the compression of blood vessels, substrate utilization, muscle fiber type, and neuromuscular activation.

Cardiovascular and Respiratory Function

When placed on a cycle ergometer, where the power output can be precisely controlled independent of body weight (50 W, as an example), women generally have a higher heart rate (HR) response for any absolute level of submaximal exercise. However, maximum heart rate (HR_{max}) is generally the same in both sexes. Stroke volume (SV) is lower in women, but cardiac output ($\dot{Q}$) is usually the same in men and women at any absolute submaximal power output. The higher submaximal HR

response in women appears to compensate for the lower SV, allowing a similar $\dot{Q}$ for the same power output, since $\dot{Q} = HR \times SV$. The lower SV results primarily from at least two factors:

- Women have smaller hearts and therefore smaller left ventricles because of their smaller body size and possibly lower testosterone concentrations.

- Women have a smaller blood volume, which also is related to their size (lower FFM).

The average woman may also be less aerobically active and therefore less aerobically conditioned.

When power output is controlled to provide the same relative level of exercise, usually expressed as a fixed percentage of maximal oxygen uptake ($\dot{V}O_{2max}$), women's heart rates are still slightly elevated compared with men's, and their stroke volumes are markedly lower. At 60% $\dot{V}O_{2max}$, for example, a woman's cardiac output, stroke volume, and oxygen consumption are generally less than a man's, and her heart rate is slightly higher. With the exception of HR_{max}, these differences also are seen at maximal levels of exercise.

These relationships between HR, SV, and $\dot{Q}$ for the same absolute power output (50 W) and the same relative power output (60% $\dot{V}O_{2max}$) are illustrated in figure 19.4. These data were derived from the HERITAGE Family Study.[37] Interestingly, when these same relationships are compared between 7- to 9-year-old boys and girls, there are no sex differences.[35]

a 50 watts

b 60% of $\dot{V}O_{2max}$

FIGURE 19.4 Comparison of submaximal heart rate (HR), stroke volume (SV), and cardiac output ($\dot{Q}$) between men and women at the same absolute power output (50 W) and the same relative power output (60% $\dot{V}O_{2max}$).

Data from Wilmore et al., 2001.

In focus

At the same submaximal exercise intensity, women generally have cardiac outputs similar to those of men. Thus, their higher heart rates appear to fully compensate for their lower stroke volumes. The lower stroke volume is largely due to a smaller left ventricle and lower blood volume, both the result of women's smaller body size.

FIGURE 19.5 Differences in maximal ventilatory volumes with age in untrained (UT) and trained (TR) females and males.

While several early studies reported $\dot{Q}$ to be higher in women at identical submaximal power outputs, possibly compensating for their lower hemoglobin concentrations, more recent studies have consistently shown no differences between sexes.[19, 37] Apparently, women can compensate for their lower hemoglobin concentrations with a steeper increase in their arterial–venous oxygen difference, or $(a-\bar{v})O_2$ difference, for a given power output.

Women also have less potential for increasing their peak $(a-\bar{v})O_2$ difference. This is likely attributable to their lower hemoglobin content, which results in lower arterial oxygen content and reduced muscle oxidative potential. Lower hemoglobin content is an important contributor to **sex-specific differences** in $\dot{V}O_{2max}$ because less oxygen is delivered to the active muscle for a given volume of blood.

Sex differences in respiratory responses to exercise are largely attributable to body size differences. Breathing frequency during exercise at the same relative power output (e.g., 60% $\dot{V}O_{2max}$) differs little between the sexes. However, at the same absolute power output, women tend to breathe more rapidly than men, probably because when men and women are working at the same absolute power output, women are working at a higher percentage of their $\dot{V}O_{2max}$.

Tidal volume and ventilatory volume are generally smaller in women at the same relative and absolute power outputs, up to and including maximal levels. Most highly trained female athletes have maximal ventilatory volumes below 125 L/min; but highly trained men have maximal values of 150 L/min and higher, some exceeding 250 L/min (figure 19.5). Again, these differences are closely associated with body size.

Metabolic Function

$\dot{V}O_{2max}$ is regarded by most exercise scientists as the single best index of a person's cardiorespiratory endurance capacity. Recall that $\dot{V}O_2$ is the product of cardiac output and $(a-\bar{v})O_2$ difference. This means that $\dot{V}O_{2max}$ represents that point during exhaustive exercise at which the subject has maximized oxygen delivery and utilization capabilities. The average female tends to

reach her peak $\dot{V}O_{2max}$ between ages 12 and 15, but the average male does not reach his peak until ages 17 to 21 (see chapter 17). Beyond puberty, the average woman's $\dot{V}O_{2max}$ is only 70% to 75% of the average man's.

$\dot{V}O_{2max}$ differences between women and men must be interpreted carefully. A classic study published in 1965 showed considerable variability in $\dot{V}O_{2max}$ within each sex and a considerable overlap of values between sexes.[21] The study involved a group of women and men 20 to 30 years of age, including

- elite female athletes,
- female nonathletes,
- elite male athletes, and
- male nonathletes.

When the subjects' physiological responses to submaximal and maximal exercise were compared, 76% of the female nonathletes overlapped 47% of the male nonathletes, and 22% of the female athletes overlapped 7% of the male athletes. The relationships are illustrated in figure 19.6. These data demonstrate the importance of looking beyond mean values to consider both the subjects' levels of physical conditioning and the extent of overlap between the groups being compared.

Although the $\dot{V}O_{2max}$ values of females and males are similar until puberty, comparisons of $\dot{V}O_{2max}$ values of normal nonathletic females and males beyond puberty might not be valid. Such data likely reflect an unfair

FIGURE 19.6 Range of $\dot{V}O_{2max}$ values (mean ±2 standard deviations from the mean) for female nonathletes, male nonathletes, elite female athletes, and elite male athletes. The mean $\dot{V}O_{2max}$ value is presented above each bar. This figure illustrates that although differences in the average $\dot{V}O_{2max}$ between groups can be substantial, there can be considerable overlap of one group with another.

Data from Hermansen and Andersen, 1965.

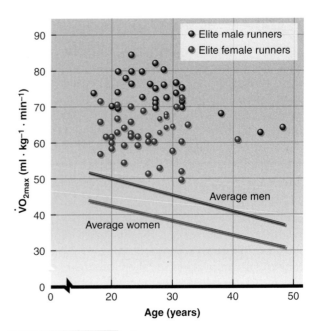

FIGURE 19.7 $\dot{V}O_{2max}$ values, compiled from the literature, for elite female and male distance runners compared with average values in untrained women and men.

comparison of relatively sedentary females with relatively active males. Thus, reported differences would reflect the level of conditioning as well as possible sex-specific differences. To overcome this potential problem, investigators began to examine highly trained female and male athletes, with the assumption that the level of training would be similar for both sexes and would allow a more accurate evaluation of true sex-specific differences.

Saltin and Åstrand[30] compared $\dot{V}O_{2max}$ values of female and male athletes from Swedish national teams. In comparable events, the women had 15% to 30% lower $\dot{V}O_{2max}$ values. However, more recent data suggest a smaller difference. The $\dot{V}O_{2max}$ values for a group of elite female distance runners are compared in figure 19.7 with values for elite male distance runners and average, nonathletic women and men. The elite female runners had substantially higher values than untrained men and women. Some women's values were even higher than a few of the elite male runners' values, but when we consider the average for each elite group, the women's values were still 8% to 12% lower than those of the elite male runners.

Several studies have been aimed at scaling $\dot{V}O_{2max}$ values relative to height, weight, FFM, or limb volume in an attempt to more objectively compare women's and men's values. Some of these studies have shown that differences between the sexes disappear when $\dot{V}O_{2max}$ is expressed relative to FFM or active muscle mass, yet others continue to demonstrate differences even when differences in body fat are adjusted for.[9]

In focus

While there are clear sex differences in the mean $\dot{V}O_{2max}$ for a given age and activity level, there is considerable overlap in the range of $\dot{V}O_{2max}$ values.

In one study, researchers used a novel approach to this problem.[8] They examined the responses to submaximal and maximal treadmill runs under various conditions in 10 women and 10 men who regularly engaged in distance running. The women exercised only under normal weight conditions, but the men exercised both at normal weight and with external weight added to their trunks so that the total percentage of excess weight, defined as the men's fat weight plus the added weight, equaled the percentage fat of the matched women. Equalizing the sexes for excess weight reduced the mean sex differences in

- treadmill run time (sex difference reduced by 32%),
- the $\dot{V}O_2$ expressed in milliliters per kilogram of FFM for running at various submaximal speeds (reduced by 38%), and
- $\dot{V}O_{2max}$ (reduced by 65%).

The researchers concluded that women's greater sex-specific essential body fat stores are major determinants of sex-specific differences in metabolic responses to running.

Women have lower hemoglobin levels than men, and this also has been proposed as a factor contributing to their lower $\dot{V}O_{2max}$ values. In one study, researchers attempted to equate the hemoglobin concentrations of a group of 10 men and 11 women who were active but not highly trained.[7] An amount of blood was withdrawn from the men to equalize their hemoglobin concentrations to those of the women. This significantly reduced the men's $\dot{V}O_{2max}$ values, but these reductions accounted for only a relatively small portion of the sex differences in $\dot{V}O_{2max}$.

It is also important to understand that a woman's lower cardiac output at maximal rates of work is a limitation to achieving a high $\dot{V}O_{2max}$ value. A woman's smaller heart size and lower plasma volume greatly limit her maximal stroke volume capacity. In fact, the results of several studies have suggested that women have limited ability to increase their maximal stroke volume capacity with high-intensity endurance training. However, more recent studies showed that young, premenopausal women were able to increase their stroke volume with training identically to men. Furthermore, in the untrained state after their plasma volume was artificially increased with a plasma volume expander and after β-blockade (which reduced the heart rate for a given rate of work, allowing more time to fill the left ventricle), women were able to increase their stroke volume to the same extent as untrained men during an acute bout of exercise.[25, 26]

If, instead of looking at $\dot{V}O_{2max}$, we consider submaximal oxygen consumption ($\dot{V}O_2$), little if any difference is found between women and men for the same absolute power output. But remember that at the same absolute submaximal work rate, women usually are working at a higher percentage of their $\dot{V}O_{2max}$. As a result, their blood lactate levels are higher, and lactate threshold occurs at a lower absolute power output. Peak blood lactate values are generally lower in active but untrained women than in active but untrained men. Also, limited data suggest that elite female middle-distance and long-distance runners have peak lactate concentrations that are approximately 45% lower than those of similarly trained elite male runners. Such sex differences in peak blood lactate values are unexpected and unexplained.

Lactate threshold appears to be similar between equally trained men and women if expressed in relative ($\%\dot{V}O_{2max}$), not absolute, terms. Lactate threshold appears to be closely related to the mode of testing and to the individual's state of training. Thus, sex-specific differences are not expected.

In focus

Compared to men, women generally have lower $\dot{V}O_{2max}$ values when expressed in $ml \cdot kg^{-1} \cdot min^{-1}$. Part of this difference in $\dot{V}O_{2max}$ values between women and men is related to the extra body fat carried by women and, to a lesser extent, to their lower hemoglobin concentration, which results in a lower oxygen content in the arterial blood.

In review

➤ The innate properties of muscle and the mechanisms of motor control are similar for women and men.

➤ Women and men do not differ in lower body strength expressed relative to body weight or to FFM. But women have less upper body strength expressed relative to body weight or FFM than men, largely because more of women's muscle mass is below the waist and women use their lower body muscles more than their upper body muscles.

➤ At submaximal exercise intensities, women have higher heart rates than men, but women's submaximal cardiac outputs are similar for the same intensity. This indicates that women have lower stroke volumes, primarily because they have smaller hearts and less blood volume.

➤ Women also have a lower capacity to increase $(a-\bar{v})O_2$ difference, probably because of their lower hemoglobin content, so less oxygen is delivered to their active muscles per unit of blood.

➤ Sex differences in respiratory responses are primarily attributable to body size differences.

➤ Beyond puberty, the average woman's $\dot{V}O_{2max}$ is only 70% to 75% that of the average man. However, some of this difference could be attributable to women's traditionally less active lifestyles. Research with highly trained athletes reveals an 8% to 15% difference, and much of this difference is attributable to women's greater fat mass, lower hemoglobin levels, and lower maximal cardiac output.

➤ Little or no difference in lactate threshold is found between the sexes, but peak lactate concentrations are generally higher in men.

Physiological Adaptations to Exercise Training

Basic physiological function both at rest and during exercise changes substantially with physical training. In this section we investigate how women adapt to chronic exercise, emphasizing areas in which their responses might differ from men's.

Body Composition

With either cardiorespiratory endurance training or resistance training, both women and men experience

- a decrease in total body mass,
- a decreased fat mass,
- a decreased relative (%) fat, and
- increased FFM.

The magnitude of the change in body composition appears to be related more to the total energy expenditure associated with the training activities than to the participant's sex. Significantly more FFM is gained in response to strength training than with endurance training, and the magnitude of these gains is similar between sexes.

Bone and connective tissue undergo alterations with training, but these changes are not well understood. In general, animal studies and limited human studies have shown an increase in the density of the weight-bearing long bones, primarily in growing animals and in children and adolescents. This adaptation appears to be independent of sex. In adults, weight-bearing exercise is critical for maintaining bone mass and density. We discuss some exceptions later in this chapter.

Connective tissue appears to be strengthened with endurance training, and sex-specific differences in this response have not been identified. Higher injury rates in women point to the possibility that women are more susceptible to injury than men while participating in physical activity and sport. This has led to concerns about sex-specific differences in joint integrity and laxity and the strength of ligaments, tendons, and bones. Unfortunately, the research literature contributes little to confirming or denying the validity of such concerns. Where differences in the rate of injury have been observed, the injury could be related more to the level of conditioning than to the participant's sex. Those who are less fit are more prone to injury. This is an extremely difficult area in which to obtain objective data but nevertheless is an important area that needs to be better researched.

Strength

Until the early 1970s, prescribing strength training programs for girls and women was considered inappropriate. Women were not believed capable of gaining strength because of their innately low levels of male anabolic hormones. Paradoxically, many people also generally feared that strength training would masculinize women. During the 1960s and 1970s, however, it became evident that many of the United States' best female athletes were not doing well in international competition, largely because they were weaker than their competitors. Slowly, research began to demonstrate that women can benefit considerably from strength training programs even though the resulting strength gains are usually not accompanied by large increases in muscle bulk.

In part because of their lower testosterone, women have less total muscle mass than men. If muscle mass is the major determinant of strength, then women have a distinct disadvantage. But if neural factors are as important as or more important than size, women's potential for absolute strength gains is considerable. Also, some women can attain significant muscle hypertrophy. This has been demonstrated in female bodybuilders who have remained free of anabolic steroids. Also, a number of studies have shown similar increases for men and women in FFM and muscle volume, as well as hypertrophy of type I, type IIa, and type IIx muscle fibers following periods of resistance training.

With all of this in mind, it is interesting to look at the men's and women's world records for weightlifting by weight classification. Figure 19.8 illustrates these world records for the total weight lifted (the sum of the snatch and the clean and jerk) as of 2011. As seen from this figure, men are considerably stronger at each weight classification. While the weight classifications differ slightly for men and women, the men generally lifted at least 75 kg (165 lb) more than the women did for an equivalent body weight. Part of the difference can be explained by the fact that, for a given body weight, men most likely have a higher FFM. Furthermore, fewer women participate in competitive weightlifting, and the greater the number of participants, the greater the likelihood of higher world records. Still, the differences are so great that there must be other factors operating.

In focus

Women can experience major increases in strength (20-40%) as a result of resistance training, and the magnitude of these changes is similar to that seen in men. These gains are attributable to both muscle hypertrophy and neural factors.

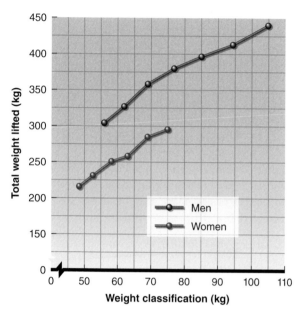

FIGURE 19.8 Men's and women's world weightlifting records as of January 2011 for the total amount of weight lifted (combination of the snatch and the clean and jerk) by weight category. The highest weight category for both men and women is not included because it is not defined by weight. The men's total weight lifted is considerably more than the women's at a similar weight classification.

International Weightlifting Federation, March 2011 (www.iwf.net/results/record_cur.php).

Cardiovascular and Respiratory Function

Major cardiovascular and respiratory adaptations result from cardiorespiratory endurance training, and these adaptations do not appear to be sex specific. Major increases in maximal cardiac output ($\dot{Q}_{max}$) accompany training. Maximum heart rate usually does not change or decreases with training, so this increase in $\dot{Q}_{max}$ is the result of a large increase in stroke volume, which results from two factors. End-diastolic volume (the amount of blood in the ventricles before contraction) increases with training because blood volume increases and venous return is more efficient. In addition, end-systolic volume (the amount of blood remaining in the ventricles after contraction) is reduced with training because the stronger myocardium produces a stronger contraction, ejecting more blood.

At submaximal work rates, cardiac output shows little or no change, although stroke volume is considerably higher for the same absolute rate of work. Consequently, heart rate for any given rate of work is reduced after training. Resting heart rate can be reduced to

50 beats/min or less. Several female distance runners have had resting heart rates below 36 beats/min. This is considered a classic training response and corresponds to an exceptionally high stroke volume.

The increases in $\dot{V}O_{2max}$ that accompany cardiorespiratory endurance training, which are discussed subsequently, result primarily from the large increases in maximal cardiac output, with only small increases in $(a-\bar{v})O_2$ difference. However, the major limitation to $\dot{V}O_{2max}$ is oxygen transport to the working muscles. Although cardiac output is important to oxygen transport, these researchers believe that the increases in $\dot{V}O_{2max}$ that accompany training are primarily attributable to increased maximal muscle blood flow and muscle capillary density. These changes are firmly established in both women and men. Although women also experience considerable increases in maximal ventilation similar to those of men, reflecting increases in both tidal volume and breathing frequency, these changes are thought to be unrelated to the increase in $\dot{V}O_{2max}$.

Metabolic Function

With cardiorespiratory endurance training, women experience the same average relative increase in $\dot{V}O_{2max}$ that has been observed in men (15% to 20% on average). The magnitude of change noted generally depends on the intensity and duration of the training sessions, the frequency of training, and the length of the study.

In focus

With aerobic training, women experience average increases in $\dot{V}O_{2max}$ (15-25%), similar to those experienced by men.

After cardiorespiratory endurance training, women's oxygen uptake at the same absolute submaximal work rate does not appear to change, although several studies have reported decreases. Women's blood lactate concentrations are reduced for the same absolute submaximal rates of work; peak lactate concentrations generally are increased; and the lactate threshold increases with training.

From this discussion, it is obvious that women respond to physical training in the same manner as men. Although the magnitudes of their adaptations to training might differ somewhat from those of their male counterparts, the overall trends appear identical. This is important to remember when one is prescribing exercise, something that is discussed in chapter 20.

In review

➤ With training, women and men experience similar changes in body composition as determined by total energy expenditure during training.

➤ Women, similar to men, gain considerable strength through resistance training, an effect associated with increases in FFM and muscle volume as well as hypertrophy of type I, type IIa, and type IIx muscle fibers.

➤ Cardiovascular and respiratory changes that accompany cardiorespiratory endurance training do not appear to be sex specific.

➤ Women experience the same relative increases in $\dot{V}O_{2max}$ as men with cardiorespiratory endurance training.

Sport Performance

Women are outperformed by men in all athletic activities in which performance can be precisely and objectively measured by distance or time. The difference is most pronounced in activities such as the shot put, where high levels of upper body strength are crucial to successful performance. In the 400 m freestyle swim, however, the winning time for women in the 1924 Olympic Games was 19% slower than that for men, but this difference decreased to 15.9% in the 1948 Olympics and to only 7.0% in the 1984 Olympics. The fastest women's 800 m freestyle swimmer in 1979 swam faster than the world record–holding man for the same distance in 1972. From these results, it would appear that the gap between the sexes is narrowing. However, as we see from table 19.1 (p. 472), the difference between the men's and women's world record for the 400 m freestyle was 10.4% in 2006. For the 1,500 m freestyle it was 8.9%. Unfortunately, making valid comparisons through the years has been difficult because the degree to which an activity has been emphasized and its popularity are not constant, and other factors—such as opportunities to participate, coaching, facilities, and training techniques—have differed considerably between the sexes over the years.

As noted at the beginning of this chapter, large numbers of girls and women did not start entering competitive sport until the 1970s. Even then, there was an initial reluctance to train women as hard as men. Once girls and women started training as hard as boys and men, their performance improved dramatically. This is illustrated in figure 19.9, which shows world records from 1960 to 2011 for women and men in six running events in track and field. For distances of 100 m

through the marathon, women's present world records are consistently 8% to 9% slower than men's. Furthermore, as can be seen in this figure, the improvement in women's records, which was initially quite dramatic, is beginning to level off and to parallel the curves for men's records.

Special Issues

Although the sexes respond to acute exercise and adapt to chronic exercise in much the same manner, several additional areas that are unique to females must be considered. Specifically, we look at

- menstruation and menstrual dysfunction,
- pregnancy,
- osteoporosis,
- eating disorders, and
- environmental factors.

Menstruation and Menstrual Dysfunction

How does the menstrual cycle or pregnancy influence exercise capacity and performance? How do physical activity and competition influence the menstrual cycle or pregnancy? These are two questions of interest to exercising women, particularly female athletes.

The three major phases of the **menstrual cycle** are illustrated in figure 19.10. The first is the menstrual (flow) phase, or **menses**, which lasts three to five days, during which time the uterine lining (endometrium) is shed and menstrual flow, or bleeding, occurs. The second is the proliferative phase, which prepares the uterus for fertilization and lasts about 10 days. During this phase, the endometrium begins to thicken, and some of the ovarian follicles that house the ova mature. These follicles secrete estrogen. The proliferative phase ends when a mature follicle ruptures, releasing its ovum (ovulation). The menstrual and proliferative phases correspond to the follicular phase of the ovarian cycle.

The third and final phase of the menstrual cycle is the secretory phase, which corresponds to the luteal phase of the ovarian cycle. This phase lasts 10 to 14 days, during which the endometrium continues to thicken, its blood and nutrient supply increases, and the uterus prepares itself for pregnancy. During this time, the remnants of the ruptured follicle (now termed a corpus luteum, hence the term *luteal phase*) secretes progesterone, and estrogen secretion also continues. The complete menstrual cycle averages 28 days; however, there is considerable variation in the cycle length among healthy women, from 23 to 36 days.

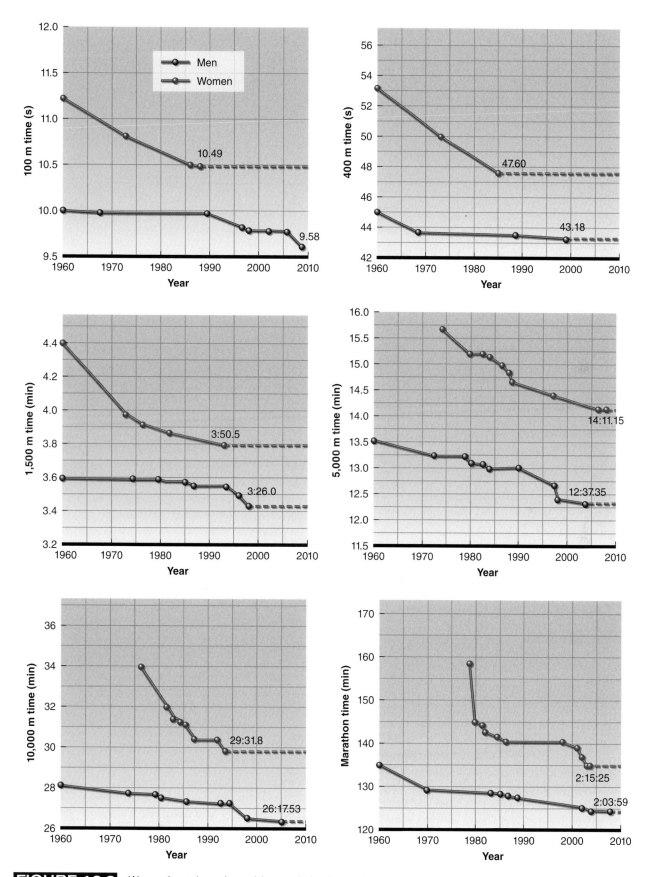

FIGURE 19.9 Women's and men's world records in six running events between 1960 and 2011.

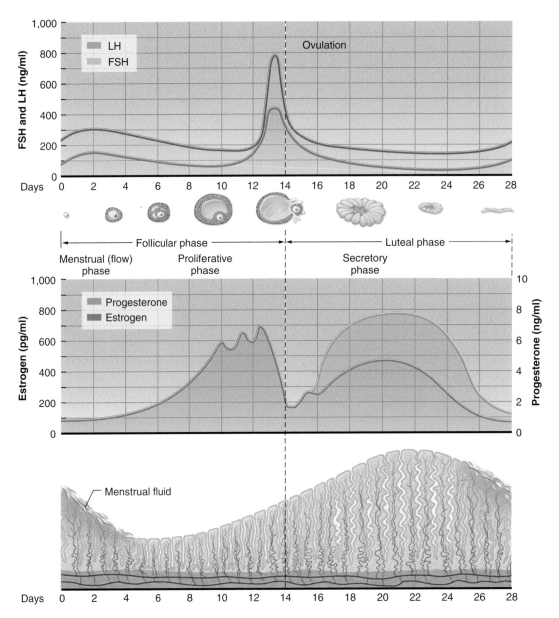

FIGURE 19.10 Phases of the menstrual cycle and the concomitant changes in progesterone and estrogen (middle) and follicle stimulating hormone (FSH) and luteinizing hormone (LH) (top). For most purposes, the cycle is divided into the follicular phase, which starts with the initiating of bleeding during menstrual flow, and the luteal phase, which starts with ovulation.

Menstruation and Performance

Alterations in athletic performance experienced during different phases of the menstrual cycle are subject to considerable individual variability. There are no reliable data that demonstrate any significant changes in athletic performance at any time during the menstrual cycle.

Several well-controlled studies have been conducted in research laboratories comparing physiological responses to exercise at various points in the menstrual cycle. These generally have shown no physiological differences across phases of the cycle. In addition, top-level performances have been achieved by female athletes during all phases of the menstrual cycle. Therefore, from data gathered both in the laboratory and during athletic competition, we can conclude that neither the physiological responses nor the performance of most women is substantially affected by the menstrual cycle.

In focus

There appears to be no general pattern concerning the ability of women to achieve their best performances during any specific phase of their menstrual cycle. World records have been set by elite female athletes during every phase of the menstrual cycle.

Menarche

Menarche, which refers to the occurrence of the first menstrual period, has been reported to be delayed in some young athletes involved in certain sports and activities, such as gymnastics and ballet. A delay in menarche is defined as a menarche after the age of 14; the median age of menarche for American girls is between 12.4 to 13.0 years, depending on the population sampled. For gymnasts, the median age appears to be closer to 14.5 years. Frisch[18] hypothesized that menarche is delayed five months for each year of training before menarche, implying that exercise training causes delayed menarche. Malina,[24] however, postulated that late-maturing girls, such as those with a later age of menarche, are more likely to be successful in sports such as gymnastics because of their small, lean bodies. This implies that those who naturally experience a later menarcheal age have an advantage in, and thus are likely to be involved in, certain sports rather than that their sport involvement delays menarche.

These opposing viewpoints can be summarized by the following questions: Does intense training to achieve the level of an elite performer delay menarche, or does a later menarche provide an advantage that contributes to the success of the elite performer? Stager and colleagues[34] used computer modeling to analyze this issue and concluded that the age of menarche in athletes naturally occurs later as opposed to being "delayed." At this time, evidence is insufficient to support the theory that training delays menarche.

In focus

Menarche appears to come later in highly trained elite athletes in certain sports such as gymnastics. However, there is no strong evidence to support the contention that the intense training for the sport delays menarche.

Menstrual Dysfunction

Female athletes can experience disruptions of their normal menstrual cycle. These disruptions are collec-tively referred to as **menstrual dysfunction**, of which there are several types. **Eumenorrhea** is the term for normal menstrual function, referring to consistent menstrual cycle lengths of 26 to 35 days. **Oligomenorrhea** refers to inconsistent and irregular menses at intervals longer than 36 days but up to 90 days. **Amenorrhea** refers to the absence of menstruation; **primary amenorrhea** refers to the absence of menarche in girls and women 15 years of age and older—women who have not yet begun menstruating. When athletes with previously normal menstrual function report the absence of menstruation for 90 days or longer, it is called secondary amenorrhea. Thus, **secondary amenorrhea** is the absence of menses for 90 days or more in girls and women who were previously menstruating. Women involved in any sport and even recreationally active women can experience amenorrhea as it occurs independent of intensity of exercise training.

The prevalence of secondary amenorrhea and oligomenorrhea among athletes is well documented and is estimated to be anywhere between 5% and 66% or higher, depending on the sport or activity, the level of competition, and the definition used for amenorrhea.[10] For example, if amenorrhea is defined as no menses for six months or more, then the prevalence rate reported tends to be lower; however, when the prevalence rate is defined as no menses for three months, the prevalence rates tend to be higher. Physicians and researchers have largely defined amenorrhea as no menses for three months because physiological consequences (discussed later in the chapter) result soon after the onset of amenorrhea. Nonetheless, the prevalence of amenorrhea in female athletes is considerably higher than the estimated 2% to 5% prevalence for amenorrhea and 10% to 12% prevalence for oligomenorrhea in the general (nonathlete) population. The prevalence appears to be greater in athletes who participate in activities that emphasize a lean physique like gymnastics and cross-country running.

Since the 1970s, scientists have been conducting experiments to determine the primary cause of secondary amenorrhea in exercising girls and women and athletes. Some of the factors that have been proposed as potential causes are the following:

- A history of menstrual dysfunction
- The acute effects of stress
- A high volume or intensity of training
- A low body weight or percent body fat
- Hormonal alterations
- An energy deficit through inadequate nutrition, disordered eating, or both

Amenorrhea and Birth Control

Many women who become amenorrheic are relieved to be free of menstruation each month. Most also assume that they have developed a simple but effective form of birth control. However, athletes have become pregnant while amenorrheic, which indicates that ovulation, and thus fertility, is not always influenced by the absence of menstruation. This point warrants further discussion. Since amenorrhea in athletes is likely caused by the failure to consume an adequate number of calories to meet energy expenditure needs, it is indeed plausible that energy status can change quickly, but menses will take longer to recover. Thus, while a woman with amenorrhea may be in the process of consuming additional calories to meet energetic needs, she may remain amenorrheic for several more months. It takes months of exposure to adequate calories for menses to recover; but during that time, the ovaries may start to produce estrogen, and ovulation may occur prior to the onset of menses. This point needs to be stressed among female athletes prone to amenorrhea to reduce the possibility of unexpected pregnancy.

Considerable research has been conducted on each of these proposed factors, and five of the six have been eliminated for consideration as the primary cause. It is tempting to surmise that high-volume or high-intensity training (or a combination of the two) leads to menstrual dysfunction, but this factor is likely not involved.

Current evidence indicates that inadequate nutrition resulting in an energy deficit is the primary cause of secondary amenorrhea. Studies have shown that inadequate intake of calories, such that the body is not matching caloric intake to caloric expenditure over an extended period of time, is the primary cause of secondary amenorrhea.

In focus

The etiology of amenorrhea in exercising women is secondary to inadequate caloric intake in the face of high exercise-related energy expenditure, resulting in a net energy deficit. The energy deficit or low energy availability, in turn, stimulates compensatory mechanisms such as weight loss and energy conservation that translate to hypothalamic suppression of ovarian function and amenorrhea.

Recent research by Dr. Anne Loucks[23, 29] at Ohio University and Dr. Nancy Williams at Penn State University[36] has clearly demonstrated that inducing an energy deficit in eumenorrheic women results in significant hormonal alterations that are associated with menstrual dysfunction, including amenorrhea. Dr. Loucks demonstrated that reducing caloric intake, with or without the added stress of increased energy expenditure resulting from a few days of exercise training, reduced LH pulse frequency and concentrations of the thyroid hormone

triiodothyronine (T_3), both of which are associated with impaired menstrual function. Dr. Williams, on the other hand, showed more specific effects of exercise on hormone profiles and menstrual function, demonstrating that failing to consume enough calories during three months of exercise training reduced estrogen and progesterone concentrations. Moreover, Dr. Williams found that the greater the energy deficit, the more severe the impact on menstrual function.[36]

Food deprivation may trigger signals that inhibit LH secretion and menstrual function. Disruption of LH pulsatility and low circulating estrogen concentrations suggest that there is a disruption of the gonadotropin-releasing hormone (GnRH) pulse generator in the hypothalamus.[29] These inhibitory signals may come from several different pathways. Potential signaling pathways include leptin secreted from adipocytes; ghrelin and peptide YY, secreted from the gut and intestine; cortisol from the adrenal gland; and other metabolic factors associated with energy deficiency.[11, 12, 33]

Thus, exercise training is likely not directly associated with menstrual dysfunction at all, other than through its contribution to an energy deficit. An energy deficit, in either the absence or presence of exercise training, is associated with these hormonal alterations. Intense or high-volume training most likely is not associated with menstrual dysfunction as long as energy intake matches or exceeds energy expenditure over days, weeks, and months.

The relationship between clinically disordered eating and menstrual dysfunction is a more recent concern; several studies have shown a strong relationship between the two. In one study, 8 of 13 amenorrheic distance runners reported disordered eating, compared with 0 of 19 eumenorrheic distance runners.[20] In another study, 7 of 9 amenorrheic elite middle- and long-distance runners were diagnosed with anorexia nervosa, bulimia nervosa, or both, compared with 0 of 5 eumenorrheic runners.[38] Eating disorders are discussed in detail later in this chapter, but we can say that eating disorders generally involve an energy deficit.

In review

➤ The consequences of competing during different phases of the menstrual cycle on performance are subject to considerable individual variability. In general, there is no research evidence that demonstrates a consistent effect of phase of the menstrual cycle on sport performance.

➤ Menarche can occur late in some young athletes in certain sports. However, the most likely explanation for this is that late maturers, because of lean body build, are more likely to participate successfully in these activities, not that these activities cause delayed menarche.

➤ Female athletes can experience menstrual dysfunction, most often in the form of secondary amenorrhea or oligomenorrhea. Current evidence implicates inadequate nutrition, or a prolonged energy deficit, as the primary cause of secondary amenorrhea.

➤ Hormonal changes associated with an energy deficit may feed back on the hypothalamus or pituitary and disrupt GnRH and LH secretion, which are needed to direct the normal cycle. This, too, is associated with a prolonged energy deficit.

Pregnancy

What are the effects of exercise during **pregnancy**? Four major physiological concerns are associated with exercise during pregnancy:

1. The acute risk associated with reduced blood flow to the uterus (blood is diverted to the mother's active muscles), leading to fetal hypoxia (insufficient oxygen).

2. Fetal hyperthermia (elevated temperature) associated with the increase in the mother's internal body temperature during prolonged aerobic-type exercise or exercise under conditions of heat stress.

3. Reduced carbohydrate availability to the fetus as the mother's body uses more carbohydrate to fuel her exercise.

4. The possibility of miscarriage and the final outcome of pregnancy.

Each of these is discussed in the following sections.

Reduced Uterine Blood Flow and Hypoxia

Uterine blood flow in both animals and humans is reduced by 25% or more during moderate to strenuous exercise, and the magnitude of the reduction is directly related to exercise intensity and duration.[39] Whether this reduction in uterine blood flow leads to fetal hypoxia is less clear. Apparently, an increase in the uterine $(a-\bar{v})O_2$ difference at least partially compensates for any reduced blood flow. Increases in fetal heart rate, although not always observed during maternal exercise, have been interpreted as an index of hypoxia in the fetus. Although increased fetal heart rate might reflect hypoxia to a certain degree, it more likely represents the fetal heart's response to increased catecholamine levels in the blood originating from both the fetus and the mother.

Hyperthermia

Fetal hyperthermia is a distinct possibility if the mother's core temperature is elevated substantially during and immediately after exercise. **Teratogenic effects** (abnormal fetal development) have been documented with chronic exposure to thermal stress in animals, and these effects have been documented with maternal fever in humans. Central nervous system defects are the most common result. Although fetal temperature has been shown to increase with exercise in animal studies, it is unclear whether this increase is sufficient to warrant concern.

Carbohydrate Availability

The potential for reduced carbohydrate availability for the fetus during exercise is also not well understood. We know that endurance athletes who train or compete for long durations reduce both liver and muscle glycogen stores and that blood glucose concentrations also can decrease. But whether this is a potential problem in pregnant women is less clear.

Miscarriage and Pregnancy Outcome

Concerns also have been expressed regarding the potential of exercise to induce miscarriage during the first trimester, to induce premature labor, and to alter the normal course of fetal development. Unfortunately, little information is available concerning the risk for miscarriage and premature labor. Regarding pregnancy outcome, data are scarce and conflicting. Although there are some indications of lighter birth weights and shorter gestation periods, most studies have shown either favorable effects of exercise (such as reduced maternal weight gain, shorter postdelivery hospital stays, and fewer cesarean sections) or no differences between the control and exercise groups.

In focus

Although there are several concerns for the health of the fetus during maternal exercise, the risk to the fetus from aerobic exercise during pregnancy appears to be low, particularly if guidelines for exercising during pregnancy are followed.

Recommendations for Exercise During Pregnancy

To summarize, exercise during pregnancy can have associated risks (see table 19.2), but the benefits far outweigh the potential risks if caution is taken in designing the exercise program. It is important that the pregnant woman coordinate her exercise program with her obstetrician so that sound medical judgment can be used to determine the most appropriate mode, frequency, duration, and intensity of activity.

The American College of Obstetricians and Gynecologists (ACOG) developed a set of guidelines in 1985, which were subsequently revised in 1994. These guidelines were summarized by Pivarnik[28] as follows:

- Pregnant women can derive health benefits from mild to moderate exercise performed at least three days per week.

- Women should avoid supine exercise and motionless standing after the first trimester because this compromises venous return, which in turn compromises cardiac output.

- Women should stop exercising when fatigued, should not exercise to exhaustion, and should modify their routines based on maternal symptoms. Weight-bearing exercises under some circumstances may be continued, but non-weight-bearing activities such as cycling or swimming are encouraged to reduce the risk of injury.

- Care should be taken not to participate in sports or exercises in which falling, a loss of balance, or blunt abdominal trauma may occur.

- Because pregnancy requires an extra 300 kcal (1,255 kJ) of energy per day, an exercising woman should pay particular attention to diet to ensure that she is receiving adequate calories.

- Heat dissipation is of particular concern in the first trimester, so an exercising woman should wear correct clothing, be sure that her fluid intake is sufficient, and select optimal environmental conditions.

TABLE 19.2 Hypothetical Risks and Postulated Benefits of Exercise During Pregnancy

Population	Hypothetical risks	Postulated benefits
Maternal	Acute hypoglycemia Chronic fatigue Musculoskeletal injury	Increased energy level (aerobic fitness) Reduced cardiovascular stress Prevention of excessive weight gain Facilitation of labor Faster recovery from labor Promotion of good posture Prevention of lower back pain Prevention of gestational diabetes Improved mood state and body image
Fetal	Acute hypoxia Acute hyperthermia Acute reduction in glucose availability Miscarriage in the first trimester Induction of premature labor Altered fetal development Shortened gestation Reduced birth weight	Fewer complications of a difficult labor

Adapted from L.A. Wolfe, P. Hall, K.A. Webb, L. Goodman, M. Monga, and M.J. McGrath, 1989, "Prescription of aerobic exercise during pregnancy," *Sports Medicine* 8: 273-301.

- A woman's regular prepregnancy exercise routine should be resumed gradually postpartum, as pregnancy-associated changes may persist four to six weeks.

In 2002, ACOG published a short "Committee Opinion" on exercise during pregnancy and the postpartum period, which essentially supported their previous guidelines.[1] In addition, they supported the current recommendation of the Centers for Disease Control and Prevention and the American College of Sports Medicine for nonpregnant individuals, which states that individuals should accumulate 30 min or more per day of moderate exercise on most, if not all, days of the week (see chapter 20). Furthermore, they state that scuba diving should be avoided throughout pregnancy because the fetus is at increased risk for decompression sickness. Also, there is an increased risk when pregnant women exercise at altitudes in excess of 6,000 ft (1,830 m).

In review

> - During exercise, major concerns for the pregnant athlete include the possible risk of fetal hypoxia, fetal hyperthermia, reduced carbohydrate supply to the fetus, miscarriage, premature labor, low birth weight, and abnormal fetal development.

> - The benefits of a properly prescribed exercise program during pregnancy outweigh the potential risks. Such an exercise program must be coordinated with the woman's obstetrician.

Osteoporosis

Maintaining a healthy lifestyle might retard one detrimental aging process that is a major health concern for women: osteoporosis. Osteoporosis is characterized by decreased bone mineral content, which causes increased bone porosity (see figure 19.11). Osteopenia, as we discovered in chapter 18, refers to a loss of bone mass that occurs with aging. Osteoporosis is a more severe loss of bone mass with deterioration of the microarchitecture of bone, leading to skeletal fragility and increased risk of bone fracture. These changes typically begin in the early 30s. The occurrence rate for fractures associated with osteoporosis increases by two to five times after the onset of menopause. Men also experience osteoporosis but to a lesser degree early in life because of a slower rate of bone mineral loss. Much remains to be learned about the etiology of osteoporosis; however, three major contributing factors common to postmenopausal women are

FIGURE 19.11 (a) Healthy bone and (b) bone showing increased porosity (decreased density, appearing darker) resulting from osteoporosis.

- estrogen deficiency,
- inadequate calcium intake, and
- inadequate physical activity.

Although the first of these is a direct result of menopause, the last two reflect dietary and exercise patterns throughout life.

In addition to postmenopausal women, women with amenorrhea and those with anorexia nervosa also suffer from low bone mass and osteoporosis attributable to insufficient calcium intake, low serum estrogen levels, or possibly both. In studies of women with anorexia, investigators found that their bone density was reduced significantly compared with those of controls. Cann and associates[5] were the first to report a substantially lower than normal bone mineral content in physically active women classified as having hypothalamic amenorrhea.

In another study, the radial and vertebral bone density of 14 athletic women (mostly runners) with amenorrhea was compared with those of 14 athletic women with normal menstruation (eumenorrhea).[15] Investigators discovered that physical activity did not protect the group with amenorrhea from significant bone density losses. The amenorrheic group's bone density values at a mean age of 24.9 were equivalent to those of normally active women at a mean age of 51.2. In a follow-up study, increases in vertebral bone mineral density were found in the women who previously had been amenorrheic but had resumed menstruation.[16] However, their bone mineral density remained well below the average for their age group, even four years after they resumed normal menses.[14]

In focus

Athletes with secondary amenorrhea are at increased risk for low bone mass. The low bone mass does not appear to be totally reversible with the resumption of normal menstrual function.

It generally is assumed that exercise is a positive factor for bone health in that it is associated with an increase in bone mass, or at least with the maintenance of bone mass in young, middle-aged, and older women. Therefore, it is confusing to learn that amenorrheic runners have reduced bone mass. Figure 19.12 is an attempt to clarify this apparent contradiction. From this figure we see that the bone mineral content of normally menstruating runners tends to be higher than that of normally menstruating non-running controls. Furthermore, female runners who are amenorrheic have higher bone mineral contents than untrained women who are amenorrheic. Thus, when we compare women

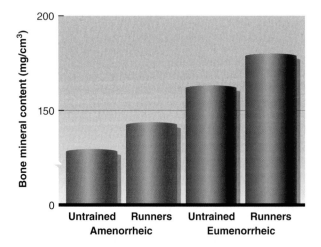

FIGURE 19.12 Bone mineral content of women runners and of untrained women who are amenorrheic (Am) and eumenorrheic (Eu). Note that when women with the same menstrual status are compared, the runners have higher bone mineral content than the untrained women.

Unpublished data from Dr. Barbara Drinkwater.

of like menstrual status, those who are exercising will have the higher bone mineral content. Caution should be used in interpreting data such as those presented in this section, because the results can be confounded by such factors as body composition, age, height, weight, and diet.

Although the precise mechanism is unknown, estrogen deficiency appears to play a major role in the development of osteoporosis. In the past, estrogen has been prescribed to peri- and postmenopausal women in an effort to reverse the degenerative effects of osteoporosis, but this therapy can have serious side effects, such as an increased risk of endometrial cancer. To reduce this risk, estrogen is given in combination with progestin (hormone therapy, or HT). However, there is an increased risk of breast cancer, strokes, and heart attacks associated with the use of HT. Bisphosphonates, considered antiresorptive medications, are also used. Increasing calcium intake to 1,200 to 1,500 mg per day and increasing vitamin D have also been proposed for decreasing the risk of osteoporosis.

Evidence certainly suggests that increased physical activity and adequate calcium intake combined with adequate caloric intake are a sensible approach to preserving the integrity of bone at any age. However, maintaining normal menstrual function is critical for those who have not reached menopause.

Eating Disorders

Eating disorders are a group of disorders that must meet specific criteria established by the American Psychiatric Association.[3] The two most commonly diagnosed eating disorders are anorexia nervosa and bulimia nervosa. **Disordered eating**, on the other hand, refers to patterns of eating that are not considered normal but don't meet the specific diagnostic criteria for a given eating disorder.

Eating disorders in girls and women became the focus of considerable attention beginning in the 1980s. Men constitute about 10% or less of the reported cases. Anorexia nervosa has been considered a clinical syndrome since the late 19th century, but bulimia nervosa was first described in 1976.

Anorexia nervosa is a disorder characterized by

- refusal to maintain more than the minimal normal weight based on age and height,
- distorted body image,
- intense fear of fatness or gaining weight, and
- amenorrhea.

Females from ages 12 to 21 are at greatest risk for this disorder. Its prevalence in this group is likely less than 1%.

Bulimia nervosa, originally termed bulimarexia, is characterized by

- recurrent episodes of binge eating;
- a feeling of lack of control during these binges; and
- purging behavior, which can include self-induced vomiting, laxative use, and diuretic use.

The prevalence of bulimia in the population at greatest risk, again adolescent and young adult females, is generally considered to be about 4% and possibly closer to 1%.

It is important to realize that a person might exhibit disordered eating and yet not meet the strict diagnostic criteria for either anorexia or bulimia. As an example, the diagnosis of bulimia requires that the individual average a minimum of two binge-eating and purging episodes a week for at least three months. What about the person who meets all the criteria, except that bingeing and purging occur only once per week? Although this person cannot technically be diagnosed as having bulimia, her or his eating is certainly disordered and is a potential cause for concern. Thus, the term "disordered eating" has been used to describe those who do not meet the strict criteria for an eating disorder but who do have abnormal eating patterns.

The prevalence of eating disorders in athletes is controversial. Numerous stud-ies have used either self-report or at least one of two inventories developed to diagnose disordered eating: the Eating Disorders Inventory (EDI) and the Eating Attitudes Test (EAT). Results have varied because not all studies used the strict standard diagnostic criteria for either anorexia or bulimia. As in the general population, female athletes are typically at a much higher risk than male athletes, and certain sports carry higher risks than others. The high-risk sports can generally be grouped into three categories:

1. Appearance sports, such as diving, figure skating, gymnastics, bodybuilding, and ballet
2. Endurance sports, such as distance running and swimming
3. Weight-classification sports, such as horse racing (jockeys), boxing, and wrestling

Self-reports or inventories do not always provide accurate results. In a study of 110 elite female athletes representing seven sports, EAT results showed that no athlete fell within the disordered eating range of the inventory. But in the subsequent two-year period, 18 of these athletes received either inpatient or outpatient treatment for eating disorders. In a second study of 14 nationally ranked middle- and long-distance runners who completed the EDI, only three were shown to have possible problems with disordered eating, and none were shown to have eating disorders.[38] In follow-up, seven subjects were subsequently diagnosed as having an eating disorder: four with anorexia nervosa, two with bulimia nervosa, and one with both. People with eating disorders, by their very nature, are secretive. We cannot realistically expect those with eating disorders to identify themselves, even when anonymity is ensured. For the athlete, this need for secrecy might be heightened by fear that a coach or a parent will learn of the eating disorder and not allow the athlete to compete.

Even though research is limited, it seems appropriate to conclude that athletes are at higher risk for eating disorders than the general population. Existing evidence likely does not reflect the seriousness of this problem in athletic populations. Although research data are not yet available, the prevalence might be as high as 60% or more in the specific high-risk athletic populations listed earlier.

In focus

Disordered eating has become a major concern in female athletes. Some researchers have estimated the prevalence to be as high as 60% or greater for elite athletes in certain sports.

Eating disorders generally are considered to be addictive disorders and are extremely difficult to treat. The physiological consequences are substantial and can even result in death. Considering this, along with the emotional distress suffered by the athlete, the extraordinary costs of treatment ($5,000-$25,000 per month for hospital inpatient treatment), and the effect on those closest to the athlete, eating disorders must be considered among the most serious problems facing female athletes today, paralleling the seriousness of anabolic steroid use in male athletes.

In 1990, the National Collegiate Athletic Association developed a list of warning signs for anorexia nervosa and bulimia nervosa (table 19.3). When an eating disorder is suspected, it is important to recognize the seriousness of the disorder and refer the athlete to a person specifically trained in dealing with eating disorders. Most athletic trainers, coaches, and even physicians are not trained to provide professional help to those with serious eating disorders. Most of the athletes who experience eating disorders are very intelligent, come from an upper-middle-class or higher socioeconomic level, and are very good at denying that they have a problem. These athletes are unfortunate victims of the unhealthy emphasis on extreme leanness promoted by the media and the challenges of attaining the optimal weight for their sport. Treating eating disorders is extremely difficult, and even the best-trained professionals are not always successful. Some extreme cases end in suicide or premature death from failure of the cardiovascular system. Immediate professional help should be sought for an athlete suspected of having an eating disorder.

Female athletes are at higher risk than nonathletes for disordered eating and eating disorders for several reasons. Perhaps most important, there is tremendous pressure on athletes, particularly female athletes, to get their weight down to very low levels, often below what is appropriate. This weight limit can be imposed by the coach, trainer, or parent or can be self-imposed by the athlete. In addition, the personality of the typical elite female athlete closely matches the profile of the female at high risk for an eating disorder (competitive, perfectionistic, and under the tight control of a parent or other significant figure such as a coach). Furthermore,

TABLE 19.3 Warning Signs for Anorexia Nervosa and Bulimia Nervosa	
Anorexia nervosa	**Bulimia nervosa**
Dramatic loss in weight	A noticeable weight loss or gain
A preoccupation with food, calories, and weight	Excessive concern about weight
Wearing baggy or layered clothing	Bathroom visits after meals
Relentless, excessive exercise	Depressed mood
Mood swings	Strict dieting followed by eating binges
Avoiding food-related social activities	Increased criticism of one's body

Note. The presence of one or two of these signs does not necessarily indicate an eating disorder. Diagnosis should be made by appropriate health professionals.

Adapted from a poster distributed by the National Collegiate Athletic Association, 1990.

Female Athlete Triad

In the early 1990s, it became apparent that there is a reasonably strong association among

- disordered eating, energy deficiency, or low energy availability;
- secondary amenorrhea; and
- low bone mass.

This group of disorders has been termed the Female Athlete Triad. The Female Athlete Triad is a syndrome of interrelated conditions that involves disordered eating or low energy availability (or both), low bone mass, and amenorrhea in physically active women and female athletes. Disordered eating, however, is not a necessary component. Rather, the common element is low energy availability that may or may not be a product of disordered eating. Either way, low energy availability or energy deficiency occurs as the athlete fails to consume an adequate volume of calories for her respective exercise energy expenditure. Over a period of time (the length of which has not been well established and might vary considerably from one athlete to another), an athlete who has low energy availability may start to experience abnormal menstrual function, which eventually may lead to secondary amenorrhea. Over time, this may lead to low bone mass.

This condition was first described by the American College of Sports Medicine in 1997 and is associated with significant health risks.[27] The condition is most common in women involved in sports that emphasize leanness, such as cross-country running, gymnastics, and figure skating; however, it may also affect recreationally active women and athletes from other sports.[13] Inadequate nutrition precedes the clinical occurrence of amenorrhea and low bone mass. Nutritional deficits are often associated with internal and external pressures on these women to maintain a low body weight.

A number of researchers have become interested in these intriguing relationships, and considerable research is now under way. In the past two decades, much has been learned about symptoms, risk factors, causes, and treatment strategies for the Female Athlete Triad and particularly amenorrhea and low bone mass, although only limited clinical guidelines are available to date. The clinical recommendation for prevention and treatment of the Female Athlete Triad is to increase caloric intake and, in some cases, reduce exercise energy expenditure. The rationale for reducing exercise training when caloric intake is already high is likely an excessively conservative approach, since exercise per se does not play a causal role in the etiology of athletic amenorrhea.

The most recent revision of the American College of Sports Medicine Position Stand on the Female Athlete Triad,[2] published in 2007, emphasizes that the three triad disorders can occur alone or in combination and should be addressed early before serious consequences develop.

Interested athletes, parents, coaches, and health professionals can visit the Female Athlete Triad Coalition website at www.femaleathletetriad.org for additional information.

the nature of the sport or activity largely dictates those at high risk. As previously mentioned, athletes in three categories are at high risk: appearance sports, endurance sports, and weight-classification sports. Added to these risks are the normal pressures imposed by the media and culture on young women, whether they are athletes or not.

Environmental Factors

Exercise in the heat, in the cold, or at altitude provides additional stress or challenge to the body's adaptive abilities (see chapters 12 and 13). Many early studies indicated that women are less tolerant to heat than men are, particularly when exercising. Much of this difference, however, was the result of lower fitness levels of the women included in these studies, because the men and women were tested at the same absolute rate of work. When the rate of work is adjusted relative to individual $\dot{V}O_{2max}$ values, women's responses are almost identical to men's. Women have a delayed onset of sweating and dilation of the skin during the luteal phase of the menstrual cycle (i.e., onset occurs at a higher core temperature). However, this should not affect performance until core temperature approaches 40 °C. Women generally have lower sweat rates for the same exercise and heat stress: Although they possess a larger number of active sweat glands than men do, women produce less sweat per gland. This is a slight disadvantage in hot, dry environments, but a slight advantage in humid conditions in which sweat evaporation is minimal.

When exposed to repeated bouts of heat stress, the body undergoes considerable adaptation (acclimatization) that enables it to withstand future heat stress more efficiently. After acclimatization, the internal temperature at which sweating and vasodilation begin is similarly lowered in women and men. Also, the sensitivity of the sweating response per unit increase in internal temperature increases by a similar amount in the two sexes following both physical training and heat acclimatization. Therefore, most differences noted between women and men in the early studies can be attributed to initial differences in their physical conditioning and acclimation status and not to their sex.

In focus

Women generally have slightly lower sweat rates than men for the same heat stress, the result of lower sweat production per sweat gland. However, this reduced maximal sweating capacity typically has a minimal effect on thermoregulation, especially in humid heat.

Women have a slight advantage over men during cold exposure because they have more subcutaneous body fat. But their smaller muscle mass is a disadvantage in extreme cold because shivering is the major adaptation for generating body heat. The greater the active muscle mass, the greater the subsequent heat generation. Muscle also provides an additional insulating layer.

Several studies have reported sex differences in response to altitude hypoxia, both at rest and during submaximal exercise. Maximal oxygen consumption decreases during hypoxic work in both sexes, but these decreases do not seem to adversely affect women's ability to work at high altitude. Studies of maximal exercise at altitude demonstrate no difference in response between the sexes.

In review

➤ Three major contributing factors to osteoporosis are estrogen deficiency, inadequate calcium intake, and inadequate physical activity.

➤ Postmenopausal women, amenorrheic women, and those who have anorexia nervosa are at greater risk of osteoporosis. Physical activity and adequate calcium and caloric intake are important to the preservation of bone at any age.

➤ Eating disorders, such as anorexia nervosa and bulimia nervosa, are much more common in women than in men and are especially common among athletes in appearance sports, endurance sports, and weight-classification sports. Athletes seem to be at a higher risk for eating disorders than the general population.

➤ When exercise intensity is adjusted relative to an individual's $\dot{V}O_{2max}$, women and men respond almost identically to heat stress. Most differences noted are likely attributable to different initial levels of conditioning.

➤ Because they have more insulating subcutaneous fat, women have a slight advantage over men during cold exposure, but their smaller muscle mass limits their ability to generate body heat.

➤ Studies indicate that maximal responses during exercise at altitude do not differ in women and men, but differences might exist at rest and during submaximal exercise.

In closing

In this chapter, we discussed sex-specific differences in performance. Most true differences between the sexes result from women's smaller body size, lower FFM, and greater relative and absolute body fat. We also considered how women's relatively more sedentary lifestyle, an artifact from a society that traditionally frowned on women's participation in physical activity, has affected research through the years. Making valid comparisons in sport performances has been difficult because an event's popularity and other factors—such as opportunities to participate, coaching, facilities, and training techniques—have differed considerably between the sexes over the years. Finally, we have found that female and male athletes are not as different as many people believe.

With this chapter, we conclude our examination of age and sex considerations in sport and exercise. In the next part of the book, we turn our attention from athletics to a different application of exercise physiology: the use of physical activity for health and fitness. We begin with an examination of exercise prescription.

Key Terms

amenorrhea

anorexia nervosa

bulimia nervosa

disordered eating

eating disorders

estrogen

eumenorrhea

lipoprotein lipase

menarche

menses

menstrual cycle

menstrual dysfunction

oligomenorrhea

pregnancy

primary amenorrhea

secondary amenorrhea

sex-specific differences

teratogenic effects

testosterone

Study Questions

1. How does the body composition of females compare with that of males? How do male and female athletes differ from male and female nonathletes?

2. What are the roles of testosterone and estrogen in the development of strength, fat-free mass, and fat mass?

3. How does women's upper body strength compare with men's? Lower body strength? Fat-free mass? Can women gain strength with resistance training?

4. What differences in $\dot{V}O_{2max}$ exist between average females and males? Between highly trained females and males? What can explain these differences?

5. What cardiovascular differences exist between females and males with respect to submaximal exercise? Maximal exercise?

6. How does the menstrual cycle influence athletic performance?

7. What is the primary reason that some female athletes in exercise training stop menstruating for intervals of several months to several years or more?

8. What risks are associated with training during pregnancy? How can these be avoided?

9. What are the effects of amenorrhea on bone mineral? How does exercise training affect bone mineral?

10. What are the two major eating disorders, and what is the level of risk for elite female athletes having these eating disorders? How does this vary by sport?

11. What is the Female Athlete Triad? What factors are involved and how does the triad develop?

12. How do women differ from men in their exercise response when exposed to intense heat and humidity? To cold? To altitude?

Study Guide Activities

In addition to the activities listed in the chapter opening outline on page 471, two other activities are available in the web study guide, located at

www.HumanKinetics.com/PhysiologyOfSportAndExercise

 The **KEY TERMS** activity reviews important terms, and the end-of-chapter **QUIZ** tests your understanding of the material covered in the chapter.

PART VII

Physical Activity for Health and Fitness

In previous parts of this book, we focused on the physiological bases of physical activity and sport performance, describing the physiological responses to an acute bout of exercise, the adaptations to chronic training, and the means of improving performance in sports. In part VII, we shift our focus from athletic performance to a special area of exercise physiology: the role of physical activity in improving and maintaining overall health and physical fitness. In chapter 20, "Prescription of Exercise for Health and Fitness," we discuss how to design an exercise program that can improve health and fitness. We consider the essential components, ways to tailor the program to an individual's specific needs, and the unique role of physical activity for rehabilitation of people who are ill. In chapter 21, "Cardiovascular Disease and Physical Activity," we examine the major types of cardiovascular disease, their physiological bases, and how physical activity can help prevent or slow the progression of these diseases. Finally, in chapter 22, "Obesity, Diabetes, and Physical Activity," we examine the causes of obesity and diabetes, the health risks associated with each, and the ways in which physical activity can be used to control both disorders.

Prescription of Exercise for Health and Fitness

20

In this chapter and in the web study guide

 ACTIVITY 20.1 Who Needs Medical Clearance? reviews the guidelines for assessing whether or not a person needs clearance before starting an exercise program.

 ACTIVITY 20.2 The Exercise Electrocardiogram reviews terms related to exercise ECGs and explains the factors that affect the accuracy of this test's results.

 ACTIVITY 20.3 Monitoring Exercise Intensity explores methods of monitoring exercise intensity (your own and someone else's).

Jason Walker, a 55-year-old executive, went in for his annual physical examination with a vow to start a long overdue exercise program. Because of his obesity, high blood pressure, and one-pack-a-day smoking habit, his physician decided to give Jason a graded exercise test to determine the normality of his electrocardiogram (ECG) during the stress of exercise. As Jason was nearing exhaustion on the treadmill, his doctor noticed changes in the ST segment of his ECG, which are considered indicative of coronary artery disease. The following week, Jason, with fear and trepidation, underwent a coronary arteriogram procedure to check for coronary artery disease. His arteriogram was abnormal, indicating that he had 85% occlusion or blockage of the circumflex coronary artery and 90% occlusion of the right coronary artery. He was immediately scheduled for coronary artery bypass surgery, and the surgery was successful. Fortunately, Jason had been shocked to the point that he stopped smoking, lost weight, and began an exercise program. He is now very fit, competing in 10 km (6.2 mi) races, and has his blood pressure under control.

Patterns of today's living have channeled the average American into an increasingly sedentary existence. Humans, however, were designed and built for movement. Physiologically, we have not adapted well to this inactive lifestyle. In fact, during what appeared to be a fitness boom in the 1970s and 1980s, fewer than 20% of adult Americans were exercising at levels that would increase or maintain their aerobic fitness and strength. Yet research had clearly determined that, for almost everyone, an active lifestyle is important for optimal health.

Health Benefits of Exercise: The Great Awakening

The 1990s will be remembered as the decade in which the medical profession formally recognized the fact that physical activity is vital to the body's health. It seems rather ironic that it took this long for clinicians and scientists to reach this conclusion, as Hippocrates (460-377 BC), a prominent physician and athlete, had strongly endorsed physical activity and proper nutrition as essential to health more than 2,000 years earlier!

The first acknowledgment from the modern medical profession came in July 1992, when the American Heart Association proclaimed physical inactivity a major risk factor for coronary artery disease, placing it alongside smoking, abnormal blood lipids, and hypertension.[10] In 1994, the Centers for Disease Control and Prevention (CDCP), in collaboration with the American College of Sports Medicine (ACSM), held a press conference to announce to the American public the importance of physical activity as a public health initiative and subsequently published the full text of a consensus statement by a panel of experts in this field in February 1995.[16] The National Institutes of Health (National Heart, Lung, and Blood Institute) released a consensus statement in December 1995, the full text of which was published in 1996, advocating physical activity as important for cardiovascular health.[15] Finally, in July

1996, coinciding with the start of the Olympic Games in Atlanta, the Surgeon General of the United States released a written report on the health benefits of physical activity.[20] This was a landmark report recognizing the importance of physical activity in reducing the risk for chronic degenerative diseases.

Much of the research supporting the benefits of physical activity in reducing the risk of developing chronic degenerative disease has come from the field of epidemiology, in which large populations are studied and associations between activity levels and disease risk determined. In the year 2000, molecular biologists and exercise physiologists started waging war on what they termed the "sedentary death syndrome" by forming an action group advocating governmental support for research into the diseases and disorders associated with a sedentary lifestyle. The group, Researchers Against Inactivity-Related Disorders, or RID, has been very effective in gaining the support of top government leaders for basic research into the role of an active lifestyle in preventing or delaying chronic degenerative diseases. Key scientific articles have been published, several of which are referenced in this chapter,[3, 4] and a website has been established (http://hac.missouri.edu/rid).

With the health benefits of an active lifestyle so clearly established, what has been the response of the U.S. population in general? We need to go back a few years to get a proper historical perspective. The seeds for a fitness revolution in the United States were planted in the late 1960s with the publication of the book *Aerobics*, written by Dr. Kenneth Cooper (figure 20.1).[7] This book provided a sound medical basis for the importance of exercise, particularly aerobic exercise, to health and fitness. The fitness movement grew throughout the 1970s, possibly peaking in the early 1980s, at which time the media declared that America was in the midst of a fitness boom.

Then, in 1983, came a penetrating article by Kirshenbaum and Sullivan,[12] published in *Sports Illustrated*, that brought everything into proper perspective. The authors questioned the existence of the fitness boom, contending that involvement was basically limited to a

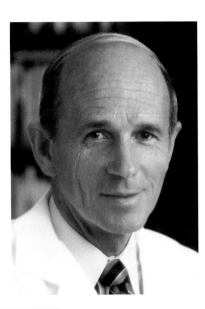

FIGURE 20.1 Dr. Kenneth H. Cooper, founder of the Cooper Institute and author of numerous books and research articles on the health-related benefits of maintaining an active lifestyle.

small but highly visible segment of the total population. They maintained that the fitness boom included mostly high-income, executive-level, white, college-educated, young to middle-aged adults. The results of several surveys confirmed this analysis.

Things have not gotten any better. According to the U.S. Department of Health and Human Services in its publication *Healthy People 2010: Understanding and Improving Health*, as of June 2004,

- nearly 40% of the U.S. population 18 years of age and older reported no leisure-time physical activity of light, moderate, or vigorous intensity for at least 10 min;

- only 22% reported engaging in vigorous physical activity sufficient to promote the development and maintenance of aerobic fitness three or more days per week for 20 or more minutes per occasion; and

- only 20% reported doing physical activities specifically designed to strengthen muscles at least twice a week.[21]

Despite these disappointing statistics, most Americans are aware that exercise is an integral part of preventive medicine. And yet people often equate exercise with jogging 8 km (5 mi) a day or lifting weights until their muscles can do no more. Many believe that high volume and intensity of exercise training are necessary to attain health-related benefits, yet this is not

true. This myth was the primary focus of the CDCP/ACSM report published in 1995,[16] which concluded that significant health benefits can be obtained if one includes a moderate amount of physical activity, such as 30 min of brisk walking, 15 min of running, or 45 min of playing volleyball, on most, if not all, days of the week. The major emphasis of this report was that through a modest increase in daily activity, most people can improve their health and quality of life. In fact, in 2006, a study of older adults (70-82 years) showed that just being more active greatly reduced the risk of mortality, independent of a formal exercise program.[14]

The Surgeon General's 1996 report,[20] however, emphasized that additional health benefits can be gained through greater amounts of physical activity. Research suggests that people who can maintain a regular regimen of activity that is of longer duration or of more vigorous intensity are likely to derive greater benefit. It is now apparent that the appropriate exercise type and intensity vary, depending on individual characteristics, current fitness level, and specific health concerns.

In 2008, the U.S. Department of Health and Human Services published the *2008 Physical Activity Guidelines for Americans*, which can be downloaded from www.health.gov/paguidelines. This publication is a rich source of information concerning the health benefits of exercise and the specific guidelines for children and adolescents, adults, older adults, and those with special needs. The sidebar summarizes the key points of the 2008 guidelines.

Knowing this, how should people begin exercise programs to improve their general health and fitness? The first step is deciding to take action. The next step is getting medical clearance.

Medical Clearance

Is a medical evaluation really necessary before starting an exercise program? Dr. Per-Olof Åstrand (see figure 20.2, p. 503), eminent Swedish physician and physiologist who has had a worldwide impact promoting physical activity for health, has suggested in jest that those individuals who elect to remain sedentary should be required to have a medical evaluation to determine if their bodies can withstand the rigors of a sedentary lifestyle. The medical evaluation is perceived as a significant barrier to starting an exercise program for many people, yet it is useful and important for the following reasons:

- Some people are considered to be at high risk for exercise and either should not exercise at all or should be restricted to exercising only under

Key Recommendations From the 2008 Physical Activity Guidelines for Americans

Substantial health benefits are gained by doing physical activity according to the following guidelines for different groups.

Children and Adolescents (aged 6–17)

- Children and adolescents should do 1 hour (60 minutes) or more of physical activity every day.
- Most of the 1 hour or more a day should be either moderate- or vigorous-intensity aerobic physical activity.
- As part of their daily physical activity, children and adolescents should do vigorous-intensity activity on at least 3 days per week. They also should do muscle-strengthening and bone-strengthening activity on at least 3 days per week.

Adults (aged 18–64)

- Adults should do 2 hours and 30 minutes a week of moderate-intensity, or 1 hour and 15 minutes (75 minutes) a week of vigorous-intensity aerobic physical activity, or an equivalent combination of moderate- and vigorous-intensity aerobic physical activity. Aerobic activity should be performed in episodes of at least 10 minutes, preferably spread throughout the week.
- Additional health benefits are provided by increasing to 5 hours (300 minutes) a week of moderate-intensity aerobic physical activity, or 2 hours and 30 minutes a week of vigorous-intensity physical activity, or an equivalent combination of both.
- Adults should also do muscle-strengthening activities that involve all major muscle groups performed on 2 or more days per week.

Older Adults (aged 65 and older)

- Older adults should follow the adult guidelines. If this is not possible due to limiting chronic conditions, older adults should be as physically active as their abilities allow. They should avoid inactivity. Older adults should do exercises that maintain or improve balance if they are at risk of falling.

For all individuals, some activity is better than none. Physical activity is safe for almost everyone, and the health benefits of physical activity far outweigh the risks. People without diagnosed chronic conditions (such as diabetes, heart disease, or osteoarthritis) and who do not have symptoms (e.g., chest pain or pressure, dizziness, or joint pain) do not need to consult with a health care provider about physical activity.

Adults With Disabilities

Follow the adult guidelines. If this is not possible, these persons should be as physically active as their abilities allow. They should avoid inactivity.

Children and Adolescents With Disabilities

Work with the child's health care provider to identify the types and amounts of physical activity appropriate for them. When possible, these children should meet the guidelines for children and adolescents—or as much activity as their condition allows. Children and adolescents should avoid being inactive.

Pregnant and Postpartum Women

Healthy women who are not already doing vigorous-intensity physical activity should get at least 2 hours and 30 minutes (150 minutes) of moderate-intensity aerobic activity a week. Preferably, this activity should be spread throughout the week. Women who regularly engage in vigorous-intensity aerobic activity or high amounts of activity can continue their activity provided that their condition remains unchanged and they talk to their health care provider about their activity level throughout their pregnancy.

Reprinted from: www.health.gov/paguidelines/factsheetprof.aspx

Exercise Is Medicine

During the first decade of the 21st century, the ACSM in collaboration with the American Medical Association (AMA) launched a major initiative to encourage health care providers to counsel their patients on the importance of physical activity in promoting and maintaining health and preventing disease. With the recognition that many physicians and associated health care workers have not had extensive education and training in this area, a formal program was developed to educate and train them in the basics of exercise prescription. Robert E. Sallis, MD, past president of the ACSM, was the task force chairman in collaboration with Ronald M. Davis, MD, past president of the AMA. Further information on this health care initiative can be obtained at the Exercise is Medicine website: www.exerciseismedicine.org. This website offers the "Health Care Providers' Action Guide," which includes the "Exercise Prescription and Referral Form."

close medical supervision. A comprehensive medical evaluation will help identify these high-risk individuals.

- The information obtained in a medical evaluation can be used to develop the exercise prescription.

- The values obtained for certain clinical measures, such as blood pressure, body fat content, and blood lipid levels, can be used to motivate the person to adhere to the exercise program.

- A comprehensive medical evaluation, particularly of healthy people, can provide a baseline against which any subsequent changes in health status can be compared.

- Children and adults should establish the habit of periodic medical evaluations because many illnesses and diseases, such as cancer and cardiovascular diseases, can be identified in their earliest stages when the chances of successful treatment are much higher.

Medical Evaluation

Although a comprehensive medical evaluation is useful and desirable before exercise is prescribed, not all people need one. Many cannot afford the cost of such an evaluation, and the medical system is not prepared to provide this service for the total population, even if money were available. Also, medical evaluation before prescribing exercise for a population presumed healthy has not been proven to reduce the medical risks associated with exercise. For these reasons, guidelines or recommendations have been established that attempt to target moderate- to high-risk individuals.[2,9] People at moderate risk are those who have no signs or symptoms of, or who have not been diagnosed with, cardiovascular, pulmonary, or metabolic disease but who have two or more risk factors for coronary artery disease (table 20.1). Those who are at high risk are people with one or more signs or symptoms of cardiovascular, pulmonary, or metabolic disease (see sidebar on page 505).

The ACSM has published specific recommendations for each phase of the medical evaluation in *ACSM's Guidelines for Exercise Testing and Prescription*.[2] This document should be consulted whenever

FIGURE 20.2 Dr. Per-Olof Åstrand, eminent Swedish physician and physiologist, bicycling through the woods.

TABLE 20.1	Coronary Artery Disease Risk Factors for Targeting At-Risk People
Positive risk factors	**Defining criteria**
Age	Men ≥45 years; women ≥55 years
Family history	Myocardial infarction, coronary revascularization, or sudden death before 55 years of age in father or other male first-degree relative or before 65 years of age in mother or other female first-degree relative
Cigarette smoking	Current cigarette smoker or those who quit within the previous six months, or exposure to environmental tobacco smoke
Hypertension	Systolic blood pressure ≥140 mmHg or diastolic blood pressure ≥90 mmHg, confirmed by measurements on at least two separate occasions, or on antihypertensive medication
Dyslipidemia	Low-density lipoprotein (LDL) cholesterol ≥130 mg/dL (3.4 mmol/L) or high-density lipoprotein (HDL) cholesterol <40 mg/dL (1.04 mmol/L), or on lipid-lowering medication. If total serum cholesterol is all that is available, use ≥200 mg/dL (5.2 mmol/L).
Impaired fasting glucose (prediabetes)	Fasting blood glucose ≥100 mg/dL (5.5 mmol/L) but <126 mg/dL (6.9 mmol/L), or impaired glucose tolerance as determined by an oral glucose tolerance test, confirmed by measurements on at least two separate occasions
Obesity	Body mass index ≥30 kg/m^2; or waist girth >102 cm (40 in.) for men and >88 cm (35 in.) for women; or waist/hip ratio ≥0.95 for men and ≥0.86 for women
Sedentary lifestyle	Persons not participating in at least 30 min of moderate-intensity (40%-60% of $\dot{V}O_2R$) physical activity on at least three days of the week for at least three months. Note: $\dot{V}O_2R$ refers to $\dot{V}O_2$ reserve, which is defined as $\dot{V}O_{2max} - \dot{V}O_{2rest}$. $\dot{V}O_{2rest}$ is assumed to be 3.5 ml · kg^{-1} · min^{-1}. See the sidebar on page 512 for more detail.
Negative risk factors	**Defining criteria**
High serum HDL-C	≥60 mg/dL (1.5 mmol/L)

Note. It is common to sum risk factors in making clinical judgments. If high-density lipoprotein cholesterol (HDL-C) is high, subtract one risk factor from the sum of positive risk factors, because high HDL-C decreases coronary artery disease risk.

Adapted, by permission, American College of Sports Medicine, 2010, *ACSM's guidelines for exercise testing and prescription*, 8th ed. (Philadelphia, PA: Lippincott, Williams, and Wilkins), p. 28.

there is any question as to what should be included. The physical examination should include discussion between the physician and patient of the proposed exercise program in case any medical contraindications are associated with the proposed activity. For example, people with hypertension should be cautioned to avoid activities that use isometric actions. Isometric actions tend to increase blood pressure considerably and usually result in the Valsalva maneuver, in which intra-abdominal and intrathoracic pressures increase to the point of restrict-ing blood flow through the vena cava, limiting venous return to the heart. Both responses can lead to serious medical complications, such as loss of consciousness or stroke. Also, even dynamic resistance training can cause a very high blood pressure response during the activity.

Graded Exercise Testing

Ideally, a comprehensive medical examination will include an exercise test, usually conducted on a motor-driven treadmill. A cycle ergometer can also be used but is not that common for clinical testing in the United States. An **exercise electrocardiogram (ECG)** (see figure 6.7, p. 147) and exercise blood pressure readings (figure 20.3, p. 506) are obtained during exercise. The ECG and blood pressure are monitored as the person progresses from low-intensity exercise, such as slow walking, up to maximal-intensity exercise. Maximal intensity might be brisk walking for an older,

In focus

Although a general medical evaluation on a regular basis is important and desirable for almost everyone, it simply is not practical to require this for all people who want to start an exercise program.

Risk Stratification Using the 2010 American College of Sports Medicine Guidelines

This logic model for risk stratification allows the health and fitness professional to determine the degree of health or medical risk for a given individual for the following purposes:

- Identification of individuals with medical contraindications for exclusion from exercise programs until those conditions have been abated or are under control
- Recognition of persons with clinically significant disease(s) or conditions who should participate in a medically supervised exercise program
- Detection of individuals at increased risk for disease because of age, symptoms, or risk factors who should undergo a medical evaluation and exercise testing before initiating an exercise program or increasing the frequency intensity, or duration of their current program
- Recognition of special needs of individuals that may affect exercise testing and programming

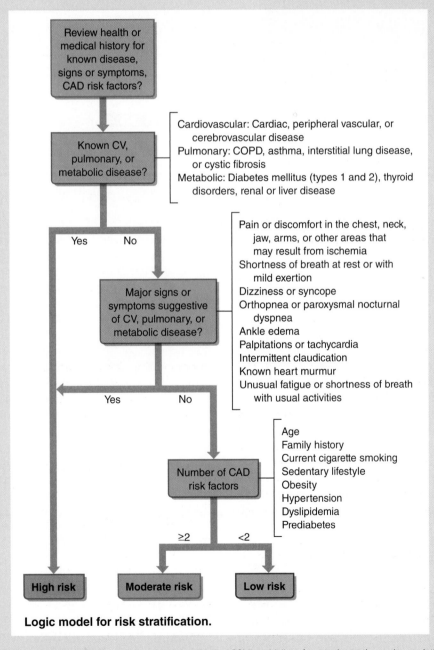

Logic model for risk stratification.

Adapted, by permission, American College of Sports Medicine, 2010, *ACSM's guidelines for exercise testing and prescription*, 8th ed. (Philadelphia, PA: Lippincott, Williams, and Wilkins), 24.

FIGURE 20.4 Illustration of (*a*) a normal ECG and (*b*) an ECG with ST-segment depression, suggestive of the presence of coronary artery disease.

deconditioned subject or running up a grade for a younger, fit individual. The rate of work is generally increased every 1 to 3 min until the maximal rate of work is achieved. This progression is referred to as a **graded exercise test (GXT)**. The ECG is monitored to detect heart rhythm and electrical conductivity abnormalities. Blood pressure is monitored to determine if there is a normal increase in systolic blood pressure and little or no change in diastolic blood pressure as the rate of work progresses from low intensity to maximal or near-maximal levels. It is also important to interact with the subject, observing signs and symptoms during and immediately following the exercise test, such as chest pain or pressure (angina), unusual shortness of breath, light-headedness or dizziness, and inappropriate heart rate response.

The exercise ECG is an important part of the medical evaluation because a small but significant percentage of the adult population have abnormalities in ECGs taken during or following exercise even though they have normal resting ECGs. These abnormalities include arrhythmias (irregular heart rhythms) and ST-segment changes. Figure 20.4 illustrates a normal ECG and an abnormal ECG in which there is depression of the ST segment. Generally a horizontal (flat) or down-sloping ST segment of 1.0 mm or greater below the isoelectric line, for at least 60 to 80 ms, is suggestive of myocardial ischemia (insufficient blood flow to the myocardium) and thus the presence of **coronary**

artery disease (CAD). Results from an exercise test are considered either positive, when there have been abnormal changes in the ECG, or negative, which implies a normal response and that no disease was detected. But some people have normal, or negative, exercise tests, yet have CAD. These are false-negative tests. Others do not have the disease, yet have positive exercise ECGs suggesting disease. These are false-positive tests.

Typically, a person with abnormal changes in his or her exercise ECG will be referred for additional tests. The presence of CAD can be determined by a coronary arteriogram, in which radiopaque dye (dye that is opaque to X rays) is injected through a catheter into the coronary arteries, allowing visualization of the insides of the arteries. Other imaging techniques, such as computed tomography (CT) and magnetic resonance imaging (MRI) scans, are also being used. If there is narrowing of one or more of the coronary arteries, the degree (%) of occlusion or narrowing can be determined.

To determine the accuracy of exercise ECG results, the sensitivity, specificity, and predictive value of an exercise test must be considered. **Sensitivity** refers to the exercise test's ability to correctly identify people who have the disease in question, such as CAD. **Specificity** refers to the test's ability to correctly identify people who do not have the disease. And the **predictive value of an abnormal exercise test** refers to the accuracy with which abnormal test results reflect presence of the disease.

Unfortunately, both the sensitivity and the predictive value of an abnormal exercise test for detecting CAD

In focus

The sensitivity and predictive value of an abnormal exercise test are generally low in a young, healthy population in which there is a low prevalence of CAD. As a result, the value of using exercise electrocardiography to screen for CAD in this population is questionable.

are relatively low in healthy populations of people who have no symptoms of this disease. Past studies reveal that sensitivity averages about 66%, indicating that 66% of those with CAD are correctly identified by exercise ECGs as having the disease. Conversely, 34% of those with the disease are incorrectly diagnosed as disease free based on exercise ECGs. Average specificity is about 84%, indicating that 84% of those without disease are correctly identified by the exercise test as being disease free. But this means that 16% are incorrectly identified as having the disease.[9]

The predictive value of an abnormal exercise test varies considerably with the prevalence of CAD in the population. Assuming an average sensitivity of 60% and an average specificity of 90%, in a population with 5% prevalence of CAD, the predictive value of an abnormal test is only 24.0%. This indicates that only 24% of those who have an abnormal exercise ECG actually have CAD. In other words, in this population, more than three of every four people identified with abnormal exercise tests will be labeled as diseased but don't actually have identifiable CAD! But if we consider a population with a prevalence of 50%, the predictive value of an abnormal exercise test is much higher—85.7%.

From this information, we can conclude that exercise testing is of limited value in screening young, apparently healthy individuals before prescribing exercise for them. The accuracy of interpreting the results of the exercise ECG is questionable, particularly in a population with such a low prevalence of disease. Also, the actual risk of death or cardiac arrest during exercise is relatively low. Another important consideration is the expense of conducting clinical exercise tests, generally around $150 to $750 per test. Finally, far too few clinical facilities are equipped to conduct these tests to accommodate testing everyone who should be involved in an exercise program. Fortunately, the ACSM and the American Heart Association have recommended this exercise test only for the at-risk groups mentioned earlier in this chapter. In fact, in 2005, the American Heart Association stated that there is insufficient evidence at this time to recommend exercise testing as a routine screening modality in asymptomatic adults.[13]

From a medical and legal perspective, however, the question must be raised as to whether a recommendation within a set of national guidelines is tantamount to a standard of practice in the medical community. The most recent guidelines of the ACSM indicate that a medical examination and exercise test are not necessary in people at low or moderate risk if low- to moderate-intensity exercise is undertaken gradually, with no competitive participation.[2] Moderate exercise is defined as exercise that is well within the individual's current capacity (e.g., 40-60% of $\dot{V}O_2$ reserve or $\dot{V}O_2R$; see sidebar on p. 512 for more detail) and can be sustained comfortably for a prolonged period of time, such as 45 min. In contrast, vigorous exercise is defined as exercise at an intensity greater than 60% of the individual's $\dot{V}O_2R$. This seems like a reasonable compromise, because moderate exercise is associated with considerable health benefits and few risks. But, as mentioned earlier, exercise testing offers benefits other than diagnosing CAD: It can provide valuable physiological data, such as a person's blood pressure response to exercise, and much of the data obtained can be used in formulating the exercise prescription.

In review

> ➤ Before beginning any exercise program, men over age 45, women over age 55, and anyone who is considered to be at a high risk for CAD should have a comprehensive medical evaluation.

> ➤ ACSM guidelines should be followed for each phase of the evaluation, and the physician should be consulted about the proposed exercise activity in case there are any medical contraindications.

> ➤ Exercise ECGs should be conducted for anyone in one of the previously mentioned high-risk categories. This test can usually detect existing undiagnosed CAD and other cardiac abnormalities.

> ➤ Test sensitivity refers to the test's ability to correctly identify people with a given disease. Test specificity refers to its ability to correctly identify people who do not have the disease. The predictive value of an abnormal exercise test refers to the accuracy with which the test reflects presence of the disease in a given population.

> ➤ The most recent ACSM guidelines state that a medical examination and exercise test might not be necessary if moderate exercise is undertaken gradually, without competition, in people without symptoms of cardiovascular, pulmonary, and metabolic disease.

Exercise Prescription

The **exercise prescription** involves four basic factors:

1. Mode or type of exercise
2. Frequency of participation
3. Duration of each exercise bout
4. Intensity of the exercise bout

In our discussion, we assume that the goal of the exercise program is to improve aerobic capacity in people who have not been exercising. Since the prescription of resistance training programs is discussed in detail in chapter 9, we will only briefly mention including resistance training as a part of a total exercise program later in this chapter. The focus of this section is on aerobic training. Also, the information contained in this section is not appropriate for designing training programs for competitive endurance athletes or for those who simply wish to gain the health-related benefits of moderate activity but do not wish to improve aerobic capacity. Aerobic and anaerobic training for competitive athletes is covered in chapter 9.

Before examining the components of the exercise prescription, we must consider how much exercise is effective. A minimum threshold for frequency, duration, and intensity of exercise must be reached before any aerobic benefits are obtained. But, as we have discussed elsewhere, individual responses to any given training program are highly variable, so the threshold required differs from one person to the next. If we use exercise intensity as an example, a position statement by the ACSM for developing and maintaining aerobic capacity recommends a training intensity of 55% or 60% to 90% of one's maximum heart rate (HR_{max}) or 40% or 50% to 85% of $\dot{V}O_{2max}$.[1] Although this recommendation is appropriate for most healthy adults, some might improve their aerobic capacities at, for example,

Parallel Careers, Lifelong Impact!

Since the early 1970s, considerable progress has been made in providing a research base for better understanding the relationship between an active lifestyle and reduced risk for chronic debilitating disease and for identifying how much and what types of activity promote health. During this time, two exercise scientists have had a particularly significant impact on helping us better understand the relationship between physical activity and disease prevention through their research, advocacy, and

Dr. William L. Haskell.   Dr. Michael L. Pollock.

professional leadership. Interestingly, both had their roots in the Los Angeles area of Southern California and both were graduate students at the same time studying for their PhD degrees in exercise physiology at the University of Illinois. Dr. William L. Haskell received his undergraduate training at the University of California, Santa Barbara, while Dr. Michael L. Pollock was playing baseball and completing his undergraduate degree at the University of Arizona. Both served in the U.S. Army before completing their PhD degrees.

During their professional careers, both were deeply committed to their research and to their primary professional organization, the ACSM, each serving as president. These men were instrumental in developing the original and subsequent ACSM Position Stands on the recommended quantity and quality of exercise needed to promote health and prevent chronic disease. These two scientists had parallel careers and a mutual impact on our understanding of the importance of an active lifestyle in promoting health and preventing chronic disease in today's sedentary society.

intensities below 40% of their $\dot{V}O_{2max}$, whereas others would have to exercise at intensities greater than 85% $\dot{V}O_{2max}$ to show improvement. Each individual's threshold for frequency, duration, and intensity must be exceeded to achieve gains in aerobic capacity, and this threshold is likely to increase as aerobic capacity improves.

In focus

A minimal threshold for frequency, duration, and intensity of exercise must be reached to provide aerobic benefits from that exercise. Furthermore, minimal thresholds vary widely, making individualized exercise prescription necessary.

Exercise Mode

The prescribed exercise program should focus on one or more **modes**, or types, of cardiovascular endurance activities. Traditionally, the activities prescribed most frequently are

- walking,
- jogging,
- running,
- hiking,
- cycling,
- rowing, and
- swimming.

Because these activities do not appeal to everyone, alternative activities have been identified that promote similar cardiovascular endurance gains. Spinning, aerobic dance, box or bench stepping, and most racket sports also have been shown to improve aerobic capacity.

For most competitive sport activities, preconditioning with one of the standard endurance activities, such as jogging/running or cycling, is advisable before one undertakes serious competition. Some researchers, clinicians, and practitioners believe that for people to successfully compete in certain sports or activities, a basic preconditioning program is essential to bring them up to the level of conditioning needed for the sport or activity and to reduce risk of injury. Rather than using the sport or activity to get in shape, people get in shape—or precondition—before participating in that sport or activity. For example, if the desired activity requires a moderate to high level of cardiovascular endurance, such as basketball, one might engage in a running or cycling program for several months until the endurance capacity increases to the necessary level.

At that time, one switches over to the sport. The sport then acts as a maintenance activity through which one maintains the desired fitness level. In some cases, with intense sports, people can continue to develop their aerobic fitness level.

In focus

Sport and recreational activities are appropriate for maintaining desirable fitness levels, but they generally are not the best for the initial development of fitness in unfit individuals. Relatively unfit individuals should use conditioning activities to reach the desired level of fitness and then switch to the sport or recreational activity.

Individuals should select activities that they enjoy and are willing to continue throughout life. Exercise must be regarded as a lifetime pursuit because, as we saw in chapter 14, the benefits are soon lost if participation stops. Motivation is probably the most important factor in a successful exercise program. Selecting an activity that is fun, provides a challenge, and can produce needed benefits is one of the most crucial tasks in exercise prescription. Having several different activities to pursue is also wise in case of inclement weather, travel, or other barriers. Other considerations include geographic location, climate, and availability of equipment and facilities. Home exercise has become more common as many people are homebound either because of responsibilities, such as child rearing, or because of weather considerations such as heat, humidity, cold, rain, ice, and snow. Exercise videos and home exercise equipment have become popular, but they need to be selected carefully to avoid inappropriate exercise or faulty equipment. Potential purchasers should seek professional advice and, when possible, use the video or equipment during a trial period before purchase.

Exercise Frequency

The frequency of exercise participation, although certainly an important factor to consider, is probably less critical than either exercise duration or intensity. Research studies conducted on exercise frequency show that three to five days per week is an optimal frequency. This does not mean that six or seven days per week won't give additional benefits; but simply for the health-related benefits, the optimal gain is achieved with a time investment of three to five days per week. Exercise initially should be limited to three or four days per week and increased up to five or more days per week only if the activity is enjoyed and physically tolerated. All too often, a person starts out with great intentions, is highly motivated, and exercises every day for the first few weeks, only to stop from utter fatigue, soreness, injury, or boredom. Obviously, additional days above the three- or four-day frequency are beneficial for weight loss, but this level should not be encouraged until the exercise habit is firmly established and the injury risk is reduced.

Exercise Duration

Several studies have demonstrated improvement in cardiovascular conditioning with endurance exercise periods as brief as 5 to 10 min per day. More recent research has indicated that 20 to 30 min per day is an optimal amount. Again, "optimal" is used here to reflect the greatest return for time invested, and the specified time refers to the time during which one is at one's appropriate exercise intensity. Exercise duration cannot be discussed appropriately without discussion of exercise intensity also. Similar improvements in aerobic capacity are gained with either a short-duration, high-intensity program or a long-duration, low-intensity program if the minimal threshold is exceeded for both duration and intensity. Similar benefits are also gained whether the daily endurance training session is conducted in multiple shorter bouts (e.g., three 10 min bouts) or a single long one (e.g., a single 30 min bout). Obviously, longer bouts will facilitate weight loss.

Exercise Intensity

The intensity of the exercise bout appears to be the most important factor. How hard must people push themselves to gain benefits? Former athletes immediately recall the exhaustive workouts they endured to condition themselves for their sport. Unfortunately, this concept also gets carried over into the exercise programs they pursue for health benefits. Evidence now suggests that a modest training effect can be accomplished in some people through training at intensities of 40% or less of their aerobic capacities and possibly could lead to health benefits. For most, however, the appropriate minimum intensity appears to be at least 50% to 60% $\dot{V}O_{2max}$. An upper level for intensity will depend on the purpose for training. Obviously, training for competition requires high intensity (see chapter 14). However, training for purposes of attaining and maintaining optimal health would seldom exceed 80% $\dot{V}O_{2max}$. A series of recent studies from McMaster University in Hamilton, Ontario (Canada), has clearly demonstrated that very high intensity, low-volume interval training can markedly increase aerobic capacity. Substantial increases in muscle oxidative capacity

and endurance performance have been obtained in a training period as short as two weeks.[11] These studies tend to seriously question the concept of specificity of training as discussed in chapter 11.

In review

➤ The four basic factors in an exercise program are exercise mode, frequency, duration, and intensity. A minimum threshold for the last three must be met to yield any aerobic benefits, and this threshold is quite variable from one individual to another.

➤ The program should include one or more cardiovascular endurance activities. If the activity involves competition, preconditioning with a standard endurance activity is recommended before sport participation begins to bring the person up to an appropriate level of fitness.

➤ Activities must be matched with individual needs and likes so that motivation can be maintained.

➤ Optimal exercise frequency is three to five days of training each week, although greater frequency might provide additional benefits. Exercise should begin with three or four sessions per week and then progress to more if desired.

➤ Exercise duration of 20 to 30 min working at the appropriate intensity is optimal, but the key is reaching the threshold for both duration and intensity.

➤ Exercise intensity appears to be the most important factor. For most people, intensity should be at least 50% to 60% $\dot{V}O_{2max}$. However, health benefits might occur at intensities lower than those needed for aerobic conditioning in some people and can also occur at very high intensities.

Monitoring Exercise Intensity

Exercise intensity can be quantified on the basis of the training heart rate (THR), the metabolic equivalent (MET), or the rating of perceived exertion (RPE). Let's examine each of these and their strengths and weaknesses in quantifying exercise intensity.

Training Heart Rate

The concept of **training heart rate (THR)** is based on the linear relationship between heart rate and $\dot{V}O_2$ with increasing rates of work, as shown in figure 20.5. When people are exercise tested, their heart rate and $\dot{V}O_2$ values are obtained each minute and plotted against each other. The THR is established through use of the heart rate that is equivalent to a set percentage of the $\dot{V}O_{2max}$. For example, if a training level of 75% $\dot{V}O_{2max}$ is desired, 75% of $\dot{V}O_{2max}$ is calculated ($\dot{V}O_{2max} \times 0.75$), and the heart rate corresponding to this $\dot{V}O_2$ then is selected as the THR. An important point is that the exercise intensity necessary to achieve a given percentage of $\dot{V}O_{2max}$ results in a much higher heart rate than that same percentage of HR_{max}. As an example, a THR that is set at 75% of the $\dot{V}O_{2max}$ represents an intensity of 87% of the HR_{max} (see figure 20.5).

FIGURE 20.5 The linear relationship between heart rate and oxygen consumption ($\dot{V}O_2$) with increasing rates of work and the heart rate equivalent to a set percentage (75%) of $\dot{V}O_{2max}$.

The Karvonen Method

One can also establish the THR by using what is known as the Karvonen concept of maximal heart rate reserve, or the **Karvonen method. Maximal heart rate reserve** is defined as the difference between HR_{max} and the resting heart rate (HR_{rest}):

$$\text{maximal heart rate reserve} = HR_{max} - HR_{rest}.$$

With this method, the THR is calculated by taking a given percentage of the maximal heart rate reserve

and adding it to the resting heart rate. Let's consider an example. For 75% of maximal heart rate reserve, the equation would be as follows:

$$THR_{75\%} = HR_{rest} + 0.75 (HR_{max} - HR_{rest}).$$

The Karvonen method adjusts the THR so that the THR as a specific percentage of the maximal heart rate reserve is nearly identical to the heart rate equivalent of that same percentage of $\dot{V}O_{2max}$ at moderate to high intensities.[8] Thus, a THR computed as 75% of maximal heart rate reserve is approximately the same as the heart rate corresponding to 75% of the $\dot{V}O_{2max}$. However, there is a substantial difference between the two at low intensities.[19]

Training Heart Rate Range

More recently, appropriate exercise intensity has been established by setting a THR range, rather than a single THR value. This is a more sensible approach because exercising at a set percentage of $\dot{V}O_{2max}$ can place people above their lactate threshold, making it difficult for them to train for an extended period of time. With the THR range concept, low and high values are established that will ensure a training response. One starts at the low end of the THR range and progresses through the range as one feels comfortable. To illustrate this, using the Karvonen method for establishing the THR, consider the following example. A 40-year-old man has a resting heart rate of 75 beats/min and a maximum heart rate of 180 beats/min, and he is advised to exercise within a THR range of 50% to 75% of his maximal heart rate reserve. His training heart rate range would be as follows:

$$THR_{50\%} = 75 + 0.50 (180 - 75) = 75 + 53$$
$$= 128 \text{ beats/min.}$$

$$THR_{75\%} = 75 + 0.75 (180 - 75) = 75 + 79$$
$$= 154 \text{ beats/min.}$$

This same THR range method can be used if one estimates HR_{max} [208 − (0.7 × age)] without losing much accuracy if true HR_{max} has not been determined.

The concept of THR is extremely valuable. Heart rate is highly correlated with the work done by the heart. Heart rate alone is a good index of myocardial oxygen consumption as well as coronary blood flow. With use of the THR method for monitoring exercise intensity, the heart works at the same rate, even though the metabolic cost of the work might vary considerably. As an example, during exercise at high altitudes or in the heat, the heart rate will be elevated significantly if the person attempts to maintain a set rate of work, such as running at a pace of 6 min/km (9 min/mi).

With the THR method, one simply trains at a lower rate of work under these extreme environmental conditions to maintain the same heart rate (THR). This is a much safer approach to monitoring exercise intensity, particularly for high-risk patients in whom the work of the heart must be closely regulated. The THR method also allows for improvement in aerobic capacity with training. As people become better conditioned, their heart rate decreases for the same rate of work, which means that they must perform at a higher rate of work to reach their THR.

It is important to come back to an important point raised in the first paragraph of this section: As one increases exercise intensity there will be a point where the rate of lactate production exceeds the rate of its clearance, resulting in increased blood lactate levels. When people are at an exercise intensity above their lactate threshold, they limit the length of time they can train comfortably at that intensity. For those just starting a training program, it is important not to exceed the lactate threshold. Having a THR range allows people to set the lower end of the range at an intensity that would be below the expected lactate threshold for an untrained individual. Obviously it would be better to actually measure the lactate threshold so that this range could be more accurately determined. However, this isn't practical because of the difficulty and expense associated with directly determining lactate threshold from multiple blood draws.

Metabolic Equivalent

Exercise intensity also has been prescribed on the basis of the **metabolic equivalent (MET)** system. The amount of oxygen the body consumes is directly proportional to the energy expended during physical activity. In this system, it is assumed that the body uses approximately 3.5 ml of oxygen per kilogram (2.2 lb) of body weight per minute ($3.5 \text{ ml} \cdot \text{kg}^{-1} \cdot \text{min}^{-1}$) at rest. However, in the sidebar, we see that this is likely not the case. The MET system, however, is based on this value, and the resting metabolic rate value of $3.5 \text{ ml} \cdot \text{kg}^{-1} \cdot \text{min}^{-1}$ is referred to as 1.0 MET. All activities can be classified by intensity according to their oxygen requirements. An activity that is rated as a 2.0 MET activity would require two times the resting metabolic rate, or $7 \text{ ml} \cdot \text{kg}^{-1} \cdot \text{min}^{-1}$; and an activity that is rated at 4.0 METs would require approximately $14 \text{ ml} \cdot \text{kg}^{-1} \cdot \text{min}^{-1}$. Some activities and their MET values are presented in table 20.2.

(In)focus

Heart rate is the preferred method for monitoring exercise intensity because it is highly correlated to the work of the heart (or stress on the heart) and allows for a progressive increase in the rate of training with improvements in fitness to maintain the same THR. When one is prescribing exercise intensity, it is appropriate to establish a training heart rate range, with exercise starting at the low end of the range and progressing to the upper end of the range over time.

Prescribing Exercise Intensity Using the $\dot{V}O_2$ Reserve Method

In the ACSM Position Stand on exercise prescription,[1] a slightly different approach to prescribing exercise intensity was proposed. Exercise intensity is prescribed based on what has been termed the $\dot{V}O_2$ reserve method ($\dot{V}O_2R$). Instead of prescribing exercise at a given percentage of $\dot{V}O_{2max}$, one bases the prescription on a given percentage of the $\dot{V}O_2R$, where $\dot{V}O_2R$ is defined as $\dot{V}O_{2max} - \dot{V}O_{2rest}$. This also can be thought of as the $\dot{V}O_{2max}$ reserve. As an example, with a $\dot{V}O_{2max}$ of $40 \text{ ml} \cdot \text{kg}^{-1} \cdot \text{min}^{-1}$ and a $\dot{V}O_{2rest}$ of $3.5 \text{ ml} \cdot \text{kg}^{-1} \cdot \text{min}^{-1}$, $\dot{V}O_2R = 40 - 3.5 \text{ ml} \cdot \text{kg}^{-1} \cdot \text{min}^{-1} = 36.5 \text{ ml} \cdot \text{kg}^{-1} \cdot \text{min}^{-1}$.

To prescribe an exercise intensity range of between 60% and 75% of $\dot{V}O_2R$, we simply multiply $\dot{V}O_2R$ by 60% and 75%: $\dot{V}O_2R_{60\%} = 36.5 \text{ ml} \cdot \text{kg}^{-1} \cdot \text{min}^{-1} \times 0.60 = 21.9 \text{ ml} \cdot \text{kg}^{-1} \cdot \text{min}^{-1}$; and $\dot{V}O_2R_{75\%} = 36.5 \text{ ml} \cdot \text{kg}^{-1} \cdot \text{min}^{-1} \times 0.75 = 27.4 \text{ ml} \cdot \text{kg}^{-1} \cdot \text{min}^{-1}$. The major advantage of using the $\dot{V}O_2R$ technique is that one now has an equivalency between the percentage of the maximal heart rate reserve and the percentage of $\dot{V}O_{2max}$ reserve. There is a potential problem with this technique, however, in that using $3.5 \text{ ml} \cdot \text{kg}^{-1} \cdot \text{min}^{-1}$ as a standard value for $\dot{V}O_{2rest}$ assumes that everyone has the same resting value. This, in fact, is not the case. Further, in one study, a large sample of women ($n = 642$) and men ($n = 127$) were found to have average $\dot{V}O_{2rest}$ values of 2.5 and $2.7 \text{ ml} \cdot \text{kg}^{-1} \cdot \text{min}^{-1}$. The range of values varied from 1.6 to $4.1 \text{ ml} \cdot \text{kg}^{-1} \cdot \text{min}^{-1}$.[6]

TABLE 20.2 Selected Activities and Their Respective MET Values

Activity	MET Value	Activity	MET Value
REST AND SELF-CARE ACTIVITIES			
Rest (supine)	1.0	Showering	2.0
Sitting	1.5	General grooming, standing	2.0
Eating	1.5	Dressing or undressing, standing	2.5
Bathing	1.5		
HOME ACTIVITY			
Knitting or hand sewing, light effort	1.3	Vacuuming (general, moderate effort)	3.3
Washing dishes	1.8	Making beds, changing linens	3.3
Ironing	1.8	Cleaning (scrubbing floor, washing car, washing windows)	3.5
Laundry, folding or hanging clothes	2.0-2.3	Sweeping, moderate effort	3.8
Cooking or food preparation	2.0-3.5	Moving furniture, carrying boxes	5.8
Machine sewing	2.8	Scrubbing floors on hands and knees, vigorous effort	6.5
OCCUPATIONAL			
Sitting tasks, office work, working at a computer	1.5	Construction (outside)	4.0
Driving a delivery truck, taxi, school bus, etc.	2.0	Hotel housekeeper	4.0
Cook, chef	2.5	Yard work	4.0
Standing tasks, light to moderate effort	3.0-4.5	Manual or unskilled labor	2.8-6.5
Custodial work	2.5-4.0	Farming, light to vigorous effort	2.0-7.8
Carpentry (general, light to moderate effort)	2.5-4.3	Fire fighter on the job	6.8-9.0
PHYSICAL CONDITIONING			
Walking			
2.5 mph, level	3.0	4.5 mph, level	7.0
3.5 mph, level	4.3	5.0 mph, level	8.3
4.0 mph, level	5.0	5.0 mph, 3% grade	9.8
Jogging or running on level surface			
4.0 mph	6.0	10.0 mph	14.5
6.0 mph	9.8	12.0 mph	19.0
8.0 mph	11.8	14.0 mph	23.0
Swimming			
Freestyle, vigorous effort	9.8	Breaststroke, recreational/training and competition	5.3/10.3
Freestyle, slow to moderate	5.8	Sidestroke, general	7.0

(continued)

TABLE 20.2 *(continued)*

Activity	MET Value	Activity	MET Value
PHYSICAL CONDITIONING			
Backstroke, recreational/training and competition	4.8/9.5		
Cycling			
Leisure, 5.5 mph	3.5	Leisure, 14.0-15.9 mph (vigorous effort)	10.0
Leisure, 10.0-11.9 mph (slow, light effort)	6.8	Racing, 16.0-19.0 mph (vigorous effort)	12.0
Leisure, 12.0-13.9 mph (moderate effort)	8.0	Racing, >20 mph (vigorous effort)	15.8
RECREATIONAL ACTIVITIES			
Aerobic dance	5.0-7.3	General resistance training	3.5-6.0
Video game activities	2.3-6.0	Rowing machines	4.8-12.0
Stationary cycle ergometer	3.5-14.0	Water aerobics	5.3
Circuit training	4.3-8.0	Video exercise workouts, light to vigorous	2.3-6.0
SPORT ACTIVITIES			
Archery	4.3	Rock or mountain climbing	5.0-8.0
Badminton	5.5-7.0	Roller skating	7.0
Basketball	6.0-9.3	Rugby	6.3-8.3
Bowling/Lawn bowling	3.0-3.8	Skateboarding	5.0-6.0
Football, flag or touch	4.0-8.0	Soccer	7.0-10.0
Golf	4.8	Softball	5.0-6.0
Handball	12.0	Squash	7.3-12.0
Hockey, field	7.8	Table tennis (ping pong)	4.0
Hockey, ice	8.0-10.0	Tennis, singles	7.3-8.0
Horseback riding	5.8-7.3	Tennis, doubles	4.5-6.0
Lacrosse	8.0	Volleyball	3.0-4.0
Orienteering	9.0	Volleyball, competitive	8.0
Racquetball	7.0-10.0	Volleyball, competitive beach	6.0

Data from Ainsworth et al. Healthy Lifestyles Research Center, College of Nursing and Health Innovation, Arizona State University. Retrieved 7/21/2011 from http://sites.google.com/site/compendiumofphysicalactivities

These values are only approximations because of the potential error in using a standard value of 3.5 ml · kg^{-1} · min^{-1} as a constant resting value. Further, metabolic efficiency varies considerably from one person to the next, and even in the same individual. Although the MET system is useful as a guideline for training, it fails to account for changes in environmental conditions, and it does not allow for changes in physical conditioning as discussed in the previous section.

Ratings of Perceived Exertion

Ratings of perceived exertion (RPE) also have been proposed for use in prescribing exercise intensity. With this method, individuals subjectively rate how hard they feel they are working. A given numerical rating corresponds to the perceived relative intensity of exercise. When the RPE scale is used correctly, this system for monitoring

In focus

One simple way of monitoring exercise intensity is referred to as the talk test, which has been used as an informal guideline for years. Scientists have now confirmed that the highest exercise intensity that just barely allows a person to talk comfortably while exercising is a very consistent method that correlates well with ventilatory threshold (see chapter 7) and is well within the THR range.[17]

exercise intensity has proven very accurate. Using the **Borg RPE scale**,[5] which is a rating scale ranging from 6 to 20, the exercise intensity should be between an RPE of 12 to 13 (somewhat hard) and an RPE of 15 to 16 (hard). Initially, this sounds too simple. However, most people can use the RPE technique very accurately. Studies have shown that when people are asked to select a pace on a treadmill, or a resistance on a cycle ergometer, at a moderate or heavy intensity of exercise (see table 20.3), they are able to select a pace or resistance that gets their heart rates into the appropriate range. This is a more natural way to prescribe exercise and very efficient if the person is able to relate perceptions of intensity accurately.

Table 20.3 compares the various methods for rating exercise intensity. Let's use them to determine a moderate exercise intensity. As the second column shows, one would want to work within a range of 60% to 79% HR_{max}. If, instead, one is monitoring intensity by $\dot{V}O_{2max}$ or using the Karvonen method, this heart rate range is equivalent to 50% to 74% of either $\dot{V}O_{2max}$ or HR_{max} reserve, as shown in the third column. With use of the rating of perceived exertion, shown in the fourth

column, this is equivalent to an RPE value of 12 to 13. All these values reflect moderate-intensity exercise.

In review

➤ Exercise intensity can be monitored on the basis of training heart rate, metabolic equivalent, or rating of perceived exertion.

➤ Training heart rate can be established through use of the heart rate equivalent to a certain percentage of $\dot{V}O_{2max}$. It can also be determined using the Karvonen method, which takes a given percentage of maximal heart rate reserve and adds it to resting heart rate. With this method, the percentage of maximal heart rate reserve used corresponds to approximately the same percentage of $\dot{V}O_{2max}$ when a person is exercising at moderate to high intensities.

➤ A sensible approach is to establish a THR range to work within, instead of a single THR, attempting to estimate the low end at an intensity below lactate threshold.

➤ The amount of oxygen consumed reflects the amount of energy expended during an activity. $\dot{V}O_2$ at rest has been assigned a value of $3.5 \ ml \cdot kg^{-1} \cdot min^{-1}$, which equals 1.0 MET. Activity intensities can be classified by their oxygen requirements as multiples of the resting metabolic rate.

➤ The RPE method requires that a person subjectively rate how difficult the work is, using a numerical scale that is related to exercise intensity. The subject looks at the standard scale to determine the appropriate number.

TABLE 20.3 **Classification of Exercise Intensity Based on 20 to 60 min of Endurance Activity: Comparing Three Methods**

| Classification of intensity | RELATIVE INTENSITY | | Rating of perceived exertion |
	HR_{max}	$\dot{V}O_{2max}$ or HR_{max} reserve	
Very light	<35%	<30%	<9
Light	35-59%	30-49%	10-11
Moderate	60-79%	50-74%	12-13
Heavy	80-89%	75-84%	14-16
Very heavy	≥90%	≥85%	>16

Data from *Exercise in health and disease: Evaluation and prescription for prevention and rehabilitation*, 2nd ed. M.L. Pollock and J.H. Wilmore, 1990.

Exercise Program

Once the exercise prescription has been determined, it is integrated into a total exercise program, which is generally only part of an overall health improvement plan. Individual exercise capacity varies widely even among people of similar ages and physical builds. For this reason, each program must be individualized, based on results of physiological and medical tests and, if possible, individual needs and interests.

The total exercise program consists of the following activities:

- Warm-up and stretching activities
- Endurance training
- Cool-down and stretching activities
- Flexibility training
- Resistance training
- Recreational activities

Generally, the first three activities are performed three or four times each week. Flexibility training can be included as part of the warm-up, cool-down, and stretching exercises, or it can be done at a separate time during the week. Resistance training is usually done on alternate days when endurance training is not done; however, the two can be combined into the same workout.

In focus

Physical activity must be considered a lifetime pursuit! The benefits of a sound exercise program are rapidly lost once that program is discontinued.

Warm-Up and Stretching Activities

The exercise session should begin with low-intensity, calisthenic-type and stretching exercises. Such a warm-up period gradually increases both heart rate and breathing, preparing the exerciser for the efficient and safe functioning of the heart, blood vessels, lungs, and muscles during the more vigorous exercise that follows. A good warm-up can reduce the amount of muscle and joint soreness experienced during the early stages of the exercise program. An acceptable warm-up would begin with 5 to 10 min of stretching, followed by 5 to 10 min of low-intensity activity using the mode of exercise selected for endurance training. For example, someone who trains by running might start with stretching and then do 5 to 10 min of light jogging before starting to run.

Endurance Training

Physical activities that develop cardiovascular endurance are the heart of the exercise program. They are designed to improve both the capacity and efficiency of the cardiovascular, respiratory, and metabolic systems. These activities also help one control or reduce body weight. Activities such as walking, jogging, running, cycling, swimming, rowing, aerobic dancing, box stepping, and hiking are good endurance activities. Sports such as handball, racquetball, tennis, badminton, and basketball also have aerobic potential if they are pursued vigorously. Activities such as golf, bowling, and softball are generally of little value for developing aerobic capacity; but they are fun, have definite recreational value, and may offer health-related benefits. For these reasons, such activities certainly have a place in the overall exercise program.

Cool-Down and Stretching Activities

Every endurance exercise session should conclude with a cool-down period. The best way to accomplish cool-down is to slowly reduce the intensity of the endurance activity during the last several minutes of a workout. After running, for example, a slow, restful walk for several minutes helps prevent blood from pooling in the extremities. Stopping abruptly after an endurance exercise bout causes blood to pool in the legs and can result in dizziness or fainting. Also, catecholamine levels might be elevated during the immediate recovery period, and this can lead to a fatal heart arrhythmia.

After the cool-down period, stretching exercises can be performed to facilitate increased flexibility.

Flexibility Training

Flexibility exercises usually supplement exercises performed during the warm-up or cool-down period and are useful for those who have poor flexibility or muscle and joint problems, such as low back pain. These exercises should be performed slowly. Quick stretching movements are potentially dangerous and can lead to muscle pulls or spasms. At one time it was recommended that these exercises be performed before the endurance conditioning period. However, recently it has been hypothesized that the muscles, tendons, ligaments, and joints are more adaptable and responsive to flexibility exercises when they are done after the endurance conditioning phase. Research has yet to confirm this hypothesis.

Resistance Training

The importance of resistance training as part of a general health and fitness exercise program has been clearly established. Many health-related benefits can be obtained from resistance training. The ACSM has included resistance training in its recommendations for a general health and fitness program.[1]

Recall from chapter 9 that the maximum amount of weight one can lift successfully only one time is the 1-repetition maximum, or 1RM. When people begin a resistance training program, they should start with a weight that is exactly one-half of their maximal strength, or 1RM, for each lift. They should attempt to lift that weight 10 consecutive times. If they can lift the weight just 10 times before reaching fatigue, this is the proper starting point. If they can do more repetitions, they should go to the next higher weight for the second set. If, instead, they were able to lift the weight fewer than eight times in the first set, the original weight was too heavy and should be reduced to the next lower weight for the second set.

When a given weight brings the exerciser to fatigue by the 8th to 10th repetition in the first set, this is the appropriate starting weight. People should try to achieve as many repetitions as possible during the second and third sets, but the number of repetitions they can complete in these last sets will probably decrease as their muscles become fatigued. As strength increases, the number of repetitions they can complete per set will increase. When one reaches 15 repetitions on the first set, one is ready to progress to the next higher weight. This training technique is referred to as progressive resistance exercise. See chapter 9 for further details.

One can perform two or three sets of each lift per day, two or three days per week, for weight control purposes. Strength gains, however, appear to be fully achieved with just one set per day in previously untrained people.[18] People should select a variety of exercises that tax most or all of the major muscle groups of the upper body, trunk, and lower body. If time is a factor, it is better to reduce the number of sets to one or two and maintain a full-body workout.

Recreational Activities

Recreational activities are important to any comprehensive exercise program. Although people engage in these activities primarily for enjoyment and relaxation, many recreational activities can also improve health and fitness. Activities such as hiking, tennis, handball, squash, and certain team sports fall into this category. Guidelines for selecting these activities include the following:

- Can you learn or perform the activities with at least a moderate degree of success?
- Do the activities include opportunities for social development, if that is desired?

- Are the costs associated with participation reasonable and within your budget?

- Are the activities varied enough to maintain your continued long-term interest?

- Given your current age and health status, is this activity safe for you?

Many excellent opportunities exist for people who have no recreational hobbies or activities but who would like to become involved. Local public recreation centers, park districts, YMCAs, YWCAs, churches and some public schools, community colleges, and universities offer instructional classes in a wide variety of activities at little or no cost. Often the entire family can participate in these classes—an added bonus to a total health improvement program! Also, the number of commercial fitness centers is rapidly growing, and most now employ trained staff who can properly prescribe exercise programs and help individuals get started.

In review

➤ An exercise session should begin with a warm-up of low-intensity, calisthenic-type and stretching exercises to prepare the cardiovascular, respiratory, and muscle systems to work more efficiently.

➤ Endurance activities should be performed three or four times each week.

➤ Each endurance session should be followed by cool-down and stretching to prevent blood pooling in the extremities and muscle soreness.

➤ Flexibility exercise should be performed slowly, and this phase of the program might be best included immediately after the endurance component.

➤ Resistance training should begin with a weight of one-half the person's 1RM. This is the proper weight if the person can lift it about 10 times. If he or she can lift it more than that, more weight is needed; if he or she can lift it fewer than eight times, less is needed.

➤ Recreational activities should be included in an exercise program for enjoyment and relaxation.

➤ Exercise is a vital part of rehabilitation for most diseases. The type and details of the rehabilitation program depend on the patient, the specific disease involved, and its extent.

Exercise and Rehabilitation of People With Diseases

Exercise has become a major component in **rehabilitation programs** for a number of diseases. Cardiopulmonary rehabilitation programs, which began in the 1950s, have become the most visible (see chapter 21). Tremendous advances in cardiopulmonary rehabilitation have led to the formation of a professional association, the American Association of Cardiovascular and Pulmonary Rehabilitation, and a professional research journal, the *Journal of Cardiopulmonary Rehabilitation and Prevention*.

Exercise is also an important part of the rehabilitation of people with

- cancer;
- obesity;
- diabetes;
- renal disease;
- osteoporosis;
- arthritis, chronic fatigue syndrome, and fibromyalgia; and
- cystic fibrosis.

Most recently, emphasis on the use of exercise in the rehabilitation of transplant patients, including those with heart transplants, liver transplants, and kidney transplants, has increased because exercise helps alleviate some drug side effects and improves general health.

In focus

Exercise training has become an extremely important part of rehabilitation programs for a number of diseases. Although the specific physiological mechanisms explaining the benefits of exercise training for each of these diseases have not been clearly defined, exercise training carries many general health benefits that appear to improve the patient's prognosis.

The manner in which exercise is used in the rehabilitation of people with disease is highly specific to the nature and extent of the disease. It is therefore beyond the scope of this chapter to present specific details for any disease; but many resources are now available that provide extensive detail about establishing exercise programs for those with specific diseases and the clinical values of these programs.[2]

In closing

In this chapter, we have seen that the medical community now regards physical activity as vital to maintaining good health and reducing the risk of disease. We looked at the importance and practicality of both a medical examination and an exercise ECG in screening previously sedentary adults before prescribing exercise. We discussed the components of an exercise prescription and methods of monitoring exercise intensity. Finally, we reviewed the components of an exercise program and the role of exercise in rehabilitating patients with disease.

Now that we have seen the importance of exercise in disease prevention, we will look more closely at physical activity as it relates to specific disease states. In the next chapter, we turn our attention to cardiovascular diseases.

Key Terms

Borg RPE scale

coronary artery disease (CAD)

exercise electrocardiogram (ECG)

exercise prescription

graded exercise test (GXT)

Karvonen method

maximal heart rate reserve

metabolic equivalent (MET)

mode

predictive value of an abnormal exercise test

rating of perceived exertion (RPE)

rehabilitation programs

sensitivity

specificity

training heart rate (THR)

Study Questions

1. How active are adult Americans today?
2. What role does the graded exercise test to exhaustion play in medical clearance? Is this test essential for all adults?
3. Discuss the concepts of sensitivity and specificity of exercise testing and the predictive value of an abnormal test. Of what value is this information in the establishment of policy mandating who should be exercise tested?
4. How can we get our population to be more active? What levels of exercise do we need to promote to help people gain the health-related benefits associated with exercise?
5. What four factors must be considered in the exercise prescription? Which of these is the most important?
6. Discuss the concept of a minimal threshold for initiating physiological changes with exercise training as it relates to the exercise prescription.
7. Discuss the various ways of monitoring exercise intensity, and give the advantages and disadvantages of each.
8. Describe the components of a good exercise program and their importance in the total program.
9. How does one effectively motivate individuals to maintain regular exercise habits?

Study Guide Activities

In addition to the activities listed in the chapter opening outline on page 499, two other activities are available in the web study guide, located at

www.HumanKinetics.com/PhysiologyOfSportAndExercise

 The **KEY TERMS** activity reviews important terms, and the end-of-chapter **QUIZ** tests your understanding of the material covered in the chapter.

Cardiovascular Disease and Physical Activity

21

In this chapter and in the web study guide

 ACTIVITY 21.1 Cardiovascular Disease describes the major cardiovascular diseases.

 ACTIVITY 21.2 Disease Pathophysiology describes the pathophysiology of atherosclerosis.

 ACTIVITY 21.3 CAD and Hypertension compares the risk factors for developing coronary heart disease and hypertension.

 ACTIVITY 21.4 Disease Prevention explains the relationships between physical activity and the risks of coronary heart disease.

On Saturday afternoon, June 22, 2002, St. Louis Cardinals pitcher Darryl Kile was found dead in his hotel room in Chicago. The Cardinals were in Chicago to play a three-game series against the Chicago Cubs. He was scheduled to start the last game of the series on Sunday night. He was considered one of the Cardinals' best pitchers and a clubhouse leader. Kile, who was only 33 years old, apparently died of a heart attack caused by coronary atherosclerosis—on autopsy, two of his three major coronary arteries were found to be narrowed by 80% to 90%. Although he had no medical history or symptoms of disease, his father had died of a stroke at the age of 44 years, and Kile had complained of shoulder pain and fatigue during dinner the previous night.

On November 2, 2007, Ryan Shay, a highly ranked distance runner (NCAA 10,000 m champion in 2001 and USA marathon champion in 2003), collapsed during the U.S. Olympic marathon trials in New York City after running just 5.5 mi. The autopsy results stated that his death was the result of "cardiac arrhythmia due to cardiac hypertrophy with patchy fibrosis of undetermined etiology." In October 2009, three men collapsed and died while running the 32nd Detroit Free Press/Flagstar Marathon, all within a span of 16 min. Several weeks earlier, two runners, a man and a women in their mid-30s, died during the Rock n' Roll San Jose Half Marathon. Autopsy reports are not available for these five runners, but it is likely that their deaths were heart related since heat stress was likely not an issue.

These tragedies illustrate the important fact that being a good or even an outstanding athlete during youth and young adulthood does not confer lifelong immunity for coronary heart disease (CHD). Although a genetic predisposition to CHD is serious, it doesn't have to result in premature death. Paying close attention to all of the CHD risk factors and knowing how to minimize the risk become extremely important.

Most of us consider ourselves to be healthy until we experience some overt sign of illness. With chronic degenerative diseases, such as heart disease, most people are unaware that the disease process is smoldering and progressing to the point that it could cause major complications, including death. Fortunately, early detection and proper treatment of various chronic diseases can substantially reduce their severity and often avert disability and death. Even more important, decreasing the risk factors for a disease often can either prevent the disease or delay its onset. In this chapter, we look at cardiovascular diseases, focusing primarily on CHD and hypertension.

Chronic and degenerative diseases of the cardiovascular system are the major cause of serious illness and death in the United States (figure 21.1). In 2006, 81.1 million Americans had one or more types of cardiovascular disease, resulting in over 831,000 deaths.[4,5]

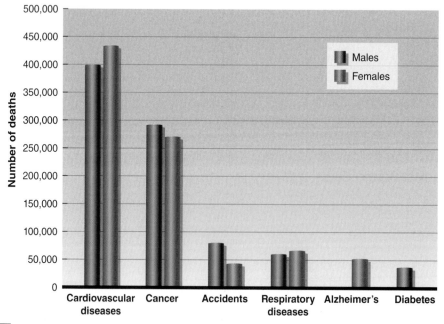

FIGURE 21.1 The leading causes of death in the United States in 2006. Diabetes was not a leading cause of death for females, and Alzheimer's was not a leading cause of death for males.

Data from American Heart Association, 2010.

It was estimated that the total costs for cardiovascular disease in the United States for the year 2010 would be $503.2 billion.[4]

From the early 1900s to the mid-1960s, the relative number of heart disease deaths, expressed per 100,000 people, increased threefold. The population of the United States more than doubled during that time, so the absolute number of heart disease deaths increased even more dramatically than the relative rate indicates. Cardiovascular diseases continue to be a major problem in the United States, accounting for one out of every 2.9 deaths, or 34.5% of all deaths in 2006.

Furthermore, it was estimated that in the United States in 2006 there were

- about 448,000 coronary artery bypass surgeries performed on 253,000 patients,
- about 1,313,000 percutaneous coronary interventions (coronary angioplasties), and
- nearly 2,200 heart transplants.

Fortunately, the death rate from heart disease and stroke has steadily decreased since its peak in the mid-1960s. Heart disease accounted for 38.2% of all deaths in the United States in 1980 but only 26.0% in 2006. Over the same time period of time, the percentages of all deaths attributable to stroke decreased from 8.6% to 5.7%.[49] The reasons for these declines have been heavily debated but likely include a greater focus on disease prevention, for example:

- Improved public awareness of risk factors and symptoms
- Increased use of preventive measures, including lifestyle changes (e.g., nutrition, exercise, stress reduction, and smoking cessation) to reduce individual risk
- Better and earlier diagnosis
- Greater awareness and use of cardiopulmonary resuscitation techniques

Another probable reason is better treatment of patients with disease, for example:

- Improved drugs for specific treatment
- Angioplasty, drug-coated stents, and bypass surgery
- Greater focus on secondary prevention

Though rates vary by country and region, cardiovascular disease continues to be a major public health concern worldwide.

In focus

In the 1970s, cardiovascular diseases accounted for well over 50% of all deaths in the United States. While cardiovascular diseases remain the number-one underlying cause of death in the United States, they accounted for only 34.5% of all deaths in 2006. Furthermore, in 1979, CHD accounted for nearly a third of all deaths in the United States, and in 2006 this had decreased to about 17%. From 1996 to 2006, death rates from all cardiovascular diseases declined 29.2%.

Table 21.1 shows rates of death from cardiovascular disease for 13 countries, many of which have rates of death that approach or exceed those seen in the United States.

Forms of Cardiovascular Disease

There are several different cardiovascular diseases. In this section, we focus primarily on those that are

TABLE 21.1 Cardiovascular Disease Deaths per 100,000 Population From Selected Countries of the World for 2007

Country	Men	Women
Argentina	406	174
Australia	196	85
Canada	212	92
China, rural/urban	413/389	279/273
England/Wales	301	138
France	183	66
Japan	170	69
Mexico	235	166
Russian Federation	1,555	659
Spain	205	79
Sweden	247	107
The Netherlands	222	102
United States	289	150

Data are for adults age 35 through 74.

Data from the American Heart Association. Available here: www.americanheart.org/downloadable /heart/1200594755071International%20Cardiovascular%20Disease%20%20Tables.pdf preventable and that affect the largest number of Americans each year; these are illustrated in figure 21.2. CHD accounts for the majority (53%) of cardiovascular disease deaths, with stroke a distant second at 17%.

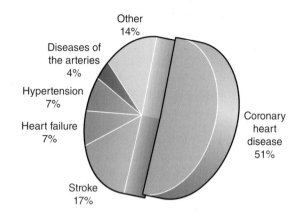

FIGURE 21.2 The leading causes of death from cardiovascular diseases.

Data from American Heart Association, 2010.

Coronary Heart Disease

As most humans age, their coronary arteries (see figure 6.4, p. 144), which supply the myocardium (heart muscle) itself, become progressively narrowed as a result of the formation of fatty **plaque** along the inner wall of the artery, as seen in figure 21.3. This progressive narrowing of the arteries in general is referred to as **atherosclerosis**; and when the coronary arteries are involved, it is termed **coronary heart disease (CHD)**, or coronary artery disease (CAD). As the disease progresses and the coronary arteries become narrower, the capacity to supply blood to the myocardium is progressively reduced. This is what happened to the baseball pitcher, and possibly the distance runners, described in the beginning of this chapter.

As the narrowing worsens, the myocardium (heart muscle) eventually can't receive enough blood to meet all of its needs. When this occurs, the portion of the myocardium that is supplied by the narrowed arteries

FIGURE 21.3 The progressive formation of plaque in a coronary artery.

becomes ischemic, meaning that it suffers a deficiency of blood, or **ischemia**. Ischemia of the heart often causes severe chest pain, referred to as *angina pectoris*. This typically is first experienced during periods of physical exertion or stress, when the demands on the heart are greatest.

When blood supply to a part of the myocardium is severely or totally restricted, ischemia can lead to a heart attack, or **myocardial infarction**, because cardiac muscle cells that are deprived of blood for several minutes are thus deprived of oxygen, which leads to irreversible damage and necrosis (cellular death). This can lead to mild, moderate, or severe disability or even death, depending on the location of the infarction and the extent of the damage. Sometimes a heart attack is so mild that the victim is unaware that it has occurred. In such cases, the heart attack is discovered weeks, months, or even years later when an electrocardiogram is obtained during a routine medical examination.

Atherosclerosis is not a disease of the aged. Rather, it is more appropriately classified as a pediatric disease because the pathological changes that lead to atherosclerosis begin in infancy and progress during childhood.[25] **Fatty streaks**, or lipid deposits, which are thought to be the precursors of atherosclerosis, commonly are found in the aortas of children by age 3 to 5. These fatty streaks start to appear in the coronary arteries during the early teens, can develop into fibrous plaques during one's 20s, and can progress to unstable or complicated lesions during one's 40s and 50s.

The rate at which atherosclerosis progresses is determined largely by genetics and lifestyle factors, including smoking history, diet, physical activity, and stress. For some people, the disease progresses rapidly, with a heart attack occurring at a relatively young age—in their 20s or 30s. For others, the disease progresses very slowly, with few or no symptoms throughout their lives. Most people fall somewhere between these two extremes.

In focus

Atherosclerosis begins in childhood and progresses at different rates, depending primarily on heredity and lifestyle choices.

To illustrate this, a study of combat fatalities from the Korean War revealed that 77% of autopsied American soldiers, average age 22.1, already had some gross evidence of coronary atherosclerosis.[19] The extent of disease ranged from fibrous thickening to complete occlusion of one or more of the main branches of the coronary arteries. The autopsied Korean soldiers, however, were free of the disease. Evidence of coronary atherosclerosis also was found in 45% of the American fatalities from the Vietnam War, and 5% exhibited severe manifestations of the disease.[33]

Hypertension

Hypertension is the medical term for high blood pressure, a condition in which blood pressure is chronically elevated above levels considered desirable or healthy for a person's age and size. Blood pressure depends primarily on body size, so children and young adolescents have much lower blood pressures than adults. For this reason, determining what constitutes hypertension in the growing child and adolescent is difficult. Clinically, hypertension in these groups is defined as blood pressure values above the 90th or the 95th percentile for the youth's age. Hypertension is uncommon during childhood but can appear during midadolescence. For adults, the Joint National Committee on Detection, Evaluation, and Treatment of High Blood Pressure has established guidelines, presented in table 21.2, for systolic blood pressure, which is the highest pressure in the arteries at any time, and for diastolic blood pressure, which is the lowest pressure in the arteries at any time.[24]

Hypertension causes the heart to work harder than normal, because it has to expel blood from the left ventricle against a greater resistance. Furthermore, hypertension places great strain on the systemic arteries and arterioles. Over time, this stress can cause the heart to enlarge and the arteries and arterioles to become scarred, hardened, and less elastic. Eventually, this can lead to atherosclerosis, heart attacks, heart failure, stroke, and kidney failure.

In 2003, at least 65 million American adults, or about 32% of the adult population, were estimated to have high blood pressure (i.e., systolic ≥140 mmHg, diastolic ≥90 mmHg, or both).[5] Prehypertension (i.e., systolic 120-139 mmHg, diastolic 80-89 mmHg, or both) accounted for an additional 28% of the adult popula-

tion. The age-adjusted death rate from hypertension increased 29.3% from 1993 to 2003, and hypertension was the primary or contributing cause of death in about 277,000 individuals. Compared with white Americans, black Americans develop high blood pressure at an earlier age, and it is more severe at any decade of life. Consequently, black Americans have a 1.3 times greater rate of nonfatal stroke, a 1.8 times greater rate of fatal stroke, a 1.5 times greater rate of heart disease deaths, and a 4.2 times greater rate of end-stage renal disease when compared with white Americans.[5] The estimated age-adjusted prevalence of high blood pressure in American adults age 20 and older was 30.6% for non-Hispanic white males, 31.0% for non-Hispanic white females, 41.8% for non-Hispanic black males, 45.4% for non-Hispanic black females, 27.8% for Hispanic males, and 28.7% for Hispanic females.[5]

(In) focus

About one of every three adult Americans has hypertension.

Stroke

Stroke is a form of cardiovascular disease that affects the cerebral arteries, those that supply the brain. Approximately 795,000 strokes occur in the United States each year, and stroke was the underlying or contributing cause of about 232,000 deaths in 2006.[5] As with CHD, the death rate from strokes also has decreased significantly in recent years—a 33.5% reduction between 1996 and 2006.

Strokes generally fall into two categories: **ischemic stroke** and **hemorrhagic stroke**. Ischemic strokes are the most common (~87% of all cases) and result from an obstruction within a cerebral blood vessel that limits the

TABLE 21.2 **Classification of Blood Pressure for Adults, Age 18 Years and Older**

Category	Systolic (mmHg)	Diastolic (mmHg)
Normal	<120	<80
Prehypertension	120-139	80-89
Hypertension		
Stage 1	140-159	90-99
Stage 2	≥160	≥100

Reprinted from the Seventh Report of the Joint National Committee on Prevention, Detection, Evaluation, and Treatment of High Blood Pressure, 2003, *Journal of the American Medical Association 289: 2560-2572.*

flow of blood to that region of the brain. Obstructions have one of two causes:

- Cerebral thrombosis, the most common, in which a thrombus (blood clot) forms in a cerebral vessel, often at the site of atherosclerotic damage to the vessel.

- Cerebral embolism, in which an embolus (an undissolved mass of material, such as fat globules, bits of tissue, or a blood clot) breaks loose from another site in the body and lodges in a cerebral artery. An irregular heartbeat, atrial fibrillation, creates conditions in which clots can form in the heart, dislodge, and then travel to the brain.[4]

In cases of ischemic stroke, blood flow beyond the blockage is restricted, and the part of the brain that relies on that supply becomes ischemic, is oxygen deficient, and can die.

Hemorrhagic strokes are of two major types:

- Intracerebral hemorrhage, in which one of the cerebral arteries ruptures in the brain

- Subarachnoid hemorrhage, in which one of the brain's surface vessels ruptures, dumping blood into the space between the brain and the skull

In both cases, blood flow beyond the rupture is diminished because the blood leaves the vessel at the site of injury. Also, as the blood accumulates outside of the vessel, it puts pressure on the fragile brain tissue, which can alter brain function. Brain hemorrhages often result from aneurysms, which arise from weak spots in the vessel wall that balloon outward; and aneurysms often arise because of hypertension or atherosclerotic damage to the vessel wall. Arteriovenous malformations, clusters of abnormally formed blood vessels, are another cause of hemorragic strokes.

As with a heart attack, a stroke results in death of the affected tissue. The consequences depend largely on the location and extent of the stroke. Brain damage from a stroke can affect the senses, speech, body movement, thought patterns, and memory. Paralysis on one side of the body is common, as is the inability to verbalize thoughts. Most effects of a stroke are indicative of the side of the brain that was damaged. One side of the brain controls the functions of the opposite side of the body. A stroke on the right side of the brain will have the following effects:

- Paralysis on the left side of the body
- Vision problems
- Quick, inquisitive behavioral style
- Memory loss[4]

A stroke on the left side of the brain will have these effects:

- Paralysis on the right side of the body
- Speech and language problems
- Slow, cautious behavioral style
- Memory loss[4]

Heart Failure

Heart failure is a chronic and progressive clinical condition in which the heart muscle (myocardium) becomes too weak to maintain an adequate cardiac output to meet the body's blood and oxygen demands. This usually results from either damage to, or overworking of, the heart. Hypertension, atherosclerosis, valvular heart disease, viral infection, and heart attack are among the possible causes of this disorder. Hypertension precedes heart failure in about 75% of all heart failure patents.[5]

When cardiac output is inadequate, blood begins to back up in the veins. This causes excess fluids to accumulate in the body, particularly in the legs and ankles. This fluid accumulation (edema) also can affect the lungs (pulmonary edema), disrupting breathing and causing shortness of breath. Heart failure can progress to the point of irreversible damage to the heart, and the patient becomes a candidate for a heart transplant.

Other Cardiovascular Diseases

Other cardiovascular diseases include peripheral vascular diseases, valvular heart diseases, rheumatic heart disease, and congenital heart disease.

Peripheral vascular diseases involve the systemic arteries and veins, as opposed to the coronary vessels. **Arteriosclerosis** refers to numerous conditions in which the walls of the arteries become thickened, hard, and less elastic. Atherosclerosis is a form of arteriosclerosis. Arteriosclerosis obliterans, in which an artery becomes completely occluded, is another form. Peripheral venous diseases include varicose veins and phlebitis. Varicose veins result from incompetency of the valves in the veins, allowing blood to back up in the veins and causing them to become enlarged, tortuous, and painful. Phlebitis is inflammation of a vein and is also very painful.

Valvular heart diseases involve one or more of the four valves that control the direction of blood flow into and out of the four heart chambers. **Rheumatic heart disease** is one form of valvular heart disease involving a streptococcal infection that has caused acute rheumatic fever, typically in children between ages 5 and 15. Rheumatic fever is an inflammatory disease of the connective tissue and commonly affects the heart, specifically the heart valves. The damage to the valves usually causes difficulty in their opening, hindering blood flow out of that chamber, or difficulty in their closing, allowing blood to flow back into the previous chamber.

Congenital heart disease includes any heart defects that are present at birth, which are also appropriately termed *congenital heart defects*. These defects occur when the heart or the blood vessels near the heart do not develop normally before birth. These include coarctation of the aorta, in which the aorta is abnormally constricted; valvular stenosis, in which one or more heart valves are narrowed; and septal defects, in which the septum or wall separating the right and left sides of the heart is defective, allowing blood from the systemic side to mix with that in the pulmonary side, and vice versa.

The remainder of this chapter focuses on the two major diseases of the heart and blood vessels: CHD and hypertension.

In review

➤ Atherosclerosis is a process in which arteries become progressively narrowed. Coronary heart disease is atherosclerosis of the coronary arteries.

➤ When blood flow to the heart is sufficiently blocked, the part of the heart supplied by the diseased artery suffers from lack of blood (ischemia), and the resulting oxygen deprivation can cause myocardial infarction, which results in tissue necrosis.

➤ Atherosclerotic changes in the arteries actually begin in young children, but the extent and progression of this disease process are quite variable.

➤ Hypertension is the clinical term for high blood pressure.

➤ Stroke affects the cerebral arteries so that the part of the brain they supply receives too little blood. Ischemic stroke is the most common form of stroke, usually resulting from cerebral thrombosis or embolism. The other cause of stroke is cerebral hemorrhage (cerebral and subarachnoid).

➤ Heart failure is a condition in which the cardiac muscle becomes too weak to maintain an adequate cardiac output, causing blood to back up in the veins.

➤ Peripheral vascular diseases involve systemic, rather than coronary, vessels and include arteriosclerosis, varicose veins, and phlebitis.

➤ Congenital heart disease includes all heart defects present at birth.

Understanding the Disease Process

Pathophysiology refers to the pathology and physiology of a specific disease process or disordered function. Understanding the pathophysiology of a disease gives us insight into how physical activity might affect or alter the disease process. In the following sections, we examine the pathophysiology of CHD and hypertension.

Pathophysiology of Coronary Heart Disease

How does atherosclerosis develop in the coronary arteries? The walls of the coronary arteries are composed of three distinct tunics or layers, as shown in figure 21.4: the tunica intima (inner layer), the tunica media (middle layer), and the tunica adventitia (outer layer). These are referred to more simply as the intima, the media, and the adventitia. The innermost layer of the intima, the **endothelium**, is formed by a thin lining of endothelial cells that provides a smooth protective coating between the blood flowing through the artery and the intimal layer of the vessel wall. The endothelium provides a protective barrier between toxic substances

Tunica adventitia

Tunica media

Tunica intima

Endothelium

FIGURE 21.4 The wall of an artery has three layers: tunica intima, tunica media, and tunica adventitia.

in the blood and the vascular smooth muscle cells. For vessels larger than 1 mm in diameter, the intima also includes a subendothelial layer, formed from connective tissue. The media consists mainly of the smooth muscle cells, which control the constriction and dilation of the vessel, and elastin. The adventitia is composed of collagen fibers that protect the vessel and anchor it to its surrounding structure.

According to an early theory of atherosclerosis, which evolved from the work of Dr. Russell Ross and his colleagues at the University of Washington, local injury to, or dysfunction of, endothelial cells appears to be an important factor in the initiation of atherosclerosis (see figure 21.5a).[43] Blood platelets and monocytes then are attracted to the site of injury and adhere to the exposed connective tissue (see figure 21.5b). These platelets release a substance referred to as **platelet-derived growth factor (PDGF)** that promotes migration of smooth muscle cells from the media into the intima. The intima normally contains few if any smooth muscle cells. A plaque, which is basically composed of smooth muscle cells, connective tissue, and debris, forms at the site of injury (see figure 21.5c). Eventually, lipids in the blood, specifically low-density lipoprotein cholesterol (LDL-C), are attracted to and deposited in the plaque (see figure 21.5d).

More recently, researchers have theorized that monocytes—white blood cells that are effector cells of the immune system—attach between endothelial cells. These monocytes differentiate into macrophages, which ingest oxidized LDL-C. They slowly become large foam cells and form fatty streaks. Smooth muscle cells then accumulate under these foam cells. The endothelial cells then separate or are sloughed off, exposing the underlying connective tissue and allowing platelets to attach to it.[43] In this modification of the original theory, endothelial injury is not always the precipitating event. Injury or disruption of the endothelium can result from high blood concentrations of the atherogenic form of cholesterol (LDL-C); free radicals caused by cigarette smoking, hypertension, and diabetes; elevated plasma homocysteine; and infectious microorganisms among other factors. In fact, atherosclerosis is now recognized as an inflammatory disease.[44]

The plaque consists of a collection of smooth muscle cells and inflammatory cells (macrophages and T lymphocytes), with both intracellular and extracellular lipid.[10] The plaque also contains a fibrous cap. It is

ⓐ A blood-borne irritant injures the arterial wall, disrupting the endothelial layer and exposing the underlying connective tissue.

ⓑ Blood platelets and circulating immune cells known as monocytes are then attracted to the site of the injury and adhere to the exposed connective tissue. The platelets release a substance referred to as platelet-derived growth factor (PDGF) that promotes migration of smooth muscle cells from the media to the intima.

ⓒ A plaque, which is basically composed of smooth muscle cells, connective tissue, and debris, forms at the site of injury.

ⓓ As the plaque grows, it narrows the arterial opening and impedes blood flow. Lipids in the blood, specifically low-density lipoprotein cholesterol (LDL-C), are deposited in the plaque.

FIGURE 21.5 Changes in the arterial wall with injury, illustrating the disruption of the endothelium and the subsequent alterations that lead to atherosclerosis.

now recognized that the composition of the plaque and its fibrous cap is critical to its stability. Unstable plaques are those that have thin fibrous caps and are heavily infiltrated by foam cells. These plaques are much more susceptible to rupture; and when rupture occurs, proteolytic enzymes are released, causing a breakdown of the cellular matrix leading to blood clotting (thrombus) as illustrated in figure 21.6. The thrombus, depending on its size, can occlude or block the artery, resulting in myocardial infarction (MI, heart attack) and even cardiac arrest. In fact, plaque rupture and thrombosis account for up to 70% of MIs and cardiac arrests. Interestingly, the plaques that do rupture are typically small, causing less than 50% stenosis or narrowing of a coronary artery.[10, 15, 44]

There is now good evidence that plaque is a dynamic structure, undergoing cycles of erosion and repair that are responsible for its growth. Ironically, smooth muscle cells are important to the stability of the plaque, and smooth muscle cell proliferation is potentially beneficial to maintaining the integrity of the plaque. Plaque rupture sites are characterized by a low density of smooth muscle cells.[10]

Pathophysiology of Hypertension

The pathophysiology of hypertension is not well understood. In fact, it is estimated that 90% to 95% of those identified with hypertension are classified as having idiopathic hypertension, or hypertension of unknown origin. Idiopathic hypertension is also referred to as essential or primary hypertension. The remaining 5% to 10% are classified as having secondary hypertension, meaning that the cause is secondary to other health issues such as kidney disease, adrenal tumors, or congenital defect of the aorta.

Plaque rupture
Triggers: physical exertion, mechanical stress due to an increase in cardiac contractility, pulse rate, blood pressure, and possibly vasoconstriction

FIGURE 21.6 Illustration of fissure or rupture of an unstable plaque in a coronary artery, releasing its contents into the bloodstream and stimulating the formation of a thrombus (clot).

In review

➤ Pathophysiology refers to the pathology and physiology of a specific disease process or disordered function.

➤ Early theories held that CHD can be initiated by damage to the smooth endothelial lining of the intimal layer of the arterial wall. This damage attracts platelets to the area, which in turn release PDGF. Platelet-derived growth factor attracts smooth muscle cells; and a plaque, composed of smooth muscle cells, connective tissue, and debris, begins to form. Eventually, lipids are deposited in the plaque.

➤ More recent research indicates that monocytes, involved with the immune system, can attach between endothelial cells in the intima and begin forming fatty streaks; this then leads to plaque formation. According to this theory, endothelial damage is not necessary for plaque formation.

➤ It is now clear that the composition of the plaque and its fibrous cap is critical with respect to myocardial infarctions and cardiac arrest. Smaller plaques, where there is typically less than 50% occlusion of the artery, that have thin fibrous caps and are heavily infiltrated with foam cells are the most dangerous.

➤ The pathophysiology of hypertension is poorly understood.

➤ More than 90% of people with hypertension have idiopathic, or essential, hypertension, meaning that its cause is unknown.

Determining Individual Risk

Over the years, scientists have attempted to determine the basic etiology, or cause, of both CHD and hypertension. Much of our understanding of these two diseases comes from the field of epidemiology, a science that studies the relationships of various factors to a specific disease or disease process. In a number of studies, selected members of various communities have been observed for extended periods of time. These observations include periodic medical examinations and clinical tests.

Eventually, some of the participants in such studies become diseased, and over time, many die. All those who develop heart disease or hypertension or who die from heart attacks or hypertension are grouped accordingly. Then their previous medical and clinical tests are analyzed to determine shared attributes or factors. Although this approach does not define the causal mechanism of the disease, it does provide researchers with valuable insights into the disease process. As identified in long-term longitudinal population studies, the factors that place individuals at risk for disease are referred to as **risk factors**. Let's examine the risk factors for heart disease and hypertension.

Risk Factors for Coronary Heart Disease

The factors associated with an increased risk for premature development of CHD can be classified into two groups: those over which a person has no control and those that can be altered through basic changes in lifestyle. Those that a person cannot control include heredity (family history of CHD), race, male sex, and advanced age. According to the American Heart Association,[4] the **primary risk factors** that can be controlled or altered include

- tobacco smoke,
- hypertension,
- abnormal blood lipids and lipoproteins,
- physical inactivity,
- obesity and overweight, and
- diabetes and insulin resistance.

Table 21.3 illustrates the ranges of values for some of these risk factors based on the categories of "desirable," "borderline," and "high." These are approximations and vary somewhat by sex and age.

In Focus

When one or more risk factors for a certain disease are present, an individual is at increased risk for morbidity (developing the disease) and mortality (dying from it).

Other CHD risk factors have been proposed, but there is not yet sufficient evidence to support their inclusion as primary risk factors as determined by the American Heart Association. The following are considered good candidates to be added to the list of primary risk factors:

- C-reactive protein (CRP): CRP is produced in the liver and smooth muscle cells within coronary arteries in response to injury or infection. C-reactive protein is a marker of inflammation.

- Fibrinogin: Fibrinogin is a blood protein integral to the process of blood clotting. Excessive concentrations can lead to abnormal clumping of blood platelets. Fibrinogin is also an indicator of inflammation.

- Homocysteine: Homocysteine is an amino acid used to make protein and build and maintain body tissues. Excessive levels are associated with increased risk for CHD and other cardiovascular diseases.

- Lipoprotein(a) [Lp(a)]: Lp(a) is similar in structure to LDL-C and might reduce the body's ability to dissolve blood clots. Its specific role in atherosclerosis has not yet been determined, but high levels are associated with increased risk for CHD.

Lipids and Lipoproteins

The inclusion of elevated **blood lipids** as a primary risk factor needs to be further defined. For many years, cholesterol and triglycerides were the only lipids observed in these epidemiological studies. The public was confused by conflicting data and opinions about the role of lipids in the development of atherosclerosis. More recently, scientists have studied the manner in which lipids are transported in the blood. Lipids by themselves are insoluble in blood, so they are packaged with a protein to allow transport through the body. **Lipoproteins** are the proteins that carry the blood lipids. Two classes of lipoproteins of major concern for CHD are low-density lipoprotein (LDL) and high-density lipoprotein (HDL). High levels of **low-density lipoprotein cholesterol (LDL-C)** and low levels of

TABLE 21.3 Level of Risk Associated With Selected Risk Factors for Coronary Heart Disease

		LEVEL OR RISK		
Risk factor		Desirable	Borderline	High
Blood pressure[a]				
Systolic	mmHg	<120	120-139	≥140
Diastolic	mmHg	<80	80-89	≥90
Blood lipids and lipoproteins[a]				
Cholesterol	mg/dL	<200	200-239	≥240
LDL-C	mg/dL	<130	130-159	≥160
HDL-C	mg/dL	≥60	40-59	<40
Triglycerides	mg/dL	<150	150-199	≥200
Overweight/Obesity (BMI)[a]	kg/m^2	18.5-24.9	25.0-29.9	≥30
Fasting plasma glucose[b]	mg/dL	<100	100-125	≥126
Physical inactivity[a, c]	min/day	30-60	15-29	<15

[a]American Heart Association, 2010.

[b]Data from the American Diabetes Association: www.diabetes.org/diabetes-basics.

[c]Moderate to vigorous exercise on most days of the week.

high-density lipoprotein cholesterol (HDL-C) place a person at extremely high risk of having a heart attack at a relatively young age—under age 60. Conversely, a high level of HDL-C and a low level of LDL-C place a person at an extremely low risk. Yet a third class of lipoproteins is called very low density lipoproteins (VLDL). **Very low density lipoprotein cholesterol (VLDL-C)** is becoming increasingly implicated as a risk factor for CHD.

Merely looking at total cholesterol is not sufficient. A person might have a moderately high level of total cholesterol (Total-C) and yet be at a relatively low risk because of a high concentration of HDL-C and low concentration of LDL-C. Conversely, a person might have a moderately low level of Total-C and yet be at a relatively high risk because of a high concentration of LDL-C and a low concentration of HDL-C.

Why are these two cholesterol carriers associated with different risk levels? Low-density lipoprotein cho-lesterol is theorized to be responsible for depositing cholesterol in the arterial wall. High-density lipoprotein cholesterol, however, is regarded as a scavenger that removes cholesterol from the arterial wall and transports it to the liver to be metabolized. Because of these very different roles, it is essential to know the specific concentrations of both of these lipoproteins when one is determining individual risk. The ratio of Total-C to HDL-C may be the best blood lipid index of risk for CHD. Values of 3.0 or less place a person at low risk, but values of 5.0 or greater place a person at high risk. As an example, a Total-C of 225 mg/dL and an HDL-C of 45 mg/dL would provide a ratio of 5.0 (225 ÷ 45 = 5.0), but with the same Total-C and an HDL-C of 75 mg/dL the ratio would be 3.0 (225 ÷ 75 = 3.0). Others have used the ratio of Total-C to LDL-C or LDL-C to HDL-C to establish the degree of risk. At this time, there is no consensus as to which ratio provides the best estimate of risk, although most use the Total-C to HDL-C ratio.

In focus

High concentrations of HDL-C and low concentrations of LDL-C place the individual at the lowest risk for CHD. Low-density lipoprotein cholesterol has been implicated in plaque formation, whereas HDL-C is likely involved in plaque regression.

In focus

The ratio of Total-C to HDL-C is possibly the most accurate lipid index of risk for CHD, with values of 5.0 or greater indicating increased risk and values of 3.0 or lower representing low risk.

The Framingham Heart Study

In July 1948, the National Institutes of Health (National Heart, Lung, and Blood Institute) started the Framingham Heart Study (FHS). The FHS was designed as a longitudinal investigation aimed at identifying those factors that influence the development of cardiovascular diseases. The original study population of 5,209 persons from Framingham, Massachusetts, was examined over a four-year period starting in September 1948 and was reexamined every two years for over 48 years. The FHS pioneered the concept of risk factors associated with the development of CHD. Inclusion of the offspring, or second generation, of the original Framingham group began in 1971, and the third generation study started in 2002. The FHS has been one of the most successful longitudinal studies in the history of medical research and resulted in over 2,090 research-based publications as of 2009. As just one example, the FHS was the first to indicate the importance of cholesterol as a risk factor for CHD, and subsequently to demonstrate that the true risk was associated with high levels of LDL-C and low levels of high-density lipoprotein cholesterol (HDL-C).

Early Detection of Risk Factors

Coronary heart disease risk factors can be identified at an early age, and the earlier they are identified, the earlier preventive treatment can begin. In a study of 96 boys ages 8 to 12,

- 19.8% had Total-C values above the suggested high-normal value of 200 mg/dL;
- 5.2% exhibited abnormal resting electrocardiograms;
- 37.5% had more than 20% relative body fat; and
- none had elevated blood pressure.[55]

Similar data were reported in a later study of 13- to 15-year-old boys.[54] Individuals with an elevated risk during childhood generally remain at elevated risk as young adults.

In addition, the results of the Bogalusa Heart Study must be considered. This was a longitudinal study of cardiovascular disease risk factor development from birth through age 39. In 204 of the subjects who died prematurely (primarily from accidents, homicides, or suicides), the scientists found a strong relationship between the risk factors and development of fatty streaks; the greater the number of risk factors, the greater the development of aortic and coronary artery fatty streaks.[6]

Risk Factors for Hypertension

The risk factors for hypertension, like those for CHD, can be classified as ones we can control and ones we cannot. Those we cannot control are heredity (family history of hypertension), sex, advanced age, and race (increased risk for people of African or Hispanic ancestry). Risk factors we can control are

- insulin resistance,
- obesity and overweight,
- diet (sodium, alcohol),
- use of tobacco products,
- use of oral contraceptives,
- stress, and
- physical inactivity.

Although heredity is a risk factor for hypertension, it probably plays a much smaller role than many of the other proposed factors. We must remember that lifestyle factors are often quite similar within a family.

Recently, scientists have shown great interest in a possible link between hypertension, obesity, type 2 diabetes, and coronary heart disease through the common pathway of insulin resistance or impaired insulin action (see sidebar). But obesity also has been established as an independent risk factor for hypertension. Numerous studies have shown substantial reductions in blood pressure with weight loss in hypertensive patients. Also, although sodium intake traditionally has been linked to hypertension, this relationship is likely limited to those who are salt sensitive.

Physical inactivity is a risk factor for hypertension. Its role has been conclusively established in epidemiological studies. Furthermore, substantial evidence indicates that increasing physical activity tends to reduce elevated blood pressure.[3]

In review

> ➤ Risk factors for CHD that we cannot control are heredity (and family history), male sex, and advanced age. Those that we can control are abnormal blood lipids and lipoproteins, hypertension, smoking, physical inactivity, obesity, diabetes, and insulin resistance.

> ➤ Low-density lipoprotein cholesterol is thought to be responsible for depositing cholesterol in the arterial walls. Very low density lipoprotein cholesterol is also implicated in the development of atherosclerosis. However, HDL-C acts as a scavenger, removing cholesterol from the vessel walls. Thus, high HDL-C levels provide some degree of protection from CHD.

> ➤ The ratio of Total-C to HDL-C might be the best indicator of personal risk for CHD. Values below 3.0 reflect a low risk; values above 5.0 reflect a high risk.

> ➤ Risk factors for hypertension that can't be controlled include heredity, advanced age, and race. Those we can control are insulin resistance, obesity, diet (sodium and alcohol), use of tobacco products and oral contraceptives, stress, and physical inactivity.

In focus

Although the pathways are complex, it is becoming increasingly clear that hypertension, CHD, abnormal blood lipids, obesity, and diabetes might be linked through the common pathway of insulin resistance. It is also possible that obesity is the trigger that starts a cascade of events leading to the metabolic syndrome.

Reducing Risk Through Physical Activity

The role that physical activity might play in preventing or delaying the onset of CHD and hypertension has been of major interest to the medical community for many years. In the following sections, we try to unravel this mystery by examining the following areas:

- Epidemiological evidence
- Physiological adaptations with training that might reduce risk
- Risk factor reduction with exercise training

Reducing the Risk of Coronary Heart Disease

Physical activity has been proven effective in reducing the risk of CHD. In the following sections, we discover what is known about this topic and what physiological mechanisms are involved.

Epidemiological Evidence

Hundreds of research papers have dealt with the epidemiological relationship between physical inactivity and CHD. Generally, studies have shown the risk of heart attack in sedentary male populations to be about two to three times that of men who are physically active in either their jobs or their recreational pursuits. The early studies of Dr. J.N. Morris (see figure 21.7) and his colleagues in England in the 1950s were among the first to demonstrate this relationship.[37] In these studies, sedentary bus drivers were compared with active bus conductors who worked on double-decker buses, and sedentary postal workers were compared with active postal carriers who walked their routes. The death rate from CHD was about twice as high in the sedentary groups as in the active groups. Many studies published

Metabolic Syndrome

Metabolic syndrome is a term that has been used to link CHD, hypertension, abnormal blood lipids, type 2 diabetes, and abdominal obesity to insulin resistance and hyperinsulinemia. This syndrome also has been referred to as syndrome X and the insulin resistance syndrome. It is not totally clear where the syndrome starts, but it has been observed that upper body obesity is associated with insulin resistance and that insulin resistance is highly correlated with increased risk for CHD, hypertension, and type 2 diabetes. It appears, however, that obesity or insulin resistance (or a combination of the two) is the trigger that initiates a cascade of events leading to the metabolic syndrome. Systemic inflammation has also been suggested as a causal factor. This became a major topic of research in the 1990s and continues to be today. The results of this research should help us better understand the pathophysiology of these diseases and their interrelationships.

over the subsequent 20 years showed essentially the same results: Those who were occupationally sedentary were at approximately twice the risk for death from CHD as those who were active.

Most of these early epidemiologic studies focused exclusively on occupational activity. Not until the 1970s did researchers start looking at leisure-time activity as well. Again, the studies by Dr. Morris and his colleagues[36, 38] were among the first to observe the relationship between leisure-time activity and the risk of CHD: The least active people were at two to three times greater risk. Subsequent studies by epidemiologists such as Drs. Paffenbarger, Leon, and Blair (figure 21.7) have provided similar results.[8, 9, 31, 32, 39] Physical inactivity approximately doubles the risk of having a fatal heart attack.[42] While most of these early studies were conducted on men, subsequent studies have demonstrated similar results in women.[13]

Scientists at the Centers for Disease Control in Atlanta conducted an extensive review of all epidemiologic studies published on physical inactivity and CHD up to the mid-1980s.[42] They used stringent criteria for including studies in their analysis, and the quality of each study also was assessed. They found that the average relative risk of CHD associated with inactivity ranged from 1.5 to 2.4, with a median value of 1.9, meaning that inactive people have about twice the risk of more active people. Additionally, these researchers found that the relative risk from physical inactivity is similar to the risk associated with other major risk factors for CHD. The results of these epidemiological studies played a major role in leading the American Heart Association in 1992 to declare physical inactivity a primary risk factor for CHD.

Another important concern was raised in the mid-1980s: What level of physical activity or fitness is necessary to reduce one's risk of CHD? It was not totally clear from the epidemiological studies what level of fitness or activity was effective. In fact, during the mid-1980s, scientists had just started to differentiate between activity level and fitness, defining aerobic or cardiovascular endurance fitness by a person's $\dot{V}O_{2max}$. In retrospect, distinguishing these two terms was crucial because a person can be active yet unfit (low $\dot{V}O_{2max}$) or fit (high $\dot{V}O_{2max}$) yet inactive. Dr. Ronald LaPorte and his colleagues at the University of Pittsburgh were instrumental in redirecting the thinking and subsequent research in this area.[28] Dr. LaPorte pointed out that, based on various epidemiological studies, the levels of activity associated with a lower risk for CHD were generally low and certainly not at the level that would be necessary to increase aerobic capacity. Subsequent studies have supported this.[9, 31, 32] Low levels of activity, such as walking and gardening, can provide considerable benefit by reducing the risk for CHD. More vigorous exercise likely provides even greater benefit.[30, 46]

In focus

From epidemiological studies, it has been established that physical inactivity doubles the risk of CHD. However, it is now equally clear that low-intensity activity is sufficient to reduce the risk of this disease. Health benefits do not require high-intensity exercise. But, more vigorous exercise likely provides even greater benefits!

FIGURE 21.7 Key exercise epidemiologists whose research activities were instrumental in leading the American Heart Association to include physical inactivity as a major risk factor for coronary heart disease: Drs. Steven Blair, Ralph Paffenbarger, Jerry Morris, and Art Leon.

Exercise Type and Intensity Are Related to CHD Risk

In 2002, a group of scientists from Harvard University reported in the *Journal of the American Medical Association* the results of their epidemiologic study of the relationship of exercise type and intensity to CHD in more than 44,000 men enrolled in the Health Professional's Follow-Up Study.[46] These men were followed every two years from 1986 through 1998 to assess potential CHD risk factors, to identify newly diagnosed cases of CHD, and to assess levels of leisure-time physical activity. Men who ran 6 mph (9.7 km/h) or faster for 1 h or more per week had a 42% risk reduction compared with men who didn't run. Men who trained with weights for 30 min or more per week had a 23% risk reduction when compared with men who didn't weight train. Brisk walking for 30 min or more per day was associated with an 18% risk reduction, as was rowing for 1 or more hours per week. Surprisingly, swimming and cycling were unrelated to risk. This study was the first to show the direct benefits of weight training on CHD risk and to indicate that exercise intensity is also a critical consideration, with higher intensities providing greater risk reduction.

Training Adaptations That Might Reduce Risk

The importance of regular physical activity in reducing the risk of CHD becomes apparent when we consider anatomical and physiological adaptations in response to exercise training. For example, as we learned in chapter 11, exercise training causes the heart to hypertrophy, primarily through an increase in left ventricular chamber size but also through increases in left ventricular wall thickness. This adaptation may be important for improved contractility and increased cardiac work capacity.

The capacity of the coronary circulation appears to increase with training. Studies have shown that the size of major coronary vessels increases, which implies an increased capacity for blood flow to all regions of the heart. In fact, several studies have demonstrated that the peak flow rate in the major coronary arteries increases following an exercise training program. An important study in the early 1980s addressed the effects of moderate exercise training on the development of CHD in monkeys.[27] The monkeys were divided into three groups: a control group that ate normal, low-fat monkey chow, a nonexercising group that ate an atherogenic (high fat) diet known to induce heart disease, and an exercising group that also ate the high-fat atherogenic diet

The sedentary group that consumed the atherogenic diet developed atherosclerosis. The coronary arteries of the exercising monkeys on this same diet had a greater internal diameter and substantially less atherosclerosis than those in the sedentary monkeys. For the exercise group, the cross-sectional area of the lumen (diameter) of all of the major coronary vessels was two

to three times larger than in the sedentary monkeys. Similar data have been shown for the coronary arteries of marathon runners versus sedentary men following a nitroglycerin challenge.[22]

Evidence is also starting to accumulate suggesting that exercise training increases endothelial function. Endothelial dysfunction is primarily the result of decreased nitric oxide bioavailability. Exercise training has been shown to increase nitric oxide bioavailability.[50] Further, exercise training has been shown to have an anti-inflammatory effect, and we have just learned that atherosclerosis is an inflammatory disease.[40]

In focus

Aerobic training produces favorable anatomical and physiological changes that decrease the risk of heart attack, including larger coronary arteries, increased heart size, and increased pumping capacity. Aerobic training also has a favorable effect on most of the other risk factors for CHD. While not studied to the same extent, resistance training appears to provide many of these same benefits.

Some evidence also suggests that the heart's collateral circulation improves with exercise training. The collateral circulation is a system of small vessels that branch off the major coronary vessels and are important in providing blood to all regions of the heart, particularly when there are blockages in the major coronary arteries. It is possible, however, that collateral circulation development is more the result of the blockages and compromised circulation than of exercise training.

Physical Activity Versus Physical Fitness: Are Both Important?

In 2001, Dr. Paul Williams of the Lawrence Berkeley National Laboratory published an important article suggesting that both physical fitness, as measured by $\dot{V}O_{2max}$, and physical activity levels are independent risk factors for CHD.[53] He conducted a meta-analysis of a number of studies that had been done on large populations. He found that both being active and having a high fitness level were independently related to the degree of risk for CHD and cardiovascular diseases (CVD) in general. This is illustrated in the figure within this sidebar. As the percentiles for both physical activity and physical fitness increase, going from the lowest percentile to the highest, above about 15% there is a reduction in risk for both CHD and cardiovascular diseases. Being physically fit had a stronger relationship to reduced risk than being physically active. These findings are controversial among epidemiologists, so there will likely be more studies and dialogue on this issue.[8]

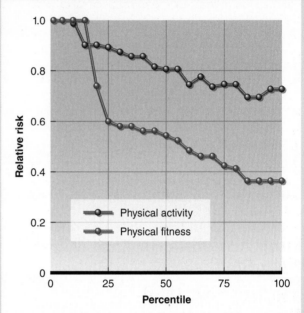

The reduction in relative risk for both coronary heart disease and cardiovascular disease with increasing levels of physical activity and physical fitness. This is a dose–response curve, indicating that the higher the dose, the better the response.

Reprinted, by permission, from P.T. Williams, 2001, "Physical fitness and activity as separate heart disease risk factors: A metaanalysis," *Medicine and Science in Sports and Exercise* 33: 754-761.

Risk Reduction With Exercise Training

Many studies have investigated the role of exercise in altering risk factors associated with heart disease. Let's consider the major risk factors and how exercise might affect them.

Little direct evidence is available to indicate that exercise leads to smoking cessation or reduces the number of cigarettes smoked. Relatively strong data support the effectiveness of exercise in reducing blood pressure in those with mild to moderate hypertension. Endurance training can reduce systolic and diastolic blood pressures by approximately 7 and 6 mmHg, respectively, in individuals who are hypertensive with blood pressures of ≥140 systolic or ≥90 mmHg diastolic.[3] Exercise even can lead to small reductions in both systolic and diastolic blood pressure (by ~2 mmHg each) in those with normal blood pressure.[3] The specific mechanisms responsible for the decreases in blood pressure with endurance training have yet to be fully determined.

Exercise possibly exerts its most beneficial effect on blood lipid levels.[17] Although the decreases in Total-C and LDL-C with endurance training are relatively small (generally less than 10%), there appear to be relatively major increases in HDL-C and major decreases in triglycerides. Cross-sectional studies of athletes and nonathletes alike show unequivocally that people with greater levels of aerobic activity or higher aerobic capacities have higher HDL-C and lower triglyceride levels. Results of longitudinal training studies, however, are much less clear. Many studies have shown increases in HDL-C and decreases in triglycerides from training, yet others have reported little or no change. Several even have shown decreased HDL-C levels. Almost all studies, however, have shown that the ratios of LDL-C to HDL-C and of Total-C to HDL-C are decreased following endurance training. This implies a reduced risk.

We must consider two confounding factors when we evaluate lipid changes with exercise training because they can have a marked independent effect on such changes. Because plasma lipids are expressed as a concentration (milligrams of lipid per deciliter of blood), any change in plasma volume will affect plasma concentrations independently of the change in total lipid. Recall that training typically increases plasma volume (chapter 11). With this plasma expansion, the absolute amount of HDL-C could increase yet the HDL-C concentration might not change or even could be lowered. In addition, plasma lipid levels are tightly coupled with changes in body weight. When we evaluate the effects of exercise training, the independent effects that a change in body weight could have on plasma lipids must be considered.

With respect to the remaining risk factors, exercise plays an important role in weight reduction and control and in the management of diabetes. These topics are discussed in detail in chapter 22. Exercise also has been reported to be effective for stress reduction and control and for reducing anxiety.[41] Some research supports the use of exercise training in the treatment of depression, although the results are not yet conclusive.[16, 35]

Reducing the Risk of Hypertension

Physical activity's role in reducing the risk of hypertension has not been as well established as its role with respect to CHD. As we saw in the last section, exercise training lowers blood pressure in those with moderate hypertension, but the precise mechanisms allowing this reduction are not yet fully known. Let's consider what is known.

Epidemiological Evidence

Very few epidemiological studies have dealt with the relationship between physical inactivity and hypertension. In the Tecumseh Community Health Study, 1,700 males (age 16 and older) completed questionnaires and interviews to provide estimates of their average daily energy expenditures, their peak daily energy expenditures, and the hours they spent in particular activities. The more active men had significantly lower systolic and diastolic blood pressures, irrespective of age.[34] Similar results were obtained when resting blood pressure was analyzed by fitness level in nearly 3,000 adult men and more than 3,900 adult women tested at the Cooper Clinic in Dallas.[14, 21] The more fit individuals exhibited lower systolic and diastolic blood pressures. In a follow-up of the participants from the Cooper Clinic study, the investigators reported a relative risk of 1.5 for the development of hypertension in people with low levels of fitness compared with highly fit people.[7] In a sample of 2,205 adults (20-49 years) from the National Health and Nutrition Examination Survey (NHANES), newly identified hypertension was associated with low fitness estimated from a submaximal treadmill test, with an odds ratio of 2.12 for women and 1.83 for men.[11] From these limited studies, active people and fit people are at reduced risk for developing hypertension. Epidemiological studies also have shown that higher physical activity levels and aerobic fitness are related to a decreased risk for stroke in both men[29] and women.[23]

Training Adaptations That Might Reduce Risk

A number of physiological adaptations that accompany endurance training could affect blood pressure both at rest and during exercise. One of the most important changes associated with endurance training is the previously mentioned plasma volume increase. We might logically assume that any increase in plasma volume would increase blood pressure, particularly because one of the first lines of drug treatment for hypertension is the prescription of a diuretic to reduce total body water and thus plasma volume. However, recall from chapter 11 that trained muscle has a notable increase in capillaries. Also, the venous system in a trained person has a greater capacity, allowing it to contain more blood. For these reasons, the increased plasma volume following exercise training does not increase blood pressure.

Specific mechanisms responsible for reductions in resting blood pressure with endurance training have not been established. Some studies show that resting cardiac output is reduced and that the body's oxygen demands are met by an increased arterial–mixed venous oxygen difference, or $(a-\bar{v})O_2$ difference. But other studies have shown cardiac output to remain unchanged. Without a decrease in cardiac output, the observed reductions in resting blood pressure that follow training must result from reductions in peripheral vascular resistance, which may be attributable to an overall reduction of sympathetic nervous system activity. Increased vasodilation and vascular remodeling of existing arteries and new vessel growth are also possible mechanisms of blood pressure reduction. Weight loss also is associated with reductions in blood pressure.[3]

In focus

Aerobic training reduces blood pressure in healthy individuals as well as in those who have hypertension. The mechanisms by which exercise reduces blood pressure have not been completely determined.

Risk Reduction With Exercise Training

In the previous section on CHD, we determined that exercise training lowers resting blood pressure in normotensives and in those with hypertension, with greater decreases in hypertensives. These reductions are unrelated to the duration of the training program but might be greater in response to low- to moderate-intensity activity compared with higher-intensity activity.

Not only does exercise reduce blood pressure itself, but it also affects other risk factors. Exercise is important in reducing body fat and can increase muscle mass, which may be important in reducing blood glucose

levels and thus assisting in better glycemic (blood sugar) control. This could partially explain the reduction in insulin resistance, another risk factor for hypertension, that has been observed in training studies. Exercise training also has been associated with stress reduction.

Risk of Heart Attack and Death During Exercise

When a person dies while exercising, the incident usually makes newspaper headlines. Deaths during exercise don't happen often, but they are highly publicized. The stories at the beginning of this chapter offer a few examples. How safe, or how dangerous, is exercise? It was estimated in a review of the scientific literature prior to 1982 that there would be approximately one death for every 7,620 middle-aged joggers per year.[48] In a study several years later, the estimate was one death for every 18,000 physically active men.[45] In a study published in 2000, the risk with vigorous exercise was one death per 1.42 million hours of exercise.[1] Most of the research results up to this time were predominantly for men. In 2006, a study in women showed a much lower risk of one death for every 36.5 million hours of moderate to vigorous exercise.[51] So, the overall risk of heart attack and death during exercise is very low. Further, although the risk of death increases during a period of vigorous exercise, habitual vigorous exercise is associated with an overall decreased risk of heart attack.[45] This is illustrated in figure 21.8. There is concern, however, with those athletes who pursue ultra-endurance exercise, that is, training bouts or competition in excess of 4 h per session. Theoretically, they are at greater risk for cardiovascular disorders because of the high oxidative stress associated with this type of training or competition.[26] Information is insufficient at this time to allow resolution of this potential issue.

When death during exercise occurs in people age 35 or older, it usually results from a cardiac arrhythmia caused by atherosclerosis of the coronary arteries. On the other hand, those under age 35 are most likely to die from hypertrophic cardiomyopathy (enlarged diseased heart, usually genetically transmitted), congenital coronary artery anomalies, an aortic aneurysm, or myocarditis (inflammation of the myocardium).

In review

- ➤ Epidemiological studies generally have found that the risk of CHD in sedentary male populations is about two to three times that of men who are physically active and that physical inactivity approximately doubles a person's risk of a fatal heart attack.

- ➤ The levels of activity associated with a reduced risk for CHD can be lower than those needed to increase aerobic capacity.

- ➤ Physical training improves the heart's contractility, work capacity, and coronary circulation.

- ➤ Exercise may have its major impact on blood lipid levels. Studies show that endurance training decreases the ratios of LDL-C to HDL-C and of Total-C to HDL-C, largely due to increases in HDL-C.

- ➤ Exercise is anti-inflammatory and appears to improve endothelial function.

- ➤ Exercise can help control blood pressure, weight, and blood glucose levels and can help alleviate stress.

- ➤ People who are active and those who are fit have reduced risk for developing hypertension.

- ➤ Increased plasma volume that accompanies physical training does not increase blood pressure because trained people have more capillaries and greater venous capacity.

- ➤ Resting blood pressure is decreased by training in people with hypertension; this is probably attributable to decreased peripheral resistance, but the actual mechanisms are unknown.

- ➤ Exercise also reduces body fat, blood glucose levels, and insulin resistance, factors related to an increased risk for hypertension.

In focus

The risk of heart attack increases during an actual period of exercise. However, over the course of a 24 h period, those who exercise regularly have a much lower risk of a heart attack than those who do not exercise.

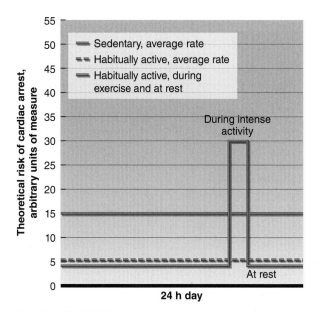

FIGURE 21.8 The risk of primary cardiac arrest during vigorous exercise and at other times throughout a 24 h period, comparing sedentary men with habitually active men.

Data from Siscovick et al., 1984.

Exercise Training and Rehabilitating Patients With Heart Disease

Can active participation in a cardiac rehabilitation program that has a strong aerobic and resistance exercise component help a heart attack survivor to either survive a subsequent attack or avoid one altogether? Endurance training leads to many physiological changes that reduce the work or oxygen demand of the heart. As we have seen, many of these are peripheral, not involving the heart directly. To recap, training increases the capillary/muscle fiber ratio and plasma volume. Because of these changes, blood flow increases to the muscles. In some cases, as mentioned earlier, this allows a reduction in cardiac output, with the body's oxygen demands being met by an increased $(a-\bar{v})O_2$ difference. Training also possibly can increase or maintain oxygen supply to the heart.

However, significant changes also may occur in the heart itself. Studies of heart disease patients at Washington University in St. Louis have provided dramatic evidence that intense aerobic conditioning not only can substantially change peripheral factors but also can alter the heart itself, possibly increasing blood flow to the heart and increasing left ventricular function.[18]

From our previous discussions in this chapter, it is clear that endurance exercise training can significantly reduce the risk of cardiovascular disease through its independent effect on the individual risk factors for CHD and hypertension. Favorable changes in blood pressure, lipid levels, body composition, glucose control, and stress have been reported in patients undergoing exercise training for cardiac rehabilitation. We have every reason to believe that these changes are just as important to the health of a patient who has had a heart attack as they are for a presumably healthy person.

Resistance training also has substantial benefits when included as a part of a comprehensive cardiac rehabilitation program.[52] Table 21.4 provides a summary comparison of the benefits of resistance training and aerobic training on various physiological and clinical markers of health and fitness. It is obvious that combining the two types of training into a comprehensive

TABLE 21.4 Comparison of the Effects of Aerobic Endurance Training and Strength Training on Health and Fitness Variables

Variable	Aerobic exercise	Resistance exercise
Bone mineral density	↑↑	↓↓
Body composition		
% fat	↓↓	↓
Lean body mass	↔	↑↑
Strength	↔	↑↑↑
Glucose metabolism		
Insulin response to glucose challenge	↓↓	↓↓
Basal insulin levels	↓	↓
Insulin sensitivity	↑↑	↑↑
Serum lipids		
HDL-C	↑↔	↑↔
LDL-C	↓↔	↓↔
Resting heart rate	↓↓	↔
Stroke volume	↑↑	↔
Blood pressure at rest		
Systolic	↓↔	↔
Diastolic	↓↔	↓↔
$\dot{V}O_{2max}$	↑↑↑	↑↑
Endurance performance	↑↑↑	↑↑
Basal metabolism	↑	↑↑

Note. HDL-C = high-density lipoprotein cholesterol; LDL-C = low-density lipoprotein cholesterol; ↑= increase; ↓= decrease; ↔ = little or no change. The more arrows, the greater the change.

Reprinted from M.L. Pollock and K.R. Vincent, 1996, "Resistance training for health," *Research Digest: Presidents' Council on Physical Fitness and Sports* 2(8): 1-6.

Major Breakthrough in Cardiopulmonary Resuscitation—Chest Compressions Only

The use of cardiopulmonary resuscitation dates back to 1740 when the Paris Academy of Sciences officially recommended mouth-to-mouth resuscitation for drowning victims. The first documented use of chest compressions in humans was in 1891. In 1960, cardiopulmonary resuscitation (CPR) was developed, and the American Heart Association formally endorsed CPR in 1963. Incredibly, Leonard Cobb established the world's first mass-citizen training in CPR in Seattle, Washington, in 1972 and trained over 100,000 people during the first two years of the program.

In the 1990s, Dr. Gordon A. Ewy, cardiologist and director of the Sarver Heart Center at the University of Arizona College of Medicine, and his colleagues started investigating new ways to improve the existing CPR techniques to increase the number of people willing to provide CPR and to improve the outcome of CPR. A significant barrier to people providing CPR was and is the requirement for mouth-to-mouth resuscitation. Dr. Ewy and his colleagues started investigating the concept of chest compressions only, excluding mouth-to-mouth resuscitation. A series of laboratory trials supported their contention, so they began advocating chest-compression-only CPR statewide in Arizona in 2004. They found that between 2005 and 2010, when chest-compression-only CPR was advocated and taught to the lay public in Arizona, the survival rates were twice as high, when compared to traditional CPR. It appears that when a rescuer stops chest compressions for any reason, including to breathe into a victim's mouth, circulation to the brain and other vital organs stops. Maintaining adequate blood flow to the heart and brain is critical to the success of CPR. Chest-compression-only CPR is for patients with primary cardiac arrest: They have been witnessed (either seen or heard) to have unexpectedly collapsed and are not responding. Standard CPR is for patients with arrest secondary to drowning or respiration arrest. The chest-compression-only CPR advance is important because 75% of out-of-hospital cardiac arrests are primary.

rehabilitation program will maximize the overall training benefits.

A comprehensive cardiac rehabilitation must consider all aspects of the patient's recovery, not just exercise and physical activity. Nutritional counseling is extremely important for those who are not eating appropriately, dealing not only with the total calories consumed in a day, but most importantly with wise food selection to minimize risk and maximize health. Psychological and sexual counseling might also be necessary in some patients. It is not unusual for patients to become anxious about their heart, and their spouses sometimes fear that having sexual relations will further harm the heart of their recovering husband or wife. Well-organized cardiac rehabilitation programs have patient support groups in which patients can openly discuss their concerns.

A number of researchers have tried to determine whether participation in a cardiac rehabilitation program reduces the risk of a subsequent heart attack or of death from a subsequent heart attack. It is nearly impossible to design a study to answer these questions, primarily because it would be necessary to enroll several thousand people into one study to have a large enough sample to prove a statistically significant effect. Consequently, several published reports have combined the results of the most highly controlled of these studies and have used meta-analysis, a special type of statistical analysis, to examine these data. The most recent report (2004), which is a composite covering the patients from previous reports plus >4,000 more patients (8,940 total), concluded that exercise rehabilitation substantially reduces total mortality (20% lower) as well as the risk of death from a subsequent heart attack (26% lower). However, the effect on reducing the risk for recurrence of a nonfatal heart attack, while substantial (21% lower), was not statistically significant.[47]

The evidence that physical activity is important in the rehabilitation of the cardiac patient is sufficiently clear. The American College of Sports Medicine issued a position stand that concluded, "most patients with CHD should engage in individually designed exercise programs to achieve optimal physical and emotional health" (p. iv).[2] The recommendation was that such programs include a comprehensive preexercise medical evaluation, including a graded exercise test, and an individualized exercise prescription. Such programs should focus on multifactorial risk factor modification by using diet, drugs, and exercise to control blood lipid disorders, diabetes, and hypertension. With an aggressive approach to rehabilitation, it is even possible to see slight regression in the disease.[20]

In review

➤ Deaths during exercise are rare, although typically highly publicized.

➤ Deaths during exercise in people over age 35 usually are caused by a cardiac arrhythmia resulting from atherosclerosis.

➤ Deaths during exercise in people under age 35 usually are caused by hypertrophic cardiomyopathy, congenital coronary artery anomalies, aortic aneurysm, or myocarditis.

➤ Cardiac rehabilitation programs are designed to facilitate recovery from heart attacks and other cardiovascular-related health problems and to reduce the risk of subsequent heart attacks or other health issues.

➤ These programs should include both aerobic and resistance training components.

➤ The benefits of comprehensive cardiac rehabilitation programs include favorable changes in body composition, glucose metabolism, plasma lipids and lipoproteins, heart function and cardiovascular dynamics, metabolism, health-related quality of life, and reduction in the risk of subsequent heart attacks and death from heart attacks.

In this chapter, we have seen how important physical activity is in reducing the risk for cardiovascular diseases, especially CHD and hypertension. We discussed the prevalence of these disorders, the risk factors associated with each, and the ways in which physical activity can help reduce our personal risks. In the next chapter, we continue examining the effects of exercise on health as we turn our attention to obesity and diabetes.

Key Terms

arteriosclerosis

atherosclerosis

blood lipids

congenital heart disease

coronary heart disease (CHD)

endothelium

fatty streaks

heart failure

hemorrhagic stroke

high-density lipoprotein cholesterol (HDL-C)

hypertension

ischemia

ischemic stroke

lipoproteins

low-density lipoprotein cholesterol (LDL-C)

metabolic syndrome

myocardial infarction

pathophysiology

peripheral vascular disease

plaque

platelet-derived growth factor (PDGF)

primary risk factors

rheumatic heart disease

risk factor

stroke

valvular heart disease

very low density lipoprotein cholesterol (VLDL-C)

Study Questions

1. What are currently the major causes of death in the United States?
2. What is atherosclerosis, how does it develop, and at what age does it begin?
3. What is hypertension, how does it develop, and at what age does it begin?
4. What is stroke? How does stroke occur? What are the results of stroke?
5. What are the basic risk factors for coronary heart disease? For hypertension?
6. What is the risk of death from coronary heart disease associated with a sedentary lifestyle as compared with an active lifestyle? How has this been established?
7. What are three basic physiological alterations resulting from exercise training that would reduce the risk of death from coronary heart disease?
8. In what ways does endurance exercise training alter risk factors for heart disease?
9. What is a sedentary individual's risk of developing hypertension compared with an active individual's?
10. What are three basic physiological alterations resulting from exercise training that would reduce the risk for developing hypertension?
11. What changes in blood pressure would be expected to result from endurance exercise training in hypertensive individuals?
12. Of what value is cardiac rehabilitation in treating a patient who has had a heart attack?
13. What is the risk of death with endurance exercise training?

Study Guide Activities

In addition to the activities listed in the chapter opening outline on page 521, two other activities are available in the web study guide, located at

www.HumanKinetics.com/PhysiologyOfSportAndExercise

 The **KEY TERMS** activity reviews important terms, and the end-of-chapter **QUIZ** tests your understanding of the material covered in the chapter.

Obesity, Diabetes, and Physical Activity

22

In this chapter and in the web study guide

 ACTIVITY 22.1 Obesity-Related Terms reviews key terms associated with obesity.

 ACTIVITY 22.2 Causes and Health Risks of Obesity explains the causes, conditions, and health risks of obesity and overweight.

 ACTIVITY 22.3 Methods of Assessing Obesity describes how to measure obesity using body mass index and waist-to-hip ratio.

 ACTIVITY 22.4 Physical Activity and Weight Control reviews physiological responses to exercise.

William "the Refrigerator" Perry,

defensive lineman for the Chicago Bears professional football team during the 1980s and early 1990s, reported to the 1988 summer training camp at a weight of 170 kg (375 lb), some 25 kg (55 lb) over his mandated playing weight. Although there was an obvious concern regarding his ability to perform on the football field at this excessive weight, of greater concern are the health risks associated with obesity. Chris Taylor, an Iowa State University wrestler and a U.S. Olympic team bronze medalist in 1972, competed at a weight between 181 and 204 kg (400-450 lb). He died in his sleep at age 29, most likely from obesity-related causes.

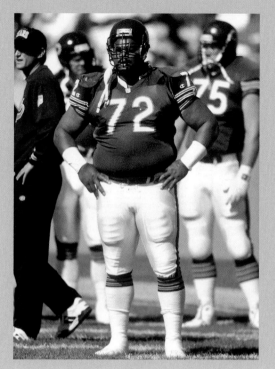

The excessive weight of some amateur and professional athletes not only affects their performance but also increases their health risks.

While millions of people are dying of starvation each year in most parts of the world, many Americans are dying as an indirect result of overconsumption of food. Billions of dollars are spent each year overfeeding the American public. This, in turn, leads to the expenditure of billions of dollars more each year on various weight loss methods and additional billions for the increase in health care costs associated with obesity. Another common disorder in America is diabetes mellitus, which affects nearly 25 million Americans. This disorder of carbohydrate metabolism centers on insulin. Interestingly, scientists have established a link between insulin resistance and obesity, coronary artery disease, hypertension, and many other diseases or disorders.

A sedentary lifestyle has been associated with an increased risk for both obesity and diabetes, and both are strongly associated with other diseases that have high mortality rates, such as coronary artery disease and cancer. Furthermore, approximately one-third of all Americans are obese, diabetic, or both. The consequences of these diseases are debilitating, and the costs associated with their treatment are exceedingly high. In this chapter we focus on obesity and diabetes, discussing their prevalence, their etiology, health problems associated with each disease, and general treatment options. Finally, we consider the role that physical activity can play in their prevention and treatment.

Obesity

The terms *overweight* and *obesity* often are used interchangeably, but technically they have different meanings, as discussed next.

Terminology and Classification

Overweight is defined as a body weight that exceeds the normal or standard weight for a particular person based on height and frame size. These standard weights were established in 1959 but are still widely used. New weight and height tables were introduced in 1983, but their introduction was controversial because many experts believed the weight allowances were too liberal. Many professional health organizations refused to accept the newer tables.

Weight values in the standard tables are based solely on population averages. For this reason, a person can be overweight according to these standards and yet have a lower than normal body fat content. For example, football players frequently are found to be overweight according to standard tables, yet many are typically much leaner than people of the same age, height, and frame size who are of normal weight or even underweight (see chapter 15). Still other people are within the normal range of body weights for their height and frame size by the standard tables and yet are obese. Use of these standard tables is now discouraged due to uncertainty about how to accurately determine frame size and to the lack of a representative data base in the original tables.

Obesity refers to the condition of having an excessive amount of body fat. This implies that the actual amount of body fat or its percentage of total weight must be assessed or estimated (see chapter 15 for assessment techniques). Exact standards for allowable fat percentages have not been established. However, men with more than 25% body fat and women with more than 35% should be considered obese. Men with relative fat values of 20% to 25% and women with values of 30% to 35% should be considered borderline obese. Allowances are higher for women due to sex-specific fat deposits such as breast tissue and hips, buttocks, and thighs, as will be discussed later in the chapter.

Body mass index (BMI) is now the most widely used clinical standard to estimate obesity. To determine a person's BMI, body weight in kilograms is divided by the square of body height in meters. As an example, a man who weighs 104 kg (230 lb) and is 1.83 m (6 ft) tall would have a BMI of 31 kg/m² [104 kg/(1.83 m)² = 104 kg/3.35 m² = 31 kg/m²]. Generally, the BMI is highly correlated with body fat and usually provides a reasonable estimate of obesity. Table 22.1 provides a simple way of determining BMI from height and weight.

In 1997, the World Health Organization proposed a classification system for underweight, overweight, and obesity based solely on BMI values.[47] This classification system was adopted by the National Institutes of Health in 1998 with several modifications and has been used widely since 2000.[30] In table 22.2, BMI values have been divided into five categories: underweight, normal weight, overweight, obesity, and extreme obesity. Within the obesity classification, there are two subclassifications, Class I and Class II. Extreme obesity is Class III. The degree of disease risk also is included and is determined by both BMI and waist circumference. A larger waist circumference increases risk for a given BMI

category. Waist circumference reflects the abdominal visceral fat, which plays a major role in increasing the risk for disease. It is now known that racial and ethnic differences affect the relationship between BMI and fatness, necessitating different BMI cut-points for overweight and obesity in specific groups. As an example, a number of studies have indicated that for the same BMI, the health risk is higher in Asian populations in that for a given BMI Asians have a higher percentage body fat. However, a BMI of 30 or higher almost always indicates excessive adiposity or obesity for all populations, independent of race or ethnicity.

This classification system has made a major contribution to our understanding of the true prevalence of overweight and obesity. Prior to the adoption of this system, there was a wide range of estimates of the percentage of adults who were overweight or obese or both. This was the result of studies using different cut-points or standards for overweight and obesity, which led to considerable confusion among scientists and the general public about the true prevalence of weight disorders. We can now better understand the true prevalence of overweight and obesity and how it has changed over time. Further, having an overweight category has been very useful, providing a buffer zone between normal weight and obesity. People falling in this category can be overweight based on an above-average fat-free body mass, such as the football players

TABLE 22.1 Body Mass Index

BMI	19	20	21	22	23	24	25	26	27	28	29	30	31	32	33	34	35
HEIGHT (ft AND in.)								**WEIGHT (lb)**									
4'10" (58")	91	96	100	105	110	115	119	124	129	134	138	143	148	153	158	162	167
5' (60")	97	102	107	112	118	123	128	133	138	143	148	153	158	163	168	174	179
5'2" (62")	104	109	115	120	126	131	136	142	147	153	158	164	169	175	180	186	191
5'4" (64")	110	116	122	128	134	140	145	151	157	163	169	174	180	186	192	197	204
5'6" (66")	118	124	130	136	142	148	155	161	167	173	179	186	192	198	204	210	216
5'8" (68")	125	131	138	144	151	158	164	171	177	184	190	197	203	210	216	223	230
5'10" (70")	132	139	146	153	160	167	174	181	188	195	202	209	216	222	229	236	243
6' (72")	140	147	154	162	169	177	184	191	199	206	213	221	228	235	242	250	258
6'2" (74")	148	155	163	171	179	186	194	202	210	218	225	233	241	249	256	264	272
6'4" (76")	156	164	172	180	189	197	205	213	221	230	238	246	254	263	271	279	287

1 lb = 0.454 kg, 1 in. = 2.54 cm

Evidence Report of Clinical Guidelines on the Identification, Evaluation, and Treatment of Overweight and Obesity in Adults, 1998. NIH/National Heart, Lung, and Blood Institute (NHLBI).

TABLE 22.2 Classification of Overweight and Obesity by BMI, Waist Circumference, and Associated Disease Risk[a]

Classification	BMI (kg/m²)	Obesity class	DISEASE RISK[b] Men ≥40 in. (102 cm) Women ≥35 in. (88 cm)	Men >40 in. (102 cm) Women >35 in. (88 cm)
Underweight	<18.5		–	–
Normal[c]	18.5-24.9		–	–
Overweight	25.0-29.9		Increased	High
Obesity	30.0-34.9	I	High	Very high
	35.0-39.9	II	Very high	Very high
Extreme obesity	≥40	III	Extremely high	Extremely high

[a]Disease risk for type 2 diabetes, hypertension, and cardiovascular disease.

[b]Relative to normal weight and waist circumference.

[c]Increased waist circumference also can be a marker for increased risk even in persons of normal weight.

Adapted, by permission, from World Health Organization, 1998, "Obesity: Preventing and managing the global epidemic." *In Report of a WHO Consultation on Obesity* (Geneva: WHO).

mentioned previously, or they can be slightly overfat. As stated earlier, all people with a BMI of 30.0 or greater are highly likely to be obese.

Prevalence of Overweight and Obesity

The prevalence of overweight and obesity in the United States has increased dramatically since the 1970s, as illustrated in figure 22.1. This figure presents data from national surveys of large numbers of men and women representative of the total U.S. population, obtained in 1960-1962, 1971-1974, 1976-1980, 1988-1994, 1999-2004, and 2007-2008.[15, 17, 31] The percentage overweight represents the percentage of the total population with a BMI of 25.0 to 29.9 (figure 22.1a), and the percentage obese is for those with BMI values of 30.0 or greater (figure 22.1b), consistent with the World Health Organization and National Institutes of Health classification system. Figure 22.1c presents the data for the combination of those who are overweight and obese (BMI ≥25.0). It is most striking to note that as of 2008 32.2% of men and 35.5% of women in the United States were obese and that nearly 72% of men and 64% of women were either overweight or obese.[17, 31] Further, the prevalence of obesity increased by 62% in men and 52% in women between the 1976-1980 and the 1988-1994 data collection periods,[15] and by an additional 34% in men and 31% in women between the 1988-1994 and 1999-2004 periods.[16, 17] Interestingly, the prevalence of overweight remained relatively constant over these same time periods, suggesting that the United States population is gradually moving from

normal weight to overweight and from overweight to obesity. Fortunately, the increase in the prevalence of overweight and obesity appears to be leveling off. When looking at these data by race, it is apparent that the problem is much more significant in Mexican American men and women and in black women (figure 22.2).

These trends are not unique to the United States. Canada, Australia, and most of Europe have seen similar increases but, with few exceptions, not to the extent seen in the United States.[47] The most recent studies are now showing that obesity is spreading to all regions of the world. Table 22.3 provides estimates of the rates of obesity in men and women for 15 representative countries around the world. These data are somewhat misleading since there is a great variation in the dates for these surveys, making direct comparison somewhat difficult. However, indications are that rates of obesity have continued to skyrocket during the past 10 years, resulting in the worldwide obesity epidemic!

Unfortunately, this same trend of increasing prevalence of overweight and obesity has been reported in U.S. children and adolescents.[31, 40] Figure 22.3 illustrates the trends in the prevalence of overweight from 1963 through 2008 in preadolescent and adolescent boys and girls. Because BMI is much less precise for estimating body fat in children and adolescents, scientists typically use the cut-point for BMI of greater than the 95th percentile, a value that likely indicates that the child is overfat. Similar to the adult trends shown in figure 22.1, the prevalence of overweight remained relatively constant from 1963 through 1980 but increased dramatically from 1980 through 2004, with indications of leveling off in 2007-2008.

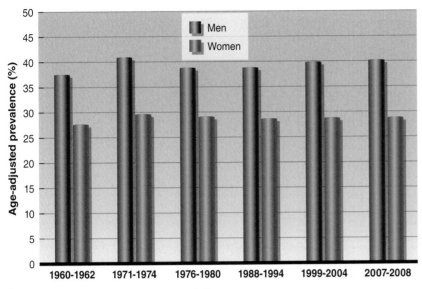

a Prevalence of overweight (BMI of 25.0-29.9)

b Prevalence of obesity (BMI of 30.0 or above)

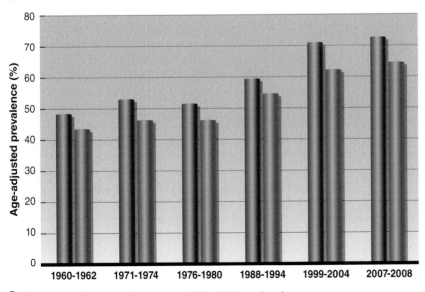

c Prevalence of overweight and obesity (BMI of 25.0 or above)

FIGURE 22.1 The increasing prevalence of overweight (body mass index [BMI] = 25.0-29.9), obesity (BMI of 30.0 and greater), and the combination of overweight and obesity (BMI of 25.0 and greater) in the United States from 1960 through 2008.

Data from Flegal et al., 1998; Flegal et al., 2002; Ogden et al., 2006; and Flegal et al., 2010.

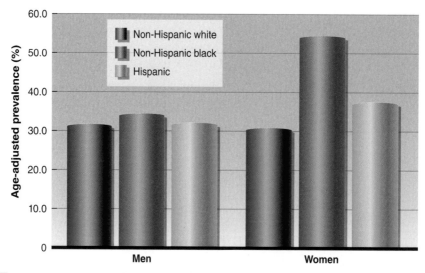

FIGURE 22.2 The prevalence of obesity in men and women by race (2008).

Data from *Morbidity and Mortality Weekly Report*, Volume 58, Number 27, July 17, 2009, pages 740-744. Department of Health and Human Services, Centers for Disease Control.

TABLE 22.3 Prevalence of Adult Obesity in Selected Countries

Country	Men	Women	Date of survey
Argentina	19.5%	17.5%	2003
Australia	25.6%	24.0%	2007-2008
Brazil	8.9%	13.1%	2002-2003
Canada	27.6%	23.5%	2007-2009
China	2.4%	3.4%	2002
England	22.1%	23.9%	2009
France	16.1%	17.6%	2006-2007
India	1.3%	2.8%	2005-2006
Japan	2.3%	3.4%	2000
Mexico	24.2%	34.5%	2006
Russian Federation	10.3%	21.6%	2000
Spain	15.7%	21.5%	1990-2001
Sweden	14.8%	11.0%	2000
The Netherlands	10.4%	10.1%	1998-2002
United States	32.2%	35.5%	2007-2008

Data from the International Obesity Taskforce: www.iaso.org/iotf/obesity/

In focus

More than 70% of men and over 64% of women in the U.S. adult population are overweight or obese, and the prevalence of overweight in children has increased at an alarming rate from 1980 through 2004. Data from 2007-2008 suggest that the prevalence is starting to plateau in both children and adults.[17, 32]

The average person in the United States will gain approximately 0.3 to 0.5 kg (0.7 to 1.1 lb) of additional weight each year after age 25. Such a seemingly small gain, however, results in 9 to 15 kg (15 to 33 lb) of excess weight by age 55. At the same time, bone and muscle mass decrease by approximately 0.1 kg (0.22 lb) per year due to reduced physical activity and normal aging. Taking this into account, an average person's fat mass actually increases by about 0.4 kg (0.9 lb) each year, which equates to a 12 kg (27 lb) fat gain over a 30-year period! It is no wonder that weight loss is an American obsession. It should be noted that these values are approximations and vary by sex and race or ethnicity.

A major concern in light of the increasing rates of obesity, coupled with the earlier onset of obesity, is the impact that this will have on individual and national health care. With the onset of obesity occurring at younger ages will come an increased accumulative exposure to excess weight, and this will likely precipitate an earlier onset of obesity-related diseases such as diabetes.

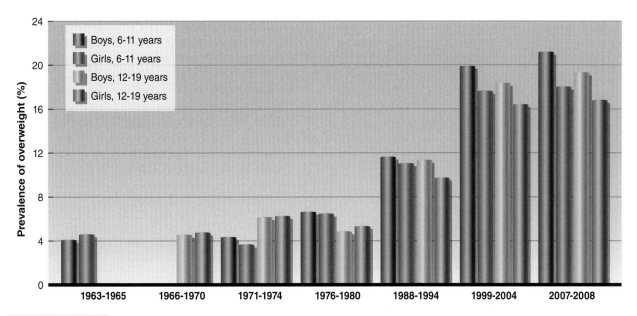

FIGURE 22.3 The increasing prevalence of overweight (95th percentile) in children and adolescents in the United States from 1963 through 2008.

Data from C.L. Ogden, K.M. Flegal, M.D. Carroll, and C.L. Johnson, 2002, "Prevalence and trends in overweight among US children and adolescents," *Journal of the American Medical Association* 288: 1728-1732; Ogden et al., 2006; and Ogden et al., 2010.

Control of Body Weight

One must understand how body weight is controlled or regulated to better understand how a person becomes obese. Body weight regulation has puzzled scientists for years. The average-sized person takes in about 2,500 kcal per day, or nearly 1 million kcal per year. An average gain of 0.4 kg (0.9 lb) of fat each year represents an imbalance of only 3,111 kcal per year between energy intake and expenditure (3,500 kcal is the energy equivalent of 0.45 kg, or 1 lb, of adipose tissue). This translates into a surplus of less than 9 kcal per day. Even with a weight gain of 1.5 lb (0.7 kg) of fat per year, the body can balance caloric intake to within one potato chip per day of what is expended, a truly remarkable example of homeostasis.

The body has the ability to balance energy intake and expenditure to within 8 to 15 kcal per day, about the equivalent of one potato chip!

The body's ability to balance its caloric intake and expenditure to within such a narrow range has led scientists to propose that body weight is regulated around a given set point, similar to the way in which body temperature is regulated. Excellent evidence for this is found in the animal research literature.[24] When animals are force-fed or starved for various periods of time, their weights respectively increase or decrease markedly. But when they go back to their normal eating patterns, they always return to their original weight or to the weight of the control animals, animals that naturally continue to gain weight throughout their life span.

Similar results have been found in humans, although the number of studies is limited due primarily to the difficulty and cost of conducting studies of this type. Subjects placed on semistarvation diets have lost up to 25% of their body weight but regained that weight within months of returning to a normal diet.[25] In a study involving Vermont prisoners, overfeeding resulted in weight gains of 15% to 25%, yet their weights returned to original levels shortly after the experiment ended.[37]

How is the body able to do this? Considering energy expenditure, the total amount of energy expended each day can be expressed as the sum of its three components (see figure 22.4):

1. Resting metabolic rate (RMR)
2. The thermic effect of a meal (TEM)
3. The thermic effect of activity (TEA)

Recall that resting metabolic rate (RMR), as discussed in chapter 5, is the body's metabolic rate early in the morning following an overnight fast and 8 h of sleep. The term basal metabolic rate (BMR) also is used but generally implies that the person fasted for 12 to 18 h and slept over in the clinical facility where the BMR

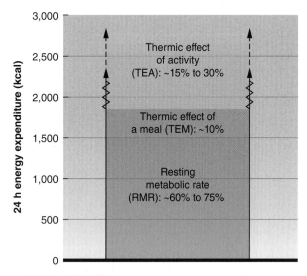

FIGURE 22.4 The three components of energy expenditure. See the text for a detailed explanation.

Adapted, by permission, from E.T. Poehlman, 1989, "A review: Exercise and its influence on resting energy metabolism in man," *Medicine and Science in Sports and Exercise* 21: 515-525.

measurement would be made. Most research today uses the RMR. This value, as we learned in chapter 5, represents the minimal amount of energy expenditure needed to support basic physiological processes. It accounts for about 60% to 75% of the total energy we expend each day.

The **thermic effect of a meal (TEM)** represents the increase in the metabolic rate that is associated with the digestion, absorption, transport, metabolism, and storage of ingested food. The TEM accounts for approximately 10% of our total energy expenditure each day. This value also includes some energy waste, because the body can increase its metabolic rate above that necessary for food processing and storage. We seldom notice the TEM; however, after a very large holiday meal with family, people will start feeling warm and sleepy, with small beads of sweat forming on the forehead. These changes indicate that the metabolic rate has increased considerably. The TEM component of metabolism might be defective in people with obesity, possibly attributable to a defect in the energy-wastage component, leading to a surplus of calories.

The **thermic effect of activity (TEA)** is simply the energy expended above the RMR to accomplish a given task or activity, whether it is combing one's hair or running a 10 km race. The TEA accounts for the remaining 15% to 30% of our energy expenditure.

The body adapts to major increases or decreases in energy intake by altering the energy expended by each of these three components—RMR, TEM, and TEA. With fasting or very low calorie diets, all three decrease. The body appears to be attempting to con-

serve its energy stores. This is dramatically illustrated by decreases in RMR of 20% to 30% or more reported within several weeks after patients begin fasting or consuming a very low calorie diet. Conversely, all three components of energy expenditure increase with overeating. In this case, the body appears to be trying to prevent unnecessary storage of the surplus calories. All of these adaptations may be under the control of the sympathetic nervous system and may play a major, if not the primary, role in maintaining weight around a given set point. This remains a most important area for future research.

In focus

The body attempts to defend its weight when overfed or underfed by increasing or decreasing the three components of energy expenditure: RMR, TEM, and TEA.

If the body has a set-point weight, how can we explain the increasing prevalence of overweight and obesity? It appears that the set point can change, at least in animals that have been most extensively studied. In several studies, the duration of overfeeding and the composition of the diet during overfeeding appear to alter the set point. For example, the set-point weight of rats maintained on a high-fat diet over a six-month period tends to increase: When the rats are placed back on a low-fat diet, they do not go back to their expected weight but stabilize at a much higher weight. For dieting periods of less than six months, they do tend to go back to their original set point. The composition of the diet is therefore a prime suspect for increasing set-point weight, providing the duration of the intervention is sufficient. The physical activity level is also a potential factor. It is quite possible that an increase in the fat content of the diet and a decrease in physical activity levels over an extended period of time would increase set-point weight. This could at least partially explain the increasing prevalence of overweight and obesity in the United States today. It is also important to note that people, like rats, generally consume more calories per day when they are placed on high-fat diets.

Another related factor evolved in the United States during the 1990s—supersizing food portions. Fast food restaurants and many restaurant food chains have started serving much larger portions of food, a trend referred to as *supersizing*. How much difference does it make when one supersizes a meal? At several fast food chains, a single meal including a supersized double cheeseburger (~900 kcal), supersized French fries (~500 kcal), and a supersized 42 oz (1.24 L) soft drink

When Did Supersizing Begin?

Supersizing food portions is supposedly a recent phenomenon, certainly starting within the past 20 years. A study published in the *International Journal of Obesity* in 2010 suggests that this might not be the case.[42] The authors proposed that if art imitates life and if food portions have been increasing with time, then this should be reflected in paintings depicting food. Taking the most commonly painted and most famous meal, Jesus Christ's Last Supper, chronicled in the New Testament, they discovered that the relative sizes of the main dish, the bread, and the plates had increased linearly over the past millennium!

(~500 kcal) provides about 1,900 kcal of energy. For smaller and less active people, this would be sufficient to supply their total daily caloric needs!

The trends of food availability and food purchasing and preparation in the United States between 1970 and 1998 revealed that the per capita energy availability increased by 15%.[20] Furthermore, Americans are eating more meals outside the home, relying more heavily on convenience foods, and consuming larger food portions. These trends point to a general increase in daily energy intake over the past 30 years.

In review

➤ Overweight is a body weight that exceeds the standard weight for a certain height and frame size. Obesity refers to excessive body fat, meaning more than 25% body fat for men and more than 35% body fat for women.

➤ To calculate a person's BMI, we divide body weight in kilograms by the square of height in meters. This value is highly correlated with relative body fat and provides a reasonable estimate of obesity. Body mass index values of 25.0 to 25.9 correspond to overweight, and values of 30.0 or higher correspond to obesity.

➤ Prevalence of obesity and overweight in the United States has increased dramatically since the 1970s.

➤ The average person gains 0.3 to 0.5 kg (0.7 to 1.1 lb) per year after age 25 but also loses 0.1 kg (0.22 lb) of fat-free mass per year, meaning a net gain of 0.4 kg (0.9 lb) of fat each year.

➤ Body weight appears to be regulated around a set point.

➤ Daily energy expenditure is reflected by the sum of the RMR, the TEM, and the TEA. The body adapts to changes in energy intake by adjusting any or all of these components.

Etiology of Obesity

At various times throughout human history, obesity has been thought to be caused by basic hormonal imbalances resulting from failure of one or more of the endocrine glands to properly regulate body weight. At other times, it has been believed that gluttony, rather than glandular malfunction, was the primary cause of obesity. In the first case, a person is perceived as having no control over the situation, and yet in the second, he or she is held directly responsible! Results of recent medical and physiological research show that obesity can be the result of any one or a combination of many factors. Its etiology, or cause, is not as simple as was once believed.

Experimental studies on animals have linked obesity to hereditary (genetic) factors. Studies of humans have also shown a direct genetic influence on height, weight, and BMI. A study from Laval University in Quebec provided possibly the strongest evidence yet of a significant genetic component for obesity.[6] The investigators took 12 pairs of young adult male monozygotic (identical) twins and housed them in a closed section of a dormitory under 24 h observation for 120 consecutive days. The subjects' diets were monitored during the initial 14 days to determine their baseline caloric intake. Over the next 100 days, the subjects were fed 1,000 kcal above their baseline consumption for six of every seven days. On the seventh day, the subjects were fed only their baseline diet. Thus, they were overfed by 1,000 kcal per day for 84 of the 100 days. Activity levels were also tightly controlled. At the end of the study period, as shown in figure 22.5, the actual individual weight gained varied widely, from 4.3 to 13.3 kg (9.5-29.3 lb)—a threefold variation in weight gain for overconsumption of the same number of calories. However, the response of both twins in any given twin pair was quite similar; the major variations occurred between different twin pairs. Similar results were found for gains in fat mass, percentage body fat, and subcutaneous fat. Subsequent studies have reported similar results, demonstrating that genetics plays a major role in determining one's susceptibility to becoming obese. But, as the individual data points in figure 22.5 demonstrate, other factors are also responsible since every subject in this study gained at least 4 kg!

Hormonal imbalances, emotional trauma, and alterations in basic homeostatic mechanisms all have been shown to be either directly or indirectly related to the onset of obesity. Environmental factors, such as cultural habits, inadequate physical activity, and improper diets, are major causes of obesity as discussed in the previous section.

Thus, obesity is of complex origin, and the specific causes undoubtedly differ from one person to the next. Recognizing this is important for treating existing obesity and for preventing its onset. To attribute obesity solely to gluttony is unfair and psychologically damaging to people who are concerned about their problem and are attempting to correct it. In fact, several studies have shown that some obese people actually eat less, although they get far less physical activity, than people of the same sex and similar age with average body fat contents.

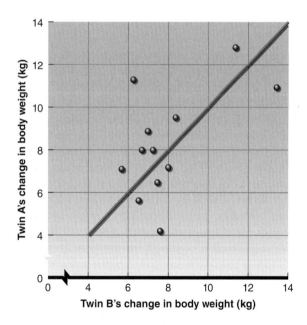

FIGURE 22.5 Similarity in weight gains between twins in response to a 1,000 kcal increase in dietary intake for 84 days of a 100-day study. A data point represents the weight gain for each twin in a pair; twin A's value is shown on the y-axis, twin B's on the x-axis. See the text for further explanation of these data.

Adapted, by permission, from C. Bouchard et al., 1990, "The response to long-term overfeeding in identical twins," *New England Journal of Medicine* 322: 1477-1482. Copyright © 1990 Massachusetts Medical Society. All rights reserved.

In focus

Recent research confirms that there is a significant genetic component in the etiology of obesity. However, it is possible to be obese, attributable basically to lifestyle choices, in the absence of a family history (genetics) of obesity. It is also possible to be relatively lean, even with a genetic predisposition to obesity, through proper diet and activity levels.

Health Problems Linked With Excessive Weight and Obesity

Before we look at the health problems associated with overweight and obesity, it is necessary to define two terms: morbidity and mortality. Morbidity refers to the presence, or rates of the presence, of a particular disease; mortality refers to deaths or rates of death related to the presence of a disease. Overweight and obesity are associated with an increased overall rate of death (general excess mortality).[7, 48] This relationship is curvilinear, as shown in figure 22.6. A major increase in risk of death occurs when the BMI exceeds 30 kg/m^2, although BMI values between 25.0 and 29.9 are associated with an increased morbidity risk for many diseases. A number of more recent studies, published between 2005 and 2010, have reported that excess mortality is primarily associated with BMI values of 35.0 and above. Excess morbidity and mortality associated with obesity and overweight are linked with the following major diseases[9]:

- Coronary heart disease
- Hypertension
- Stroke
- Type 2 diabetes
- Certain types of cancer—endometrial, breast, and colon
- Liver and gallbladder disease
- Osteoarthritis
- Sleep apnea and respiratory problems

With the large increase in the prevalence of obesity in the United States since the 1970s, it is not surprising to also see a very high prevalence of the metabolic syndrome (chapter 21) in U.S. adults. Data from the National Health and Nutrition Examination Survey (2003-2006) revealed that 34% of all adults in the United States met the criteria for the metabolic syndrome. In adults over 60 years of age, the prevalence

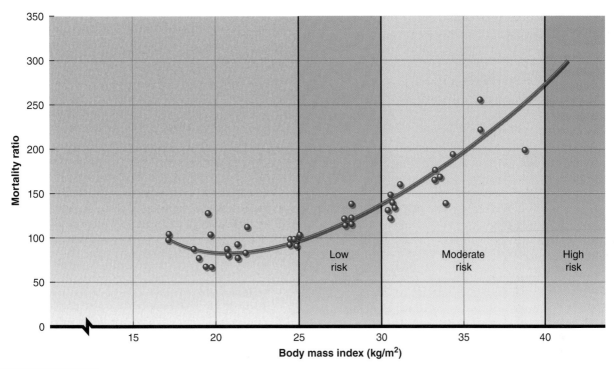

FIGURE 22.6 The relationship of body mass index to excess mortality. A mortality ratio of 100 represents average mortality. The lowest portion of the curve (body mass indexes under 25) indicates very low risk.

Bray, G.A. "Obesity: Definition, diagnosis and disadvantages." MJA 1985; 142: S2-S8. © Copyright 1985. *The Medical Journal of Australia*—reproduced with permission.

was 54%. Consistent with the trends in obesity, the prevalence was highest in Hispanic men and women.[14, 18] Furthermore, obesity has been directly related to changes in normal body function, increased risk for certain diseases, detrimental effects on established diseases, and adverse psychological reactions.

Changes in Normal Body Function

The prevalence and extent of changes in body function vary with the individual and with the degree of obesity. Respiratory problems are common among people with obesity, including sleep apnea. These can lead to other common consequences of obesity, such as lethargy (sluggishness), because of increased carbon dioxide levels in the blood, and polycythemia (increased red blood cell production) in response to lower arterial blood oxygenation. These conditions can lead to abnormal blood clotting (thrombosis), enlargement of the heart, and congestive heart failure. Those with obesity typically have a lower exercise tolerance because of these respiratory problems and also because of the increased body mass that must be moved during exercise. Additional weight gains further reduce activity levels, and exercise tolerance decreases even more.

Increased Risk for Certain Diseases

An increased risk of developing certain chronic degenerative diseases also is associated with obesity. Both hypertension and atherosclerosis have been directly linked to obesity (see chapter 21). So have various metabolic and endocrine disorders, such as impaired carbohydrate metabolism and diabetes. Obesity is a problem associated particularly with the onset of type 2 diabetes (non-insulin-dependent diabetes). A major research breakthrough has enabled us to better understand the role of obesity as a risk factor for most of these diseases. Since the 1940s, major sex differences in the way in which fat is stored or patterned on the body have been recognized. As shown in figure 22.7, males tend to store fat in the upper body, particularly the abdominal area, whereas females tend to store fat in the lower body, particularly the hips, buttocks, and thighs. Obesity that follows the male pattern is referred to as **upper body (android) obesity**, or apple-shaped obesity, and the female pattern is referred to as **lower body (gynoid) obesity**, or pear-shaped obesity.

Research beginning in the late 1970s and early 1980s established upper body obesity as a risk factor for the following conditions:

- Coronary artery disease
- Hypertension
- Stroke
- Elevated blood lipids
- Diabetes

Furthermore, upper body obesity appears to be more important than total-body fatness as a risk factor for these diseases. Waist and hip circumference, or girth, measurements can be used to identify people with increased risk. A waist/hip girth ratio greater than 0.90 for men and greater than 0.85 for women indicates increased risk. With upper body obesity, the increased risk may result from the close proximity of visceral fat depots to the portal circulatory system (circulation to the liver). Figure 22.8 shows a young woman being placed into a computed tomography (CT) scanner in order to assess visceral abdominal fat (figure 22.8a) and CT scans at the level of the fourth lumbar vertebra of two men (figure 22.8, b and c).[36] The subject in figure 22.8c has considerably more visceral (deep) abdominal fat than subcutaneous abdominal fat.

ⓐ Upper body (android) obesity **ⓑ** Lower body (gynoid) obesity

FIGURE 22.7 Obesity patterns tend to differ by sex.

In focus

Obesity places an individual at significantly increased risk for hypertension; stroke; diabetes; coronary artery disease; various cancers; and various respiratory, metabolic, and digestive diseases. The health risks associated with obesity are most likely associated with the manner in which the fat is distributed on the body, with upper body obesity (representing high levels of visceral fat) posing significantly greater risk.

FIGURE 22.8 (a) A subject in the process of undergoing a computed tomography (CT) scan. (b and c) CT scans at the level of the fourth lumbar vertebra in two people. The subject in c has considerably more visceral abdominal fat (light areas) than subcutaneous abdominal fat. The subject in b is leaner but also has considerably less of his fat as visceral fat.

Detrimental Effects on Established Diseases

The effects of obesity on existing diseases are not clear. Obesity can contribute to further development of certain diseases and medical conditions, and weight reduction usually is prescribed as an integral part of treatment. Conditions that generally benefit from weight reduction include

- angina pectoris,
- hypertension,
- congestive heart disease,
- myocardial infarction (reduced risk of recurrence),
- varicose veins,
- diabetes, and
- orthopedic problems.

Adverse Psychological Reactions

Emotional or psychological problems might be the cause of obesity in some people. Furthermore, emotional or psychological problems can arise from the condition itself. In many societies, obesity carries a social stigma that contributes substantially to the problems of those who have it. Media, especially Western media, typically glamorize only people with extremely lean bodies. Consequently, some obese people might need professional counseling assistance in their efforts to lose weight. On the other hand, as the population as a whole becomes more obese, what the general public once considered "fat" becomes the norm.

General Treatment of Obesity

In theory, weight control seems to be a simple matter. To maintain weight, the energy consumed by the body in the form of food must equal the total energy expended, which is the sum of the RMR, TEM, and TEA. Ideally, the body normally maintains a balance between caloric intake and caloric expenditure, but when this balance is upset, weight will be lost or gained. Both weight losses and weight gains appear to depend largely on just two factors: dietary intake and physical activity. This is now recognized as an oversimplification, considering the results of the overfeeding study of monozygotic twins discussed earlier, in which considerable variation in weight gains occurred for the same amount of overfeeding.[6] Not everyone responds to the same intervention in the same way. This difference in response must be considered when one is designing treatment programs for individuals attempting to lose weight, and people

trying to lose weight must understand this difference so that those who are low responders won't be discouraged. In the past, we have tended to label the low responders as *noncompliers*, but we know now that this is generally not accurate.

Weight loss generally should not exceed 0.45 to 0.9 kg (1-2 lb) per week. Losses greater than this should not be attempted without direct medical supervision. Losing just 0.45 kg (1 lb) of fat a week will result in the loss of 23.4 kg (52 lb) of fat in only a year! Few people become obese that rapidly. Weight loss also should be considered a long-term project. Research and experience have proven that rapid weight losses are usually short-lived, and the lost weight is usually quickly regained because rapid weight losses are generally the result of large losses of body water. The body has built-in safety mechanisms to prevent an imbalance in body fluid levels, so the lost water eventually will be replaced. Thus, a person wishing to lose 9 kg (20 lb) of fat is advised to attempt to attain this goal in a minimum of three to five months.

Many special diets have achieved popularity over the years, such as the Drinking Man's Diet, the Beverly Hills Diet, the Cambridge Diet, the California Diet, Dr. Stillman's Diet, The Zone, the South Beach Diet, the Weight Watchers diet, the Mediterranean diet, the Pritikin diet, and the Atkins Diet. Each claims to be the ultimate in effective and comfortable weight loss. Some diets have been developed for use either in the hospital or at home under the supervision of a physician. These are often referred to as very low calorie diets, as they allow only 350 to 500 kcal of food per day. Most of these have been formulated with a certain amount of protein and carbohydrate to minimize the loss of fat-free body mass. Research has shown that many of these are effective, but no single diet has been shown to be more effective than any other. Again, the important factor is the development of a caloric deficit while maintaining a complete, balanced diet that meets the body's nutritional needs. The best diet is the one that meets these criteria and is best suited to individual comfort and personality.

Generally, improper eating habits are at least partially responsible for most weight problems, so no diet should be viewed as a quick fix. A person should learn to make permanent changes in dietary habits, especially reducing the intake of fat and simple sugars. For most people, simply eating a low-fat diet will gradually reduce weight to a desirable level. One potential problem with low-fat diets, however, is that people mistakenly assume that low fat means low calorie, and this often is not the case. For most people, simply reducing total caloric intake by 250 to 500 kcal per day, combined with a selection of low-fat and low simple sugar foods, would be sufficient to accomplish their desired weight loss goals.

A Genetic Predisposition to Obesity: The Pima Indians

It has been clearly established that genetics is a major factor in the development of obesity. Dr. Claude Bouchard, former director of the Pennington Biomedical Research Center (Louisiana State University system), has conducted many studies on the heritability of obesity and has concluded that the heritability of fat mass or relative body fat (percentage fat) is about 25% of the age- and sex-adjusted variance.[5] Does having a genetic predisposition to obesity mean that someone is destined to be obese? The answer is no! We have learned a great deal from the study of the Pima Indians, who have lived for at least 2,000 years near the Gila River in the Sonora Desert in what is now southern Arizona. Until the early 1900s, the Pima Indians apparently were lean and healthy people who were

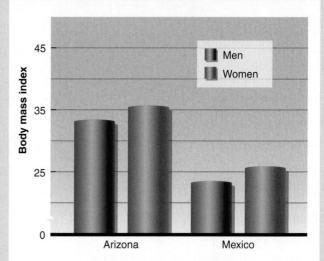

The BMI for Pima Indian men and women living in Arizona and in northern Mexico (2006).

Data from Schulz et al., 2006.

physically active and ate a healthy diet. As they have moved onto reservations, stopped farming, and started eating a westernized high-fat diet and consuming alcohol, they have become very obese, with a prevalence of obesity of 64% for men and 75% for women.[35] Associated with their obesity is a high prevalence of diabetes—34% of men and 41% of women. The Pima Indians are such an unusual population with respect to their extremely high prevalence of obesity and diabetes that the National Institutes of Health established a special research center in Phoenix, Arizona, just to study them and their health issues.

Interestingly, another group of Pima Indians who live in northern Mexico have stayed active working on farms but use no motorized equipment. Furthermore, they consume a diet high in carbohydrates and low in fat. They have managed to stay relatively lean. The BMIs for these two groups of Indians are seen in the figure in this sidebar. The bottom line is that a person can have a genetic predisposition for obesity but with proper diet and exercise can maintain a relatively normal body weight. The Pima Indians have taught us an important lesson.

Hormones and drugs also have been used to assist patients in weight loss by decreasing their appetite or increasing their RMR. Unfortunately, a number of side effects are associated with these drugs, some of which are very serious and can be life threatening. Surgical techniques also are used to treat extreme obesity, but only as a last resort when other treatment procedures have failed and the obesity is life threatening. Intestinal bypass surgery involves surgically bypassing a large segment of the small intestine, thus reducing food absorption. While popular at one time, this procedure is now seldom used because of potential complications. Today, gastric bypass surgery and gastric banding are the most common procedures, both restricting the amount of food that can enter the stomach. While highly effec-

tive, these procedures are very expensive and do have associated risk, although the average mortality rate is under 1% to 2%. Generally, surgery should be reserved for those who have extreme obesity or obesity with significant risk factors.

Behavior modification has been proposed as one of the most effective techniques for helping people with weight problems. Major weight losses have been achieved through changes in basic behavior patterns associated with eating. Furthermore, these weight losses appear much more permanent. This approach appeals to most people because the techniques seem to make sense and are often easy to incorporate into a normal daily routine. For example, an individual might not have to consciously reduce the amount of food eaten but simply agree that all eating will be done in one location, which often cuts down on snacking. Or an individual might be allowed to take as much food as desired with the first helping, but no second helpings. Many such simple changes can help regulate eating behavior and result in substantial weight loss.

In review

➤ The etiology of obesity is not simple; the cause can be any one or a combination of many factors.

➤ Studies of animals and humans indicate that there is a genetic component to obesity. The disorder also has been linked to hormonal imbalances, emotional trauma, homeostatic imbalances, cultural influences, physical inactivity, and improper diets.

➤ Overweight and obesity are associated with increased risk of general excess mortality.

➤ Respiratory problems are quite common among people with obesity. These, in turn, can lead to lethargy and polycythemia.

➤ Obesity increases the risk of certain chronic degenerative diseases. Upper body (visceral) obesity increases the risk of developing coronary artery disease, hypertension, stroke, elevated blood lipids, diabetes, and the metabolic syndrome. Also, obesity can worsen preexisting health conditions and diseases.

➤ Emotional or psychological problems may contribute to obesity; and the disorder itself, with the stigma it carries, can be psychologically damaging.

➤ In the treatment of obesity, it is important to remember that people respond differently to the same intervention. Some have large losses of weight in a relatively short period of time, while others appear to be resistant to weight loss and see only small losses.

➤ Weight loss generally should not exceed 0.45 to 0.9 kg (1-2 lb) per week. Simple diet modification, reducing intake of fat and simple sugars, is sufficient to help most people lose weight. Behavior modification is also an effective method of weight loss.

➤ The use of drugs or surgery in the treatment of obesity is generally not recommended unless deemed necessary for the patient's health by a physician.

Role of Physical Activity in Weight Control

Inactivity is a major cause of obesity in the United States. In fact, a sedentary lifestyle may be just as important in the development of obesity as overeating! Thus, increasing physical activity levels must be recognized as an essential component in any program of weight reduction or control.

Changes in Body Composition With Exercise Training

Physical training can alter body composition. Many people have believed that physical activity has little or no influence on changing body composition and that even vigorous exercise burns too few calories to lead to substantial body fat reductions. Yet research has conclusively demonstrated the effectiveness of exercise training in promoting moderate alterations in body composition.

A person who jogs three days a week for 30 min each day at an 11 km/h (7 mph) pace (slightly over 5.4 min/km, or 8.5 min/mi) will expend about 14.5 kcal/min, or 435 kcal, for the 30 min run each day. This results in a total expenditure per week of about 1,305 kcal, the equivalent loss of about 0.15 kg (0.33 lb) of adipose tissue (fat plus connective tissue and water) each week just from the exercise period alone. This might lead some people to believe that exercise is a painfully slow way to significantly reduce body fat levels and that there are better and easier ways to lose fat. However, in 52 weeks, providing energy intake remained constant, this person would lose 7.8 kg (17 lb)!

In estimations of an activity's energy cost, typically the average or steady-state rate of energy expenditure for that activity is multiplied by the number of minutes the activity is performed. For example, if the steady-state rate for shoveling snow is 7.5 kcal/min, 1 h of shoveling would require a total of 450 kcal. This would result in an approximate loss of 0.06 kg (0.13 lb) of adipose tissue (450 kcal ÷ 3,500 kcal × 0.45 kg of adipose tissue = 0.06 kg, or 450 kcal ÷ 3,500 kcal × 1 lb = 0.13 lb).

But examining the energy expended only during exercise does not give us the full picture. Metabolism remains temporarily elevated after exercise ends. This phenomenon was at one time referred to as the oxygen debt but, as mentioned in chapter 5, is now referred to as the excess postexercise oxygen consumption (EPOC). Returning the metabolic rate back to its preexercise level can require several minutes following light exercise, such as walking; several hours following very heavy exercise, such as playing a football game; and up to 12 to 24 h or even longer for prolonged, exhaustive exercise, such as running a marathon in a hot and humid environment.

In focus

Physical activity is important in both weight maintenance and weight loss. In addition to the calories that are expended during exercise, a substantial expenditure of calories occurs during the postexercise period (EPOC).

The EPOC can require a substantial energy expenditure when considered over the entire recovery period. If, for example, the oxygen consumption following exercise remains elevated by an average of only 0.05 L/min, this will amount to approximately 0.25 kcal/min or 15 kcal/h. If the metabolism remained elevated for 5 h, this would provide an additional expenditure of 75 kcal that would not normally be included in the calculated total energy expenditure for that particular activity. This additional energy expenditure is ignored in most calculations of the energy costs of various activities. The person in this example, if exercising five days per week, would expend 375 kcal, or lose the equivalent of about 0.05 kg (0.1 lb) of fat in one week, or 0.45 kg (1.0 lb) in 10 weeks, from the additional caloric expenditure during the recovery period alone!

Studies have shown relatively small, but significant changes in both weight and body composition with both aerobic and resistance training, which include

- total weight decrease,
- fat mass and relative body fat decrease, and
- either maintained or increased fat-free mass.

Overall, these changes are not large. In a summary of hundreds of individual studies that monitored body composition changes with aerobic training, the expected changes from a typical one-year exercise training program (three times per week, 30-45 min per day, at 55-75% of $\dot{V}O_{2max}$) would be as follows: −3.2 kg (−7.1 lb) total body mass, −5.2 kg (−11.5 lb) fat mass, and +2.0 kg (+4.4 lb) fat-free mass.[44] Furthermore, relative body fat would decrease by nearly 6% (e.g., from 30% body fat to 24% body fat).

To help put this into perspective, table 22.4 presents a hypothetical example of the expected weight loss in six months and weight loss retained one year after the completion of the intervention for an overweight and obese man who has a starting weight of 90 kg (198 lb), comparing the expected weight loss from (1) low and very low calorie diets only, (2) behavior modification only, (3) exercise only, and (4) a combination of low and very low calorie diets, behavior modification, and formal exercise. The expected weight loss for a six-month intervention is estimated from averages derived from the available research studies in each area.

TABLE 22.4 Expected Weight Loss Following Six Months (26 Weeks) of Treatment, and the Retention of Weight Loss After One Year of Follow-Up Resulting From Various Weight Loss Interventions in an Overweight, 90 kg Man

Variables	Low or very low calorie diet only	Behavior modification only	Exercise only	Combination of all three interventions
Rate of weight loss, kg/wk	0.90	0.40	0.06	1.00
Weight loss in six months, kg/lb	23.4/51.6	10.4/22.9	1.6/3.5[a]	26.0
Weight loss, % of initial weight	26.0	11.6	1.8	28.9
Weight loss retention one year after intervention completion, %	25	68	70[b]	75

[a]This figure is misleading: The total fat loss would be 2.6 kg (5.7 lb) since there would be a gain in the fat-free mass of 1.0 kg (2.2 lb) during this six-month exercise program.

[b]Very limited data from exercise-only studies are available.

Although most weight loss studies have used aerobic training, a number of studies have used resistance training and have shown impressive decreases in body fat and increases in fat-free mass. The evidence shows that exercise is an important part of any weight loss program. But to maximize losses in body weight and body fat, it is necessary to combine exercise with decreased caloric intake.

(In) focus

Attempts to lose weight are much more successful when one loses only 0.45 to 0.9 kg (1-2 lb) per week and when dietary restriction is combined with moderate exercise (300-500 kcal per day). This combination minimizes the loss of fat-free mass and maximizes the loss of fat mass.

Since the 1990s, abdominal visceral fat (figure 22.8 on p. 556) has been identified as a major independent risk factor for cardiovascular diseases and obesity. There is now substantial evidence that physical activity reduces the rate of accumulation of visceral fat and that exercise training actually reduces visceral fat stores.[38] This could be one of the most important health benefits of an active lifestyle!

Mechanisms for Change in Body Weight and Composition

When attempting to explain how exercise causes such changes in body weight and composition, it is necessary to consider both sides of the energy-balance equation. Evaluating energy expenditure requires that we consider each of the three components of energy expenditure: RMR, TEM, and TEA. Evaluating energy intake requires that we also consider the energy that is lost in the feces (energy excreted), which is generally less than 5% of the total caloric intake. Keeping this balance in mind, in the next section we examine some of the possible mechanisms through which exercise might affect body weight and body composition.

(In) focus

The energy-balance equation:

$$\text{energy intake} - \text{energy excreted} = RMR + TEM + TEA.$$

Exercise and Appetite Some believe that exercise stimulates the appetite to such an extent that food intake is unconsciously increased to at least equal that expended during exercise. In 1954, Jean Mayer, a world-famous nutritionist, reported that animals exercising for periods of from 20 min to 1 h per day had a lower food intake than nonexercising control animals.[29] He concluded from this and other studies that when activity is less than a certain minimal level, food intake does not decrease correspondingly and the animal (or human) begins to accumulate body fat. This led to the theory that a certain minimal level of physical activity is necessary for the body to precisely regulate food intake to balance energy expenditure. A sedentary lifestyle may reduce this regulatory ability, resulting in a positive energy balance and weight gain.

Exercise does, in fact, appear to be a mild appetite suppressant, at least for the first few hours following

How Much Activity Is Necessary for Weight Control?

In 2005, a group of scientists from the Mayo Clinic made an important discovery with respect to activity levels and weight control.[27] Ten lean and 10 obese sedentary subjects were instrumented with a very sophisticated system for monitoring even minor changes in body position over a 10-day period. Although both groups were sedentary, the lean subjects spent 350 kcal per day more in postural changes and movement when compared to the obese subjects. As just one example, obese subjects were seated 2 h longer per day than lean subjects. These results point to the importance of simply being active, independent of a formal exercise program.

intense exercise training. Furthermore, studies have shown that the total number of calories consumed per day does not change when a person begins a training program. Although some people interpret this as evidence that exercise does not affect appetite, a more accurate conclusion might be that appetite was affected, in fact suppressed, because caloric intake did not increase in proportion to the additional caloric expenditure from the exercise program. In studies conducted on rats, male rats appear to reduce food intake with exercise training, whereas female rats tend to eat the same or even more than nonexercising control rats.[33] There is no obvious explanation for this sex difference. Also, it is unclear whether this sex difference is present in humans.

The decrease in appetite might occur only with intense levels of exercise, in which the resulting increased catecholamine (epinephrine and norepinephrine) levels might suppress the appetite. The increased body temperature that accompanies either high-intensity activity or almost any activity performed under hot and humid conditions also might suppress appetite. We all know from experience that we desire less food when the weather is hot or when our body temperatures are elevated because of illness. This also might explain why a hard running workout in the heat results in little or no desire to eat, yet a hard swimming workout in a cool swimming pool elicits a relatively strong craving for food. In the pool, provided that the water temperature is well below body core temperature, the heat generated by exercise is lost very effectively, so core temperature typically is not elevated to the same extent.

In focus

Regular physical activity may assist in controlling appetite so that caloric intake balances caloric expenditure.

Exercise and Resting Metabolic Rate The effects of exercise on the components of energy expenditure became a major topic of interest among researchers in the late 1980s and early 1990s. Of obvious interest is how exercise training might affect the RMR, since RMR represents 60% to 75% of the total calories expended each day. For example, if a 25-year-old man's total daily caloric intake was 2,700 kcal and his RMR accounted for just 60% of that total ($0.60 \times 2,700 = 1,620$ kcal RMR), a mere 1% increase in his RMR would require an extra 16 kcal expenditure each day, or 5,840 kcal per year. This small increase in RMR alone would account for the equivalent of a 0.8 kg (1.7 lb) fat loss per year!

The role of physical training in increasing RMR has not been totally resolved. Several cross-sectional studies have shown that highly trained runners have higher RMRs than untrained people of similar age and size. But other studies have not been able to confirm this.[34] Few longitudinal studies have been conducted to determine the change in RMR in untrained people who undergo training for a period of time. Some of these suggest that RMR might increase following training.[8] However, in a study of 40 men and women 17 to 62 years of age (HERITAGE Family Study), a 20-week aerobic training program (three times per week, 35-55 min per day, at 55-75% of $\dot{V}O_{2max}$) failed to increase RMR even though $\dot{V}O_{2max}$ increased by nearly 18%.[47] Because RMR is closely related to the fat-free mass of the body (fat-free tissue is more metabolically active), interest has increased in the use of resistance training to increase fat-free mass in an attempt to increase RMR.[8]

Exercise and the Thermic Effect of a Meal Several studies have examined the role of individual bouts of exercise and exercise training in increasing the TEM. A single bout of exercise, either before or after a meal, increases the thermic effect of that meal. Less clear is the role of exercise training on the TEM. Some studies have shown increases; others have shown decreases; and yet others have shown no effect at all. As with measuring changes in RMR accompanying exercise training, measurement of the TEM must be timed carefully with the last exercise bout. When measurements are made within 24 h of the last bout, the TEM is typically lower than it is three days afterward.[39]

Exercise and Mobilization of Body Fat During exercise, fatty acids are freed from their storage sites to be used for energy. Several studies suggest that human growth hormone may be responsible for this increased fatty acid mobilization. Growth hormone levels increase sharply with exercise and remain elevated for up to several hours in the recovery period. Other research has suggested that, with exercise, the adipose tissue is more sensitive to either the sympathetic nervous system or the increasing levels of circulating catecholamines. Either situation would increase lipid mobilization. More recent research suggests that this mobilization occurs in response to a specific fat-mobilizing substance that is highly responsive to elevated levels of activity. Thus, we cannot state with certainty which factors are of greatest importance in mediating this response.

Spot Reduction

Many people, including athletes, believe that exercising a specific area of the body will use the fat in that area, reducing the locally stored fat. Results of several early research studies tended to support this concept of **spot reduction**. But later research suggests that spot reduction is a myth and that exercise, even when localized, draws from almost all of the fat stores of the body, not just from local depots.

One such study used outstanding tennis players, theorizing that they would be ideal subjects for studying spot reduction because they could act as their own controls: Their dominant arms exercise vigorously for several hours every day, whereas their nondominant arms are relatively inactive.[19] Researchers postulated that if spot reduction is a reality, the nondominant (inactive) arm should have substantially more fat than the dominant (active) arm. The players' dominant arms had substantially greater girths attributable to exercise-

induced muscle hypertrophy. But the subcutaneous skinfold fat thicknesses in the active and inactive arms showed absolutely no differences.

Another study examined the localized effects of a 27-day intense sit-up training program. Researchers found no difference in the rate at which fat cell diameter changed in the abdomen, the subscapular region, and the gluteal region.[23] This indicates a lack of specific adaptation at the site of the exercise training (the abdomen). Researchers now theorize that fat is mobilized during exercise either mostly from those areas of highest concentration or equally from all areas, thus negating the spot-reduction theory. Decreases in girth can occur with exercise training, but these primarily result from increased muscle tone, not fat loss.

Low-Intensity Aerobics

As we have discussed in earlier chapters, the higher the exercise intensity, the greater the body's reliance on carbohydrate as an energy source. With high-intensity aerobic exercise, carbohydrate can supply up to 90% or more of the body's energy needs. During the late 1980s, various professional exercise groups promoted **low-intensity aerobic exercise** to increase the loss of body fat. These groups theorized that low-intensity aerobic training would allow the body to use more fat as the energy source, hastening the loss of body fat. Indeed, the body uses a higher percentage of fat for energy at lower exercise intensities. However, the total calories expended does not necessarily change as a result of the body's use of fat.

This is illustrated in table 22.5. In this hypothetical example, a 23-year-old woman with a $\dot{V}O_{2max}$ of 3.0 L/min exercises for 30 min at 50% of her $\dot{V}O_{2max}$ on one day and for 30 min at 75% of her $\dot{V}O_{2max}$ on another. The total calories from fat do not differ between the low- and high-intensity aerobic workouts: In both cases she burns about 110 kcal of fat during 30 min. Most important, however, for the higher-intensity workout she expends about 50% more total calories for the same time period.

TABLE 22.5 **Estimation of Kilocalories Used From Fat and Carbohydrate for a Low- and High-Intensity Aerobic Training Bout of 30 min Duration**

Exercise intensity	Average $\dot{V}O_2$ (L/min)	Average RER	% kcal CHO	% kcal fat	kcal for 30 min CHO	kcal for 30 min fat	kcal for 30 min total
Low, 50%	1.50	0.85	50	50	110	110	220
High, 75%	2.25	0.90	67	33	222	110	332

Note. RER = respiratory exchange ratio; CHO = carbohydrate. Subject was a fit but not highly trained 23-year-old woman ($\dot{V}O_{2max}$ = 3.0 L/min).

Scientists have determined that there is an optimal zone where rates of fat oxidation are at their highest. The Fat$_{max}$ zone, defined as that zone where fat oxidation rates are within 10% of the peak rate, was found to vary from between 55% and 72% of $\dot{V}O_{2max}$.[1] This is illustrated in figure 22.9.

FIGURE 22.9 Rate of fat oxidation at various exercise intensities, expressed as a percentage of $\dot{V}O_{2max}$.

Reprinted, by permission, from J. Acten, M. Gleeson, and A.E. Jeukendrup, 2002, "Determination of the exercise intensity that elicits maximal fat oxidation," *Medicine and Science in Sports and Exercise* 34: 92-97.

In focus

Low-intensity aerobic activity does not necessarily lead to a greater expenditure of calories from fat. More important, the total caloric expenditure for a given period of time is much less than with high-intensity aerobic activity.

Exercise Gadgets

We seldom get something for nothing. An effortless exercise program would be ideal, of course, but such a program would result in no significant changes in fitness, body composition, or physical dimensions. With the increasing popularity of exercise, many gimmicks and gadgets have appeared on the market. Some of these are legitimate and effective, but unfortunately many are of no practical value for either exercise conditioning or weight loss. Three such devices were evaluated to determine the legitimacy of their claims: the Mark II Bust Developer, the Astro-Trimmer Exercise Belt, and Slim-Skins Vacuum Pants. The first device, marketed to women, was claimed to add 2 to 3 in. (5-8 cm) to the bust within three to seven days, and the other two devices were claimed to take inches off of the abdomen, hips, buttocks, and thighs in a matter of minutes. All three devices failed to produce any changes whatsoever when evaluated in highly controlled scientific studies.[45, 46]

Those who are considering weight reduction often cringe at the thought of increasing their physical activity, and who wouldn't prefer immediate results over waiting for a payoff? But reality must be addressed. To gain the benefits from exercise, it is necessary to actually do the work!

Physical Activity and Health Risk Reduction

An important relationship was discovered during the 1990s that suggests another substantial benefit of fitness and activity. For those who are overweight or obese, their overall risk of death from disease is greatly reduced if they are physically active and fit.[2, 43] This is good news for those who seem destined to remain obese or overweight: An active lifestyle and moderate to high fitness levels can greatly reduce their risk of dying from chronic degenerative diseases, such as coronary artery disease and diabetes.

Obesity and Diabetes Risk Among Athletes

In the chapter opening, William "Refrigerator" Perry was used as an example of the possible health risks associated with being a big, oversized athlete. A study published in 2009 estimated the prevalence of the metabolic syndrome, insulin resistance, and their associated risk factors in a group of 90 NCAA Division I collegiate football players.[4] The prevalence of obesity, insulin resistance, and metabolic syndrome was 21%, 21%, and 9% respectively. The linemen included 19 of the 19 obese subjects, 13 of the 19 subjects with insulin resistance, and all subjects with the metabolic syndrome. The authors concluded that linemen are at a significant risk for metabolic syndrome and insulin resistance compared with other positions, largely related to their obesity!

> Inactivity is a major cause of obesity in the United States, perhaps as important as overeating.

> The energy expended by activity includes the steady-state rate of energy expenditure during the activity and also the energy expended after the exercise because the metabolic rate remains elevated for some time after the activity ends, a phenomenon known as excess postexercise oxygen consumption (EPOC).

> Diet alone causes fat loss, but fat-free mass is also lost. With exercise, either alone or with diet, fat is lost, but fat-free mass is either maintained or increased. Possibly the greatest benefit of physical activity and formal exercise is their role in attenuating the accumulation of visceral fat, or in reducing visceral fat stores.

> Simply being active, independent of a formal exercise program, is important in the prevention of obesity.

> Energy intake – energy excreted = RMR + TEM + TEA when one is in energy balance.

> A certain amount of activity appears to be needed for the body to precisely balance energy intake and expenditure.

> Research indicates that exercise can suppress appetite.

> Resting metabolic rate might increase slightly following training, and even a single bout of exercise increases the TEM.

> Exercise increases lipid mobilization from adipose tissue.

> Spot reduction is a myth.

> Low-intensity aerobic exercise burns no more fat than more vigorous exercise, and more total calories are spent in a more strenuous workout.

Diabetes

Diabetes mellitus, also called simply diabetes, is a disease characterized by high blood glucose levels (hyperglycemia) resulting from either an inadequate production of insulin by the pancreas, an inability of insulin to facilitate the transport of glucose into the cells, or both. Recall from chapter 4 that insulin is a hormone that reduces the amount of glucose circulating in the blood by facilitating its transport into the cells. We first address the terminology used in defining diabetes and disordered blood sugar control (glycemic control) and then look at the prevalence of diabetes in the United States.

Terminology and Classification

Historically, diabetes mellitus was classified into two major categories: juvenile-onset diabetes, or insulin-dependent diabetes mellitus (now known as type 1 diabetes), and adult-onset diabetes (now known as type 2 diabetes). This classification was based on the age of onset of diabetes.[28] Unfortunately, there has been an epidemic of type 2 diabetes in children, which largely can be attributed to the increased rates of obesity in children; thus use of the term "adult-onset" is no longer appropriate.

Type 1 diabetes is caused by the inability of the pancreas to produce sufficient insulin as a result of failure of the β-cells in the pancreas, which is the result of the body's immune system destroying pancreatic β-cells. Thus, this type has also been referred to as **insulin-dependent diabetes mellitus (IDDM)**. Type 1 diabetes accounts for only 5% to 10% of all cases of diabetes.

Type 2 diabetes is the result of the ineffectiveness of insulin to facilitate the transport of glucose into the cells and is a result of insulin resistance. It had previously been referred to as **non-insulin-dependent diabetes mellitus (NIDDM)**. Type 2 diabetes accounts for 90% to 95% of all cases of diabetes. **Insulin resistance** refers to the condition in which a "normal" insulin concentration in the blood produces a less than normal biological response. Insulin's primary function is to facilitate the transport of glucose from the blood into the cell, across the cell membrane. With insulin resistance, the body needs more insulin to transport a given amount of glucose across the cell membrane into the cell. **Insulin sensitivity** is a related term and provides an index of the effectiveness of a given insulin concentration in the blood. As insulin sensitivity increases, insulin resistance decreases.

A third type of diabetes, gestational diabetes, is a form of diabetes that develops in pregnant women and their fetuses in about 4% of all pregnancies. Fortunately, it usually disappears in both mother and baby after delivery. Unfortunately, when gestational diabetes is present, there can be complications during pregnancy.

Another category, **prediabetes**, refers to the condition in those who have impaired fasting glucose, impaired glucose tolerance, or both. Both type 1 and type 2 diabetes are diagnosed on the basis of a plasma glucose level of greater than 125 mg/dL following an 8 h fast. **Impaired fasting glucose** is defined as a plasma glucose level of between 100 and 125 mg/dL, again following an 8 h fast. **Impaired glucose tolerance** is determined by a glucose tolerance test. This involves drinking a solution in which 75 g of anhydrous glucose is dissolved in water. Plasma glucose levels are measured 2 h later. A glucose value of 200 mg/dL or higher is diagnostic of diabetes. Values of 140 to 199 mg/dL represent impaired glucose tolerance, and values below 140 mg/dL are considered normal.

For research studies, an intravenous glucose tolerance test (IVGTT) often is used. A catheter is placed in each arm, with the glucose solution injected into a vein in one arm, and blood samples are withdrawn from a vein in the other arm over the course of 3 h. Samples are taken more frequently during the first 15 to 45 min of the test and less frequently from 60 to 180 min. This allows one to develop a curve of both the glucose and insulin responses to the injected glucose load. The IVGTT is more precise than the oral glucose tolerance test.

There are also symptoms of diabetes that can be used to identify those at risk for diabetes. These include the following:

Type 1 Diabetes

- Frequent urination
- Excessive or unusual thirst
- Unusual, unexplained weight loss
- Extreme hunger
- Extreme fatigue and irritability

Type 2 Diabetes

- Any of the type 1 symptoms
- Frequent infections
- Blurred vision or sudden vision changes
- Tingling or numbness in hands or feet
- Cuts, bruises, or sores that are slow to heal
- Recurring skin, gum, or bladder infections

Prevalence of Diabetes

Approximately 17.9 million Americans have been diagnosed with diabetes. An estimated 5.7 million people likely have diabetes that has not been diagnosed, and an estimated 57 million people are prediabetic. The prevalence of diabetes increased from 4.9% of the U.S. population in 1990 to 7.0% in 2005—a 43% increase. Between 1990 and 1998, the largest increase (76%) occurred in 30- to 39-year-olds. Over 23% of people 60 years or older have diabetes. Further, the prevalence of diagnosed diabetes in those aged 20 years or older is 6.6% for non-Hispanic whites, 7.5% for Asian Americans, 11.8% for non-Hispanic blacks, and 10.4% for Hispanics. These racial differences in diabetes rates parallel the racial differences in obesity rates discussed earlier in this chapter.

The true prevalence of type 2 diabetes in children has not been established through national epidemiological studies. However, prevalence has been estimated to have increased by as much as 10-fold over the last 20 years. Furthermore, in 10- to 19-year-olds, studies have shown that type 2 diabetes accounted for between 33% and 46% of all diabetes.[28] This is very disturbing, considering that not too long ago we referred to type 2 diabetes as adult-onset diabetes.

Etiology of Diabetes

Heredity appears to play a major role in both type 1 and type 2 diabetes. With type 1 diabetes, the β-cells (insulin-secreting cells) of the pancreas are destroyed. This destruction may be caused by the body's immune system, increased β-cell susceptibility to viruses, or β-cell degeneration.

Type 1 diabetes generally has a sudden onset during childhood or young adulthood. This leads to almost total insulin deficiency, and daily injections of insulin usually are required to control the disease.

The Diabetes Prevention Program Study

During the mid-1990s, the National Institute of Diabetes and Digestive and Kidney Diseases (NIDDK) of the National Institutes of Health designed and initiated a study to determine whether either a lifestyle intervention (diet and exercise) or an oral diabetes drug (metformin) could prevent or delay the onset of type 2 diabetes in people with impaired glucose tolerance (prediabetics). A total of 3,234 persons 25 years of age and older were randomly assigned to placebo, drug, or lifestyle groups. The goals for the lifestyle modification program were to lose at least 7% of body weight and to participate in physical activity for at least 150 min per week. Both interventions proved very successful, delaying the development of diabetes by 11 years in the lifestyle group and by three years in the metformin group.[26] This landmark study clearly illustrates the power of lifestyle intervention in reducing the risk of a major debilitating disease and its serious consequences.

In type 2 diabetes, the onset of the disease is more gradual, and the causes are more difficult to establish. Type 2 diabetes often is characterized by any of the following three major metabolic abnormalities: delayed or impaired insulin secretion; impaired insulin action (insulin resistance) in the insulin-responsive tissues of the body, including muscle; or excessive glucose output from the liver.

Obesity plays a major role in the development of type 2 diabetes. With obesity, the β-cells of the pancreas often become less responsive to the stimulation of increased blood glucose concentrations. Furthermore, the target cells throughout the body, including muscle, often undergo a reduction in the number or activation of their insulin receptors, so the insulin in the blood is less effective in transporting glucose into the cell.

Health Problems Linked With Diabetes

Considerable health risks are associated with diabetes. People with this disease have a relatively high mortality rate. Diabetes places a person at increased risk for[10]

- coronary artery or heart disease;
- cerebrovascular disease and stroke;
- hypertension;
- peripheral vascular disease;
- kidney disease;
- nervous system disease;
- eye disorders, including blindness;
- dental disease;
- amputations; and
- complications during pregnancy.

During the late 1980s, scientists made the important associations among coronary artery disease, hypertension, obesity, and type 2 diabetes. **Hyperinsulinemia** (high blood levels of insulin) and insulin resistance appear to be the important threads linking these disorders, possibly through insulin-mediated sympathetic

(In)focus

Diabetes increases the risk of coronary artery disease, cerebrovascular disease and stroke, hypertension, and peripheral vascular disease. Coronary artery disease, hypertension, obesity, and diabetes may be linked through the common pathway of increased insulin levels in the blood or to target cells' becoming insulin resistant. However, obesity appears to be the trigger.

nervous system stimulation (increased insulin levels cause increased sympathetic nervous system activity).[12] Again, obesity seems to be the trigger setting off this reaction.

General Treatment of Diabetes

The major modes of treatment for type 1 diabetes are insulin administration, diet, and exercise. The dosage of insulin is adjusted to allow normal carbohydrate, fat, and protein metabolism. The type of insulin injected— short acting or intermediate acting—and the time of day at which the injections are administered are also individualized to maintain glycemic control throughout the day.

For type 2 diabetes, the focus has traditionally been on three factors: weight loss, diet, and exercise. However, during the mid to late 1990s, a number of new drugs were introduced that effectively treat type 2 diabetes. Two types of drugs have been particularly effective: sulfonylureas, to lower blood sugar levels, and biguanides, to reduce hepatic glucose production. Metformin, a biguanide, has been particularly successful in obese patients.

A well-balanced diet generally is prescribed for people with diabetes. In the past, patients were prescribed a low-carbohydrate diet to better control blood sugar levels. However, a low-carbohydrate diet necessitates an increase in dietary fat, which can have a major negative effect on blood lipid levels. Because people with diabetes are already at greater risk for coronary artery disease, this is not desirable. Maintenance of proper blood sugar levels is difficult in patients with obesity, so a reduced-calorie diet is necessary for them to achieve major body fat losses. For many people with type 2 diabetes, weight loss alone can bring blood sugar levels back into the normal range. This can be the most important aspect of the treatment plan for the overweight or obese person with diabetes.

Role of Physical Activity in Diabetes

Indirect scientific evidence has clearly established that a physically active lifestyle reduces the risk of developing type 2 diabetes,[3] but there is little evidence to support this for type 1 diabetes. However, most physicians and scientists agree that physical activity is an important part of the treatment plan for either type of diabetes. Because there is such a disparity between the characteristics and responses of those with type 1 and type 2 diabetes, we discuss each separately.

Type 1 Diabetes

The role of regular exercise and physical training in improving glycemic control (regulation of blood sugar levels) in patients with type 1 diabetes has not been clearly defined and is controversial. The most distinguishing feature differentiating type 1 and type 2 diabetes is that those with type 1 have low blood insulin levels attributable to the inability or reduced ability of the pancreas to produce insulin. Those with type 1 diabetes are prone to hypoglycemia (low blood sugar levels) during and immediately after exercise because the liver fails to release glucose at a rate that can keep up with glucose utilization. For these people, exercise can lead to excessive swings in plasma glucose levels that are unacceptable for the management of the disease. The degree of glycemic control during exercise varies tremendously among individuals with type 1 diabetes. As a result, exercise and exercise training can improve glycemic control in some patients, mainly those who are less prone to hypoglycemia, but not in others.[41]

Although glycemic control is generally not improved by exercise in most people with type 1 diabetes, there are other potential benefits of exercise for these patients. Because these patients have a two to three times greater risk for coronary artery disease, exercise may help reduce this risk. Exercise may also help reduce the risk for cerebrovascular and peripheral arterial diseases.

People with uncomplicated type 1 diabetes do not have to restrict physical activity, providing that blood sugar levels are controlled appropriately. A number of athletes who have type 1 diabetes have trained and competed successfully. Monitoring blood sugar levels in an exercising person with type 1 diabetes is important so that diet and insulin dosages can be altered accordingly. This is particularly important for those who are competing at high intensities or for extended periods of time.

Special attention also should be given to the feet of people with diabetes, as it is common for them to experience peripheral neuropathy (diseased nerves) with some loss of sensation in the feet. Peripheral vascular disease is also more common in patients with diabetes, so the circulation to the extremities, especially the feet, often is impaired significantly. Ulcerations and other lesions on the feet account for more than half of all hospitalizations of diabetic patients.[11] Because weight-bearing exercise places additional stress on the feet, the proper selection of footwear and appropriate preventive foot care are important.

Type 2 Diabetes

Exercise plays a major role in glycemic control for people with type 2 diabetes. Insulin production is generally not of concern in this group, particularly during the early stages of the disease, so the major problem with this form of diabetes is the lack of target cell response to insulin (insulin resistance). Because the cells become resistant to insulin, the hormone cannot perform its function of facilitating glucose transport across the cell membrane, resulting in decreased insulin sensitivity. Muscle contraction has an insulin-like effect.[22] Membrane permeability to glucose increases with muscular contraction, likely attributable to an increase in the number of GLUT-4 glucose transporters associated with the plasma membrane.[21] Thus, acute bouts of exercise decrease insulin resistance and increase insulin sensitivity. This reduces the cells' requirements for insulin, which means that people taking insulin must reduce their dosages. Resistance and aerobic training appear to produce similar effects,[49] although some evidence suggests that combining resistance and aerobic exercise is the optimal exercise strategy for reducing insulin resistance.[13] This decrease in insulin resistance and increase in insulin sensitivity may primarily be a response to each individual bout of exercise rather than the result of a long-term change associated with training. Some studies have shown that the effect dissipates within 72 h.

In focus

Physical activity has many desirable effects for people with diabetes, particularly those with type 2 diabetes. Glycemic control is improved, primarily in people with type 2 diabetes, possibly attributable to the insulin-like effect of muscle contraction on transporting glucose from the plasma into the cell.

➤ Diabetes is a disorder of carbohydrate metabolism characterized by hyperglycemia. It develops from inadequate insulin secretion or utilization.

➤ Type 1 diabetes involves destruction of the β-cells in the pancreas, and it typically has a sudden, early onset. Type 2 diabetes typically involves impaired insulin secretion or action, excessive liver glucose output, or a combination of these.

➤ Major modes of treatment for diabetes are drugs (if needed), diet, and exercise.

➤ In people with type 1 diabetes, glycemic control might or might not be improved with exercise. But these people have a greater risk for coronary artery disease, and exercise certainly can decrease that risk.

➤ Blood glucose levels must be carefully monitored during exercise, particularly in people with type 1 diabetes, so that diet and insulin dosage can be altered as needed.

➤ The feet of people with type 1 diabetes deserve special attention, because peripheral neuropathy causes loss of sensation, and impaired peripheral circulation decreases blood flow. These people might not be aware of injuries to their feet, but these injuries can be very serious.

➤ Type 2 diabetes responds well to exercise. Membrane permeability to glucose improves with exercise, likely associated with an increase in GLUT-4 receptors, which decreases the person's insulin resistance and increases insulin sensitivity.

In these last two chapters, we have concluded our look at the role of physical activity in preventing and treating coronary artery disease, hypertension, obesity, and diabetes. We have seen that exercise can decrease individual risk and also can be an integral part of treatment, improving overall health as well as alleviating some symptoms.

With this chapter, we also conclude our journey to an understanding of exercise and sport physiology. We began this book by reviewing how various body systems function during exercise and how they respond to chronic training. We saw how physical activity and performance are affected by the environment, such as extremes of heat, cold, and barometric pressure. We turned our attention to ways in which athletes can, or can attempt to, optimize their performance. Then we considered the unique differences among older and younger participants and male and female participants in sport and exercise. And finally, we examined the role of exercise in the maintenance of health and the development of fitness.

It has been a long journey from cover to cover, but we hope that you will close this book with a new appreciation of physical activity. Perhaps you will leave this book with a new awareness of how your body performs physical activity. Maybe, if you have not yet done so, you will feel compelled to commit to a personal exercise program. And we hope that you now feel some excitement about exercise and sport physiology, realizing that these areas of study affect so many aspects of our lives.

Key Terms

β-cells

body mass index (BMI)

diabetes mellitus

hyperinsulinemia

impaired fasting glucose

impaired glucose tolerance

insulin-dependent diabetes mellitus (IDDM)

insulin resistance

insulin sensitivity

lower body (gynoid) obesity

low-intensity aerobic exercise

non-insulin-dependent diabetes mellitus (NIDDM)

obesity

overweight

prediabetes

spot reduction

thermic effect of activity (TEA)

thermic effect of a meal (TEM)

type 1 diabetes

type 2 diabetes

upper body (android) obesity

Study Questions

1. What is the difference between overweight and obesity?
2. What is ideal body weight, and how is it determined?
3. What is body mass index? What is its significance?
4. What is the prevalence of obesity in the United States today? Is there a difference in the prevalence between men and women? Children and adults? Blacks and whites?
5. What are several health-related problems associated with obesity?
6. What is the association among obesity, coronary artery disease, hypertension, insulin resistance, and diabetes?
7. Describe several methods for treating obesity. Which are the most effective?
8. What role does exercise play in the prevention and treatment of obesity?
9. By what mechanisms might exercise effect losses in total weight and fat weight?
10. How effective is spot reduction? Low-intensity aerobic exercise?
11. Describe the two major types of diabetes. How are they caused?
12. What health risks are associated with diabetes?
13. Describe the role of exercise in preventing diabetes and in treating patients with type 1 and type 2 diabetes.

Study Guide Activities

In addition to the activities listed in the chapter opening outline on page 545, two other activities are available in the web study guide, located at

www.HumanKinetics.com/PhysiologyOfSportAndExercise

 The **KEY TERMS** activity reviews important terms, and the end-of-chapter **QUIZ** tests your understanding of the material covered in the chapter.

1-repetition maximum (1RM)—The maximal amount of weight that can be lifted just one time.

1RM—*See* 1-repetition maximum.

acclimation—Physiological adaptation to repeated environmental stresses, occurring over a relatively brief period of time (days to weeks). Acclimation often occurs in a laboratory environment.

acclimatization—Physiological adaptation to repeated environmental stress in a natural environment, occurring over months and years of living and exercising in that environment.

ACE—*See* angiotensin converting enzyme.

acetylcholine—A primary neurotransmitter that transmits impulses across the synaptic cleft.

acetyl CoA—*See* acetyl coenzyme A.

acetyl coenzyme A (acetyl CoA)—The compound that forms the common entry point into the Krebs cycle for the oxidation of carbohydrate and fat.

actin—A thin protein filament that acts with myosin filaments to produce muscle action.

action potential—A rapid and substantial depolarization of the membrane of a neuron or muscle cell that is conducted through the cell.

activation energy—The initial energy required to start a chemical reaction or chain of reactions.

acute altitude (mountain) sickness—Illness characterized by headache, nausea, vomiting, dyspnea, and insomnia. It typically begins 6 to 96 h after one reaches high altitude and lasts several days.

acute exercise—A single bout of exercise.

acute muscle soreness—Soreness or pain felt during and immediately after an exercise bout.

acute overload—an "average" training load, whereby the athlete is stressing the body to the extent necessary to improve both physiological function and performance.

adenosine diphosphate (ADP)—A high-energy phosphate compound from which ATP is formed.

adenosine triphosphatase (ATPase)—An enzyme that splits the last phosphate group off ATP, releasing a large amount of energy and reducing the ATP to ADP and P_i.

adenosine triphosphate (ATP)—A high-energy phosphate compound from which the body derives its energy.

ADH—*See* antidiuretic hormone.

adolescence—The period of life between the end of childhood and the beginning of adulthood. The onset of puberty marks the beginning of adolescence.

ADP—*See* adenosine diphosphate.

adrenaline—A chemical compound that serves as a neurotransmitter throughout the body. Also called epinephrine.

adrenergic—Refers to norepinephrine or epinephrine (also called noradrenaline and adrenaline, respectively).

aerobic capacity—*See* maximal oxygen uptake.

aerobic endurance—*See* cardiorespiratory endurance.

aerobic metabolism—A process occurring in the mitochondria that uses oxygen to produce energy (ATP). Also known as cellular respiration.

aerobic power—Another name for maximal oxygen uptake, or $\dot{V}O_{2max}$.

aerobic training—Training that improves the efficiency of the aerobic energy-producing systems and can improve cardiorespiratory endurance.

afferent nerves—The sensory division of the peripheral nervous system, carrying impulses toward the central nervous system.

afterload—The pressure against which the heart must pump blood, determined by the peripheral resistance in the large arteries.

air plethysmography—A procedure for assessing body composition by using air displacement to measure body volume, allowing the calculation of body density.

aldosterone—A mineralocorticoid hormone secreted by the adrenal cortex that prevents dehydration by promoting renal absorption of sodium.

alveolar capillary membrane—*See* respiratory membrane.

alveolus—Terminal air sac at the end of the bronchial tree in the lungs, where gas exchange takes place with the capillaries (plural, alveoli).

amenorrhea—The absence (primary amenorrhea) or cessation (secondary amenorrhea) of normal menstrual function.

amino acids—The chief components of proteins that are synthesized by living cells or are obtained by the body through the diet.

α-motor neuron—A neuron innervating extrafusal skeletal muscle fibers.

amphetamine—A central nervous system stimulant proposed to have ergogenic properties.

anabolic steroids—Prescription drugs with the anabolic (growth stimulating) characteristics of testosterone, taken by some athletes to increase body size, muscle mass, and strength.

anabolism—The building up of body tissue; the constructive phase of metabolism.

anaerobic—In the absence of oxygen.

anaerobic glycolysis—The anaerobic breakdown of glucose to lactic acid in the production of energy (ATP).

anaerobic metabolism—The production of energy (ATP) in the absence of oxygen.

anaerobic power—Mean or peak power output in exercise lasting 30 s or less.

anaerobic threshold—The point at which the metabolic demands of exercise can no longer be met by available aerobic sources and at which an increase in anaerobic metabolism occurs, reflected by an increase in blood lactate concentration.

anaerobic training—Training that improves the efficiency of the anaerobic energy-producing systems and can increase muscular strength and tolerance for acid–base imbalances during high-intensity effort.

angiotensin converting enzyme (ACE)—An enzyme that converts angiotensin I to angiotensin II.

anorexia nervosa—A clinical eating disorder characterized by distorted body image, intense fear of fatness or weight gain, amenorrhea, and refusal to maintain more than the minimal normal weight based on age and height.

antidiuretic hormone (ADH)—A hormone secreted by the pituitary gland that regulates fluid and electrolyte balance in the blood by reducing urine production.

arginine vasopressin—See antidiuretic hormone.

arterial–mixed venous oxygen difference, or a-v̄O₂ difference—The difference in oxygen content between arterial and mixed venous blood, which reflects the amount of oxygen removed by the whole body.

arterial–venous oxygen difference, or (a-v)O₂ difference—The difference in oxygen content between arterial and venous blood at the tissue level.

arteries—Blood vessels that transport blood away from the heart.

arterioles—The smallest arteries that transport blood from larger arteries to the capillaries.

arteriosclerosis—A condition that involves loss of elasticity, thickening, and hardening of the arteries.

atherosclerosis—A form of arteriosclerosis that involves changes in the lining of the arteries and plaque accumulation, leading to progressive narrowing of the arteries.

athlete's heart—A nonpathological enlarged heart, often found in endurance athletes, that results primarily from left ventricular hypertrophy in response to training.

ATP—See adenosine triphosphate.

ATPase—See adenosine triphosphatase.

ATP-PCr system—The short-term anaerobic energy system that maintains ATP levels. Breakdown of phosphocreatine (PCr) frees P_i, which then combines with ADP to form ATP.

atrioventricular (AV) node—The specialized mass of conducting cells in the heart located at the atrioventricular junction.

atrophy—Loss of size, or mass, of body tissue, such as muscle atrophy with disuse.

autogenic inhibition—Reflex inhibition of a motor neuron in response to excessive tension in the muscle fibers it supplies, as monitored by the Golgi tendon organs.

autoregulation—Local control of blood distribution (through vasodilation) in response to a tissue's changing needs.

AV node—See atrioventricular node.

(a-v)O₂ difference—See arterial–venous oxygen difference.

(a-v̄)O₂ difference—See arterial–mixed venous oxygen difference.

axon hillock—A part of the neuron, between the cell body and the axon, that controls traffic down the axon through summation of excitatory and inhibitory postsynaptic potentials.

axon terminal—One of numerous branched endings of an axon. Also known as a terminal fibril.

barometric pressure—The total pressure exerted by the atmosphere at a given altitude.

baroreceptor—Stretch receptor located within the cardiovascular system that senses changes in blood pressure.

basal metabolic rate (BMR)—The lowest rate of body metabolism (energy use) that can sustain life, measured after an overnight sleep in a laboratory under optimal conditions of quiet, rest, and relaxation and after a 12 h fast. See also resting metabolic rate.

β-blockers—A class of drugs that block transmission of neural impulses from the sympathetic nervous system, proposed to have ergogenic properties.

BCAA—*See* branched-chain amino acids.

β-cells—Cells in the islets of Langerhans in the pancreas that secrete insulin.

bicarbonate loading—Ingesting bicarbonate to elevate blood pH with hopes of delaying fatigue by increasing the capacity to buffer acids.

bioelectric impedance—A procedure for assessing body composition in which an electrical current is passed through the body. The resistance to current flow through the tissues reflects the relative amount of fat present.

bioenergetics—Term given to the study of metabolic processes that yield or consume energy.

blood doping—Any means by which a person's total volume of red blood cells is increased, typically via transfusion of red blood cells or use of erythropoietin.

blood lipids—Bloodborne fats, such as triglycerides and cholesterol.

BMI—*See* body mass index.

BMR—*See* basal metabolic rate.

body composition—The chemical composition of the body. The model used in this book considers two components: fat-free mass and fat mass.

body density (D_{body})—Body weight divided by body volume.

body mass index (BMI)—A measurement of body overweight or obesity determined by dividing weight (in kilograms) by height (in meters) squared. BMI is highly correlated with body composition.

bone mineral density—The mass of bone per unit volume. Decreased bone mineral density increases the risk of fractures.

Borg RPE scale—A numerical scale for rating perceived exertion.

β-oxidation—The first step in fatty acid oxidation, in which fatty acids are broken into separate two carbon units of acetic acid, each of which is then converted to acetyl CoA.

Boyle's gas law—Law stating that at a constant temperature, the number of gas molecules in a given volume depends on the pressure.

bradycardia—A resting heart rate lower than 60 beats/min.

branched-chain amino acids (BCAA)—Specific amino acids—leucine, isoleucine, and valine—that have been postulated to work in combination with L-tryptophan to delay fatigue, primarily through central nervous system mechanisms.

buffer—A substance that combines with either an acid or a base to maintain a constant acid–base (pH) balance.

bulimia nervosa—A clinical eating disorder characterized by recurrent episodes of binge eating, a feeling of lack of control during these binges, and purging behavior, which may include self-induced vomiting and use of laxatives and diuretics. Sometimes the disorder also includes fasting or excessive exercise behaviors.

CAD—*See* coronary artery disease.

caffeine—A central nervous system stimulant believed by some athletes to have ergogenic properties.

calcitonin—A hormone secreted by the thyroid gland that assists in the control of calcium ion concentrations in the blood.

calorie (cal)—A unit of measure of energy in biological systems, where 1.0 calorie is equal to the amount of heat energy needed to raise the temperature of 1.0 g of water 1 °C, from 15 to 16 °C.

calorimeter—A device for measuring the heat produced by the body (or by specific chemical reactions).

cAMP—*See* cyclic adenosine monophosphate.

capillaries—The smallest vessels transporting blood from the heart to the tissues and the actual sites of exchange between the blood and tissue.

capillary-to-fiber ratio—The number of capillaries per muscle fiber.

carbohydrate—An organic compound formed from carbon, hydrogen, and oxygen; includes starches, sugars, and cellulose.

carbohydrate loading—Increased dietary consumption of carbohydrates, a process used by athletes to increase carbohydrate stores in the body prior to prolonged endurance exercise.

cardiac cycle—The period that includes all events between two consecutive heartbeats.

cardiac hypertrophy—Enlargement of the heart by increases in muscle wall thickness or chamber size or both.

cardiac output ($\dot{Q}$)—The volume of blood pumped out by the heart per minute. $\dot{Q}$ = heart rate × stroke volume.

cardiorespiratory endurance—The ability of the body to sustain prolonged exercise.

cardiovascular deconditioning—A decrease in the cardiovascular system's ability to deliver sufficient oxygen and nutrients.

cardiovascular drift—An increase in heart rate during exercise to compensate for a decrease in stroke volume. This compensation helps maintain a constant cardiac output.

cardiovascular endurance training—*See* aerobic training.

catabolism—The tearing down of body tissue; the destructive phase of metabolism.

catecholamines—Biologically active amines (organic compounds derived from ammonia), such as epinephrine and norepinephrine, that have powerful effects similar to those of the sympathetic nervous system.

central command—Information originating in the brain that is transmitted to the cardiovascular, muscular, or pulmonary systems.

central nervous system (CNS)—System consisting of the brain and spinal cord.

chemoreceptor—A sensory organ capable of reacting to a chemical stimulus.

Cheyne-Stokes breathing—Alternating periods of rapid breathing and slow, shallow breathing including periods in which breathing may actually cease temporarily. A symptom of acute mountain sickness.

childhood—The period of life between the first birthday and the onset of puberty.

chronic adaptation—A physiological change that occurs when the body is exposed to repeated exercise bouts over weeks or months. These changes generally improve the body's efficiency at rest and during exercise.

chronic fatigue syndrome—A syndrome that appears to involve immune system dysfunction. Patients have incapacitating fatigue, sore throat, muscle tenderness or pain, and cognitive dysfunction; the symptoms may vary in severity over time but generally last for months or years.

chronic hypertrophy—An increase in muscle size that results from repeated long-term resistance training.

CK—*See* creatine kinase.

CNS—*See* central nervous system.

cold acclimatization—*See* acclimatization.

cold habituation—A response to repeated cold exposure, often to the hands and face, in which skin vasoconstrictor and shivering responses are blunted.

concentric contraction—Muscle shortening.

conduction—(1) Transfer of heat through direct molecular contact with a solid object; (2) movement of an electrical impulse, such as through a neuron.

congenital heart disease—A heart defect present at birth that occurs from abnormal prenatal development of the heart or associated blood vessels. Also known as congenital heart defect.

continuous training—Training at a moderate to high intensity without stopping to rest.

contractile velocity (V_o)—The speed of action associated with specific muscle fiber types.

control group—In an experimental design, the nontreated group to which the experimental group is compared.

convection—The transfer of heat or cold via the movement of a gas or liquid across an object, such as the body.

coronary artery disease (CAD)—Progressive narrowing of the coronary arteries.

coronary heart disease (CHD)—A disease characterized by pathological changes in the coronary arteries that supply blood flow to the myocardium.

cortisol—A corticosteroid hormone released from the adrenal cortex that stimulates gluconeogenesis, increases mobilization of free fatty acids, decreases use of glucose, and stimulates catabolism of protein. Also known as hydrocortisone.

creatine—A substance found in skeletal muscles most commonly in the form of PCr. Creatine supplements are often used as ergogenic aids because they are theorized to increase PCr levels, thus enhancing the ATP-PCr energy system by better maintaining muscle ATP levels.

creatine kinase (CK)—The enzyme that facilitates the breakdown of PCr to creatine and P_i.

critical temperature theory—Theory that prolonged exercise in hot environments is limited by attainment of a fixed elevated core temperature.

crossover design—Experimental design in which the control group becomes the experimental group after the first experimental period, and vice versa.

cross-sectional research design—A research design in which a cross section of a population is tested at one specific time and then data from groups within that population are compared.

cross-training—Training for more than one sport at the same time, or training multiple fitness components (such as endurance, strength, and flexibility) within the same period.

cycle ergometer—An exercise device that uses cycling to measure physical work.

cyclic adenosine monophosphate (cAMP)—Intracellular second messenger that mediates hormone action.

cytochrome—A series of iron-containing proteins that facilitate the transport of electrons within the electron transport chain.

Dalton's Law of Partial Pressures—A principle that states that the total pressure exerted by a mixture of gases is equal to the sum of the partial pressures of those individual gases.

DBP—*See* diastolic blood pressure.

dehydration—Loss of body fluids.

delayed-onset muscle soreness (DOMS)—Muscle soreness that develops a day or two after a heavy bout of exercise and that is associated with actual injury within the muscle.

densitometry—The measurement of body density.

dependent variable—The physiological factor that is allowed to vary as another factor (the independent variable) is manipulated. Usually plotted on the *y*-axis.

depolarization—A decrease in the electrical potential across a membrane, as when the inside of a neuron becomes less negative relative to the outside.

detraining—Changes in physiological function in response to a reduction or cessation of regular physical training.

development—Changes that occur in the body starting at conception and continuing through adulthood; differentiation along specialized lines of function, reflecting changes that accompany growth.

diabetes mellitus—A disorder of carbohydrate metabolism characterized by hyperglycemia (high blood sugar levels) and glycosuria (presence of sugar in the urine). The disease develops when there is inadequate production of insulin by the pancreas or inadequate utilization of insulin by the cells.

diastolic blood pressure (DBP)—The lowest arterial pressure, resulting from ventricular diastole (the resting phase).

direct calorimetry—A method that gauges the body's rate and quantity of energy production by direct measurement of the body's heat production.

direct gene activation—The method of action of steroid hormones. They bind to receptors in the cell, and then the hormone–receptor complex enters the nucleus and activates certain genes.

disordered eating—Abnormal eating behavior that ranges from excessive restriction of food intake to pathological behaviors, such as self-induced vomiting and laxative abuse. Disordered eating can lead to clinical eating disorders, such as anorexia nervosa and bulimia nervosa.

diuretics—Substances that promote water excretion.

diurnal variation—Fluctuations in physiological responses that occur during a 24 h period.

DOMS—*See* delayed-onset muscle soreness.

dose–response relation—A relationship between two variables in which one changes predictably as the other increases or decreases.

downregulation—Decreased cellular sensitivity to a hormone, likely the result of a decreased number of cell receptors available to bind with the hormone.

dry heat exchange—Heat transfer by the combined avenues of convection, conduction, and radiation.

dual-energy X-ray absorptiometry (DEXA)—A technique used to assess both regional and total-body composition through the use of X-ray absorptiometry.

dynamic contraction—Any muscle action that produces joint movement.

dyspnea—Labored or difficult breathing.

eating disorders—A group of clinical disorders involving eating. *See* anorexia nervosa, bulimia nervosa.

eccentric contraction—Any muscle action in which muscle lengthens.

eccentric training—Training that involves eccentric action.

eccrine sweat glands—Simple sweat glands dispersed over the body surface that respond to increases in core or skin temperature (or both) and facilitate thermoregulation.

ECG—*See* electrocardiogram and exercise electrocardiogram.

EDV—*See* end-diastolic volume.

EF—*See* ejection fraction.

efferent nerves—The motor division of the peripheral nervous system, carrying impulses from the CNS toward the periphery.

EIAH—*See* exercise-induced arterial hypoxemia.

ejection fraction (EF)—The fraction of blood pumped out of the left ventricle with each contraction, determined by dividing stroke volume by end-diastolic volume and expressed as a percentage.

electrical stimulation training—Stimulation of a muscle via passing an electrical current through it.

electrocardiogram (ECG)—A recording of the heart's electrical activity.

electrocardiograph—A machine used to obtain an electrocardiogram.

electrolyte—A dissolved substance that can conduct an electrical current.

electron transport chain—A series of chemical reactions that convert the hydrogen ion generated by glycolysis and the Krebs cycle into water and produce energy for oxidative phosphorylation.

end branches—Branches coming off the ends of the axons leading to the axon terminals.

end-diastolic volume (EDV)—The volume of blood inside the left ventricle at the end of diastole, just before contraction.

endomysium—A sheath of connective tissue that covers each muscle fiber.

endothelial dysfunction—Negative changes in the cells that line the lumen of blood vessels resulting in a relative inability of those vessels to constrict or dilate.

endothelium—Layer of thin cells that line the lumen of the blood vessels.

end-systolic volume (ESV)—The volume of blood remaining in the left ventricle at the end of systole, just after contraction.

endurance—The ability to resist fatigue; includes muscular endurance and cardiorespiratory endurance.

energy—The capability of producing force, performing work, or generating heat.

environmental physiology—Study of the effects of the environment (heat, cold, altitude, hyperbaria, etc.) on the function of the body.

enzyme—Protein molecules that speed up reactions by lowering their energy of activation.

ephedrine—a sympathomimetic amine derived from ephedra herbs (also known as ma huang) and is used as a decongestant and as a bronchodilator in the treatment of asthma.

epimysium—The outer connective tissue that surrounds an entire muscle, holding it together.

epinephrine—A catecholamine released from the adrenal medulla that, along with norepinephrine, prepares the body for a fight-or-flight response. It is also a neurotransmitter. *See* catecholamines.

EPOC—*See* excess postexercise oxygen consumption.

EPSP—*See* excitatory postsynaptic potential.

ergogenic—Able to improve work or performance.

ergogenic aid—A substance or phenomenon that can improve work or athletic performance.

ergolytic—Able to impair work or performance.

ergometer—An exercise device that allows the amount and rate of a person's physical work to be controlled (standardized) and measured.

erythropoietin (EPO)—The hormone that stimulates erythrocyte (red blood cell) production.

essential amino acids—The eight or nine amino acids necessary for human growth that the body cannot synthesize and are thus essential parts of our diets.

estrogen—A female sex hormone.

ESV—*See* end-systolic volume.

eumenorrhea—Normal menstrual function.

evaporation—Heat loss through the conversion of water (such as in sweat) to vapor.

excessive training—Training in which volume, intensity, or both are too great or are increased too quickly without proper progression.

excess postexercise oxygen consumption (EPOC)—Elevated oxygen consumption above resting levels after exercise; at one time referred to as oxygen debt.

excitation-contraction coupling—The sequence of events by which a nerve impulse reaches the muscle membrane and leads to cross-bridge activity and thus muscle contraction.

excitatory postsynaptic potential (EPSP)—A depolarization of the postsynaptic membrane caused by an excitatory impulse.

exercise-associated muscle cramps (EAMCs)—Painful prolonged contractions of muscles that accompany or result from muscle contractions.

exercise electrocardiogram (ECG)—A recording of the heart's electrical activity during exercise.

exercise-induced arterial hypoxemia (EIAH)—A decline in arterial PO_2 and arterial oxygen saturation during maximal or near-maximal exercise.

exercise physiology—The study of how body structure and function are altered by exposure to acute and chronic bouts of exercise.

exercise prescription—Individualization of the prescription of exercise duration, frequency, intensity, and mode.

exhaustion—Inability to continue exercise.

expiration—The process by which air is forced out of the lungs through relaxation of the inspiratory muscles and elastic recoil of the lung tissue, which increase the pressure in the thorax.

external respiration—The process of bringing air into the lungs and the resulting exchange of gas between the alveoli and the capillary blood.

extracellular fluid—The 35% to 40% of the water in the body that is outside the cells, including interstitial fluid, blood plasma, lymph, cerebrospinal fluid, and other fluids.

extrinsic neural control—Redistribution of blood at the system or body level through neural mechanisms.

Fartlek training—Developed in the 1930s; the term comes from the Swedish for "speed play." This type of training combines continuous and interval training and stresses both the aerobic and anaerobic energy pathways.

fasciculus—A small bundle of muscle fibers wrapped in a connective tissue sheath within a muscle.

fat—A class of organic compounds with limited water solubility that exists in the body in many forms, such as triglycerides, free fatty acids, phospholipids, and steroids.

fat-free mass—The mass (weight) of the body that is not fat, including muscle, bone, skin, and organs.

fatigue—General sensations of tiredness and accompanying decrements in muscular performance.

fat mass—The absolute amount or mass of body fat.

fatty streaks—Early lipid deposits within blood vessels.

female athlete triad—Three interrelated disorders—disordered eating, menstrual dysfunction, and bone mineral disorders—to which some female athletes are prone.

$FEV_{1.0}$—*See* forced expiratory volume in 1 s.

FFA—*See* free fatty acids.

fiber hyperplasia—An increase in the number of muscle fibers.

fiber hypertrophy—An increase in the size of existing individual muscle fibers.

fibromyalgia syndrome—A chronic syndrome that includes muscle pain as its dominant symptom but is also characterized by muscle weakness, migraine type headaches, and depression.

Fick equation—$\dot{V}O_2 = \dot{Q} \times (a\text{-}\bar{v})O_2$ difference.

Fick's law—Law stating that the net diffusion rate of a gas across a fluid membrane is proportional to the difference in partial pressure, proportional to the area of the membrane, and inversely proportional to the thickness of the membrane.

force—Strength or energy exerted or brought to bear.

forced expiratory volume in 1 s ($FEV_{1.0}$)—The volume of air exhaled in the first second after maximal inhalation.

Frank-Starling mechanism—The mechanism by which an increased amount of blood in the ventricle causes a stronger ventricular contraction to increase the amount of blood ejected.

free fatty acids (FFA)—The components of fat that are used by the body for metabolism.

free radicals—Univalent (unpaired) oxygen intermediates that leak out of the electron transport chain during metabolic processes and may damage tissues.

free weights—Traditional resistance training modality that uses only barbells, dumbbells, and so on to provide resistance.

frostbite—Tissue damage that occurs during cold exposure because circulation to the skin decreases, in an attempt to retain body heat, to the point that the tissue receives insufficient oxygen and nutrients.

gastric emptying—The movement of food mixed with gastric secretions from the stomach into the duodenum.

glucagon—A hormone released by the pancreas that promotes increased breakdown of liver glycogen to glucose (glycogenolysis) and increased gluconeogenesis.

glucocorticoid—A family of steroid hormones produced by the adrenal cortex that help maintain homeostasis through a variety of effects throughout the body.

gluconeogenesis—The conversion of protein or fat into glucose.

glucose—Six-carbon sugar that is the primary form of carbohydrate used for metabolism.

glycogen—The form of carbohydrate stored in the body, found predominantly in the muscles and liver.

glycogenesis—The conversion of glucose to glycogen.

glycogen loading—The manipulation of exercise and diet to optimize the body's glycogen storage.

glycogenolysis—The conversion of glycogen to glucose.

glycogen sparing—Increased reliance on fats for energy production during endurance activity, rather than stores of glycogen.

glycolysis—The breakdown of glucose to pyruvic acid.

glycolytic enzymes—Enzymes that are specific to the glycolytic energy system.

glycolytic system—A system that produces energy through glycolysis.

Golgi tendon organ—A sensory receptor in a muscle tendon that monitors tension.

graded exercise test (GXT)—An exercise test in which the rate of work is increased gradually in 1 to 3 min increments, usually to the point of fatigue or exhaustion.

graded potential—A localized change (depolarization or hyperpolarization) in the membrane potential.

growth—An increase in the size of the body or any of its parts.

growth hormone—An anabolic agent that stimulates fat metabolism and promotes muscle growth and hypertrophy by facilitating amino acid transport into the cells.

GXT—*See* graded exercise test.

habituation—Short-term adaptation to a stress.

HACE—*See* high-altitude cerebral edema.

Haldane transformation—An equation allowing one to calculate the inspired air volume from expired air volume, or expired air volume from inspired air volume.

HAPE—*See* high-altitude pulmonary edema.

HDL—*See* high-density lipoprotein.

HDL-C—*See* high-density lipoprotein cholesterol.

heart failure—A clinical condition in which the myocardium becomes too weak to maintain adequate cardiac output to meet the body's oxygen demands; heart failure usually results from the heart's being damaged or overworked.

heart rate recovery period—The time it takes for heart rate to return to the resting rate following exercise.

heat acclimation—*See* acclimation.

heat cramp—Cramping of the skeletal muscles as a result of excessive dehydration and the associated salt loss.

heat exhaustion—A heat disorder resulting from an inability of the cardiovascular system to meet all the body tissues' needs while also shifting blood to the periphery for cooling, characterized by elevated body temperature, breathlessness, extreme tiredness, dizziness, and rapid pulse.

heatstroke—The most serious heat disorder, resulting from failure of the body's thermoregulatory mechanisms. Heatstroke is characterized by body temperature above 40.5 °C (105 °F), cessation of sweating, and total confusion or unconsciousness and can lead to death.

hematocrit—The percentage of cells or formed elements in the total blood volume. More than 99% of the cells or formed elements are red blood cells.

hematopoiesis—Increased red blood cell concentration by increased production of cells.

hemoconcentration—A relative (not absolute) increase in the cellular content per unit of blood volume, resulting from a reduction in plasma volume.

hemodilution—An increase in blood plasma, resulting in a dilution of the blood's cellular contents.

hemoglobin—The iron-containing pigment in red blood cells that binds oxygen.

hemoglobin saturation—The amount of oxygen bound by each molecule of hemoglobin.

hemorrhagic stroke—Involves bleeding within the brain, which damages nearby brain tissue.

Henry's law—Law stating that gases dissolve in liquids in proportion to their partial pressures, depending also on their solubilities in the specific fluids and on the temperature.

hGH—*See* human growth hormone.

high-altitude cerebral edema (HACE)—A condition of unknown cause in which fluid accumulates in the cranial cavity at altitude; characterized by mental confusion that can progress to coma and death.

high-altitude pulmonary edema (HAPE)—A condition of unknown cause in which fluid accumulates in the lungs at altitude, interfering with ventilation, resulting in shortness of breath and fatigue, and characterized by impaired blood oxygenation, mental confusion, and loss of consciousness.

high-density lipoprotein (HDL)—A cholesterol carrier regarded as a scavenger; theorized to remove cholesterol from the arterial wall and transport it to the liver to be metabolized.

high-density lipoprotein cholesterol (HDL-C)—The cholesterol carried by HDL.

high-intensity interval training—Training that uses short bursts of very intense exercise interspersed with only a few minutes of rest or low-intensity exercise.

high responders—Those individuals within a population that show clear or exaggerated responses or adaptations to a stimulus.

homeostasis—Maintenance of a constant internal environment.

hormonal agents—A group of hormones proposed to have ergogenic properties.

hormone—A chemical substance produced or released by an endocrine gland and transported by the blood to a specific target tissue.

HR$_{max}$—*See* maximum heart rate.

human growth hormone (hGH)—A hormone that promotes anabolism and is believed by some athletes to have ergogenic properties.

hydrocortisone—*See* cortisol.

hydrostatic pressure—The pressure exerted by a stationary column of fluid in a tube.

hydrostatic weighing—A method of measuring body volume in which a person is weighed while submerged underwater. The difference between the scale weight on land and the underwater weight (corrected for water density) equals body volume. This value must be further corrected to account for any air trapped in the lungs and other parts of the body.

hyperglycemia—An elevated blood glucose level.

hyperinsulinemia—High levels of insulin in the blood.

hyperplasia—An increase in the number of cells in a tissue or organ. *See also* fiber hyperplasia.

hyperpolarization—An increase in the electrical potential across a membrane.

hypertension—Abnormally high blood pressure. In adults, hypertension is usually defined as a systolic pressure of 140 mmHg or higher or a diastolic pressure of 90 mmHg or higher.

hyperthermia—Elevated body temperature; any temperature above a person's normal resting body temperature.

hypertrophy—Increase in the size or mass of an organ or body tissue. *See also* fiber hypertrophy.

hyperventilation—A breathing rate or tidal volume greater than necessary for normal function.

hypobaric—Referring to an environment, such as that at high altitude, involving low atmospheric pressure.

hypoglycemia—A low blood glucose level.

hyponatremia—A blood sodium concentration below the normal range of 136 to 143 mmol/L.

hypothermia—Low body temperature; any temperature below the given person's normal temperature.

hypoxemia—A decreased oxygen content or concentration within the blood.

hypoxia—A decreased availability of oxygen to the tissues.

hypoxic vasoconstriction—The constriction of blood vessels in response to low levels of oxygen.

IDDM—*See* insulin-dependent diabetes mellitus.

immune function—The body's normal ability to fight infection and illness with antibodies and lymphocytes.

impaired fasting glucose—A plasma glucose level between 110 and 125 mg/dL following an 8 h fast.

impaired glucose tolerance—An abnormal glucose response to an oral glucose load (glucose tolerance test), sometimes seen as a precursor to diabetes.

independent variable—In an experiment, the variable that is manipulated by the experimenter to determine the response of the dependent variable. Usually plotted on the *x*-axis.

indirect calorimetry—A method of estimating energy expenditure by measuring respiratory gases.

infancy—The first year of life.

inhibiting factors—Hormones transmitted from the hypothalamus to the anterior pituitary that inhibit release of some other hormones.

inhibitory postsynaptic potential (IPSP)—A hyperpolarization of the postsynaptic membrane caused by an inhibitory impulse.

inspiration—The active process involving the diaphragm and the external intercostal muscles that expands the thoracic dimensions and thus the lungs. The expansion decreases pressure in the lungs, allowing outside air to rush in.

insulation—Resistance to dry heat loss.

insulative acclimation—A pattern of cold acclimation in which enhanced skin vasoconstriction increases peripheral insulation and minimizes heat loss.

insulin—A hormone produced by the β-cells in the pancreas that assists glucose entry into cells.

insulin-dependent diabetes mellitus (IDDM)—One of two major categories of diabetes mellitus that is caused by the inability of the pancreas to produce sufficient insulin as a result of failure of the β-cells in the pancreas. This is also known as type 1 diabetes.

insulin resistance—A deficient target cell response to insulin.

insulin sensitivity—An index of the effectiveness of a given insulin concentration on the disposal of glucose.

intercalated disks—Specialized cell junctions in the myocardium where one muscle cell connects to the next.

internal respiration—The exchange of gases between the blood and tissues.

interval-circuit training—Training program that involves rapid movement from one exercise to another around a "circuit" or established set of exercises.

interval training—Repeated, brief, fast-paced exercise bouts with short rest intervals between bouts.

intracellular fluid—The approximately 60% to 65% of total-body water that is contained in the cells.

IPSP—*See* inhibitory postsynaptic potential.

ischemia—A temporary deficiency of blood to a specific area of the body.

ischemic stroke—Brain tissue damage resulting from insufficient oxygen supply to an area of the brain. May be caused by narrowing or blockage of blood vessels supplying the area.

isokinetic training—Resistance training in which the rate of movement is kept constant through the range of motion.

isometric training—Resistance training involving a static action.

Karvonen method—The calculation of training heart rate in which a given percentage of the maximal heart rate reserve is added to the resting heart rate. This method gives an adjusted heart rate that is approximately equivalent to the desired percentage of $\dot{V}O_{2max}$.

kilocalorie (kcal)—The equivalent of 1,000 calories. *See* calorie.

Krebs cycle—A series of chemical reactions that involve the complete oxidation of acetyl CoA and produce 2 mol of ATP (energy) along with hydrogen and carbon, which combine with oxygen to form H_2O and CO_2.

lactate—A salt formed from lactic acid.

lactate dehydrogenase (LDH)—A key glycolytic enzyme involved in the conversion of pyruvate to lactate.

lactate threshold—The point during exercise of increasing intensity at which blood lactate begins to accumulate above resting levels, where lactate clearance is no longer able to keep up with lactate production.

L-carnitine—A substance important for fatty acid metabolism because it assists in the transfer of fatty acids from the cytosol (the fluid portion of the cytoplasm, exclusive of organelles) across the inner mitochondrial membrane for β-oxidation.

LDH—*See* lactate dehydrogenase.

LDL—*See* low-density lipoprotein.

LDL-C—*See* low-density lipoprotein cholesterol.

lean body mass—The sum of the body's fat-free mass and essential fat. This is not to be confused with fat-free mass.

lipogenesis—The process of converting protein into fatty acids.

lipolysis—The process of breaking down triglyceride to its basic units to be used for energy.

lipoprotein lipase—The enzyme that breaks down triglycerides to free fatty acids and glycerol, allowing the free fatty acids to enter the cells for use as a fuel or for storage.

lipoproteins—The proteins that carry the blood lipids.

longevity—The length of a person's life.

longitudinal research design—A research design in which subjects are tested initially and then one or more times later to directly measure changes over time resulting from a given intervention.

low-density lipoprotein (LDL)—A cholesterol carrier theorized to be responsible for depositing cholesterol in the arterial wall.

low-density lipoprotein cholesterol (LDL-C)—The cholesterol carried by LDL.

lower body (gynoid) obesity—Obesity that follows the typically female pattern of fat storage, in which fat is stored primarily in the lower body, particularly in the hips, buttocks, and thighs.

low-intensity aerobic exercise—Aerobic exercise performed at low intensity, theoretically to cause the body to burn a higher percentage of fat.

low responders—Those individuals within a population that show little or no response or adaptation to a stimulus.

LSD training—Endurance training involving long, slow distances.

L-tryptophan—An essential amino acid that has been proposed to increase aerobic endurance performance through its effects on the central nervous system. It theoretically acts as an analgesic and delays fatigue.

macrominerals—Those minerals of which the body needs more than 100 mg per day.

MAP—*See* mean arterial pressure.

maturation—The process by which the body takes on the adult form and becomes fully functional. It is often defined by the system or function being considered.

maximal expiratory ventilation ($\dot{V}_{Emax}$)—The highest ventilation that can be achieved during exhaustive exercise.

maximal heart rate reserve—The difference between maximal heart rate and resting heart rate.

maximal oxygen uptake ($\dot{V}O_{2max}$)—The maximal capacity for oxygen consumption by the body during maximal exertion. It is also known as aerobic power, maximal oxygen intake, maximal oxygen consumption, and cardiorespiratory endurance capacity.

maximal voluntary ventilation—The maximal capacity to move air into and out of the lungs, usually measured for 12 s and extrapolated to a per-minute value.

maximum heart rate (HR_{max})—The highest heart rate value attainable during an all-out effort to the point of exhaustion.

mean arterial pressure (MAP)—The average pressure exerted by the blood as it travels through the arteries. It is estimated as follows: MAP = DBP + [0.333 × (SBP − DBP)].

mechanoreceptors—An end organ that responds to changes in mechanical stress, such as stretch, compression, or distension.

menarche—The onset of menstruation; the first menses.

menses—The menstrual or flow phase of the menstrual cycle.

menstrual cycle—The cycle of uterine changes, averaging 28 days and consisting of the menstrual (flow) phase, the proliferative phase, and the secretory phase.

menstrual dysfunction—Disruption of the normal menstrual cycle; includes oligomenorrhea, primary amenorrhea, and secondary amenorrhea.

metabolic acclimation—A pattern of cold acclimation involving increased metabolic heat production through enhanced nonshivering and shivering thermogenesis.

metabolic equivalent (MET)—A unit used to estimate the metabolic cost (oxygen consumption) of physical activity. One MET equals the resting metabolic rate of approximately 3.5 ml of $O_2 \cdot kg^{-1} \cdot min^{-1}$.

metabolic syndrome—A term that has been used to link coronary artery disease, hypertension, type 2 diabetes, and upper body obesity to insulin resistance and hyperinsulinemia. This syndrome has also been referred to as syndrome X and the civilization syndrome.

metabolism—All energy-producing and energy-using processes within the body.

microgravity—An environment in which the body experiences a reduced gravitational force.

microminerals (trace elements)—The minerals of which the body needs less than 100 mg per day.

mineralocorticoids—Steroid hormones released from the adrenal cortex that are responsible for electrolyte balance within the body, for example aldosterone.

mitochondria—Cellular organelles that generate ATP through oxidative phosphorylation.

mitochondrial oxidative enzymes—Oxidative enzymes located in the mitochondria.

MK—*See* myokinase.

mode—Type of exercise.

morphology—The form and structure of the body.

motor nerves—Efferent nerves that carry impulses to skeletal muscle.

motor reflex—An involuntary motor response to a given stimulus.

motor unit—The motor nerve and the group of muscle fibers it innervates.

mountain sickness—*See* acute altitude sickness.

muscle buffering capacity—The muscles' ability to tolerate the acid that accumulates in them during anaerobic glycolysis.

muscle fiber—An individual muscle cell.

muscle pump—The rhythmic mechanical compression of the veins that occurs during skeletal muscle contraction in many types of movement and exercise, for example during walking and running, and assists the return of blood to the heart.

muscle spindle—A sensory receptor located in the muscle that senses how much the muscle is stretched.

muscular endurance—The ability of a muscle to resist fatigue.

musculoskeletal system—Body system composed of the skeleton and skeletal muscles that allows, supports, and helps control human movement.

myelination—The process of acquiring a myelin sheath.

myelin sheath—The outer covering of a myelinated nerve fiber, formed by a fatlike substance called myelin.

myocardial infarction—Death of heart tissue that results from insufficient blood supply to part of the myocardium.

myocardium—The muscle of the heart.

myofibril—The contractile element of skeletal muscle.

myoglobin—A compound similar to hemoglobin, but found in muscle tissue, that carries oxygen from the cell membrane to the mitochondria.

myokinase (MK)—A key enzyme in the ATP-PCr energy system.

myosin—One of the proteins that form filaments that produce muscle action.

myosin cross-bridge—The protruding part of a myosin filament. It includes the myosin head, which binds to an active site on an actin filament to produce a power stroke that causes the filaments to slide across each other.

nebulin—A giant protein that coextends with actin and appears to play a regulatory role in mediating actin and myosin interactions.

needs analysis—An assessment of factors that determine the specific training program appropriate for an individual.

negative feedback system—The primary mechanism through which the endocrine system maintains homeostasis. Some change in the body upsets homeostasis, which triggers release of a hormone to correct the change. Once that correction is accomplished, the hormone is no longer needed, so its secretion decreases.

nerve impulse—The electrical signal conducted along a neuron, which can be transmitted to another neuron or an end organ such as a group of muscle fibers.

neuromuscular junction—The site at which a motor neuron communicates with a muscle fiber.

neuron—A specialized cell in the nervous system responsible for generating and transmitting nerve impulses.

neurotransmitter—A chemical used for communication between a neuron and another cell.

NIDDM—*See* non-insulin-dependent diabetes mellitus.

nonessential amino acids—The 11 or 12 amino acids that the body synthesizes.

non-insulin-dependent diabetes mellitus (NIDDM)—One of two major categories of diabetes mellitus that is caused by the ineffectiveness of insulin to facilitate the transport of glucose into the cells and is a result of insulin resistance. This is also known as type 2 diabetes.

nonresponders—Individuals who show little or no improvement compared with others who undergo the same training program.

nonshivering thermogenesis—The stimulation of metabolism by the sympathetic nervous system to generate more metabolic heat.

nonsteroid hormones—Hormones derived from protein, peptides, or amino acids that cannot easily cross cell membranes.

norepinephrine—A catecholamine released from the adrenal medulla that, along with epinephrine, prepares the body for a fight-or-flight response. It is also a neurotransmitter. *See* catecholamines.

nutritional agents—Nutritional substances proposed to have ergogenic benefits.

obesity—An excessive amount of body fat, generally defined as more than 25% in men and 35% in women; a body mass index of 30 or greater.

oligomenorrhea—Abnormally infrequent or scant menstruation.

oncotic pressure—The pressure exerted by the concentration of proteins in a solution, drawing water from regions with lower oncotic pressures.

oral contraceptives—Drugs used for birth control and other medical purposes, which are believed by some female athletes to have ergogenic properties.

osmolarity—The number of solutes (such as electrolytes) dissolved in a fluid divided by the volume of that fluid; usually expressed in units of osmols (or milliosmols) per liter.

osmolality—The number of solutes (such as electrolytes) dissolved in a fluid divided by the weight of that fluid; usually expressed in units of osmols (or milliosmols) per kg.

osmotic pressure—The pressure exerted by the concentration of electrolytes in a solution, drawing water from regions with lower osmotic pressures.

ossification—The process of bone formation.

osteopenia—The loss of bone mass with aging.

osteoporosis—Decreased bone mineral content that increases bone porosity.

overload—To train at a level above that at which one normally trains; for example, to impart more physiological strain on muscle during training than the muscle is normally exposed to.

overreaching—A systematic attempt to intentionally overstress the body, allowing the body to adapt even more to the training stimulus above and beyond the adaptation attained during a period of acute overload.

overtraining—The attempt to do more work than can be physically tolerated.

overtraining syndrome—A condition brought on by overtraining and characterized by performance decrements and a general breakdown in physiological function.

overweight—Body weight that exceeds the normal or standard weight for a particular individual based on sex, height, and frame size; a BMI of 25.0 to 29.9.

oxidative capacity of muscle ($\dot{Q}O_2$)—A measure of the muscle's maximal capacity to use oxygen.

oxidative phosphorylation—Mitochondrial process that uses oxygen and high-energy electrons to produce ATP and water.

oxidative system—The body's most complex energy system, which generates energy by disassembling fuels with the aid of oxygen and has a very high energy yield.

oxygen diffusion capacity—The rate at which oxygen diffuses from one place to another.

oxygen supplementation—The breathing of supplemental oxygen, which is proposed to have ergogenic properties.

oxygen transport system—The components of the cardiovascular and respiratory systems involved in transporting oxygen.

parathyroid hormone (PTH)—The hormone released by the parathyroid gland to regulate plasma calcium concentration and plasma phosphate.

partial pressure—The pressure exerted by an individual gas in a mixture of gases.

partial pressure of oxygen (PO_2)—The pressure exerted by oxygen in a mixture of gases.

pathophysiology—The physiology of a specific disease or disorder.

PCr—*See* phosphocreatine.

PDGF—*See* platelet-derived growth factor.

peak oxygen uptake ($\dot{V}O_{2peak}$)—The highest oxygen uptake achieved during a graded exercise test when a subject reaches volitional fatigue before a plateau occurs in the $\dot{V}O_2$ response (the criterion for a true $\dot{V}O_{2max}$).

pericardium—A double-layered outer covering of the heart.

perimysium—The connective tissue sheath surrounding each muscle fasciculus.

peripheral blood flow—Blood flow to the extremities and the skin.

peripheral nervous system (PNS)—That section of the nervous system through which motor nerve impulses are transmitted from the brain and spinal cord to the periphery and sensory nerve impulses are transmitted from the periphery to the brain and spinal cord.

peripheral vascular disease—Diseases of the systemic arteries and veins, especially those to the extremities, that impede adequate blood flow.

peripheral vasoconstriction—*See* vasoconstriction.

PFK—*See* phosphofructokinase.

pharmacological agents—A group of drugs proposed to have ergogenic properties.

phosphate loading—The practice of ingesting sodium phosphate, which has been proposed to have ergogenic properties.

phosphocreatine (PCr)—An energy-rich compound that plays a critical role in providing energy for muscle action by maintaining ATP concentration.

phosphofructokinase (PFK)—A key rate-limiting enzyme of the anaerobic glycolytic energy system.

phosphorylase—A key enzyme of the anaerobic glycolytic energy system.

phosphorylation—The addition of a phosphate group (PO_4) to a molecule.

physical maturity—The point at which the body has attained the adult physical form.

physiological agents—A group of agents normally present in the body that have been proposed to have ergogenic properties.

physiology—The study of the function of organisms.

placebo—An inactive substance, usually provided in a manner identical to that for an active substance, typically to test for real results produced by the test substance versus equally real results of psychological origin.

placebo effect—An effect produced by the subject's expectations after administration of an inactive substance (placebo).

placebo group—The group in an intervention study that receives a placebo rather than a test substance.

plaque—A buildup of lipids, smooth muscle cells, connective tissue, and debris that forms at the site of injury to an artery.

plasmalemma—Plasma membrane, the selectively permeable lipid bilayer coated by proteins that composes the outer layer of a cell.

platelet-derived growth factor (PDGF)—A substance released by blood platelets that promotes the migration of smooth muscle cells from the media of an artery into the intima.

plyometrics—A type of dynamic-action resistance training based on the theory that use of the stretch reflex during jumping will recruit additional motor units.

PNS—*See* peripheral nervous system.

PO₂—*See* partial pressure of oxygen.

POAH—*See* preoptic-anterior hypothalamus.

polycythemia—Increased red blood cells.

power—The rate of performing work; the product of force and velocity. The rate of transformation of metabolic potential energy to work or heat.

power stroke—The tilting of the myosin head, caused by a strong intermolecular attraction between the myosin cross-bridge and the myosin head, that causes the actin and myosin filaments to slide across each other.

prediabetes—a term used to define those who have impaired fasting glucose, impaired glucose tolerance, or both, but are not truly diabetic.

predictive value of an abnormal exercise test—The accuracy with which abnormal test results reflect the presence of a disease.

pregnancy—The state of carrying an embryo or fetus in the body.

preload—The degree to which the myocardium is stretched before it contracts, determined by factors such as central blood volume.

premature ventricular contraction (PVC)—A common cardiac arrhythmia that results in the feeling of skipped or extra beats caused by impulses originating outside the SA node.

preoptic-anterior hypothalamus (POAH)—The area of the midbrain that is the primary controller of thermoregulatory function.

primary amenorrhea—The absence of menarche (the beginning of menstruation) beyond age 18.

primary risk factors—Risk factors that have been conclusively shown to have a strong association with a certain disease. Primary risk factors for coronary artery disease include smoking, hypertension, high blood lipid levels, obesity, and physical inactivity.

principle of individuality—The theory that any training program must consider the specific needs and abilities of the individual for whom it is designed.

principle of orderly recruitment—The theory that motor units generally are activated on the basis of a fixed order of recruitment, in which the motor units within a given muscle appear to be ranked according to the size of the motor neuron.

principle of periodization—The gradual cycling of specificity, intensity, and volume of training to achieve peak levels of fitness for competition (also called the *principle of variation*).

principle of progressive overload—The theory that, to maximize the benefits of a training program, the training stimulus must be progressively increased as the body adapts to the current stimulus.

principle of reversiblity—The theory that a training program must include a maintenance plan to ensure that the gains from training are not lost.

principle of specificity—The theory that a training program must stress the physiological systems critical for optimal performance in a given sport to achieve desired training adaptations in that sport.

principle of variation—The systematic process of changing one or more variables in an exercise training program—mode, volume, or intensity—over time to allow for the training stimulus to remain challenging and effective (also called the *principle of periodization*).

progesterone—A hormone secreted by the ovaries that promotes the luteal phase of the menstrual cycle.

prostaglandins—Substances derived from a fatty acid that act as hormones at the local level.

protein—A class of nitrogen-containing compounds formed by amino acids.

pseudoephedrine—a sympathomimetic amine that is used in over-the-counter medications primarily as a decongestant and in the illicit manufacturing of methamphetamine.

PTH—*See* parathyroid hormone.

puberty—The point at which a person becomes physiologically capable of reproduction.

pulmonary diffusion—The exchange of gases between the lungs and the blood.

pulmonary ventilation—The movement of gases into and out of the lungs.

Purkinje fibers—The terminal branches of the AV bundle that transmit impulses through the ventricles six times faster than through the rest of the cardiac conduction system.

PVC—*See* premature ventricular contraction.

$\dot{Q}$—*See* cardiac output.

$\dot{Q}O_2$—*See* oxidative capacity of muscle.

radiation—The transfer of heat through electromagnetic waves.

rate coding—Refers to the frequency of impulses sent to a muscle. Increased force can be generated through increase in either the number of muscle fibers recruited or the rate at which the impulses are sent. Also called frequency coding.

rating of perceived exertion (RPE)—A person's subjective assessment of how hard he or she is exercising.

rate-limiting enzyme—An enzyme found early in a metabolic pathway that determines the rate of the pathway.

rate–pressure product (RPP)—The mathematical product of heart rate × systolic blood pressure. Also called the double product.

rehabilitation programs—Programs designed to reestablish health or fitness following a disability or illness.

relative body fat—The ratio of fat mass to total-body mass, expressed as a percentage.

releasing factors—Hormones transmitted from the hypothalamus to the anterior pituitary that promote release of some other hormones.

renin—An enzyme formed by the kidneys to convert a plasma protein called angiotensinogen into angiotensin II. *See also* renin–angiotensin-aldosterone mechanism.

renin–angiotensin-aldosterone mechanism—The mechanism involved in renal control of blood pressure. The kidneys respond to decreased blood pressure or blood flow by forming renin, which converts angiotensinogen into angiotensin I, which is finally converted to angiotensin II. Angiotensin II constricts arterioles and triggers aldosterone release.

RER—*See* respiratory exchange ratio.

residual volume (RV)—The amount of air that cannot be exhaled from the lungs.

resistance training—Training designed to increase strength, power, and muscular endurance.

respiratory alkalosis—A condition in which increased carbon dioxide clearance allows blood pH to increase.

respiratory centers—Autonomic centers located in the medulla oblongata and the pons that establish breathing rate and depth.

respiratory exchange ratio (RER)—The ratio of carbon dioxide expired to oxygen consumed at the level of the lungs.

respiratory membrane—The membrane separating alveolar air and blood, composed of the alveolar wall, the capillary wall, and their basement membranes.

respiratory pump—Passive movement of blood through the central circulation as a function of pressure changes during breathing.

resting heart rate (RHR)—The heart rate at rest, averaging 60 to 80 beats/min.

resting membrane potential (RMP)—The potential difference between the electrical charges inside a cell and outside the cell, caused by a separation of charges across the membrane.

resting metabolic rate (RMR)—The body's metabolic rate early in the morning following an overnight fast and 8 h of sleep. Determining RMR does not require sleeping overnight in a laboratory or clinical facility. *See also* basal metabolic rate.

retraining—Recovery of conditioning after a period of inactivity.

rhabdomyolysis—Breakdown of muscle fibers resulting in protein buildup in the blood and often in the urine.

rheumatic heart disease—A form of valvular heart disease involving a streptococcal infection that has caused acute rheumatic fever, typically in children between ages 5 and 15.

RHR—*See* resting heart rate.

risk factor—A predisposing factor statistically linked to the development of a disease, such as coronary artery disease.

RMP—*See* resting membrane potential.

RMR—*See* resting metabolic rate.

RPE—*See* rating of perceived exertion.

RV—*See* residual volume.

saltatory conduction—The means of rapid nerve impulse conduction along myelinated neurons.

SA node—*See* sinoatrial node.

sarcolemma—A muscle fiber's cell membrane.

sarcomere—The basic functional unit of a myofibril.

sarcopenia—The loss of muscle mass associated with aging.

sarcoplasm—The gelatin-like cytoplasm in a muscle fiber.

sarcoplasmic reticulum (SR)—A longitudinal system of tubules that is associated with the myofibrils and that stores calcium for muscle action.

satellite cells—Immature cells that can develop into mature cell types, such as myoblasts.

SBP—*See* systolic blood pressure.

SDH—*See* succinate dehydrogenase.

secondary amenorrhea—The cessation of menstruation in a woman with previously normal menstrual function.

second messenger—A substance inside a cell that acts as a messenger after a nonsteroid hormone binds to receptors outside the cell.

sensitivity—A test's ability to correctly identify subjects who fit the criteria being tested, such as coronary artery disease.

sensory nerves—Afferent nerves that carry impulses toward the central nervous system from the periphery.

sensory-motor integration—The process by which the sensory and motor systems communicate and coordinate with each other.

sex-specific differences—True physiological differences between females and males.

shivering—A rapid, involuntary cycle of contraction and relaxation of skeletal muscles that generates heat.

single-fiber contractile velocity (V$_o$)—The rate at which an individual muscle cell can shorten and develop tension.

sinoatrial (SA) node—A group of specialized myocardial cells, located in the wall of the right atrium, that control the heart's rate of contraction; the pacemaker of the heart.

size principle—Principle asserting that the size of the motor neuron dictates the order of motor unit recruitment, with small-sized motor neurons being recruited first.

skinfold fat thickness—The most widely applied field technique used to estimate body density, relative body fat, and fat-free mass. It involves measurement with calipers of the skinfold fat at one or more sites.

sliding filament theory—A theory explaining muscle action: A myosin cross-bridge attaches to an actin filament, and then the power stroke drags the two filaments past one another.

sodium-potassium pump—An enzyme called Na$^+$-K$^+$-ATPase, which maintains the resting membrane potential in disequilibrium at –70 mV.

specificity—A test's ability to correctly identify subjects who do not fit the criteria being tested.

specificity of training—The principle that physiological adaptations in response to physical training are highly specific to the nature of the training activity. To maximize benefits, training should be carefully matched to an athlete's specific performance needs.

spirometry—The measurement of lung volumes and capacities.

sport physiology—The application of the concepts of exercise physiology to training athletes and enhancing sport performance.

spot reduction—The practice of exercising a specific area of the body, theoretically to reduce locally stored fat.

sprint training—A form of anaerobic training involving very brief, intense training bouts.

SR—*See* sarcoplasmic reticulum.

ST—*See* slow-twitch fiber.

static-contraction resistance training—Resistance training that emphasizes static muscle action. Also known as isometric resistance training.

static (isometric) muscle contraction—Action in which the muscle contracts without moving, generating force while its length remains static (unchanged). Also known as isometric action.

steady-state heart rate—A heart rate that is maintained constant at submaximal levels of exercise when the rate of work is held constant.

steroid hormones—Hormones with chemical structures similar to cholesterol that are lipid soluble and that diffuse through cell membranes.

strength—The ability of a muscle to exert force—generally the maximal ability.

stroke—A cerebral vascular accident, a condition in which blood supply to some part of the brain is impaired, typically caused by infarction or hemorrhage, so that the tissue is damaged.

stroke volume (SV)—The amount of blood ejected from the left ventricle during contraction; the difference between the end-diastolic volume and the end-systolic volume.

submaximal endurance capacity—The average absolute power output a person can maintain during a fixed period of time on a cycle ergometer, or the average speed or velocity a person can maintain during a fixed period of time. Generally, these tests will last at least 30 min but usually not more than 90 min.

submaximal exercise—All intensities of exercise below maximal exercise intensity.

substrate—Basic fuel source, such as carbohydrates, proteins, and fats.

succinate dehydrogenase (SDH)—A key enzyme of the oxidative enzyme system.

summation—The summing of all individual changes in a neuron's membrane potential.

SV—*See* stroke volume.

swimming flume—A device that uses propeller pumps to circulate water past a swimmer, who attempts to maintain body position by swimming against the current.

sympatholysis—The process by which locally-released vasodilating substances in exercising muscle compete with, and dominate, the vasoconstrictor influence of sympathetic stimulation.

synapse—The junction between two neurons.

systolic blood pressure (SBP)—The greatest arterial blood pressure, resulting from systole (the contracting phase of the heart).

T_3—*See* triiodothyronine.

T_4—*See* thyroxine.

tachycardia—A resting heart rate greater than 100 beats/min.

tapering—A reduction in training intensity prior to a major competition to give the body and mind a break from the rigors of intense training.

taper period—A period during which training intensity is reduced, allowing time for tissue damage from intense training to heal and for the body's energy reserves to be fully replenished.

target cells—Cells that possess specific hormone receptors.

TEA—*See* thermic effect of activity.

TEM—*See* thermic effect of a meal.

teratogenic effects—Effects that cause abnormal fetal development.

testosterone—The predominant male sex hormone.

test specificity—Matching the type of ergometer used in testing to the type of activity an athlete usually performs to ensure the most accurate results.

tetanus—Highest tension developed by a muscle in response to stimulation of increasing frequency.

tethered swimming—A method of monitoring a swimmer in which the swimmer is attached to a harness connected to a rope, a series of pulleys, and a pan that contains weights, which allows the swimmer to swim while maintaining a constant position in the pool.

therapeutic use exemption—An exemption granted by the governing body of a sport that allows an athlete to use an otherwise banned substance if it is needed to treat a medical condition.

thermal stress—Stress imposed on the body by external temperature.

thermic effect of activity (TEA)—The energy expended in excess of the resting metabolic rate to accomplish a given task or activity.

thermic effect of a meal (TEM)—The energy expended in excess of resting metabolic rate associated with digestion, absorption, transport, metabolism, and storage of ingested food.

thermoreceptors—Sensory receptors that detect changes in body temperature and external temperature and relay this information to the hypothalamus (also called thermoceptors).

thermoregulation—The process by which the thermoregulatory center, located in the hypothalamus, readjusts body temperature in response to small deviations from the set point.

thermoregulatory center—An autonomic nervous center located in the hypothalamus that is responsible for maintaining normal body temperature.

thirst mechanism—A neural mechanism that triggers thirst in response to dehydration.

THR—*See* training heart rate.

threshold—A minimum amount of stimulus needed to elicit a response. Also, the minimum depolarization required to produce an action potential in neurons.

thyrotropin (TSH)—A hormone secreted by the anterior lobe of the pituitary gland that promotes the release of thyroid hormones.

thyroxine (T_4)—A hormone secreted by the thyroid gland that increases the rate of cellular metabolism and the rate and contractility of the heart.

tidal volume—The amount of air inspired or expired during a normal breathing cycle.

titin—A protein that positions the myosin filament to maintain equal spacing between actin filaments.

TLC—*See* total lung capacity.

total lung capacity (TLC)—The sum of vital capacity and residual volume.

total peripheral resistance (TPR)—The resistance to the flow of blood through the entire systemic circulation.

trace elements—*See* microminerals.

training effect—Physiological adaptation to repeated bouts of exercise.

training heart rate (THR)—A heart rate goal established by using the heart rate equivalent of a desired percentage of $\dot{V}O_{2max}$. For example, if a training level of 75% $\dot{V}O_{2max}$ is desired, 75% of $\dot{V}O_{2max}$ is calculated, and the heart rate corresponding to this $\dot{V}O_2$ is selected as the THR.

transient hypertrophy—The "pumping up" of muscle that happens during a single exercise bout, resulting mainly from fluid accumulation in the interstitial and intracellular spaces of the muscle.

transverse tubules (T-tubules)—Extensions of the sarcolemma (plasma membrane) that pass laterally through the muscle fiber, allowing nutrients to be transported and nerve impulses to be transmitted rapidly to individual myofibrils.

treadmill—An ergometer in which a motor and pulley system drive a large belt that a person can either walk or run on.

triglycerides—The body's most concentrated energy source and the form in which most fats are stored in the body.

triiodothyronine (T_3)—A hormone released by the thyroid gland that increases the rate of cellular metabolism and the rate and contractility of the heart.

tropomyosin—A tube-shaped protein that twists around actin strands, fitting into the groove between them.

troponin—A complex protein attached at regular intervals to actin strands and tropomyosin.

TSH—*See* thyrotropin.

twitch—The smallest contractile response of a muscle fiber or a motor unit to a single electrical stimulus.

type 1 diabetes—A type of diabetes mellitus that generally has a sudden onset during childhood or young adulthood and leads to almost total insulin deficiency, usually requiring daily insulin injections. Also known as insulin-dependent diabetes mellitus (IDDM) or juvenile-onset diabetes.

type 2 diabetes—A type of diabetes mellitus in which disease onset is more gradual and the causes are more difficult to establish than in type 1 diabetes. Type 2 diabetes is characterized by impaired insulin secretion, impaired insulin action, or excessive glucose output from the liver. Also known as non-insulin-dependent diabetes mellitus (NIDDM).

type I (slow-twitch) muscle fiber—A type of muscle fiber that has a high oxidative and a low glycolytic capacity, associated with endurance-type activities.

type II (fast-twitch) muscle fiber—A type of muscle fiber with a low oxidative capacity and a high glycolytic capacity; associated with speed or power activities.

undertraining—The type of training an athlete would undertake between competitive seasons or during active rest. Generally, physiological adaptations will be minor, and there will be no improvement in performance.

upper body (android) obesity—Obesity that follows the typically male pattern of fat storage, in which fat is stored primarily in the upper body, particularly in the abdomen.

upregulation—An increased cellular sensitivity to a hormone, often caused by increased hormone receptors.

Valsalva maneuver—The process of holding the breath and attempting to compress the contents of the abdominal and thoracic cavities, causing increased intra-abdominal and intrathoracic pressure.

valvular heart disease—A disease involving one or more of the heart valves. Rheumatic heart disease is one example.

variable-resistance training—A technique that allows variation in the resistance applied throughout the range of motion in an attempt to match the ability of the muscle or muscle groups to apply force at any specific point in the range of motion.

vasoconstriction—The constriction or narrowing of blood vessels.

vasodilation—The dilation or widening of blood vessels.

vasopressin—*See* antidiuretic hormone.

VC—*See* vital capacity.

$\dot{V}CO_2$—The volume of CO2 produced per minute.

$\dot{V}_E$—The volume of air expired per minute.

$\dot{V}_{Emax}$—*See* maximal expiratory ventilation.

$\dot{V}_E/\dot{V}CO_2$—*See* ventilatory equivalent for carbon dioxide.

$\dot{V}_E/\dot{V}O_2$—*See* ventilatory equivalent for oxygen.

veins—Blood vessels that transport blood back to the heart.

ventilatory equivalent for carbon dioxide ($\dot{V}_E/\dot{V}CO_2$)—The ratio of the volume of air ventilated ($\dot{V}_E$) to the amount of carbon dioxide produced ($\dot{V}CO_2$).

ventilatory equivalent for oxygen ($\dot{V}_E/\dot{V}O_2$)—The ratio between the volume of air ventilated ($\dot{V}_E$) and the amount of oxygen consumed ($\dot{V}O_2$); indicates breathing economy.

ventilatory threshold—Older name for the ventilator breakpoint.

ventricular fibrillation—A serious cardiac arrhythmia in which the contraction of the ventricular tissue is uncoordinated, affecting the heart's ability to pump blood. *See also* ventricular tachycardia.

ventricular tachycardia—A serious cardiac arrhythmia consisting of three or more consecutive premature ventricular contractions. *See also* premature ventricular contraction and ventricular fibrillation.

venules—Small vessels that transport blood from the capillaries to the veins and then back to the heart.

very low density lipoprotein (VLDL)—A lipoprotein carrier of cholesterol.

very low density lipoprotein cholesterol (VLDL-C)—The cholesterol carried by VLDL.

vital capacity (VC)—The maximal volume of air expelled from the lungs after maximal inhalation.

vitamin—One of a group of unrelated organic compounds that perform specific functions to promote growth and to maintain health. Vitamins act primarily as catalysts in chemical reactions.

VLDL—*See* very low density lipoprotein.

VLDL-C—*See* very low density lipoprotein cholesterol.

$\dot{V}O_2$—The volume of oxygen consumed per minute.

$\dot{V}O_2$ **drift**—A slow increase in $\dot{V}O_2$ during prolonged submaximal exercise at a constant power output.

$\dot{V}O_{2max}$—*See* maximal oxygen uptake.

wet-bulb globe temperature (WBGT)—A measurement of temperature that simultaneously accounts for conduction, convection, evaporation, and radiation, providing a single temperature reading to estimate the cooling capacity of the surrounding environment. The apparatus for measuring WBGT consists of a dry bulb, a wet bulb, and a black globe.

windchill—A chill factor created by the increase in the rate of heat loss via convection and conduction caused by wind.

work—Force expressed through distance, or a displacement, independent of time.

Introduction

1. Åstrand, P.-O. (1991). Influence of Scandinavian scientists in exercise physiology. *Scandinavian Journal of Medicine and Science in Sports, 1,* 3-9.

2. Åstrand, P.-O., & Rhyming, I. (1954). A nomogram for calculation of aerobic capacity (physical fitness) from pulse rate during submaximal work. *Journal of Applied Physiology, 7,* 218-221.

3. Bainbridge, F.A. (1931). *The physiology of muscular exercise.* London: Longmans, Green.

4. Brown, R.C., & Kenyon, G.S. (1968). *Classical studies on physical activity.* Englewood Cliffs, NJ: Prentice Hall.

5. Buskirk, E.R. (1996). Early history of exercise physiology in the United States: Part I. A contemporary historical perspective. In J.D. Messengale & R.A. Swanson (Eds.), *History of exercise and sport science* (pp. 55-74). Champaign, IL: Human Kinetics.

6. Buskirk, E.R., & Taylor, H.L. (1957). Maximal oxygen uptake and its relation to body composition, with special reference to chronic physical activity and obesity. *Journal of Applied Physiology, 11,* 72-78.

7. Cooper, K.H. (1968). *Aerobics.* New York: Evans.

8. Dill, D.B. (1938). *Life, heat, and altitude.* Cambridge, MA: Harvard University Press.

9. Dill, D.B. (1968). Historical review of exercise physiology science. In R. Warren & R.E. Johnson (Eds.), *Science and medicine of exercise and sports* (2nd ed., pp. 42-48). New York: Harper.

10. Dill, D.B. (1985). *The hot life of man and beast.* Springfield, IL: Charles C Thomas.

11. Fletcher, W.M., & Hopkins, F.G. (1907). Lactic acid in amphibian muscle. *Journal of Physiology, 35,* 247-254.

12. Flint, A., Jr. (1871). On the physiological effects of severe and protracted muscular exercise; with special reference to the influence of exercise upon the excretion of nitrogen. *New York Medical Journal, 13,* 609-697.

13. Foster, M. (1970). *Lectures on the history of physiology.* New York: Dover.

14. Horvath, S.M., & Horvath, E.C. (1973). *The Harvard Fatigue Laboratory: Its history and contributors.* Englewood Cliffs, NJ: Prentice Hall.

15. LaGrange, F. (1889). *Physiology of bodily exercise.* London: Kegan Paul International.

16. McArdle, W.D., Katch, F.I., & Katch, V.L. (2001). *Exercise physiology: Energy, nutrition, and human performance* (5th ed.). Baltimore: Williams & Wilkins.

17. Robinson, S. (1938). Experimental studies of physical fitness in relation to age. *Arbeitsphysiologie, 10,* 251-327.

18. Séguin, A., & Lavoisier, A. (1793). Premier mémoire sur la respiration des animaux. *Histoire et Mémoires de l'Academie Royale des Sciences, 92,* 566-584.

19. Taylor, H.L., Buskirk, E.R., & Henschel, A. (1955). Maximal oxygen intake as an objective measure of cardiorespiratory performance. *Journal of Applied Physiology, 8,* 73-80.

20. Tipton, C.M. (2003). *Exercise physiology: People and ideas.* New York: Oxford University Press.

21. Zuntz, N., & Schumberg, N.A.E.F. (1901). *Studien Zur Physiologie des Marches* (p. 211). Berlin: A. Hirschwald.

Chapter 1

1. Brooks, G.A., Fahey, T.D., & Baldwin, K.M. (2005). *Exercise physiology: Human bioenergetics and its applications* (4th ed.). New York: McGraw-Hill.

2. Close, R. (1967). Properties of motor units in fast and slow skeletal muscles of the rat. *Journal of Physiology* (London), *193,* 45-55.

3. Costill, D.L., Daniels, J., Evans, W., Fink, W., Krahenbuhl, G., & Saltin, B. (1976). Skeletal muscle enzymes and fiber composition in male and female track athletes. *Journal of Applied Physiology, 40,* 149-154.

4. Costill, D.L., Fink, W.J., Flynn, M., & Kirwan, J. (1987). Muscle fiber composition and enzyme activities in elite female distance runners. *International Journal of Sports Medicine, 8,* 103-106.

5. Costill, D.L., Fink, W.J., & Pollock, M.L. (1976). Muscle fiber composition and enzyme activities of elite distance runners. *Medicine and Science in Sports, 8,* 96-100.

6. MacIntosh, B.R., Gardiner, P.F., & McComas, A.J. (2006). *Skeletal muscle form and function* (2nd ed.). Champaign, IL: Human Kinetics.

Chapter 3

1. Edstrom, L., & Grimby, L. (1986). Effect of exercise on the motor unit. *Muscle and Nerve, 9,* 104-126.

2. Guyton, A.C., & Hall, J.E. (2000). *Textbook of medical physiology* (10th ed.). Philadelphia: Saunders.

3. Marieb, E.N. (1995). *Human anatomy and physiology* (3rd ed.). New York: Benjamin/Cummings.

4. Petajan, J.H., Gappmaier, E., White, A.T., Spencer, M.K., Mino, L., & Hicks, R.W. (1996). Impact of aerobic training on fitness and quality of life in multiple sclerosis. *Annals of Neurology, 39,* 432-441.

5. Pette, D., & Vrbova, G. (1985). Neural control of phenotypic expression in mammalian muscle fibers. *Muscle and Nerve, 8,* 676-689.

Chapter 5

1. Bar-Or, O. (1987). The Wingate Anaerobic Test: An update on methodology, reliability and validity. *Sports Medicine, 4,* 381-394.

2. Barstow, T.J., Jones, A.M., Nguyen, P.H., & Casaburi, R. (1996). Influence of muscle fiber type and pedal frequency on oxygen uptake kinetics of heavy exercise. *Journal of Applied Physiology, 81,* 1642-1650.

3. Costill, D.L. (1986). *Inside running: Basics of sports physiology.* Indianapolis: Benchmark Press.

4. Gaesser, G.A., & Poole, D.C. (1996). The slow component of oxygen uptake kinetics in humans. *Exercise and Sport Sciences Reviews, 24,* 35-70.

5. Galloway, S.D.R., & Maughan, R.J. (1997). Effects of ambient temperature on the capacity to perform prolonged cycle exercise in man. *Medicine and Science in Sports and Exercise, 29,* 1240-1249.

6. Hawley, J.A. (1997). Carbohydrate loading and exercise performance: An update. *Sports Medicine, 24,* 73-81.

7. Hill, D.W. (1993). The critical power concept: A review. *Sports Medicine, 16,* 237-254.

8. Medbø, J.I., Mohn, A.C., Tabata, I., Bahr, R., Vaage, O., & Sejersted, O.M. (1988). Anaerobic capacity determined by maximal accumulated O2 deficit. *Journal of Applied Physiology, 64,* 50-60.

9. Westerblad, H., Allen, D.G., & Lännergren, J. (2002). Muscle fatigue: Lactic acid or inorganic phosphate the major cause? *News in the Physiological Sciences, 17,* 17-21.

10. Zuntz, N., & Hagemann, O. (1898). *Untersuchungen uber den Stroffwechsel des Pferdes bei Ruhe und Arbeit.* Berlin: Parey.

Chapter 8

1. Hermansen, L. (1981). Effect of metabolic changes on force generation in skeletal muscle during maximal exercise. In R. Porter & J. Whelan (Eds.), *Human muscle fatigue: Physiological mechanisms* (pp. 75-88). London: Pitman Medical.

2. McKirnan, M.D., Gray, C.G., & White, F.C. (1991). Effects of feeding on muscle blood flow during prolonged exercise in miniature swine. *Journal of Applied Physiology, 70,* 1097-1104.

3. Poliner, L.R., Dehmer, G.J., Lewis, S.E., Parkey, R.W., Blomqvist, C.G., & Willerson, J.T. (1980). Left ventricular performance in normal subjects: A comparison of the responses to exercise in the upright and supine position. *Circulation, 62,* 528-534.

4. Powers, S.K., Martin, D., & Dodd, S. (1993). Exercise-induced hypoxaemia in elite endurance athletes: Incidence, causes and impact on $\dot{V}O_{2max}$. *Sports Medicine, 16,* 14-22.

5. Tanaka, H., Monahan, D.K., & Seals, D.R. (2001). Age-predicted maximal heart rate revisited. *Journal of the American College of Cardiology, 37,* 153-156.

6. Turkevich, D., Micco, A., & Reeves, J.T. (1988). Noninvasive measurement of the decrease in left ventricular filling time during maximal exercise in normal subjects. *American Journal of Cardiology, 62,* 650-652.

7. Wasserman, K., & McIlroy, M.B. (1964). Detecting the threshold of anaerobic metabolism in cardiac patients during exercise. *American Journal of Cardiology, 14,* 844-852.

Chapter 9

1. American College of Sports Medicine. (2009). ACSM position stand: Progression models in resistance training for healthy adults. *Medicine and Science in Sports and Exercise, 41,* 687-708.

2. Behm, D.G., Drinkwater, E.J., Willardson, J.M., & Cowley, P.M. (2010). The use of instability to train the core musculature. *Applied Physiology Nutrition and Metabolism 35,* 91-108.

3. Fleck, S.J., & Kraemer, W.J. (2004). *Designing resistance training programs* (3rd ed.). Champaign, IL: Human Kinetics.

4. Fox, E.L., & Mathews, D.K. (1974). Interval training conditioning for sports and general fitness. Philadelphia: Saunders.

5. Willardson, J.M. (2007). Core stability training: applications to sports conditioning programs. *Journal of Strength and Conditioning Research 21,* 979-985.

Chapter 10

1. Armstrong, R.B., Warren, G.L., & Warren, J.A. (1991). Mechanisms of exercise-induced muscle fibre injury. *Sports Medicine, 12,* 184-207.

2. Bergeron, M.F. (2008) Muscle cramps during exercise—Is it fatigue or electrolyte deficit? *Current Sports Medicine Reports, 7,* S50-55.

3. Duchateau, J., & Enoka, R.M. (2002). Neural adaptations with chronic activity patterns in able-bodied humans. *American Journal of Physical Medicine and Rehabilitation, 81*(11 Suppl.), 517-527.

4. Cheung, K., Hume, P.A., Maxwell, L. (2003). Delayed onset muscle soreness treatment strategies and performance factors. *Sports Medicine, 33,* 145-164.

5. Enoka, R.M. (1988). Muscle strength and its development: New perspectives. *Sports Medicine, 6,* 146-168.

6. Gonyea, W.J. (1980). Role of exercise in inducing increases in skeletal muscle fiber number. *Journal of Applied Physiology, 48,* 421-426.

7. Gonyea, W.J., Sale, D.G., Gonyea, F.B., & Mikesky, A. (1986). Exercise induced increases in muscle fiber number. *European Journal of Applied Physiology, 55,* 137-141.

8. Graves, J.E., Pollock, M.L., Leggett, S.H., Braith, R.W., Carpenter, D.M., & Bishop, L.E. (1988). Effect

of reduced training frequency on muscular strength. *International Journal of Sports Medicine, 9,* 316-319.

9. Green, H.J., Klug, G.A., Reichmann, H., Seedorf, U., Wiehrer, W., & Pette, D. (1984). Exercise-induced fibre type transitions with regard to myosin, parvalbumin, and sarcoplasmic reticulum in muscles of the rat. *Pflugers Archiv: European Journal of Physiology, 400,* 432-438.

10. Hagerman, F.C., Hikida, R.S., Staron, R.S., Sherman, W.M., & Costill, D.L. (1984). Muscle damage in marathon runners. *Physician and Sportsmedicine, 12,* 39-48.

11. Hakkinen, K., Alen, M., & Komi, P.V. (1985). Changes in isometric force and relaxation-time, electromyographic and muscle fibre characteristics of human skeletal muscle during strength training and detraining. *Acta Physiologica Scandinavica, 125,* 573-585.

12. Hawke, T.J., & Garry, D.J. (2001). Myogenic satellite cells: Physiology to molecular biology. *Journal of Applied Physiology, 91,* 534-551.

13. Koopman, R., Saris, W.H.M., Wagenmakers, A.J.M., & van Loon, L.J.C. (2007). Nutritional interventions to promote post-exercise muscle protein synthesis. *Sports Medicine, 37,* 895-906.

14. Kraemer, W.J. (2000). Physiological adaptations to anaerobic and aerobic endurance training programs. In T.R. Baechle & R.W. Earle (Eds.), *Essentials of strength training and conditioning* (2nd ed., p. 150). Champaign, IL: Human Kinetics.

15. McCall, G.E., Byrnes, W.C., Dickinson, A., Pattany, P.M., & Fleck, S.J. (1996). Muscle fiber hypertrophy, hyperplasia, and capillary density in college men after resistance training. *Journal of Applied Physiology, 81,* 2004-2012.

16. Schwane, J.A., Johnson, S.R., Vandenakker, C.B., & Armstrong, R.B. (1983). Delayed-onset muscular soreness and plasma CPK and LDH activities after downhill running. *Medicine and Science in Sports and Exercise, 15,* 51-56.

17. Schwane, J.A., Watrous, B.G., Johnson, S.R., & Armstrong, R.B. (1983). Is lactic acid related to delayed-onset muscle soreness? *Physician and Sportsmedicine, 11*(3), 124-131.

18. Shepstone, T.N., Tang, J.E., Dallaire, S., Schuenke, M.D., Staron, R.S., & Phillips, S.M. (2005). Short-term high- vs. low-velocity isokinetic lengthening training results in greater hypertrophy of the elbow flexors in young men. *Journal of Applied Physiology, 98,* 1768-1776.

19. Sjöström, M., Lexell, J., Eriksson, A., & Taylor, C.C. (1991). Evidence of fibre hyperplasia in human skeletal muscles from healthy young men? A left-right comparison of the fibre number in whole anterior tibialis muscles. *European Journal of Applied Physiology, 62,* 301-304.

20. Staron, R.S., Karapondo, D.L., Kraemer, W.J., Fry, A.C., Gordon, S.E., Falkel, J.E., Hagerman, F.C., & Hikida, R.S. (1994). Skeletal muscle adaptations during early

phase of heavy resistance training in men and women. *Journal of Applied Physiology, 76,* 1247-1255.

21. Staron, R.S., Leonardi, M.J., Karapondo, D.L., Malicky, E.S., Falkel, J.E., Hagerman, F.C., & Hikida, R.S. (1991). Strength and skeletal muscle adaptations in heavy-resistance-trained women after detraining and retraining. *Journal of Applied Physiology, 70,* 631-640.

22. Staron, R.S., Malicky, E.S., Leonardi, M.J., Falkel, J.E., Hagerman, F.C., & Dudley, G.A. (1990). Muscle hypertrophy and fast fiber type conversions in heavy-resistance-trained women. *European Journal of Applied Physiology, 60,* 71-79.

23. Tidball, J.G. (1995). Inflammatory cell response to acute muscle injury. *Medicine and Science in Sports and Exercise, 27,* 1022-1032.

24. Warren, G.L., Ingalls, C.P., Lowe, D.A., & Armstrong, R.B. (2001). Excitation-contraction uncoupling: Major role in contraction-induced muscle injury. *Exercise and Sport Sciences Reviews, 29,* 82-87.

Chapter 11

1. Armstrong, R.B., & Laughlin, M.H. (1984). Exercise blood flow patterns within and among rat muscles after training. *American Journal of Physiology, 246,* H59-H68.

2. Bouchard, C. (1990). Discussion: Heredity, fitness, and health. In C. Bouchard, R.J. Shephard, T. Stephens, J.R. Sutton, & B.D. McPherson (Eds.), *Exercise, fitness, and health* (pp. 147-153). Champaign, IL: Human Kinetics.

3. Bouchard, C., An, P., Rice, T., Skinner, J.S., Wilmore, J.H., Gagnon, J., Pérusse, L., Leon, A.S., & Rao, D.C. (1999). Familial aggregation of $\dot{V}O_{2max}$ response to exercise training: Results from the HERITAGE Family Study. *Journal of Applied Physiology, 87,* 1003-1008.

4. Bouchard, C., Dionne, F.T., Simoneau, J.-A., & Boulay, M.R. (1992). Genetics of aerobic and anaerobic performances. *Exercise and Sport Sciences Reviews, 20,* 27-58.

5. Bouchard, C., Lesage, R., Lortie, G., Simoneau, J.A., Hamel, P., Boulay, M.R., Pérusse, L., Theriault, G., & Leblanc, C. (1986). Aerobic performance in brothers, dizygotic and monozygotic twins. *Medicine and Science in Sports and Exercise, 18,* 639-646.

6. Costill, D.L., Coyle, E.F., Fink, W.F., Lesmes, G.R., & Witzmann, F.A. (1979). Adaptations in skeletal muscle following strength training. *Journal of Applied Physiology: Respiratory Environmental Exercise Physiology, 46,* 96-99.

7. Costill, D.L., Fink, W.J., Ivy, J.L., Getchell, L.H., & Witzmann, F.A. (1979). Lipid metabolism in skeletal muscle of endurance-trained males and females. *Journal of Applied Physiology, 28,* 251-255.

8. Dempsey, J.A. (1986). Is the lung built for exercise? *Medicine and Science in Sports and Exercise, 18,* 143-155.

9. Ehsani, A.A., Ogawa, T., Miller, T.R., Spina, R.J., & Jilka, S.M. (1991). Exercise training improves left ventricular systolic function in older men. *Circulation, 83,* 96-103.

10. Ekblom, B., Goldbarg, A.M., & Gullbring, B. (1972). Response to exercise after blood loss and reinfusion. *Journal of Applied Physiology, 33,* 175-180.

11. Fagard, R.H. (1996). Athlete's heart: A meta-analysis of the echocardiographic experience. *International Journal of Sports Medicine, 17,* S140-S144.

12. Gibala, M.J., Little, J.P., van Essen, M., Wilkin, G.P., Burgomaster, K.A., Safdar, A., Raha, S., & Tarnopolsky, M.A. (2006). Short-term sprint interval *versus* traditional endurance training: Similar initial adaptations in human skeletal muscle and exercise performance. *Journal of Physiology, 575,* 901-911.

13. Hagberg, J.M., Ehsani, A.A., Goldring, D., Hernandez, A., Sinacore, D.R., & Holloszy, J.O. (1984). Effect of weight training on blood pressure and hemodynamics in hypertensive adolescents. *Journal of Pediatrics, 104,* 147-151.

14. Hermansen, L., & Wachtlova, M. (1971). Capillary density of skeletal muscle in well-trained and untrained men. *Journal of Applied Physiology, 30,* 860-863.

15. Holloszy, J.O., Oscai, L.B., Mole, P.A., & Don, I.J. (1971). Biochemical adaptations to endurance exercise in skeletal muscle. In B. Pernow & B. Saltin (Eds.), *Muscle metabolism during exercise* (pp. 51-61). New York: Plenum Press.

16. Jacobs, I., Esbjörnsson, M., Sylvén, C., Holm, I., & Jansson, E. (1987). Sprint training effects on muscle myoglobin, enzymes, fiber types, and blood lactate. *Medicine and Science in Sports and Exercise, 19,* 368-374.

17. Jansson, E., Esbjörnsson, M., Holm, I., & Jacobs, I. (1990). Increase in the proportion of fast-twitch muscle fibres by sprint training in males. *Acta Physiologica Scandinavica, 140,* 359-363.

18. Klissouras, V. (1971). Adaptability of genetic variation. *Journal of Applied Physiology, 31,* 338-344.

19. MacDougall, J.D., Hicks, A.L., MacDonald, J.R., Mc Kelvie, R.S., Green, H.J., & Smith, K.M. (1998). Muscle performance and enzymatic adaptations to sprint interval training. *Journal of Applied Physiology, 84,* 2138-2142.

20. Martino, M., Gledhill, N., & Jamnik, V. (2002). High $\dot{V}O_{2max}$ with no history of training is primarily due to high blood volume. *Medicine and Science in Sports and Exercise, 34,* 966-971.

21. McCarthy, J.P., Pozniak, M.A., & Agre, J.C. (2002). Neuromuscular adaptations to concurrent strength and endurance training. *Medicine and Science in Sports and Exercise, 34,* 511-519.

22. McGuire, D.K., Levine, B.D., Williamson, J.W., Snell, P.G., Blomqvist, C.G., Saltin, B., & Mitchell, J.H. (2001). A 30-year follow-up of the Dallas Bedrest and Training Study: II. Effect of age on cardiovascular adaptation to exercise training. *Circulation, 104,* 1358-1366.

23. Milliken, M.C., Stray-Gundersen, J., Peshock, R.M., Katz, J., & Mitchell, J.H. (1988). Left ventricular mass as determined by magnetic resonance imaging in male endurance athletes. *American Journal of Cardiology, 62,* 301-305.

24. Pirnay, F., Dujardin, J., Deroanne, R., & Petit, J.M. (1971). Muscular exercise during intoxication by carbon monoxide. *Journal of Applied Physiology, 31,* 573-575.

25. Prud'homme, D., Bouchard, C., LeBlanc, C., Landrey, F., & Fontaine, E. (1984). Sensitivity of maximal aerobic power to training is genotype-dependent. *Medicine and Science in Sports and Exercise, 16,* 489-493.

26. Rico-Sanz, J., Rankinen, T., Joanisse, D.R., Leon, A.S., Skinner, J.S., Wilmore, J.H., Rao, D.C., & Bouchard, C. (2003). Familial resemblance for muscle phenotypes in The Heritage Family Study. *Medicine and Science in Sports and Exercise, 35*(8): 1360-1366.

27. Saltin, B., Nazar, K., Costill, D.L., Stein, E., Jansson, E., Essen, B., & Gollnick, P.D. (1976). The nature of the training response: Peripheral and central adaptations to one-legged exercise. *Acta Physiologica Scandinavica, 96,* 289-305.

28. Saltin, B., & Rowell, L.B. (1980). Functional adaptations to physical activity and inactivity. *Federation Proceedings, 39,* 1506-1513.

29. Sawka, M.N., Convertino, V.A., Eichner, E.R., Schnieder, S.M., & Young, A.J. (2000). Blood volume: Importance and adaptations to exercise training, environmental stresses, and trauma/sickness. *Medicine and Science in Sports and Exercise, 32,* 332-348.

30. Strømme, S.B., Ingjer, F., & Meen, H.D. (1977). Assessment of maximal aerobic power in specifically trained athletes. *Journal of Applied Physiology, 42,* 833-837.

31. Wilmore, J.H., Stanforth, P.R., Gagnon, J., Rice, T., Mandel, S., Leon, A.S., Rao, D.C., Skinner, J.S., & Bouchard, C. (2001). Cardiac output and stroke volume changes with endurance training: The HERITAGE Family Study. *Medicine and Science in Sports and Exercise, 33,* 99-106.

32. Wilmore, J.H., Stanforth, P.R., Hudspeth, L.A., Gagnon, J., Daw, E.W., Leon, A.S., Rao, D.C., Skinner, J.S., & Bouchard, C. (1998). Alterations in resting metabolic rate as a consequence of 20 wk of endurance training: The HERITAGE Family Study. *American Journal of Clinical Nutrition, 68,* 66-71.

Chapter 12

1. American College of Sports Medicine. (2006). Prevention of cold injuries during exercise. *Medicine and Science in Sports and Exercise, 38*(11), 2012-2029.

2. King, D.S., Costill, D.L., Fink, W.J., Hargreaves, M., & Fielding, R.A. (1985). Muscle metabolism during exercise in the heat in unacclimatized and acclimatized humans. *Journal of Applied Physiology, 59,* 1350-1354.

3. Rowell, L.B. (1974). Human cardiovascular adjustments to heat stress. *Physiological Reviews, 54,* 75-159.

4. Young, A.J. (1996). Homeostatic responses to prolonged cold exposure: Human cold acclimation. In M.J. Fregley & C.M. Blatteis (Eds.), *Handbook of physiology: Section 4. Environmental physiology* (pp. 419-438). New York: Oxford University Press.

Chapter 13

1. Bartsch, P. & Saltin, B. (2008). General introduction to altitude adaptation and mountain sickness. *Scandinavian Journal of Medicine and Science in Sports*, **18** (suppl. 1), 1-10.

2. Bonetti, D.L. & Hopkins, W.G. (2009). Sea-level exercise performance following adaptation to hypoxia: a Meta-analysis. *Sports Medicine*, **39**, 107-127.

3. Brooks, G.A., Wolfel, E.E., & Groves, B.M. (1992). Muscle accounts for glucose disposal but not blood lactate appearance during exercise after acclimatization to 4,300 m. *Journal of Applied Physiology*, **72**, 2435-2445.

4. Brosnan, M.J., Martin, D.T., Hahn, A.G., Gore, C.J., & Hawley, J.A. (2000). Impaired interval exercise responses in elite female cyclists at moderate simulated altitude. *Journal of Applied Physiology*, **89**, 1819-1824.

5. Buskirk, E.R., Kollias, J., Piconreatigue, E., Akers, R., Prokop, E., & Baker, P. (1967). Physiology and performance of track athletes at various altitudes in the United States and Peru. In R.F. Goddard (Ed.), *The effects of altitude on physical performance* (pp. 65-71). Chicago: Athletic Institute.

6. Daniels, J., & Oldridge, N. (1970). Effects of alternate exposure to altitude and sea level on world-class middledistance runners. *Medicine and Science in Sports*, **2**, 107-112.

7. Forster, P.J.G. (1985). Effect of different ascent profiles on performance at 4200 m elevation. *Aviation, Space, and Environmental Medicine*, **56**, 785-794.

8. Levine, B.D., & Stray-Gundersen, J. (1997). "Living high–training low": Effect of moderate-altitude acclimatization with low-altitude training on performance. *Journal of Applied Physiology*, **83**, 102-112.

9. Norton, E.G. (1925). *The fight for Everest: 1924*. London: Arnold.

10. Pugh, L.C.G.E., Gill, M., Lahiri, J., Milledge, J., Ward, M., & West, J. (1964). Muscular exercise at great altitudes. *Journal of Applied Physiology*, **19**, 431-440.

11. Stray-Gundersen, J., Chapman, R.F., & Levine, B.D. (2001). "Living high–training low" altitude training improves sea level performance in male and female elite runners. *Journal of Applied Physiology*, **91**, 1113-1120.

12. Sutton, J., & Lazarus, L. (1973). Mountain sickness in the Australian Alps. *Medical Journal of Australia*, **1**, 545-546.

13. Sutton, J.R., Reeves, J.T., Wagner, P.D., Groves, B.M., Cymerman, A., Malconian, M.K., Rock, P.B., Young, P.M., Walter, S.D., & Houston, C.S. (1988). Operation Everest II: Oxygen transport during exercise at extreme simulated altitude. *Journal of Applied Physiology*, **64**, 1309-1321.

14. Ward, M.P., Milledge, J.S., & West, J.B. (1989). *High altitude medicine and physiology*. Philadelphia: University of Pennsylvania Press.

15. West, J.B., Peters, R.M., Aksnes, G., Maret, K.H., Milledge, J.S., & Schoene, R.B. (1986). Nocturnal periodic breathing at altitudes of 6300 and 8050 m. *Journal of Applied Physiology*, **61**, 280-287.

Chapter 14

1. Armstrong, L.E., & VanHeest, J.L. (2002). The unknown mechanism of the overtraining syndrome. *Sports Medicine*, **32**, 185-209.

2. Costill, D.L. (1998). Training adaptations for optimal performance. Paper presented at the VIII International Symposium on Biomechanics and Medicine of Swimming, June 28, University of Jyväskylä, Finland.

3. Costill, D.L., Fink, W.J., Hargreaves, M., King, D.S., Thomas, R., & Fielding, R. (1985). Metabolic characteristics of skeletal muscle during detraining from competitive swimming. *Medicine and Science in Sports and Exercise*, **17**, 339-343.

4. Costill, D.L., King, D.S., Thomas, R., & Hargreaves, M. (1985). Effects of reduced training on muscular power in swimmers. *Physician and Sportsmedicine*, **13**(2), 94-101.

5. Costill, D.L., Maglischo, E., & Richardson, A. (1991). *Handbook of sports medicine: Swimming*. London: Blackwell.

6. Costill, D.L., Thomas, R., Robergs, R.A., Pascoe, D.D., Lambert, C.P., Barr, S.I., & Fink, W.J. (1991). Adaptations to swimming training: Influence of training volume. *Medicine and Science in Sports and Exercise*, **23**, 371-377.

7. Coyle, E.F., Martin, W.H., III, Sinacore, D.R., Joyner, M.J., Hagberg, J.M., & Holloszy, J.O. (1984). Time course of loss of adaptations after stopping prolonged intense endurance training. *Journal of Applied Physiology*, **57**, 1857-1864.

8. Fitts, R.H., Costill, D.L., & Gardetto, P.R. (1989). Effect of swim-exercise training on human muscle fiber function. *Journal of Applied Physiology*, **66**, 465-475.

9. Fleck, S.J., & Kraemer, W.J. (2004). *Designing resistance training programs* (3rd ed.). Champaign, IL: Human Kinetics.

10. Hickson, R.C., Foster, C., Pollock, M.L., Galassi, T.M., & Rich, S. (1985). Reduced training intensities and loss of aerobic power, endurance, and cardiac growth. *Journal of Applied Physiology*, **58**, 492-499.

11. Houmard, J.A., Costill, D.L., Mitchell, J.B., Park, S.H., Hickner, R.C., & Roemmish, J.N. (1990). Reduced training maintains performance in distance runners. *International Journal of Sports Medicine*, **11**, 46-51.

12. Houmard, J.A., Scott, B.K., Justice, C.L., & Chenier, T.C. (1994). The effects of taper on performance in distance runners. *Medicine and Science in Sports and Exercise,* **26,** 624-631.

13. Kraemer, W.J., & Ratamess, N.A. (2003). Endocrine responses and adaptations to strength and power training. In P.V. Komi (Ed.), *Strength and power in sport* (pp. 379-380). Oxford: Blackwell Scientific.

14. Krivickas, L.S. (2006). Recurrent rhabdomyolysis in a collegiate athlete: A case report. *Medicine and Science in Sports and Exercise,* **38,** 407-410.

15. Lemmer, J.T., Hurlbut, D.E., Martel, G.F., Tracy, B.L., Ivey, F.M., Metter, E.J., Fozard, J.L., Fleg, J.L., & Hurley, B.F. (2000). Age and gender responses to strength training and detraining. *Medicine and Science in Sports and Exercise,* **32,** 1505-1512.

16. Mujika, I., & Padilla, S. (2003). Scientific bases for precompetition tapering strategies. *Medicine and Science in Sports and Exercise,* **35,** 1182-1187.

17. Nieman, D.C. (1994). Exercise, infection, and immunity. *International Journal of Sports Medicine,* **15,** S131-S141.

18. Saltin, B., Blomqvist, G., Mitchell, J.H., Johnson, R.L., Jr., Wildenthal, K., & Chapman, C.B. (1968). Response to submaximal and maximal exercise after bed rest and training. *Circulation,* **38**(Suppl. 7).

19. Selye, H. (1956). *The stress of life.* New York: McGraw-Hill.

20. Shepherd, R.J. (2001). Chronic fatigue syndrome: An update. *Sports Medicine,* **31,** 167-194.

21. Smith, L.L. (2000). Cytokine hypothesis of overtraining: A physiological adaptation to excessive stress? *Medicine and Science in Sports and Exercise,* **32,** 317-331.

22. Springer, B.L., & Clarkson, P.M. (2003). Two cases of exertional rhabdomyolysis precipitated by personal trainers. *Medicine and Science in Sports and Exercise,* **35,** 1499-1502.

23. Trappe, T., Trappe, S., Lee, G., Widrick, J., Fitts, R., & Costill, D. (2006). Cardiorespiratory responses to physical work during and following 17 days of bed rest and spaceflight. *Journal of Applied Physiology,* **100,** 951-957.

24. Watenpaugh, D.E., & Hargens, A.R. (1996). The cardiovascular system in microgravity. In M.J. Fregly & C.M. Blatteis (Eds.), *Handbook of physiology: Environmental physiology* (Vol. 1, pp. 631-674). New York: Oxford University Press.

Chapter 15

1. American College of Sports Medicine. Nattiv A, Loucks AB, Manore MM, Sanborn CF, Sundgot-Borgen J, Warren MP (2007). The female athlete triad. Position Stand. Med Sci Sports Exerc **39**(10):1867-82.

2. American College of Sports Medicine, American Dietetic Association, and Dietitians of Canada. (2000). Nutrition and athletic performance. Joint position statement. *Medicine and Science in Sports and Exercise,* **32,** 2130-2145.

3. Armstrong, L.E., Costill, D.L., & Fink, W.J. (1985). Influence of diuretic-induced dehydration on competitive running performance. *Medicine and Science in Sports and Exercise,* **17,** 456-461.

4. Åstrand, P.-O. (1967). Diet and athletic performance. *Federation Proceedings,* **26,** 1772-1777.

5. Beaton, L.J., Allan, D.A., Tarnopolsky, M.A., Tiidus, P.M., & Phillips, S.M. (2002). Contraction-induced muscle damage is unaffected by vitamin E supplementation. *Medicine and Science in Sports and Exercise,* **34,** 798-805.

6. Cheuvront, S.N. (1999). The Zone diet and athletic performance. *Sports Medicine,* **27,** 213-228.

7. Coombes, J.S., & Hamilton, K.L. (2000). The effectiveness of commercially available sports drinks. *Sports Medicine,* **29,** 181-209.

8. Costill, D.L., Bowers, R., Branam, G., & Sparks, K. (1971). Muscle glycogen utilization during prolonged exercise on successive days. *Journal of Applied Physiology,* **31,** 834-838.

9. Costill, D.L., & Saltin, B. (1974). Factors limiting gastric emptying during rest and exercise. *Journal of Applied Physiology,* **37,** 679-683.

10. De Souza MJ, Toombs RJ, Scheid JL, O'Donnell E, West SL, Williams NI 2010 High prevalence of subtle and severe menstrual disturbances in exercising women: confirmation using daily hormone measures. Hum Reprod **25**(2):491-503.

11. De Souza MJ, Williams NI 2004 Physiological aspects and clinical sequelae of energy deficiency and hypoestrogenism in exercising women. Hum Reprod Update **10**(5):433-48.

12. Dougherty, K.A., Baker, L.B., Chow, M., & Kenney, W.L. (2006). Two percent dehydration impairs and six percent carbohydrate drink improves boys basketball skills. *Medicine and Science in Sports and Exercise,* **38,** 1650-1658.

13. Fairchild, T.J., Fletcher, S., Steele, P., Goodman, C., Dawson, B., & Fournier, P.A. (2002). Rapid carbohydrate loading after a short bout of near maximal-intensity exercise. *Medicine and Science in Sports and Exercise,* **34,** 980-986.

14. Foster-Powell, K., Holt, S.H.A., & Brand-Miller, J.C. (2002). International table of glycemic index and glycemic load values: 2002. *American Journal of Clinical Nutrition,* **76,** 5-56.

15. Frizzell, R.T., Lang, G.H., Lowance, D.C., & Lathan, S.R. (1986). Hyponatremia and ultramarathon running. *Journal of the American Medical Association,* **255,** 772-774.

16. Gollnick, P.D., Piehl, K., & Saltin, B. (1974). Selective glycogen depletion pattern in human muscle fibres after exercise of varying intensity and at varying pedaling rates. *Journal of Physiology,* **241,** 45-57.

17. Ivy, J.L. (2004). Regulation of muscle glycogen repletion, muscle protein synthesis and repair following exercise. *Journal of Sports Science and Medicine,* **3,** 131-138.

18. Ivy, J.L., Katz, A.L., Cutler, C.L., Sherman, W.M., & Coyle, E.F. (1988). Muscle glycogen synthesis after exercise: Effect of time of carbohydrate ingestion. *Journal of Applied Physiology, 64,* 1480-1485.

19. Ivy, J.L., Lee, M.C., Brozinick, J.T., Jr., & Reed, M.J. (1988). Muscle glycogen storage after different amounts of carbohydrate ingestion. *Journal of Applied Physiology, 65,* 2018-2023.

20. Jeukendrup, A., & Gleeson, M. (2010). *Sport nutrition: An introduction to energy production and performance* (2nd ed.). Champaign, IL: Human Kinetics.

21. Rigotti NA, Neer RM, Skates SJ, Herzog DB, Nussbaum SR. The clinical course of osteoporosis in anorexia nervosa. A longitudinal study of cortical bone mass. JAMA. (1991) 265(9):1133-1138.

22. Schabort, E.J., Bosch, A.N., Weltan, S.M., & Noakes, T.D. (1999). The effect of a preexercise meal on time to fatigue during prolonged cycling exercise. *Medicine and Science in Sports and Exercise, 31,* 464-471.

23. Sears, B. (1995). *The Zone.* New York: HarperCollins.

24. Sears, B. (2000). The Zone diet and athletic performance [letter]. *Sports Medicine, 29,* 289-291.

25. Sundgot-Borgen J 2002 Weight and eating disorders in elite athletes. Scand J Med Sci Sports **12**(5):259-60.

26. Sundgot-Borgen J 1999 Eating disorders among male and female elite athletes. Br J Sports Med **33**(6):434.

27. Wade GN, Schneider JE 1992 Metabolic fuels and reproduction in female mammals. Neuroscience and biobehavioral reviews **16**(2):235.

28. Wilmore, J.H., Brown, C.H., & Davis, J.A. (1977). Body physique and composition of the female distance runner. *Annals of the New York Academy of Sciences, 301,* 764-776.

29. Wilmore, J.H., Morton, A.R., Gilbey, H.J., & Wood, R.J. (1998). Role of taste preference on fluid intake during and after 90 min of running at 60% of $\dot{V}O_{2max}$ in the heat. *Medicine and Science in Sports and Exercise, 30,* 587-595.

30. Wolfe, R.R. (2006). Skeletal muscle protein metabolism and resistance exercise. *Journal of Nutrition, 136,* 525S-528S.

Chapter 16

1. Alvois, L., Robinson, N., Saudan, D., Baume, N., Mangin, P., & Saugy, M. (2006). Central nervous system stimulants and sport practice. *British Journal of Sports Medicine, 40*(Suppl. I), i16-i20.

2. American College of Sports Medicine Consensus Statement. (2000). The physiological and health effects of oral creatine supplementation. *Medicine and Science in Sports and Exercise, 32,* 706-717.

3. American College of Sports Medicine Position Stand. (1996). The use of blood doping as an ergogenic aid. *Medicine and Science in Sports and Exercise, 28*(6), i-xii.

4. Ariel, G., & Saville, W. (1972). Anabolic steroids: The physiological effects of placebos. *Medicine and Science in Sports and Exercise, 4,* 124-126.

5. Bannister, R.G., & Cunningham, D.J.C. (1954). The effects on respiration and performance during exercise of adding oxygen to the inspired air. *Journal of Physiology, 125,* 118-137.

6. Berning, J.M., Adams, K.J., & Stamford, B.A. (2004). Anabolic steroid usage in athletics: Facts, fiction, and public relations. *Journal of Strength and Conditioning Research, 18,* 908-917.

7. Bhasin, S., Storer, T.W., Berman, N., Callegari, C., Clevenger, B., Phillips, J., Bunnell, T.J., Tricker, R., Shirazi, A., & Casaburi, R. (1996). The effects of supraphysiologic doses of testosterone on muscle size and strength in normal men. *New England Journal of Medicine, 335,* 1-7.

8. Birkeland, K.I., Stray-Gundersen, J., Hemmersbach, P., Hallén, J., Haug, E., & Bahr, R. (2000). Effect of rhEPO administration on serum levels of sTfR and cycling performance. *Medicine and Science in Sports and Exercise, 32,* 1238-1243.

9. Broeder, C.E., Quindry, J., Brittingham, K., Panton, L., Thomson, J., Appakondu, S., Bruel, K., Byrd, R., Douglas, J., Earnest, C., Mitchell, C., Olson, M., Roy, T., & Yarlagadda, C. (2000). The Andro Project: Physiological and hormonal influences of androstenedione supplementation in men 35 to 65 years old participating in a high-intensity resistance training program. *Archives of Internal Medicine, 160,* 3093-3104.

10. Bronson, F.H., & Matherne, C.M. (1997). Exposure to anabolic-androgenic steroids shortens life span of male mice. *Medicine and Science in Sports and Exercise, 29,* 615-619.

11. Brown, G.A., Vukovich, M.D., Sharp, R.L., Reifenrath, T.A., Parsons, K.A., & King, D.S. (1999). Effect of oral DHEA on serum testosterone and adaptations to resistance training in young men. *Journal of Applied Physiology, 87,* 2274-2283.

12. Buick, F.J., Gledhill, N., Froese, A.B., Spriet, L., & Meyers, E.C. (1980). Effect of induced erythrocythemia on aerobic work capacity. *Journal of Applied Physiology, 48,* 636-642.

13. Calfee, R., & Fadale, P. (2006). Popular ergogenic drugs and supplements in young athletes. *Pediatrics, 117,* e577-e589.

14. Costill, D.L., Dalsky, G.P., & Fink, W.J. (1978). Effects of caffeine ingestion on metabolism and exercise performance. *Medicine and Science in Sports, 10,* 155-158.

15. Costill, D.L., Verstappen, F., Kuipers, H., Janssen, E., & Fink, W. (1984). Acid-base balance during repeated bouts of exercise: Influence of HCO-3. *International Journal of Sports Medicine, 5,* 228-231.

16. Davis, J.M. (1995). Carbohydrates, branched-chain amino acids, and endurance: The central fatigue hypothesis. *International Journal of Sport Nutrition, 5,* S29-S38.

17. Eichner, E.R. (1989). Ergolytic drugs. *Sports Science Exchange,* **2**(15), 1-4.

18. Ekblom, B., & Berglund, B. (1991). Effect of erythropoietin administration on maximal aerobic power. *Scandinavian Journal of Medicine and Science in Sports,* **1,** 88-93.

19. Ekblom, B., Goldbarg, A.N., & Gullbring, B. (1972). Response to exercise after blood loss and reinfusion. *Journal of Applied Physiology,* **33,** 175-180.

20. Evans, N.A. (2004). Current concepts in anabolic-androgenic steroids. *American Journal of Sports Medicine,* **32,** 534-542.

21. Forbes, G.B. (1985). The effect of anabolic steroids on lean body mass: The dose response curve. *Metabolism,* **34,** 571-573.

22. Gledhill, N. (1985). The influence of altered blood volume and oxygen transport capacity on aerobic performance. *Exercise and Sport Sciences Reviews,* **13,** 75-93.

23. Goforth, H.W., Jr., Campbell, N.L., Hodgdon, J.A., & Sucec, A.A. (1982). Hematologic parameters of trained distance runners following induced erythrocythemia [abstract]. *Medicine and Science in Sports and Exercise,* **14,** 174.

24. Graham, T.E. (2001). Caffeine and exercise: Metabolism, endurance and performance. *Sports Medicine,* **31,** 785-807.

25. Hartgens, F., & Kuipers, H. (2004). Effects of androgenic-anabolic steroids in athletes. *Sports Medicine,* **34,** 513-554.

26. Heinonen, O.J. (1996). Carnitine and physical exercise. *Sports Medicine,* **22,** 109-132.

27. Hervey, G.R., Knibbs, A.V., Burkinshaw, L., Morgan, D.B., Jones, P.R.M., Chettle, D.R., & Vartsky, D. (1981). Effects of methandienone on the performance and body composition of men undergoing athletic training. *Clinical Science,* **60,** 457-461.

28. Ivy, J.L., Costill, D.L., Fink, W.J., & Lower, R.W. (1979). Influence of caffeine and carbohydrate feedings on endurance performance. *Medicine and Science in Sports and Exercise,* **11,** 6-11.

29. Juhn, M.S. (2003). Popular sports supplements and ergogenic aids. *Sports Medicine,* **33,** 921-939.

30. King, D.S., Sharp, R.L., Vukovich, M.D., Brown, G.A., Reifenrath, T.A., Uhl, N.L., & Parsons, K.A. (1999). Effect of oral androstenedione on serum testosterone and adaptations of resistance training in young men: A randomized controlled trial. *Journal of the American Medical Association,* **281,** 2020-2028.

31. Linderman, J., & Fahey, T.D. (1991). Sodium bicarbonate ingestion and exercise performance: An update. *Sports Medicine,* **11,** 71-77.

32. Magkos, F., & Kavouras, S.A. (2004). Caffeine and ephedrine: Physiological, metabolic and performanceenhancing effects. *Sports Medicine,* **34,** 871-889.

33. Nissen, S.L., & Sharp, R.L. (2003). Effect of dietary supplements on lean mass and strength gains with resistance exercise: A meta-analysis. *Journal of Applied Physiology,* **94,** 651-659.

34. Pärssinen, M., & Seppälä, T. (2002). Steroid use and long-term health risks in former athletes. *Sports Medicine,* **32,** 83-94.

35. Roth, D.A., & Brooks, G.A. (1990). Lactate transport is mediated by a membrane-bound carrier in rat skeletal muscle sarcolemmal vesicles. *Archives of Biochemistry and Biophysics,* **279,** 377-385.

36. Rudman, D., Feller, A.G., Nagraj, H.S., Gergans, G.A., Lalitha, P.Y., Goldberg, A.F., Schlenker, R.A., Cohn, L., Rudman, I.W., & Mattson, D.E. (1990). Effects of human growth hormone in men over 60 years old. *New England Journal of Medicine,* **323,** 1-6.

37. Smith-Rockwell, M., Nickols-Richardson, S.M., & Thye, F.W. (2001). Nutrition knowledge, opinions, and practices of coaches and athletic trainers at a division 1 university. *International Journal of Sports Nutrition and Exercise Metabolism,* **11,** 174-185.

38. Spriet, L.L. (1991). Blood doping and oxygen transport. In D.R. Lamb & M.H. Williams (Eds.), *Ergogenics: Enhancement of performance in exercise and sport* (pp. 213-242). Dubuque, IA: Brown & Benchmark.

39. Spriet, L.L., & Gibala, M.J. (2004). Nutritional strategies to influence adaptations to training. *Journal of Sports Sciences,* **22,** 127-141.

40. Tamaki, T., Uchiyama, S., Uchiyama, Y., Akatsuka, A., Roy, R.R., & Edgerton, V.R. (2001). Anabolic steroids increase exercise tolerance. *American Journal of Physiology: Endocrinology and Metabolism,* **280,** E973-E981.

41. Tokish, J.M., Kocher, M.S., & Hawkins, R.J. (2004). Ergogenic aids: A review of basic science, performance, side effects, and status in sports. *American Journal of Sports Medicine,* **32,** 1543-1553.

42. van Hall, G., Raaymakers, J.S.H., Saris, W.H.M., & Wagenmakers, A.J.M. (1995). Ingestion of branched-chain amino acids and tryptophan during sustained exercise in man: Failure to affect performance. *Journal of Physiology,* **486,** 789-794.

43. Villareal, D.T., & Holloszy, J.O. (2006). DHEA enhances effects of weight training on muscle mass and strength. *American Journal of Physiology: Endocrinology and Metabolism.* **291,** E1003-1008.

44. Williams, M.H. (Ed.). (1983). *Ergogenic aids in sport.* Champaign, IL: Human Kinetics.

45. Williams, M.H., Wesseldine, S., Somma, T., & Schuster, R. (1981). The effect of induced erythrocythemia upon 5-mile treadmill run time. *Medicine and Science in Sports and Exercise,* **13,** 169-175.

46. Winter, F.D., Snell, P.G., & Stray-Gundersen, J. (1989). Effects of 100% oxygen on performance of professional soccer players. *Journal of the American Medical Association,* **262,** 227-229.

47. Yarasheski, K.E. (1994). Growth hormone effects on metabolism, body composition, muscle mass, and strength. *Exercise and Sport Sciences Reviews*, **22**, 285-312.

48. Yesalis, C.E. (Ed.). (2000). *Anabolic steroids in sport and exercise* (2nd ed.). Champaign, IL: Human Kinetics.

Chapter 17

1. Bar-Or, O. (1983). *Pediatric sports medicine for the practitioner: From physiologic principles to clinical applications.* New York: Springer-Verlag.

2. Bar-Or, O. (1989). Temperature regulation during exercise in children and adolescents. In C.V. Gisolfi & D.R. Lamb (Eds.), *Perspectives in exercise science and sports medicine: Youth, exercise and sport* (pp. 335-362). Carmel, IN: Benchmark Press.

3. Beneke, R., Hütler, M., Jung, M., & Leithäuser, R.M. (2005). Modeling the blood lactate kinetics at maximal short-term exercise conditions in children, adolescents, and adults. *Journal of Applied Physiology*, **99**, 499-504.

4. Burgeson, C.R., Wechsler, H., Brener, N.D., Young, J.C. & Spain, C.G. (2001). Physical education and activity: results from the School Health Policies and Programs 2000. *Journal of School Health*, **71**, 279-293.

5. Clarke, H.H. (1971). *Physical and motor tests in the Medford boys' growth study.* Englewood Cliffs, NJ: Prentice Hall.

6. Cureton, K.J., Sloniger, M.A., Black, D.M., McCormack, W.P., & Rowe, D.A. (1997). Metabolic determinants of the age-related improvement in one-mile run/walk performance in youth. *Medicine and Science in Sports and Exercise*, **29**, 259-267.

7. Daniels, J., Oldridge, N., Nagle, F., & White, B. (1978). Differences and changes in VO2 among young runners 10 to 18 years of age. *Medicine and Science in Sports and Exercise*, **10**, 200-203.

8. Eriksson, B.O. (1972). Physical training, oxygen supply and muscle metabolism in 11-13-year-old boys. *Acta Physiologica Scandinavica* (Suppl. 384), 1-48.

9. Falk, B., & Eliakim, A. (2003). Resistance training, skeletal muscle and growth. *Pediatric Endocrinology Reviews*, **1**, 120-127.

10. Fleck, S.J., & Kraemer, W.J. (2004). *Designing resistance training programs* (3rd ed.). Champaign, IL: Human Kinetics.

11. Froberg, K., & Lammert, O. (1996). Development of muscle strength during childhood. In O. Bar-Or (Ed.), *The child and adolescent athlete* (p. 28). London: Blackwell.

12. Gunter, K., Baxer-Jones, A.D., Mirwald, R.L., Almstedt, H., Fuller, A., Durski, S. & Snow, C. (2008). Jump starting skeletal health: a 4-year longitudinal study assessing the effects of jumping on skeletal development in pre and circum pubertal children. *Bone*, **4**, 710-718.

13. Halpern, A., Mancini, M.C., Magelhaes, M.E.C., Fisbert, M., Radominski, R., Berolami, M.C., Bertolami, A., de Melo, M.E., Zanella, M.T., Queiroz, M.S. & Nery, M. (2010). Metabolic syndrome, dyslipidemia, hypertension and type 2 diabetes in youth: from diagnosis to treatment. *Diabetology & Metabolic Syndrome*, **2**, 55-75.

14. Kaczor, J.J., Ziolkowski, W., Popinigis, J., Tarnopolsky, M.A. (2005). Anaerobic and aerobic enzyme activities in human skeletal muscle from children and adults. *Pediatric Research*, 57(3), 331-5.

15. Kraemer, W.J., & Fleck, S.J. (2005). *Strength training for young athletes* (2nd ed.). Champaign, IL: Human Kinetics

16. Krahenbuhl, G.S., Morgan, D.W., & Pangrazi, R.P. (1989). Longitudinal changes in distance-running performance of young males. *International Journal of Sports Medicine*, **10**, 92-96.

17. Lobstein, T., Baur, L. & Uauy, R. (2004). Obesity in children and young people: a crisis in public health. *Obesity Reviews*, **1**, 4-104.

18. Mahon, A.D., & Vaccaro, P. (1989). Ventilatory threshold and $\dot{V}O_{2max}$ changes in children following endurance training. *Medicine and Science in Sports and Exercise*, **21**, 425-431.

19. Malina, R.M. (1989). Growth and maturation: Normal variation and effect of training. In C.V. Gisolfi & D.R. Lamb (Eds.), *Perspectives in exercise science and sports medicine: Youth, exercise and sport* (pp. 223-265). Carmel, IN: Benchmark Press.

20. Malina, R.M., Bouchard, C., & Bar-Or, O. (2004). *Growth, maturation, and physical activity* (2nd ed.). Champaign, IL: Human Kinetics.

21. Pitukcheewanont, P., Punyasavatsut, N. & Feuille, M. (2010). Physical activity and bone health in children and adolescents. *Pediatric Endocrinology Reviews*, **7**, 275-82.

22. Ramsay, J.A., Blimkie, C.J.R., Smith, K., Garner, S., MacDougall, J.D., & Sale, D.G. (1990). Strength training effects in prepubescent boys. *Medicine and Science in Sports and Exercise*, **22**, 605-614.

23. Riddell, M.C. (2008). The endocrine response and substrate utilization during exercise in children and adolescents. *Journal of Applied Physiology*, **105**, 725-733.

24. Robinson, S. (1938). Experimental studies of physical fitness in relation to age. *Arbeitsphysiologie*, **10**, 251-323.

25. Rogers, D.M., Olson, B.L., & Wilmore, J.H. (1995). Scaling for the VO2-to-body size relationship among children and adults. *Journal of Applied Physiology*, **79**, 958-967.

26. Rowland, T.W. (1985). Aerobic response to endurance training in prepubescent children: A critical analysis. *Medicine and Science in Sports and Exercise*, **17**, 493-497.

27. Rowland, T.W. (1989). Oxygen uptake and endurance fitness in children: A developmental perspective. *Pediatric Exercise Science*, **1**, 313-328.

28. Rowland, T.W. (1991). "Normalizing" maximal oxygen uptake, or the search for the holy grail (per kg). *Pediatric Exercise Science*, **3**, 95-102.

29. Rowland, T.W. (2005). *Children's exercise physiology* (2nd ed.). Champaign, IL: Human Kinetics.

30. Rowland, T.W. (2007). Evolution of maximal oxygen uptake in children. *Medicine Sport Science*, **50,** 200-209.

31. Rowland, T.W. (2008). Thermoregulation during exercise in the heat in children: old concepts revisited. *Journal of Applied Physiology*, **105,** 718-724.

32. Santos, A.M.C., Welsman, J.R., De Ste Croix, M.B.A., & Armstrong, N. (2002). Age- and sex-related differences in optimal peak power. *Pediatric Exercise Science*, **14,** 202-212.

33. Sjödin, B., & Svedenhag, J. (1992). Oxygen uptake during running as related to body mass in circumpubertal boys: A longitudinal study. *European Journal of Applied Physiology*, **65,** 150-157.

34. Small, E.W., McCambridge, T.M., Benjamin, H.J., Bernhardt, D.T., Brenner, J.S., Cappetta, C.T., Congeni, J.A., Gregory, A.J., Griesemer, B.A., Reed, F.E., Rice, S.G., Gomez, J.E., Gregory, D.B., Stricker, P.R., Le Blanc, C.M., Raynor, J., Bergeron, M.F. & Emanuel, A. (2008). Strength training by children and adolescents. *Pediatrics*, **121,** 835-840.

35. Turley, K.R., & Wilmore, J.H. (1997). Cardiovascular responses to treadmill and cycle ergometer exercise in children and adults. *Journal of Applied Physiology*, **83,** 948-957.

Chapter 18

1. Buskirk, E.R., & Hodgson, J.L. (1987). Age and aerobic power: The rate of change in men and women. *Federation Proceedings*, **46,** 1824-1829.

2. Connelly, D.M., Rice, C.L., Roos, M.R., & Vandervoort, A.A. (1999). Motor unit firing rates and contractile properties in tibialis anterior of young and old men. *Journal of Applied Physiology*, **87,** 843-852.

3. Costill, D.L. (1986). *Inside running: Basics of sports physiology.* Indianapolis: Benchmark Press.

4. Dill, D.B., Robinson, S., & Ross, J.C. (1967). A longitudinal study of 16 champion runners. *Journal of Sports Medicine and Physical Fitness*, **7,** 4-27.

5. DeGroot, D.W., Havenith, G., & Kenney, W.L. (2006). Responses to mild cold stress are predicted by different individual characteristics in young and older subjects. *Journal of Applied Physiology*, **101,** 1607-1615.

6. Doherty, T.J., Vandervoort, A.A., Taylor, A.W., & Brown, W.F. (1993). Effects of motor unit losses on strength in older men and women. *Journal of Applied Physiology*, **74,** 868-874.

7. Fitzgerald, M.D., Tanaka, H., Tran, Z.V., & Seals, D.R. (1997). Age-related declines in maximal aerobic capacity in regularly exercising vs. sedentary women: A metaanalysis. *Journal of Applied Physiology*, **83,** 160-165.

8. Frontera, W.R., Meredith, C.N., O'Reilly, K.P., Knuttgen, W.G., & Evans, W.J. (1988). Strength conditioning in older men: Skeletal muscle hypertrophy and improved function. *Journal of Applied Physiology*, **64,** 1038-1044.

9. Goodrick, C.L. (1980). Effects of long-term voluntary wheel exercise on male and female Wistar rats: 1. Longevity, body weight and metabolic rate. *Gerontology*, **26,** 22-33.

10. Hagerman, F.C., Walsh, S.J., Staron, R.S., Hikida, R.S., Gilders, R.M., Murray, T.F., Toma, K., & Ragg, K.E. (2000). Effects of high-intensity resistance training on untrained older men. I. Strength, cardiovascular, and metabolic responses. *Journals of Gerontology Series A: Biological Sciences and Medical Sciences*, **55,** B336-B346.

11. Häkkinen, K., Kraemer, W.J., Pakarinen, A., Triplett-McBride, T., McBride, J.M., Häkkinen, A., Alen, M., McGuigan, M.R., Bronks, R., & Newton, R.U. (2002). Effects of heavy resistance/power training on maximal strength, muscle morphology, and hormonal response patterns in 60-75-year-old men and women. *Canadian Journal of Applied Physiology*, **27,** 213-231.

12. Häkkinen, K., Pakarinen, A., Kraemer, W.J., Häkkinen, A., Valkeinen, H., & Alen, M. (2001). Selective muscle hypertrophy, changes in EMG and force, and serum hormones during strength training in older women. *Journal of Applied Physiology*, **91,** 569-580.

13. Hameed, M., Harridge, S.D.R., & Goldspink, G. (2002). Sarcopenia and hypertrophy: A role for insulin-like growth factor-1 and aged muscle? *Exercise and Sport Sciences Reviews*, **30,** 15-19.

14. Hawkins, S.A., Marcell, T.J., Jaque, S.V., & Wiswell, R.A. (2001). A longitudinal assessment of change in $\dot{V}O_{2max}$ and maximal heart rate in master athletes. *Medicine and Science in Sports and Exercise*, **33,** 1744-1750.

15. Hikida, R.S., Staron, R.S., Hagerman, F.C., Walsh, S., Kaiser, E., Shell, S., & Hervey, S. (2000). Effects of highintensity resistance training on untrained older men. II. Muscle fiber characteristics and nucleocytoplasmic relationships. *Journals of Gerontology Series A: Biological Sciences and Medical Sciences*, **55,** B347-B354.

16. Holloszy, J.O. (1997). Mortality rate and longevity of food-restricted exercising male rats: A reevaluation. *Journal of Applied Physiology*, **82,** 399-403.

17. Jackson, A.S., Beard, E.F., Wier, L.T., Ross, R.M., Stuteville, J.E., & Blair, S.N. (1995). Changes in aerobic power of men, ages 25-70 yr. *Medicine and Science in Sports and Exercise*, **27,** 113-120.

18. Jackson, A.S., Wier, L.T., Ayers, G.W., Beard, E.F., Stuteville, J.E., & Blair, S.N. (1996). Changes in aerobic power of women, ages 20-64 yr. *Medicine and Science in Sports and Exercise*, **28,** 884-891.

19. Janssen, I., Heymsfield, S.B., Wang, Z., & Ross, R. (2000). Skeletal muscle mass and distribution in 468 men and women aged 18-88 yr. *Journal of Applied Physiology*, **89,** 81-88.

20. Johnson, M.A., Polgar, J., Weihtmann, D., & Appleton, D. (1973). Data on the distribution of fiber types in

thirty-six human muscles: An autopsy study. *Journal of Neurological Science, 1*, 111-129.

21. Kenney, W.L. (1997). Thermoregulation at rest and during exercise in healthy older adults. *Exercise and Sport Sciences Reviews,* **25,** 41-77.

22. Kohrt, W.M., Malley, M.T., Coggan, A.R., Spina, R.J., Ogawa, T., Ehsani, A.A., Bourey, R.E., Martin, W.H., III, & Holloszy, J.O. (1991). Effects of gender, age, and fitness level on response of $\dot{V}O_{2max}$ to training in 60-71 yr olds. *Journal of Applied Physiology,* **71,** 2004-2011.

23. Kohrt, W.M., Malley, M.T., Dalsky, G.P., & Holloszy, J.O. (1992). Body composition of healthy sedentary and trained, young and older men and women. *Medicine and Science in Sports and Exercise,* **24,** 832-837.

24. uk, J.L., Saunders, T.J., Davidson, L.E., & Ross, R. (2009) Age-related changes in total and regional fat distribution. *Ageing Research Reviews.* **4,** 339-48.

25. Lexell, J., Taylor, C.C., & Sjostrom, M. (1988). What is the cause of the aging atrophy? Total number, size, and proportion of different fiber types studied in whole vastus lateralis muscle from 15- to 83-year-old men. *Journal of Neurological Science,* **84,** 275-294.

26. Marcell, T.J., Hawkins, S.A., Tarpenniing, K.M., Hyslop, D.M., Wiswell, R.A. (2003). Longitudinal analysis of lctate theresbold in male and female masters athletes. Medicince Sciences in Sports and Exercise. 35(5), 810-17.

27. Meredith, C.N., Frontera, W.R., Fisher, E.C., Hughes, V.A., Herland, J.C., Edwards, J., & Evans, W.J. (1989). Peripheral effects of endurance training in young and old subjects. *Journal of Applied Physiology,* **66,** 2844-2849.

28. Proctor, D.N., Shen, P.H., Dietz, N.M., Eickhoff, T.J., Lawler, L.A., Ebersold, E.J., Loeffler, D.L., & Joyner, M.J. (1998). Reduced leg blood flow during dynamic exercise in older endurance-trained men. *Journal of Applied Physiology,* **85,** 68-75.

29. Robinson, S. (1938). Experimental studies of physical fitness in relation to age. *Arbeitsphysiologie,* **10,** 251-323.

30. Saltin, B. (1986). The aging endurance athlete. In J.R. Sutton & R.M. Brock (Eds.), *Sports medicine for the mature athlete* (pp. 59-80). Indianapolis: Benchmark Press.

31. Seals, D.R., Walker, A.E., Pierce, G.L., & Lesniewski, L.A. (2009) Habitual exercise and vascular aging. *Journal of Physiology.* 5541-5549. Shephard, R.J. (1997). *Aging, physical activity, and health.* Champaign, IL: Human Kinetics.

32. Shibata, S., Hastings, J.L. Prasad, A., Fu, Q., Palmer, M.D., & Levine, B.D. (2008) 'Dynamic' starling mechanisms; effects of ageing and physical fitness on ventricular-arterial coupling. *Journal of Physiology.* 586 (7), 1951-62.

33. Spirduso, W.W. (2005). *Physical dimensions of aging.* Champaign, IL: Human Kinetics.

34. Tanaka, H., Monahan, K.D., & Seals, D.R. (2001). Agepredicted maximal heart rate revisited. *Journal of the American College of Cardiology,* **37,** 153-156.

35. Tanaka, H. & Seals, D.R. (2008) Endurance exercise performance in Masters athletes: age-associated changes and underlying physiological mechanisms. Journal of Physiology. 55-63.

36. Trappe, S.W., Costill, D.L., Fink, W.J., & Pearson, D.R. (1995). Skeletal muscle characteristics among distance runners: A 20-yr follow-up study. *Journal of Applied Physiology,* **78,** 823-829.

37. Trappe, S.W., Costill, D.L., Goodpaster, B.H., & Pearson, D.R. (1996). Calf muscle strength in former elite distance runners. *Scandinavian Journal of Medicine and Science in Sports,* **6,** 205-210.

38. Trappe, S.W., Costill, D.L., Vukovich, M.D., Jones, J., & Melham, T. (1996). Aging among elite distance runners: A 22-yr longitudinal study. *Journal of Applied Physiology,* **80,** 285-290.

39. Wiswell, R.A., Jaque, S.V., Marcell, T.J., Hawkins, S.A., Tarpenning, K.M., Constantino, N., & Hyslop, D.M. (2000). Maximal aerobic power, lactate threshold, and running performance in master athletes. *Medicine and Science in Sports and Exercise,* **32,** 1165-1170.

40. Chodzko-Zajko, W.J., Proctor, D.N., Fiatarone Singh, M.A., Minson, C.T., Nigg C.R., Salem, G.J., & Skinner, J.S. (2009) American College of Sports Medicine Position Stand: Exercise and physical activity for older adults. Medicine and Science in Sports and Exercise. 1515-1530.

Chapter 19

1. American College of Obstetricians and Gynecologists Committee Opinion. (2002). Exercise during pregnancy and the postpartum period. *Obstetrics and Gynecology,* **99,** 171-173.

2. American College of Sports Medicine. (2007). The female athlete triad. *Medicine and Science in Sports and Exercise* **39,** 1867-1882.

3. American Psychiatric Association. (1994). *Diagnostic and statistical manual of mental disorders* (4th ed.). Washington, DC: American Psychiatric Association.

4. Åstrand, P.-O., Rodahl, K., Dahl, H.A., & Strømme, S.B. (2003). *Textbook of work physiology: Physiological bases of exercise* (4th ed.). Champaign, IL: Human Kinetics.

5. Cann, C.E., Martin, M.C., Genant, H.K., & Jaffe, R.B. (1984). Decreased spinal mineral content in amenorrheic women. *Journal of the American Medical Association,* **251,** 626-629.

6. Costill, D.L., Fink, W.J., Flynn, M., & Kirwan, J. (1987). Muscle fiber composition and enzyme activities in elite female distance runners. *International Journal of Sports Medicine,* **8**(Suppl. 2), 103-106.

7. Cureton, K., Bishop, P., Hutchinson, P., Newland, H., Vickery, S., & Zwiren, L. (1986). Sex differences in maximal oxygen uptake: Effect of equating haemoglobin concentration. *European Journal of Applied Physiology,* **54,** 656-660.

8. Cureton, K.J., & Sparling, P.B. (1980). Distance running performance and metabolic responses to running in men and women with excess weight experimentally equated. *Medicine and Science in Sports and Exercise, 12,* 288-294.

9. Davis, J.A., Wilson, L.D., Caiozzo, V.J., Storer, T.W., & Pham, P.H. (2006). Maximal oxygen uptake at the same fat-free mass is greater in men than women. *Clinical Physiology and Functional Imaging, 26,* 61-66.

10. De Souza MJ, Toombs RJ, Scheid JL, O'Donnell E, West SL, Williams NI 2010 High prevalence of subtle and severe menstrual disturbances in exercising women: confirmation using daily hormone measures. Hum Reprod 25(2):491-503.

11. De Souza, MJ, Lee, D.K., VanHeest, J.L., Scheid, J.L., West, S.L., & Williams, N.L. (2007). Severity of energy-related menstrual disturbances increases in proportion to indices of energy conservation in exercising women. *Fertility and Sterility* 88, 971-975.

12. De Souza M.J,. & Williams, N.I. (2004). Physiological aspects and clinical sequelae of energy deficiency and hypoestrogenism in exercising women. *Hum Reprod Update* 10, 433-48.

13. De Souza, M.J., Miller, B.E., Loucks, A.B., Luciano, A.A., Pescatello, L.S., Campbell, C.G. & Lasley, B.L. (1998). High frequency of luteal phase deficiency and anovulation in recreational woman runners: blunted elevation in follicle-stimulating hormone observed during luteal-follicular transition. *Journal of Clinical Endocrinology and Metabolism* 83, 4220-4232.

14. Drinkwater, B.L., Bruemner, B., & Chesnut, C.H., III. (1990). Menstrual history as a determinant of current bone density in young athletes. *Journal of the American Medical Association, 263,* 545-548.

15. Drinkwater, B.L., Nilson, K., Chesnut, C.H., III, Bremner, W.J., Shainholtz, S., & Southworth, M.B. (1984). Bone mineral content of amenorrheic and eumenorrheic athletes. *New England Journal of Medicine, 311,* 277-281.

16. Drinkwater, B.L., Nilson, K., Ott, S., & Chesnut, C.H., III. (1986). Bone mineral density after resumption of menses in amenorrheic athletes. *Journal of the American Medical Association, 256,* 380-382.

17. Fink, W.J., Costill, D.L., & Pollock, M.L. (1977). Submaximal and maximal working capacity of elite distance runners: Part II. Muscle fiber composition and enzyme activities. *Annals of the New York Academy of Sciences, 301,* 323-327.

18. Frisch, R.E. (1983). Fatness and reproduction: Delayed menarche and amenorrhea of ballet dancers and college athletes. In P.E. Garfinkel, P.L. Darby, & D.M. Garner (Eds.), *Anorexia nervosa: Recent developments in research* (pp. 343-363). New York: Liss.

19. Fu, Q., & Levine, B.D. (2005). Cardiovascular response to exercise in women. *Medicine and Science in Sports and Exercise, 37,* 1433-1435.

20. Gadpaille, W.J., Sanborn, C.F., & Wagner, W.W. (1987). Athletic amenorrhea, major affective disorders, and eating disorders. *American Journal of Psychiatry, 144,* 939-942.

21. Hermansen, L., & Andersen, K.L. (1965). Aerobic work capacity in young Norwegian men and women. *Journal of Applied Physiology, 20,* 425-431.

22. Janssen, I., Heymsfield, S.B., Wang, Z., & Ross, R. (2000). Skeletal muscle mass and distribution in 468 men and women aged 18-88 yr. *Journal of Applied Physiology, 89,* 81-88.

23. Loucks, A.B., & Thuma, J.R. (2003). Luteinizing hormone pulsatility is disrupted at a threshold of energy availability in regularly menstruating women. *Journal of Clinical Endocrinology and Metabolism, 88,* 297-311.

24. Malina, R.M. (1983). Menarche in athletes: A synthesis and hypothesis. *Annals of Human Biology, 10,* 1-24.

25. Mier, C.M., Domenick, M.A., Turner, N.S., & Wilmore, J.H. (1996). Changes in stroke volume and maximal aerobic capacity with increased blood volume in men and women. *Journal of Applied Physiology, 80,* 1180-1186.

26. Mier, C.M., Domenick, M.A., & Wilmore, J.H. (1997). Changes in stroke volume with ⊠-blockade before and after 10 days of exercise training in men and women. *Journal of Applied Physiology, 83,* 1660-1665.

27. Otis, C.L., Drinkwater, B., Johnson, M., Loucks, A., & Wilmore, J. (1997). The female athlete triad. *Medicine and Science in Sports and Exercise, 29*(5), i-ix.

28. Pivarnik, J.M. (1994). Maternal exercise during pregnancy. *Sports Medicine, 18,* 215-217.

29. Redman, L.M., & Loucks, A.B. (2005). Menstrual disorders in athletes. *Sports Medicine, 35,* 747-755.

30. Saltin, B., & Åstrand, P.-O. (1967). Maximal oxygen uptake in athletes. *Journal of Applied Physiology, 23,* 353-358.

31. Saltin, B., Henriksson, J., Nygaard, E., & Andersen, P. (1977). Fiber types and metabolic potentials of skeletal muscles in sedentary man and endurance runners. *Annals of the New York Academy of Sciences, 301,* 3-29.

32. Schantz, P., Randall-Fox, E., Hutchison, W., Tyden, A., & Åstrand, P.-O. (1983). Muscle fibre type distribution, muscle cross-sectional area and maximal voluntary strength in humans. *Acta Physiologica Scandinavica, 117,* 219-226.

33. Scheid, J.L., Williams, N.I., West, S.L., VanHeest, J.L., & De Souza, M.J.. (2009). Elevated PYY is associated with energy deficiency and indices of subclinical disordered eating in exercising women with hypothalamic amenorrhea. *Appetite* 52, 184-192.

34. Stager, J.M., Wigglesworth, J.K., & Hatler, L.K. (1990). Interpreting the relationship between age of menarche and prepubertal training. *Medicine and Science in Sports and Exercise, 22,* 54-58.

35. Turley, K.R., & Wilmore, J.H. (1997). Cardiovascular responses to submaximal exercise in 7- to 9-yr old boys

and girls. *Medicine and Science in Sports and Exercise*, **29**, 824-832.

36. Williams, N.I., McConnell, H.J., Gardner, J.K., Frye, B.R., Richard, E.L., Snook, M.L., Dougherty, K.L., Parrott, T.S., Albert, A., & Schukert, M. (2004). Exercise-associated menstrual disturbances: dependence on daily energy deficit, not body composition or body weight changes. *Medicine and Science in Sports and Exercise*, **36**(5), S280.

37. Wilmore, J.H., Stanforth, P.R., Gagnon, J., Rice, T., Mandel, S., Leon, A.S., Rao, D.C., Skinner, J.S., & Bouchard, C. (2001). Cardiac output and stroke volume changes with endurance training: The HERITAGE Family Study. *Medicine and Science in Sports and Exercise*, **33**, 99-106.

38. Wilmore, J.H., Wambsgans, K.C., Brenner, M., Broeder, C.E., Paijmans, I., Volpe, J.A., & Wilmore, K.M. (1992). Is there energy conservation in amenorrheic compared to eumenorrheic distance runners? *Journal of Applied Physiology*, **72**, 15-22.

39. Wolfe, L.A., Brenner, I.K.M., & Mottola, M.F. (1994). Maternal exercise, fetal well-being and pregnancy outcome. *Exercise and Sport Sciences Reviews*, **22**, 145-194.

Chapter 20

1. American College of Sports Medicine. (1998). The recommended quantity and quality of exercise for developing and maintaining cardiorespiratory and muscular fitness, and flexibility in healthy adults. *Medicine and Science in Sports and Exercise*, **30**, 975-991.

2. American College of Sports Medicine. (2010). *ACSM's guidelines for exercise testing and prescription* (8th ed.). Philadelphia: Lippincott Williams & Wilkins.

3. Booth, F.W., Chakravarthy, M.V., Gordon, S.E., & Spangenburg, E.E. (2002). Waging war on physical inactivity: Using modern molecular ammunition against an ancient enemy. *Journal of Applied Physiology*, **93**, 3-30.

4. Booth, F.W., Gordon, S.E., Carlson, C.J., & Hamilton, M.T. (2000). Waging war on modern chronic disease: Primary prevention through exercise biology. *Journal of Applied Physiology*, **88**, 774-787.

5. Borg, G.A.V. (1998). *Borg's perceived exertion and pain scales.* Champaign, IL: Human Kinetics.

6. Byrne, N.M., Hills, A.P., Hunter, G.R., Weinsier, R.L., & Schutz, Y. (2005). Metabolic equivalent: One size does not fit all. *Journal of Applied Physiology*, **99**, 1112-1119.

7. Cooper, K.H. (1968). *Aerobics.* New York: Evans.

8. Davis, J.A., & Convertino, V.A. (1975). A comparison of heart rate methods for predicting endurance training intensity. *Medicine and Science in Sports*, **7**, 295-298.

9. Fletcher, G.F., Balady, G.J., Amsterdam, E.A., Chaitman, B., Eckel, R., Fleg, J., Froelicher, V.F., Leon, A.S., Piña, I.L., Rodney, R., Simons-Morton, D.G., Williams, M.A., & Bazzarre, T. (2001). Exercise standards for testing and training: A statement for healthcare professionals from the American Heart Association. *Circulation*, **104**, 1694-1740.

10. letcher, G.F., Blair, S.N., Blumenthal, J., Caspersen, C., Chaitman, B., Epstein, S., Falls, H., Froelicher, E.S.S., Froelicher, V.F., & Piña, I.L. (1992). Statement on exercise: Benefits and recommendations for physical activity programs for all Americans. *Circulation*, **86**, 340-344.

11. Gibala, M.J., & McGee, S. (2008). Metabolic adaptations to short-term high-intensity interval training: A little pain for a lot of gain? *Exercise and Sports Science Reviews*, **36**, 58-63.

12. Kirshenbaum, J., & Sullivan, R. (1983). Hold on there, America. *Sports Illustrated*, **58**(5), 60-74.

13. Lauer, M., Sivarajan Froelicher, E., Williams, M., & Kligfield, P. (2005). Exercise testing in asymptomatic adults. *Circulation*, **112**, 771-776.

14. Manini, T.M., Everhart, J.E., Patel, K.V., Schoeller, D.A., Colbert, L.H., Visser, M., Tylavsky, F., Bauer, D.C., Goodpaster, B.H., & Harris, T.B. (2006). Daily activity energy expenditure and mortality among older adults. *Journal of the American Medical Association*, **296**, 171-179.

15. National Institutes of Health, Consensus Development Panel on Physical Activity and Cardiovascular Health. (1996). Physical activity and cardiovascular health. *Journal of the American Medical Association*, **276**, 241-246.

16. Pate, R.R., Pratt, M., Blair, S.N., Haskell, W.L., Macera, C.A., Bouchard, C., Buchner, D., Ettinger, W., Heath, G.W., King, A.C., Kriska, A., Leon, A.S., Marcus, B.H., Morris, J., Paffenbarger, R.S., Patrick, K., Pollock, M.L., Rippe, J.M., Sallis, J., & Wilmore, J.H. (1995). Physical activity and public health: A recommendation from the Centers for Disease Control and Prevention and the American College of Sports Medicine. *Journal of the American Medical Association*, **273**, 402-407.

17. Persinger, R., Foster, C., Gibson, M., Fater, D.C.W., & Porcari, J.P. (2004). Consistency of the talk test for exercise prescription. *Medicine and Science in Sports and Exercise*, **36**, 1632-1636.

18. Pollock, M.L., Franklin, B.A., Balady, G.J., Chaitman, B.L., Fleg, J.L., Fletcher, B., Limacher, M., Piña, I.L., Stein, R.A., Williams, M., & Bazzarre, T. (2000). Resistance exercise in individuals with and without cardiovascular disease: Benefits, rationale, safety and prescription. *Circulation*, **101**, 828-833.

19. Swain, D.P., & Leutholtz, B.C. (1997). Heart rate reserve is equivalent to %VO2 reserve, not to %VO2max. *Medicine and Science in Sports and Exercise*, **29**, 410-414.

20. U.S. Department of Health and Human Services. (1996). *Physical activity and health: A report of the Surgeon General.* Atlanta: U.S. Department of Health and Human Services, Centers for Disease Control and Prevention, National Center for Chronic Disease Prevention and Health Promotion.

21. U.S. Department of Health and Human Services. (2000, November). *Healthy people 2010: Understanding and improving health* (2nd ed.). Washington, DC: U.S. Government Printing Office.

Chapter 21

1. Albert, C.M., Mittleman, M.A., Chae, C.U., Lee, I.-M., Hennekens, C.H., & Manson, J.E. (2000). Triggering of sudden death from cardiac causes by vigorous exertion. *New England Journal of Medicine,* **343,** 1355-1361.

2. American College of Sports Medicine Position Stand. (1994). Exercise for patients with coronary artery disease. *Medicine and Science in Sports and Exercise,* **26**(3), i-v.

3. American College of Sports Medicine Position Stand. (2004). Exercise and hypertension. *Medicine and Science in Sports and Exercise,* **36,** 533-553.

4. American Heart Association. (2010). *Heart Disease and Stroke Statistics – 2010 Update.* Dallas, TX: American Heart Association.

5. American Heart Association. (2010). Heart Disease and Stroke Statistics—2010 Update. *Circulation,* **121,** e46-e215.

6. Berenson, G.S., Srinivasan, S.R., Bao, W., Newman, W.P., Tracy, R.E., & Wattigney, W.A. (1998). Association between multiple cardiovascular risk factors and atherosclerosis in children and young adults. The Bogalusa Heart Study. *New England Journal of Medicine,* **338,** 1650-1656.

7. Blair, S.N., Goodyear, N.N., Gibbons, L.W., & Cooper, K.H. (1984). Physical fitness and incidence of hypertension in healthy normotensive men and women. *Journal of the American Medical Association,* **252,** 487-490.

8. Blair, S.N., & Jackson, A.S. (2001). Guest editorial: Physical fitness and activity as separate heart disease risk factors: A meta-analysis. *Medicine and Science in Sports and Exercise,* **33,** 762-764.

9. Blair, S.N., Kohl, H.W., Paffenbarger, R.S., Clark, D.G., Cooper, K.H., & Gibbons, L.W. (1989). Physical fitness and all-cause mortality: A prospective study of healthy men and women. *Journal of the American Medical Association,* **262,** 2395-2401.

10. Braganza, D.M., & Bennett, M.R. (2001). New insights into atherosclerotic plaque rupture. *Postgraduate Medical Journal,* **77,** 94-98.

11. Carnethon, M.R., Gulati, M., & Greenland, P. (2005). Prevalence and cardiovascular disease correlates of low cardiorespiratory fitness in adolescents and adults. *Journal of the American Medical Association,* **294,** 2981-2988.

12. Caspersen, C.J. (1987). Physical inactivity and coronary heart disease. *Physician and Sportsmedicine,* **15**(11), 43-44.

13. Conroy, M.B., Cook, N.R., Manson, J.E., Buring, J.E., & Lee, I-M. (2005). Past physical activity, current physical activity, and risk of coronary heart disease. *Medicine and Science in Sports and Exercise,* **37,** 1251-1256.

14. Cooper, K.H., Pollock, M.L., Martin, R.P., White, S.R., Linnerud, A.C., & Jackson, A. (1976). Physical fitness levels vs. selected coronary risk factors: A cross-sectional study. *Journal of the American Medical Association,* **236,** 166-169.

15. Corti, R., Hutter, R., Badimon, J.J., & Fuster, V. (2004). Evolving concepts in the triad of atherosclerosis, inflammation and thrombosis. *Journal of Thrombosis and Thrombolysis,* **17,** 35-44.

16. Dunn, A.L., & Dishman, R.K. (1991). Exercise and the neurobiology of depression. *Exercise and Sport Sciences Reviews,* **19,** 41-98.

17. Durstine, J.L., Grandjean, P.W., Cox, C.A., & Thompson, P.D. (2002). Lipids, lipoproteins, and exercise. *Journal of Cardiopulmonary Rehabilitation,* **22,** 385-398.

18. Ehsani, A.A. (1987). Cardiovascular adaptations to endurance exercise training in ischemic heart disease. *Exercise and Sport Sciences Reviews,* **15,** 53-66.

19. Enos, W.F., Holmes, R.H., & Beyer, J. (1953). Coronary disease among United States soldiers killed in action in Korea. *Journal of the American Medical Association,* **152,** 1090-1093.

20. Franklin, B.A., & Kahn, J.K. (1996). Delayed progression or regression of coronary atherosclerosis with intensive risk factor modification: Effects of diet, drugs, and exercise. *Sports Medicine,* **22,** 306-320.

21. Gibbons, L.W., Blair, S.N., Cooper, K.H., & Smith, M. (1983). Association between coronary heart disease risk factors and physical fitness in healthy adult women. *Circulation,* **67,** 977-983.

22. Haskell, W.L., Sims, C., Myll, J., Bortz, W.M., St. Goar, F.G., & Alderman, E.L. (1993). Coronary artery size and dilating capacity in ultradistance runners. *Circulation,* **87,** 1076-1082.

23. Hu, F.B., Stampfer, M.J., Colditz, G.A., Ascherio, A., Rexrode, K.M., Willett, W.C., & Manson, J.E. (2000). Physical activity and risk of stroke in women. *Journal of the American Medical Association,* **283,** 2961-2967.

24. Joint National Committee on Prevention, Detection, Evaluation, and Treatment of High Blood Pressure. (2003). The seventh report of the Joint National Committee on Prevention, Detection, Evaluation, and Treatment of High Blood Pressure. *Journal of the American Medical Association,* **289,** 2560-2572.

25. Kannel, W.B., & Dawber, T.R. (1972). Atherosclerosis as a pediatric problem. *Journal of Pediatrics,* **80,** 544-554.

26. Knez, W.L., Coombes, J.S., & Jenkins, D.G. (2006). Ultraendurance exercise and oxidative damage: Implications for cardiovascular health. *Sports Medicine,* **36,** 429-441.

27. Kramsch, D.M., Aspen, A.J., Abramowitz, B.M., Kreimendahl, T., & Hood, W.B. (1981). Reduction of coronary atherosclerosis by moderate conditioning exercise in monkeys on an atherogenic diet. *New England Journal of Medicine,* **305,** 1483-1489.

28. LaPorte, R.E., Adams, L.L., Savage, D.D., Brenes, G., Dearwater, S., & Cook, T. (1984). The spectrum of physical activity, cardiovascular disease and health: An epidemiologic perspective. *American Journal of Epidemiology,* **120,** 507-517.

29. Lee, C.D., & Blair, S.N. (2002). Cardiorespiratory fitness and stroke mortality in men. *Medicine and Science in Sports and Exercise, 34,* 592-595.

30. Lee, I.-M., & Paffenbarger, R.S., Jr. (1996). Do physical activity and physical fitness avert premature mortality? *Exercise and Sport Sciences Reviews, 24,* 135-171.

31. Leon, A.S., & Connett, J. (1991). Physical activity and 10.5 year mortality in the Multiple Risk Factor Intervention Trial (MRFIT). *International Journal of Epidemiology, 20,* 690-697.

32. Leon, A.S., Connett, J., Jacobs, D.R., & Rauramaa, R. (1987). Leisure-time physical activity levels and risk of coronary heart disease and death. *Journal of the American Medical Association, 258,* 2388-2395.

33. McNamara, J.J., Molot, M.A., Stremple, J.F., & Cutting, R.T. (1971). Coronary artery disease in combat casualties in Vietnam. *Journal of the American Medical Association, 216,* 1185-1187.

34. Montoye, H.J., Metzner, H.L., Keller, J.B., Johnson, B.C., & Epstein, F.H. (1972). Habitual physical activity and blood pressure. *Medicine and Science in Sports and Exercise, 4,* 175-181.

35. Morgan, W.P. (1994). Physical activity, fitness and depression. In C. Bouchard, R.J. Shephard, & T. Stephens (Eds.), *Physical activity, fitness, and health* (pp. 851-867). Champaign, IL: Human Kinetics.

36. Morris, J.N., Adam, C., Chave, S.P.W., Sirey, C., Epstein, L., & Sheehan, D.J. (1973). Vigorous exercise in leisuretime and the incidence of coronary heart disease. *Lancet, 1,* 333-339.

37. Morris, J.N., Heady, J.A., Raffle, P.A.B., Roberts, C.G., & Parks, J.W. (1953). Coronary heart-disease and physical activity of work. *Lancet, 265,* 1053-1057, 1111-1120.

38. Morris, J.N., Pollard, R., Everitt, M.G., Chave, S.P.W., & Semmence, A.M. (1980). Vigorous exercise in leisuretime: Protection against coronary heart disease. *Lancet, 2,* 1207-1210.

39. Paffenbarger, R.S., Hyde, R.T., Wing, A.L., & Hsieh, C.-C. (1986). Physical activity, all-cause mortality, and longevity of college alumni. *New England Journal of Medicine, 314,* 605-613.

40. Petersen, A.M.W., & Pedersen, B.K. (2005). The antiinflammatory effect of exercise. *Journal of Applied Physiology, 98,* 1154-1162.

41. Petruzzello, S.J., Landers, D.M., Hatfield, B.D., Kubitz, K.A., & Salazar, W. (1991). A meta-analysis on the anxiety-reducing effects of acute and chronic exercise: Outcomes and mechanisms. *Sports Medicine, 11,* 143-182.

42. Powell, K.E., Thompson, P.D., Caspersen, C.J., & Kendrick, J.S. (1987). Physical activity and the incidence of coronary heart disease. *Annual Reviews in Public Health, 8,* 253-287.

43. Ross, R. (1986). The pathogenesis of atherosclerosis—an update. *New England Journal of Medicine, 314,* 488-500.

44. Ross, R. (1999). Atherosclerosis—an inflammatory disease. *New England Journal of Medicine, 340,* 115-126.

45. Siscovick, D.S., Weiss, N.S., Fletcher, R.H., & Lasky, T. (1984). The incidence of primary cardiac arrest during vigorous exercise. *New England Journal of Medicine, 311,* 874-877.

46. Tanasescu, M., Leitzmann, M.F., Rimm, E.B., Willett, W.C., Stampfer, M.J., & Hu, F.B. (2002). Exercise type and intensity in relation to coronary heart disease in men. *Journal of the American Medical Association, 288,* 1994-2000.

47. Taylor, R.S., Brown, A., Ebrahim, S., Jolliffe, J., Noorani, H., Rees, K., Skidmore, B., Stone, J.A., Thompson, D.R., & Oldridge, N. (2004). Exercise-based rehabilitation for patients with coronary heart disease: Systematic review and meta-analysis of randomized controlled trials. *American Journal of Medicine, 116,* 682-692.

48. Thompson, P.D. (1982). Cardiovascular hazards of physical activity. *Exercise and Sport Sciences Reviews, 10,* 208-235.

49. U.S. Department of Health and Human Services. (2010). Deaths: Leading causes for 2006. National Vital Statistics Reports. *58,* Number 14, p. 8.

50. Walther, C., Gielen, S., & Hambrecht, R. (2004). The effect of exercise training on endothelial function in cardiovascular disease in humans. *Exercise and Sport Sciences Reviews, 32,* 129-134.

51. Whang, W., Manson, J.E., Hu, F.B., Chae, C.U., Rexrode, K.M., Willett, W.C., Stampfer, M.J., & Albert, C.M. (2006). Physical exertion, exercise, and sudden cardiac death in women. *Journal of the American Medical Association, 295,* 1399-1403.

52. Williams, M.A., Haskell, W.L., Ades, P.A., Amsterdam, E.A., Bittner, V., Franklin, B.A., Gulanick, M., Laing, S.T., & Stewart, K.J. (2007). Resistance exercise in individuals with and without cardiovascular disease: 2007 update. *Circulation, 116,* 572-584.

53. Williams, P.T. (2001). Physical fitness and activity as separate heart disease risk factors: A meta-analysis. *Medicine and Science in Sports and Exercise, 33,* 754-761.

54. Wilmore, J.H., Constable, S.H., Stanforth, P.R., Tsao, W.Y., Rotkis, T.C., Paicius, R.M., Mattern, C.M., & Ewy, G.A. (1982). Prevalence of coronary heart disease risk factors in 13- to 15-year-old boys. *Journal of Cardiac Rehabilitation, 2,* 223-233.

55. Wilmore, J.H., & McNamara, J.J. (1974). Prevalence of coronary heart disease risk factors in boys 8 to 12 years of age. *Journal of Pediatrics, 84,* 527-533.

Chapter 22

1. Achten, J., Gleeson, M., & Jeukendrup, A.E. (2002). Determination of the exercise intensity that elicits maximal fat oxidation. *Medicine and Science in Sports and Exercise, 34,* 92-97.

2. Barlow, C.E., Kohl, H.W., III, Gibbons, L.W., & Blair, S.N. (1995). Physical fitness, mortality and obesity. *International Journal of Obesity,* **19**(Suppl. 4), 41-44.

3. Bassuk, S.S., & Manson, J.E. (2005). Epidemiological evidence for the role of physical activity in reducing risk of type 2 diabetes and cardiovascular disease. *Journal of Applied Physiology,* **99,** 1193-1204.

4. Borchers, J.R., Clem, K.L., Habash, D.L., Nagaraja, H.N., Stokley, L.M. & Best, T.M. (2009). Metabolic syndrome and insulin resistance in Divisions 1 collegiate football players. *Medicine and Science in Sports and Exercise,* **41,** 2105-2110.

5. Bouchard, C. (1991). Heredity and the path to overweight and obesity. *Medicine and Science in Sports and Exercise,* **23,** 285-291.

6. Bouchard, C., Tremblay, A., Després, J.-P., Nadeau, A., Lupien, P.J., Theriault, G., Dussault, J., Moorjani, S., Pinault, S., & Fournier, G. (1990). The response to longterm overfeeding in identical twins. *New England Journal of Medicine,* **322,** 1477-1482.

7. Bray, G.A. (1985). Obesity: Definition, diagnosis and disadvantages. *Medical Journal of Australia,* **142,** S2-S8.

8. Broeder, C.E., Burrhus, K.A., Svanevik, L.S., & Wilmore, J.H. (1992). The effects of either high intensity resistance or endurance training on resting metabolic rate. *American Journal of Clinical Nutrition,* **55,** 802-810.

9. Centers for Disease Control and Prevention. (March 3, 2011). *Obesity and overweight for professionals: Health consequences.* Retrieved from www.cdc.gov/obesity/causes/health.html.

10. Centers for Disease Control and Prevention. (2008). *National diabetes fact sheet: general information and national estimates on diabetes in the United States, 2007.* Atlanta, GA: U.S. Department of Health and Human Services, Centers for Disease Control and Prevention.

11. Chisholm, D.J. (1992). Diabetes mellitus. In J. Bloomfield, P.A. Fricker, & K.D. Fitch (Eds.), *Textbook of science and medicine in sport* (pp. 555-561). Boston: Blackwell Scientific.

12. Daly, P.A., & Landsberg, L. (1991). Hypertension in obesity and NIDDM: Role of insulin and sympathetic nervous system. *Diabetes Care,* **14,** 240-248.

13. Davidson, L.E., Hudson, R., Kilpatrick, K., Kuk, J.L., McMillan, K., Janiszewski, P.M., Lee, S., Lam, M., & Ross, R. (2009). Effects of exercise modality on insulin resistance and functional limitation in older adults: a randomized controlled trial. *Archives of Internal Medicine,* **169,** 122-131.

14. Ervin R.B. (2009). Prevalence of metabolic syndrome among adults 20 years of age and over, by sex, age, race and ethnicity, and body mass index: United States, 2003–2006. *National health statistics reports* (no. 13). Hyattsville, MD: National Center for Health Statistics.

15. Flegal, K.M., Carroll, M.D., Kuczmarski, R.J., & Johnson, C.L. (1998). Overweight and obesity in the United States: Prevalence and trends, 1960-1994. *International Journal of Obesity,* **22,** 39-47.

16. Flegal, K.M., Carroll, M.D., Ogden, C.L., & Johnson, C.L. (2002). Prevalence and trends in obesity among US adults, 1999-2000. *Journal of the American Medical Association,* **288,** 1723-1727.

17. Flegal, K.M., Carroll, M.D., Ogden, C.L., & Curtin, L.R. (2010). Prevalence and trends in obesity among US adults, 1999-2008. *Journal of the American Medical Association* **303,** 235-241.

18. Ford, E.S., Giles, W.H., & Dietz, W.H. (2002). Prevalence of the metabolic syndrome among US adults. *Journal of the American Medical Association,* **287,** 356-359.

19. Gwinup, G., Chelvam, R., & Steinberg, T. (1971). Thickness of subcutaneous fat and activity of underlying muscles. *Annals of Internal Medicine,* **74,** 408-411.

20. Harnack, L.J., Jeffery, R.W., & Boutelle, K.N. (2000). Temporal trends in energy intake in the United States: An ecologic perspective. *American Journal of Clinical Nutrition,* **71,** 1478-1484.

21. Holloszy, J.O. (2005). Exercise-induced increase in muscle insulin sensitivity. *Journal of Applied Physiology,* **99,** 338-343.

22. Ivy, J.L. (1987). The insulin-like effect of muscle contraction. *Exercise and Sport Sciences Reviews,* **15,** 29-51.

23. Katch, F.I., Clarkson, P.M., Kroll, W., McBride, T., & Wilcox, A. (1984). Effects of sit up exercise training on adipose cell size and adiposity. *Research Quarterly for Exercise and Sport,* **55,** 242-247.

24. Keesey, R.E. (1986). A set-point theory of obesity. In K.D. Brownell & J.P. Foreyt (Eds.), *Handbook of eating disorders: Physiology, psychology, and treatment of obesity, anorexia, and bulimia* (pp. 63-87). New York: Basic Books.

25. Keys, A., Brozek, J., Henschel, A., Mickelsen, O., & Taylor, H.L. (1950). *The biology of human starvation.* Minneapolis: University of Minnesota Press.

26. Knowler, W.C., Barrett-Connor, E., Fowler, S.E., Hamman, R.F., Lachin, J.M., Walker, E.A., & Nathan, D.M. (2002). Reduction in the incidence of type 2 diabetes with lifestyle intervention or metformin. *New England Journal of Medicine,* **346,** 393-403.

27. Levine, J.A., Lanningham-Foster, L.M., McCrady, S.K., Krizan, A.C., Olson, L.R., Kane, P.H., Jensen, M.D., & Clark, M.M. (2005). Interindividual variation in posture allocation: Possible role in human obesity. *Science,* **307,** 584-586.

28. Ludwig, D.S., & Ebbeling, C.B. (2001). Type 2 diabetes mellitus in children. *Journal of the American Medical Association,* **286,** 1426-1430.

29. Mayer, J., Marshall, N.B., Vitale, J.J., Christensen, J.H., Mashayekhi, M.B., & Stare, F.J. (1954). Exercise, food intake, and body weight in normal rats and genetically obese adult mice. *American Journal of Physiology,* **177,** 544-548.

30. National Institutes of Health. (2000). *The practical guide: Identification, evaluation, and treatment of overweight and obesity in adults* (NIH Publication No. 00-4084). Washington, DC: U.S. Department of Health and Human Services.

31. Ogden, C.L., Carroll, M.D., Curtin, L.R., McDowell, M.A., Tabak, C.J., & Flegal, K.M. (2006). Prevalence of overweight and obesity in the United States, 1999-2004. *Journal of the American Medical Association, 295,* 1549-1555.

32. Ogden, C.L., Carroll, M.D., Curtin, L.R., Lamb, M.M., & Flegal, K.M. (2010). Prevalence of high body mass index in US children and adolescents, 2007-2008. *Journal of the American Medical Association, 303,* 242-249

33. Oscai, L.B. (1973). The role of exercise in weight control. *Exercise and Sport Sciences Reviews, 1,* 103-123.

34. Poehlman, E.T. (1989). A review: Exercise and its influence on resting energy metabolism in man. *Medicine and Science in Sports and Exercise, 21,* 515-525.

35. Schulz, L.O., Bennett, P.H., Ravussin, E., Kidd, J.R., Kidd, K.K., Esparza, J., & Valencia, ME. (2006). Effects of traditional and western environments on prevalence of type 2 diabetes in Pima Indians in Mexico and the U.S. *Diabetes Care, 29,* 1866-1871.

36. Seidell, J.C., Deurenberg, P., & Hautvast, J.G.A.J. (1987). Obesity and fat distribution in relation to health—current insights and recommendations. *World Review of Nutrition and Dietetics, 50,* 57-91.

37. Sims, E.A.H. (1976). Experimental obesity, dietaryinduced thermogenesis and their clinical implications. *Clinics in Endocrinology and Metabolism, 5,* 377-395.

38. Slentz, C.A., Aiken, L.B., Houmard, J.A., Bales, C.W., Johnson, J.L., Tanner, C.J., Duscha, B.D., & Kraus, W.E. (2005). Inactivity, exercise and visceral fat. STRRIDE: A randomized, controlled study of exercise intensity and amount. *Journal of Applied Physiology, 99,* 1613-1618.

39. Tremblay, A., Nadeau, A., Fournier, G., & Bouchard, C. (1988). Effect of a three-day interruption of exercise-training on resting metabolic rate and glucose-induced thermogenesis in trained individuals. *International Journal of Obesity, 12,* 163-168.

40. Troiano, R.P., Flegal, K.M., Kuczmarski, R.J., Campbell, S.M., & Johnson, C.L. (1995). Overweight prevalence and trends for children and adolescents: The National Health and Nutrition Examination Surveys, 1963 to 1991. *Archives of Pediatric Adolescent Medicine, 149,* 1085-1091.

41. Vitug, A., Schneider, S.H., & Ruderman, N.B. (1988). Exercise and type I diabetes mellitus. *Exercise and Sport Sciences Reviews, 16,* 285-304.

42. Wansink, B., & Wansink, C.S. (2010). The largest Last Supper: Depictions of food portions and plate size increased over the millennium. *International Journal of Obesity, 34,* 943-944.

43. Welk, G.J., & Blair, S.N. (2000). Physical activity protects against the health risks of obesity. *Research Digest: President's Council on Physical Fitness and Sports, 3*(12), 1-6.

44. Wilmore, J.H. (1996). Increasing physical activity: Alterations in body mass and composition. *American Journal of Clinical Nutrition, 63,* 456S-460S.

45. Wilmore, J.H., Atwater, A.E., Maxwell, B.D., Wilmore, D.L., Constable, S.H., & Buono, M.J. (1985). Alterations in body size and composition consequent to Astro-Trimmer and Slim-Skins training programs. *Research Quarterly for Exercise and Sport, 56,* 90-92.

46. Wilmore, J.H., Atwater, A.E., Maxwell, B.D., Wilmore, D.L., Constable, S.H., & Buono, M.J. (1985). Alterations in breast morphology consequent to a 21-day bust developer program. *Medicine and Science in Sports and Exercise, 17,* 106-112.

47. Wilmore, J.H., Stanforth, P.R., Hudspeth, L.A., Gagnon, J., Daw, E.W., Leon, A.S., Rao, D.C., Skinner, J.S., & Bouchard, C. (1998). Alterations in resting metabolic rate as a consequence of 20-wk of endurance training: The HERITAGE Family Study. *American Journal of Clinical Nutrition, 68,* 66-71.

48. World Health Organization. (1998). *Obesity: Preventing and managing the global epidemic. Report of a WHO consultation on obesity.* Geneva: WHO.

49. Yaspelkis, B.B. (2006). Resistance training improves insulin signaling and action in skeletal muscle. *Exercise and Sport Sciences Reviews, 34,* 42-46.

Index

Note: The f and t following page numbers refer to figures and tables, respectively.

W. Larry Kenney, PhD,

is a professor of physiology and kinesiology at Pennsylvania State University in University Park, Pennsylvania. He received his PhD in physiology from Penn State in 1983. Working at Noll Laboratory, Kenney is currently researching the effects of aging and elevated cholesterol on the control of blood flow in human skin and has been continuously funded by NIH since 1983. He has also studied the effects of heat, cold, and dehydration on various aspects of health, exercise, and athletic performance as well as the biophysics of heat exchange between humans and the environment. He is the author of some 200 papers, books, book chapters, and other publications.

Kenney served as president of the American College of Sports Medicine from 2003 to 2004. He is a fellow of the American College of Sports Medicine and of the American Academy of Kinesiology and Physical Education and a member of the American Physiological Society.

For his service to the university and his field, Kenney has been awarded Penn State University's Faculty Scholar Medal, the Evan G. and Helen G. Pattishall Distinguished Research Career Award, and the Pauline Schmitt Russell Distinguished Research Career Award. He was awarded the American College of Sports Medicine's New Investigator Award in 1987 and the Citation Award in 2008.

Kenney has been a member of the editorial and advisory boards for several journals, including *Medicine and Science in Sports and Exercise, Current Sports Medicine Reports* (inaugural board member), *Exercise and Sport Sciences Reviews,* the *Journal of Applied Physiology, Human Performance, Fitness Management,* and *ACSM's Health & Fitness Journal* (inaugural board member). He is also an active grant reviewer for the National Institutes of Health and many other organizations. He and his wife, Patti, have three children, all of whom are or were college athletes: Matt (Cornell football), Alex (Penn State football and track), and Lauren (Penn State track).

Jack H. Wilmore, PhD,

is the Margie Gurley Seay Centennial professor emeritus in the department of kinesiology and health education at the University of Texas at Austin. He retired in 2003 from Texas A&M University as a distinguished professor in the department of health and kinesiology. From 1985 to 1997, Wilmore was the chair of the department of kinesiology and health education and the Margie Gurley Seay Endowed Centennial professor at the University of Texas at Austin. Prior to that, he served on the faculties at the University of Arizona, the University of California, and Ithaca College. Wilmore earned his PhD in physical education from the University of Oregon in 1966.

Wilmore has published 53 chapters, more than 320 peer-reviewed research papers, and 15 books on exercise physiology. He was one of five principal investigators for the HERITAGE Family Study, a large multicenter clinical trial investigating the possible genetic basis for the variability in the responses of physiological measures and risk factors for cardiovascular disease and type 2 diabetes to endurance exercise training. Wilmore's research interests have included determining the role of exercise in the prevention and control of both obesity and coronary heart disease, determining the mechanisms accounting for alterations in physiological function with training and detraining, and factors limiting the performance of elite athletes.

A former president of the American College of Sports Medicine, Wilmore was the recipient of the American College of Sports Medicine's Honor Award in 2006. In addition to serving as chair for many ACSM organizational committees, Wilmore served on the United States Olympic Committee's Sports Medicine Council and chaired their Research Committee. He is currently a member of the American Physiological Society and a fellow and former president of the American Academy of Kinesiology and Physical Education. Wilmore has served as a consultant for several professional sports teams, the California Highway Patrol, the President's Council on Physical Fitness and Sport, NASA, and the U.S. Air Force. He has also served on several journal editorial boards.

In his free time, Wilmore enjoys Bible study, running, walking, and playing with his grandchildren. He and his wife, Dottie, have three daughters (Wendy, Kristi, and Melissa) and sons-in-law and seven grandchildren.

David L. Costill, PhD, is the emeritus John and Janice Fisher chair in exercise science at Ball State University in Muncie, Indiana. He established the Ball State University Human Performance Laboratory in 1966 and served as its director for over 32 years.

Costill has written and coauthored more than 425 publications over the course of his career, including 6 books and articles in both peer-reviewed and lay publications. He served as the original editor in chief of the *International Journal of Sports Medicine* for 12 years. Between 1971 and 1998, he averaged 25 U.S. and international lecture trips each year. He was president of the ACSM from 1976 to 1977, a member of its board of trustees for 12 years, and a recipient of ACSM Citation and Honor Awards. Many of his former students are now leaders in the field of exercise physiology.

Costill received his PhD in physical education and physiology from Ohio State University in 1965. He and his wife, Judy, have two daughters, Jill and Holly. In his leisure time, Costill is a private pilot, auto and experimental airplane builder, competitive masters swimmer, and ex-marathon runner.